Advances in Intelligent Systems and Computing

Volume 697

Series editor

Janusz Kacprzyk, Polish Academy of Sciences, Warsaw, Poland
e-mail: kacprzyk@ibspan.waw.pl

The series "Advances in Intelligent Systems and Computing" contains publications on theory, applications, and design methods of Intelligent Systems and Intelligent Computing. Virtually all disciplines such as engineering, natural sciences, computer and information science, ICT, economics, business, e-commerce, environment, healthcare, life science are covered. The list of topics spans all the areas of modern intelligent systems and computing such as: computational intelligence, soft computing including neural networks, fuzzy systems, evolutionary computing and the fusion of these paradigms, social intelligence, ambient intelligence, computational neuroscience, artificial life, virtual worlds and society, cognitive science and systems, Perception and Vision, DNA and immune based systems, self-organizing and adaptive systems, e-Learning and teaching, human-centered and human-centric computing, recommender systems, intelligent control, robotics and mechatronics including human-machine teaming, knowledge-based paradigms, learning paradigms, machine ethics, intelligent data analysis, knowledge management, intelligent agents, intelligent decision making and support, intelligent network security, trust management, interactive entertainment, Web intelligence and multimedia.

The publications within "Advances in Intelligent Systems and Computing" are primarily proceedings of important conferences, symposia and congresses. They cover significant recent developments in the field, both of a foundational and applicable character. An important characteristic feature of the series is the short publication time and world-wide distribution. This permits a rapid and broad dissemination of research results.

Advisory Board

Chairman

Nikhil R. Pal, Indian Statistical Institute, Kolkata, India
e-mail: nikhil@isical.ac.in

Members

Rafael Bello Perez, Universidad Central "Marta Abreu" de Las Villas, Santa Clara, Cuba
e-mail: rbellop@uclv.edu.cu

Emilio S. Corchado, University of Salamanca, Salamanca, Spain
e-mail: escorchado@usal.es

Hani Hagras, University of Essex, Colchester, UK
e-mail: hani@essex.ac.uk

László T. Kóczy, Széchenyi István University, Győr, Hungary
e-mail: koczy@sze.hu

Vladik Kreinovich, University of Texas at El Paso, El Paso, USA
e-mail: vladik@utep.edu

Chin-Teng Lin, National Chiao Tung University, Hsinchu, Taiwan
e-mail: ctlin@mail.nctu.edu.tw

Jie Lu, University of Technology, Sydney, Australia
e-mail: Jie.Lu@uts.edu.au

Patricia Melin, Tijuana Institute of Technology, Tijuana, Mexico
e-mail: epmelin@hafsamx.org

Nadia Nedjah, State University of Rio de Janeiro, Rio de Janeiro, Brazil
e-mail: nadia@eng.uerj.br

Ngoc Thanh Nguyen, Wroclaw University of Technology, Wroclaw, Poland
e-mail: Ngoc-Thanh.Nguyen@pwr.edu.pl

Jun Wang, The Chinese University of Hong Kong, Shatin, Hong Kong
e-mail: jwang@mae.cuhk.edu.hk

More information about this series at http://www.springer.com/series/11156

Hasmat Malik · Smriti Srivastava
Yog Raj Sood · Aamir Ahmad
Editors

Applications of Artificial Intelligence Techniques in Engineering

SIGMA 2018, Volume 2

Editors
Hasmat Malik
Department of Instrumentation
and Control Engineering
Netaji Subhas Institute of Technology
New Delhi, Delhi, India

Smriti Srivastava
Department of Instrumentation
and Control Engineering
Netaji Subhas Institute of Technology
New Delhi, Delhi, India

Yog Raj Sood
National Institute of Technology
Puducherry, India

Aamir Ahmad
Perceiving Systems Department
Max Planck Institute for Intelligent
Systems
Tübingen, Germany

ISSN 2194-5357 ISSN 2194-5365 (electronic)
Advances in Intelligent Systems and Computing
ISBN 978-981-13-1821-4 ISBN 978-981-13-1822-1 (eBook)
https://doi.org/10.1007/978-981-13-1822-1

Library of Congress Control Number: 2018949630

© Springer Nature Singapore Pte Ltd. 2019
This work is subject to copyright. All rights are reserved by the Publisher, whether the whole or part of the material is concerned, specifically the rights of translation, reprinting, reuse of illustrations, recitation, broadcasting, reproduction on microfilms or in any other physical way, and transmission or information storage and retrieval, electronic adaptation, computer software, or by similar or dissimilar methodology now known or hereafter developed.
The use of general descriptive names, registered names, trademarks, service marks, etc. in this publication does not imply, even in the absence of a specific statement, that such names are exempt from the relevant protective laws and regulations and therefore free for general use.
The publisher, the authors and the editors are safe to assume that the advice and information in this book are believed to be true and accurate at the date of publication. Neither the publisher nor the authors or the editors give a warranty, express or implied, with respect to the material contained herein or for any errors or omissions that may have been made. The publisher remains neutral with regard to jurisdictional claims in published maps and institutional affiliations.

This Springer imprint is published by the registered company Springer Nature Singapore Pte Ltd.
The registered company address is: 152 Beach Road, #21-01/04 Gateway East, Singapore 189721, Singapore

Preface

This Conference Proceedings Volume 2 contains the written versions of most of the contributions presented at the International Conference SIGMA 2018. The conference was held at Netaji Subhas Institute of Technology (NSIT), New Delhi, India, during February 23–25, 2018. NSIT is an autonomous institute under the Government of NCT of Delhi and affiliated to University of Delhi, India. The International Conference SIGMA 2018 aimed to provide a common platform to the researchers in related fields to explore and discuss various aspects of artificial intelligence applications and advances in soft computing techniques. The conference provided excellent opportunities for the presentation of interesting new research results and discussion about them, leading to knowledge transfer and the generation of new ideas.

The conference provided a setting for discussing recent developments in a wide variety of topics including power system, electronics and communication, renewable energy, tools and techniques, management and e-commerce, motor drives, manufacturing process, control engineering, health care and biomedical, cloud computing, image processing, environment and robotics. This book contains broadly 13 areas with 59 chapters that will definitely help researchers to work in different areas.

The conference has been a good opportunity for participants coming from all over the globe (mostly from India, Qatar, South Korea, USA, Singapore, and so many other countries) to present and discuss topics in their respective research areas.

We would like to thank all the participants for their contributions to the conference and for their contributions to the proceedings. Many thanks go as well to the NSIT participants for their support and hospitality, which allowed all foreign participants to feel more at home. Our special thanks go to our colleagues for their devoted assistance in the overall organization of the conference.

It is our pleasant duty to acknowledge the financial support from Defence Research and Development Organisation (DRDO), Gas Authority of India Limited (GAIL), MARUTI SUZUKI, CISCO, Power Finance Corporation (PFC) Limited, CommScope, Technics Infosolutions Pvt. Ltd., Bank of Baroda, and Jio India.

We hope that it will be interesting and enjoying at least as all of its predecessors.

New Delhi, India — Hasmat Malik
New Delhi, India — Smriti Srivastava
Puducherry, India — Yog Raj Sood
Tübingen, Germany — Aamir Ahmad

Contents

About the Editors

Hasmat Malik (M'16) received his B.Tech. degree in electrical and electronics engineering from GGSIP University, New Delhi, India, and M.Tech. degree in electrical engineering from NIT Hamirpur, Himachal Pradesh, India, and he is currently doing his Ph.D. degree in Electrical Engineering Department, Indian Institute of Technology Delhi, New Delhi, India.

He is currently Assistant Professor in the Department of Instrumentation and Control Engineering, Netaji Subhas Institute of Technology, New Delhi, India. His research interests include power systems, power quality studies, and renewable energy. He has published more than 100 research articles, including papers in international journals, conferences, and chapters. He was a Guest Editor of Special Issue of *Journal of Intelligent & Fuzzy Systems*, 2018 (SCI Impact Factor 2018:1.426) (IOS Press), and Special Issue of *International Journal of Intelligent Systems Design and Computing* (IJISDC) (Inderscience—three times).

He received the POSOCO Power System Award (PPSA-2017) for his Ph.D. work on research and innovation in the area of power system in 2017. His interests are in artificial intelligence/soft computing applications to fault diagnosis, signal processing, power quality, renewable energy, and microgrids.

He is a life member of the Indian Society for Technical Education (ISTE); International Association of Engineers (IAENG), Hong Kong; International Society for Research and Development (ISRD),

London; and he is a member of the Institute of Electrical and Electronics Engineers (IEEE), USA, and MIR Labs, Asia.

Smriti Srivastava received her B.E. degree in electrical engineering and her M.Tech. degree in heavy electrical equipment from Maulana Azad College of Technology [now Maulana Azad National Institute of Technology (MANIT)], Bhopal, India, in 1986 and 1991, respectively, and Ph.D. degree in intelligent control from the Indian Institute of Technology Delhi, New Delhi, India, in 2005. From 1988 to 1991, she was Faculty Member at MANIT, and since August 1991, she has been with the Department of Instrumentation and Control Engineering, Netaji Subhas Institute of Technology, University of Delhi, New Delhi, India, where she is working as Professor in the same division since September 2008 and as Dean of Undergraduate Studies. She also worked as Associate Head of the Instrumentation and Control Engineering Division at NSIT, New Delhi, India, from April 2004 to November 2007 and from September 2008 to December 2011. She was Dean of Postgraduate Studies from November 2015 to January 2018. She is Head of the division since April 2016. She is the author of a number of publications in transactions, journals, and conferences in the areas of neural networks, fuzzy logic, control systems, and biometrics. She has given a number of invited talks and tutorials mostly in the areas of fuzzy logic, process control, and neural networks. Her current research interests include neural networks, fuzzy logic, and hybrid methods in modeling, identification, and control of nonlinear systems and biometrics.

She is Reviewer of *IEEE Transactions on Systems, Man and Cybernetics* (SMC), Part-B, *IEEE Transactions on Fuzzy Systems*, *International Journal of Applied Soft Computing* (Elsevier), *International Journal of Energy, Technology and Policy* (Inderscience).

Her paper titled 'Identification and Control of a Nonlinear System using Neural Networks by Extracting the System Dynamics' was selected by the Institution of Electronics and Telecommunication Engineers for IETE K S Krishnan Memorial Award for the best system-oriented paper. She was also nominated for

'International Engineer of the Year 2008' by International Biographical Center of Cambridge, England, in 2008. Her name appeared in Silver Jubilee edition of 'Marquis Who's Who in the World' in November 2007, 2008, and 2009. Her paper titled 'New Fuzzy Wavelet Neural Networks for System Identification and Control' was the second most downloadable paper of the year 2006 from the list of journals that come under 'ScienceDirect'.

Prof. Yog Raj Sood (SM'10) is a member of DEIS. He obtained his B.E. degree in electrical engineering with 'Honors' and M.E. degree in power system from Punjab Engineering College, Chandigarh (UT), in 1984 and 1987, respectively. He obtained his Ph.D. degree from Indian Institute of Technology Roorkee in 2003. He is Director of the National Institute of Technology Puducherry, Karaikal, India. He joined Regional Engineering College, Kurukshetra, in 1986. Since 2003, he has been working as Professor in the Department of Electrical Engineering, National Institute of Technology Hamirpur, Himachal Pradesh, India. He has published a number of research papers. He has received many awards, prizes, and appreciation letters for his excellence in research academic and administration performance. His research interests are deregulation of power system, power network optimization, condition monitoring of power transformers, high-voltage engineering, nonconventional sources of energy.

Aamir Ahmad is Research Scientist at the Perceiving Systems Department, Max Planck Institute for Intelligent Systems, Tübingen, Germany. He received his Ph.D. degree in electrical and computer engineering from the Institute for Systems and Robotics, Instituto Superior Técnico, Lisbon, Portugal. His main research interests lie in Bayesian sensor fusion and optimization-based state estimation. The core application focus of his research is vision-based perception in robotics. He has worked with several ground robots in the recent past. Currently, he is focusing on aerial robots and active perception in multiple aerial vehicle systems. He has published 5 peer-reviewed journal articles, 2 chapters, 11 conference papers, and 2 workshop papers.

A New Approach for Power Loss Minimization in Radial Distribution Networks

Sarfaraz Nawaz, Manish Sharma and Ankush Tandon

Abstract A new methodology is presented in this paper to reduce active power losses achieved in the distribution system. The multiple DG units are placed to solve the problem. A new mathematical expression, Power Voltage Sensitivity Constant (PVSC), has been formulated to solve the DG placement problem. The PVSC determines the site and size of multiple DG units. The size of DG units is also restricted up to 50% of total system load. IEEE 69-bus reconfigured distribution system at three different loading conditions is considered to validate the results. The obtained results are compared to the results of the latest approaches to evaluate its robustness.

Keywords Radial distribution system (RDS) · Real power loss
Distributed generation (DG)

1 Introduction

In India, the distribution losses are around 20–25%. Therefore, myriads of efforts have been made for the reduction of losses in distribution networks like network reconfiguration, optimal allocation of distributed generation. Network reconfiguration is done by modifying the feeder topologies by proper handling of operation of sectionalizing and tie-switches and taking care of their open/close status at the time of emergency or normal operation. Modification of feeder topologies is an adequate and effective technique to minimize losses, enhance voltage stability, and better load balance. The authors suggested various techniques to solve the problem of reconfiguration of existing network [1–20]. Merlin et al. [1] were the first to evolve a scheme to minimize feeder loss using distribution network reconfiguration. They solved the problem using discrete branch and bound technique by formulating

S. Nawaz · M. Sharma (✉) · A. Tandon
Swami Keshvanand Institute of Technology, Management and Gramothan,
Jaipur, India
e-mail: manishpandya67@gmail.com

© Springer Nature Singapore Pte Ltd. 2019

H. Malik et al. (eds.), *Applications of Artificial Intelligence Techniques in Engineering*, Advances in Intelligent Systems and Computing 697,
https://doi.org/10.1007/978-981-13-1822-1_1

it as mixed-integer nonlinear optimization problem. Civanlar et al. [2] estimated the loss reduction by deriving a simple formula and performing switching operation between two feeders and adopted a branch exchange procedure for this. Lin and Chin [3] suggested a modern technique to resolve the problem of reconfiguration of distribution feeder through which an effective network configuration could be developed to minimize losses. Shirmohammadi et al. [4] introduced a technique that depends on an optimal flow pattern in which switches were made open one after another starting from a complete meshed system. Amanulla et al. [5] reformed the pattern of switches in the system using PSO technique.

In DG technology, the small generating units (1 kW–50 MW) are connected near the load side. Both renewable and nonrenewable energy source can be served as DG units. As fossil fuels are getting depleted day by day, the outcome of this has lead to incorporate renewable-based distribution generating units, which is the topmost priority. The biggest asset of using DG device is that it minimizes both real and reactive power losses and it also revamps the voltage profile while conserving reliability and efficiency of the power system. To utilize the benefits of DG technology, it is required to determine optimal place and size of DG units in RDS; otherwise, it causes some unfavorable effects like increased real power losses, poor voltage profile, increased cost, etc. Various approaches have been proposed in the literature to solve the DG allocation problem. DG units were determined by particle swarm optimization (PSO) in [6]. Acharya et al. [7] incorporated an analytical methodology to decipher DG placement problem in the distribution system. Solve. DG placement problem is solved using analytical approach by Gozel and Hocaoglu [8]. In [9, 10], the author suggested GA-based method to get the optimal position and amount of DG units. Kean and Omalley [11] proposed the constrained linear programming (LP) approach to solve the problem in Irish system. Zhang et al. [12] proposed a novel approach to solve the above problem. In [13], Injeti and Kumar formulated an objective function to decrease active power losses and to get better voltage stability. Loss sensitivity factor (LSF) is the key to determine optimal location of DG and the size is determined by simulated annealing (SA) technique. Nawaz et al. [14] presented a sensitivity analysis technique and tested it on a standard 33 bus test system under different loading conditions. Viral et al. [15] proposed an analytical technique to get the best place and size of DG units in the balanced distribution system to reduce real power. The various types of DG units are [16]:

Type-I: Generate active power
Type-II: Generate reactive power
Type-III: Conjure both kW and kVAr
Type-IV: Consume reactive power by conjure active power.

In this paper, multiple DG (Type-I) units are placed for boosting voltage profile and maximizing the percentage of active power loss reduction. A new method is projected to determine site and size of multiple DG units. A pristine mathematical expression is investigated that is called PVSC (Power Voltage Sensitivity Constant).

The constant evaluates size and location of any type of DGs at the same time. DG Type-I (PV solar module) is used to solve the problem. Upto 50% penetration level of DG units is also taken into consideration, so that less size of DG units produces maximum loss reduction. Standard IEEE 69 bus reconfigured distribution system is considered as a test system. It has been observed that the proposed approach gives optimum results than other approaches mentioned in this paper.

The rest of the paper is organized as follows: Sect. 2 states problem formulation with constraints. Then, the proposed approach is discussed in Sect. 3. The comparative analysis of the numerical results of the proposed approach on 69 bus test system is reported in Sect. 4 and finally, the conclusion is given in Sect. 5.

2 Problem Formulation

The DG's placement problem can be mathematically expressed as [17] (Fig. 1):

$$\mathrm{Min} f = W_{\mathrm{Loss}} \tag{1}$$

The distribution network power loss is calculated by using

$$P_{\mathrm{Loss}} = \sum_{i=1}^{n}\sum_{j=1}^{n} R\frac{|V_i|^2 + |V_j|^2 - 2|V_i||V_j|\cos\delta_{ij}}{Z^2} \tag{2}$$

Subjected to:

(i) Network total power balancing.

(ii) $P_{\mathrm{DG}j}^{\min} \le P_{\mathrm{DG}j} \le P_{\mathrm{DG}j}^{\max}$

$Q_{\mathrm{DG}j}^{\min} \le Q_{\mathrm{DG}j} \le Q_{\mathrm{DG}j}^{\max}$

(iii) Bus voltage limits $0.95\,\mathrm{pu} \le V_i \le 1.0\,\mathrm{pu}$

(iv) Thermal limit constraint of line.

where

R	Line section resistance;
X	Line section reactance;
Z	Line section impedance;
V_i, V_j	voltage at bus i and j;

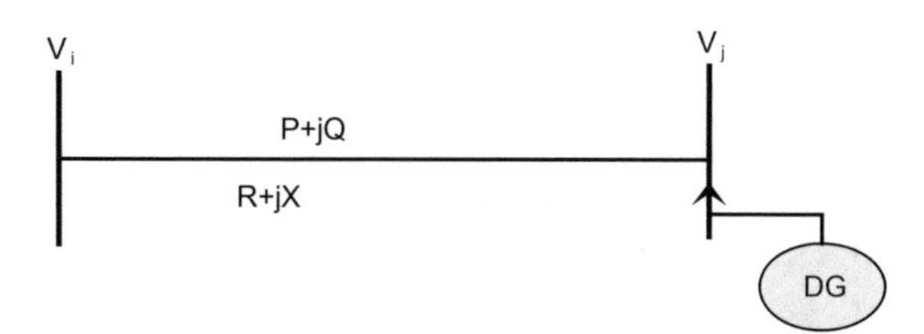

Fig. 1 Single line diagram of DG connected network

δ_i, δ_j Voltage angle at bus *i* and *j*;
P and Q kW and kVAr flow in the section.

3 Approach

A new method is proposed here for the allocation of multiple DG units. The Power Voltage Sensitivity Constant (PVSC) determines the size and position of DG units.

$$\mathrm{PVSC} = \frac{V_{\mathrm{rated}}}{V_{\mathrm{min}}} + \frac{P_{\mathrm{dgloss}}}{P_{\mathrm{realloss}}} \tag{3}$$

where

P_{realloss} active real power loss (base case).
P_{dgloss} power loss after DG placement.
V_{rated} rated bus voltage in pu (always be 1 pu).
V_{min} minimum bus voltage after DG placement.

The following constraints are to be met for optimal allocation of DG units: (i) V_{min} should be maximum (ii) P_{dgloss} should be minimum and (iii) PVSC should be minimum. The algorithm of the proposed technique is as follows:

Step 1: Calculate P_{realloss} by executing load flow program.
Step 2: Start with 5% DG size of total load and run load flow program.
Step 3: Compute P_{dgloss} and "PVSC" values.
Step 4: Proliferate DG size in small steps and calculate P_{dgloss}.
Step 5: Restore the DG size which corresponds to least value of P_{dgloss}.
Step 6: Minimum value of "PVSC" yields optimal location of DG unit.
Step 7: Next location of DG is identified by repeating steps from 3.

4 Results

The proposed methodology is tested on IEEE 69-bus (12.66 kV and 100 MVA base value) distribution system. The real power loss of 69-bus system 225 kW (before reconfiguration) and minimum bus voltage is 0.9092 pu. After reconfiguration [20] the loss at nominal load is 98.6 kW. Three different loading conditions light (50% load), nominal (100% load), and heavy (160% load) load levels are used here.

The following cases are considered here:

I: Base Case
II: After Network Reconfiguration
III: DGs placement after network reconfiguration.

Table 1 Results of the proposed method for 69-bus system at various load levels

Case	Items	Load level		
		Light (50%)	Nominal (100%)	Heavy (160%)
Base case	Tie switches	69, 70, 71, 72, 73	69, 70, 71, 72, 73	69, 70, 71, 72, 73
	Power loss (kW)	51.61	225	652.47
	Minimum bus voltage (pu)	0.956	0.9092	0.845
After feeder reconfiguration (FR) [19]	Tie switches	14, 58, 61, 69, 70	14, 58, 61, 69, 70	14, 58, 61, 69, 70
	Power loss (kW)	23.60	98.6	264.6
	Minimum bus voltage (pu)	0.9753	0.9497	0.917
	% Loss reduction (from base case)	54.24%	56.17%	59.44%
DG placement after FR (proposed method)	DG size (location)	600 (61) 180 (64) 140 (27)	1350 (61) 310 (64) 210 (27)	1800 (61) 650 (64)
	Size of DG (kW)	920	1870	2450
	P_{Loss} (kW)	9.8	37	108
	$V_{\min}$ (pu)	0.99	0.98	0.956
	% Loss reduction	81%	83.55%	83.3%

The results of 69 bus system after DG allocation is shown in Table 1. The optimal location is found at bus no. 27, 61, and 64. At light, nominal and heavy load level, the real power losses are reduced to 9.8, 37, and 108 kW, respectively. The bus voltage profile is also enhanced at each load level as shown in Fig. 2.

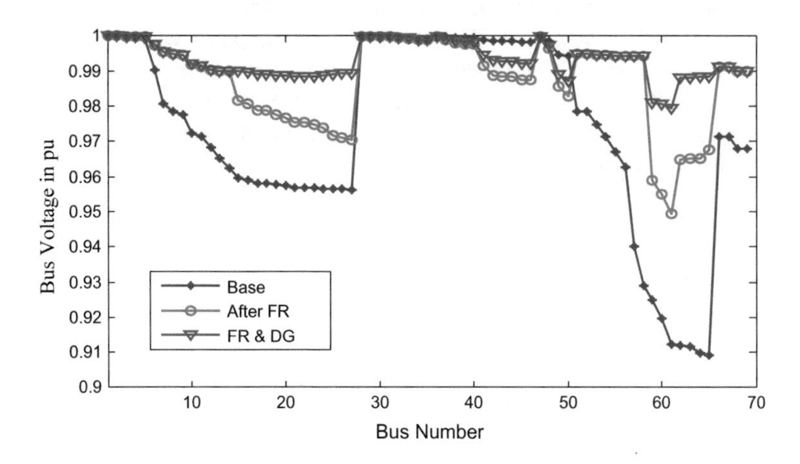

Fig. 2 Comparison of bus voltage at nominal load for 69 bus system

Table 2 Comparison of results for 69-bus systems at nominal load

Technique	Tie switches	DG size in kW (location)	Total DG size in kW	Real Power loss in kW	% Loss reduction from base case	Min. bus voltage (pu)
Integrated approach [20]	69, 18, 13, 56, 61	1066 (61) 355 (60) 425 (58)	1846	51.3	77	0.9619
UVDA [20]	14, 58, 61, 69, 70	1378 (61) 620 (11) 722 (64)	2720	37.84	83.18	0.9801
Proposed	14, 58, 61, 69, 70	1350 (61) 310 (64) 210 (27)	1870	37	83.55	0.98

The overall size of DG unit is tethered upto 50% of system load. The obtained results are compared with the results of the latest optimization approaches as mentioned in Table 2. It is obvious from Table 2 that the proposed analytical approach yields minimum losses while optimizing DG size also.

5 Conclusion

This paper presents a new approach for minimizing losses by placing of DG unit at the optimal location. Solar PV module (Type-I DG) is used for placement. A new mathematical expression, PVSC, is proposed which yields both optimal size and location of DG unit. The developed approach is tested on IEEE 69 bus reconfigured system at three diverse loading of total load and heavy conditions mainly nominal similar to base case, light load with a decrement of 50% of total load and heavy loading which an increment of 50% of the total load. The overall size of DG unit is tethered upto 50% of system load. The results are compared with the latest algorithms. It can be justified by Table 2 that the proposed methodology gives maximum loss reduction while considering DG size also. The voltage profile is also equal to or higher than other mentioned methods. The proposed approach can be easily implemented in the real distribution system.

References

1. Merlin, A., Back, H.: Search for a minimal-loss operating spanning tree configuration in urban power distribution systems. In: Proceedings of 5th Power Systems Computer Conference, Cambridge, U.K., 1–5 Sept 1975
2. Civanlar, S., Grainger, J.J., Yin, H., Lee, S.H.: Distribution feeder reconfiguration for loss reduction. IEEE Trans. Power Deliv. **3**(3), 1217–1223 (1988)

3. Lin, M.W., Chin, H.-C.: A new approach for distribution feeder reconfiguration for loss reduction and service restoration. IEEE Trans. Power Deliv. **13**(3), 870–875 (1988)
4. Shirmohammadi, D., Hong, H.W.: Reconfiguration of electric distribution networks for resistive losses reduction. IEEE Trans. Power Deliv. **4**, 1402–1498 (1989)
5. Amanulla, B., Chakrabarti, S.: Reconfiguration of power distribution systems considering reliability and power loss. IEEE Trans. Power Deliv. **27**(2) (April 2012)
6. Katsigiannis, Y.A., Georgilakis, P.S.: Optimal sizing of small isolated hybrid power systems using Tabu search. J. Optoelectron. Adv. Mater. **10**(5), 1241–1245 (2008)
7. Acharya, N., Mahat, P., Mithulananthan, N.: An analytical approach for DG allocation in primary distribution network. Int. J. Electr. Power Energy Syst. **28**, 669–678 (2006)
8. Gozel, T., Hocaoglu, M.H.: An analytical method for the sizing and siting of distributed generators in radial systems. Electr. Power Syst. Res. **79**, 912–918 (2009)
9. Kim, K.H., Lee, Y.J., Rhee, S.B., Lee, S.K.: Dispersed generator placement using fussy-GA in distribution systems. In: Proceeding of IEEE Power Engineering Society Summer Meeting, USA vol. 2, pp. 1148–1153 (2002)
10. Kim, J.O., Nam, S.W., Park, S.K., Singh, C.: Dispersed generation planning using improved Hereford Ranch algorithm. Electr. Power Syst. Res. **47**(1), 47–55 (1998)
11. Kean, A., Omalley, M.: Optimal allocation of embedded generation on distribution networks. IEEE Trans. Power Syst. (2006)
12. Zhang, X., Karady, G.G., Ariaratnam, S.T.: Optimal allocation of CHP-based distributed generation on urban energy distribution networks. IEEE Trans. Sustain. Energy **5**(1), 246–253 (2014)
13. Injeti, S.K., Kumar, N.P.: A novel approach to identify optimal access point and capacity of multiple DGs in a small, medium and large scale radial distribution systems. Int. J. Electr. Power Energy Syst. **45**, 142–151 (2013)
14. Nawaz, S., Imran, M., Sharma, A., Jain, A.: Optimal feeder reconfiguration and DG placement in distribution network. Int. J. Appl. Eng. Res. **11**(7), 4878–4885 (2016)
15. Viral, R., Khatod, D.K.: An analytical approach for sizing and siting of DGs in balanced radial distribution networks for loss minimization. Electr. Power Energy Syst. **67**, 191–201 (2015)
16. Devi, S., Geethanjali, M.: Application of modified bacterial foraging optimization algorithm for optimal placement and sizing of distributed generation. Expert Syst. Appl. **41**, 2772–2781 (2014)
17. El-Fergany, Attia: Study impact of various load models on DG placement and sizingusing backtracking search algorithm. Appl. Soft Comput. **30**, 803–811 (2015)
18. Savier, J.S., Das, D.: Impact of network reconfiguration on loss allocation of radial distribution systems. IEEE Trans. Power. Deliv. **2**(4), 2473–2480 (2007)
19. Tandon, A., Saxena, D.: A comparative analysis of SPSO and BPSO for power loss minimization in distribution system using network reconfiguration. In: IEEE Conference on Computational Intelligence on Power, Energy and Controls with their impact on Humanity (CIPECH), 28–29 Nov 2014
20. Bayat, A., Bagheri, A., Noroozian, R.: Optimal siting and sizing of distributed generation accompanied by reconfiguration of distribution networks for maximum loss reduction by using a new UVDA-based heuristic method. Electr. Power Energy Syst. **77**, 360–371 (2016)

Security-Constrained Unit Commitment for Joint Energy and Reserve Market Based on MIP Method

Pranda Prasanta Gupta, Prerna Jain, Suman Sharma and Rohit Bhakar

Abstract In this work, a Security-Constrained Unit Commitment (SCUC) is proposed for day-ahead scheduling with joint energy and reserve markets. Independent System Operator (ISO) executes SCUC to optimize reserve requirements in restructured power system. Though, SCUC structure determines reserve requirements for simulating UC necessities and line contingency with DC optimal power flow (DCOPF) for adequate purpose of energy market. In this context, proposed SCUC problem is formulated using Mixed Integer Programming (MIP) in which schedule and dispatch of generating units is considered to be deterministic. The overall objective of this paper, ISO is to minimize the cost of supply energy and reserve requirement over the optimization horizon (24 h) subject to satisfying all the operating and network security constraints. However, comprehensive Benders decomposition (BD) is new to explain SCUC formulation and simulation results are compare without and by inclusion of network security. In order to show that effective reach of the proposed model, it is executed on a transmission system of modified IEEE-30 bus system with seven generating units on Zone A and two generating units on Zone B. Moreover various, test cases are investigated and compared, which shows that the proposed optimization model is promising.

Keywords Security-constrained unit commitment · Benders decomposition
Mixed integerprogramming · MIP Method · IEEE-30 bus system
DCOPF

P. P. Gupta (✉) · P. Jain · S. Sharma · R. Bhakar
Department of Electrical Engineering, Malaviya National Institute
of Technology, Jaipur, India
e-mail: prandaprasantagupta@gmail.com

P. Jain
e-mail: pjain.ee@mnit.ac.in

S. Sharma
e-mail: sharma_sumi@yahoo.comand

R. Bhakar
e-mail: rohitbhakar@gmail.com

© Springer Nature Singapore Pte Ltd. 2019

H. Malik et al. (eds.), *Applications of Artificial Intelligence Techniques in Engineering*, Advances in Intelligent Systems and Computing 697,
https://doi.org/10.1007/978-981-13-1822-1_2

1 Introduction

Provision of energy and reserve markets over the restructured situation would not immediately promote as it is described in vertically integrated utilities [1]. Though a lot of reasons would be to figure elsewhere, the major motive is that entity provided that reserve market might not be over-straight managed for the operator. The issues are emphasized by two examples. Generator in the ready-for-action market is dispatched as for each bids provided by those to the market operator. The markets were unoccupied consecutively in a series determined by the rate of response of the service. Energy is the major product which can be traded in electricity market. There are different types of products in competitive bidding, for example regulation reserve, spinning reserve, and non-spinning reserve and other types of ancillary markets [2]. The thermal units must have condition for extra generation all through contingencies similar to outage of generator. The purpose of this paper is for joint optimization of cross commodities dispatch energy and reserve decision taking into consideration the availability of resources and the overall cost. Another product is ancillary reserve which is single reserve requirements that is crucial for ensuring the reliability process for electric grid [3, 4]. Here in few markets, simultaneous application is to apply for procure energy and spinning reserve (SR), even as new as one of these two markets are clear independently and successively [5, 6]. In [3], the above literature, it is clear that both energy and reserve markets could be called by ISO of different amount based on benchmark system under consideration and there in after obtaining the order of SR market. Moreover, probabilistic and deterministic scale is executed with ISO to determine the desired stage of SR. During deterministic perspective, SR ability is regularly presumed to be approximately equivalent on a particular fraction of the generator load hourly, or else ability of the enormous existence generator [7–9]. In [10], present a MILP intended for reimbursement of the ancillary service and energy markets on which the essential regulation down (REGD), regulation up (REGU), automatic generation control (AGC), 30-minute operating reserve (TMOR), 10-minute non-spinning reserve (TMNR) and 10-minute spinning reserve (TMSR), and are assumed to be set and predefined.

In deregulated power system, security is superior to supply adequate SR on the system [1, 11]. SCUC define to UC explicitly considering security constraints in equally normal operation and contingency situation to guarantee that transmission flows are within limits. The straight method for considering network constraints establishes Lagrangian relaxation (LR) which is used to explain SCUC [12, 13]. With additional SCUC model which prominence to concurrent the optimization of ancillary service and energy market is presented in [14]. In [8], the probabilistic approach is used, while taking into account of transmission and line contingency to optimize the generation schedule. A lot of work in these part utilities for solving optimization technique in SCUC problem efficiently (such as mixed integer linear programming (MILP)-based method, Lagrangian relaxation method [15–17]. In [18] UC derived from uncertainty and an amalgamation of BD and proposed the outer approximation technique. BD is a popular optimization method for solving SCUC problem presented, example in [15]. The BD moves toward the master problem in UC and subproblem in network security separately. The job for deterministic UC apparatus signifies the network classically incorporating DC power flow equations as an alternative of AC constraint [19]. Novel contribution of this paper is incorporation of contingency line in the model of SCUC by network constraint and solve by operating DCOPF. The approximation of DC model is operating as a simplification in the computation of power flow [20]. This paper also suggests contingency analysis ranking based as is performed with DCOPF solution [23]. The main contribution of this research work, with respect to previous work in this region, proposed SCUC model for joint optimization of energy and reserve markets as a MIP method and solved using CPLEX solver in General Algebraic Modeling System (GAMS). To achieve this objective, constraints are introduced to alleviate the effect of variable power generation such as power flow limits on preferred line, bus voltage limits, etc. The problem is solved for a benchmark of IEEE 30-bus system such that simulation results are found to be inspiring.

Organization of this work is as follows: Sect. 2 explains the mathematical modeling of SCUC problem based on MIP formulation; Sect. 3 describe the proposed SCUC framework; Sect. 4 describes the simulation portion to show the effect of SCUC on mutually technical and economical feature and Sect. 5 represents termination and future work.

2 Mathematical Formulation of SCUC Based on MIP

Considering the above literature facts, in this paper, the SCUC model are formulated as optimization problem that minimizes the total cost of supply ancillary services and energy requirements. The objective function (OF) is formulated as follows:

$$\begin{aligned} \min \mathrm{OF} = & \sum_{g=1}^{N_G}\sum_{t=1}^{N_T}\begin{bmatrix} FC^h_{g,t}(P_{g,t}) + FC^d_{g,t}(R^d_{g,t}) + FC^u_{g,t}(R^u_{g,t}) \\ FC^s_{g,t}(R^s_{g,t}) \end{bmatrix} U_{g,t} \\ & \sum_{g=1}^{N_G}\sum_{t=1}^{N_T}\left[FC^n_{g,t}(R^n_{g,t}) + FC^o_{g,t}(R^o_{g,t})\right] + \sum_{g=1}^{N_G}\sum_{t=1}^{N_T} S_{g,t} \end{aligned} \tag{1}$$

In Eq. (1), the first term indicates the costs of supplying REGU, REGD, TMSR, and energy of the generating unit is on; second term indicates the costs of supplying TMOR and TMNR. On the other hand, last third term indicates shutdown and startup costs, which incur if the unit is turned on or off. Where, $F^x_{g,t}(.)$ indicates the generator bidding cost of unit g at time t for product x. $P_{g,t}$ indicates energy of generator unit g at time t, $S_{g,t}$ represents shutdown/startup cost of generator unit g at time t, and $U_{g,t}$ represents commitment status of generator unit g at time t. In the above Eq. (1) the constraints are included as

Constraints

Power energy balance constraint

$$\sum_{g=1}^{N_G} P_{g,t} = PD_t + PL_t, \quad \forall t \tag{2}$$

In Eq. (2), the power balance constraint of the system is represented. Here, PD_t is the system load demand at time t, and PL_t represents the system losses at time t.

Ancillary service requirement

$$\sum_{g=1}^{N_G} R_{g,t}^d \leq D_t^d, \ \forall t \tag{3}$$

$$\sum_{g=1}^{N_G} R_{g,t}^u \geq D_t^u, \ \forall t \tag{4}$$

$$\sum_{g=1}^{N_G} R_{g,t}^u + R_{g,t}^s \geq D_t^u + D_t^s, \ \forall t \tag{5}$$

$$\sum_{g=1}^{N_G} R_{g,t}^u + R_{g,t}^s + R_{g,t}^n \geq D_t^u + D_t^s + D_t^n, \ \forall t \tag{6}$$

$$\sum_{g=1}^{N_G} R_{g,t}^u + R_{g,t}^s + R_{g,t}^n + R_{g,t}^o \geq D_t^u + D_t^s + D_t^n + D_t^o, \ \forall t \tag{7}$$

Ancillary service supplies for REGD, TMSR, REGU, TMOR, and TMNR are specified in (3)–(7), correspondingly. For illustration, (7) show that the additional REGU might be used to assure TMSR supplies.

2.1 *Formulation of Thermal UC Model*

In this sector, SCUC model proposed MIP method for thermal unit commitment problem considering joint energy and reserve markets. Normally, the thermal unit OF can be spoken as follows:

$$\sum_{g=1}^{N_G}\sum_{t=1}^{N_T}\left[FC_{g,t}^{h}(P_{g,t})\right]U_{g,t}+\sum_{g=1}^{N_G}\sum_{t=1}^{N_T}S_{g,t} \tag{8}$$

A generation output of each unit in Eq. (9), ought to be among its maximum and minimum limit which is subsequent inequality used for all generator is fulfilled.

$$P_{g,\min}U_{g,t}\leq P_{g,t}\leq P_{g,\max}U_{g,t},\ \ \forall g,\ \forall t \tag{9}$$

The above Eq. (8) indicates the constraints and OF (10–13) for a MIP solution. Generator constraints scheduled subsequently contain ramp down/up limits (10), (11) minimum ON/OFF limits (12), (13).

$$P_{g,t}-P_{g,t-1}\leq \mathrm{UR}_g,\quad \forall g,\ \forall t \tag{10}$$

$$P_{g,t-1}-P_{g,t}\leq \mathrm{DR}_g,\quad \forall g,\ \forall t \tag{11}$$

$$\left[UM_{g,t-1}^{\mathrm{on}}-T_g^{\mathrm{on}}\right]\left[U_{g,t-1}-U_{g,t}\right]\geq 0,\quad \forall g,\ \forall t \tag{12}$$

$$\left[UM_{g,t-1}^{\mathrm{off}}-T_g^{\mathrm{off}}\right]\left[U_{g,t}-U_{g,t-1}\right]\geq 0,\quad \forall g,\ \forall t \tag{13}$$

Wherever DR_g and UR_g indicates ramp down and up rate limit of unit g. $UM_{g,t}^{\mathrm{on}}$ and $UM_{g,t}^{\mathrm{off}}$ indicates is time duration for which unit g has been ON and OFF at time t, T_g^{on} and T_g^{off} indicates is minimum ON and OFF time of unit g. To reach this OF expressed in (8) is subjected to following constraint.

System spinning and operating reserve requirement

$$\sum_{g=1}^{N_G}\mathrm{OR}_{g,t}U_{g,t}\geq R_{ot},\ \forall t \tag{14}$$

$$\sum_{g=1}^{N_G}\mathrm{SR}_{g,t}U_{g,t}\geq R_{st},\ \forall t \tag{15}$$

Security constraints

$$-\mathrm{FL}_{ij}^{\max}\leq \mathrm{FL}_{ij,t}\leq \mathrm{FL}_{ij}^{\max} \tag{16}$$

DC power flow constraints

$$\mathrm{FL}_{ij,t}=\frac{\theta_{j,t}-\theta_{i,t}}{X_{ij}} \tag{17}$$

In Eqs. (14), (15), (16), and (17) $\mathrm{OR}_{g,t}$, $\mathrm{SR}_{g,t}$ denoting operating and spinning reserve of unit g at time t. $\mathrm{FL}_{ij,t}$ are the line flow at node i and j at hour t. X_{ij} are the branch connecting line reactance of node j and i. Further $F_{ij}^{\max}$ represent maximum flow of line at node j and i correspondingly.

3 Proposed Algorithm

In this part, the proposed algorithm and strategies have been discussed and explained for optimum SCUC among energy and reserve market in power system. In short to explain the MIP formulation in Portion 2, a BD has been adopting to answer this problem. BD concert should be relevant to integrate that optimization problem expressing master problem for solving ED and UC, network constraint subproblem are revealed in Fig. 1. The master problem of UC, which includes Eqs. (9–15), of (8) give the schedule and dispatch result to minimize operation cost apart from energy and reserve market. Subsequent to the answer of master problem, the network constraint check is applied for every hour, which is shown in Fig. 1. The vector outputs Eq. (18) are existing dispatch and commitment solution acquire since the problem are check with the transmission line violation. If any line violation appears on the subproblem, the benders cuts created and associated to the

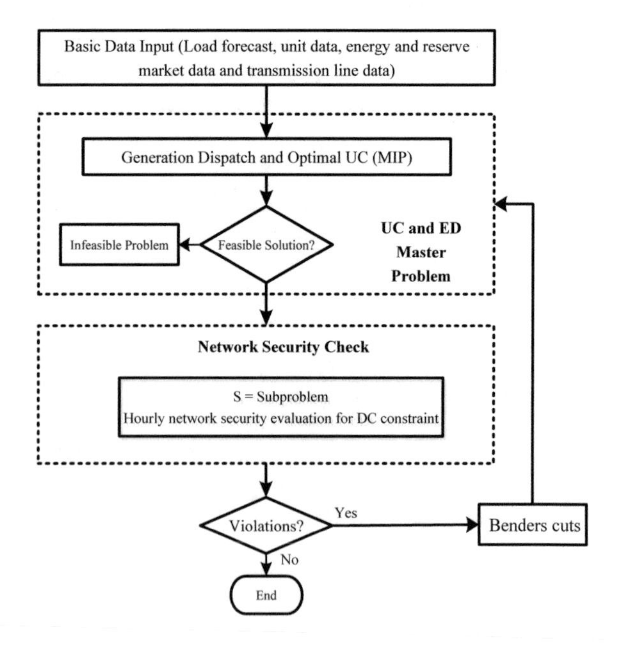

Fig. 1 SCUC algorithm

master problem in the consequently iteration of SCUC. The iteration among the SCUC and network constraints will remain till no more development in hourly price is succeeded. Moreover Benders cuts are created established on subproblem results.

$$w(\hat{U},\hat{P}_g) = \hat{w} + \sum_{g=1}^{N_G} \pi^n_{g,t}(P_{g,t}U_{g,t} - \hat{P}_{g,t}\hat{U}_{g,t}) \tag{18}$$

In Eq. (18), where $\hat{w}$, $\pi^n_{g,t}$, $\hat{P}_{g,t}$ and $\hat{U}_{g,t}$ represents the optimal solution for the master problem, simplex vector, vector output those are obtained in commitment state and real power generation of unit g at time t of the master problem.

4 Simulation Results

In sort to edge a main issue, IEEE-30 bus systems are adapted to solve in this paper. In view of two zones take part with A and B. Seven generator at Zone A, although remaining two generators in Zone B, providing detailed data in website [21–23]. In the section, the test cases are investigated to shows its efficiency: (I) Base case with/without network security, energy and reserve markets. (II) Impact of SCUC with energy and reserve markets on branch contingency. (III) Impact of LMP on energy and reserve markets.

Case 1 Energy and reserve market in SCUC

In Case 1, UC is solve by MIP problem. The SCUC algorithm solves simply the master problem. Successively the optimization method with every thermal unit as of Eqs. (1–15) is fulfilled. Apart from network security, energy and reserve market the problem of UC prove the commitment status and dispatch of units precise in Tables 1 and 3 (i.e., Method S1), respectively. Extra difference is experimental in commitment outcome and power flow shall be analyzed in after that. Units G1, G5, G8, and G15 are more inexpensive unit although they are committed for entire period (Tables 1 and 2). Units G13, G24, and G30 are does not dispatched all during the whole period, though G2 be dispatch 40% of hour, thus a large amount of costly and cheap units to position the generator OFF somewhere additional unit supply the demand Tables 2 and 3 (i.e. Method S2). Whole production cost without energy and AS markets is \$92,237.432. When there is energy and AS markets, the additional cheap units would not offer the demand at peak hour (hour 22) and valuable unit G2 is dispatch. As the entire operational cost increase to \$101,783.342 that its \$9545.91 (9.37%). Table 3 illustrates the dispatch result without and with reserve and energy markets. When a reserve and energy market is calculated, the valuable units G11 should to be dispatch in control to influence the load at peak hour. The extra commitment shall be essential in the case of cheaper generating assets cannot be completely utilized, due to the energy and reserve markets. The dispatch result of unit G2 in Table 3, say method S1 considering without energy and reserve markets committed from 2 to 9 h and from 14 to 24 h. With energy and reserve markets the unit is committed say method S2 from 2 to 9 h and from 16 to 21 h. At hour 9, G2 is generating 25.0 MW that is within energy and reserve markets. Since G1 is highest power thermal unit (100 MW), G8 supplies be dispatched to constancy the reserve construction and operation. G15 is least power thermal unit (10 MW), since energy required decreasing the generation stage of G1 to balance the phase of energy, since reserve and energy region of G1, fairly make 100 MW. Although G1 is a cheap unit, except it would not be use cost-effectively because of energy and AS markets.

Case 2 SCUC effect on branch contingency

In this case, the implementation of an OPF model which includes the contingency that makes the system $N - 1$ secure. The resulting implementation is referred as a DCOPF [15–17]. Table 4 shows the very large critical contingency rank select on outage of line 23–24 to show the particular effects on the thermal values and the operational price to secure the line network. Zone B, units G13 was require taking on for the period of the absent of the line since the remaining line plus peak the exacting generator G15 is unable to distribute the area without violations. Besides at 19 and 8 h, all units are committed. This contingency which proves to be incredibly costly and the operational costs of energy and reserve markets turn to be $124,751.169.

Case 3 Impact of LMP on energy and reserve markets

Figure 2, indicates the day-ahead LMP at bus 2. Once congestion happens on branch 5 through peak hours, cheap generator might be additional schedule and dispatch in event to fulfill the demand at bus 2. The extra commitment will be compulsory in these sections subsequently simple generating units cannot be entirely operated, in line for the congestion of bus 2, energy and reserve markets region of economical generator, to fulfill the demand on bus 2. It can be observed that considering energy and reserve markets increases congestion cost and LMP significantly.

Table 1 SCUC without energy and reserve market

Total costs = $92,237.432		
Zone	Bus no	Hours (1–24)
A	30	0 0 0 0 0 0 0 0 0 0 0 0 0 0 0 0 0 0 0 0 0 0 0 0
	24	0 0 0 0 0 0 0 0 0 0 0 0 0 0 0 0 0 0 0 0 0 0 0 0
	11	0 0 0 0 0 0 0 0 0 0 0 0 0 0 0 0 0 0 0 0 0 0 0 0
	2	1 0 1 1 1 1 1 1 1 0 0 0 0 1 1 1 1 1 1 1 1 1 1 1
	8	1 1 1 1 1 1 1 1 1 1 1 1 1 1 1 1 1 1 1 1 1 1 11
	5	1 1 1 1 1 1 1 1 1 1 1 1 1 1 1 1 1 1 1 1 1 1 1 1
	1	1 1 1 1 1 1 1 1 1 1 1 1 1 1 1 1 1 1 1 1 1 1 1 1
B	13	0 0 0 0 0 0 0 0 0 0 0 0 0 0 0 0 0 0 0 0 0 0 0 0
	15	1 1 1 1 1 1 1 1 1 1 1 1 1 1 1 1 1 1 1 1 1 1 1 1

Table 2 SCUC with energy and reserve market

Total costs = $101,783.342		
Zone	Bus no	Hours (1–24)
A	30	0 0 0 0 0 0 0 0 0 0 0 0 0 0 0 0 0 0 0 0 0 0 0 0
	24	0 0 0 0 0 0 0 0 0 0 0 0 0 0 0 0 0 0 0 0 0 0 0 0
	11	1 1 1 1 1 1 1 1 1 1 1 1 1 1 1 1 1 1 1 1 1 1 1 1
	2	1 0 1 1 1 1 1 1 1 0 0 0 0 0 0 1 1 1 1 1 1 0 0 0
	8	1 1 1 1 1 1 1 1 1 1 1 1 1 1 1 1 1 1 1 1 1 1 11
	5	1 1 1 1 1 1 1 1 1 1 1 1 1 1 1 1 1 1 1 1 1 1 1 1
	1	1 1 1 1 1 1 1 1 1 1 1 1 1 1 1 1 1 1 1 1 1 1 1 1
B	13	0 0 0 0 0 0 0 0 0 0 0 0 0 0 0 0 0 0 0 0 0 0 00
	15	1 1 1 1 1 1 1 1 1 1 1 1 1 1 1 1 1 1 1 1 1 1 1 1

Table 3 Generation dispatch without/with energy and reserve markets

Hour	Method	A							B	
		30	24	11	2	8	5	1	13	15
1	S1	0.0	0.0	0.0	36.0	25.0	25.0	50.0	0.0	35.0
	S2	0.0	0.0	20.0	40.0	20.0	25.0	50.0	0.0	35.0
2	S1	0.0	0.0	0.0	0.0	46.0	50.0	100.0	0.0	26.0
	S2	0.0	0.0	5.0	0.0	36.0	50.0	100.0	0.0	26.0
3	S1	0.0	0.0	0.0	29.0	50.0	50.0	100.0	0.0	22.0
	S2	0.0	0.0	5.0	36.2	47.0	50.0	100.0	0.0	22.0
4	S1	0.0	0.0	0.0	49.6	50.0	50.0	100.0	0.0	57.0
	S2	0.0	0.0	5.0	67.0	50.0	50.0	100.0	0.0	57.0
5	S1	0.0	0.0	0.0	57.4	50.0	50.0	100.0	0.0	70.0
	S2	0.0	0.0	5.0	73.4	50.0	50.0	100.0	0.0	70.0
6	S1	0.0	0.0	0.0	57.4	50.0	50.0	100.0	0.0	70.0
	S2	0.0	0.0	20.0	72.0	50.0	50.0	100.0	0.0	70.0
7	S1	0.0	0.0	0.0	50.0	50.0	50.0	100.0	0.0	70.0
	S2	0.0	0.0	20.0	46.0	50.0	50.0	100.0	0.0	70.0
8	S1	0.0	0.0	0.0	42.0	50.0	50.0	100.0	0.0	53.0
	S2	0.0	0.0	5.0	13.0	50.0	50.0	100.0	0.0	53.0
9	S1	0.0	0.0	0.0	25.0	50.0	47.0	100.0	0.0	32.0
	S2	0.0	0.0	5.0	10.0	50.0	47.0	100.0	0.0	32.0
10	S1	0.0	0.0	0.0	0.0	50.0	50.0	100.0	0.0	10.0
	S2	0.0	0.0	5.0	0.0	50.0	50.0	100.0	0.0	10.0
11	S1	0.0	0.0	0.0	0.0	50.0	37.0	100.0	0.0	10.0
	S2	0.0	0.0	5.0	0.0	50.0	50.0	100.0	0.0	10.0
12	S1	0.0	0.0	0.0	0.0	50.0	50.0	100.0	0.0	10.0
	S2	0.0	0.0	5.0	0.0	50.0	50.0	100.0	0.0	10.0

(continued)

Table 3 (continued)

Hour	Method	A							B	
		30	24	11	2	8	5	1	13	15
13	S1	0.0	0.0	0.0	0.0	50.0	50.0	100.0	0.0	10.0
	S2	0.0	0.0	5.0	0.0	50.0	50.0	100.0	0.0	10.0
14	S1	0.0	0.0	0.0	0.0	50.0	50.0	100.0	0.0	25.0
	S2	0.0	0.0	5.0	10.0	50.0	50.0	100.0	0.0	25.0
15	S1	0.0	0.0	0.0	0.0	50.0	50.0	100.0	0.0	48.0
	S2	0.0	0.0	5.0	10.0	50.0	50.0	100.0	0.0	48.0
16	S1	0.0	0.0	0.0	27.4	48.0	50.0	100.0	0.0	70.0
	S2	0.0	0.0	5.0	32.0	42.0	50.0	100.0	0.0	70.0
17	S1	0.0	0.0	0.0	36.5	41.0	50.0	100.0	0.0	70.0
	S2	0.0	0.0	5.0	36.0	36.0	50.0	100.0	0.0	70.0
18	S1	0.0	0.0	0.0	47.2	36.2	50.0	100.0	0.0	70.0
	S2	0.0	0.0	5.0	41.0	25.0	50.0	100.0	0.0	70.0
19	S1	0.0	0.0	0.0	50.0	15.0	50.0	100.0	0.0	70.0
	S2	0.0	0.0	11.0	36.0	10.0	50.0	100.0	0.0	70.0
20	S1	0.0	0.0	0.0	36.0	10.0	50.0	100.0	0.0	65.0
	S2	0.0	0.0	5.0	15.0	10.0	50.0	100.0	0.0	65.0
21	S1	0.0	0.0	0.0	22.0	10.0	50.0	100.0	0.0	44.0
	S2	0.0	0.0	5.0	10.0	10.0	50.0	100.0	0.0	44.0
22	S1	0.0	0.0	0.0	10.0	10.0	50.0	100.0	0.0	22.0
	S2	0.0	0.0	5.0	0.0	10.0	50.0	100.0	0.0	22.0
23	S1	0.0	0.0	0.0	10.0	10.0	41.0	100.0	0.0	10.0
	S2	0.0	0.0	5.0	0.0	10.0	41.0	100.0	0.0	10.0
24	S1	0.0	0.0	0.0	10.0	10.0	16.0	95.0	0.0	10.0
	S2	0.0	0.0	0.0	0.0	10.0	16.0	95.0	0.0	10.0

S1 = without energy and AS markets; S2 = with energy and AS markets

Table 4 Outage of line 23–24 on SCUC

Total costs = $124,751.169		
Zone	Bus no	Hours (1–24)
A	30	0 0 0 0 0 0 0 0 0 0 0 0 0 0 0 0 0 0 0 1 1 0 0 0
	24	0 0 0 0 0 0 0 0 0 0 0 0 0 0 0 0 0 0 0 1 1 1 0 0
	11	1 0 0 0 0 1 1 1 1 1 1 1 1 1 1 1 1 1 1 1 1 1 1 1
	2	1 0 1 1 1 1 1 1 1 1 1 1 1 1 1 1 1 1 1 1 1 1 1 1
	8	1 1 1 1 1 1 1 1 1 1 1 1 1 1 1 1 1 1 1 1 1 1 11
	5	1 1 1 1 1 1 1 1 1 1 1 1 1 1 1 1 1 1 1 1 1 1 1 1
	1	1 1 1 1 1 1 1 1 1 1 1 1 1 1 1 1 1 1 1 1 1 1 1 1
B	13	1 1 1 1 1 1 1 1 1 1 1 1 1 1 1 1 1 1 1 1 1 1 11
	15	1 1 1 1 1 1 1 1 1 1 1 1 1 1 1 1 1 1 1 1 1 1 1 1

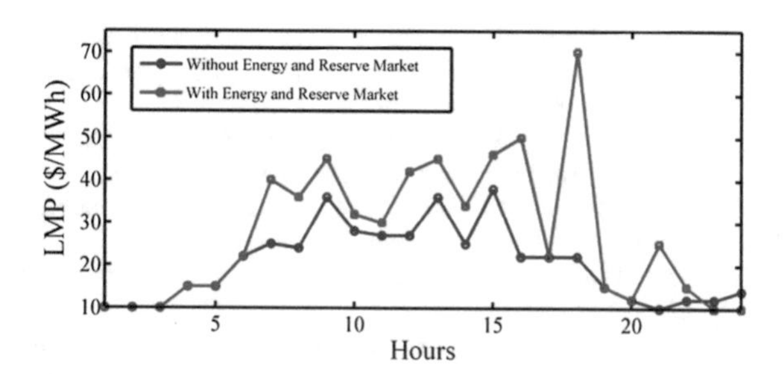

Fig. 2 LMP on Bus 2 without and with energy and reserve markets

5 Conclusions and Future Work

This paper proposed for joint optimization of energy and reserve markets excluding DC security using MIP method. Suggested method is implemented on modified IEEE-30 and explained for UC and SCUC. A significant improvement in the fast computational requirement can be applied by using the CPLEX solver in GAMS. Moreover a variation of proposed is presented to obtain better optimal solution. The BD is used to explain SCUC problem in case of any violations of network constraint, and owing to its capability to find the global or close to global optima solution. The proposed method is more realistic for practical power systems due to incorporation of DCOPF in SCUC solution. In future, the comprehensive proposed work is used to investigate AC-based SCUC problems by integrating renewable generation considering uncertainty of renewable and load demand.

References

1. Ma, X., Sun, D.I.: Energy and ancillary service dispatch in a competitive pool. To appear in IEEEPES Power Engineering Letters (1998)
2. Wu, Z., Zeng, P., Zhang, X.P., Zhou, Q.: A solution to the chance on strained two-stage stochastic program for unit commitment with wind energy integration. IEEE Trans. Power Syst. **31**(6), 4185–4196 (2016)
3. Billinton, R., Allan, R.N.: Reliability Evaluation of Power Systems, 2nd edn. Plenum Press, USA (1996)
4. Wood, A.J., Wollenberg, B.F.: Power Generation, Operation and Control. Wiley, New York (1996)
5. Hejazi, H.A., Mohabati, H.R., Hosseinian, S.H., Abedi, M.: Differential evolution algorithm for security-constrained energy and reserve optimization considering credible contingencies. IEEE Trans Power System, **26**, 1145–1155 (2011)
6. Singh, H., Papalexopoulos, A.: Competitive procurement of ancillary services by an independent system operator. IEEE Trans. Power Syst. **14**, 498–504 (1999)
7. Li, Z., Shahidehpour, M.: Security-constrained unit commitment for simultaneous clearing of energy and ancillary services markets. IEEE Trans. Power Syst. **20**(2) (2005)
8. Soleymani, S., Ranjbar, A.M., Shirani, A.R.: New approach for strategic bidding of GENCOs in energy and spinning reserve markets. Energy Convers. Manage. **48**, 2044–2052 (2007)

9. Nasr Azadani, E., Hosseinian, S.H., Moradzadeh, B.: Generation and reserve dispatch in a competitive market using constrained particle swarm optimization. Int. J. Electr. Power Energy Syst. **32**, 79–86 (2010)
10. Fu, Y., Li, Z., Wu, L.: Modeling and solution of the large-scale security constrained unit commitment. IEEE Trans. Power Syst. **28**(4), 3524–3533 (2013)
11. Shahidehpour, M., Marwali, M.: Maintenance Scheduling in Restructured Power Systems. Kluwer Academic Publishers, London (2000)
12. Shahidehpour, M., Yamin, H., Li, Z.: Market Operations in Electric Power Systems. Wiley, Chichester (2002)
13. Amjady, N., Dehghan, S., Attarha, A., Conejo, A.J.: Adaptive robust network-constrained AC unit commitment. IEEE Trans. Power Syst. **32**(1) (2017)
14. Ye, H., Ge, Y., Shahidehpour, M.: Uncertainty marginal price, transmission reserve, and day-ahead market clearing with robust unit commitment. IEEE Trans. Power Syst. **31**(2) (2016)
15. Shafie-khah, M., Moghaddam, M.P., Sheikh-El-Eslami, M.K.: Unified solution of a non-convex SCUC problem using combination of modified branch-and-bound method with quadratic programming. Energy Convers. Manage. 3425–3432 (2011)
16. Nikoobakht, A., Mardaneh, M., Aghaei, J., Guerrero-Mestre, V., Contreras, J.: Flexible power system operation accommodating uncertain wind power generation using transmission topology control: an improved AC SCUC model. IET GTD, pp. 142–153 (2017)
17. Gupta, P.P., Jain, P., Sharma, S., Bhakar, R.: Reliability-security constrained unit commitment based on BD and mixed integer nonlinear programming. In: International Conference on Computer, Communication, and Electronics (Comptelix) (2017)
18. Hedman, K.W., O'Neill, R.P., Fisher, E.B., Oren, S.S.: Optimal transmission switching with contingency analysis. IEEE Trans. Power Syst. **24**, 1577–1586 (2009)
19. Liu, C., Shahidehpour, M., Wu, L.: Extended benders decomposition for two-stage SCUC. IEEE Trans. Power Syst. **25**(2), 1192–1194 (2010)
20. Ding, T., Bo, R., Yang, Y., Blaabjerg, F.: Impact of negative reactance on definiteness of B-matrix and feasibility of DC power flow. IEEE Trans. Power Syst. 1949–3053 (2017)
21. Khanabadi, M., Fu, Y., Gong, L.: A fully parallel stochastic multi-area power system operation considering large-scale wind power integration. IEEE Trans. Sustain. Energy **9**(1) (2018)
22. The GAMS Software Website: [Online]. Available: http://www.gams.com/dd/docs/solvers/cplex.pdf (2016)
23. http://www.cesr.tntech.edu/PaperSupplements/SolvingSecurityConstrainedUnitCommietment/LineLoadData.xls

In-Depth Analysis of Charge Leakage Through Vegetation in Transmission and Distribution Lines

Hari Shankar Jain, Swati Devabhaktuni and T. Sairama

Abstract For power transmission and distribution lines, the impact of plant and vegetation has been simulated by researchers and reported for redistribution of field stresses and leakage of charge. Both plants and vegetation (creepers) are simulated as grounded electrodes (equivalent to a metallic surface) in the reported literature. However, on critical review of the physical structure of these, one notices that considering them as grounded objects is erroneous and an approximation to simplify the simulation. Reported simulation also suggests their presence as supplementing reactive kVAr on line and better voltage profile during service. In this work, the vegetation (plants, creepers, etc.) is considered as dielectrics, with partial conductivity for charge, dependent on their type, and the age.

Keywords FEM · Finite element method · Charge leakage · Capacitive loading Plant and vegetation · Transmission · Distribution

1 Introduction

Presence of plant and vegetation along power transmission and distribution lines is common to tropical countries [1]. The T&D lines in these countries present a typical site, where the creepers happily extend themselves up to live wires, and encircle it. Interestingly, these creepers achieve a critical balance and it is found that initial portion of the creepers dries up to improve insulation and support rest of the creeper body, analogous to a permanent high impedance fault [2–4].

Not only creepers, one can notice plants of various sizes, heights, and ages along the line. Their presence not only shortens the design clearances but also modifies the line parameters like ground capacitance, impacting the voltage regulation and stability of the system [5].

H. S. Jain · S. Devabhaktuni (✉) · T. Sairama
Department of Electrical and Electronics Engineering,
Vardhaman College of Engineering, Hyderabad 501218, India
e-mail: swatikjm@gmail.com

© Springer Nature Singapore Pte Ltd. 2019

H. Malik et al. (eds.), *Applications of Artificial Intelligence Techniques in Engineering*, Advances in Intelligent Systems and Computing 697,
https://doi.org/10.1007/978-981-13-1822-1_3

Addition of virtual shunt capacitance through such growth while visible and beneficial is a reliability threat and need to be mitigated in time through periodic line maintenance [3].

The simulation is carried out using FEM model for an 11 kV, 3-φ distribution feeder, PCC poles, metallic cross arms, and 160 kN pin insulators. A pole span of 100 m is considered for subject simulation for four different configurations [1].

2 Problem Statement

A plant consists of stem and the branches, the maximum spread of branches and leaves form the surface area for the virtual ground in reported simulations. Charges transported through this system (the plant) necessarily travel through the stem and shall return to system through roots and earth. This conductor is recognized as a system with three conducting constituents, each with different moisture dependent conductivity.

The assumptions considered in this study include the following:

1. The plant is exogenous (hard core, soft sap outside),
2. Length of branch and roots is identical to that of the stem,
3. The cumulative cross section of branches is equal to the cross section of the stem, and
4. The roots are in harmony with ground, and represent ground.

The complete system together is considered as a series RC element connected across the line. The simulation considers number of elements in proportion to the plant's linear density. Only plants are considered here, creepers are planned to be reported next in part-2 of this work.

As plants grow older, the formation of annual rings is natural. The core thickness for a young plant is smaller compared to the sap; however, the sap thickness over a period saturates and remains constant for rest of the life of the plant. The two parts of the plant provide different conductivities as the core is considered to be with minimum or no moisture. The sap on other hand is always moist in a living plant and is the medium providing charge conductivity.

The bark gets dried identical to the root due to exposure to atmosphere, and thus conduction through the bark and core is neglected or considered as a function of plant age.

The conductivity (resistance) of the sap is calculated for the simulation using the following equation:

$$R_s = \rho \frac{l}{a} \quad \text{Ohms} \tag{1}$$

where Rs is the resistance offered by plant sap, ρ is the specific resistance of the sap, l is the length of the stem, and a is the annular area of the sap.

The conductivity (resistance) of the core is similarly calculated for the simulation using the following equation:

$$R_c = 0.9\rho\frac{l}{a} \quad \text{Ohms} \tag{2}$$

where Rc is the resistance of the core, ρ is the specific resistance of the core, l is the length of the core (=stem), and a is the area of the core.

The factor 0.9 is used here to compensate for the bark (considered equally hard as the core).

Both of these elements are considered in parallel to calculate combined resistance of the plant, as

$$R_p = \frac{1}{R_s} + \frac{1}{R_c} \quad \text{Ohms} \tag{3}$$

The line capacitance per unit length is related to the gap (d) between line conductor and the ground and the conductor area (A) as

$$C = k\frac{A}{d} \text{ Farads} \tag{4}$$

Any reduction in distance d causes an increase in line capacitance, and hence higher charge leakage is eminent. Correction (reduction) of this leakage with addition of a series impedance is simulated using the above methodology.

The distribution system chosen is a rural section fed by a 33/11 kV Narkuda substation in Shamshabad Mandal of Telangana State, India. The section consists of cement poles and cross arms, Fig. 1a, b, respectively.

The 11 kV distribution line construction strand is shown in Fig. 1b.

Figure 1a, b is reconstructed in FEM for the simulation purpose and electrical stresses are simulated under the following two conditions:

1. A clear distribution line with single growing tree, below the line, and
2. A clear distribution line with single dead and aging tree, simulation with the variation $\epsilon_{r.}$

Variation of the voltage with respect to the height of the tree for a prefixed/identified location is studied along with changes in electrical stress (average), in the medium.

The analysis has been carried out using Finite Element Method (FEMM-2.04) software. The variable density mesh has been used for simulation with optimal number of nodes (11,320). Simulation time accordingly is optimized.

Care has been taken to study the effect of only the tree and the conductor, and thus the location for the tree has been selected at a distance away from poles. The properties assigned are considered to be affected by the moisture content of the plant sap. As the plant ages, the moisture content in sap is assumed to be decline. Accordingly the permittivity of the material (ϵ_r) is modified in simulation.

(a)

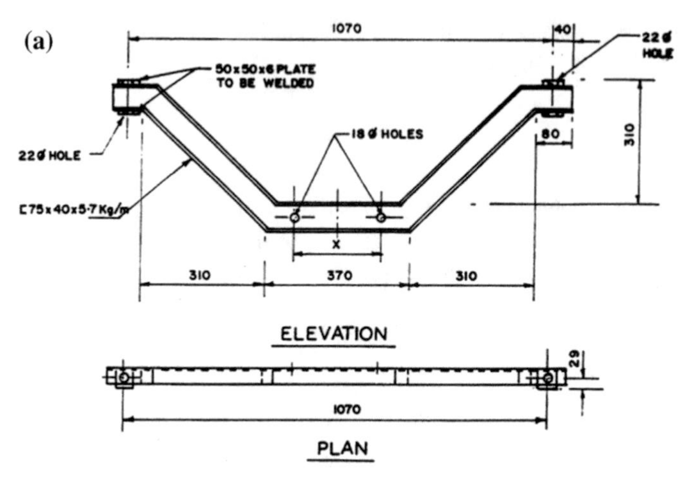

(b)

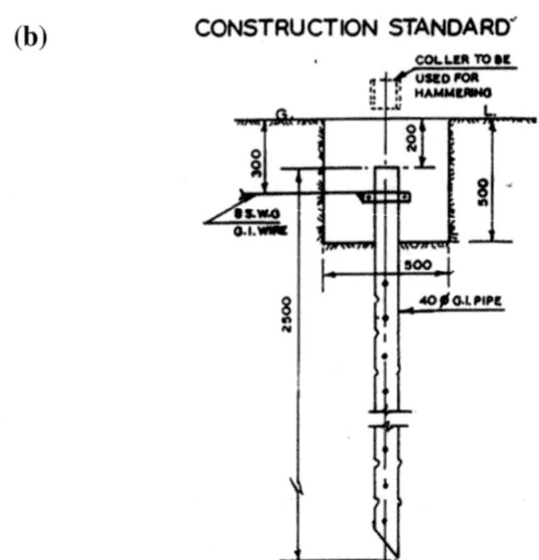

Fig. 1 **a** 11 kV distribution line insulator design. **b** 11 kV distribution construction strand design

3 Results and Discussions

In this section, the results were presented for the above-mentioned problems.

3.1 Electrical Stress Variation with and Without Tree

The proposed model for the 11 kV distribution line shown in Fig. 1a, b has been modeled in FEM and the layout for the smooth distribution assuming that there may

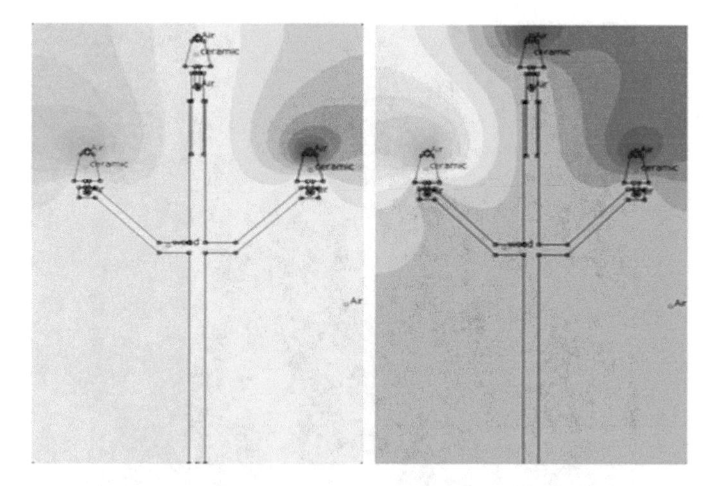

Fig. 2 Variation of the electric stress around the conductor with the change in the conductor voltage

be no tree/creeper surrounding the pole. The variation of the electric stress on the conductor in FEM is shown in Fig. 2.

Figure 2 shows that the electric stress is distributed as per the variations of the voltage on the conductor and also shows that the stress distribution is undisturbed.

After analyzing the variation of the electric stress as mentioned above, the author now introduced a tree below the conductor and observed the variation of the electric stress as shown in Fig. 3. As the height of the tree is increasing, the stress between the conductor and the tree is also increasing.

The variation of the stress with the increase in the height of the tree is shown in Fig. 4.

With the increase in the height of the tree, it has been observed that the electrical stress on the conductor of the line is increasing leading to the increase in the value of the capacitance and thereby increase in the charge leakage current.

3.2 *Electrical Stress Variation with the Variation of Tree Characteristics*

The characteristics of the tree as per the UFEI (Urban Forest Ecosystem Institute) are broadly classified as: Growth Rate and Tree Shape.

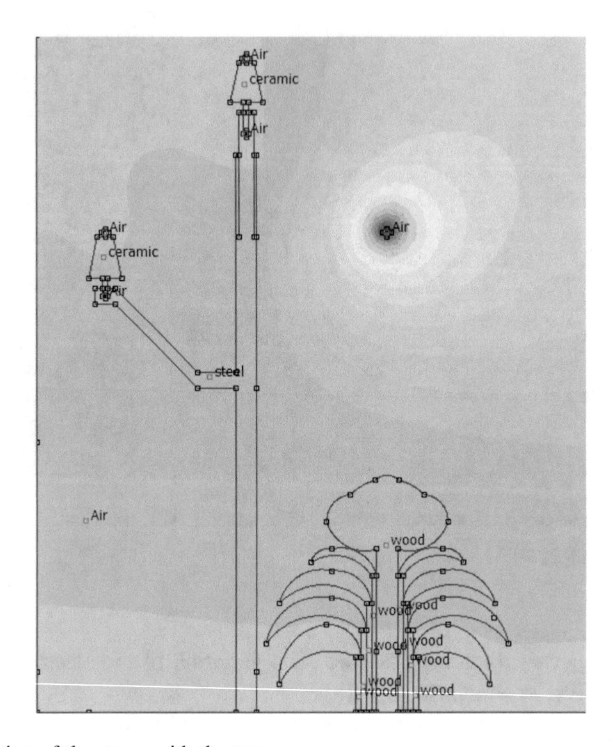

Fig. 3 Variation of the stress with the tree

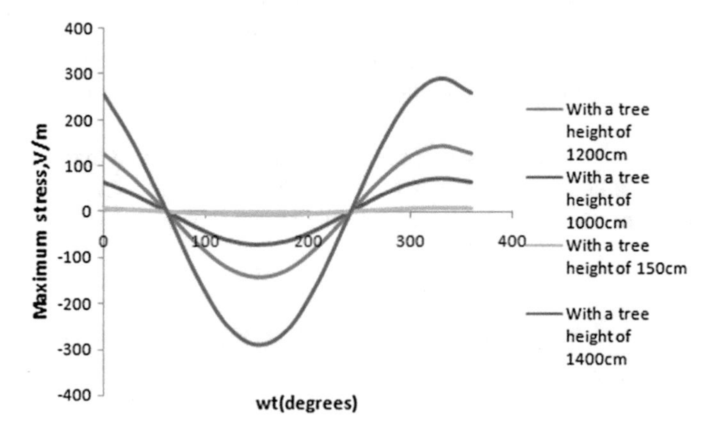

Fig. 4 Variation of the stress with the increase in the height of the tree

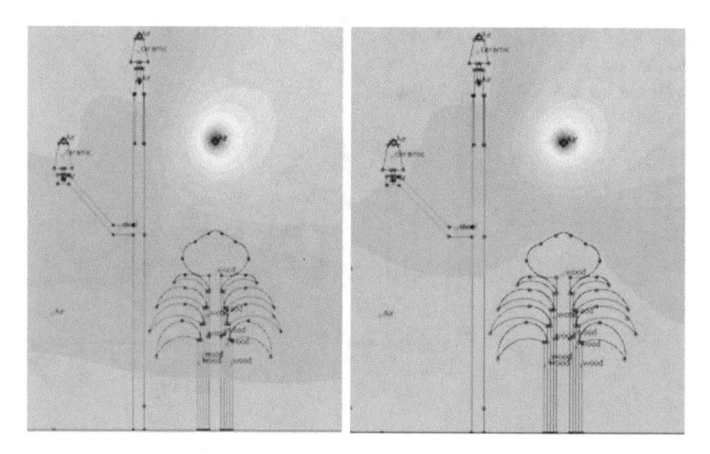

Fig. 5 FEM analysis of the system with tree height = 12 m. **a** With $\epsilon_r = 5$. **b** With $\epsilon_r = 20$

- With the increase in the growth rate of the tree, the moisture content in the tree is also varying. The moisture content in the tree represents the relative permittivity (ϵ_r) of the tree. At the Juvenile Stage of a tree, the trees have ample moisture, hence ϵ_r of the tree is very low, i.e., the tree have the less resistance. As the life of the tree increases the moisture content in the tree is reducing and hence the ϵ_r is increasing, and the tree is acting like more insulated material.
- This has been analyzed with the FEM by varying the values of the ϵ_r. The FEM model for the $\epsilon_r = 5$ and $\epsilon_r = 80$ are shown in Fig. 5a, b, respectively.

From Fig. 5a, b it can be observed that with the change in the life cycle of a tree the electric stress on the conductor is also increasing.

Figure 6 shows the variation of the electric stress on the conductor with the variation of the ϵ_r. From Fig. 6, it can be concluded that with the change of the life cycle the tree has becoming a conductor to an insulator. As the tree has become an insulator, there the ground has been shifted to the height of the tree which increases the electrical stress on the tree.

Average stress variation on the conductor with the increase in the ϵ_r of the tree as shown in Fig. 7.

The variation of the voltage on the conductor with the increase in the age and height of the tree is as shown in Fig. 8.

From Fig. 8, it can be observed that with the increase in the height of the tree the leakage current in the capacitance of the line has been increased resulting in the current leading the voltage. The increase in the leakage current also increases the transmission line losses. It has been observed that the effective capacitance is varying between 1.75 and 3.04 μF as the height of the tree is increasing.

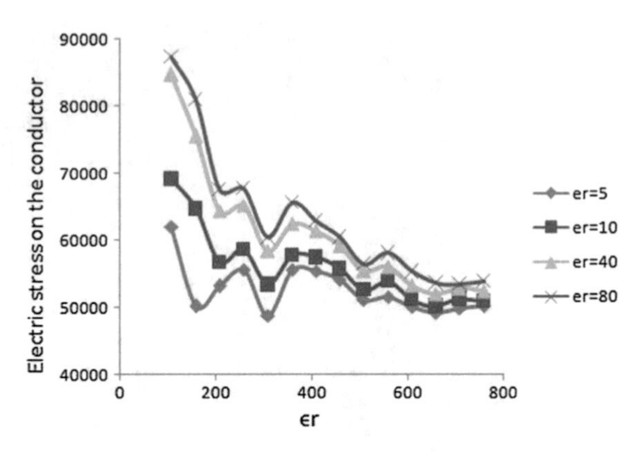

Fig. 6 Variation of the electric stress on the conductor with the increase of the age of the tree

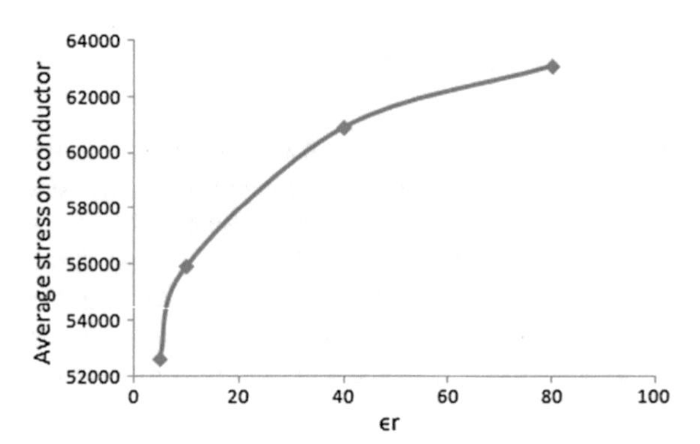

Fig. 7 Variation of the average stress ϵr

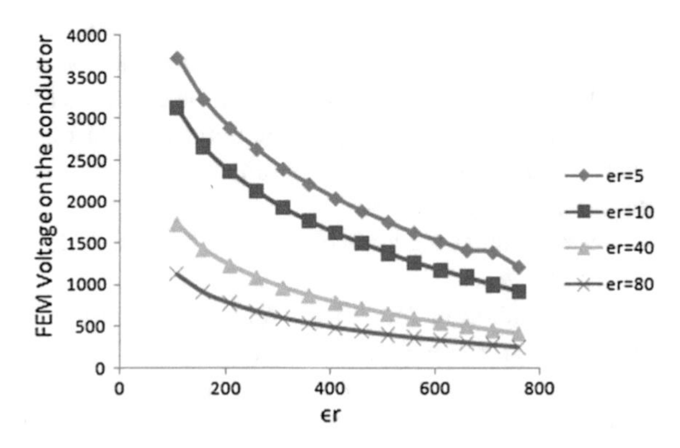

Fig. 8 Variation of the voltage with respect to height and age of the tree

4 Conclusions

This study, first of its kind, clearly brings out fetal potential prevailing on plants growing in the vicinity of T&D lines. It also brings out the fact that it is unsafe to approach such vegetation by the linesman using metal objects like ladder, etc.

The second important information which this study confirms is the fact that the leakage load on the system gets multiplied with the closing gap between conductor and the vegetation. The study also suggests a marginal improvement in line capacitance which would supplement system power factor in isolated cases.

Acknowledgements The authors are thankful to the management and staff of Vardhaman College of Engineering for their whole-hearted support in the laboratory as well as the field work for reported work. The dedicated efforts made for successfully completing this study are thankfully acknowledged.

References

1. Devabhaktuni, S., Jain, H.S., Pramod Kumar, V.: Impact of shrubbery in rural transmission and distribution networks. In: IEEE-ICCIC, 978-1-5090-6621-6/17/$31.00 ©2017 IEEE
2. Devabhaktuni, S., Jain, H.S., Kiran Kumar, P.: A study on RC element based current limiter reactor in AC power systems, pp. 3172–3176. In: IEEE-ICEEOT-2016, 978-1-4673-9939-5/16/$31.00 ©2016 IEEE
3. Devabhaktuni, S., Jain, H.S., Kiran Kumar, P.: A comparative study of CLR and CLCR as Current Limiters in AC Power Systems. IEEE-ICCPEIC-2016, 978-1-5090-0901-5/16/$31.00 ©2016 IEEE
4. Devabhaktuni, S., Jain, H.S., Kiran Kumar, P.: Optimum selection of inductor as a current limiter in AC power systems. In: IEEE-TENSYMP-2016, IEEE proceedings, P. No. 88, IEEE explore, pp. 416–419
5. Devabhaktuni, S., Jain, H.S., Pramod Kumar, V.: Statistically generalized rating for CLR in power transmission systems. In: IEEE-PIICON-2016, Bikaner, India

Comparative Study of Different Neural Networks for 1-Year Ahead Load Forecasting

Hasmat Malik and Manisha Singh

Abstract Load forecasting is a technique used to forecast the amount of energy needed to meet the demand of load side. Load forecasting comes under energy forecasting, which also include generation forecasting, price forecasting, demand response forecasting, and so on. Demand on load side affects from different parameters which cannot be controlled like weather, temperature, humidity, etc, so in order to fulfill the demand of load side, some measures should be considered in advance to prevent the consequences of sudden increment of load, outages, etc, Here, in this paper 1-year ahead load is predicted from previous years load. The techniques most popularly used in today's century are Artificial Neural Network (ANN), Artificial Intelligence (AI), Fuzzy Logic, etc. In these techniques neurons are trained with information to perform a certain task and they are trained up to a level that they work intelligently in situation they are meant for. Thus, by using artificial intelligence in this paper, load forecasting will be done.

Keywords ANN (Artificial neural network) · AI (Artificial intelligence) STLF (Short-term load forecasting) · RBE (Exact radial basis) · GRNN (General regression neural network)

1 Introduction

The demand at load side affects from various factors such as sudden peak load, sudden drop in load, etc. thus there is a need to assume the demand at load side at the time of load unbalance situation [9–12]. Here, in this paper various techniques are used to find the most accurate way to assume the demand required.

H. Malik
Electrical Engineering Department, IIT Delhi, Hauz Khas, New Delhi 110016, India
e-mail: hmalik.iitd@gmail.com

M. Singh (✉)
Instrumentation and Control Engineering Department, NSIT, New Delhi 110078, Delhi, India
e-mail: manishaei0292@gmail.com

© Springer Nature Singapore Pte Ltd. 2019

H. Malik et al. (eds.), *Applications of Artificial Intelligence Techniques in Engineering*, Advances in Intelligent Systems and Computing 697,
https://doi.org/10.1007/978-981-13-1822-1_4

Let us consider an example of weather, earlier only the lighting is the demand of generation, but as technology develops, there comes air conditioning which requires more electricity than earlier. The air conditioning is used in summers thus keeping this factor only in mind, the demand at load side will be more in summers as compared to winters. The various cooling systems are also used with the invention of new instruments, thus another factor comes into play which requires more electricity at load side. Thus, if large amount of air conditioning will use at same time like in day time when its too sunny, then the demand at load side with increase at tremendous rate will change. Thus, outage will happen at some other place in order to fulfill demand at this side. This can also happen in some emergency situations also where it will be necessary to provide electricity at such a high rate and thus these all factors should be kept in mind and forecasting or prediction is necessary to tackle this situation.

2 Material and Methodology for Load Forecasting

2.1 Datasets Used

In this paper, data has been taken from GEFcom2012 [1] in which there were four input variables, i.e., year, month, day, hour, temperature, and one-output variable, i.e., load are included. In this paper, training file is created from 2 years data, i.e., (2004–2005) with four input variables in one file and load in other file. Then testing files are created in the same manner of 1 year each, i.e., 2006 year, 2007 year, and 2008 year and load is forecasted 1-year ahead (Table 1).

2.2 Methodology

A comparative study has been done in this paper in which different neural network models are used to obtain the best approximate model. The various methods that can be used are as follows: (a) Regression analysis, (b) Time series analysis, (c) Artificial neural network, (d) Fuzzy regression, and (e) Support vector machine.

Table 1 Dataset used for the study

Dataset	Source	Year	Open source
Data 1	GEFcom2012	2004–2008	https://www.dropbox.com/s/epj9b57eivn79j7/GEFCom2012.zip?dl=0
Data 2	GEFcom2014	2005–2011	https://www.dropbox.com/s/pqenrr2mcvl0hk9/GEFCom2014.zip?dl=0

2.2.1 Generalized Regression Neural Network [2, 3]

Basically GRNN consists of four layers, i.e., input layer, pattern layer, summation layer, and output layer. Let the inputs in the layer are denoted by $X = [X_1, X_2, X_3 \ldots\ldots X_m]^T$ and the outputs to be $Y = [Y_1, Y_2, Y_3 \ldots Y_k]^T$

Input layer neurons consists of 'm' variables on which load demand depends, i.e., 4 and the pattern layers also consists of same no of neurons as of input layer. The transfer function of pattern layer neurons is

$$p_i = \exp\left[-\frac{D_i^2}{2\sigma^2}\right], \quad i = 1, \ldots, n. \tag{1}$$

Here, $D_i^2 = (X - X_i)^T(X - X_i)$

X is the input variables and X_i is the learning sample corresponding to ith neuron, where D_i is the Euclidean distance between X and X_i.

In the summation layer, two computational formulas are used, first one is as follows:

$$\sum_{i=1}^{n} \exp\left[-\frac{D_i^2}{2\sigma^2}\right] = \sum_{i=1}^{n} \exp\left[-\frac{(X - X_i)^T(X - X_i)}{2\sigma^2}\right]. \tag{2}$$

It gives the summation of output neurons in pattern layer and the connecting weights or synapses of each neurons is 1. Hence, the transfer function is

$$S_D = \sum_{i=1}^{n} P_i = \sum_{i=1}^{n} \exp\left[-\frac{D_i^2}{2\sigma^2}\right] \tag{3}$$

Second computational formula used is

$$\sum_{i=1}^{n} Y_i \exp\left[-\frac{(X - X_i)^T(X - X_i)}{2\sigma^2}\right] \tag{4}$$

The transfer function is the summation of all the neurons in pattern layer and summation layer. The ith neuron from pattern layer and jth neuron from summation layer is the jth element from ith output sample Y_i

$$S_{Nj} = \sum_{i=1}^{n} y_{ij} p_i, j = 1, \ldots, k \tag{5}$$

The output neurons consist of k neurons of learning sample. Each neuron should be divided by summation layer output, i.e., S_D. Therefore, the estimated result should be

$$y_i = \frac{S_{Nj}}{S_D}, \quad j = 1, \ldots, k \tag{6}$$

2.2.2 Radial Basis Neural Network [3, 4]

Basically Exact Radial Basis (RBE) consists of three layers, i.e., input layer, hidden layer, and the output layer. Gaussian approximation is used as activation function in hidden neurons. The values for Gaussian functions are the distances between input value x and the midpoint of Gaussian function.

$$g_j(s) = \exp\left(-\frac{\|x - c_j\|^2}{2\sigma_j^2}\right) \tag{7}$$

Here is c_j the midpoint position of jth neuron and σ_j is the spread factor, which is usually taken as 1. The synaptic weights (w_j) are taken as 1 between input layer and hidden layer. The weights are adjusted based on the adaptive rule $(\dot{w}_j)$

$$y = \sum_{j=1}^{n} w_j g_j(s) = \sum_{j=1}^{n} w_j \exp\left(-\frac{\|x - c_j\|^2}{2\sigma_j^2}\right) \tag{8}$$

The adaptive rule can be calculated by minimized with respect to w_j

$$\dot{w}_j = -\eta \frac{\partial E(t)}{\partial w_j(t)} = -\eta \frac{\partial x(t)\dot{x}(t)}{\partial w_j(t)} \tag{9}$$

η is the adaptive rate parameter, where $0 < \eta < 1$.

2.2.3 Multi-layer Perceptron Neural Network [5–8]

There are three layers in MLP, i.e., input layer, hidden layer, and the output layer. The neurons used in input layer are 4 neurons. The below-mentioned approach is used here in this algorithm with Levenberg–Marquadt technique. For more detail of MLP network, reference [5–8] can be referred.

2.2.4 Proposed Approach

The proposed approach for load forecasting problem used in this study is represented in Fig. 1, which includes basically eight steps. Dataset used in this study has been collected from available open source of GEFcom2012 and GEFcom2014.

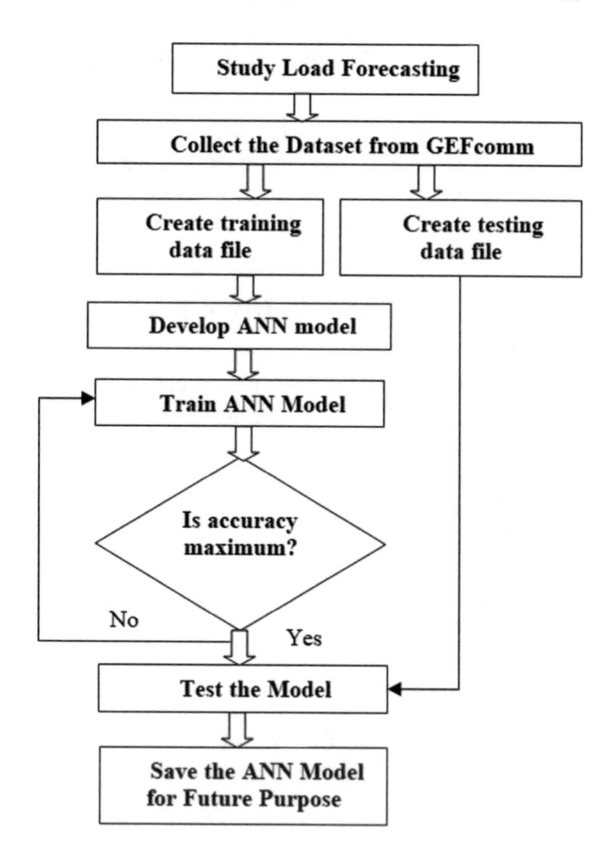

Fig. 1 Proposed approach

By using this dataset, a comparative study has been presented using three different neural networks named MLP, GRNN, and RBF to found out the best suitable network for this problem so that it may be used for future prospective. To forecast 1-year ahead load, dataset is divided into training and testing dataset. Models have been implemented by using training dataset and validation has been done through testing dataset, which is shown in subsequence section.

3 Results and Discussion

In this paper, three models are considered to predict best approximate load. Namely GRNN, RBF, and MLP, out of which RBF model is best. It gives output very near to the given. Error in this method is negligible.

3.1 Load Forecasting Using GRNN

Under this neural network, the regression is obtained is not too good. The training state gives 98.09% whereas testing states give maximum 90%; thus, this method is least preferred. Following are the graphs obtained as yearly load: (Figs. 2, 3, 4, and 5).

3.2 Load Forecasting Using RBF

Radial Basis Function gives approximate values that can be considered as good results. This model of Neural Network gives the training state as 91.10% and testing

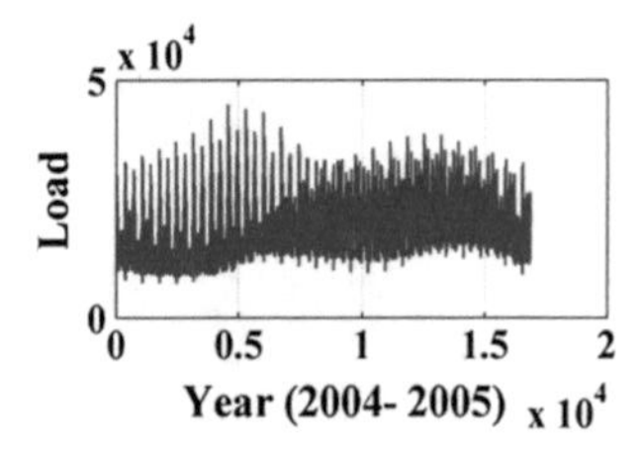

Fig. 2 Training graphs—years versus load

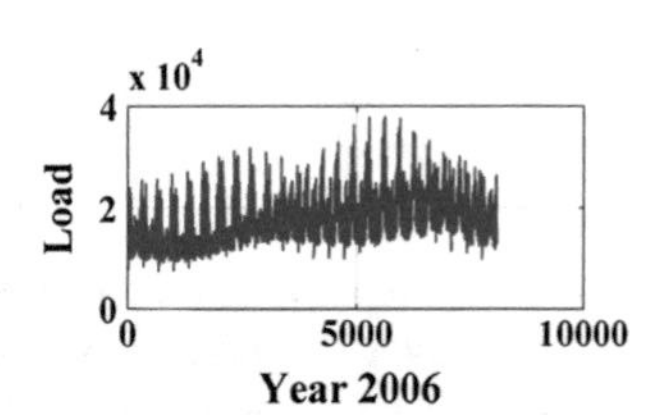

Fig. 3 Testing graph—year 2006 versus load

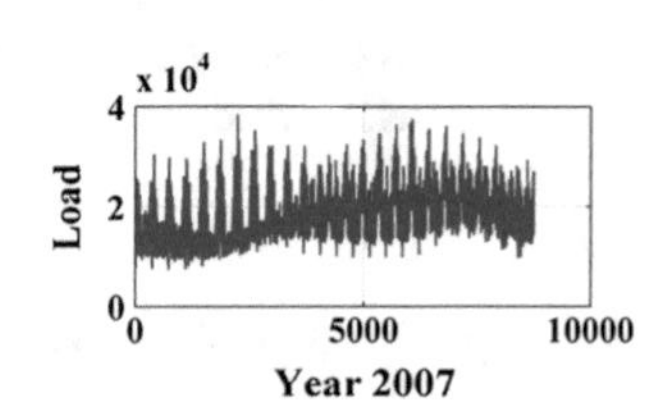

Fig. 4 Testing graph—year 2007 versus load

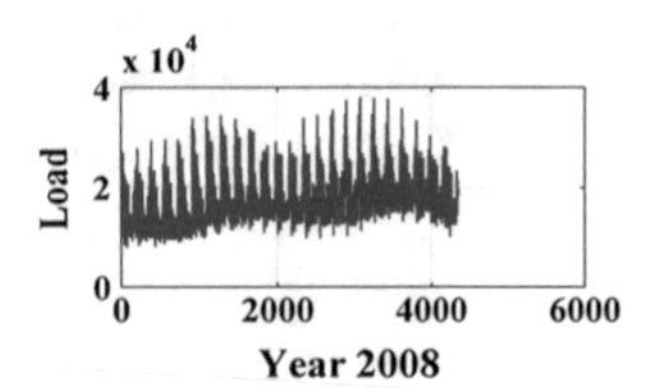

Fig. 5 Testing graph—year 2008 versus load

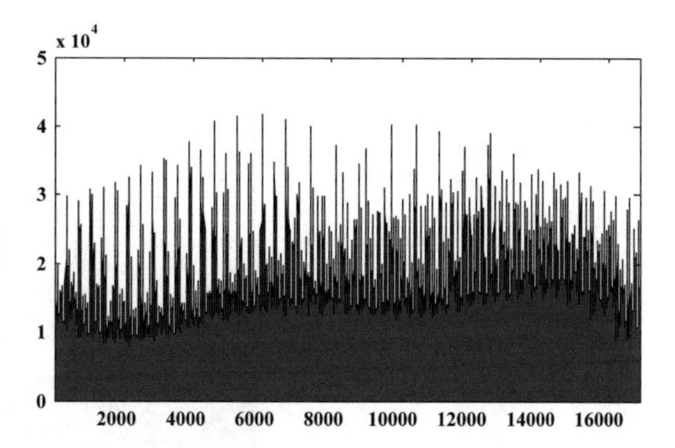

Fig. 6 Training area graph

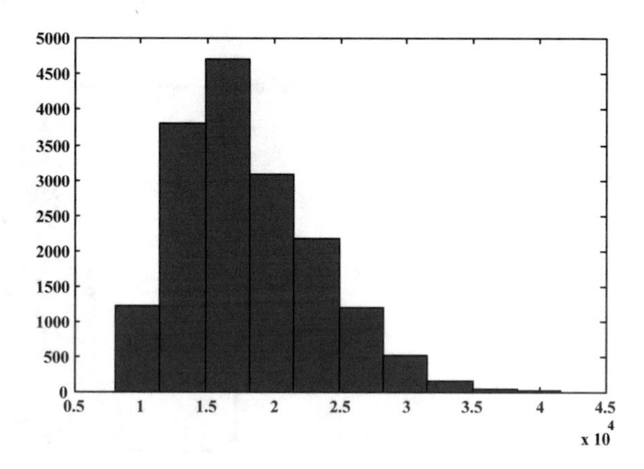

Fig. 7 Training histogram

state as 90.01%. Thus, it can be considered as a best approach in prediction field out of three used in this paper. Following are the results obtained from this model (Figs. 6, 7, 8, and 9).

3.3 Load Forecasting Using MLP

Three types of methods are applied here to obtain best possible method for load forecasting. Multi-layer perceptron gives accuracy of about 88.9%. Training regression graphs are discussed in Figs. 10, 11, 12, 13, 14, and 15.

In MLP, the testing file gives the accuracy 88.9%, which is good in comparison to GRNN. The following graphs are shown in Figs. 16 and 17.

Fig. 8 Testing area graph

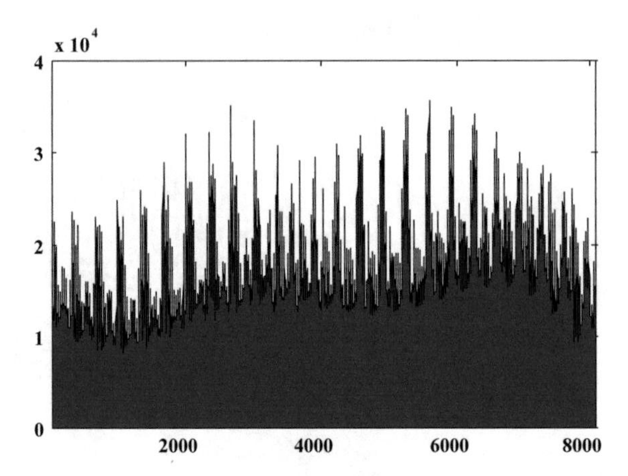

Fig. 9 Testing histogram

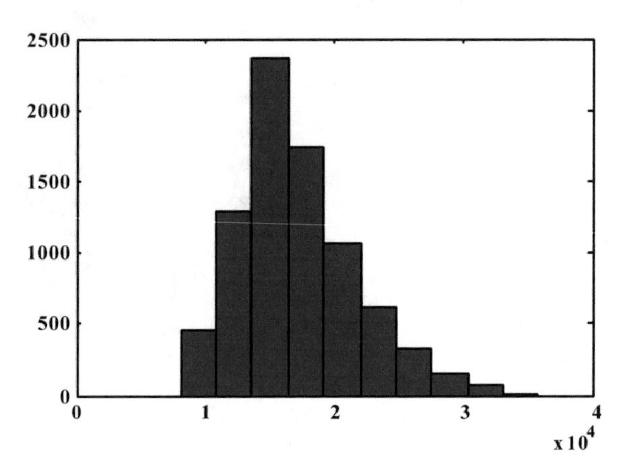

Fig. 10 Training phase: Regression plot for training

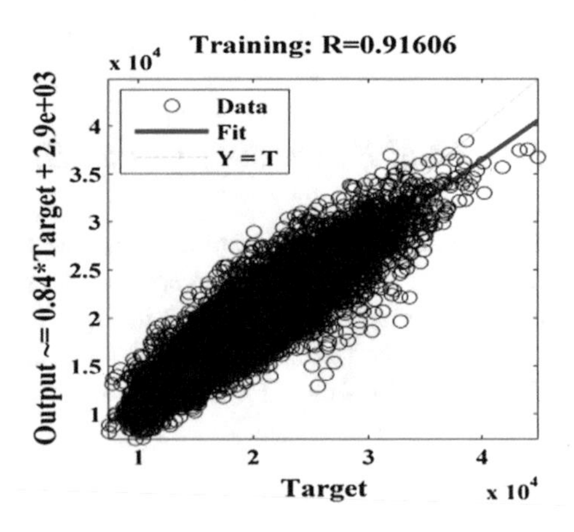

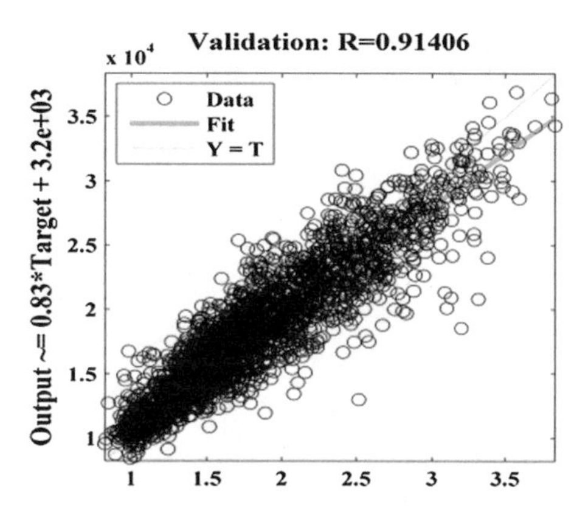

Fig. 11 Training phase: Regression plot for validation

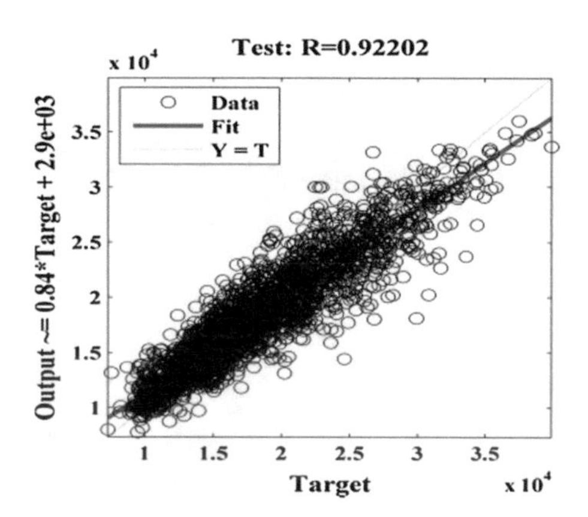

Fig. 12 Training phase: Regression plot for test

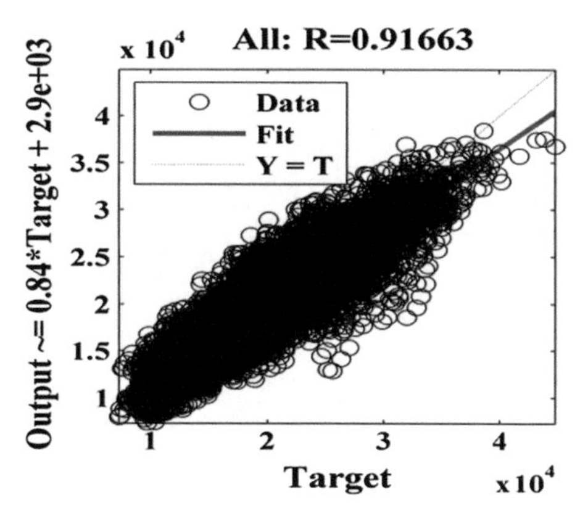

Fig. 13 Training phase: Overall regression plot

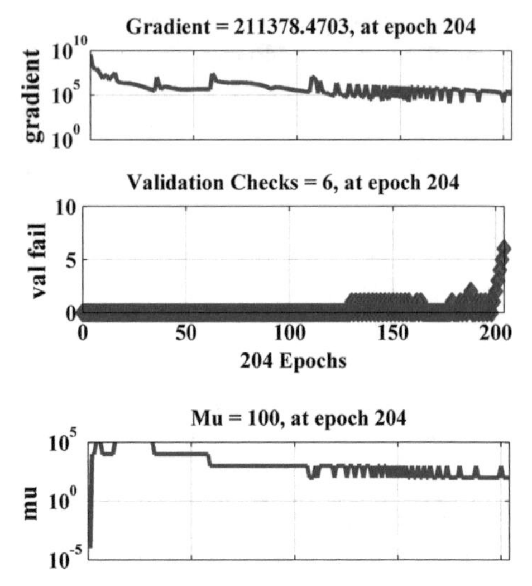

Fig. 14 Training phase: Training state

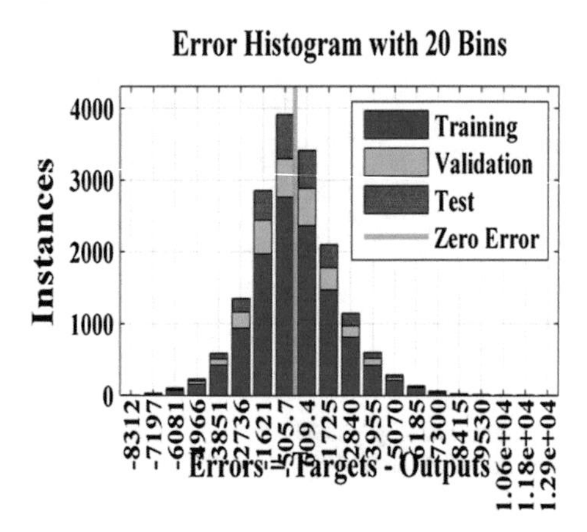

Fig. 15 Training phase:Error Hostogram

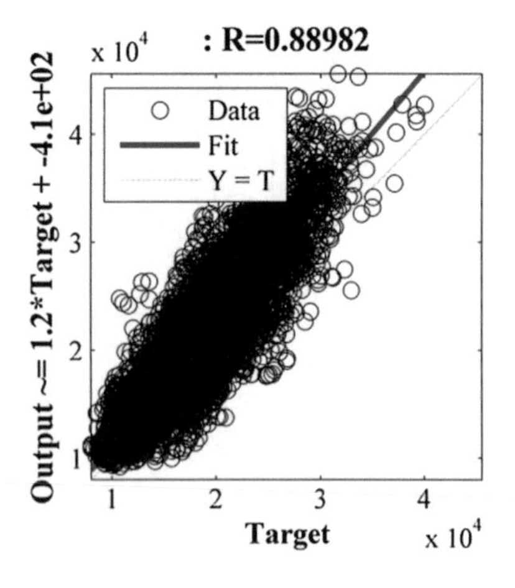

Fig. 16 Testing phase regression graph

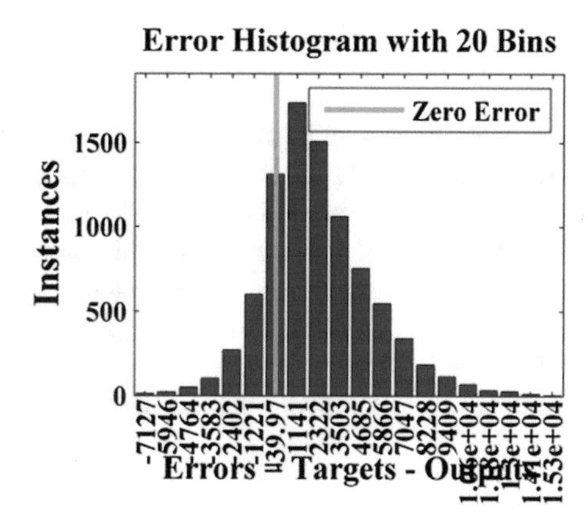

Fig. 17 Testing phase Error histogram

4 Comparative Study

With the development in technologies, load demand rises and to provide the accurate demand is the most needed thing in today's world. To obtain the approximate load demand, prediction plays an important role. Here, in this study, three methods are applied to obtain the best possible demanded load. These are RBF, GRNN, and MLP. Out of the three, the best method for load forecasting is RBF method with training as 91.10% and testing as 90.01% (Fig. 18; Table 2).

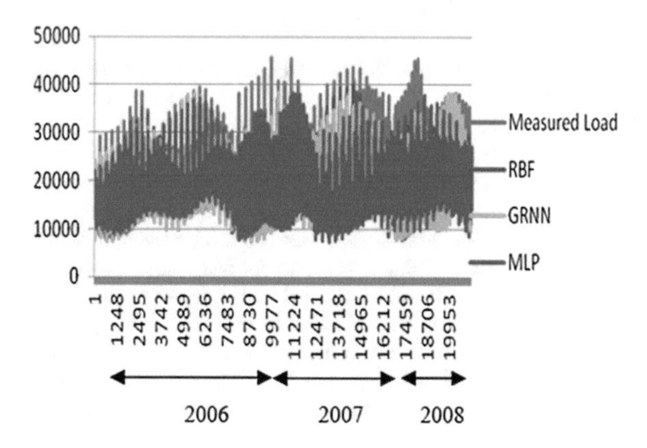

Fig. 18 Years versus load

Table 2 Comparision of various models

S. no.	Technique (%)	Training (%)	Testing (%)
1	GRNN	98	90
2	RBF	91.10	90.01
3	MLP	91.6	88.9

5 Conclusion

Artificial neural network gives great value to statisticians for load forecasting as they produce very accurate results. The only thing needed is that the problem of load forecasting should be modeled carefully with all the factors that may affect the demand at load side. Here in this paper three techniques are discussed, i.e., Multi-Layer Perceptron (MLP), General Regression Neural Network (GRNN), Exact Radial Basis Function (RBF). The best accuracy obtained in RBF after that MLP. So for forecasting purpose, RBF should be used widely. Results showed that artificial neural network can be used for large datasets effectively.

References

1. GEFcom2012 Dataset, https://www.dropbox.com/s/epj9b57eivn79j7/GEFCom2012.zip?dl=0 . Accessed on 26 Jan 2018
2. Savita, Ansari, M.A., Pal, N.S., Malik, H.: Wind speed and power prediction of prominent wind power potential states in India using GRNN. In: Proceedings of IEEE ICPEICES-2016, pp. 1–6 (2016). https://doi.org/10.1109/icpeices.2016.7853220
3. Kumar, G., Malik, H.: Generalized regression neural network based wind speed prediction model for western region of India. Elsevier Procedia Comput. Sci. **93**, 26–32 (2016). https://doi.org/10.1016/j.procs.2016.07.177
4. Yadav, A.K., Malik, H., Chandel, S.S.: Application of rapid miner in ANN based prediction of solar radiation for assessment of solar energy resource potential of 76 sites in Northwestern India. Renew. Sustain. Energy Rev. **52**, 1093–1106 (2015). https://doi.org/10.1016/j.rser.2015.07.156
5. Malik, H., Mishra, S.: Artificial neural network and empirical mode decomposition based imbalance fault diagnosis of wind turbine using TurbSim, FAST and Simulink. IET Renew. Power Gener. **11**(6), 889–902. https://doi.org/10.1049/iet-rpg.2015.0382
6. Malik, H., Sharma, R.: EMD and ANN based intelligent fault diagnosis model for transmission line. J. Intell. Fuzzy Syst. **32**(4), 3043–3050 (2017). https://doi.org/10.3233/JIFS-169247
7. Yadav, A.K., Malik, H., Chandel, S.S.: Selection of most relevant input parameters using WEKA for artificial neural network based solar radiation prediction models. Renew. Sustain. Energy Rev. **31**, 509–519 (2014). https://doi.org/10.1016/j.rser.2013.12.008
8. Saad, S., Malik, H.: Selection of most relevant input parameters using WEKA for artificial neural network based concrete compressive strength prediction model. In: Proceedings of IEEE PIICON-2016, pp. 1–6, 25–27 Nov 2016. https://doi.org/10.1109/poweri.2016.8077368
9. Luo, J., Hong, T., Fang, S.-C.: Benchmarking robustness of load forecasting models under data integrity attacks. Int. J. Forecast. **34**(1), 89–104
10. Hong, T., Fan, S.: Probabilistic electric load forecasting: a tutorial review. Int. Forecast. **32**(3), 914–938
11. Wang, P., Liu, B., Hong, T.: Electric load forecasting with recency effect: a big data approach. Int. J. Forecast. **32**(3), 585–597
12. Hong, T., Pinson, P., Fan, S.: Global energy forecasting competition 2012. Int. J. Forecast. **30**(2), 357–363

Reliability of PMU Using Fuzzy Markov Dynamic Method

Pramod Kumar, Poonam Juneja and Rachna Garg

Abstract The reliability of phase measurement unit (PMU) considering uncertainties in various modules is computed using the fuzzy logic tool. The Markov reliability model for phase measurement unit is implemented and logic gate representation is developed for the same. The theory and configuration of PMU for synchronization detection is also presented for the compilation and operation. PMU is an important device for intelligent SCADA and smart grid operation and management. It is also useful for the detection of islanding, isolation, and protection of microgrid. This study is useful for designer and protection engineer.

Keywords Phase measurement unit(PMU) · SCADA · Fuzzy logic
Markov's reliability model

1 Introduction

Power systems are one of the most complex man-made systems. They are, generally, spread over a wide area. The integration of distributed power generation, nonconventional energy resources, with power system grid has made its operation very cumbersome. This needs high accuracy in data measurement of electrical quantities in magnitude and phase. Further, each measurement needs to be time stamped for traceability. Phase measurement unit, an intelligent electronic device, can meet these requirements. It captures the real-time synchronized measurements in power system, having synchronized accuracy better than one microsecond. It measures the dynamics state picture of the power system, whereas SCADA system

P. Kumar (✉) · P. Juneja
Department of Electrical Engineering, Maharaja Agarsen
Institute of Technology, New Delhi, India
e-mail: pramodk2003@yahoo.co.in

R. Garg
Department of Electrical Engineering, Delhi Technological University,
New Delhi, India

© Springer Nature Singapore Pte Ltd. 2019

H. Malik et al. (eds.), *Applications of Artificial Intelligence Techniques in Engineering*, Advances in Intelligent Systems and Computing 697,
https://doi.org/10.1007/978-981-13-1822-1_5

offers the steady-state picture of power system, and measure the magnitude of electrical quantities only. Its failure may affect the operation of DG (distributed generation)-based grid, and efficiency of the integrated grid. This leads to computing the reliability of PMU as the key characterization. In the present paper, the authors have computed the reliability of PMU using fuzzy-dynamic logic gates.

Nikita, Rachna, Pramod [1] have computed the reliability of PV system connected to the grid. They have computed the sensitivity of PV system with respect to the parameter of interest considering the stress developed due to environmental effects. Aminifer et al. [2] have computed the reliability model for PMU for wide area measurement. They have considered the uncertainties using fuzzy logic functions for various phase measurement components. Ghosh et al. [3] have computed the reliability model for PMU using fuzzy logic. Here, they have also calculated failure and repair rate using symmetrical triangular membership function. Murthy et al. [4] have developed the reliability model for PMU using Hidden Markov model (HMM). Wang et al. [5] have computed the reliability of PMU considering data uncertainties using fuzzy logic model. They have also developed an index based on fuzzy logic to determine the various uncertainties. Zhang et al. [6] have done reliability evaluation of PMU using Monte Carlo Dynamic fault tree method. In this paper, the redundancy of GPS and CPU hardware have also been taken into account to improve the reliability of PMU.

2 Theory of Phase Measurement Unit (PMU)

The synchronous detector of PMU receives a delay as time standard. Let e_1 be delayed time standard signal.

$$e_1(t-\tau) = E_1 \cos \omega(t-\tau) \tag{1}$$

and measured signal e_2 is given by

$$e_2(t) = E_2 \cos(\omega t + \emptyset), \text{ then product} \tag{2}$$

$$\begin{aligned} P(t,\tau) &= e_1(t-\tau) \cdot e_2(t) \\ &= E_1 \cdot E_2\{\cos(\omega t + \tau) \cos \omega(t-\tau)\} \end{aligned} \tag{3}$$

$$= \frac{E}{2}\{\cos(\varphi + \omega\tau) + \cos(2\omega t + \psi)\}, \quad \text{here } \psi = 2\omega t + (\varphi - \omega t) \tag{4}$$

For the same frequency, product signal, $m(\tau)$ is given by

$$m(\tau) = \frac{1}{2}\text{Ecos}(\varphi + \omega t) = \frac{1}{T}\int_0^T p(t,\tau)\text{d}t \tag{5}$$

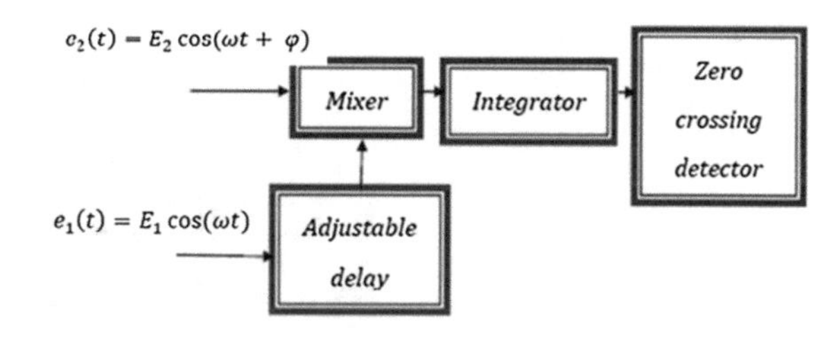

Fig. 1 Syncronisation of phase measurement using delay function

Delay time is reduced until $m(\tau_0) = 0$, which enables that
$0 = \varphi + \omega\tau_0 = n(2K + 1)$, where $n = \pm 1$ and K is an integer ≥ 0
Let $K = 0$ and $n = +1$, then
$\varphi = (\pi/2 - \omega\tau_0)$, where $\omega = 2\pi f$ and τ_0 = delay angle.
Therefore,

$$\frac{\mathrm{d}m(\tau)}{\mathrm{d}t} = \frac{1}{2}E.\cos\varphi', \quad \text{where } \varphi' = \omega(t - \tau_0) \tag{6}$$

For good sensitivity, the peak slope is obtained for $\varphi' = 0$ (Fig. 1)
Therefore,

$$\left|\frac{\mathrm{d}m(\tau)}{\mathrm{d}\varphi'}\right| = \frac{1}{2}E \tag{7}$$

3 Components and Configuration

Phase measurement unit broadly needs to capture the data from the field in time synchronization with respect to other PMU, located at a far distance and communicate with central energy control center. Thus, it shall have a time-synchronized traceability device, called geographical position system (GPS) receiver. Also, it shall have modem and transmitter to communicate with other unit or energy control center, power system data mostly are of analog nature. They need to be converted to digital format after filtering out the noise. Also, the digital data need to be time stamped to ensure the time of occurrence of an event. Thus, a PLL-based oscillator is added with GPS receiver. Processing and managing of all these activities are regulated and controlled by microcontroller or (FPGA, or digital signal controller—DSC). A block diagram of PMU thus can be drawn as shown in Fig. 2. Here, anti-aliasing filter is used to enable that all signals shall have the same phase shift and attenuation. GPS system is used in determining the coordinates of the receiver. Usually, one pulse per second is received by any receiver on earth is coincident with all other received pulses within 1 μs.

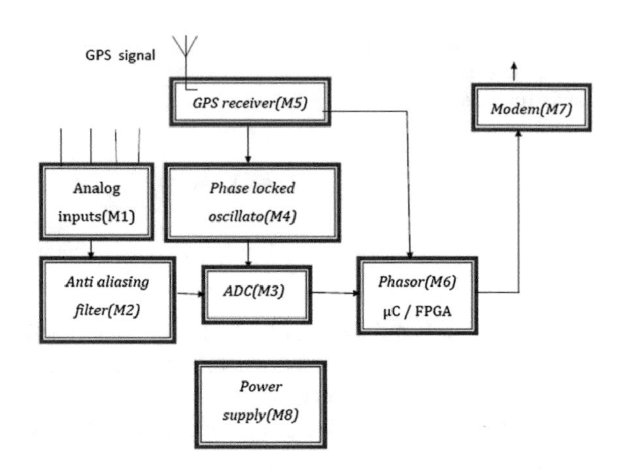

Fig. 2 Functional block diagram of the phase measurement unit (PMU)

PMU consists of eight functional blocks. Module M1 transforms the analog inputs to the standard signal of 4–20 mA or 1–10 V range. This transformed signal is fed to the anti-aliasing filter, M2. This eliminates the high-frequency noise. The filtered analog signal is then converted to digital signal by ADC module, M3. The synchronous sampling pulse is supplied to ADC by PLL module (M4), which receives the synchronous pulse from the GPS module, M5. CPU module, M6 computes the phasor voltages and currents from the digitized signals and time stamped them with coordinated and supplied by GPS module, M5. Time-stamped signal is sent to communication module (MODEM), M7. All the modules (M1–M7) are energized by the power supply module, M8. This module is duplicated and remains in hot standby mode.

4 Dynamic Reliability Model

The dynamic reliability of PMU is the function of its operating time and repairing period. The satisfactory operation of the complete unit depends upon the submodule, which is a further function of various parameters. These parameters may go under stresses during operation of the complete submodules. Thus, the characterization of PMU may be defined as its availability in a defined period and traceability.

$$\begin{aligned}\text{Availability} &= \frac{\text{Operational time of PMU}}{\text{Operational time of PMU } + \text{ Down Time}} * 100\\ &= \frac{\text{mean time between failures (MTBF)}}{\text{Mean cycle time}}\end{aligned} \tag{8}$$

Mean time between failures is the function of failure rate (λ) of sun modules can be written as

$$\text{MTBF} = \lambda^{-1} \tag{9}$$

Thus, the reliability can be computed using

$$\text{Reliability} = e^{-\lambda t} \tag{10}$$

If σ_i represented the stress due to operating factors, then the failure rate can be computed using the relation

$$\lambda = \lambda_b \left(\prod_{i=1}^{n} \sigma_i \right) \tag{11}$$

where σ_i is the product of all the stress factors applicable to a particular functional block of PMU. The different stress factors $\sigma_a, \sigma_{e,}, \sigma_q, \sigma_s, \sigma_t$ and σ_v correspond to device power rating, operational environment, quality of the device, reverse voltage, index factor of device, and temperature, respectively. The total failure rate of a PMU λ_{PMU} can be computed by the summation of the failure rate of all the individual functional blocks.

$$\lambda_{\text{system}} = \sum \lambda_{i(\text{functional block})} \tag{12}$$

Based on the frequency balance technique, the steady-state probability equation can be written as

$$p_0 \lambda_i = p_i \psi_i \tag{13}$$

Here, p_0, and p_i is the probability of PMU in working state and failure state, respectively.

Hence,

$$p_i = \frac{\lambda_i}{\psi_i} p_0 \tag{14}$$

Since, the sum of probability failure rate + sum of probability repair rate = 1 i.e., $p_0 + \sum p_i = 1$, substituting the value of p_i, we get

$$p_0 + \sum p_0 \frac{\lambda_i}{\psi_i} = 1 \quad \Rightarrow p_0\{1 + \sum (\lambda_i/\psi_i)\} = 1 \quad \Rightarrow p_0 = [1 + \sum (\lambda_i/\psi_i)]^{-1} \tag{15}$$

Thus, in a two-state model of PMU, the probability of working of PMU, P_w

$$P_w = p_0, \tag{16}$$

and the probability of not working, P_D

$$P_D = 1 - P_w = \left\{ \sum \left(\frac{\lambda_i}{\psi_i} \right) \right) [1 + \sum (\lambda_i / \psi_i)]^{-1} \tag{17}$$

and

$$\psi = (\sum \lambda_i) \left[\sum \left(\frac{\lambda_i}{\psi_i} \right) \right]^{-1} \tag{18}$$

$$\lambda = \sum \lambda_i \tag{19}$$

This results in the uncertainty of the reliability parameters of PMU's functional blocks. For accurate results, a range of reliability parameters are estimated instead of single value. α-cut fuzzy logic is an alternative tool to restructure the fuzzy functions. The authors of this paper, thus, have computed the reliability of PMU using fuzzy logic gates relation. Logic gates state the dynamic response of PMU, whereas uncertainty in data of failure rate and repair rate is taken care of by fuzzy logic—α cut evaluation. Figure 2 shows the functional block diagram f PMU. It consists of eight functional blocks. Power supply, anti-alias filter, and ADC blocks are redundant blocks. Each of these three modules forms two parallel circuits due to redundancy in blocks. It is assumed here that one PS is functional, while other in cold standby mode. These three blocks are connected in series.

4.1 Failure Module of Data Acquisition

Figure 3 shows the reliability model in terms of failure rate using dynamic logic gates. The combined three modules are called a digital input module.

Let $\lambda_{MiX,}$ and ϱ_{MiY} (i = 1, 3) are failure rate, and repair rate of analog input module, anti-alias filter module, and ADC module, respectively.

4.2 Failure Model of Synchronizing GPS Module

GPS receiver (M4) and phase-locked oscillator (M5) forms series circuit, while combination of two forms a parallel circuit. Figure 4 shows the synchronizing GPS failure reliability model.

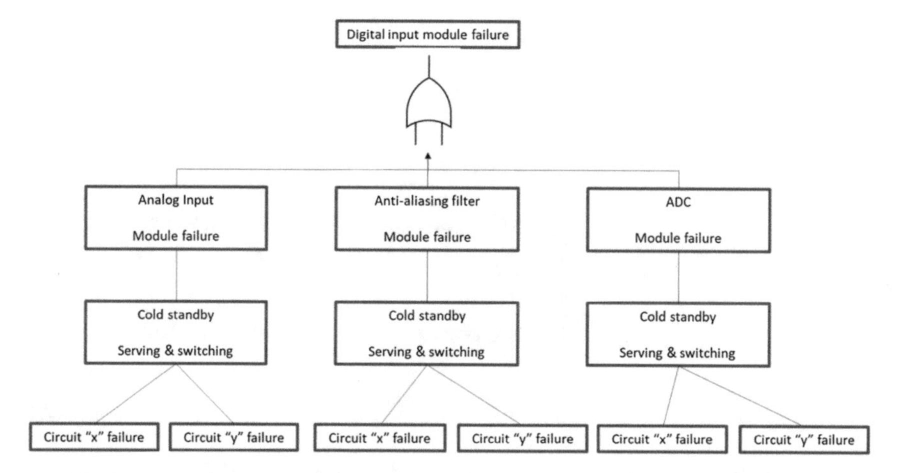

Fig. 3 Reliablity of data acquisition model

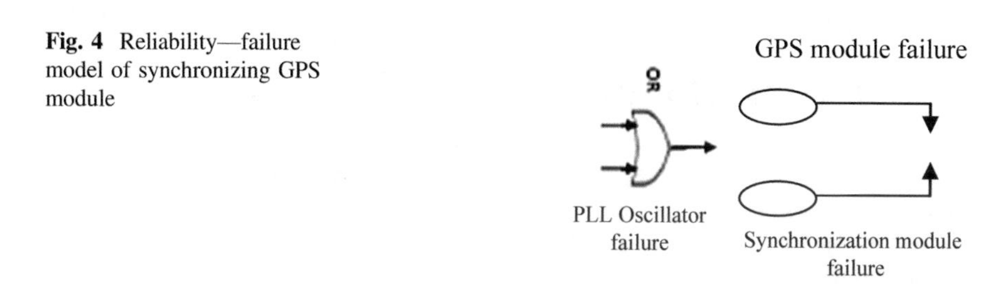

Fig. 4 Reliability—failure model of synchronizing GPS module

In synchronization GPS module, PLL oscillator supply the sampling clock pulse for the ADC in case GPS receiver fails. It keeps a track on the error between PLL crystal oscillator frequency and pulse per second (PPS) supplied by GPS receiver. This error shall always be less than 1 μs/h, when the sampling pulse is supplied by PLL oscillator.

4.3 Failure Model of Microcontroller

Microcontroller comprises of hardware and software. Process and computational functions such as phasor, frequency estimation, etc., are carried out by a software algorithm. The computation of software reliability is different from hardware. Hardware reliability is evaluated from the physical failure of hardware of processor, whereas software reliability is computed based on the functional failure, may be due to improper logic, code error, etc. Software reliability is computed using Eq. (20), which is as follows:

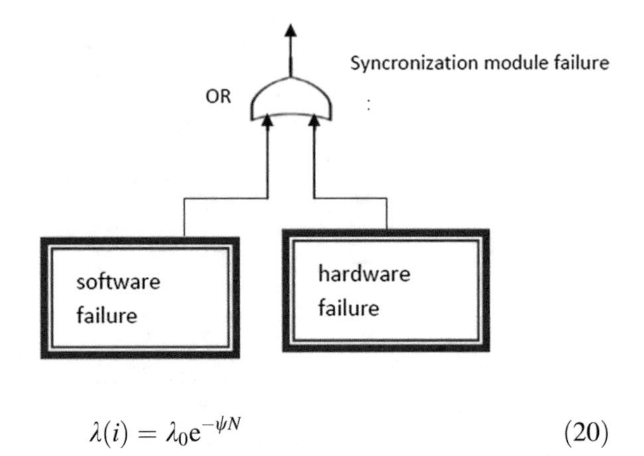

Fig. 5 Reliability—failure model of microcontroller

$$\lambda(i) = \lambda_0 e^{-\psi N} \tag{20}$$

where λ_0 is the initial failure rate, ψ is the failure decay, and N is the number of failures observed (Fig. 5).

4.4 Failure Model of MODEM and Power Supply

There are at least two communication ports in MODEM, which are dual to each other. Power supply module is also in dual, standby mode. Figure 6 shows the reliability—failure module of MODEM and power supply. It is similar to the data acquisition module (Fig. 7).

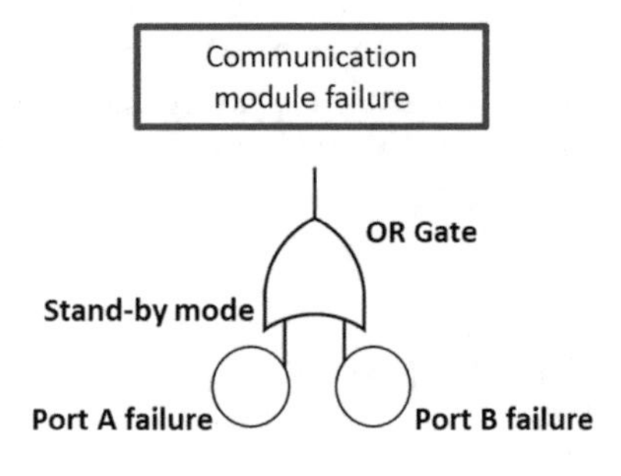

Fig. 6 Reliability—failure model communication

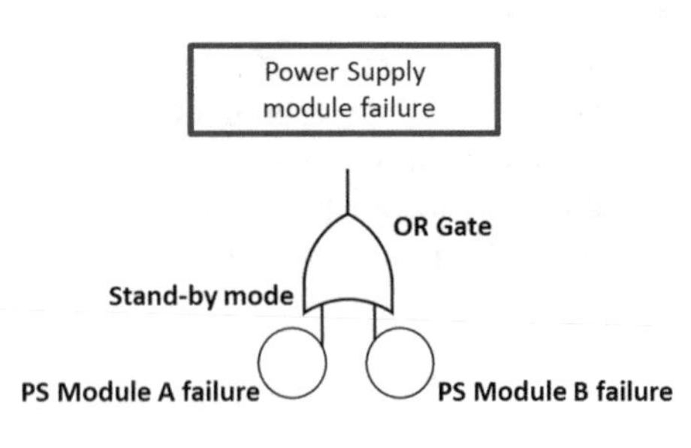

Fig. 7 Reliability—failure model power supply

5 Complete Reliability Model of PMU

Failure of any functional module results in failure of PMU. Such logic is called series circuit. Thus, OR logic is used to integrate all the eight reliabilities—failure modules associated with eight functional modules of PMU to compute the dynamic model of PMU, as shown in Fig. 8:

6 Fuzzy Model of Reliability Based on α-Cut

The reliability of each constituent functional module of PMU depends on the stress on various components. The stress on the components of functional modules varies due to electrical and environmental conditions. The variation in stress result in variation of reliability values of functional modules. The reliability value, thus, is not a constant. It is based on the principle that any fuzzy set can be reconstructed from a family of nested, assuming that they satisfy consistency constraint

$$\propto_1 \; > \sup \propto_i A_i(x) \quad \text{where } \propto \in [0,1] \tag{21}$$

Steps to determine the fuzzy reliability of PMU:

- Select the symmetrical triangular fuzzy membership functions for failure and repair rates.
- α-cut sets of failure and repair rates are computed.
- The equivalent reliability values are computed for α-cut sets.
- Fuzzy out function is reconfigured from α-cut outputs.

Application to PMU:

The computation of the reliability of PMU using the fuzzy logic technique, we consider the mean value of upper and lower bounds as in [2] and presented in Table 1.

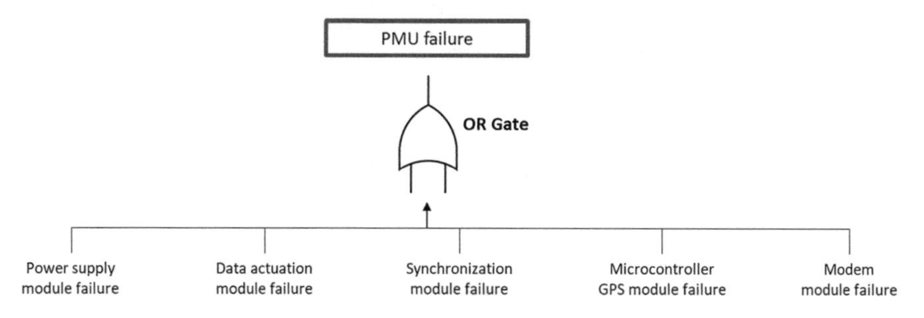

Fig. 8 Reliability model of PMU based on the failure rate

Table 1 Failure and repair rates of PMU module

i	1	2	3	4	5	6	7
λ_i	0.24	0.275	0.875	0.34	1.4	1.2	3.584×10^{-4}
ψ_i	438	438	876	438	876	876	370.79

Table 2 Available and unavailable of PMU

A	P_{w1}	P_{w2}	P_{D1}	P_{D2}
1.0	0.9939	0.9939	0.0056	0.0056
0.9	0.9933	0.9946	0.0050	0.0065
0.6	0.9902	0.9960	0.0028	0.0091
0.3	0.9862	0.9973	0.0019	0.0132
0.0	0.9798	0.9987	0.0009	0.021

Table 3 Failure and repair rate of PMU

α	λ_1	λ_2	ψ_1	ψ_2
1.0	4.2843	4.2843	730.8743	730.8743
0.9	3.9881	4.5823	689.4336	761.3276
0.6	3.0213	5.5168	579.0936	878.3265
0.3	2.08754	6.4830	470.0017	992.0117
0.0	1.17	7.4002	361.8737	1115.101

Given the failure and repair rates, fuzzy reliability values are computed for a two-state Markov model of PMU. Tables 2 and 3 give the upper and lower limits of available and unavailable PMU, and failure and repair rates of PMU.

Tables 2 and 3 can be drawn graphically. This shows that reliability lies in a range for PMU with different values of α (Figs. 9 and 10).

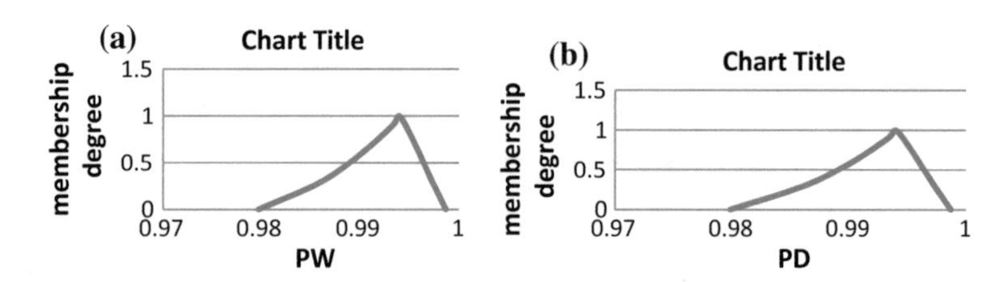

Fig. 9 **a** Reliability versus membership for working PMU. **b** Reliability versus membership for non-working PMU

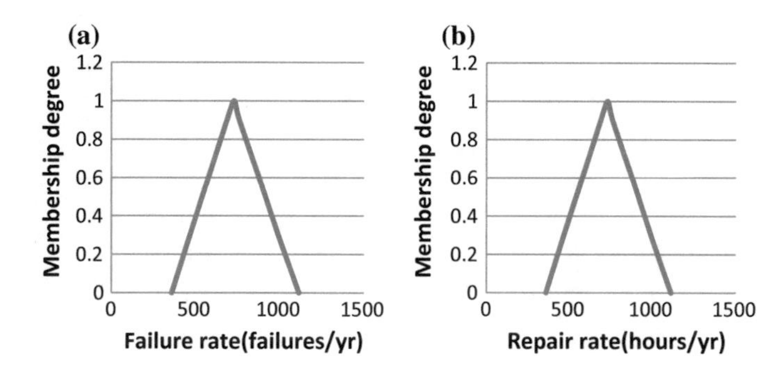

Fig. 10 **a** Fuzzy member function for failure rate. **b** Fuzzy member function for repair rate

7 Conclusion

Reliability of PMU is computed using fuzzy logic α-cut method. The stresses developed in components of submodules, due to environmental effect, change the failure rate of submodule. This has been considered in computing the fuzzy membership of each module. The model is converted into two state fuzzy Markov models which, in turn, gives a more realistic result for reliability. The redundancy analysis is also taken into account. The study is important for digital power systems operation and control

References

1. Gupta, N., Garg, R., Kumar, P.: Sensitivity and reliability models of a PV system connected to grid. Renew. Sustain. Energy Rev. www.elsevier.com/locate/rser
2. Aminifar, F., Bagheri-Shouraki, S., Fotuhi-Firuzabad, M., Shahidehpour, M.: Reliability modeling of PMUs using fuzzy sets. IEEE Trans. Power Deliv. **25**(4), 2384–2390 (2010)
3. Ghosh, S., Das, S., Ghosh, D., Mohanta, D.K.: Fuzzy approach for reliability analysis of PMU. In: 2014 14th International Conference on Environment and Electrical Engineering
4. Murthy, C., Mishra, A., et al.: Reliability analysis of phase measurement unit using hidden Markov model. IEEE Syst. J. **8**(4), 1293–1130 (2014)
5. Wang, Y., Li, W., et al.: Reliability analysis of phasor measurement unit considering data uncertainity. IEEE Trans. Power Syst. **27**(3), 1503–1510 (2012)
6. Zhang, P., Chan, K.W.: Reliability evaluation of phasor measurement using Monte Carlo dynamic fault tree method. IEEE Trans. Smart Grid **3**(3), 1235–1243 (2012)

A Hybrid Intelligent Model for Power Quality Disturbance Classification

Hasmat Malik, Paras Kaushal and Smriti Srivastava

Abstract With the abundant demand for power and increasing number of loads causes a number of power quality disturbances. The detection and classification of these PQ disturbances have become a pressing concern. This paper presents a novel technique for disturbances classification in power distribution line using empirical mode decomposition of raw data and multi-layer perceptron method for classification. The electrical distribution model is designed over MATLAB/Simulink environment to create PQ disturbances. This proposed method successfully classifies five types of PQ disturbances, i.e., sag, swell, interruption, transients, harmonics, and one healthy for comparison, and obtains 98.9% correct classification rate for tested events.

Keywords Empirical mode decomposition · Classifier · PQ disturbances
Feature extraction · Multi-layer perceptron · Distribution system

1 Introduction

Considering different power quality assessment methods, one of the most important methodologies is power disturbance (PQD) classification. Generally, disturbances occur in distribution and transmission lines, in distribution lines disturbances are more complex with more frequent, whereas the most well-known transmission line

H. Malik
Electrical Engineering Department, IIT Delhi, New Delhi 110016, India
e-mail: hmalik.iitd@gmail.com

P. Kaushal (✉) · S. Srivastava
Instrumentation and Control Engineering Department, NSIT, Dwarka,
New Delhi 110078, India
e-mail: parask2903@gmail.com

S. Srivastava
e-mail: smriti.nsit@gmail.com

© Springer Nature Singapore Pte Ltd. 2019

H. Malik et al. (eds.), *Applications of Artificial Intelligence Techniques in Engineering*, Advances in Intelligent Systems and Computing 697,
https://doi.org/10.1007/978-981-13-1822-1_6

disturbances are voltage sags and swells, interruption, flickers, and harmonic generation due to nonlinear loads.

Although a large number of disturbances occur and are recorded [1], it is prohibitively expensive to classify all of them manually. Thus, the automatic classification of PQ disturbances is highly desirable. However, relatively some amount of work has been done on automatic classification but correct classification rates for the actual events are not as high as in the areas such as pattern recognition, speech recognition. So, there is broad scope for improvement of PQD classification.

Over the few decades, the demand for power has been increased tremendously with the interest of clean energy. Good power quality (PQ) is essential for the healthy operation of power systems. Main causes of PQ degradation include faults, switching heavy loads, capacitor banks, solid-state power converter devices, arc furnaces, and energized transformers. Mitigation of these disturbances requires fast and accurate classification [2]. Signal processing with instrumentation at the source end and feature extraction at load end plays a key role. PQDs are essentially nonstationary in behavior and need instantaneous time–frequency analysis.

2 Model Description

A comprehensive model of real power distribution and transmission system is created by using a Simulink toolbox in MATLAB. Simulink is one of the most powerful simulation tools for analyzing and modeling a real-time system. Voltage and current analysis corresponding to one healthy and five faulty conditions, naming, sag, swell, interruption, transients, and harmonics are simulated on a three-phase, 25 kV, 50 Hz transmission line of length 20 km connected between three-phase source and three-phase nonlinear load is generated. Various disturbances are created by using faults, nonlinear loads, and capacitor banks at different locations and respective responses are recorded to classify them automatically using the neural network technique [1].

All the parameters for the above Simulink model are considered as the parameter of the actual distribution system. The parameters are listed in Table 1.

Table 1 Parameters of the proposed method

Phase source		3 Phase load	
Voltage: 25 kV	Impedance: 5 + j0.3 Ω	Voltage ($V_{p\text{-}p}$): 1000 V	P_{Active}: 10 KW
Transformer (Δ → Y)		*Universal bridge: 3 Arm (diode)*	
Primary winding	Secondary winding	Series R-L load	
$V_{p\text{-}p}$: 25 kV	$V_{p\text{-}p}$: 600 V	V_n: 1000 V	
R_p: 0.002 p.u.	R_s: 0.002 p.u.	P_{Active}: 10 KW	
L_p: 0.08 p.u.	L_s: 0.08 p.u.	$P_{Reactive}$: 100 p.u.	
Transmission line		*Capacitor bank*	
20 K.M.		$V_{Nominal}$: 600 V	$P_{Reactive}$: 0.8 MW

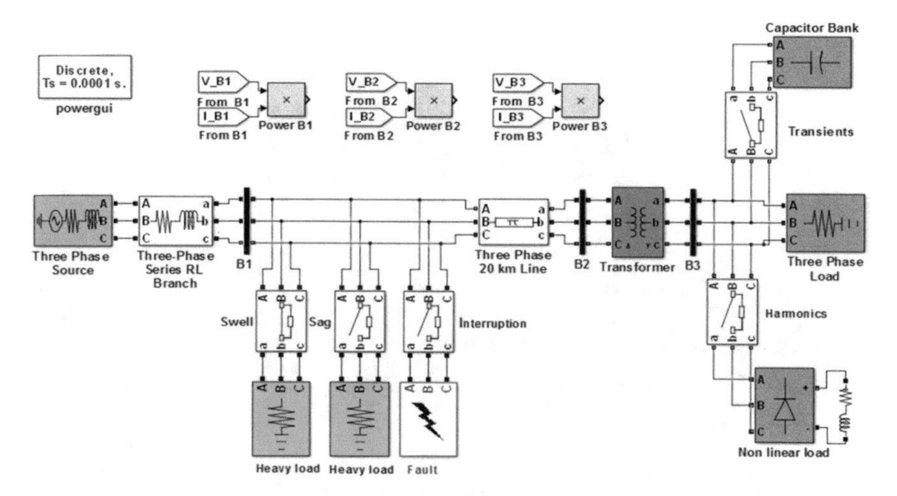

Fig. 1 Simulink model of the electrical power distribution

Bus B1, B2, and B3 are placed to capture the effect of disturbance on voltage and current waveform at different locations of 20 km transmission lines. Furthermost, the there effect on net power is also formulated. PQ waveform is taken for one second with frequency of 50 Hz and a sampling frequency of 10 kHz. The disturbance is created for 0.1 s or 5 cycles of three-phase sinusoidal input alternatively with normal condition. The normal condition is a clean pure sinusoidal waveform, which is free from all disturbances (Fig. 1).

Switching of three-phase series RLC load causes voltage swell and dip while short circuit fault leads to interruption type of PQ disturbance. Capacitor banks are used to create transients and harmonics which are generated by high-frequency solid-state switching devices or nonlinear loads. The disturbances flow less toward the source due to transients and harmonics, most of the effect is seen at load side (Fig. 2).

In this model, Eempirical mode decomposition (EMD) technique is used for feature extraction while the artificial neural network (ANN) is used to classify PQ disturbances [3]. For classification of disturbances, a multi-layer perceptron (MLP) network is used. It is a nonlinear learning and modeling method which uses for the precise classification of disturbances. In the presented approach, post disturbance currents, voltages, and power have been predicted. The proposed scheme is accurately evaluated on a 25 kV, 50 Hz and transmission line of 25 km length using MATLAB [4].

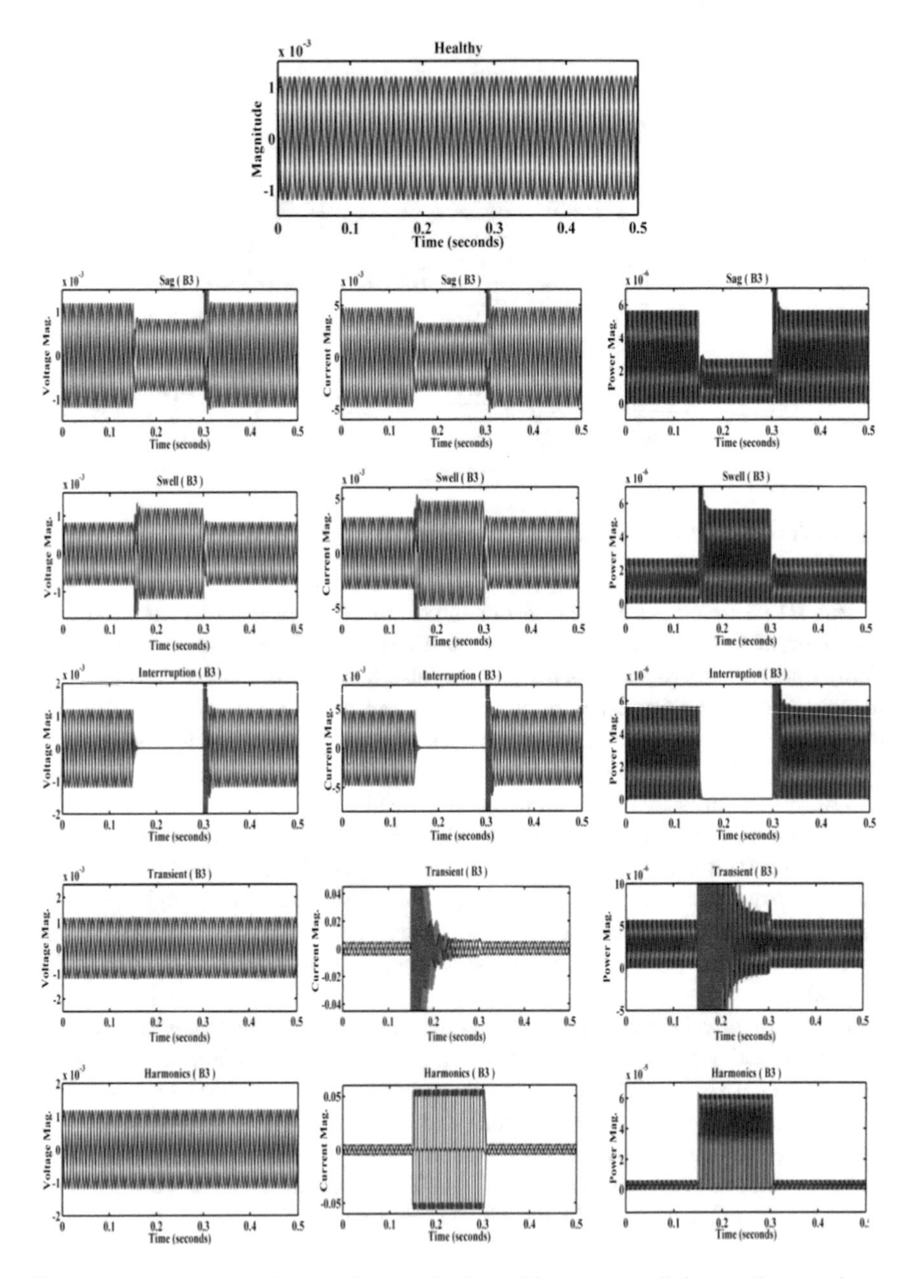

Fig. 2 Voltage, current and power signature for 5 conditions (sag, swell, interruption, transients, harmonics) and one healthy condition at B3

3 Empirical Mode Decomposition (EMD) [3, 5, 6]

EMD method decomposes a nonstationary signal into multiple stationary sub signals called as intrinsic mode function (IMF). The EMD process decomposes *x(t)* to

$$\sum_{n=1}^{N} \mho n(t) + \Phi N(t) \tag{1}$$

where N is number of IMFs; $\mho n(t)$ is nth IMF and $\phi N(t)$ is final residue.

Two conditions for each IMF have to satisfy the following:

1. In the complete dataset, only one extreme (maxima or minima) between zero crossings should exist.
2. At any point in process, the mean value between maxima or minima should be zero.

First, upper and lower envelopes are generated then averaging is performed on these envelopes [3]. The final value is obtained by subtracting average from the original value. The above result is examined to see whether it satisfies above two IMFs conditions. If the difference signal does not satisfy any of the two conditions, it is considered as the original signal's envelope and calculations for difference signal are repeated. This procedure is continued until the resultant signal satisfies conditions that make it an IMF.

4 Multi-layer Perceptron (MLP) Classifier

MLP is a network of neurons called perceptron, which is used to describe any general feed forward network [7, 8–16]. Feed forward network is a network in which each layer has to feed their output forward to the next layer until it generates the final output from the neural network. Such computations for feed forward network with a hidden layer, nonlinear activation functions and an output layer can be represented mathematically as [7]:

$$Y = \mathrm{X}(s) = B\Phi(As + b) + c \tag{2}$$

where Y is output vector, S is input vector, A is weight matrix of the first layer, B is weight matrix, b is vector of bias for the first layer and c is bias vector of the second layer. NN learning is terminated when a specified value of mean square error (MSE) is achieved [17]. Any number of hidden layers can be present in a feed forward network. For best result, the number of neurons can be calculated as

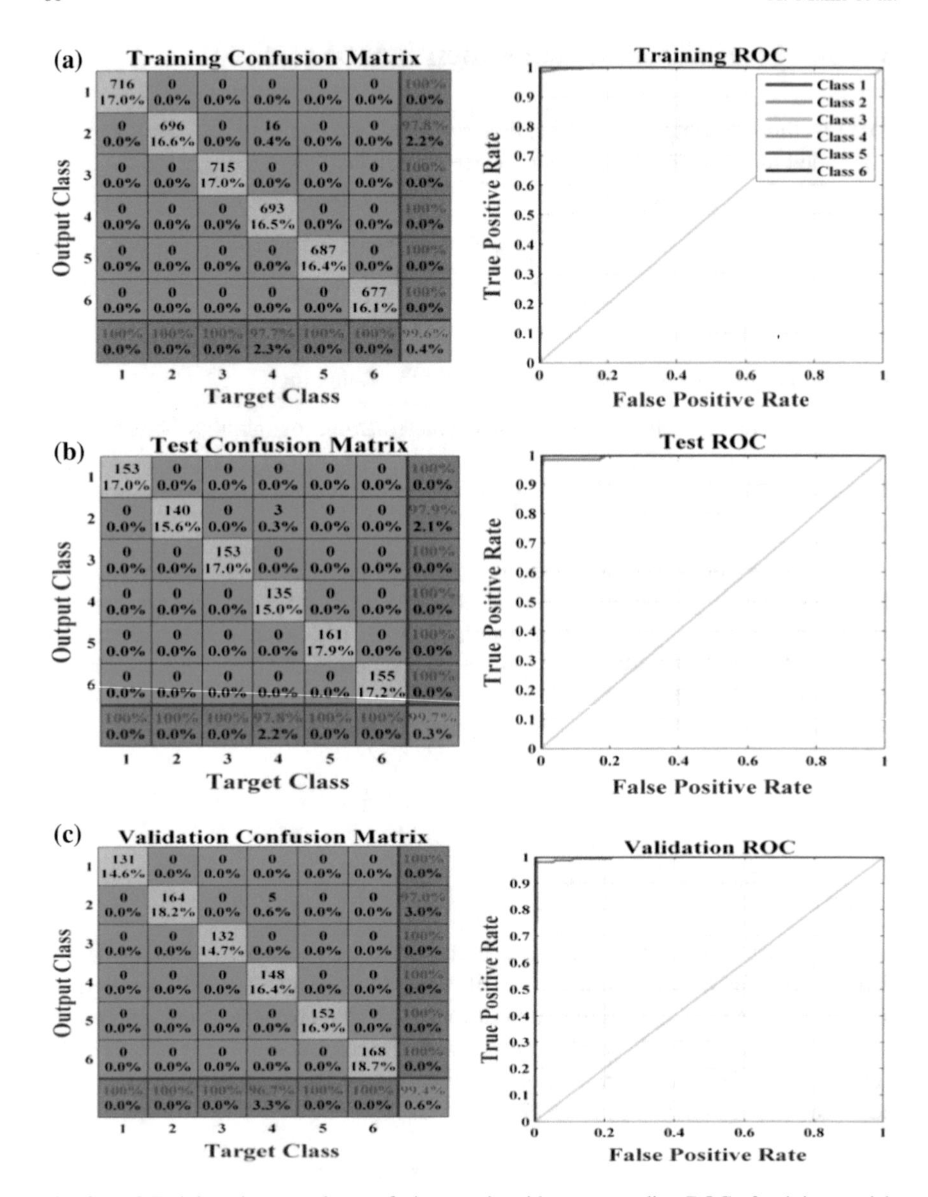

Fig. 3 **a–d** Training phase results: confusion matrix with corresponding ROC of training model. **e** Gradient plot and validation checks plot

$$\text{Neurons} = \left[\left(\frac{\text{Input} + \text{Output}}{2}\right) + \sqrt{\text{Sample}}\right] \pm 10\% \tag{3}$$

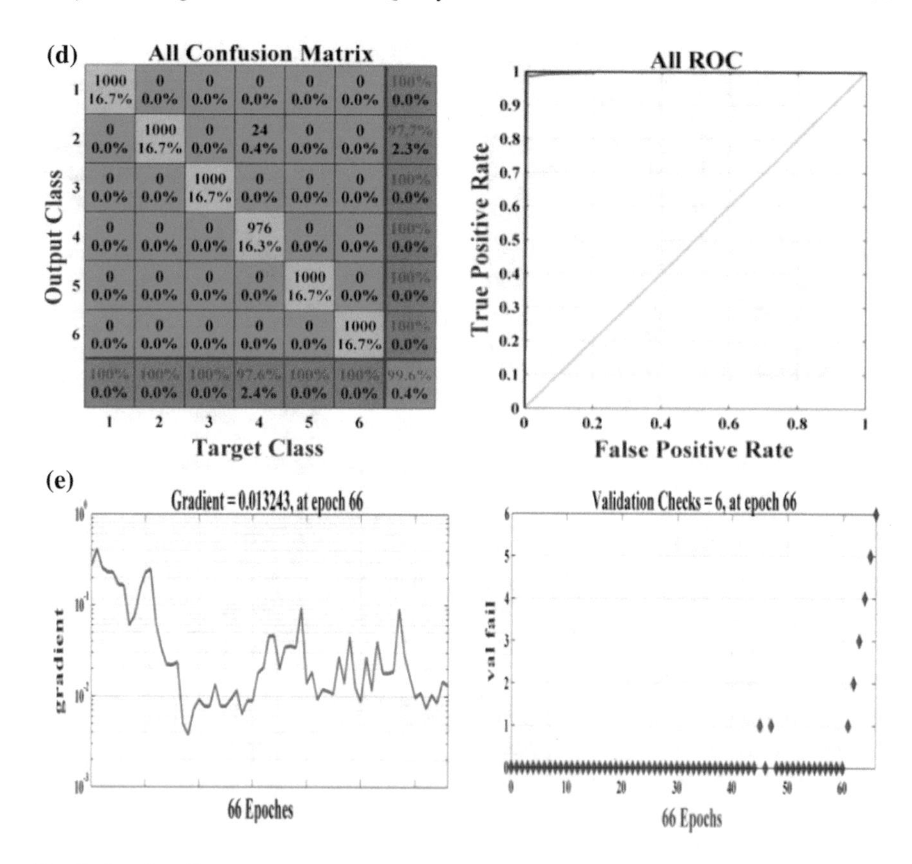

Fig. 3 (continued)

5 Results and Discussion

Simulation of the model generates 1 out of 10,000 data samples for each case. This data set is decomposed into IMFs, which are used as input variables to the ANN. The first 1000 IMFs samples from each case are used for analysis (Figs. 3 and 4; Table 2).

Out of this data, 75% randomly chosen samples are fed to classifier during the training phase and the rest of the data is used for testing and validation. The developed neural network structure is 3-N-6, where N is the number of hidden layer neuron from 1 to 90 calculated from Eq. (3) gives results obtained in terms of voltage, current and their combination at B3 measuring point of the above Simulink model.

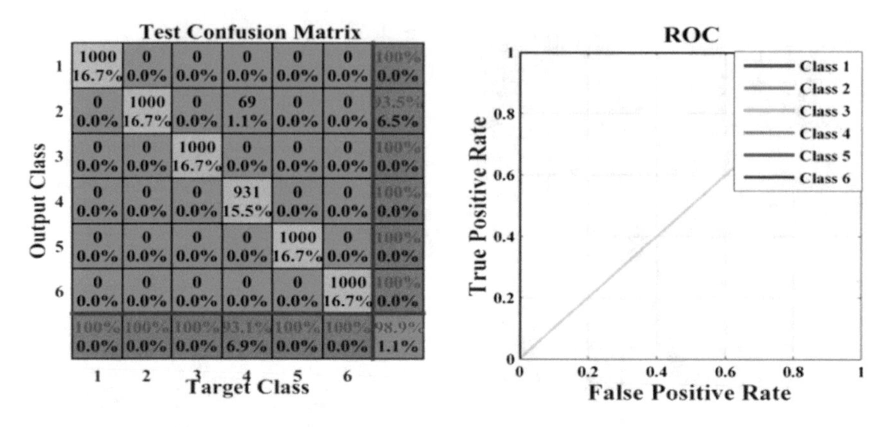

Fig. 4 Testing phase results: confusion matrix with corresponding ROC of test model

Table 2 Percent accuracy of training and testing

Measurement at B3	% Accuracy training	% Accuracy testing
Voltage (raw data)	67.08	58.75
Current (raw data)	79.67	68.92
Voltage + current (raw data)	84.91	70.75
Voltage (IMF data)	99.98	92.77
Current (IMF data)	99.60	98.90

6 Conclusion

This paper presented a combined EMD and ANN technique for power quality classification for the distribution system. Simulink model is designed and disturbance such as sag, swell, interruption, transient and harmonics are created at a time to extract relevant data. IMFs are obtained using EMD and MLP is used for classification of PQ disturbances. The technique achieved a training accuracy of 99.60% and testing accuracy of 98.9% for the current at B3.

References

1. Khokhar, S., Mohd Zin, A.A., Mokhtar, A.S., Ismail, N.A.M.: MATLAB/simulink based modeling and simulation of power quality disturbances, IEEE conference (2014), 4799–4848
2. Huang, J., Negnevitsky, M., Thong Nguyen, D.: A neural-fuzzy classifier for recognition of power quality disturbances. IEEE Trans. Power Deliv. **97**, 0885–8977 (2002)
3. Malik, H., Sharma, R.: EMD and ANN based intelligent fault diagnosis model for transmission line. J. Intell. Fuzzy Syst. **32**, 3043–3050 (2017)

4. Liu, Z., Cui, Y., Li, W.: A classification method for complex power quality disturbances using EEMD and rank wavelet SVM. IEEE Trans. Smart Grid (2015). https://doi.org/10.1109/TSG.2015.2397431
5. Malik, H., Mishra, S.: Artificial neural network and empirical mode decomposition based imbalance fault diagnosis of wind turbine Using TurbSim, FAST and Simulink. IET Renew. Power Gener. **11**(6), 889–902 (2017). https://doi.org/10.1049/iet-rpg.2015.0382
6. Malik, H., Sharma, R.: Transmission line fault classification using modified fuzzy Q learning. IET Gener. Transm. Distrib. **11**(16), 4041–4050 (2017). https://doi.org/10.1049/iet-gtd.2017.0331
7. Yadav, A.K., Malik, H., Chandel, S.S.: Application of rapid miner in ANN based prediction of solar radiation for assessment of solar energy resource potential of 76 sites in Northwestern India. Renew. Sustain. Energy Rev. **52**, 1093–1106 (2015). https://doi.org/10.1016/j.rser.2015.07.156
8. Yadav, A.K., Malik, H., Chandel, S.S.: Selection of most relevant input parameters using WEKA for artificial neural network based solar radiation prediction models. Renew. Sustain. Energy Rev. **31**, 509–519 (2014). https://doi.org/10.1016/j.rser.2013.12.008
9. Yadav, A.K., Sharma, V., Malik, H., Chandel, S.S.: Daily array yield prediction of grid-interactive photovoltaic plant using relief attribute evaluator based radial basis function neural network. Renew. Sustain. Energy Rev. **81**(Part 2), 2115–2127. https://doi.org/10.1016/j.rser.2017.06.023
10. Yadav, A.K., Malik, H., Mittal, A.P.: Artificial neural network fitting tool based prediction of solar radiation for identifying solar power potential. J. Electr. Eng. **15**(2), 25–29 (2015). (University "POLITEHNICA" Timisoara, Romania)
11. Yadav, A.K., Singh, A., Malik, H., Azeem, A., Rahi, O.P.: Application research based on artificial neural network (ANN) to predict no load loss for transformer design. In: Proceedings of IEEE International Conference on Communication System's Network Technologies, pp. 180–183 (2011). https://doi.org/10.1109/csnt.2011.45
12. Yadav, A.K., Malik, H., Chandel, S.S.: ANN based prediction of daily global solar radiation for photovoltaics applications. In Proceedings of IEEE India Annual Conference (INDICON), pp. 1–5 (2015). https://doi.org/10.1109/indicon.2015.7443186
13. Malik, H., Savita, Application of artificial neural network for long term wind speed prediction. In: Proceedings of IEEE CASP-2016, pp. 217–222, 9–11 June 2016. https://doi.org/10.1109/casp.2016.7746168
14. Azeem, A., Kumar, G., Malik, H.: Artificial neural network based intelligent model for wind power assessment in India. In: Proceedings of IEEE PIICON-2016, pp. 1–6, 25–27 Nov. 2016. https://doi.org/10.1109/poweri.2016.8077305
15. Saad, S., Malik, H.: Selection of most relevant input parameters using WEKA for artificial neural network based concrete compressive strength prediction model. In: Proceedings of IEEE PIICON-2016, pp. 1–6, 25–27 Nov 2016. https://doi.org/10.1109/poweri.2016.8077368
16. Azeem, A., Kumar, G., Malik, H.: Application of waikato environment for knowledge analysis based artificial neural network models for wind speed forecasting. In: Proceedings of IEEE PIICON-2016, pp. 1–6, 25–27 Nov 2016. https://doi.org/10.1109/poweri.2016.8077352
17. Biswal, B., Mishra, S., Jalaja, R.: Automatic classification of power quality events using balanced neural tree. IEEE Trans. Ind. Electron. (2014). https://doi.org/10.1109/TIE.2013.2248335

Voltage Stability Enhancement of Primary Distribution System by Optimal DG Placement

Vani Bhargava, S. K. Sinha and M. P. Dave

Abstract Voltage stability of a power system is the ability of the same to maintain a steady voltage at all the system buses even after the system is subjected to some disturbance. This steady voltage is compared from some initial system state when there was no disturbance in the system. By far shunt capacitors were found to be the simplest and most economical method for correcting the distribution system voltage. But they do suffer from certain limitations which are inherent, as far as voltage stability is concerned. The paper studies the voltage stability of a distribution system and proposes placement of DG units with sufficient capacity and at suitable sites so as to improve the voltage stability of the distribution system.

Keywords Primary distribution system · Stability · Voltage stability
Distributed generation · Voltage profile · DG site · DG size · Optimal placement

1 Introduction

In general terms, the stability, in the context of electrical power system, refers to the system's ability of returning back to the normal state after it is subjected to some disturbances. The stability can be divided into steady-state and transient stability. Steady-state stability refers to the variations in system state after small and gradual changes take place in system state, whereas transient stability is referred to the variations in system state after system is subjected to sudden and large disturbances. In steady-state stability studies, the prime consideration is on the system bus

V. Bhargava (✉) · S. K. Sinha
Amity University, Noida, Uttar Pradesh, India
e-mail: vanigarg03@gmail.com

S. K. Sinha
e-mail: sinha.sanjay66@gmail.com

M. P. Dave
Shiv Nadar University, Greater Noida, India
e-mail: davemp2003@yahoo.com

© Springer Nature Singapore Pte Ltd. 2019

H. Malik et al. (eds.), *Applications of Artificial Intelligence Techniques in Engineering*, Advances in Intelligent Systems and Computing 697,
https://doi.org/10.1007/978-981-13-1822-1_7

voltages, to restrict them close to their nominal/supposed values and also to make sure that phase angle between two system buses is within prescribed limits. The study also ensures that in any case overloading of the line and apparatuses do not take place. Usually, these all are ascertained using load/power flow studies of the system. In the transient stability study, power system is studied against sudden and large changes. The objective is basically to study the behavior of load angle, to know whether it is returning to a steady-state value after the disturbance is cleared or not. Another one is the stability of the system against continuous small disturbances which is referred as dynamic stability [1].

2 Distribution System and Voltage Stability

2.1 *Voltage Stability*

The distribution system is mainly associated with the voltage stability. The voltage stability is defined as the power system's ability to restore nominal system voltage after the occurrence of a disturbance in the system. The main factor responsible for voltage instability is the mismatch in reactive power demand and supply [2–4]. A distribution system enters the instability state if the occurrence of a disturbance, uncontrollably, drops the system voltage. One more criterion for system to be voltage stable is that, when reactive power injection at a certain bus at nominal frequency is increased, the voltage magnitude at that bus should get increased. If for a single bus in the distribution system, increasing reactive power injection at any particular bus, causes a reduction in voltage at that bus, the DS is said to be voltage unstable. In more technical terms, it can be said that if the V-Q sensitivity of system is positive, system is voltage stable otherwise if V-Q sensitivity is negative, even for a single bus, the system is voltage unstable. In a distribution system, voltage stability is a local phenomena but it may lead to system voltage collapse which is a widespread phenomena. Voltage collapse is defined as a phenomenon where a sequence of events caused by voltage instability, results in abnormally low voltage and may result in a total blackout in some parts of the system. Line impedance voltage drop is the main contributor to voltage instability. After the occurrence of some disturbance, the reactive power demand in the system may rise to a higher value, causing an increased current and an increased line impedance drop.

2.2 *Causes of Voltage Instability*

The instability in voltage is caused by a number of reasons, it may be due to increment in loading, the operation of synchronous generators, and condensers near limiting value of the reactive power, operation of tap changing transformer, load recovery dynamics, feeder outage, or generating unit outage. All these reasons affect voltage stability as they affect the reactive power consumption profile of the system. For prevention of voltage instability various measures like shunt capacitor switching, tap changing transformer blocked operation, generation re-dispatch, scheduling the load, permitting the temporary reactive power overloading of the generators, can be taken [1, 5].

Voltage stability analysis is based upon examining two main aspects of the system, where voltage stability is to be examined. They are: what is the proximity of the system to voltage instability? And the second one is the voltage instability mechanism. Voltage instability mechanism is basically associated with the factors responsible for voltage instability, system areas where system voltage is weak and the measures that can be taken to enhance voltage stability. Most of the aspects of the voltage stability problem can be easily analyzed by employing static methods. The static methods inspect the violation of system equilibrium point under some specified system operating conditions [1].

3 Methodology

In this work, the voltage stability of an IEEE 33 bus system is accesses based upon the PV characteristics of the system and then placement of DG is proposed to enhance the voltage stability. The selection of size and site for DG placement in the distribution system is a crucial concern, as if they are not placed in proper capacity and at proper site, their placement will result in increased system losses. In this paper, from the viewpoint of voltage stability analysis of the distribution system, the DGs are placed optimally in such a manner that the overall system losses are reduced along with the improvement in system voltage profile and stability enhancement [6]. The optimal DG size is obtained by an analytical method using exact loss formula [7–9].

$$P_l = \sum_{i=1}^{N} \sum_{j=1}^{N} \left[\alpha_{ij}\left(P_iP_j + Q_iQ_j\right) + \beta_{ij}\left(Q_iP_j - P_iQ_j\right)\right] \tag{1}$$

where

$$\begin{aligned} \alpha_{ij} &= \frac{r_{ij}}{V_iV_j}\cos(\delta_i - \delta_j), \\ \beta_{ij} &= \frac{r_{ij}}{V_iV_j}\sin(\delta_i - \delta_j), \end{aligned} \tag{2}$$

and

$$r_{ij} + jx_{ij} = Z_{ij}, \tag{3}$$

being *ij*th element of Z-bus matrix [Zbus] also [Zbus] = $[\text{Ybus}]^{-1}$. P_i, P_j, Q_i, and Q_j are the real and reactive power injected at *i*th bus and the *j*th bus, respectively, and N is the total number of buses. If DG power factor is known, then

$$\begin{aligned} Q_{\text{DG}i} &= \left[\pm \tan\left\{\cos^{-1}(PF_{\text{DG}})\right\}P_{\text{DG}i}\right] \\ Q_{\text{DG}i} &= a(P_{\text{DG}i}) \\ a &= \pm \tan\left\{\cos^{-1}(PF_{\text{DG}})\right\} \end{aligned} \tag{4}$$

DG injects reactive power if 'a' is positive and DG absorbs the same if the sign of 'a' is negative [8, 10].

The above equations are combined to get the real power loss equation as

$$\begin{aligned} P_L = \sum_{i=1}^{N}\sum_{j=1}^{N} &\left[\alpha_{ij}\left[(P_{\text{DG}i} - P_{Di})P_j + (aP_{\text{DG}i} - Q_{Di})Q_j\right]\right. \\ &\left. + \beta_{ij}\left[(P_{\text{DG}i} - P_{Di})P_j + (aP_{\text{DG}i} - Q_{Di})Q_j\right]\right] \end{aligned} \tag{5}$$

Now, if the partial derivative of this equation with respect to real power of *i*th bus is reduced to zero, it will give the optimal size of DG to minimize these losses. In this work, two types of DGs are employed; type 2 and type 3. For type 3 DG, the optimal DG size is

$$P_{\text{DG}i} = \frac{\alpha_{ij}(P_{Di} + aQ_{Di}) + \beta_{ii}(aP_{Di} - Q_{Di}) - X_i - aY_i}{a^2\alpha_{ii} + \alpha_{ii}} \tag{6}$$

For real power and for reactive power it is

$$\begin{aligned} Q_{\text{DG}i} &= a(P_{\text{DG}i}) \\ a &= \pm \tan\left\{\cos^{-1}(PF_{\text{DG}})\right\} \end{aligned} \tag{7}$$

For type 2 DG, optimal DG size is given as

$$Q_{\text{DG}i} = Q_{Di} + \frac{1}{\alpha_{ii}}\left[\beta_{ii}P_{Di} - \sum_{\substack{j=1 \\ j \neq i}} (\alpha_{ij}Q_j + \beta_{ij}P_j)\right] \tag{8}$$

After getting optimal DG size, the optimal place for their location is determined considering the total system losses placing each optimally sized DG one at a time at respective buses. The location which gives minimum system losses is selected as the optimal site and the system is now studied from the viewpoint of voltage stability by running load flow [1, 11].

4 Test System and Results

In this paper, for the purpose of voltage stability enhancement, the optimal placement of DG is proposed. In fact, here, it is investigated that DG when placed for the purpose of loss reduction can also serve the purpose of voltage stability improvement. In this work, type 2 and type 3 DG's are investigated for voltage stability enhancement of the primary distribution system.

The system is tested for two cases from the viewpoint of voltage stability. The first case is without DG and the second one is with DG. The procedure is followed once with DG type 2 and once with DG type 3. The DG systems can be classified into four categories as [7, 8]

1. Type 1: Which injects only the real power into the system.
2. Type 2: This injects only the reactive power into the system.
3. Type 3: Which injects real as well as reactive power into the system, and
4. Type 4: This injects real power but at the same time withdraws reactive power from the system.

The algorithm for DG placement is tested on IEEE 33 bus system having a 3.72 MW of real power and 2.3 MVAR reactive power [7]. The system is a radial distribution system and is studied considering the following constraints:

- The DG capacity should not exceed the total load demand of the system, and
- The allowable voltage range is taken to be between 0.95 and 1.05 pu.

The 33-bus test distribution system is shown in Fig. 1.

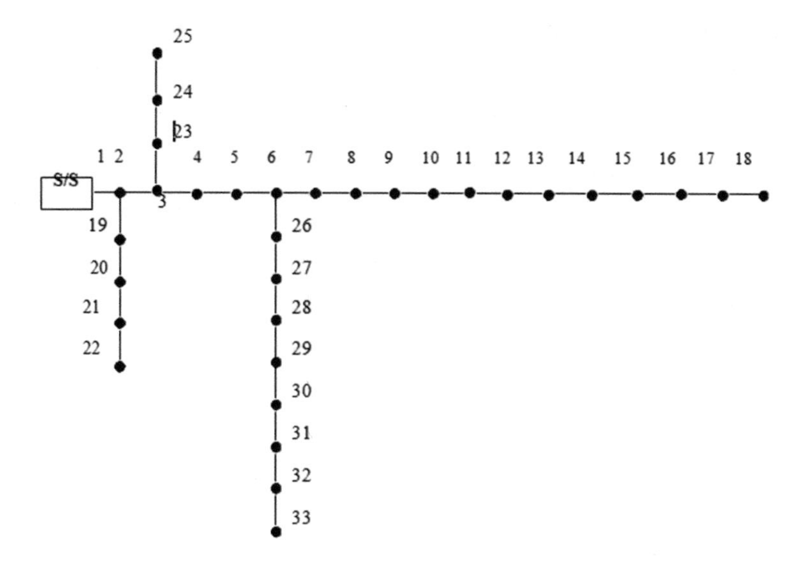

Fig. 1 33-Bus test distribution system

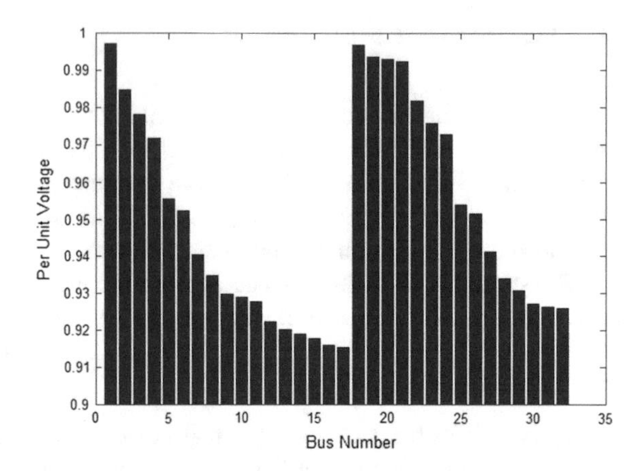

Fig. 2 Base case voltage profile

The program for determination of DG size and voltage profile of the system is written in MATLAB environment and results are obtained, first, for determination of system voltage profile without any DG penetration and then for obtaining DG size and DG site. The optimized DG size and site is obtained keeping in mind the distribution system loss reduction. The voltage profile of the test system is recorded in two cases, i.e., without and with DG penetration. System load flow is run for voltage profile determination. The loading factor is increased by 0.1 and the system performance in terms of bus voltage is recorded to draw the system PV curve.

The voltage profile of the system without DG in Fig. 2 shows that the voltage is minimum at bus no. 17.

The minimum voltage is 0.9156 pu and is occurring at bus no. 17 and the total distribution system losses are 180.1369 kW. The PV curve for the same bus is given in Fig. 3.

The variation of bus voltages with loading factor varying from 1 to 2.0 (as for distribution system overloading up to twice the system load is permitted for 1 hour) is shown in Table 1.

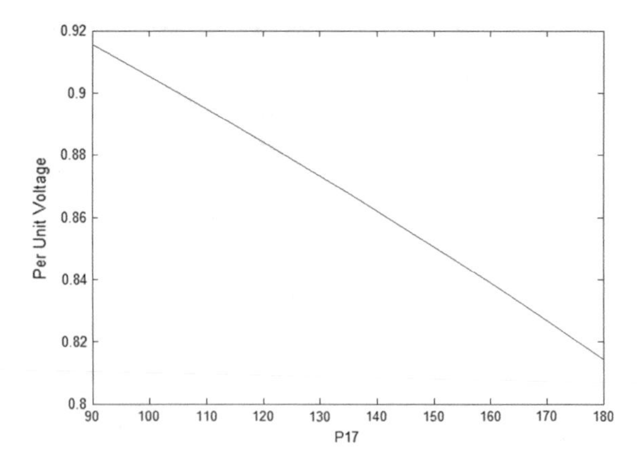

Fig. 3 P-V curve for bus No. 17

Table 1 Voltage variation with loading factor increment by 0.1 up to 2 without DG

Bus No.	Vpu at 1.0	Vpu at 1.1	Vpu at 1.2	Vpu at 1.3	Vpu at 1.4	Vpu at 1.5	Vpu at 1.6	Vpu at 1.7	Vpu at 1.8	Vpu at 1.9	Vpu at 2.0
1	0.9974	0.9971	0.9968	0.9965	0.9962	0.9959	0.9957	0.9953	0.9950	0.9947	0.9944
2	0.9850	0.9834	0.9817	0.9801	0.9784	0.9767	0.9750	0.9732	0.9715	0.9696	0.9678
3	0.9784	0.9761	0.9737	0.9713	0.9689	0.9664	0.9639	0.9614	0.9588	0.9561	0.9534
4	0.9719	0.9688	0.9658	0.9626	0.9595	0.9562	0.9530	0.9496	0.9462	0.9427	0.9391
5	0.9555	0.9507	0.9458	0.9409	0.9358	0.9307	0.9254	0.9201	0.9146	0.9090	0.9033
6	0.9524	0.9472	0.9420	0.9366	0.9312	0.9257	0.9200	0.9143	0.9084	0.9024	0.8963
7	0.9404	0.9339	0.9274	0.9207	0.9138	0.9069	0.8998	0.8926	0.8852	0.8776	0.8698
8	0.9349	0.9278	0.9206	0.9133	0.9058	0.8982	0.8904	0.8824	0.8743	0.8660	0.8575
9	0.9299	0.9222	0.9145	0.9066	0.8985	0.8903	0.8319	0.8733	0.8645	0.8556	0.8464
10	0.9291	0.9214	0.9136	0.9056	0.8974	0.8891	0.8806	0.8719	0.8631	0.8540	0.8447
11	0.9278	0.9199	0.9120	0.9038	0.8955	0.8870	0.8784	0.8695	0.8605	0.8512	0.8417
12	0.9224	0.9139	0.9053	0.8966	0.8876	0.8785	0.8692	0.8596	0.8499	0.8399	0.8296
13	0.9204	0.9117	0.9029	0.8939	0.8847	0.8753	0.8657	0.8559	0.8459	0.8356	0.8251
14	0.9191	0.9103	0.9013	0.8922	0.8829	0.8733	0.8636	0.8536	0.8435	0.8330	0.8223
15	0.9179	0.9090	0.8999	0.8906	0.8811	0.8714	0.8615	0.8514	0.8411	0.8305	0.8196
16	0.9161	0.9070	0.8977	0.8882	0.8785	0.8686	0.8585	0.8481	0.8375	0.8267	0.8156
17	0.9156	0.9064	0.8970	0.8874	0.8777	0.8677	0.8575	0.8471	0.8365	0.8256	0.8144
18	0.9969	0.9966	0.9963	0.9959	0.9956	0.9952	0.9949	0.9945	0.9942	0.9938	0.9935
19	0.9937	0.9931	0.9924	0.9918	0.9911	0.9905	0.9898	0.9891	0.9884	0.9878	0.9871
20	0.9931	0.9924	0.9917	0.9910	0.9903	0.9895	0.9888	0.9881	0.9873	0.9866	0.9858
21	0.9925	0.9918	0.9910	0.9902	0.9895	0.9887	0.9879	0.9871	0.9863	0.9855	0.9847
22	0.9818	0.9799	0.9779	0.9759	0.9739	0.9719	0.9698	0.9678	0.9656	0.9635	0.9613
23	0.9759	0.9734	0.9708	0.9682	0.9656	0.9629	0.9603	0.9575	0.9548	0.9520	0.9491

(continued)

Table 1 (continued)

Bus No.	Vpu at 1.0	Vpu at 1.1	Vpu at 1.2	Vpu at 1.3	Vpu at 1.4	Vpu at 1.5	Vpu at 1.6	Vpu at 1.7	Vpu at 1.8	Vpu at 1.9	Vpu at 2.0
24	0.9730	0.9701	0.9673	0.9644	0.9614	0.9585	0.9555	0.9524	0.9494	0.9462	0.9431
25	0.9538	0.9488	0.9438	0.9386	0.9333	0.9280	0.9225	0.9170	0.9113	0.9055	0.8996
26	0.9516	0.9463	0.9410	0.9356	0.9301	0.9245	0.9187	0.9129	0.9069	0.9008	0.8946
27	0.9414	0.9350	0.9286	0.9220	0.9153	0.9085	0.9015	0.8944	0.8872	0.8798	0.8722
28	0.9340	0.9269	0.9196	0.9122	0.9047	0.8970	0.8892	0.8812	0.8730	0.8647	0.8561
29	0.9309	0.9234	0.9153	0.9080	0.9001	0.8921	0.8339	0.8755	0.8669	0.8581	0.8492
30	0.9272	0.9193	0.9113	0.9031	0.8948	0.8863	0.8777	0.8688	0.8598	0.8505	0.8411
31	0.9264	0.9184	0.9103	0.9020	0.8936	0.8850	0.8763	0.8673	0.8582	0.8489	0.8393
32	0.9262	0.9182	0.9100	0.9017	0.8933	0.8846	0.8759	0.8669	0.8577	0.8483	0.8387

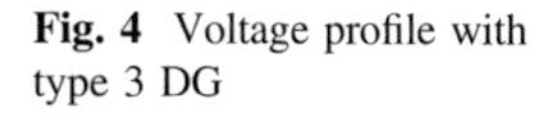

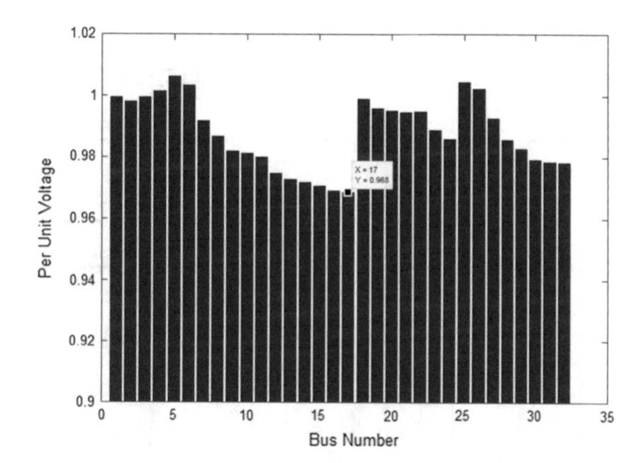

Fig. 4 Voltage profile with type 3 DG

The DG3 placed at bus 5 results in the voltage profile given in Fig. 4, which indicates that per unit voltage at bus 17 is still minimum but now it got improved to 0.968 which is well in the acceptable range. Total system losses are reduced to 57.86124 kW with DG placement of 2.93 MW and 1.82 MVAr at bus 5 (Table 2; Fig. 5).

With loading factor increased from 1 to 2 with a successive difference of 0.1, shows the following PV characteristic for weakest bus 17 as follows:

With the PV curve it can be seen that loading characteristics in respect of voltage stability are enhanced by connecting DG type 3. The same procedure can be implemented by placing DG type 2. Type 2 DG's are basically the suppliers of reactive power only. Following figures shows the implementation with the help of Type 2 DG. The losses are reduced to 138.66 kW with a DG size of 1.8175 MVAr at bus 5 (Table 3; Figs. 6 and 7).

Based on the observation from table, it is found that bus no. 17 is the weakest bus and hence for the same PV curve is drawn which renders the variation of its voltage with increased total power withdrawal at the bus. The minimum voltage at the bus for a loading of 2 times of the allocated load is 0.8339, which is less as compared to that obtained with DG 3 placement.

Table 2 Voltage variation with loading factor increment by 0.1 up to 2 with DG type 3

Bus No.	Vpu at 1.0	Vpu at 1.1	Vpu at 1.2	Vpu at 1.3	Vpu at 1.4	Vpu at 1.5	Vpu at 1.6	Vpu at 1.7	Vpu at 1.8	Vpu at 1.9	Vpu at 2.0
1	0.9994	0.9992	0.9989	0.9986	0.9984	0.9981	0.9981	0.9976	0.9973	0.9970	0.9967
2	0.9979	0.9965	0.9950	0.9935	0.9919	0.9903	0.9904	0.9873	0.9856	0.9840	0.9824
3	0.9994	0.9973	0.9952	0.9930	0.9908	0.9884	0.9886	0.9841	0.9818	0.9794	0.9770
4	1.0013	0.9986	0.9958	0.9930	0.9901	0.9870	0.9872	0.9813	0.9783	0.9752	0.5721
5	1.0062	1.0019	0.9975	0.9931	0.9886	0.9835	0.5840	0.9746	0 9698	0.9650	0.9600
6	1.0032	0.9986	0.9939	0.9891	0.9842	0.9788	0.9793	0.9692	0.9641	0.9588	0.9535
7	0.9919	0.9860	0.9801	0.9740	0.9679	0.9611	0.9616	0.9489	0.9423	0.9356	0.9288
8	0.9867	0.9802	0.9737	0.9670	0.9603	0.929	0.9534	0.9394	0.9322	0.9248	0.9174
9	0.9819	0.9750	0.9679	0.9607	0.9535	0.9455	0.9461	0.9309	0.9231	0.9151	0.9070
10	0.9812	0.9742	0.9670	0.9598	0.9524	0.9444	0.9449	0.9296	0.9217	0.9136	0.9054
11	0.9800	0.9728	0.9655	0.9581	0.9506	0.9425	0.9430	0.9273	0.9193	0.9111	0.9027
12	0.9749	0.9671	0.9593	0.9513	0.9432	0.9344	0.9350	0.9181	0.9094	0.9005	0.8915
13	0.9730	0.9650	0.9570	0.9488	0.9404	0.9315	0.9320	0.9146	0.9057	0.8966	0.8873
14	0.9718	0.9637	0.9555	0.9472	0.9387	0.9296	0.9301	0.9125	0.9034	0.8541	0.8847
15	0.9706	0.9614	0.9541	0.9457	0.9371	0.9273	0.9283	0.9104	0.9012	0.8918	0.8821
16	0.9689	0.9606	0.9520	0.9434	0.9346	0.9251	0.9257	0.9073	0.8979	0.8882	0.8784
17	0.9684	0.9600	0.9514	0.9427	0.9339	0.9243	0.9249	0.9064	0.8969	0.8872	0.8773
18	0.9990	0.9987	0.9983	0.9980	0.9977	0.9974	0.9974	0.9968	0.9964	0.9961	0.9958
19	0.9958	0 9952	0.9945	0.9939	0.9933	0.9926	0.9926	0.9914	0.9907	0.9901	0.9894
20	0.9952	0.9945	0.9938	0.9931	0.9924	0.9917	0.9917	0.9903	0.9896	0.9889	0.9881
21	0.9946	0.9939	0.9931	0.9924	0.9916	0.9908	0.9908	0.9893	0.9885	0.9878	0.9870
22	0.9948	0.9930	0.9912	0.9894	0.9875	0.9855	0.9857	0.9818	0.9799	0.9779	0.9759
23	0.9890	0.9866	0.9842	0.9818	0.9793	0.9767	0.9768	0.9713	0.9692	0.9666	0.9640

(continued)

Table 2 (continued)

Bus No.	Vpu at 1.0	Vpu at 1.1	Vpu at 1.2	Vpu at 1.3	Vpu at 1.4	Vpu at 1.5	Vpu at 1.6	Vpu at 1.7	Vpu at 1.8	Vpu at 1.9	Vpu at 2.0
24	0.9861	0.9834	0.9807	0.9780	0.9752	0.9723	0.9724	0.9667	0.9639	0.9610	0.9580
25	1.0046	1.0001	0.9955	0.9909	0.9862	0.9810	0.9815	0.9718	0.9668	0.9617	0.9565
26	1.0025	0.9977	0.9929	0.9881	0.9832	0.9776	0.9782	0.9679	0.9627	0.9573	0.9519
27	0.9928	0.9870	0.9812	0.5753	0.9692	0.9626	0.5631	0.9506	0.9442	0.9377	0.9310
28	0.9859	0.9793	0.9727	0.9660	0.9592	0.9518	0.9523	0.9382	0.9309	0.9235	0.9160
29	0.9829	0.9760	0.9691	0.9621	0.9549	0.9472	0.9477	0.9328	0.9252	0.9174	0.9095
30	0.9794	0.9722	0.9649	0.9575	0.9499	0.9417	0.9423	0.9266	0.9185	0.9103	0.9020
31	0.9786	0.9714	0.9640	0.9564	0.9488	0.9405	0.9411	0.9252	0.9170	0.9087	0.9003
32	0.9784	0.9711	0.9637	0.9561	0.9485	0.9402	0.9407	0.9248	0.9166	0.9083	0.8993

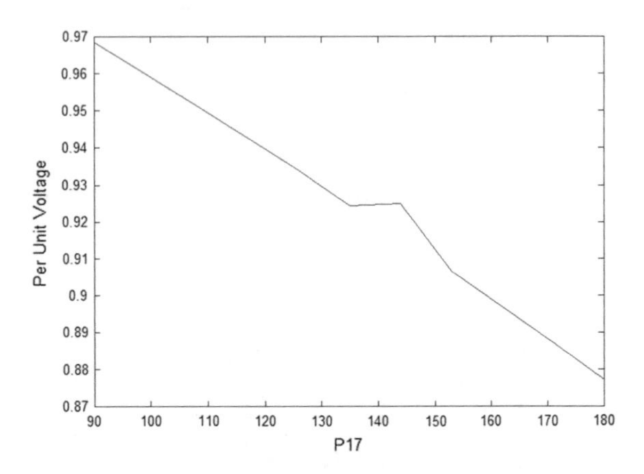

Fig. 5 PV curve for bus 17 after DG 3 placement

Table 3 Voltage variation with loading factor increment by 0.1 up to 2 with DG type 2

Bus No.	Vpu at 1.0	Vpu at 1.1	Vpu at 1.2	Vpu at 1.3	Vpu at 1.4	Vpu at 1.5	Vpu at 1.6	Vpu at 1.7	Vpu at 1.8	Vpu at 1.9	Vpu at 2.0
1	0.9979	0.9976	0.9974	0.9971	0.9968	0.9965	0.9962	0.9959	0.9956	0.9953	0.9950
2	0.9383	0.9868	0.9852	0.9836	0.9819	0.9803	0.9786	0.9769	0.9752	0.9734	0.9717
3	0.9838	0.9816	0.9793	0.9770	0.9746	0.9722	0.9698	0.9673	0.9648	0.9623	0.9597
4	0.9795	0.9765	0.9736	0.9705	0.9675	0.9643	0.9612	0.9579	0.9547	0.9513	0.9479
5	0.9710	0.9664	0.9617	0.9569	0.9520	0.9471	0.9420	0.9369	0.9316	0.9263	0.9208
6	0.9679	0.9629	0.9579	0.9527	0.9475	0.9422	0.9368	0.9312	0.9256	0.9199	0.9140
7	0.9562	0.9499	0.9435	0.9370	0.9305	0.9237	0.9169	0.9099	0.9028	0.8955	0.8881
8	0.9507	0.9439	0.9369	0.9298	0.9225	0.9152	0.9077	0.9000	0.8922	0.8842	0.8761
9	0.9458	0.9384	0.9309	0.9232	0.9154	0.9075	0.8994	0.8911	0 8826	0.8740	0.8652
10	0.9451	0.9376	0.9300	0.9222	0.9143	0.9063	0.8981	0.8897	0.8812	0.8725	0.8635
11	0.9438	0.9362	0.9284	0.9205	0.9125	0.9043	0.8959	0.8874	0.8787	0.8698	0.8607
12	0.9385	0.9303	0.9219	0.9134	0.9047	0.8959	0.8869	8.777	0.8683	0.8587	0.8488
13	0.9365	0.9281	0.9195	0.9107	0.9018	0.8928	0.8835	0.8741	0.8644	0.8545	0.8444
14	0.9353	0.9267	0.9180	0.9091	0.9001	0.8908	0.8814	0.8718	0.8620	0.8520	0.8417
15	0.9341	0.9254	0.9165	0.9075	0.8983	0.8890	0.8794	0.8697	0.8597	0.8495	0.8391
16	0.9323	0.9234	0.9144	0.9051	0.8958	0.8862	0.8764	0.8664	0.8562	0.8458	0.8351
17	0.9318	0.9228	0.9137	0.9044	0.8950	0.8853	0.8755	0.8655	0.8552	0.8447	0.8339
18	0.9974	0.9971	0.9968	0.9965	0.9961	0.9958	0.9955	0.9951	0.9948	0.9944	0.9941
19	0.9943	0.9936	0.9930	0.9923	0.9917	0.9910	0.9904	0.9897	0.9890	0.9884	0.9877
20	0.9936	0.9929	0.9922	0.9915	0.9908	0.9901	0.9894	0.9886	0.9879	0.9872	0.9864
21	0.9931	0.9923	0.9916	0.9905	0.9900	0.9892	0.9885	0.9877	0.9869	0.9861	0.9853
22	0.9852	0.9833	0.9814	0.9794	0.9775	0.9755	0.9735	0.9715	0.9694	0.9673	0.9652
23	0.9793	0.9768	0.9743	0.9717	0.9692	0 9666	09,639	0.9613	0.9586	0.9558	0.9531
24	0.9764	0.9736	0.9707	0.9679	0.9650	0.9621	0.9592	0.9562	0.9532	0.9501	0.9470
25	0.9694	0.9645	0.9596	0.9547	0.9496	0.9444	0.9392	0.9339	0.9284	0.9229	0.9172
26	0.9671	0.9621	0.9569	0.9517	0.9464	0.9410	0.9355	0.9298	0.9241	0.9183	0.9123

(continued)

Table 3 (continued)

Bus No.	Vpu at 1.0	Vpu at 1.1	Vpu at 1.2	Vpu at 1.3	Vpu at 1.4	Vpu at 1.5	Vpu at 1.6	Vpu at 1.7	Vpu at 1.8	Vpu at 1.9	Vpu at 2.0
27	0.9571	0.9510	0.9447	0.9383	0.9319	0.9253	0.9186	0.9118	0.9048	0.8977	0.8904
28	0.9499	0.9430	0.9359	0.9238	0.9215	0.9140	0.9065	0.8988	0.8909	0.8829	0.8747
29	0.9468	0.9395	0.9321	0.9246	0.9170	0.9092	0.9013	0.8932	0.8849	0.8765	0.8679
30	0.9432	0.9355	0.9277	0.9198	0.9118	0.9035	0.8952	0.8867	0.8780	0.8691	0.8600
31	0.9424	0.9347	0.9268	0.9188	0.9106	0.9023	0.8938	0.8852	0.8764	0.8674	0.8582
32	0.9422	0.9344	0.9265	0.9184	0.9102	0.9019	0.8934	0.8848	0.8759	0.8669	0.8577

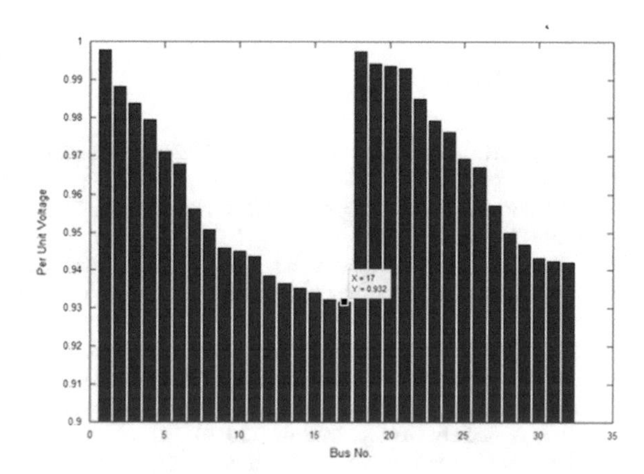

Fig. 6 Voltage profile with type DG at bus 5

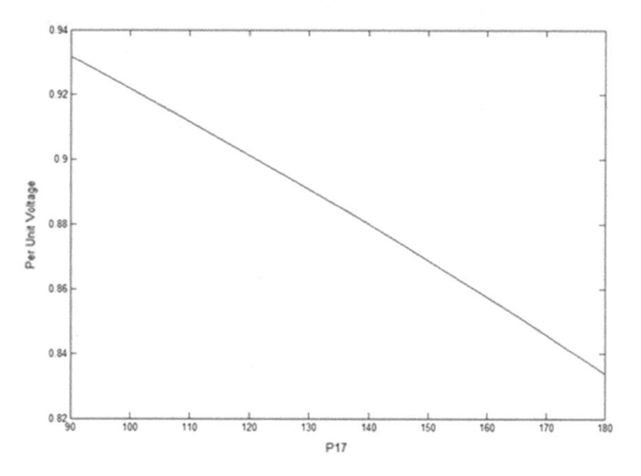

Fig. 7 PV curve for bus 17 after DG 2 placement

5 Conclusion

The work presented here proposes the use of DG for enhancement of the distribution system's voltage stability. But as the placement and size of DG in distribution system are very important from viewpoint of system I^2R losses, so they must be placed optimally to minimize these losses. In the paper, two types of DG's are placed for the voltage stability enhancement and they are placed at optimal locations and their size is also taken optimally to minimize losses. Placing DG for voltage profile improvement is advantageous in respect of reduction of system losses and also their placement relieves the substation capacity, reducing substation size and hence cost. The era is going to be of renewable energy which by its nature is distributed. The system can further be studied with two other types of DG which provide only active power and those which inject real power but absorb reactive power.

References

1. Teng, J.H.: A direct approach for distribution system load flow solutions. IEEE Trans. Power Delivery **18**(3), 882–887 (2003)
2. Kundur, P.: Power System Stability and Control. Tata McGraw-Hill Edition, New York (1994)
3. Georgilakis, P.S., Hatziargyriou, N.D.: Optimal distributed generation placement in primary distribution networks: Models, methods, and future research. IEEE Trans. Power Syst. **28**(3), 3420–3428 (2013)
4. Hoff, T.E., Wenger, H.J., Farmer, B.K.: Distributed generation: an alternative to electric utility investments in system capacity. Energ. Policy **24**(2), 137–147 (1996)
5. Zhang, Y., Rajagopalan, S., Conto, J.: Practical voltage stability analysis. In: IEEE Power and Energy Society General Meeting (2010)
6. Hien, N.C., Mithulananthan, N., Bansal, R.C.: Location and sizing of distributed generation units for loadabilty enhancement in primary feeder. IEEE Syst. J. **7**(4), 797–806 (2013)
7. Hung, D.Q., Mithulananthan, N., Bansal, R.C.: Analytical expressions for DG allocation in primary distribution networks. IEEE Trans. Energ. Convers. **25**(3), 814–820 (2010)
8. Hung, D.Q., Mithulananthan, N., Bansal, R.C.: Analytical expressions for DG allocation in primary distribution networks. IEEE Trans. Energ. Convers. **25**(3), 814–820 (2010)
9. Kothari, D.P., Dhillo, J.S.: Power System Optimization. Prentice-Hall of India, NJ, USA (2004)
10. Rani, P.S., Devi, A.L.: Optimal sizing of DG units using exact loss formula at optimal power factor. Int. J. Eng. Sci. Tech. **4**(9), 4043–4050 (2012)
11. Kumaraswamy, I., Tarakalyani, S., VenkataPrasanth, B.: Comparison of Voltage Stability Indices and its Enhancement Using Distributed Generation. In: Proceedings of World Congress on Engineering, 2016, vol I, WEC 2016, June 29–July 1 2016, London, UK. Comparison of Voltage Stability Indices and its Enhancement Using Distributed Generation

Demand Forecasting Using Artificial Neural Networks—A Case Study of American Retail Corporation

Aditya Chawla, Amrita Singh, Aditya Lamba, Naman Gangwani and Umang Soni

Abstract Artificial neural networks (ANNs) provide a way to make intelligent decisions while leveraging on today's processing power. In this paper, an attempt has been made to use ANN in demand forecasting by modeling it mathematically. MATLAB and R software are used to create the neural networks. Data has been organized and results are compared using Python. The complete analysis has been done using demand forecasting of American multinational retail corporation, Walmart. What we have managed to achieve in the end is almost perfect accuracy in forecasting demand of Walmart by ensuring that the set of inputs are complete enough to provide an output and then further ensuring that we do obtain an output. In compliance with the same, average sales of each Walmart store in question was calculated from training data and normalized. A correction factor was used to compensate for the effect of seasonality which is an external factor. By doing this, the model is saved from the trouble of having to map an extra factor which can otherwise be easily compensated for. The method used is a multi-layered perceptron in all cases. Iterations were done to find the best parameters to build the model.

Keywords Supply chain · Demand forecasting · Artificial neural networks Case study · Feature selection · Bullwhip effect

---

A. Chawla (✉) · A. Singh · A. Lamba · N. Gangwani · U. Soni
Netaji Subhas Institute of Technology, Sector-3 Dwarka,
New Delhi 110078, Delhi, India
e-mail: aditya21196@gmail.com

A. Singh
e-mail: amritacholia@gmail.com

A. Lamba
e-mail: adityalamba007@gmail.com

N. Gangwani
e-mail: naman.gangwani@gmail.com

U. Soni
e-mail: umang.soni.iitd@gmail.com

© Springer Nature Singapore Pte Ltd. 2019

H. Malik et al. (eds.), *Applications of Artificial Intelligence Techniques in Engineering*, Advances in Intelligent Systems and Computing 697,
https://doi.org/10.1007/978-981-13-1822-1_8

1 Introduction

Demand forecasting plays a key role in inventory management and optimization and determining how profitable a business is. All of this, in turn, affects supply chain management. A supply chain (SC) has a dynamic structure involving the constant flow of information, product, and funds between different stages [1]. Accurate demand forecasting could achieve a lot in terms of maximizing profit, increased sales, efficient production planning, etc. It is central to the planning and operation of retail business at micro and macro level [2]. All in all, a forecast of demand is a decision so crucial that an error committed while calculating it might cost an organization too much.

As the market grows, it becomes more diverse and factors affecting demand also become more dynamic. Each factor affects the demand for a product in its own complicated way and thus, cannot be mapped that easily. The advantage of using neural networks in such a scenario is that it maps the dependence of each variable involved accurately without us having to worry about the details of what is happening inside as long as we ensure that our artificial neural network learns as well as possible. But to use a neural network model for a practical problem, one of the most important tasks is to fit the problem well and properly model it mathematically. To do that, each of the factors which could affect the demand should be considered and be converted to a mathematical, physical quantity.

These factors should be independent of each other or else our neural network will get redundant inputs. Sometimes these factors could be random as well. For example, climate changes may affect the sales for the day and since climate for a far-off time cannot be predicted accurately, we cannot predict the sales for that day. But fortunately enough, even climate follows a set pattern and if we consider a longer period of time, say a week—we might be able to predict the sales for that week meanwhile assuming that climate acts the way it is not expected to only on a specific day of the week. Here, we are using the fact that random errors follow a Gaussian curve pattern to our advantage. Other factors may include region. There is no way, we can obtain a physical, mathematical quantity corresponding to a specific region which will help in forecasting demand of a product in that region. However, mapping the effect of how a region affects demand is not entirely impossible. In fact, it can be modeled very accurately using neural networks.

2 Literature Review

2.1 Supply Chain Management and Demand Forecasting

Supply chain management (SCM)—It is the management of flow of goods and services. It involves the movement and storage of raw materials, of work in process inventory, and of finished goods from point of origin to point of consumption. In short, it can be described as design, planning, execution, control, and monitoring of supply chain activities with the objective of creating net value, building a competitive infrastructure, leveraging worldwide logistics, synchronizing supply with demand, and measuring performance globally. With increasing globalization and easier access to alternative products in today's markets, the importance of product design to generating demand is more significant than ever [1]. Some points to be noted about supply chain management are as follows [3, 4]:

1. Supply chain initial decisions (supply chain design decisions) are important because once taken, they are very expensive to alter on a short notice. For e.g., a company should plan the locations of its warehouses and its proximity to the consumer base before starting production of a product. The more flexible is the supply chain design, the more likely it is that the business will be a successful one. That is because more opportunities will open up during supply chain planning later.
2. Sharing of information plays a key role in supply chain management. So setting up information systems to support supply chain operations can be quite valuable.
3. Effective forecasting helps a great deal in effective supply chain management.

Demand forecasting plays a pivotal role in SCM. If we know the demand in advance, we can counteract the bullwhip effect effectively. Mistakes can be made like overinvesting in overproducing due to increased estimates or under producing goods of one kind and not being able to meet the demand of the next stage in the supply chain, which, in turn, scales exponentially and upsets the entire supply chain (the bullwhip effect). In either of the cases, the use of capital available is less than satisfactory and this showcases the obvious need of demand forecasting.

1. There are many forecasting techniques that can be classified into four main groups:
2. Qualitative methods are primarily subjective; they rely on human judgment and opinion to make a forecast.
3. Time-series methods use historical data to make a forecast.
4. Causal methods involve assuming that the demand forecast is highly correlated with certain factors in the environment (e.g., the state of the economy, interest rate).
5. Simulation methods imitate the consumer choices that give rise to demand to arrive at a forecast [1].

2.2 *Artificial Neural Networks (ANNs)*

Any problem can be modeled mathematically and Artificial Neural Network (ANN) is one of the attempts of doing so [5–10]. Neural network, similar to the real human brain, has the required ability for learning and are able to utilize the acquired new experiences from new and similar affairs. Although ANNs are not comparable to the real human brain system, these networks equipped special features which make them privileged in some applications abilities such as separation of patterns and its amenability to learn the networks by linear and nonlinear mapping wherever the learning is required [4]. Just like the human brain, a neural network can learn techniques by various methods which could include learning by memory, by correction of parameters after randomly predicting results, by classification, and so on. But to use a neural network effectively in practical application, we must make a mathematical model of it as accurate as possible.

The idea of neural networks has been inspired from the brain. The brain consists of a large number (approximately 10^{11}) of highly connected elements (approximately 10^4 connections per element) called neurons. These neurons are connected in the sense that they pass electric signals among themselves on receiving input from the sensory organs and thus coming to various "output nodes" whence the decision is taken. This can be modeled mathematically as shown.

Figure 1: The first layer is that of input factors. In biological terms, that would be our 5 senses: sight, taste, skin sensations, sound, and smell. What we see through our eyes are images, colors, shapes all of which can be said to be input data. The hidden layers are where the decisions are taken. Each branch between two nodes has a certain "synaptic weight" which determines what input will be fed to the next node. We model that mathematically as a multiplication factor. The next layer is the hidden layer which in the case of our brain exists in thousands. Each node of this layer receives input from each connection to it and based on the cumulative result, fires itself. We model the hidden layer as shown in the illustration above. Each input received is multiplied by its synaptic weight and summed. A bias is further added to this sum. The role of the bias is to improve learning. A bias may be seen as an individual property of a neuron independent of input factors. The net input is fed through an activation function and then this neuron "fires", i.e., feeds this output to all the nodes it is connected to, as input. But then the question arises—how to decide the weights?

Fig. 1 Basic model of a single-layer neural network

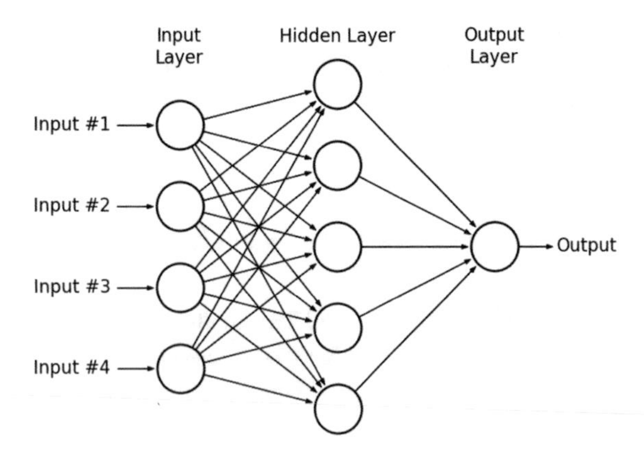

Our brain learns from experience or logic. But the trick behind our model is to learn from experience, while we make sure that the logic behind learning is sound. For that, we need data of the past and what were outputs corresponding to it. At first, we expose the neural network to the output along with input and let the network perform its learning. Then we use it to make a prediction. If the error is too much, the weights are changed. The weights are decided by backpropagation (BP) algorithm.

3 Problem Formulation

In 2014, Walmart had held an online contest to choose its employees on Kaggle. Walmart provided its sales data from 2010–2012 and asked the participants to predict the data for 2012–2013. The problem was approached by most statisticians using generic forecasting techniques. Today, in 2017, we attempt to solve the same problem using artificial neural networks. The data provided (Fig. 1) by Walmart consists of about 4.2 lakh entries. We separated out the data and used it for modeling different neural networks.

3.1 Devising a Neural Network for Each Department of Each Store

The input factors provided were in Excel format and contained all sorts of factors like CPI, discount markers, oil prices, holiday markers, etc., for each corresponding week. Out of these, we segregated the data for each store and each department and used these inputs for training our model. The model gave decent results. But there were overall 45 stores and 99 departments each. So, clearly, this model though accurate would be too tedious to work out for a single data scientist in MATLAB as it would require us to train 4455 neural networks. But then, the problem that arose was that if we use entries from all the departments and stores, there would be colossal values of error in our predictions because the effect of region was not mapped. To correct that, we decided a new input factor in our model called rank.

3.2 Devising a Neural Network for All the Stores but a Single Department

Since we are now modeling a neural network which can specifically predict sales of any store on any given day for a specific department, we tried to introduce as many corrective factors as possible. The first one was rank.

Rank—For that, we took the average sales of each store over the training period and ranked them. We then used the store rank as an input factor.

Null values—The Walmart data provides us with the data of those days on which there had been a sale and how much sales of discounted items occurred on those days due to sale. But when there had been no sale, their values are given as null. In the neural network, null inputs do not affect the transfer function, i.e., they are seen by the neural network as a 0. So, we carefully picked values to replace these null values to finally come up with these inputs for training.

The input factors (Table 1 shows a sample) were all decided as they affected the sales by intuition. An attempt was made to cover for all the random factors which could affect the sales. For example, the temperature for the day could affect which kinds of products would be sold that day and in what amount. The correlation seems weak but since the temperature could affect the sales in many ways, it was taken into consideration. In fact, the beauty of the neural networks lies in its ability to map factors such as temperature and CPI to the sales.

The data provided was too numerous to be processed on MATLAB. Therefore, details of 10 stores were first separated out. In the given set provided by Walmart, there were 143 dates and each date represented the first day of a week. So in total, the data was of 143 weeks. We divided the data using regex expressions in Python into 104 dates for training and rest 39 dates for testing purposes. The results of those 39 dates were withheld from the model and later compared to the values predicted by the models (Graphs 1, 2, 3, 4 and 5). All of the data separations were done using NumPy and Pandas libraries of python.

4 Solution

4.1 Use of MATLAB in Creating ANN

MATLAB's NN Toolbox was used to create the neural network. First the training, simulation, and target files were imported into the tool. Then similar to Efendigil [11], various combinations of activation function, no. of hidden layers, no. of neurons in the hidden layer and transfer function of the hidden layer were used to make different networks and their simulation results were analyzed to find the combination that gives best results, i.e., least error in the forecast. The type of network (feed forward backpropagation), adaption learning function (LEARNGDM—gradient descent with momentum weight and bias learning function) and performance function (MSE—mean squared error or mean squared deviation) were kept the same for all these combinations.

Different Activation functions and Transfer used were the following.

The neural network which gave the best results for 11 inputs was a network with 1 hidden layer having 22 neurons and transfer function TANSIG—sigmoid tangent (Table 3 (1)) and activation function TRAINLM—Levenberg–Marquardt

Table 1 Sample interpretation of input factors after data processing

Store	Date	Temperature	Fuel price index	Discount marker 1	Discount marker 2	Discount marker 3	Discount marker 4	CPI	Unemployment index	Is holiday	Rank
1	2/5/2010	42.31	2.572	−2000	−500	−100	−500	0.964	8.106	0	13
1	2/12/2010	38.51	2.548	−2000	−500	−100	−500	0.2422	8.106	1	13
1	2/19/2010	39.93	2.514	−2000	−500	−100	−500	0.2891	8.106	0	13
1	2/26/2010	46.63	2.561	−2000	−500	−100	−500	0.3196	8.106	0	13
1	3/5/2010	46.5	2.625	−2000	−500	−100	−500	0.3501	8.106	0	13
1	3/12/2010	57.79	2.667	−2000	−500	−100	−500	0.3806	8.106	0	13
1	3/19/2010	54.58	2.72	−2000	−500	−100	−500	0.2156	8.106	0	13

Table 2 Different activation functions used

Sr. No.	Function name	Algorithm
1	TRAINLM	Levenberg–Marquadt backpropagation
2	TRAINSCG	Scaled conjugate gradient backpropagation
3	TRAINRP	Resilient backpropagation
4	TRAINBFG	BFGS quasi-Newton backpropagation
5	TRAINBR	Bayesian regularization

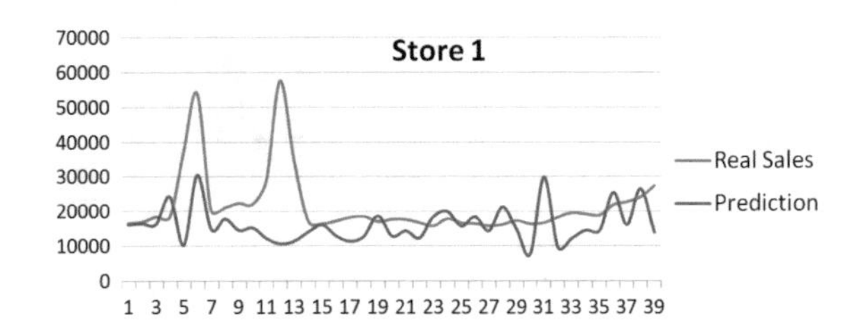

Graph 1 Comparison of predictions with real sales for store 1

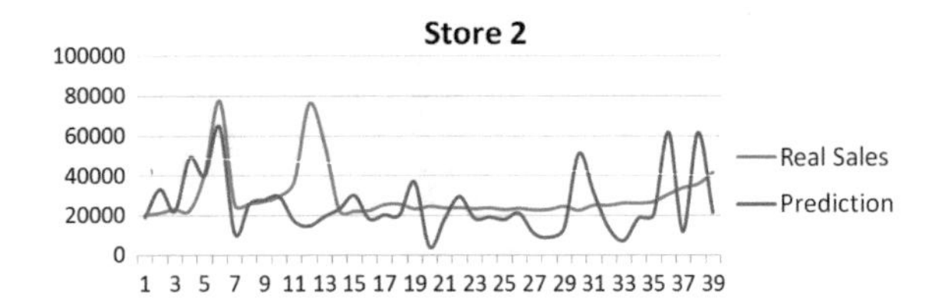

Graph 2 Comparison of predictions with real sales for store 2

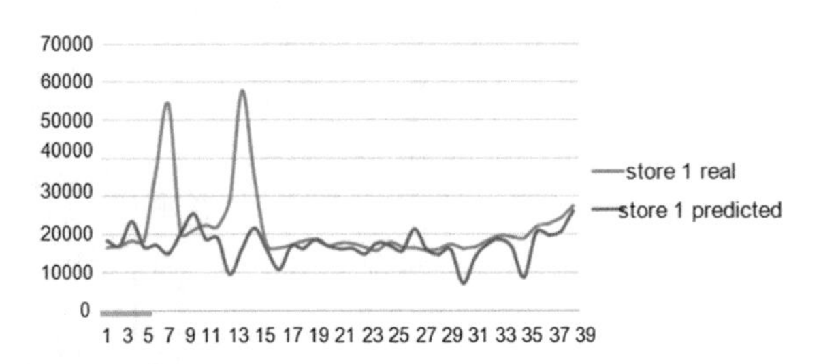

Graph 3 Comparison of real sales versus predicted sales by ANNs in R

backpropagation (Table 2 (1)). This network was then used for rest of the demand forecasting.

As we can see in Graphs 1 and 2, no model fits the graph perfectly. This is because the factors incorporated were not sufficient to map all factors. Certain festive occasions such as the Thanksgiving Day and Christmas were leading to increased sales. To map this, the festive seasons were identified and if the sales were far lower than expected, we assumed that our neural network was producing an outlier and multiplied the sales of that by a carefully chosen factor.

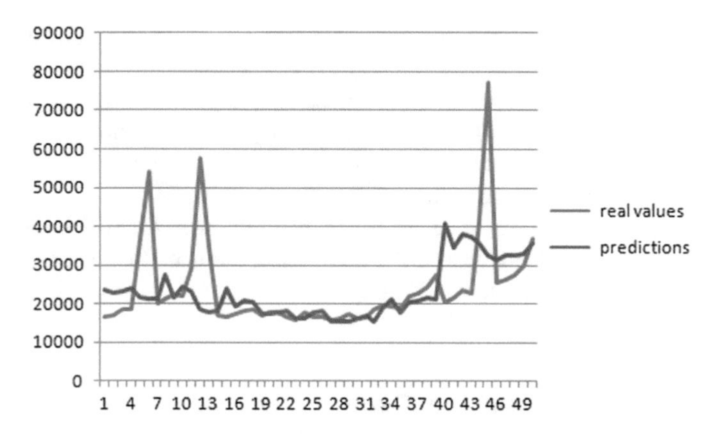

Graph 4 Comparison of predictions with real sales using random forest regression with fully optimized algorithm from scikit learn

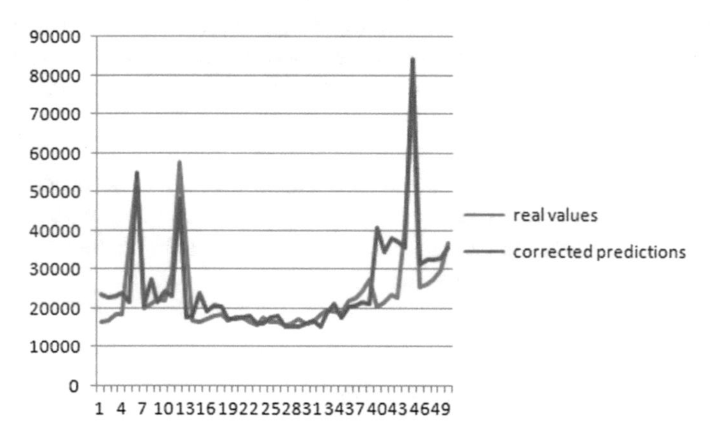

Graph 5 Comparison of random forest regression after applying corrections

Table 3 Transfer functions used

Sr. No.	Function name
1	Tansig (sigmoid tangent)
2	Logsig (logarithmic sigmoid)

4.2 Use of R in Creating ANN

R is not a name of software, but it is a language and environment for data management, graphic plotting, and statistical analysis [12]. R is freely available and is an open-source environment that is supported by the world research community

Neuralnet is a very flexible package. The backpropagation algorithm and three versions of resilient backpropagation are implemented and it provides a custom choice of activation and error function. An arbitrary number of covariates and response variables as well as of hidden layers can theoretically be included [6].

Neural networks are applicable to almost all problems where the goal is to represent a relation between predictor variables (independent or input variables) and predicted variables (dependent or output variables), especially when this relation is too complex to be considered as association or clustering 2].

The results obtained (Graph 3) are better than the ones obtained in MATLAB, but at the cost of processing power. As we can see, the neural network is far more complex than the one used in MATLAB.

4.3 Results After Applying Correction Factor

Here, Graph 3 shows the results after training the data with all the features accounted for (except seasonality) on Python random forest regression. The reason for such results is that there is no dearth of computational power available. But some peaks exist for which the results are inconsistent with actual values. The reason for those is occasional festivities like Christmas, Super Bowl, etc. They need to be corrected still.

Here, in Graph 4, the correction makes the result fit almost perfectly. As compared to MATLAB, even with limited processing power, neural networks were able to produce very good results as well. This shows that if all the independent inputs are used and the forecasting technique is recursive enough, the challenge of demand forecasting can be reduced to just data sampling.

5 Limitations and Future Scope

The main drawback lies in the fact that MATLAB was unable to iterate through the data when the number of layers was increased. This leads to improper mapping of input factors to the output. Despite the same, efforts were made to choose the technique which was reliable enough to compensate for this drawback. What could be done to bypass this is to set up an environment (preferably GPU based) which could handle more perceptron layers. Alternatively, RNNs (Recursive neural networks) could be used atop this model for better results.

6 Conclusion

Artificial neural networks offer great value to statisticians for demand forecasting as they are capable of producing very accurate results. The only pre-requisite is that the problem of demand forecasting is modeled carefully. All factors that might affect the demand should be taken into perspective and should be independent of each other as much as possible. The feature optimized results of MATLAB and R produced here can be leveraged work on larger datasets to optimize hyper-parameters of neural networks working on GPUs to produce best results meanwhile providing immense accuracy in demand forecasting which could further help in Supply Chain Management. In this study, an effort was made to showcase how a practical problem could be modeled to fit the application of ANN's as accurately as possible. The results have shown that ANN's can be used on very large datasets as well and still produce very less errors. While R provides better accuracy, it consumes more time in doing so. In MATLAB, the neural networks produce results faster in comparison and also allow for better parameter optimization. Though these software cannot be utilized in industry application as they cannot process bulk data, they provide a way to find the parameters which in turn could be used to carry out demand forecasting using neural networks on an industry scale.

References

1. Chen, C.H.: Neural networks for financial market prediction. In: IEEE World Congress Computational Intelligence (1994). Chopra, S., Meindl, P.: Supply Chain Management: Strategy, Planning and Operation. Prentice Hall, NJ, (2001)
2. Stalidis, G., Karapistolis, D., Vafeiadis, A.: Marketing decision support using artificial intelligence and knowledge modeling: Application to tourist destination and management. Proc. Soc. Behav. Sci. **175**, 106–113 (2015)
3. Zhang, G., Patuwo, B., Hue, M.: Forecasting with artificial neural networks: The state of the art. Int. J. Forecast. **14**, 35–62 (1998). Horton, N.J., Kleinman, K.: Using R For Data Management, Statistical Analysis, and Graphics. CRC Press, Clermont (2010)
4. Sultan, J.A., Jasim, R.M.: Demand forecasting using artificial neural networks optimized by artificial bee colony. Int J. Manage, Inf. Technol. Eng. **4**(7), 77–88 (2016)
5. MathWorks, Inc: Using MATLAB, Version 6. MathWorks, Inc, MA, USA (2000)
6. Günther, F., Fritsch, S.: Neuralnet: Training of Neural Networks. R J. **2**(1), 30–38 (2010)
7. Hagan, M.T., Demuth, H.B.: Neural Networks Design, 2nd edn. PWS Publication, Boston (1996)
8. Sengupta, S.: NPTEL lectures on Neural networks and Applications. http://nptel.ac.in/courses/117105084/
9. Chang, P.C., Wang, Y.W.: Fuzzy Delphi and backpropagation model for sales forecasting in PCB industry. Expert Syst. Appl. **30**(4), 715–726 (2006)
10. Trippi, R.R., Turban, E. (eds.): Neural Networks in Finance and Investing: Using Artificial Intelligence Toimprove Real—World Performance. Probus, Chicago (1993)
11. Efendigil, T.: a decision support system for demand forecasting with artificial neural networks and neuro fuzzy models: A comparative analysis. Expert Syst. Appl. **36**(3), 6697–6707 (2009)
12. Kabacoff, R.: R in Action. Manning Publications Co, Shelter Island (2011). Lander, J.P.: R for Everyone: Advanced Analytics and Graphics. Addison-Wesley Professional, Boston (2014)

Comparative Analysis of Reliability Optimization Using a New Approach of Redundancy Optimization Under Common Cause Failure

Kalpana Hazarika and G. L. Pahuja

Abstract Redundancy Optimization (RO) remained as a regular practice for reliability optimization of any systems since a long time ago. This practice is being carried out either at component level or system level with identical elements. This demands selection of right component or subsystem to improve the overall reliability. For component selection of a simple system, existing heuristic approach does not consume ample time, but it increases the time and computational complexity once the system becomes complex. Hence, a new approach has been proposed here for redundancy optimization. In this approach, the algorithm is developed to search the highest reliable path for optimization, rather than searching for weak components in all possible paths unlike in existing heuristic methods. The highest reliable path is now considered as main path and second highest reliable path as back up path. Further to mitigate the root cause of Common Cause Failures (CCF), design divergent components have been introduced at all subsystem levels instead of identical one. For validation of the proposed approach, an existing heuristic component selection criterion is implemented using this new approach for reliability optimization of a complex system represented by a reliability block diagram. To compare the effects of CCF, system reliability is evaluated with identically and nonidentically distributed components in the presence of CCF.

Keywords Common cause failures · Main path and backup path
Redundancy optimization · Complex network

K. Hazarika (✉) · G. L. Pahuja
Department of Electrical Engineering, National Institute of Technology, Kurukshetra, Haryana, India
e-mail: hazarika.kalpana@gmail.com

© Springer Nature Singapore Pte Ltd. 2019

H. Malik et al. (eds.), *Applications of Artificial Intelligence Techniques in Engineering*, Advances in Intelligent Systems and Computing 697,
https://doi.org/10.1007/978-981-13-1822-1_9

1 Introduction

Search for a main and back up path in complex communication network has become an important and essential task for achieving the more reliable communication under any circumstances to maximize the customer's satisfaction and to remain on the top priority in today's market. Generally, optimization of any networks with s-identical component is remained as a very common practice till date [1–3]. When critically analyzed, it has been observed that, it does not yield the benefits of redundancy under common cause failures. Because, when any common cause failures occur, identical redundant components fail concurrently inside the group [2, 4, 5]. Although metaheuristic approaches are used for reliability estimation [6, 7], but reliability optimization under common cause failures is still remained as a rarely studied area. Hence, by introducing the design of divergent components in a complex system, the impact of common cause failure is tried to reduce here [5]. The assumption of identically distributed components represents that the reliability of each component is similar. Whereas nonidentical indicates the components of different design or technology and hence having a different component reliability. On the other hand, different components cause failure processes that also involve different number of components failure, hence, different failure rates are considered here. The complete analysis is discussed under two major headings such as reliability optimization under common cause failures, where problem assumptions, process to implement the proposed approach, and evaluation are done on a complex network. In the next successive section, results are compared with respective cases to observe the effects.

Notations

$R_s(x_i)$	System reliability consisting of x_i components
$R_{IC}(t)$	System reliability under CCFs at the instant t
r_i	Reliability of subsystem i
$R_{ICi}(t)$	Reliability of subsystem i under CCFs at the instant t
q_i	Unreliability of subsystem i
F_{ICi}	Unreliability of the subsystem i under CCFs at the instant t
p_i	Reliability of single component of individual subsystem
λ_{ij}	Failure rate of subsystem i involving j no of components
R_{P_m}	Reliability of main path
R_{P_b}	Reliability of backup path
R_{ddP_m}	Reliability of main path with design divergent component
R_{ddP_b}	Reliability of backup path with design divergent component.

2 Reliability Optimization Under Common Cause Failures

A typical complex network encompassing below-mentioned assumptions is considered here for analysis and comparison of the existing and the proposed improved heuristic approach for redundancy optimization. To reduce the adverse effects of CCFs caused in presence of identical redundant components, they are being replaced by design divergent components. To incorporate common cause failures in system, conditional probability is applied here [5].

2.1 *Assumptions*

I. Identical components of individual subsystems will have identical failure rates. On the contrary, design divergent components of all individual subsystems will have different failure rates.
II. All components are of bi-states, i.e., either good or bad.
III. Components are subjected to different failure groups.

2.2 *Procedure for Redundancy Optimization as per the Proposed Approach*

I. Select highest reliable path of the existing network as main path
II. Select second highest reliable path of the same network as backup path
III. Apply component selection criteria [3, 8]
IV. Determine the optimum configuration
V. Determine reliability of main path and back up path for s-identical components with and without CCF
VI. Determine reliability of above for non-*s*-identical components with and without CCF.

2.3 *Problem*

(A) Applying old approach [3] reliability of optimized network $\{x^* = 3, 2, 2, 1, 1\}$ without considering CCFs is (Fig. 1; Table 1)

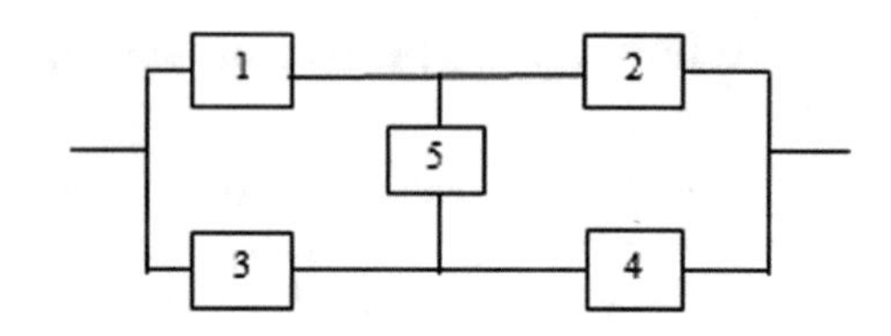

Fig. 1 Complex network

Table 1 Reliability and cost of single identical components of subsystems are

x_i	1	2	3	4	5
P_i	0.7	0.85	0.75	0.80	0.9
C_i	2	3	2	3	1

$$\begin{aligned} R_s = r_1r_2 + r_3r_4 + r_1r_4r_5 + r_2r_3r_5 - r_1r_2r_3r_4 - r_1r_2r_4r_5 \\ - r_1r_3r_4r_5 - r_1r_2r_3r_5 - r_2r_3r_4r_5 + 2r_1r_2r_3r_4r_5 \end{aligned} \quad (1)$$

Including CCF effects using conditional probability, the network reliability is

$$\begin{aligned} R_{\mathrm{IC}}(t) = {} & R_{\mathrm{IC1}}R_{\mathrm{IC2}} + R_{\mathrm{IC3}}R_{\mathrm{IC4}} + R_{\mathrm{IC1}}R_{\mathrm{IC4}}R_{\mathrm{IC5}} + R_{\mathrm{IC2}}R_{\mathrm{IC3}}R_{\mathrm{IC5}} - R_{\mathrm{IC1}}R_{\mathrm{IC2}}R_{\mathrm{IC3}}R_{\mathrm{IC4}} \\ & - R_{\mathrm{IC1}}R_{\mathrm{IC2}}R_{\mathrm{IC4}}R_{\mathrm{IC5}} - R_{\mathrm{IC1}}R_{\mathrm{IC3}}R_{\mathrm{IC4}}R_{\mathrm{IC5}} - R_{\mathrm{IC1}}R_{\mathrm{IC2}}R_{\mathrm{IC3}}R_{\mathrm{IC5}} \\ & - R_{\mathrm{IC2}}R_{\mathrm{IC3}}R_{\mathrm{IC4}}R_{\mathrm{IC5}} + 2R_{\mathrm{IC1}}R_{\mathrm{IC2}}R_{\mathrm{IC3}}R_{\mathrm{IC4}}R_{\mathrm{IC5}} \end{aligned} \quad (2)$$

(B) (i) Considering the proposed new approach, configurations of main path and backup path of above network optimizing with *s*-identical components are as follows:

$$\begin{aligned} P_{\mathrm{m}} &= \{\mathrm{X}_3,\ \mathrm{X}_4\} \Rightarrow \{4, 2\} \\ P_{\mathrm{b}} &= \{\mathrm{X}_1, \mathrm{X}_2\} \Rightarrow \{1, 1\} \end{aligned}$$

Hence, overall reliability of optimized network is

$$R_s(x_i) = (Rp_{\mathrm{m}}) + (Rp_{\mathrm{b}}) - (Rp_{\mathrm{m}}xRp_{\mathrm{b}}) \quad (3)$$

Now incorporating CCFs to above, reliability of this network is

$$R_{\mathrm{IC}}(t) = R_{\mathrm{IC}P_{\mathrm{m}}}(t) + R_{\mathrm{IC}P_{\mathrm{b}}}(t) - R_{\mathrm{IC}P_{\mathrm{m}}}(t)xR_{\mathrm{IC}P_{\mathrm{b}}}(t) \quad (4)$$

(ii) Considering proposed approach, configurations of main path and backup path of above network optimizing with non-*s*-identical components, are as follows:

Table 2 Reliability and cost of multiple design divergent components of subsystems are

Subsystems	X_1			X_2			X_3			X_4			X_5
Component types	1	1′	1″	2	2′	2″	3	3′	3″	4	4′	4″	5
P_i	0.7	0.75	0.69	0.85	0.87	0.83	0.75	0.77	0.73	0.8	0.82	0.78	1
C_i	2	2.5	1.8	3	3.2	2.9	2	2.4	1.9	3	3.2	2.9	1

Table 3 Comparative reliability results of optimized network under CCFs with *s*-identical and design divergent components

Configurations	Type of components	Reliability with CCFs
(B) As per the new approach (i) $X^* = \{2, 1, 4, 2\}$ (ii) $X^* = \{2, 1, 3, 2\}$	(i) With identical components	0.9822815
	(ii) With design divergent components	0.9891059
(A) As per "Shi" approach $X^* = \{3, 2, 2, 1, 1\}$	With identical components	0.930623
	With design divergent components	0.9723322

$$P_{\mathrm{m}} = \{X_3, X_4\} \Rightarrow \{3, 2\}$$
$$P_{\mathrm{b}} = \{X_1, X_2\} \Rightarrow \{2, 1\}$$

Now incorporating CCFs for divergent components (Table 2) in above Eq. (3), overall reliability of optimized network is

$$R_{\mathrm{dds}}(x_{\mathrm{idd}}) = (R_{\mathrm{dd}P_{\mathrm{m}}}) + (R_{\mathrm{dd}P_{\mathrm{b}}}) - (R_{\mathrm{dd}P_{\mathrm{m}}} x R_{\mathrm{dd}P_{\mathrm{b}}}) \tag{5}$$

2.4 Comparative Results

See Table 3.

3 Observation and Conclusion

There is a large difference observed, when comparing the results of same network with *s*-identical and non-identical components under common cause failures. To improve on that, when the reliability optimization by redundancy allocation is done as per existing heuristic method [9], the configuration found based on new approach is showing better performance as compared to "Shi's" configuration under CCFs. Hence, it may be concluded that for the search of optimized configuration;

diversified components should be opted for redundancy optimization in component level for improved reliability achievement. As the network size increases, the no. of paths increases. As per "Shi's" Algorithm, the selection factors must be checked for all the paths which increase the computational time. Instead, based on the new approach, if redundancy allocation is done only by checking the main and backup paths selection factor, the reliability can be improved within less computational time and complexity. The results may be even better than this if all divergent components would be of superior design with higher reliability as compared to existing network components. Here, in this example, we have considered both high- and low-reliable components as divergent components.

References

1. Bar Ness, Y., Livni, H.: Reliability optimization in the design of telephone networks. IEEE Trans. Reliab. **R-27**(5) (1978)
2. Chae, K.C., Clark, G.M.: System reliability in the presence of common-cause failures. IEEE Trans. Reliab. **R-35**(1) (1986)
3. Dighua, S.: A new heuristic algorithm for constrained redundancy optimization in complex system. IEEE Trans. Reliab. **R-36**(5) (1987)
4. Yaun, J.: Pivotal decomposition to find availability and failure frequency of systems with common cause failures. IEEE Trans. Reliab. **R-36**(1) (1987)
5. Page, L.B., Perry, J.E.: A model for system reliability with common cause failure. IEEE Trans. Reliab. **R-33**(4) (1989)
6. Bhardwaj, A.K., Gajpal, Y., Singh, M.: Fuzzy reliability analysis of integrated network traffic visualization system. J. Intell. Fuzzy Syst. **31**(3), 1941–1953 (2016) https://doi.org/10.3233/jifs-16128
7. Chen, X.G.: Research on reliability of complex network for estimating network reliability. J. Intell. Fuzzy Syst. **32**(5), 3551–3560 (2017). https://doi.org/10.3233/JIFS-169291
8. Hazarika, K., Pahuja, G.L.: Effect of common cause failures on redundancy optimization. Int. J. Appl. Mech. Mater. **592–594**, 2491–2495 (SCOPUS Indexed) (2014) ISSN 1662-7482
9. Remirez-Marquez, J.E., Coit, D.W.: Optimization of system reliability in the presence of common cause failures. Reliab. Eng. Syst. Saf. **92**, 1421–1434 (2007)

Brain Health Assessment via Classification of EEG Signals for Seizure and Non-seizure Conditions Using Extreme Learning Machine (ELM)

Nuzhat Fatema

Abstract In this paper, we are going to propose a new method for brain health assessment via classification of EEG signals for seizure and non-seizure conditions using Extreme Learning Machine (ELM). EEG signals are collected by Brain–Computer Interface (BCI) and are utilized to distinguish between ictal and seizure-free signals. Here, we used the empirical mode decomposition (EMD) method to obtain IMFs from EEG signals for classification of seizure and seizure-free signals. These IMFs are combined with statistical parameters and are used as an input features set to ELM. This method is accurate and provides a better classification of highly densed and complex EEG signals, which are utilized for classification of seizure and seizure-free signals.

Keywords ELM · EMD · EEG signals · Seizure · Classification

1 Introduction

The human brain is a boon for the human. It helps in coordination of the several parts of the body; and entire organs of the body are regulated by our brain. If some disorders occur in it, the entire functioning of the body will be affected. Some of diseases are caused due to an imbalance of nervous system called neurological disorder. According to statistics (1 out of 6 people suffer from neurological disorder). Various neurological diseases are there but our main concern in this paper is toward epileptic seizure. Basically, abnormal activities of the electrical signal cause seizure due to which movements of body, functioning of body, and the human behavior changes. Various diseases are there due to which seizure is induced in the human brain and these are [1]: (1) Encephalotrigeminal Angiomatosis, (2) Early

N. Fatema (✉)
International Institute of Health Management Research, New Delhi, India
e-mail: nuzi62@gmail.com

© Springer Nature Singapore Pte Ltd. 2019

H. Malik et al. (eds.), *Applications of Artificial Intelligence Techniques in Engineering*, Advances in Intelligent Systems and Computing 697,
https://doi.org/10.1007/978-981-13-1822-1_10

Infantile Epileptic Encephalopathy or Ohtahara Syndrome, (3) Acquired Epileptiform Aphasia, and (4) Macrencephaly.
Common causes of seizures vary by age of onset, which are as follows:

(i) Up to the age of 2 years: fever, birth, or development injuries.
(ii) From the age 2–14 years: idiopathic seizure disorder.
(iii) Adults: cerebral trauma, alcohol withdrawal.
(iv) Elderly: tumors and strokes.

Neuron is a backbone of the nervous system as shown in Fig. 1, and work as a transporter. All information such as touching of hot or cold object (i.e., sensation) obtained from the surrounding is sensed by specialized tips of nerve cells called dendrite. Dendrite transfer information from the brain to body in the form of neurons and if some misbalance occurs in it then the entire coordinating system of the body will be affected, which is known as neurological disorder.

For detecting neurological disorders like seizure, EEG signals can be obtained from the brain by using brain–computer interface (BCI) technique. There are two methods for collecting brain signals, i.e., invasive and noninvasive methods. An invasive BCI is obtained by the surgical method and it will be dangerous for a human being. So, we prefer noninvasive BCI methods. Several BCI methods are magneto encephalography, functional magnetic resonance imaging (FMRI).

EEG signals are the recording of the spontaneous electrical activity of the brain for a short period of time. These are highly complex signals and contain important data related to human behavior and emotions. The EEG signals are recorded by placing multiple electrodes over the scalp. Electrodes used in recording signals, usually have small metal disc shapes and made up of atomic elements like stainless steel, tin, gold, or silver having a coating of silver chloride. These are placed on the scalp using 10/20 international system and marked by letter or numerical. Letter represents the area of brain covered by the electrodes such as F represents frontal lobe and T-Temporal lobe. The numerical represents the side of the head such as even number denotes the right-hand side of scalp and odd number denotes the left-hand side of scalp.
Figure 2 represents the placement of electrodes according to the international 10/20 system.

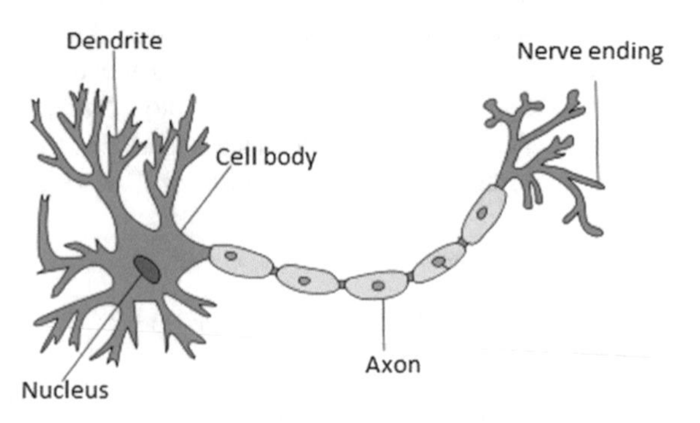

Fig. 1 Transferring of neurons

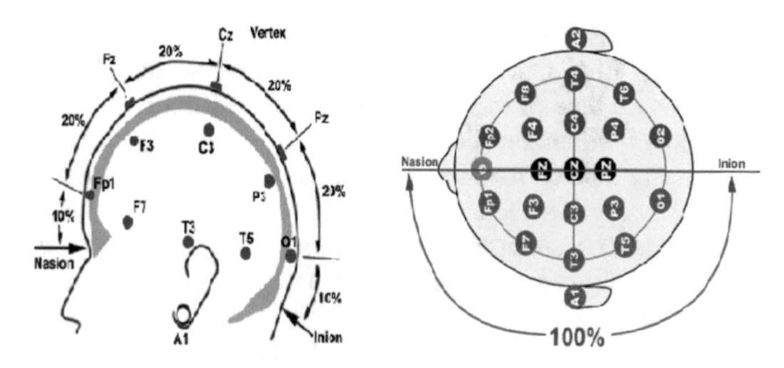

Fig. 2 Electrodes placement as per international 10/20 system

EEG signal consists of various frequencies, i.e., Delta: it has a frequency of 3 Hz or below. It is normal in infants up to 1 year (deep sleep), Theta: frequency ranges from 3.5 to 7.5 Hz and classified as slow activity (deep meditation and dreaming), Alpha: frequency ranges from 7.5 to 14 Hz (visualization and meditation) and Beta: frequency above 14 Hz and classified as fastest activity (Wakefulness condition).

The classification of sleep stages using EEG signals is represented in Table 1.

EEG signals classification is very useful in the detection of various neurological diseases such as: Alzheimer's disease (AD), Parkinson's disease, Epileptic Seizure.

The dataset of EEG signals used in this study is presented in EDF form. Seizure can be detected using EEG signal. Seizures are of various types: (a) Tonic seizure, (b) Myoclonic seizure, (c) Partial seizure, and (d) Febrile seizure. Our main aim in this paper is toward the classification of ictal and seizure-free signals.

Epilepsy (also called epileptic seizure disorder) is characterized by recurrent electrical discharge of neurons of the cerebral cortex. Almost 1–2% of worlds people suffering from this disorder. Fairly out of 60,000 people's stadium, about 500 will have epilepsy. Epilepsy is categorized into two subcategories: (a) Generalized Epilepsy and (b) Focal Epilepsy. During medication of epilepsy, patients develop a tendency to resist the action of drugs and in records, it is approximately twenty percent for first subcategory and sixty percent for the second

Table 1 Classification of sleep stages using EEG signals

Rechtschaffen and Kales (R&K) standard	American Academy of Sleep Medicine (AASM)
Wakefulness (awake)	Wakefulness (awake)
Slow eye movement (SEM) sleep stage 1 (S1)	Slow eye movement (SEM) sleep stage 1 (S1)
SEM sleep stage 2 (S2)	SEM sleep stage 2 (S2)
SEM sleep stage 3 (S3)	Slow wave sleep (SWS) or DEEP SLEEP. S3 and S4 are combined together to form SWS
SEM sleep stage 4 (S4)	
Fast or instantaneous eye movement (FEM)	Fast or instantaneous eye movement (FEM)

subcategory. Epilepsy is the caused because of brain injuries and due to chemical imbalances. Initially, EEG signals can detect seizure only up to thirty–fifty percent but if we increase the time of recording and use serial EEG tests, the chances of detection of epilepsy will automatically increase up to ninety percent. Apart from BCI, there is another method for detection of epileptic seizures by visual scanning of EEG signal by an experienced neurophysiologist. The visual scanning method of EEG signal is a time consuming and may have some possible human error. Also, there may be some disagreement among the neurophysiologists because of the subjective nature of analysis.

EMD technique is applied to the EEG signal and it gives IMF's. These IMF's can be used for features extraction for discriminating in between ictal and non-seizure signals. These IMF's are combined with statistical parameters and set as an input features to ELM. ELM is used as a classifier. The result and discussion is presented in the following section.

2 Electroencephalogram (EEG) Datasets

The material in the form of EEG dataset used in the study is publicly available online on website: http://epileptologie-bonn.de/cms/front_content.php [1]. The dataset consists of usually five databases which are denoted by X, Y, Z, W, and V all these databases contain 100 single-channel EEG signals. The duration of obtaining the signals are 23.6 s (epoch). The database X and Y have been recorded extra cranially. Whereas Z, W, and V are obtained intracranially. The database X obtained from a healthy person with an eye open, the database Y obtained from a healthy person with eye closed, database Z obtained from a hippocampal formation of opposite hemisphere, database W obtained from the epileptogenic zone and the database V obtained from the person having an epileptic seizure. In this paper, usually X, Y, Z, and W databases are not having seizure symptoms but V have seizure symptoms.

Here, we represent (Z, O, N, F, and S) by X, Y, Z, W, and V, respectively, and are tabulated in Table 2.

Table 2 Used data sets for study

Data representation	X = Z and Y = O	Z = N and W = F	V = S
Electrode type	Extra cranially	Intracranially	Intracranially
Electrode position	International 10/20 system	Epileptogenic zone	Epileptogenic zone
Affected by seizure	No	No	Yes
Duration of epoch (s)	23.6	23.6	23.6

Summary of EEG dataset

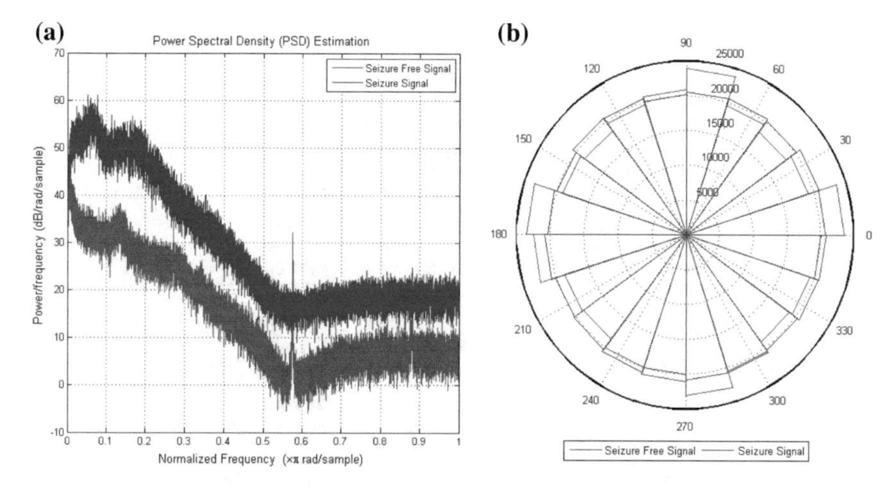

Fig. 3 Seizure EEG and seizure-free EEG signal representation by using **a** power spectral density and **b** rose graph

In this study, two EEG signals (Z and S) have been utilized for the study in which dataset Z is considered as normal EEG signals, whereas dataset S is considered as epileptic seizure EEG signals. To analyze the difference between seizure and seizure-free signal, Fig. 3 has been represented.

3 Methodology

3.1 Proposed Approach

The proposed approach for ictal and non-seizure EEG signal classification is represented in Fig. 4. The classification model includes 12 basic steps which are used to identify the 2 states: seizure and non-seizure EEG signals. First, EEG signals have been captured under these two conditions; thereafter, recorded signals have been preprocessed by using EMD technique to decompose the signals into IMFs. Obtained IMFs are utilized as an input variable in the designed ELM technique based classifier model. Two different ELM-based classifier models have been implemented under two distinct scenarios, i.e., by using raw data as input and by using IMFs as input. Thereafter, both ELM models have been trained and tested 20 times with variation of hidden layer neuron from 1 to 364 at each instance. The standard deviation has been evaluated to analyze the variation in classification accuracy, which is explained in detail in subsequent sections.

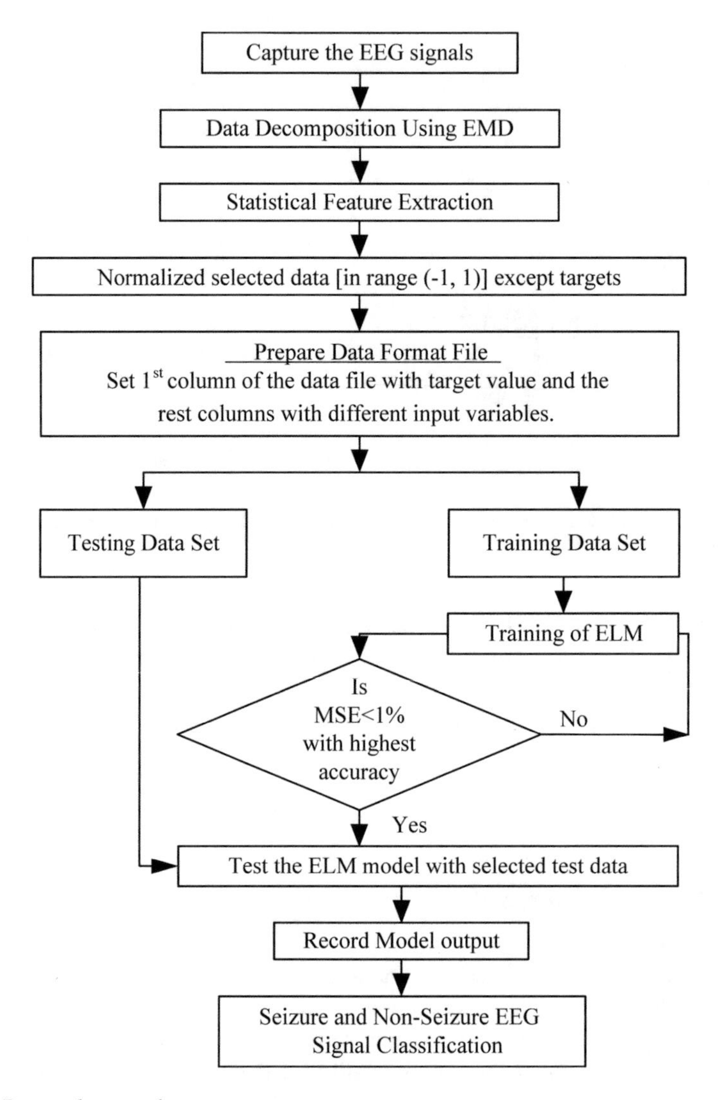

Fig. 4 Proposed approach

3.2 Data Decomposition by Using EMD

It is pioneered by Malik and Mishra [2], as the name suggests, EMD's work's to decompose the nonlinear and nonstationary EEG signal into multiple sets of IMFs. This method is a suitable and adaptive method for forming IMFs. Each IMF formed must satisfy the two basic conditions:

(1) There must be no difference in number of extremum and shifting of value on the positive and negative side of number system or it may differ by only one.

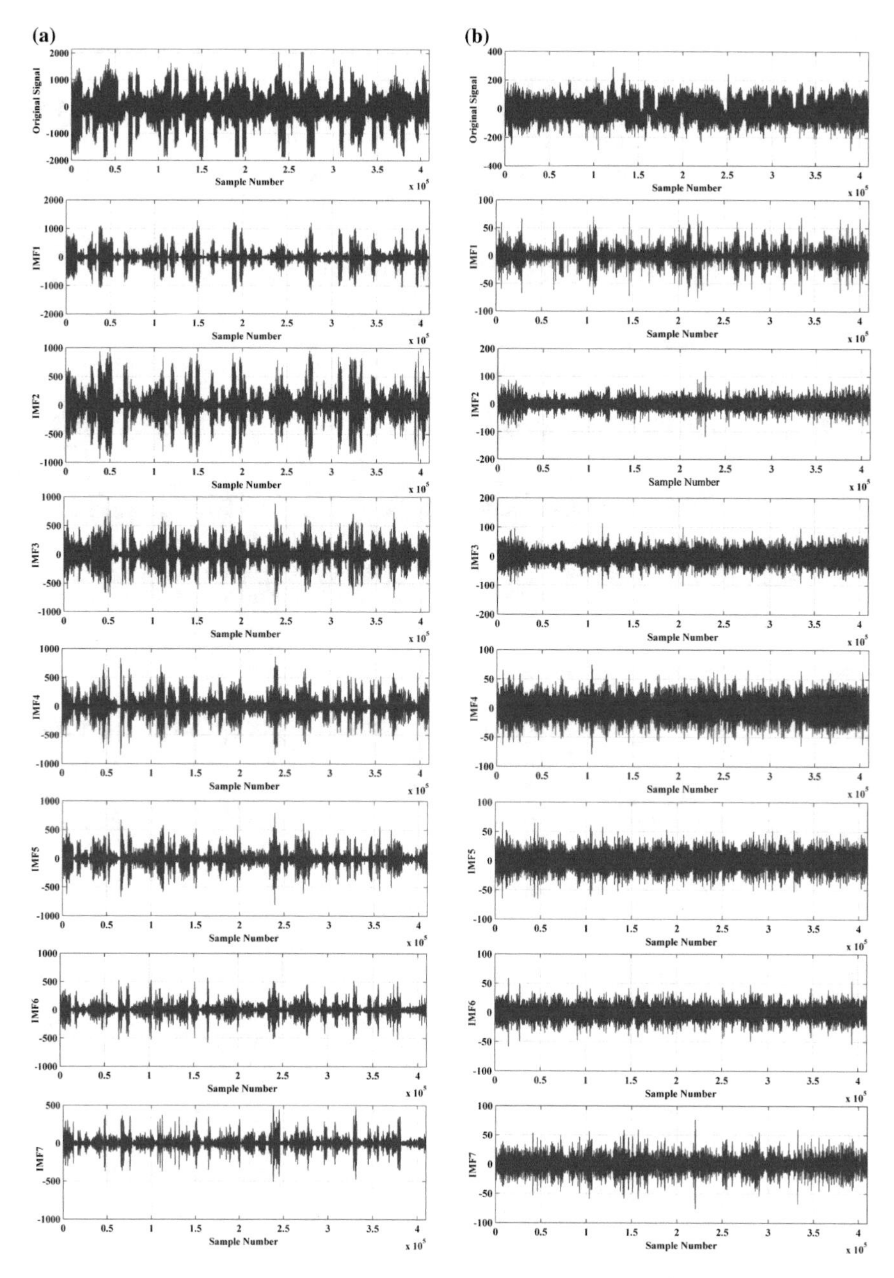

Fig. 5 **a** IMFs of S dataset (epileptic seizure EEG signals). **b** IMFs of Z dataset (seizure-free EEG signals)

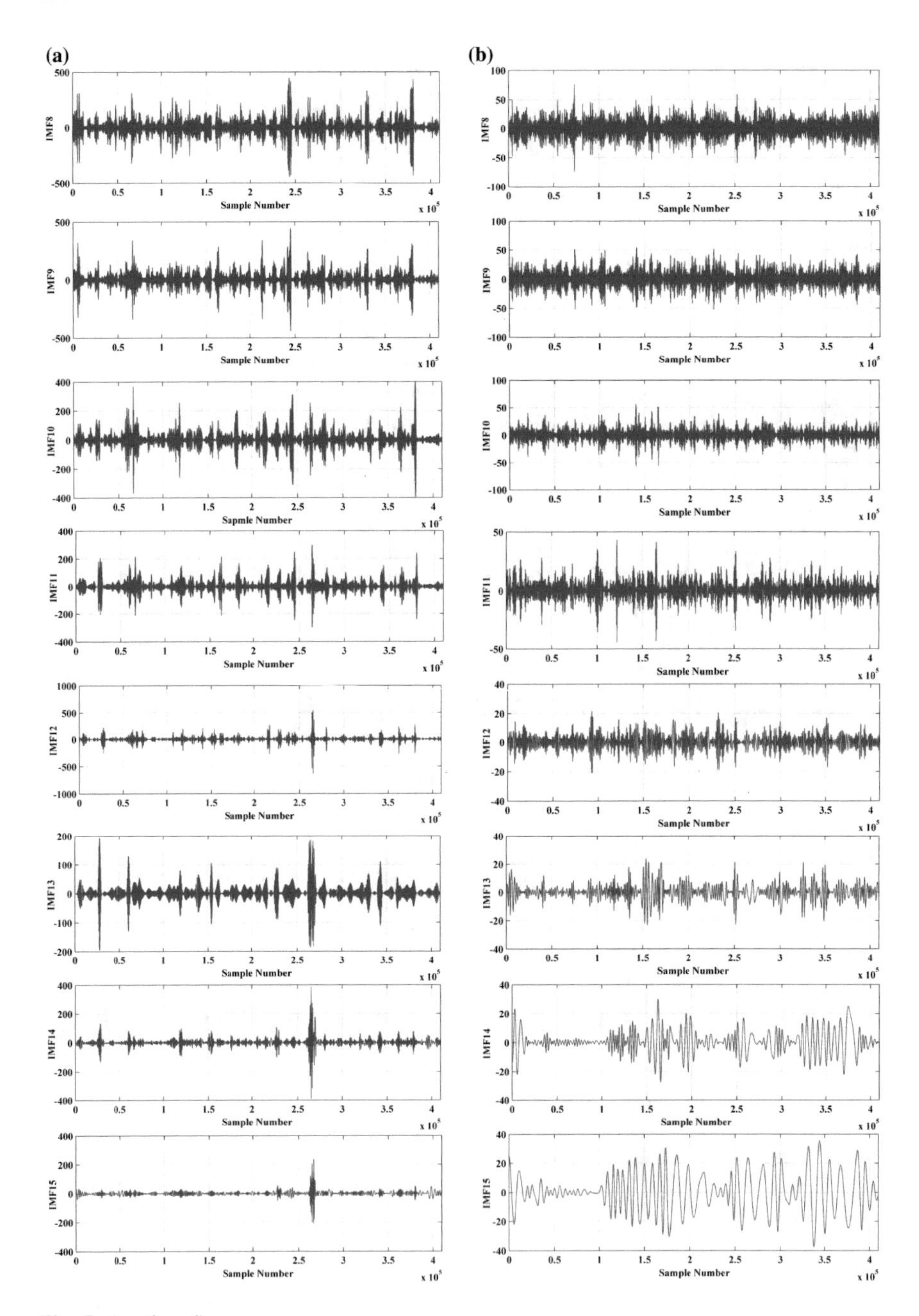

Fig. 5 (continued)

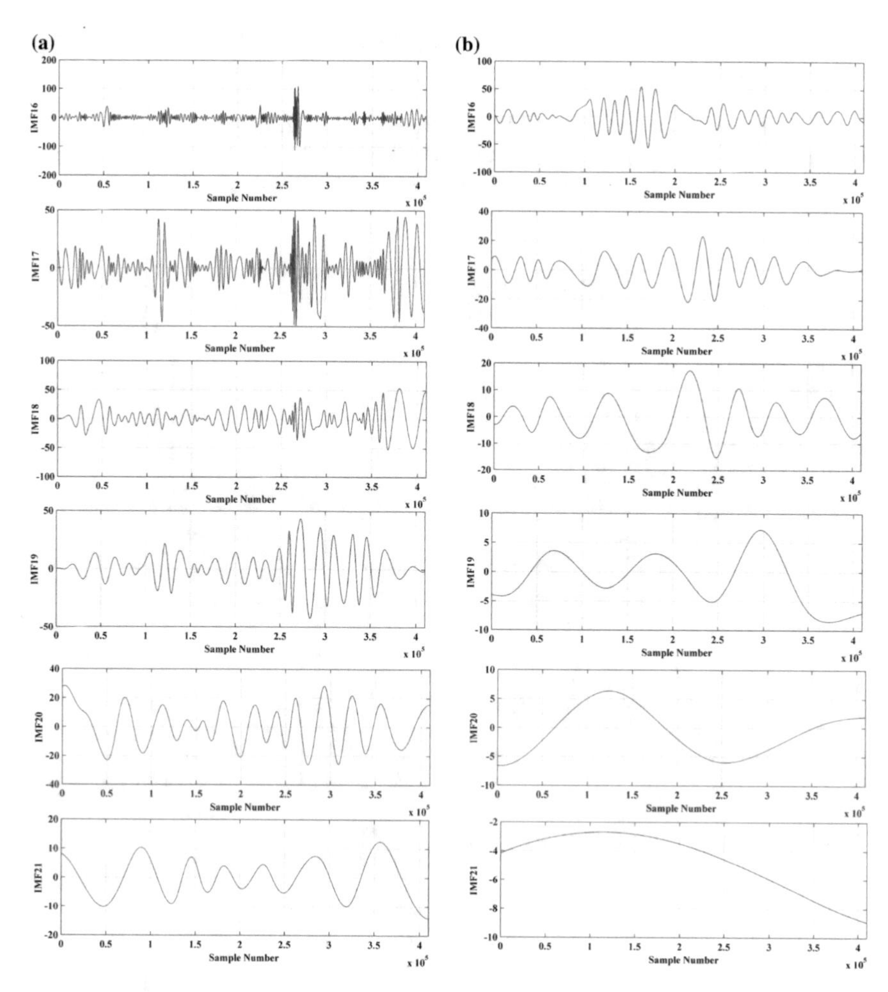

Fig. 5 (continued)

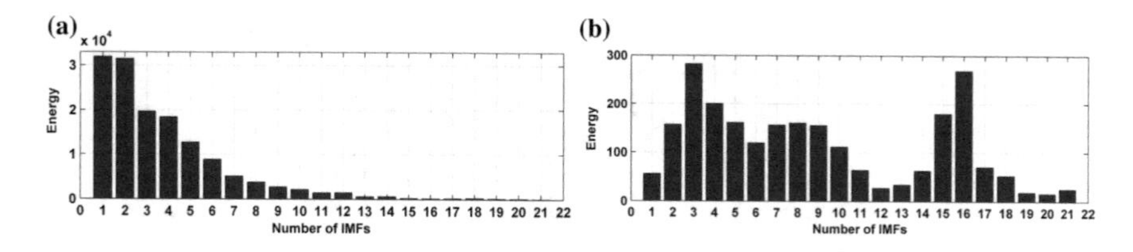

Fig. 6 **a** Energy magnitude of IMFs of S dataset. **b** Energy magnitude of IMFs of Z dataset

(2) The value obtained from the mean of local maxima and local minima must be zero at all points.

All steps used for applying EMD on signal S and Z is given in [2–5]. After decomposition, the signals by EMD method, obtained first 21 IMFs are represented in Fig. 5 and its energy distribution in Fig. 6. To visualize the difference between

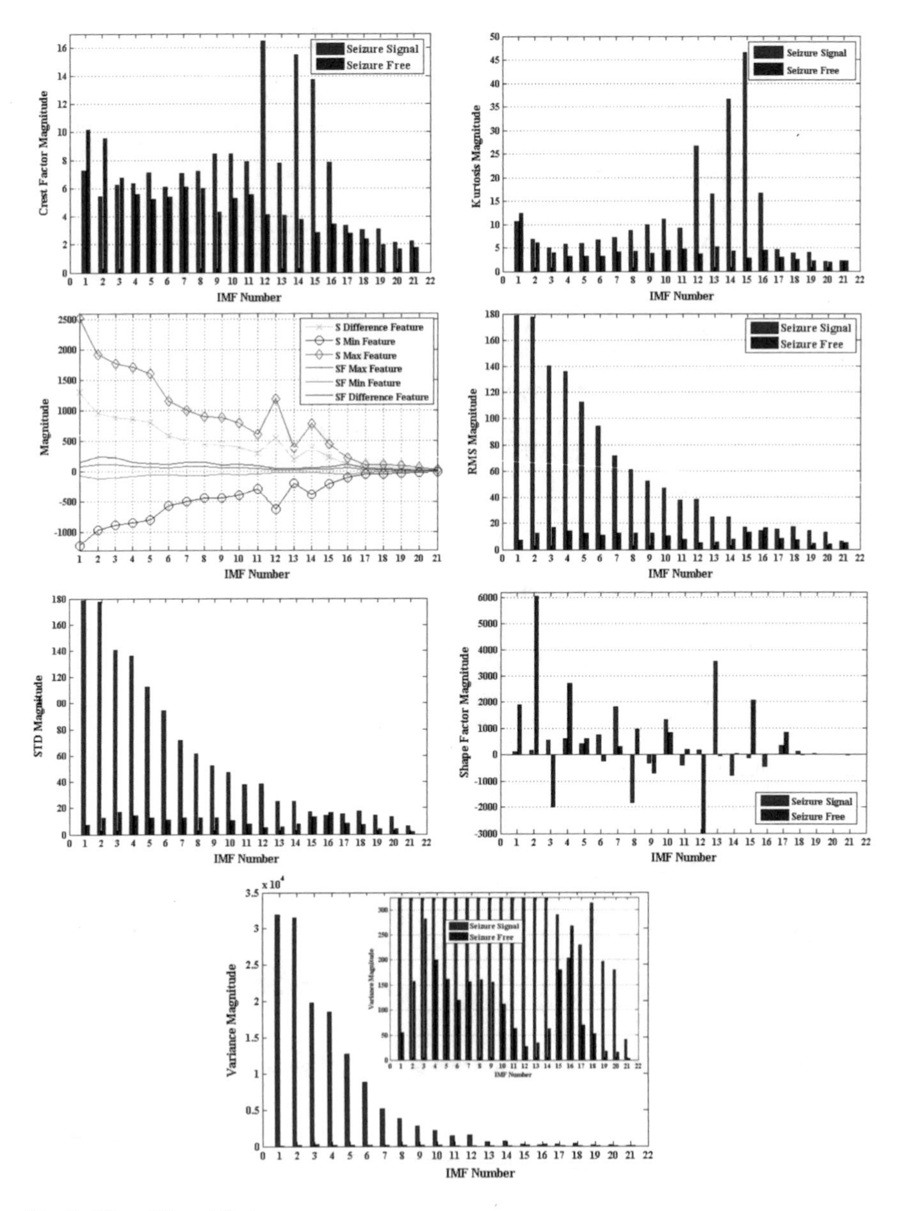

Fig. 7 Nine different features

seizure and non-seizure signals, nine statistical parameters have also been computed and represented in Fig. 7. The detailed analysis of these features has been given in [6].

4 Seizure and Non-seizure EEG Signals Classification Using ELM

4.1 Extreme Learning Machine (ELM) [7–10]

ELM approach is the recent advanced approach developed by Guang-Bin Huang, Qin-Yu Zhu and Chee-Kheong Siew in 2004 [7–10]. This approach is based on feed forward NN as represented in Fig. 8, which includes three layers (i.e., input layer, hidden layer, and output layer). The neurons at input layers are equal to the number of input of the problem and output layer neurons are equal to the number of output decided by the user. Whereas hidden layer neurons vary from 1 to nth number, it will be dependent on the problem. ELM approach may be utilized for classification, regression as well as prediction/forecasting problem. Here, it is utilized for classification of two conditions. It can be utilized for multi-class also. In the healthcare domain, it is a very recent application for classification of EEG signals.

The mathematical modeling of ELM-based classifier is presented in [7–10] in detail. Let us assume the training data specimens H(Yn, Tn) utilized for designing the ELM classifier then activation function $\left(g(x) = 1/1 + \mathrm{e}^{-\lambda x}\right)$ will be utilized and is represented as

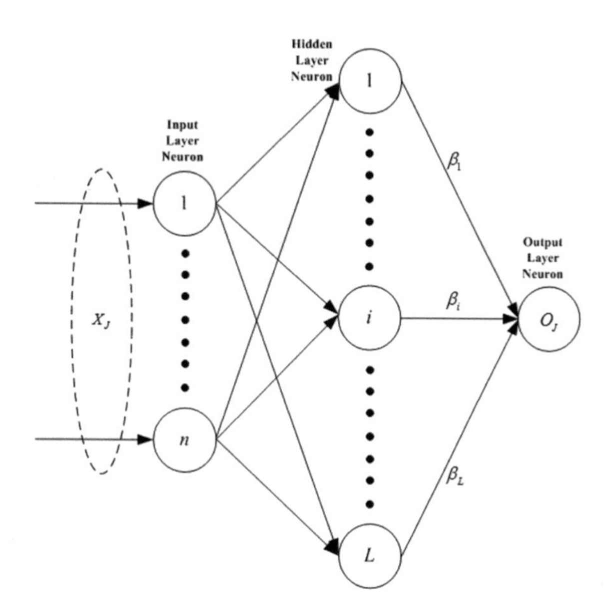

Fig. 8 Schematic diagram of ELM structure

$$f_i(x_{\text{proc}}) = \sum_{i=1}^{h} \beta_i g(w_i \cdot x_{\text{proc}} + b_i) \tag{1}$$

where

$$g(w_i \cdot x_{\text{proc}} + b_i) = \frac{1}{1 + \mathrm{e}^{(-(w_i \cdot x_{\text{proc}} + b_i))}}; \tag{2}$$

For h hidden nodes with G training data specimens and error = zero, then relation of β_i, w_i and b_i:

$$f_i(x_n) = \sum_{i=1}^{h} \beta_i g(w_i \cdot x_n + b_i) = H\beta \tag{3}$$

where

$$H = \begin{bmatrix} h(x_1) \\ \vdots \\ h(x_G) \end{bmatrix} = \begin{bmatrix} g(w_1 \cdot x_1 + b_1) & \cdots & g(w_h \cdot x_1 + b_h) \\ \vdots & \ddots & \vdots \\ g(w_1 \cdot x_G + b_1) & \cdots & g(w_h \cdot x_G + b_h) \end{bmatrix} \tag{4}$$

and β is given by

$$\beta = \begin{bmatrix} \beta_{1,1} & \ldots & \beta_{1,m} \\ \vdots & \ddots & \vdots \\ \beta_{h,1} & \cdots & \beta_{h,m} \end{bmatrix} \tag{5}$$

The output weight vector β can be calculated by

$$\beta = H^{\psi} T \tag{6}$$

where H^{ψ} = the Moore–Penrose pseudo-inverse of the hidden layer output matrix H

T = the output matrix is given as follows:

$$T = \begin{bmatrix} t_{1,1} & \ldots & t_{1,m} \\ \vdots & \ddots & \vdots \\ t_{h,1} & \cdots & t_{h,m} \end{bmatrix} \tag{7}$$

Based on this learning algorithm, the ELM training time can be extremely fast because only three calculation steps are required:

Step 1: Randomly assign the input weight w_i and bias b_i according to any continuous sampling distribution, $i = 1, 2, \ldots, h$
Step 2: Calculate the hidden layer output matrix H and
Step 3: Calculate the output weight

$$\beta(\beta = H^{\psi}T) \tag{8}$$

The final classification result for a multi-class problem can be expressed as

$$\text{Label}(x_{\text{proc}}) = \arg\max f_j(x_{\text{proc}}) \quad \text{for } j \in \{1, \ldots, m\} \tag{9}$$

Based on these basic steps, ELM has been implemented to classify the two conditions of EEG signals, and after implementation, the classification performances (in terms of MSE, RMSE, and correct identification accuracy) have been evaluated for each model under different two conditions (i.e., training and testing phase).

5 Results and Discussion

The ELM model is implemented under two distinct conditions. ELM Model#1 is utilized raw data as an input, whereas ELM Model#2 is utilized and evaluated IMFs of EMD from raw data specimens. Each model has been trained and then tested twenty times at each hidden layer neuron. The hidden layer neurons are varied from 1 to 364 and check the prediction accuracy at each time. The learning curve for both ELM models during training and testing phase has been represented in Fig. 9, which shows the variation in classification accuracy at each hidden layer neuron. After successful training and testing of both models, the obtained results are summarized in Table 3 which shows that ELM Model#2 has highest classification accuracy during training as well as a testing phase as compared with ELM

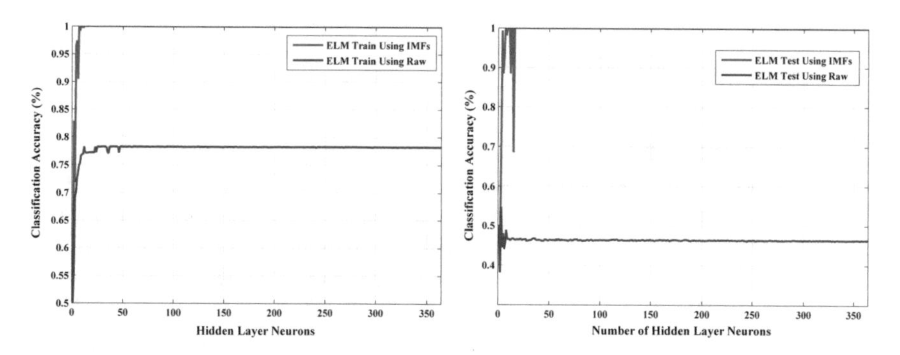

Fig. 9 ELM learning curve for training and testing phase

Table 3 ELM-based accuracy analysis of EEG signal classification model

	ELM model operational mode	Maximum accuracy (%)	Minimum accuracy (%)	Average accuracy of 1–364 neurons (%) ± STD
Raw data input	Training phase	78.312	50.0	77.956 ± 19.98
	Testing phase	54.725	38.156	46.388 ± 2.51
IMFs data input	Training phase	100.0	50.0	99.692 ± 35.35
	Testing phase	100.0	39.06	99.432 ± 35.35

Table 4 ELM results comparison with ANN using calculated input variables

Used online database	ANN (CP/TS)	ELM (CP/TS)
90 cases	82/90	89/90

CP/TS correct prediction/total samples

Model#1. The testing data is not included with training data set. The standard deviation for the training and testing phase has been evaluated which shows the maximum variation of classification by the ELM model. The Model#1, STD is lower but accuracy is also lower and highest accuracy may be achieved up to 78.31% which is not desirable. In Model#2, STD is 35.35 during both training as well as testing phase but average classification accuracy of 20 runs is 99.69 and 99.43% for training and testing phase, respectively, which are nearer to desirable values and may achieve up to 100%.

The maximum MSE of training and testing phase for ELM models are found to be 0.0555 and 0.0754, respectively, showing that the classification accuracy is higher up to 100%.

To validate the proposed ELM Model#2, ANN has also been implemented by using same input variables as IMFs of EMD and then classification accuracy for some specimens have been represented in Table 4 which shows the better results as compared with ANN.

6 Conclusions

In this study, the seizure and non-seizure EEG signals are decomposed by using empirical mode decomposition (EMD) method to form intrinsic mode functions (IMFs). Then, statistical features are extracted based on obtained IMFs to visualize the difference between seizure and seizure-free signals. Thereafter, calculated IMFs are utilized as an input variable to extreme learning machine (ELM)-based classification model. The classification accuracy of the proposed model is compared with the artificial neural network (ANN), which reveals that ELM approach performs much better than ANN.

Future work is focused on the proposed approach application for online classification of seizure and non-seizure EEG signals.

References

1. Andrzejak, R.G., Lehnertz, K., et al.: Indications of nonlinear deterministic and finite-dimensional structures in time series of brain electrical activity dependence on recording region and brain state. Phys. Rev. **E.64** (Article ID 061907) (2001)
2. Malik, H., Mishra, S.: Application of GEP to investigate the imbalance faults in direct-drive wind turbine using generator current signals. IET Renew. Power Gener. **11**(6), 889–902 (2017). https://doi.org/10.1049/iet-rpg.2016.0689
3. Malik, H., Sharma, R.: EMD and ANN based intelligent fault diagnosis model for transmission line. J. Intell. Fuzzy Syst. **32**, 3043–3050 (2017)
4. Malik, H., Mishra, S.: Artificial neural network and empirical mode decomposition based imbalance fault diagnosis of wind turbine using TurbSim, FAST and Simulink. IET Renew. Power Gener. **11**(6), 889–902 (2017). https://doi.org/10.1049/iet-rpg.2015.0382
5. Malik, H., Sharma, R.: EMD and ANN based intelligent fault diagnosis model for transmission line. J. Intell. Fuzzy Syst. **32**(4), 3043–3050 (2017). https://doi.org/10.3233/JIFS-169247
6. Shen, Z., Chen, X., Zhang, X., He, Z.: A novel intelligent gear fault diagnosis model based on EMD and multi-class TSVM. Measurement **45**, 30–40 (2012). https://doi.org/10.1016/j.measurement.2011.10.008
7. Sharma, S., Malik, H., Khatri, A.: External fault classification experienced by three-phase induction motor based on multi-class ELM. Elsevier Procedia Comput. Sci. **70**, 814–820 (2015). https://doi.org/10.1016/j.procs.2015.10.122
8. Malik, H., Mishra, S.: Application of extreme learning machine (ELM) in paper insulation deterioration estimation of power transformer. In: Proceedings of the International Conference on Nanotechnology for Better Living, vol. 3, no. 1, p. 209 (2016). https://doi.org/10.3850/978-981-09-7519-7nbl16-rps-209
9. Malik, H., Mishra, S.: Selection of most relevant input parameters using principle component analysis for extreme learning machine based power transformer fault diagnosis model. Int. J. Electr. Power Compon. Syst. **45**(12), 1–13 (2017). https://doi.org/10.1080/15325008.2017.1338794
10. Malik, H., Mishra, S.: Extreme learning machine based fault diagnosis of power transformer using IEC TC10 and its related data. In: Proceedings of the IEEE India Annual Conference (INDICON), pp. 1–5 (2015). https://doi.org/10.1109/indicon.2015.7443245

An Optimization Algorithm for Unit Commitment Economic Emission Dispatch Problem

Jitendra Kumar, Ashu Verma and T. S. Bhatti

Abstract The emission rate of greenhouse gases results in global warming and having some other environment-related issues and uncertainties. This work presents a new approach to handle the effect of uncertainties and economic generation. An optimization technique Lagrangian relaxation with priority list (LR-PL) has been used to handle Unit Commitment and Economic Emission Dispatch (UCEED) problem with the inclusion of renewable energy source such as solar energy. An optimization technique has been developed to determine optimal solution and verified. The proposed approach has been applied at standard IEEE 69-bus 11 electrical generators system. An integration of the generator units with solar system is implemented. After integration with solar system, two cases of peak and off-peak hours have been studied. It is found that the developed approach is capable of handling the uncertainties in the solar radiation in an efficient manner. The compared results show the better performance of the developed approach over the existing ones. The compared results show the effectiveness of the developed approach.

Keywords Unit commitment · Economic emission dispatch · Solar photovoltaic system · Lagrangian relaxation method

J. Kumar (✉) · A. Verma · T. S. Bhatti
Centre for Energy Studies, Indian Institute of Technology,
Hauz Khas, New Delhi 110016, India
e-mail: jitu.ee.iitd@gmail.com

A. Verma
e-mail: ashu.ee.iitd@gmail.com

T. S. Bhatti
e-mail: tsb@ces.iitd.ac.in

© Springer Nature Singapore Pte Ltd. 2019

H. Malik et al. (eds.), *Applications of Artificial Intelligence Techniques in Engineering*, Advances in Intelligent Systems and Computing 697,
https://doi.org/10.1007/978-981-13-1822-1_11

1 Introduction

Environment is a major concern these days due to the increment in hazardous conditions and being the main source for various dangerous diseases in living organisms. Therefore, lots of efforts have been placed in order to reduce the greenhouse gases (GHG) emission and climate change impacts [1]. To reduce the discussed problems associated with environment, it is very important to dig out the causes and work around them. Transportation and electricity generation are two major sectors which contribute largely to the air pollution. In electricity generation, coal-based power plants plays an eminent role in the power industry [2] and also have higher contribution in emissions such as SO_x, NO_x, CO_2, and particulate matter [3–5]. Thermal power plants are more harmful when situated near to city or town; but, it can be utilized in corporation with renewable energy sources (RES) such as solar, wind, etc., which may help to reduce the emission rate [6]. One of the main challenges among these utilities is unpredictable and intermittent nature of these resources. Hence, proper scheduling of these resources with conventional sources may help in optimum utilization [7].

In view of this, a combined unit commitment economic dispatch problem has been solved for a system having a solar photovoltaic (SPV) utility with conventional (thermal) electrical generators. Reduction of the emissions like NO_x and SO_x are the key concerns considered in this paper. In the past few decades, various techniques and methodologies have been discussed by the researchers to solve the unit commitment problem (UCP), such as dynamic programming [8], truncated dynamic programming [9], genetic algorithm [10], extended priority list [11], Lagrangian relaxation with differential evolution algorithm [12], dynamic programming integrating priority list based on best per unit cost [13], Gaussian harmony search, and jumping gene transposition algorithm [14]. Worldwide the maximum electricity generation depends on the fossil fuels and among these more than 70% electrical energy is generated by the coal. Power produced from the fossil fuels may bring adverse environmental effects and emissions. To achieve lower emission and pollutant, RES has attracted more attention [15]. Considering environmental issues, results in combined unit commitment economic emission dispatch.

The unit commitment economic emission dispatch is a challenging issue in power system which turned up as an optimal problem [16]. Therefore, an optimal solution is required which would consider economic and emission impacts on power system. Several different algorithms have been applied in the field of UCEED to determine the optimal value, such as tribe-modified differential evolution [17, 18], incremental artificial bee colony algorithm with local search [19], modified harmony search algorithm [20], a surrogate differential evolution [21].

The integration of thermal and solar systems has been investigated in the literature, as in [22], Euclidean affine flower pollination algorithm and binary flower pollination algorithm have been used for combined economic emission dispatch for photovoltaic plants and thermal power generation units. A new strategy for the optimal scheduling problem considering the impacts of uncertainties in SPV, wind, and load demand forecasts has been proposed in [23], the effectiveness of the proposed scheme has been tested on IEEE 30 and 300 bus systems. The obtained results have also been compared with genetic algorithm and two-point estimate methods.

In [24], cost-effectiveness and the potential of a SPV power plant has been analyzed. In [25], various energy payback time analysis and life cycle cost are performed for a distributed 2.7 kW_p grid-connected monocrystalline PV system operating in Singapore. A model capable of comparing several mature and emerging PV technologies has been given in [26]. An economic analysis of photovoltaic and diesel hybrid system with flywheel energy storage has been reported in [27]. In [28], the authors measured GHG effect using fuzzy-logic-based strategy. Results show that in view of GHG emission, the SPV system is a satisfactory choice to exhort present environmental issues.

Figure 1 shows the sub-problems of short-term solar thermal coordination problem (STCP), UCP, and economic emission dispatch problem (EEDP).

It has been observed from the literature that many researchers have included SPV system with conventional electrical generators. However, a continuous effort is being done to focus on integration of RES with thermal generation units to reduce emission. Hence, in this paper a LR-PL-based integration of thermal units and SPV is done to evade complete dependency on thermal units and to reduce emission and for economic operation. The proposed system reduces the effect of uncertainty of RES. It also reduces the total operating cost as compared to previous work presented and comparison is shown in the result section.

Therefore, to handle emissions-related problem, a new algorithm with the inclusion of SPV system has been proposed to reduce the emission and total generation cost. On-site and off-site power plants have been taken into account to analyze the total cost. The combination of SPV power plants, thermal power plants, priority list, and Lagrangian relaxation is used to determine the optimum solution. The main contributions of this work are as follows:

- A LR-PL-based UCEED is developed to operate the thermal units in optimum manner with minimum fuel cost.
- To reduce the emission level an SPV is integrated with thermal generation units.

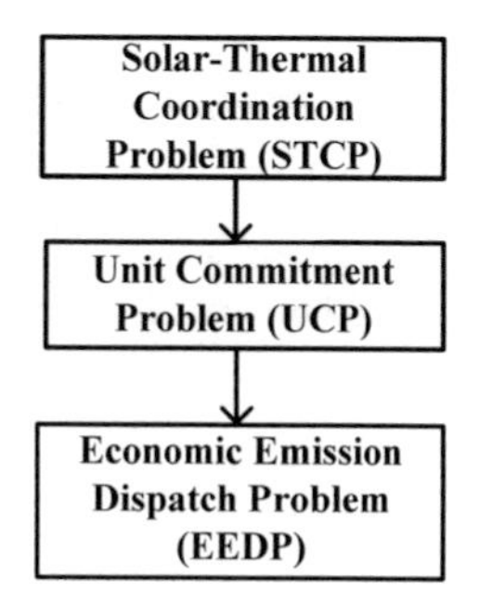

Fig. 1 Solar thermal coordination system

- An incorporation of on-site and off-site SPV power plants are presented for cost analysis.
- In the integration peak hours and off-peak hours are considered, and optimum results are obtained.

The paper is organized as follows: Sect. 1 is the introduction part of the paper. Section 2 presents the objective synthesis with the several constraints. In Sect. 3, proposed algorithm is presented. Section 4 investigated the results and discussion. Finally, Sect. 5 concludes the work.

2 Objective Synthesis

The combined UCEED problem can be treated as the minimization of total operating cost (T_{OC}—\$) and the emission (T_E—kg) given in (1) and (2) [29].

$$\text{Min}\, T_{OC} = \sum_{i=1}^{N_{TGU}} \sum_{t=1}^{T} \text{CG}_{ei}\left(P_{gi}(t)\right) + \left(C_{\text{sup},i}(t) + C_{\text{sdc},i}(t)\right) \tag{1}$$

$$\text{Min}\, T_E = \sum_{i=1}^{N_{TGU}} \sum_{t=1}^{T} \text{EG}_{ei}\left(P_{gi}(t)\right) + \left(E_{\text{sue},i}(t) + E_{\text{sde},i}(t)\right) \tag{2}$$

where $\text{CG}_{ei}\left(\text{P}_{gi}(t)\right) = \alpha_i\; P_{gi}^2(t) + \beta_i P_{gi}(t) + \gamma_i$ \$/h represents the fuel cost of ith generator, ($P_{gi}(t)$) is the generated output power of the ith generator, α_i, β_i, γ_i are the fuel cost coefficients of ith generator, T is the time horizon, and N_{TGU} is the number of thermal generator units respectively. $C_{\text{sup},i}(t)$ and $C_{\text{sdc},i}(t)$ represents the start-up and shut down cost for ith generation unit [11].

Similar to (1), total emission is given by (2). Where, emission rate $\text{EG}_{ei}\left(P_{gi}(t)\right) = d_i\, P_{gi}^2(t) + e_i\, P_{gi}(t) + f_i$ kg/h is given in [16], EG_{ei} ($P_{gi}(t)$) is the emission function, d_i, e_i, f_i are the emission coefficients of the ith generator. $E_{\text{sue},i}$ (t) and $E_{\text{sde},i}$ (t) represent the start-up and shut down emission for unit i.

Constraints: Various constraints for solar–thermal systems are given below:

2.1 Power Generation Capacity

The generated output power capacity of each thermal power plant unit must be lying between its minimum and maximum values as given in (3) [11].

$$P_{gi,\min} \leq P_{gi} \leq P_{gi,\max} \quad \text{for } i = 1, 2\ldots, N_{\mathrm{TGU}} \tag{3}$$

where $P_{gi,\min}$ and $P_{gi,\max}$ are the minimum and maximum output generated powers of the electrical generators units.

2.2 *Power Balance Satisfaction for Each Hour*

Total load demand should be satisfied by the total power output generated from the solar system and electrical generators for each hour as expressed in (4) [28].

$$\sum_{i=1}^{N} P_{gi}(t) + P_{si}(t) = P_{\mathrm{dem}}(t) \tag{4}$$

where $P_{\mathrm{dem}}(t)$ is the total load demand of end users for each hour, $P_{gi}(t)$ and $P_{si}(t)$ are the generated power from the thermal and solar generating units.

2.3 *Spinning Reserve*

The maximum output power of the generating units available on the system should be able to supply load and required spinning reserve given in (5) [11].

$$\sum_{i=1}^{N} P_{gi}(t) \geq P_{\mathrm{dem},t} + \mathrm{SR}_t \quad t = 1, 2, \ldots, T \tag{5}$$

where SR_t is spinning reserve requirement at tth hour.

2.4 *Minimum Up and Down Time*

In the operation of units, once the unit is running, it should not be turned off immediately. That means a unit requires a certain amount of time before it is shut down. Similarly, if a unit is decommitted, then there is a minimum time require before it be recommitted [29].

$$\begin{aligned} T_i^{\mathrm{up}} &\leq X_i^{\mathrm{up}} \\ T_i^{\mathrm{down}} &\leq X_i^{\mathrm{down}} \end{aligned} \tag{6}$$

where $T_i^{\mathrm{up}}/T_i^{\mathrm{down}}$, minimum up and down time at hour ‘t’ $X_i^{\mathrm{up}}/X_i^{\mathrm{down}}$, duration during units on and off continuously.

2.5 Power Output Limit/Battery Constraint

The power output from solar PV system is given in (7) [28].

$$P_{\mathrm{PV}}(G(t)) - P_{\mathrm{sb}}(t) - P_{\mathrm{s}}(t) = 0 \tag{7}$$

where $G(t)$ is the solar radiation at hour t (in W/m^2), P_{PV} is the conversion function of solar radiation to energy and P_{sb} is the total output power of the batteries at hour t.

The solar radiation to energy conversion function P_{PV} of the SPV system is given in (8)

$$P_{\mathrm{PV}}(G(t)) = \begin{cases} P_{\mathrm{sn}}\frac{(G(t))^2}{G_{\mathrm{std}}R_{\mathrm{C}}} & 0 < G(t) < R_{\mathrm{C}} \\ P_{\mathrm{sn}}\frac{(G(t))}{G_{\mathrm{std}}} & G(t) > R_{\mathrm{C}} \end{cases} \quad t = 1, 2\ldots, T \tag{8}$$

where G_{std} is the solar radiation for standard environment and set as 1000 w/m^2 and R_{C} is the cut in radiation point set as 150 W/m^2. P_{sn} is the rated power output of SPV power plants as the solar system is associated with batteries, the battery constraints are as follows:

Power limit: The output power of battery is limited as given in (9)

$$P_{\mathrm{sbi,min}} \le P_{\mathrm{sbi}} \le P_{\mathrm{sbi,max}} \tag{9}$$

where $P_{\mathrm{sbi,min}}$ and $P_{\mathrm{sbi,max}}$ are the minimum and maximum generated battery capacity range for the SPV system.

State of charge: the overcharge and undercharge both conditions are harmful. Thus, the state of charge is limited within a range described in (10).

$$\mathrm{LSb}_i \le \mathrm{Sb}_i \le \mathrm{USb}_i \tag{10}$$

where $\mathrm{USb}_i/\mathrm{LSb}_i$, respectively, represents the upper and lower limit of battery’s state of charge (SOC). Sbi represents the current SOC of ith battery i at tth hour.

3 Proposed Algorithm

In this work, a Lagrangian relaxation method with priority list has been implemented to determine the optimal solution, of UCEED problem subjected to various equality and inequality constraints.

An hourly load demand as shown in Fig. 2 is considered to provide the optimal solution. As the load demand is given in Fig. 2 is scheduled; still it is possible that it may change with the time as per end user demand. In the conventional UC the priority list is prepared based on calculations which are time consuming and limited with given load demand. Therefore, a versatile algorithm is required to be developed to run the generating units in an economic manner to reduce the overall cost and emission of generation.

Hence, in the present work an effort has been made to make the system automatic and more reliable by eliminating calculation-based priority selection method. To do so, a Lagrangian relaxation with priority list (LR-PL)-based unit commitment has been implemented to improve the reliability and robustness against unexpected change in load rather than scheduled load.

In the proposed algorithm, the total available electrical generator units are selected according to the priority list order in ascending order of the heat rate at maximum power output as defined in (11).

$$\text{Heat Rate (HR)} = \frac{\alpha_i P_{gi\max}^2(t) + \beta_i P_{gi\max}(t) + \gamma_i}{P_{gi\max}(t)} \tag{11}$$

where α_i, β_i, γ_i are the fuel cost coefficients of ith generator, $P_{gi,\max}$ is the generated output power of the ith generator.

The efficient units are assigned first for each hour for the scheduling period. If load increases in any set of time interval than the less efficient or more costly units will commit at last. Thus, all the units will operate at minimum operating cost subject to satisfying several constraints.

Thermal units running or not, have been estimated by the UCP [11]; however the distribution of load demand is necessary to solve the EED problem. The EED problem finds the optimum sharing of the load to reduce the emission which is

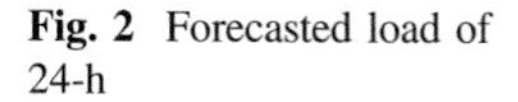

Fig. 2 Forecasted load of 24-h

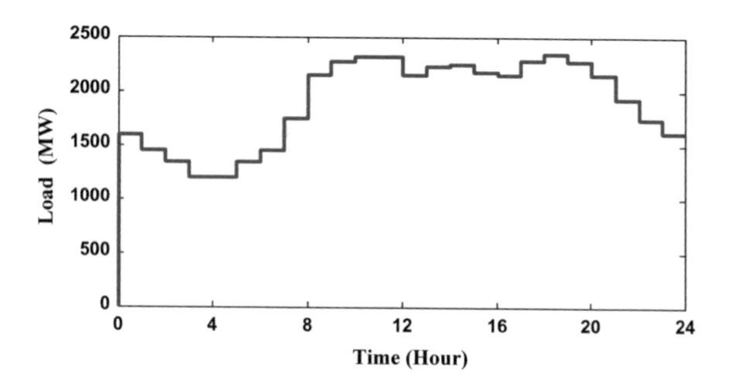

produced from the thermal units. The Lagrangian relaxation uses Lagrangian multiplier (λ) for the system constraints and obtained optimal solution for the system model. To reduce the emission rate, it is better to run the efficient units in the off-peak hours or when load demand is low.

From the heat rate expression first the most economic units are identified, these units have highest priority to meet any given load demand.

In addition, the SPV system is added intelligently with the electrical generator to determine the optimal emission and dispatch during scheduled time of period. This work focuses on the SPV system scheduling with the thermal units in peak hours and off-peak hours.

SPV power plants can be classified into two parts (i) On-site SPV power plant and (ii) Off-site SPV power plant [24]. The estimated output power considered [16] from the SPV power plants are 20, 20, and 45 MW. Remaining demand will be fulfilled by the electrical generator units. By using this approach the emission rate for peak hours and off-peak hours reduced in a significant amount. The flowchart of the used algorithm is shown in Fig. 3.

3.1 On-Site Solar Photovoltaic Power Plant

The on-site SPV power plant consists of PV array module, inverter, an inbuilt maximum power point tracker (MPPT) system, battery bank and land as per requirement of load demand as given in [24].

3.2 Off-Site Solar Photovoltaic Power Plant

The main limitation to implement on-site SPV system is the high (Land and Batteries) cost. Due to which it is not possible to set up a big MW solar power plant near the cities/towns as land is not available or may be costly. Therefore, near the cities/towns off-site SPV power plant proposal is considered.

The off-site SPV power plant is similar as the on-site SPV power plant except the cost of land and battery. Total operation cost is lower in case of off-site SPV power plant in comparison with the on-site SPV power plant. The levelized cost of energy (LCOE) for on-site and off-site SPV power plant is \$0.0846/kWh and \$0.0538/kWh at 10% discount rate for 25 years life of plant is considered (Exchange rate 1 US\$ = INR 64) [24, 30–32].

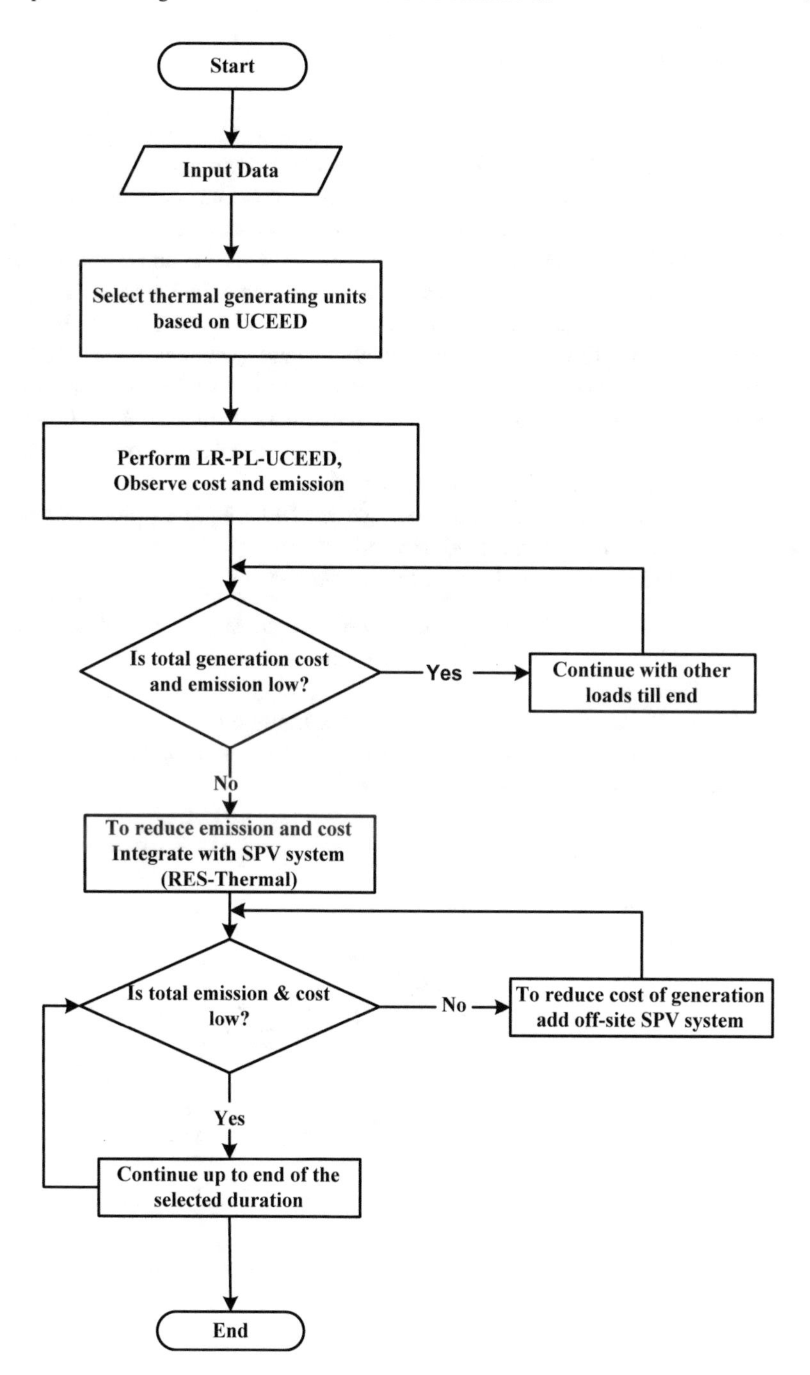

Fig. 3 The complete flow diagram of UCEED with SPV system

4 Result and Discussion

The LR-PL-based integration of RES and thermal power generation units is done in MATLAB/Simulink R2012b on i72600CPU 3.40 GHz computer.

The system with 11 electrical generators and SPV units has been considered for this work as shown in Fig. 4. The effectiveness of the proposed approach is presented on the basis of reduction in the cost of fuel and emission and compared with data reported in available literature [16]. The generation scheduling of the given system delivers the load demand ranges from 1200 to 2345 MW. Table 1 shows the load demand for 24-h time period, where peak demand occurs at 2320, 2320, and 2345 MW at 10:00–11:00 am, 11:00–12:00 am and 18:00–19:00 pm, respectively, and minimum demand occurs at 1200 MW at 03:00–04:00 am and 04:00–05:00 am respectively. The fuel cost coefficients and emission coefficients data for the 11 electrical generators are given in Tables 2 and 3.

The forecast solar radiation data is shown in Table 4. The spinning reserve requirement is set to 10% of the load demand.

Three case studies are carried out in this work:

- Case I: Only thermal generating units are used to supply the load demand (UCEED without SPV system) for scheduling time period.
- Case II: Thermal generating units and solar energy system are connected to (RES-Thermal) supply load demands for peak hours only.
- Case III: All generating units and solar energy system connected (RES-Thermal) to supply load demands for off-peak hours.

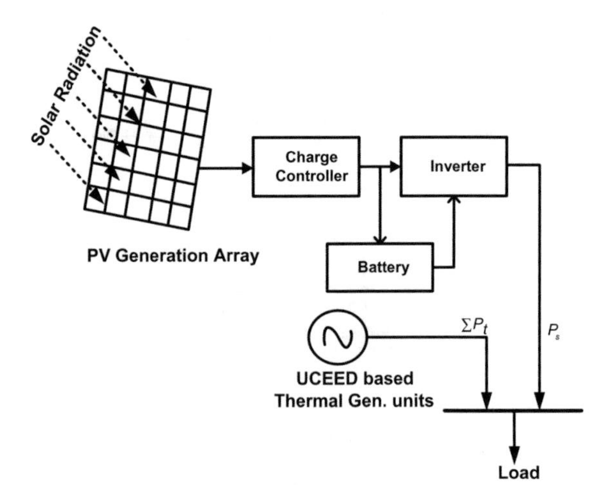

Fig. 4 Generation system under work

Table 1 The load demand for 24 h period

Time (h)	Load (MW)	Time (h)	Load (MW)
0–1	1590	12–13	2150
1–2	1450	13–14	2225
2–3	1345	14–15	2250
3–4	1200	15–16	2170
4–5	1200	16–17	2150
5–6	1345	17–18	2280
6–7	1450	18–19	2345
7–8	1750	19–20	2270
8–9	2150	20–21	2150
9–10	2270	21–22	1925
10–11	2320	22–23	1740
11–12	2320	23–24	1615

Table 2 Fuel cost coefficients

Units	Fuel cost coefficients			$P_{\min}$	$P_{\max}$
	α_i	β_i	γ_i		
1	0.00762	1.92699	387.85	20	250
2	0.00838	2.11969	441.62	20	210
3	0.00523	2.19196	422.57	20	250
4	0.00140	2.01983	552.50	60	300
5	0.00154	2.22181	557.75	20	210
6	0.00177	1.91528	562.18	60	300
7	0.00195	2.10681	568.39	20	215
8	0.00106	1.99138	682.93	100	455
9	0.00117	1.99802	741.22	100	455
10	0.00089	2.12352	617.83	110	460
11	0.00098	2.10487	674.61	110	465

Table 3 Emission coefficients

Units	NO_x emission coefficients		
	d_i	e_i	f_i
1	0.00419	−0.67767	33.93
2	0.00461	−0.69044	24.62
3	0.00419	−0.67767	33.93
4	0.00683	−0.54551	27.14
5	0.00751	−0.40006	24.15
6	0.00683	−0.54551	27.14
7	0.00751	−0.40006	24.15
8	0.00355	−0.51116	30.45
9	0.00417	−0.56228	25.59
10	0.00355	−0.41116	30.45
11	0.00417	−0.56228	25.59

Table 4 Forecast solar radiation

Time (h)	1	2	3	4	5	6
G_t (W/m^2)	0	0	0	0	13	218
Time (h)	7	8	9	10	11	12
G_t (W/m^2)	421	606	763	880	948	964
Time (h)	13	14	15	16	17	18
G_t (W/m^2)	927	838	704	535	341	136
Time (h)	19	20	21	22	23	24
G_t (W/m^2)	0	0	0	0	0	0

4.1 Case Study I: UCEED Without SPV System

In this case, only thermal generating units (11) are employed to supply load demand. LR-PL approach is applied to solve UCEED problem. Table 5 shows the obtained results. The determined results have also been compared with the results of existing method given in [16], for different conditions such as pure economic dispatch (ECD), pure emission dispatch (EMD) and for both (ECD/EMD), over the 24-h schedule period.

Figures 5 and 6 show that using proposed algorithm total electrical generators meet the load demands in an efficient manner. The applied algorithm's obtained average cost of generation is 0.0411 $/kWh and has been compared with pure ECD 0.0537 $/kWh, pure EMD 0.0597 $/kWh and ECD/EMD 0.0554, which is computationally efficient in terms of total cost and emission.

Figure 5 shows the generation cost for scheduled load demand of 24-h. The cost has been reduced significantly in peak and off-peak hours. As the emission is a vital component, that is to be considered while dealing with power generation using thermal power plants.

The proposed method combining UCEED gives better result in terms of low emission as compared to other methods given in literature [16] as shown in Fig. 6. The same method can be used to reduce the SO_x and CO_2 for an environment-friendly power generation at reduced cost of generation. Thermal power plants are largest power generation method worldwide, and they are the second largest pollution source. Due to these reasons the analysis and reduction of emission from the thermal power plants is topic of interest of researchers in recent trends.

Table 5 UCEED without SPV system for 24-h period

Methods	For 24-h period		
	Total operating cost $	Total NO_x emission kg	Generation cost $/kWh
Pure ECD [16]	245,055	35,890	0.0537
Pure EMD [16]	272,515	22,031	0.0597
ECD/EMD [16]	252,964	32,214	0.0554
Proposed UCEED	188,082	21,480	0.0411

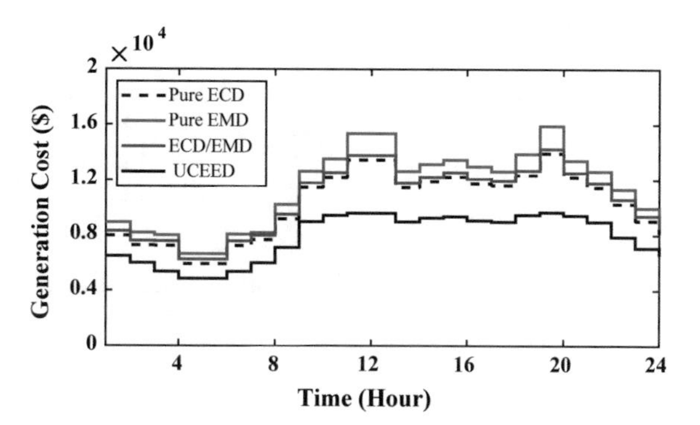

Fig. 5 Generation cost without SPV

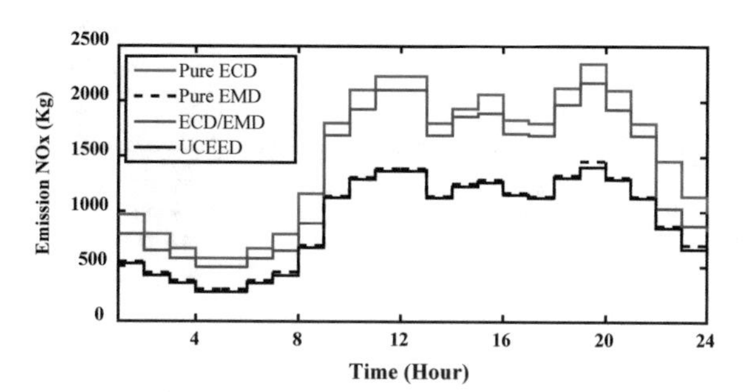

Fig. 6 Emission NO_x without SPV system for 24 h time period

4.2 Case Study II: UCEED-RES-Thermal Integration for Peak Hours Only

In this case, SPV units have also been considered in addition to 11 thermal units to check the performance of the proposed algorithm.

In order to schedule the on-site SPV power plants in peak hours and off-peak hours, total cost of generation and emission has been investigated. The determine results show the reduction in total cost for on-site and off-site is $195,028 and $192,408 for peak hours only, while the total emission is equal to 21,429 kg.

Figures 7 and 9 show the total cost and emissions at the peak hours. Tested results have been given in Table 6. The obtained results have also been compared with RES and energy storage system (ESS) [16] for peak hours only. It is seen that the peak load and emissions can be reduced by adding SPV power plants. That means the total fuel cost for the electrical generators units can be saved with the inclusion of SPV systems.

The saving in fuel cost is given as follows: The total fuel cost for on-site peak hours ($62,085 for RES, $65,836 for battery, $66,939 for flywheel and $66,371 for SMES) and off-site peak hours ($64,705 for RES, $68,456 for battery, $69,559 for flywheel and $68,991 for SMES) is saved.

The total emission for peak hours 10,701 kg is reduced. The total emission from case II is smaller than that from the case I. The air pollution is also lower.

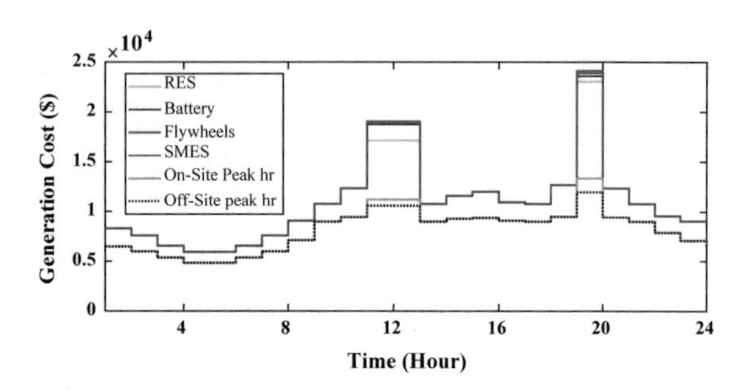

Fig. 7 Generation cost with solar photovoltaic system for peak hours during 24-h period

Table 6 SPV-thermal integration for peak hours only

Methods	For 24-h period		
	Total operating cost ($)	Total NO_x emission (kg)	Cost of generation ($/kWh)
With RES [16]	257,113	32,130	0.0563
With ESS [16] (i) Battery (ii) Flywheels (iii) SMES	260,864 261,967 261,399	32,130 32,130 32,130	0.0571 0.0574 0.0572
On-site SPV (peak hours) (RES-thermal)	195,028	21,429	0.0427
Off-site SPV (peak hours) (RES-thermal)	192,408	21,429	0.0421

4.3 Case Study III: UCEED-RES-Thermal Integration for Off-Peak Hours Only

In this case, all generating units are considered to supply the load demands. The obtained result evident that the total cost for on-site and off-site $211,242 and $202,473 is for off-peak hours while the total emission is equal to 21,335 kg.

Total cost and emission for off-peak hours only is shown in Figs. 8 and 9. The total saving in fuel cost is given as: The total fuel cost for on-site off-peak hours ($45,871 for RES, $49,622 for battery, $50,725 for flywheel and $50,157 for

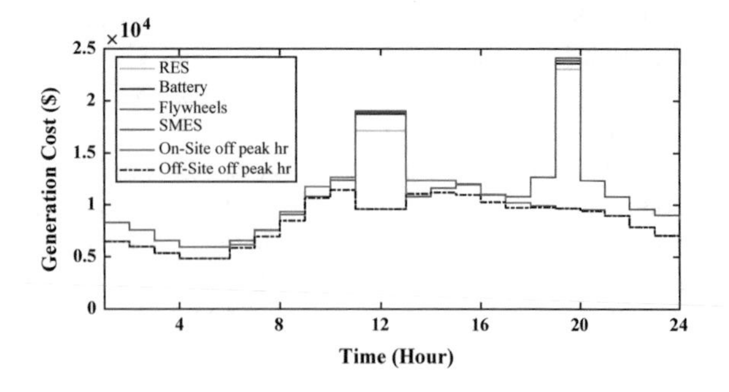

Fig. 8 Generation cost with solar photovoltaic system for off-peak hours during 24-h period

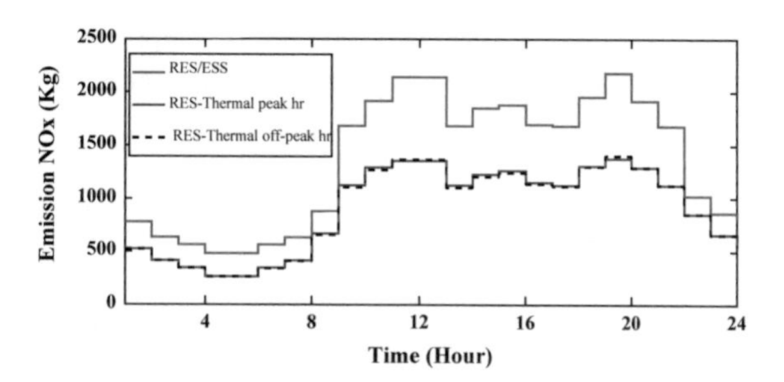

Fig. 9 Emissions with SPV system for peak/off-peak hours during 24-h

SMES) and off-site off-peak hours ($54,640 for RES, $58,391 for battery, $59,494 for flywheel and $58,926 for SMES) is saved. The total emission for off-peak hours 10,795 kg is reduced.

4.4 Comparison Between On-Site and Off-Site SPV System

Figures 7 and 8 show the generation cost of on-site/off-site peak and off-peak hours. The obtained results have been given in Tables 6 and 7. It is seen from the results that by proposing off-site SPV power plant, the total operation cost $2620 for peak hours and $8769 for off-peak hours is reduced. The emission comparison for peak/off-peak hours is shown in Fig. 9 that is also significantly low. Therefore, it can be concluded that, off-site SPV power plant is the better option for the economic operation and emission.

The generation cost of energy with thermal and solar units using the proposed approach is presented in Figs. 7 and 8.

It is seen from results that RES-Thermal approach performed better in terms of both cost and emission. The total operating cost using proposed approach is reduced in significant amount with increase in load demand. Among on-site and off-site, the

Table 7 SPV-thermal integration for off-peak hours only for 24-h period

Methods	For 24-h period		
	Total operating cost ($)	Total O_x emission (kg)	Cost of generation ($/kWh)
With RES [16]	257,113	32,130	0.0563
With ESS [16] (i) Battery (ii) Flywheels (iii) SMES	260,864 261,967 261,399	32,130 32,130 32,130	0.0571 0.0574 0.0572
On-site SPV (off-peak hours) (RES-thermal)	211,242	21,335	0.0462
Off-site SPV (off-peak hours) (RES-thermal)	202,473	21,335	0.0443

off-site approach gives good quality of supply free from sag and interruption with reduced cost as compared to on-site where cost of land and batteries avoided.

5 Conclusion

A new LR-PL method has been implemented for the combined unit commitment and economic dispatch with reduction in emissions. The thermal units are integrated with solar system with proposed approach RES-Thermal. Two case studies, UCEED without SPV system and UCEED with SPV system has been considered and presented. In the present work only NO_x emission has been focused, as the SO_x and CO_2 emissions can be calculated in the same manner. The proposed algorithm discussed on-site and off-site SPV system for the optimal solution with the various constraints. It has been found that, the off-site SPV power plant is the better option for the economic operation and emission. Solar photovoltaic power plants are environmental friendly. Due to these reasons, the analysis and reduction of emission from the thermal power plants is a topic of interest of researchers in recent trends. Scheduling of RES such as: SPV system not only useful for the society; but, also benefited the electricity energy market.

References

1. Hammons, T.J.: Impacts of electric power generation on green house gas emissions in Europe: Russia, Greece, Italy and views of the EU power plant supply industry-a critical analysis. Electr. Power Energy Syst. **28**, 548–564 (2006)
2. Mishra, U.C.: Environmental impact of coal industry and thermal power plants in India. J. Environ. Radioact. **72**, 35–40 (2004)
3. Bellhouse, G.M., Whittington, H.W.: Simulation of gaseous emissions from electricity generating plant. Electr. Power Energy Syst. **18**(8), 501–507 (1996)
4. Chakraborty, N., Mitra, A.P., Sharma, C., Bhattacharya, S., Santra, A.K., Chakraborty, S., Mukherjee, I., Chowdhury, S.: Measurement of CO_2, CO, SO_2, and NO emissions from coal-based thermal power plants in India. Atmos. Environ. **42**, 1073–1083 (2008)
5. Liu, C.H., Lewis, C., Lin, S.J.: Evaluation of NO_x, SO_x and CO_2 Emissions of Taiwan's thermal power plants by data envelopment analysis. Aerosol Air Qual. Res. **13**, 1815–1823 (2013)
6. Kuo, Y.M., Fukushima, Y.: Greenhouse gas air pollutant emission reduction potentials of renewable energy-case studies on photovoltaic and wind power introduction considering interactions among technologies in Taiwan. J. Air Waste Manag. Assoc. **59**, 360–372 (2009)
7. Lu, B., Shahidehpour, M.: Short-term scheduling of battery in a grid-connected PV/battery system. IEEE Trans. Power Syst. **20**(2), 1053–1061 (2005)
8. Lowery, P.G.: Generating unit commitment by dynamic programming. IEEE Trans. Power Apparatus Syst. **PAS-85**(5), 422–426 (1966)
9. Pang, C.K., Chen, H.C.: Optimal short-term thermal unit commitment. IEEE Trans. Power Apparatus Syst. PAS **95**(4), 1336–1346 (1976)
10. Kazarlis, S.A., Petridis, V., Bakirtzis, A.G.: A genetic algorithm solution to the unit commitment problem. IEEE Trans. Power Syst. **11**(1), 83–92 (1996)

11. Senjyu, T., Funabashi, T., Uezato, K., Shimabukuro, K.: A fast technique for unit commitment problem by extended priority list. IEEE Trans. Power Syst. **18**(2), 882–888 (2003)
12. Sum-Im, T.: Lagrangian relaxation combined with differential evolution algorithm for unit commitment problem. IEEE Emerg. Technol. Factory Autom (2014)
13. Kazemi, M., Siano, P., Sarno, D., Goudarzi, A.: Evaluating the impact of sub-hourly unit commitment method on spinning reserve in presence of intermittent generators. Energy **113**, 338–354 (2016)
14. Kumar, N., Panigrahi, B.K., Singh, B.: A solution to the ramp rate and prohibited operating zone constrained unit commitment by GHS-JGT evolutionary algorithm. Electr. Power Energy Syst. **81**, 193–203 (2016)
15. Carlos, J., Culp, C., Gilman, D.R., Haberl, J.S.: Development of a web-based emission reduction calculator for solar-thermal and solar-photovoltaic installations. In: International Conference for Enhanced Building Operations, Pittsburgh, Pennsylvania, pp. 1–10 (2005)
16. Palanichamy, C., Babu, N.S.: Day-night weather-based economic power dispatch. IEEE Trans. Power Syst. **17**(2), 469–475 (2002)
17. Liang, Y.C., Juarez, J.R.C.: A normalization method for solving the combined economic and emission dispatch problem with meta-heuristic algorithms. Electr. Power Energy Syst. **54**, 163–186 (2014)
18. Niknam, T., Firouzi, B.B., Mojarrad, H.D.: A new optimization algorithm for multi-objective economic/emission dispatch. Electr. Power Energy Syst. **46**, 283–293 (2013)
19. Aydin, D., Liao, T., Yasar, C., Ozyon, S.: Artificial bee colony algorithm with dynamic population size to combined economic and emission dispatch problem. Electr. Power Energy Syst. **54**, 144–153 (2014)
20. Jeddi, Babak, Vahidinasab, Vahid: A modified harmony search method for environmental/economic load dispatch of real-world power systems. Energy Convers. Manag. **78**, 661–675 (2014)
21. Glotic, Arnel, Zamuda, Aleš: Short-term combined economic and emission hydrothermal optimization by surrogate differential evolution. Appl. Energy **141**, 42–56 (2015)
22. Shilaja, C., Ravi, K.: Optimization of emission/economic dispatch using euclidean affine flower pollination algorithm (eFPA) and binary FPA (BFPA) in solar photo voltaic generation. Renew. Energy **107**, 550–566 (2017)
23. Reddy, S.S.: Optimal scheduling of thermal-wind-solar power system with storage. Renew. Energy **101**, 1357–1368 (2017)
24. Chandel, M., Mathur, S., Mathur, A., Aggarwal, G.D.: Techno-economic analysis of solar photovoltaic power plant for garment zone of Jaipur city. Case Stud. Thermal Eng. **2** (2014)
25. Kannan, R., Tso, C.P., Ho, H.K., Osman, R., Leong, K.C.: Life cycle assessment study of solar PV systems: an example of a 2.7 kWp distributed solar PV system in Singapore. Sol. Energy **80**, 555–563 (2006)
26. Sandwell, P., Chan, N.L.A., Foster, S., Nagpal, D., Emmott, C.J.M., Candelise, C., Buckle, S. J., Ekins-Daukes, N., Gambhir, A., Nelson, J.: Off-grid solar photovoltaic systems for rural electrification and emissions mitigation in India. Solar Energy Mater. Solar Cells **156**, 147–156 (2016)
27. Ramli, M.A.M., Hiendro, A., Twaha, S.: Economic analysis of PV/diesel hybrid system with flywheel energy storage. Renew. Energy **78**, 398–405 (2015)
28. Chakraborty, S., Senjyu, T., Ito, T.: Fuzzy logic-based thermal generation scheduling strategy with solar-battery system using advanced quantum evolutionary method. IET Gener. Trans. Distrib. **8**(3), 410–420 (2014)
29. Wood, A.J., Wollenberg, B.F.: Power Generation Operation and Control. Wiley, New York (1996)
30. International Energy Agency (IEA): Projected Costs of Generating Electricity, 2015 Edition
31. World Energy Council: World Energy Resources Solar (2016)
32. Is Solar Power Cheaper Than Coal? Curr. Sci. **109**(12) (2015)

Comparison of Throughput Metrics Based on Transmission Content in Cognitive Radio Networks

Atul Verma, Navpreet Kaur and Inderdeep Kaur Aulakh

Abstract The principle of cognitive radio technology is to enhance the proficiency of the wireless systems by using the frequency band among unlicensed users also. A cognitive radio system has been designed with two incumbents and four secondary transmitters. The secondary user transmits subjectively with the end goal that it doesn't hinder the execution of the primary node. There are a few variables which are utilized to enhance the cognitive radio system quality. Modulation techniques are one of them. In this paper we differentiate throughput between the two applications e-mail and video, modulation techniques such as 64-QAM, 16-QAM and QPSK are compared. Network throughput of all these three techniques is depicted for e-mail and video applications. For network design and simulation, the NetSim software tool has been used.

Keywords Cognitive radio networks · Primary users · Secondary users
Quadrature amplitude modulation · Quadrature phase shift keying

1 Introduction

Frequency spectrum is limited and the interest in it is expanding continuously every single day. Radio spectrum is distributed to the authorized radio administrations like the licensed system in a command-and-control way [1–3]. The radio spectrum is squandered when not utilized by the authorized or licensed users regularly. CRNs technology includes some adjustments in control of Radio Spectrum. That secondary users additionally utilizes spectrum when it is not utilized by authorized

A. Verma (✉) · N. Kaur · I. K. Aulakh
Panjab University, Chandigarh, India
e-mail: atulvermad360@gmail.com

I. K. Aulakh
e-mail: ikaulakh@pu.ac.in

© Springer Nature Singapore Pte Ltd. 2019

H. Malik et al. (eds.), *Applications of Artificial Intelligence Techniques in Engineering*, Advances in Intelligent Systems and Computing 697,
https://doi.org/10.1007/978-981-13-1822-1_12

users. For Secondary users, sensing is essential in order to maintain to avoid collisions, and packet losses [4–7].

CR technology is used to manage the spectrum shortage problem. As in CR Networks, the spectrum is fixed. Cognitive radios are intended for proficiently utilizing the spectrum on the mutual premise and not making destructive interference to licensed radio systems [8–10]. In CR, Secondary Users (SUs) use the available spectrum by utilizing spectrum-sensing procedure to identify the unused parts of the spectrum and utilize them in an opportunistic way, [11]. When SU is utilizing the band and distinguishes the presence of the Primary User, SUs must abandon it for incumbent users [12]. Spectrum sensing entails the determination of the presence of primary user's signal. It measures the level of a primary user's signal on a channel and hence makes the decision of the channel being available or not. It is the foundation of cognitive radio networks [13, 14]. CRNs allow SUs to utilize the frequency channels allocated to the licensed users when they are not accessing those frequency bands. But when licensed users need those frequency bands then the SUs utilizing that band have to immediately vacate it for PUs or incumbent users. It is of utmost importance that the licensed users must remain undisturbed by the activities of SUs [15]. SUs are always seeking free bands which can be used. If any free band is detected, it can be used by the secondary users in a cooperative manner because of the modulation techniques. Main feature of CR is coordination among the SUs. This coordination ensures efficient utilization of available spectrum in the network. In CRNs, every framework can sense its environment [16]. Parameters that enable cognitive radios to modify their operations are modulation and power of transmission [17].

1.1 Modulation

For the transmission of a signal from one place to another, it has to be strengthened. The signal can travel longer distances after it has been strengthened. Modulation is nothing but, the variation of the carrier signal in accordance with the information/ voice/data signal. Signal characteristics are changed using modulation techniques.

1.1.1 Two Types of Modulation

a. Analog Modulation.
b. Digital Modulation.

Advantages of Digital Over Analog Modulation

a. Digital communication provides additional security to our information signal.
b. Digital systems have more immunity to noise and external interference.
c. In digital systems, information can be saved and retrieved easily when needed, and it is not possible in analog systems.

1.2 Digital Modulation

Figure 1 depicts different types of modulation in which digital modulation is more powerful than analog, and hence for efficient communication and better quality digital modulation technique is used. It also increases the range of communication. Digital modulation has more advantages when compared to analog modulation that include high noise immunity, available bandwidth, etc. Digital modulation entails the conversion of the information from analog to digital form and then it modulates the carrier wave.

1.3 Multiple Access (OFDMA)

According to [18], orthogonal frequency division multiplexing (OFDM) has better air access. According to the notion of orthogonality guard bands are removed in OFDM technique, so the concept of orthogonality also provides better spectrum efficiency in OFDM, and one subcarrier does not affect another subcarrier in this technique. OFDM partitions the frequency channel into sub-bands which are orthogonal to each other and form the subcarriers. These sub-bands are then modulated by the same or different modulation technique like QAM, PSK, etc. [18]. In PSK technique, the modulating signal changes the phase of the carrier [19]. QPSK technique modulates 2 bits at one time, choosing 1 out of 4 phase shifts of the carrier, i.e., 0°, 90°, 180°, or 270°. Because of the QPSK technique, the signal can carry double the data carried by PSK using the same speed of transfer. QAM modulation is generally utilized for modulating the high-frequency carrier signal by

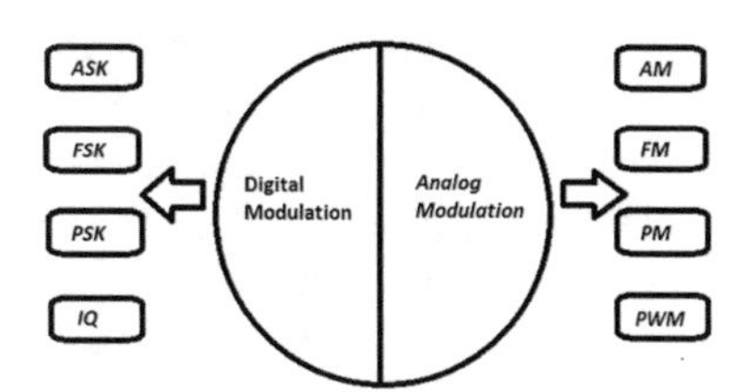

Fig. 1 Types of modulation

the information signal in communication. In QAM, 2 carriers that are out of phase by 90° are modulated and the resulting signal is both in phase and amplitude modulation.

2 Literature Review

As per [20], wireless channels are liable to a scope of mutilations because of rain constriction, multipath, and frequency fading. Information rate can be changed by utilizing modulation. For instance, the system bandwidth efficiency is better when utilizing 16-QAM rather than QPSK modulation.

In [21], the author examines the throughput execution against sensing time utilizing with OFDM as input information. OFDM strategy with 16-QAM is utilized as PU. Numerical simulation in MATLAB demonstrates that throughput is better.

In CRN, modulation is one of the variables which is useful to upgrade network performance. There are different metrics available to improve network quality like modulation, transmission power, etc. According to [22], 64-QAM performs well as compared to the 16-QAM and QPSK modulation in NetSim.

3 Network Model and Simulation

In CRNs, modulation is one of the parameters which is useful to upgrade network performances [23, 24]. In Fig. 2 There are four secondary users and two base stations, one router and two incumbents. As shown in Fig. 2, the fourth node ('D') is sending information or packet to eighth node ('H'), the relative examination has been inspected in terms of modulation, for example, QPSK, 16-QAM and 64-QAM for e-mail and video applications. We upgrade the network performance by utilizing QPSK, 16-QAM, and 64-QAM modulation. As in 64-QAM it transmits 6 bits/symbol and have total 64 symbols with 6/1 bit or baud rate, 16-QAM modulation transmits 4 bits/symbols and have 16 symbols with 4/1 bit or baud rate, and in QPSK it transmits 2 bits/symbols and have 4 symbols with 2/1 bit or baud rate, network scenario has been made in NetSim. All details are as follows.

The link is established between fourth node ('D') to eighth node ('H') by setting different properties of BS and application in the network, and after setting all parameters, simulation has been checked. The same network is used for e-mail and video applications to examine the network performance and metrics as below:

Different performance metrics are depicted when QPSK, 16-QAM, and 64-QAM modulations are used in NetSim as network metrics, packets errored, packet transmitted, bytes transmitted, payload transmitted and overhead transmitted.

The first three tables (Tables 1, 2, and 3) show the network performances of each modulation technique for E-mail application and other three tables

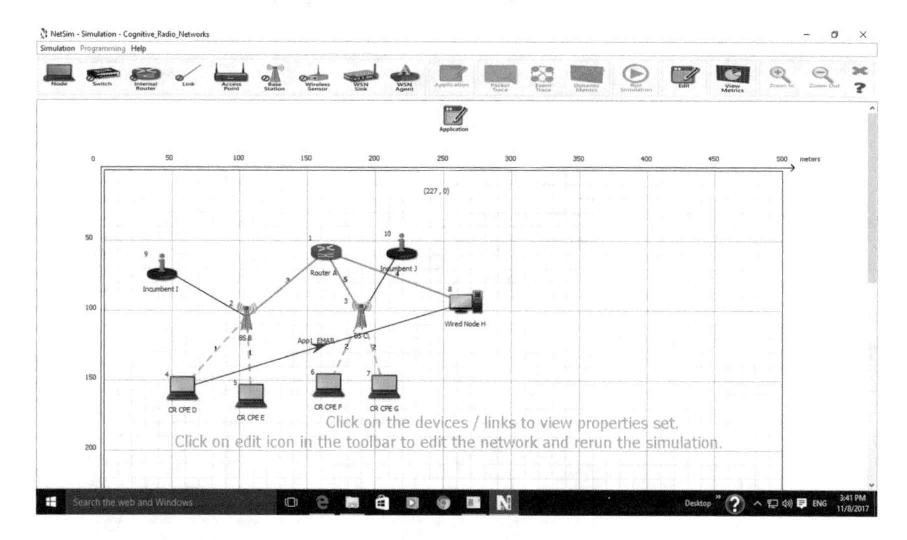

Fig. 2 Cognitive radio networks setup utilizing NetSim

(Tables 4, 5, and 6) show the network performance for video application, which shows payload transmitted, overhead transmitted, number of packets errored, bytes transmitted and compare all three modulation techniques. Now, we analyze the overall throughput of the design from fourth to eighth node for E-mail and video applications, respectively, for all three modulation techniques. According to these Tables, 64-QAM shows more packets errored in e-mail application and in case of video application packets errored are nearly same for all modulation techniques as shown below in the graphs.

Figures 3, 4, and 5 depict the mean and standard deviation of e-mail and video applications of QPSK, 16-QAM, and 64-QAM, respectively, in the throughput graphs.

All these three techniques give different mean values and SD, each graph shows different values, and in e-mail application, it shows a large drop in throughput as compared to the video application. Throughput is better in case of video application as compared to e-mail application.

Table 1 QPSK network metrics for e-mail

Metric	Value
Simulation time (ms)	100,000.00
Packets transmitted	121,082
Packets errored	4
Packets collided	0
Bytes transmitted (bytes)	3,530,998.00
Payload transmitted (bytes)	1,211,330.00
Overhead transmitted (bytes)	2,319,668.00

Table 2 16-QAM network metrics for e-mail

Metric	Value
Simulation time (ms)	100,000.00
Packets transmitted	121,053
Packets errored	4
Packets collided	0
Bytes transmitted (bytes)	3,535,258.00
Payload transmitted (bytes)	1,211,220.00
Overhead transmitted (bytes)	2,324,038.00

Table 3 64-QAM network metrics for e-mail

Metric	Value
Simulation time (ms)	100,000.00
Packets transmitted	117,959
Packets errored	6
Packets collided	0
Bytes transmitted (bytes)	3,506,632.00
Payload transmitted (bytes)	1,211,220.00
Overhead Transmitted (Bytes)	2,295,412.00

Table 4 QPSK network metrics for video

Metric	Value
Simulation time (ms)	100,000.00
Packets transmitted	115,361
Packets errored	7
Packets collided	0
Bytes transmitted (bytes)	3,443,333.00
Payload transmitted (bytes)	1,746,540.00
Overhead transmitted (bytes)	1,696,793.00

Table 5 16-QAM network metrics for video

Metric	Value
Simulation time (ms)	100,000.00
Packets transmitted	115,287
Packets errored	8
Packets collided	0
Bytes transmitted (bytes)	3,443,630.00
Payload transmitted (bytes)	1,746,994.00
Overhead transmitted (bytes)	1,696,636.00

Table 6 64-QAM network metrics for video

Metric	Value
Simulation time (ms)	100,000.00
Packets transmitted	115,261
Packets errored	7
Packets collided	0
Bytes transmitted (bytes)	3,443,379.00
Payload transmitted (bytes)	1,747,028.00
Overhead transmitted (bytes)	1,696,351.00

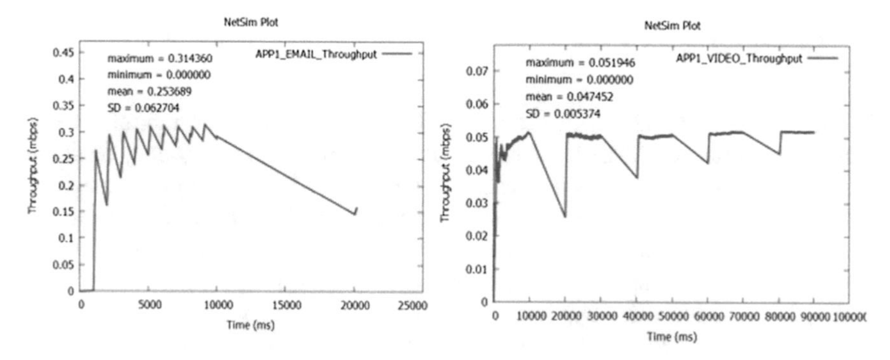

Fig. 3 Illustration of throughput for e-mail and video applications (QPSK)

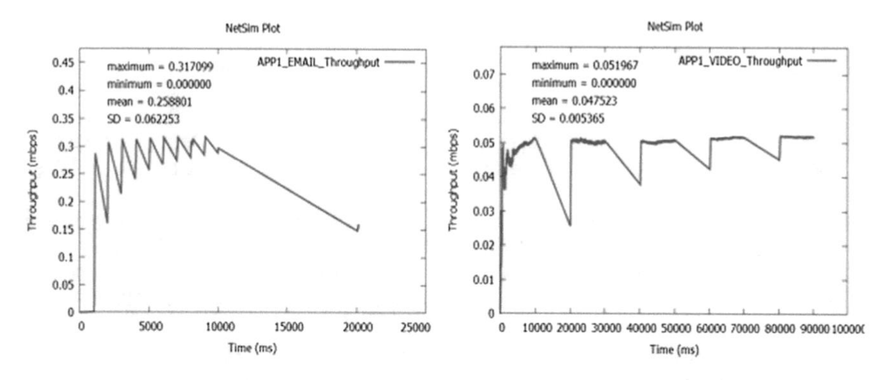

Fig. 4 Illustration of throughput for e-mail and video applications (16-QAM)

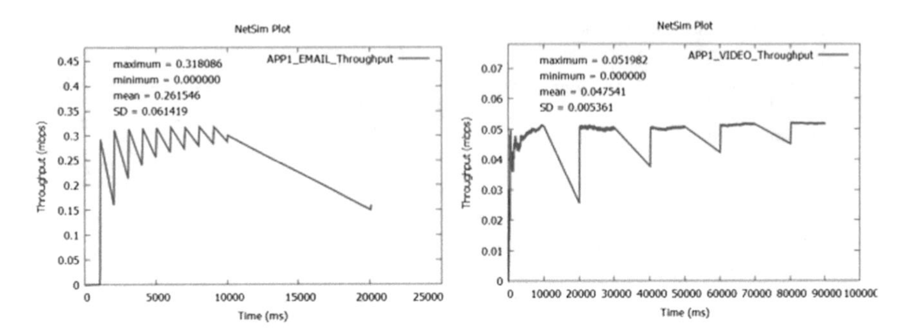

Fig. 5 Illustration of throughput for e-mail and video applications (64-QAM)

4 Conclusion and Results

In these e-mail and video applications, OFDMA is used as the multiple access technique because OFDMA has better air access. It divides the frequency band into narrow orthogonal parts, called subcarriers. Figures. 3, 4, and 5 show the graphs of throughput (mbps) versus time (ms) of all three techniques of the designed network from node 4 to node 8 for e-mail and video applications. This paper describes the overall throughput graphs between e-mail and video applications and after comparing these graphs video application shows better throughput as compared to the e-mail application. In these designed applications, like e-mail and video, throughput graphs are obtained and based on that 64-QAM shows better throughput as compared to the 16-QAM and QPSK modulation techniques

Bibliography

1. Fette, B.A.: Cognitive Radio Technology, 2nd edn. Elsevier Inc., New York, USA (2009)
2. Berlemann, L., Mangold, S.: Cognitive Radio and Dynamic Spectrum Access. Wiley, Hoboken, (2009)
3. Lai, L., Gamal, H.E., Jiang, H., Poor, H.V.: Cognitive medium access: exploration, exploitation and competition. IEEE Trans. Mob. Comput. **10**(2), 239–253 (2011)
4. Aulakh, I.K., Vig, R.: Secondary user sensing time optimization in sensing-transmission scheduling for dynamic spectrum access in cognitive radio. J. Comput. Sci. **11**(8), 880–891 (2015)
5. Aulakh, I.K., Vig, R.: Optimization of SU's probability of false alarm for dynamic spectrum access in Cognitive Radio. In: International Conference on Computing for sustainable Global Development (IndiaCom), pp. 710–715 (2014)
6. Aulakh, I.K., Vig, R.: Secondary user transmission protection optimisation in sensing-transmission scheduling under varying channel noise in cognitive radio networks. Int. J. Sys. Control Commun. **7**(2), 97–115 (2016)
7. Aulakh, I.K., Vig, R.: Secondary user aggressiveness optimization in sensing-transmission scheduling for cognitive radio networks. J. Netw. **10**(10), 543–551 (2015)

8. Aulakh, I.K., Kaur, N.: Optimal sensing simulation in cognitive radio networks under shadow–fading environments. In: International Conference on Computing for sustainable Global Development (INDIACom), pp. 901–905 (2016)
9. Aulakh, I.K., Singh, S., Kaur, N., Vig, R.: Optimization of BER in cognitive radio networks using FHSS and energy-efficient OBRMB spectrum sensing. East J. Electron. Commun. **16** (2), 297–324 (2016)
10. Aulakh, I.K., Vig, R.: Optimization of secondary user access in cognitive radio networks. In: Engineering and Computational Sciences (RAECS), pp. 6–8 (2014)
11. Chen, Z., Cooklev, T., Chen, C., Pomalaza-Raez, C.: Modeling primary user emulation attacks and defenses in cognitive radio networks. In: IEEE International Performance Computing and Communications Conference, pp. 208–215 (2009)
12. Jin, F., Varadharajan, V., Tupakula, U.: Improved detection of primary user emulation attacks in cognitive radio networks. In: International telecommunication Networks and Applications Conference, pp. 274–279 (2015)
13. Malhotra, M., Aulakh, I.K., Vig, R.: A review on energy based spectrum sensing in cognitive radio networks. In: IEEE International Conference on Furistic Trends on Computational analysis and Knowledge Management (Ablaze), 3(Ablaze), pp. 561–565 (2015)
14. Malhotra, M., Aulakh, I.K.: Secure spectrum leasing in cognitive radio networks via secure primary-secondary user interaction. In: Artificial Intelligence and Evolutionary Computations in Engineering systems, pp. 735–741. Springer, Switzerland (2016)
15. Yuan, Z., Niyato, D., Li, H., Han, Z.: Defense against primary user emulation attacks using belief propagation of location information in cognitive radio. Networks **30**(10), 599–604 (2011)
16. Maric, S., Reisenfeld, S., Goratti, L.: A single iteration belief propagation algorithm to minimize the effect of primary user emulation attacks. In: Intelligent Signal Processing and communication systems, pp. 1–6 (2016)
17. Kaur, N., Aulakh, I.K., Vig, R.: Analysis of spread spectrum techniques in cognitive radio networks. Int. J. Appl. Eng. Res. **11**(8), 5641–5645 (2016)
18. Acharya, S., Kabiraj, P., De, D.: Comparative analysis of different modulation techniques of LTE networks. In: Proceedings of the 2015 Third International Conference on Computer Communication Control and Information Technology (C3IT) Hooghly, pp. 1–6 (2015)
19. Koshti, R., Jangalwa, M.: Performance comparison of WRAN over AWGN & RICIAN channel using BPSK and QPSK modulation with convolution coding. In: International Conference on Communication Systems and Network Technologies, pp. 124–126 (2013)
20. Fong, B., Hong, G.Y., Fong, A.C.M.: A modulation scheme for broadband wireless access in high capacity networks. IEEE Trans. Consum. Electron. **48**(3), 457–462 (2002)
21. Armi, N., Chaeriah, B.A.W., Suratman, F.Y., Wijaya, A.: Sensing time-based throughput performance in OFDM cognitive radio systems. In: International Conference on Wireless and Telematics, pp. 88–91 (2017)
22. Kanti, J. Bagwari, A., Tomar, G.S.: Quality analysis of cognitive radio networks based on modulation techniques. In: International Conference on Computational Intelligence and communication Networks, pp. 566–569 (2016)
23. Pappas, N., Kountouris, M.: Throughput of a cognitive radio network under congestion constraints: a network level study. In: 9th International Conference on Cognitive Radio Oriented Wireless Networks, pp. 162–166 (2014)
24. Saifuddin, K.M., Ahmed, A.S., Reza, K.F., Alam, S.S., Rahman, S.: Performance analysis of cognitive radio: Netsim viewpoint. In: 2017 3rd International Conference on Electrical Information and Communication Technology (EICT), Khulna, pp. 1–6 (2017). https://doi.org/10.1109/eict.2017.8275155

A New Electronically Tunable CM/VM Oscillator Using All Grounded Components

Soumya Gupta, Manpreet Sandhu, Manish Gupta and Tajinder Singh Arora

Abstract With the use of two Voltage Differencing Current Conveyers (VDCC) as active building blocks, and three passive elements, i.e., one resistor and two capacitors, an electronically tunable sinusoidal oscillator has been designed. The proposed configuration offers advantages like low component count, use of grounded elements, and attainment of an independent condition of oscillation (C.O.) and frequency of oscillation (F.O.). Dual-mode operation and efficient integrated circuit implementation due to use of all grounded capacitors improve the performance of the design. The graphical results using PSPICE simulation software, for verifying the functionality of the designed oscillator, have been obtained by employing CMOS structure of VDCC as well as off-the-shelf realization using OPA860.

Keywords Voltage mode circuits · Current mode oscillator · Voltage differencing current conveyer

1 Introduction

The approach for designing reliable, good performance and wide scope circuits which can be employed in communication, instrumentation, control engineering, etc., comes from Analog Signal Processing (ASP). Analog circuits like filters, oscillators, converters, integrators, and many more have been designed in the past employing a variety of analog devices such as Operational Transconductance

S. Gupta · M. Sandhu
Department of Electronics and Communication, MSIT, Janakpuri, New Delhi, India

M. Gupta
Department of EC, Indraprastha Engineering College, Ghaziabad, New Delhi, India

T. S. Arora (✉)
Department of Electronics Engineering, National Institute of Technology, Uttarakhand, Srinagar 246174, Uttarakhand, India
e-mail: tsarora@nituk.ac.in

© Springer Nature Singapore Pte Ltd. 2019

H. Malik et al. (eds.), *Applications of Artificial Intelligence Techniques in Engineering*, Advances in Intelligent Systems and Computing 697,
https://doi.org/10.1007/978-981-13-1822-1_13

Amplifier (OTA) [1], Current Differencing Buffered Amplifier (CDBA) [2], Current Feedback Operational Amplifier (CFOA) [3], Differential Voltage Current Conveyer (DVCC) [4], Current Conveyers (CC) [5], Voltage Differencing Current Conveyer (VDCC) [6] to name a few.

Oscillator, which is the application explored in this manuscript, refers to an electronic arrangement, which provides a periodic output through feedback and amplification. Researchers in the past have made worthy attempts in designing oscillators using variant analog means [7]. However, the active building block (ABB) utilized here has not been studied much when it comes to devising a sinusoidal oscillator with features like (i) dual-mode operation, (ii) independent C. O. and F.O., (iii) high oscillation frequency, (iv) low/acceptable total harmonic distortion and (v) all grounded passive elements, hence easy integrated circuit implementation. The authors have made an attempt of introducing one such oscillator arrangement, which provides all the above-mentioned features and have justified the circuit in all possible ways.

A tabulated comparison between presented oscillator network and previously designed oscillator circuits have been provided in Table 1. Since VDCC is a device which has not been utilized much in the generation of sinusoidal oscillators, hence the comparison has been drawn among oscillators designed using different ABBs. Some of the noticeable features on the basis of which oscillators have been compared are: (i) type and number of active components, (ii) number of passive components, (iii) employment of grounded elements, and (iv) mode of operation.

The presented manuscript has been divided into following sections: the forthcoming section, i.e., Sect. 2, gives an introduction to the active device, VDCC. Section 3 presents the proposed circuit along with the transfer function, while Sect. 4 gives the nonideal and sensitivity analysis. Simulation results justifying the workability of the design are included in Sect. 5 whereas the conclusion is provided in Sect. 6.

Table 1 Comparison of previous works and proposed work

Ref. No.	Active components	Passive components	All grounded passive elements	Output in CM as well as VM
[8]	1 CDBA	5	No	No
[9]	2 CDBA	5	No	No
[10]	2 CCIII	6	No	No
[11]	3 CCII	5	Yes	No
[12]	1 CFOA	6/7	No	No
[13]	2 CFOA	4	No	No
[14]	2 VDCC	4	Yes	Yes
[15]	2 VDCC	4	Yes	Yes
Proposed Circuit	2 VDCC	3	Yes	Yes

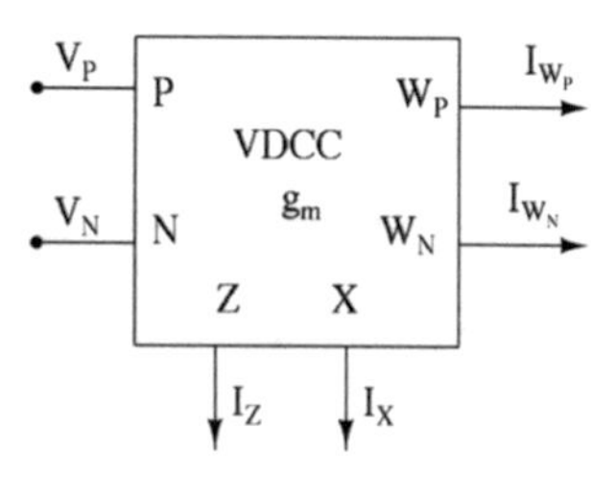

Fig. 1 Circuit symbol of VDCC

2 Description of VDCC

VDCC is a six-terminal device, as shown in Fig. 1, which is capable of providing both current mode and voltage mode output. Among the six terminals available, W_P, W_N, Z, P, and N are high impedance ports, whereas X is a low impedance port. Also, this device has an electronically controllable gain value which is represented by g_m.

The nonideal characteristic equations of VDCC are given in (1).

$$\begin{gathered} I_N = I_P = 0 \\ I_Z = \alpha g_m (V_P - V_N) \\ I_{WP} = \gamma I_X \\ I_{WN} = -\gamma I_X \\ V_X = \beta V_Z \end{gathered} \tag{1}$$

where α is the port transfer ratio of P and N terminals, and β and γ are the port transfer ratios of Z and X terminals, respectively. The ideal value of α, β, and γ is unity, substituting which we get the ideal characteristic equations of VDCC.

3 Proposed Circuit

The arrangement of active and passive elements resulting in pure sinusoidal oscillator configuration is shown in Fig. 2. In total, two active and three passive components have been employed, and all passive elements are grounded.

On analyzing the circuit shown above using the ideal characteristic equations of VDCC, which is obtained from (1) by replacing α, β, and γ with unity, the ideal transfer function of the oscillator is derived as shown in (2). Using the same equation, ideal F.O. and C.O. are obtained as given in (3) and (4), respectively.

$$s^2 + s\left(\frac{1}{R_1 C_1} - \frac{g_3}{C_2}\right) + \frac{g_2 g_3}{C_1 C_2} = 0 \tag{2}$$

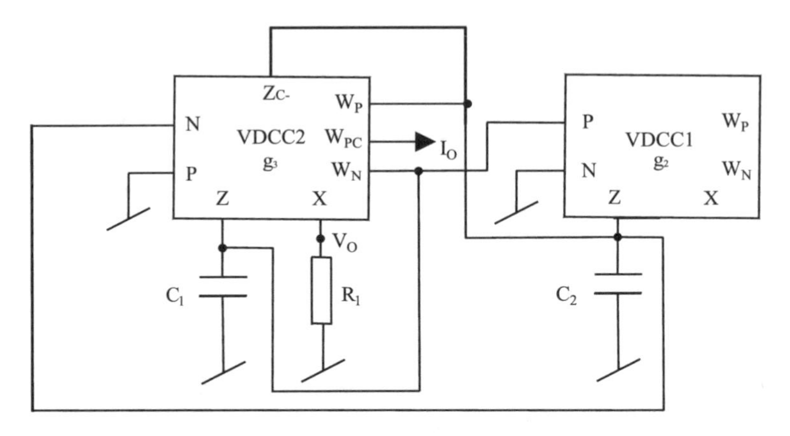

Fig. 2 Proposed oscillator configuration

$$\text{F.O.}\,\omega_0 = \sqrt{\frac{g_2 g_3}{C_1 C_2}} \tag{3}$$

$$\text{C.O.}\,\frac{1}{R_1 C_1} - \frac{g_3}{C_2} = 0 \tag{4}$$

The above-derived equations indicate that F.O. and C.O. can be independently tuned. This implies that C.O. can be tuned by varying R_1, which does not affect F. O., whereas F.O. can be adjusted electronically using g_2, which does not affect C. O., hence the design enjoys simultaneous adjustment of both the parameters to their desired values.

4 Non-ideal and Sensitivity Analysis

Analysis of the proposed circuit of Fig. 2 has also been done using the equations given in (1). These equations take into consideration the non-idealities of the active device, i.e., VDCC, thereby giving us the non-ideal characteristic equation as displayed in (5) and F.O. and C.O. as shown in (6) and (7), respectively.

$$s^2 + s\left(\frac{\gamma_1 \beta_1}{R_1 C_1} - \frac{\alpha_1 g_3}{C_2}\right) + \frac{\alpha_1 \alpha_2 g_2 g_3}{C_1 C_2} = 0 \tag{5}$$

$$\text{F.O.}\,\varpi_0 = \sqrt{\frac{\alpha_1 \alpha_2 g_2 g_3}{C_1 C_2}} \tag{6}$$

$$\text{C.O.}\,\frac{\gamma_1 \beta_1}{R_1 C_1} - \frac{\alpha_1 g_3}{C_2} = 0 \tag{7}$$

where α_1, β_1, γ_1 are the non-ideal port transfer ratios of VDCC1 while α_2, β_2, γ_2 represent the non-idealities of VDCC2.

The sensitivity of the circuit toward the passive components, for ideal and non-ideal frequency, as well as the non-ideal port transfer ratios, for non-ideal frequency, has been given in (8) and (9), respectively.

$$S_{C_1}^{\omega_0} = S_{C_2}^{\omega_0} = S_{C_1}^{\varpi_0} = S_{C_2}^{\varpi_0} = -\frac{1}{2} \tag{8}$$

$$S_{\alpha_1}^{\varpi_0} = S_{\alpha_2}^{\varpi_0} = \frac{1}{2} \tag{9}$$

As desired, a very low sensitivity of 0.5 has been obtained which indicates the efficient performance of the proposed circuit.

5 Simulation Results

For practically verifying the devised oscillator and generating its graphical results, Cadence ORCAD PSPICE simulation software has been used. Initially, all the results have been obtained by using the CMOS structure of VDCC [16] given in Fig. 3. The CMOS arrangement functions on the use of TSMC 0.18 μm process parameters [16], with a supply voltage of ±0.9 V and bias current I_{B1} being 50 μA and I_{B2} being 100 μA. The corresponding value of transconductance gain (g_m) is 277.83 μA/V.

For supporting hardware realization of the devised oscillator, VDCC has also been implemented using commercially available ICs OPA860 [17], and the PSPICE simulation results for the same are also presented later.

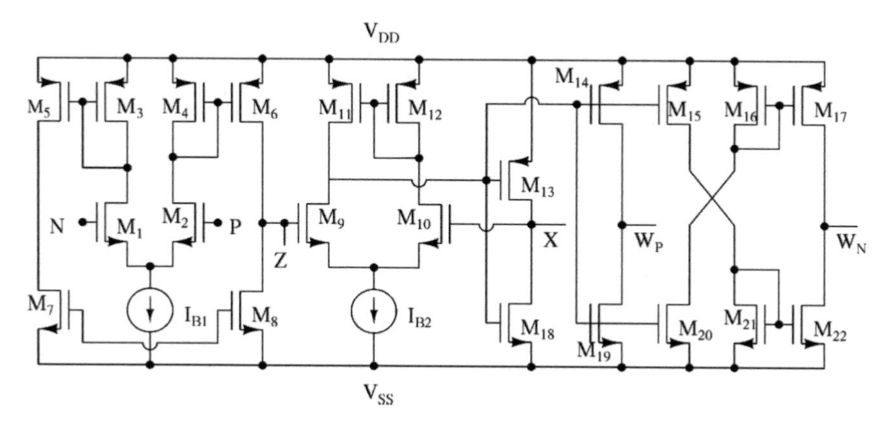

Fig. 3 CMOS structure of employed VDCC [16]

All necessary and regular simulations, such as the transient, steady state, fast Fourier transform (FFT) and total harmonic distortion (THD) analysis have been included. Presenting first the simulation results obtained by employing CMOS structure of VDCC, Fig. 4 shows the transient response of the oscillator when operated in current mode. The frequency at which oscillations have been obtained is 1 MHz. To obtain this frequency of oscillation, the values of the passive elements were chosen to be $C_1 = C_2 = 47$ pF, $R_1 = 3.6$ K. The steady-state response, obtained by focusing on few cycles of transient response, is given in Fig. 5.

FFT and THD results are given in Figs. 6 and 7, respectively. Former figures show the presence of very low magnitude harmonics at frequencies other than F.O. and play a vital role in signifying the efficiency of the oscillator. Latter figure shows

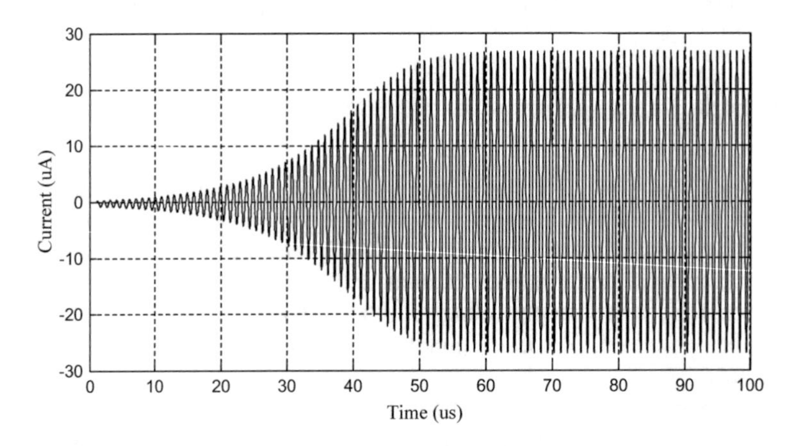

Fig. 4 Transient response of designed oscillator when operated in current mode

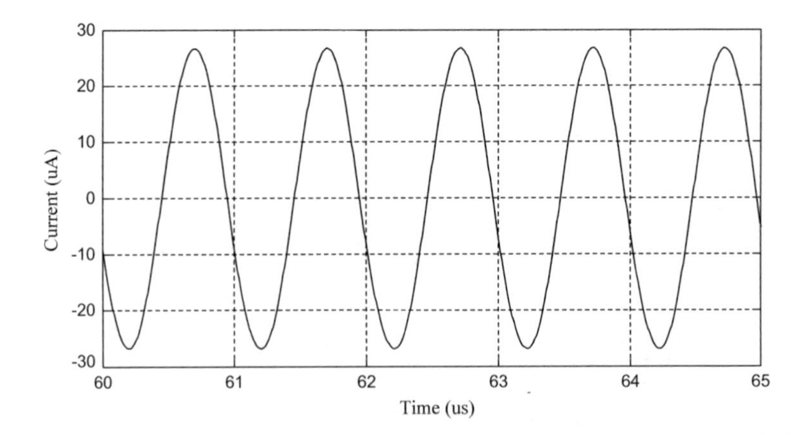

Fig. 5 Steady-state response generated using transient response of oscillator

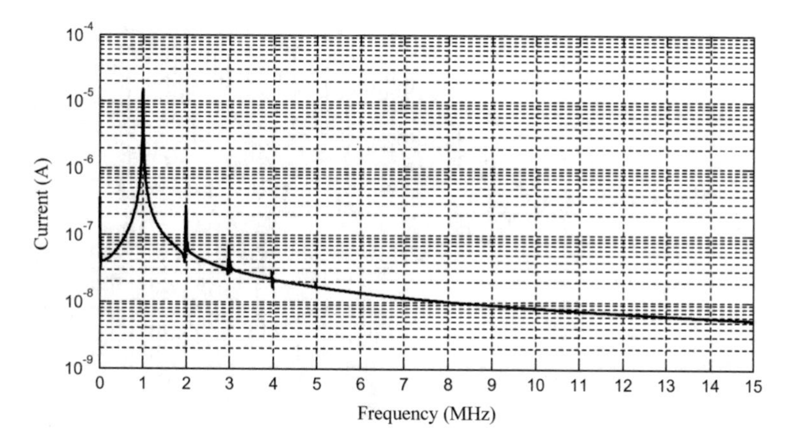

Fig. 6 FFT response with main harmonic at F.O

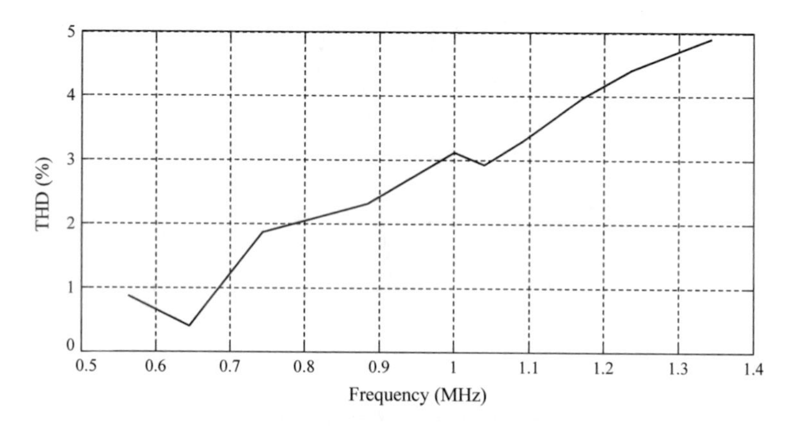

Fig. 7 THD response of the proposed oscillator

the percentage of distortion the designed oscillator will provide at different frequencies.

As stated earlier, the designed oscillator configuration operates both in current mode as well as voltage mode. The above-shown curves are for current mode operation, and explicit current output has been achieved. With the purpose of justifying its voltage mode functioning, transient response as well as steady-state response of the proposed sinusoidal oscillator, the frequency being 1 MHz, has been included in Figs. 8 and 9, respectively.

The proposed circuit has also been simulated using the OPA860 realization of VDCC. Figure 10 depicts the transient response of an oscillator whereas Fig. 11 shows the steady-state response, which has been obtained by focusing on a few oscillations of transient response. FFT results are given in Fig. 12. The frequency

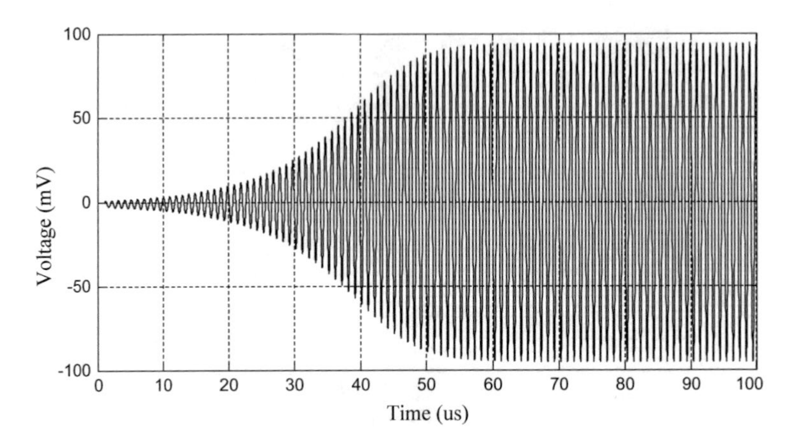

Fig. 8 Transient response of designed oscillator when operated in voltage mode

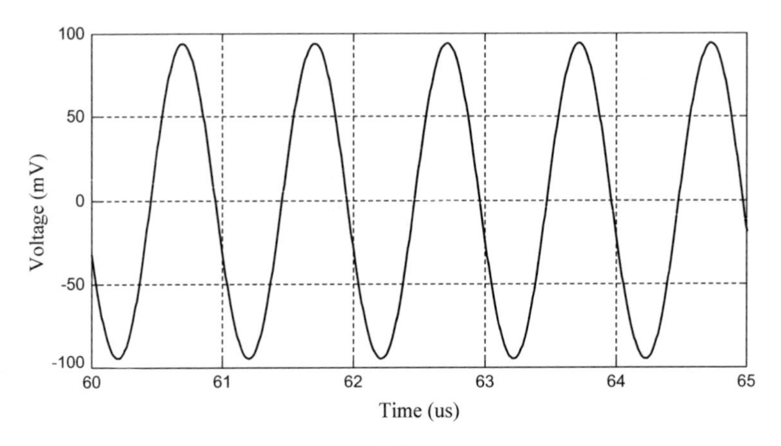

Fig. 9 Steady-state response generated using transient response of oscillator

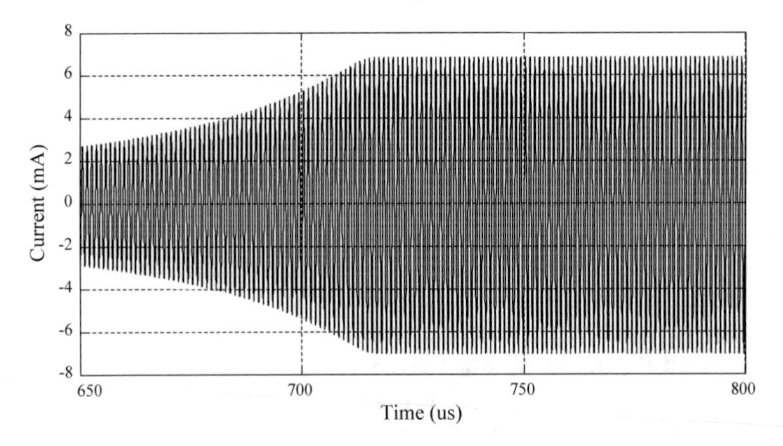

Fig. 10 Transient response obtained using OPA860

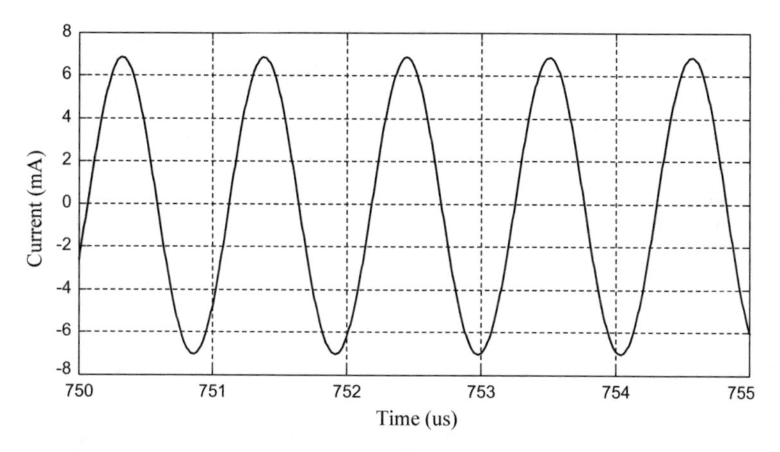

Fig. 11 Steady-state response obtained using OPA860

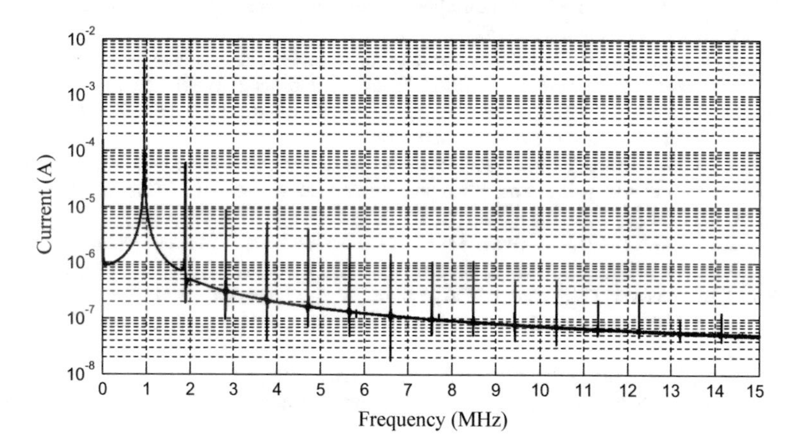

Fig. 12 FFT response obtained using OPA860

obtained in this case is 1 MHz, with an error percentage of 5.8. The values of the passive elements were chosen to be $C_1 = C_2 = 470$ pF and $R_1 = 360$ ohm which are standard values of capacitors and resistors. Hence, hardware results can be obtained by using exactly the same values.

6 Conclusion

The paper focused on introducing an oscillator network, using VDCC as an active device, which offered various advantageous features like low component count, efficient integrated circuit implementation due to all grounded capacitors and independent tunability of C.O. and F.O. For theoretically justifying the sinusoidal oscillator, all regular analysis such as ideal, non-ideal, and sensitivity were presented. The practical validation had been done using PSPICE simulations, in which the CMOS structure of VDCC was used initially and the high frequency of oscillation, i.e., 1 MHz, was achieved. The proposed oscillator configuration was also tested using off-the-shelf IC OPA860 and results had been presented in the paper. All regular analysis such as transient, steady state and FFT and THD were included for justifying its workability in both the stated modes.

References

1. Senani, R., Bhaskar, D.R., Gupta, M., Singh, A.K.: Canonic OTA-C sinusoidal oscillators: generation of new grounded-capacitor versions. American J. Electr. Electron. Eng. **3**, 137–146 (2015)
2. Arora, T.S., Rana, U.: Multifunction filter employing current differencing buffered amplifier. J. Circuits Sys. **7**, 543–550 (2016)
3. Senani, R., Sharma, R.K.: Explicit-current-output sinusoidal oscillators employing only a single current-feedback op-amp. J. IEICE Electron. Express **2**, 14–18 (2005)
4. Minaei, S., Yuce, E.: All-grounded passive elements voltage-mode DVCC-based universal filters. Circuits Sys. Sig. Process. **29**, 295–309 (2010)
5. Sharma, R.K., Arora, T.S., Senani, R.: On the realization of canonic single-resistance-controlled oscillators using third generation current conveyers. J IET Circuits Devices Sys. **11**, 10–20 (2017)
6. Prasad, D., Bhaskar, D.R., Srivastava, M.: New single VDCC-based explicit current-mode SRCO employing all grounded passive components. Electronics **18**(2), 81–88 (2014)
7. Senani, R., Bhaskar, D.R., Singh, V.K., Sharma, R.K.: Sinusoidal Oscillators and Waveform Generators Using Modern Electronic Circuit Building Blocks. Springer, New Delhi, India (2016)
8. Ozcan, S., Toker, A., Acar, C., Kuntman, H., Cicekoglu, O.: Single resistance-controlled sinusoidal oscillators employing current differencing buffered amplifier. J. Microelectronics **31**, 169–174 (2000)
9. Kalra, D., Gupta, S., Arora, T. S.: Single-resistance-controlled quadrature oscillator employing two current differencing buffered amplifier. In: IEEE International Conference on Contemporary Computing and Informatics (IC3I), pp. 688–692 (2016)
10. Un, M., Kacar, F.: Third generation current conveyor based current-mode first order all-pass filter and quadrature oscillator. J. Electr. Electron. Eng. **8**, 529–535 (2008)
11. Abdalla, K.K., Bhaskar, D.R., Senani, R.: Configuration for realizing a current-mode universal filter and dual-mode quadrature single resistor controlled oscillator. J IET Circuits Devices Sys. **6**, 159–167 (2012)
12. Toker, A., Cicekoglu, O., Kuntman, H.: On the oscillator implementations using a single current feedback op-amp. J. Comput. Electr. Eng. **28**, 375–389 (2002)
13. Chen, H.P., Wang, S.F., Ku, Y.T., Hsieh, M.Y.: Quadrature oscillators using two CFOAs and four passive components. J. IEICE Electron. Express **12**, 1–8 (2015)
14. Srivastava, M., Prasad, D.: VDCC based dual-mode quadrature sinusoidal oscillator with outputs at appropriate impedance levels. Adv. Electr. Electron. Eng. **14**(2), 168–177 (2016)

15. Gupta, M., Arora, T.S.: Realisation of current mode universal filter and a dual-mode single resistance controlled quadrature oscillator employing VDCC and grounded passive elements. Adv. Electr. Electr. Eng. **15**, 833–845 (2017)
16. Kaçar, F., Yeşil, A., Minaei, S., Kuntman, H.: Positive/negative lossy/lossless grounded inductance simulators employing single VDCC and only two passive elements. Int. J. Electron. Commun. **68**, 73–78 (2014)
17. Kaçar, F., Yeşil, A., Gürkan, K.: Design and experiment of VDCC-based voltage mode universal filter. Indian J. Pure Appl. Phy. (IJPAP) **53**(5), 341–349 (2015)

An Advance Forward Pointer-Based Routing in Wireless Mesh Network

Abhishek Hazra and Prakash Choudhary

Abstract Mobility Management in Wireless Network has been considered as a great potential innovation in recent times. A specific problem in Wireless Mesh Network is to supply a smooth Internet connectivity among Mesh Nodes. Several mobility management schemes like iMesh, MEMO, WMM, Mesh networks with Mobility management has been implemented to improve the performance of the Mesh Network. The drawback of these schemes is the high communication cost. In this paper, we have proposed an adaptive mobility management scheme called Forward Pointer-based Routing (FPBR). With this scheme Mesh Router reduces the frequent location update message to the gateway. Moreover, it uses forward pointer and shortest path for decreasing the Signaling Cost. The selection of forwarding chain on Handoff Cost, Packet Delivery Cost, and Total Communication cost are observed in this paper. Improvement of Performance is also carried out with the help of numerical analysis.

Keywords Wireless mesh network · Handoff · Forward pointer
Internet traffic

1 Introduction

Wireless Mesh Networks (WMN) [1–4] slowly supersedes the wired access networks and traditional wireless networks. Likewise, they require special attention to the resources of a mesh network in order to provide infrastructure and quality of service. Wireless Networks [5, 6] are a communication network that is often

A. Hazra (✉) · P. Choudhary
National Institute of Technology, Imphal, Manipur, India
e-mail: abhishek.hazra1@nitmanipur.ac.in

P. Choudhary
e-mail: choudharyprakash@nitmanipur.ac.in

© Springer Nature Singapore Pte Ltd. 2019

H. Malik et al. (eds.), *Applications of Artificial Intelligence Techniques in Engineering*, Advances in Intelligent Systems and Computing 697,
https://doi.org/10.1007/978-981-13-1822-1_14

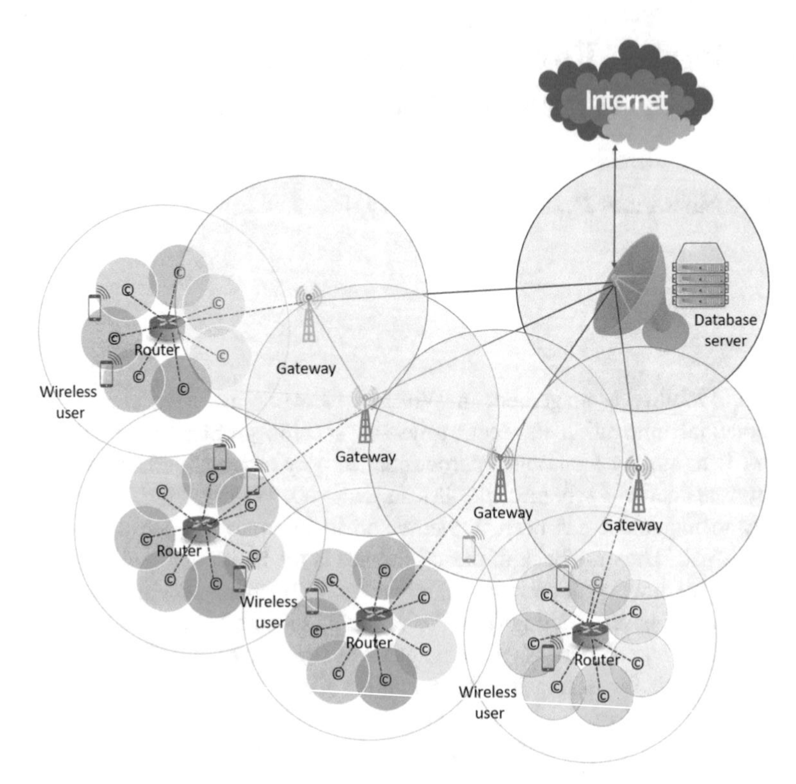

Fig. 1 The infrastructure of a wireless mesh network

constructed by three basic types of Mesh nodes Mesh Client (MC), Mesh Router (MR), and Gateway(GW). Mesh clients are highly mobile devices because of its dynamic movement. Mesh routers are the backbone of wireless network which connects the MC with the network. There is a special kind of router in the network called Gateway, which keeps track of all the information in the network. MR and Gateways are the stationary wireless devices used for routing the packets. Providing uninterrupted Internet connectivity is one of the incriminating issues in a wireless mesh network. Whenever any MC comes out of the continuity of one MR to another Handoff [7] management becomes essential for forwarding the packets. Figure 1 shows the Backbone of a Wireless Network.

The rest of the paper is organized as follows. Section 2 discusses the different mobility management scheme. In Sect. 3, the proposed FPBR scheme is introduced. Sections 4, 5, and 6 show the System model and assumptions, Numerical Analysis, and Performance Analysis. Finally, Conclusion and Feature Work are discussed in Sect. 7.

2 Related Work

Several strategies have been introduced for supporting the mobility in the field of WMN. Among them, some techniques such as iMesh [8], MEMO [9], and M^3 [10] have been discussed briefly in this section.

iMesh is an infrastructure-based network. It follows IEEE 802.11 base standard. The common problem with IEEE 802.11i and 802.11 s are that in many situations, it fails to provide fast re-authentication or fast hand off. In iMesh, whenever a Mesh Client steps from the proximity of one MR to another, serving MR broadcast encodes HNA (Host and Network Association) information in the wireless network. The main disadvantage of this scheme is the broadcasting of HNA message, whenever any MC associated with MR. In MEMO, AODV-MEMO [1] protocol was used for integrating both mobility and routing information. According to this scheme, whenever a node wakes up, it first addresses a special Route Request information to the gateway. After getting the information from the GW, serving MR broadcasts this message in the network. In a situation where the mobility of an MC is high, this scheme fails to reduce the communication cost. Huang et al. proposed a different kind of stable mobility management technique called Wireless Mesh Mobility Management (WMM) [11]. According to WMM, each router maintains two table routing table and proxy table [12]. Routing table maintains the handoff information in the wireless network and proxy table keeps the MCs location update information. This technique somehow cut down the frequent location update cost, but in a situation, where MCs frequent location update is high and location of the destination MR is not known, then this scheme also fails to reduce the total communication cost in the wireless network. Per-hop communication cost for MCs is in general the problem with MEMO, iMesh, and WMM.

3 Proposed Scheme

In this section, we proposed a Pointer Forwarding [13, 14] scheme for reducing the number of route update messages sent by the MC to minimize signaling costs for Mobility Management in MC. The Gateway maintains a database and keeps track of information about the routing in different MRs. Then, it sends an ART message to the network as shown in Table 1, which will be propagated through all the nodes until all the MRs creates an Adjacent Relationship Table in the network. When the MC joins a WMN, it first gets associated with a nearby MR and sets it as serving MR. Then, it sends location update message to the GW. The GW maintains a database recording the serving MRs of all corresponding MCs alive inside the Network. On receiving the location update message, GW first checks its database whether an entry of the MC is present or not. If there exists an entry corresponding to the MC, the GW updates the information. Else it will create a new entry for the

Table 1 Adjacent relationship table

Next Hop IP	Relation	Sequence No.	Bandwidth	....
....	....	....	....	....

MC. For handling the Intranet traffic, each MR maintains a database of the serving MRs of corresponding MCs.

The proposed scheme is based on a tree structure and mainly consists of two parts: Node Registration and Route Updation.

3.1 Node Registration

When an MC wakes up in the network, it sends an Alive message to the Corresponding MR. Then, the serving MR sends a registration request message to the Gateway and the Gateway update the entry of the corresponding MC in the Routing Table. Finally, the gateway sends a confirmation message to the MR. Whenever any MC plod from the continuity of one MR to a new MR, it deputizes two message. One is a Registration Request (RR) message to the new MR for updating the entry in the ART and second message to the current serving MR, which contains the Handoff information. After receiving the handoff information, it will forward the message to the newly serving MR. On receiving the handoff information from old serving MR and RR message from child, the new serving MR Register the entry of the MC in the Routing Table. After registration the new MR sends a Handoff confirmation to the new MR and Registration Confirmation to the Corresponding MC. In the final step, the MC sends an RSM message to the new serving MR and the old MR adds a Pointer towards the new serving MR so that packets can forward to the destination MC. With this, the registration of MCs is complete and the process of Path Update will start.

3.2 Route Update

In Wireless Network, the movements of MCs is dynamic in nature. When any MC comes into the vicinity of new MR it leads to many location update message to the GW. But with the proposed FPBR scheme, whenever any MC moves in Wireless network it will send location update message to the Gateway only when the New MR is the Parent of Old MR or the New MR receives NACK from all the child's. This phase is divided into three parts, phase one explains the Route Update process by Serving MR, Phase two explains the connection establishments of intermediate MRs.

3.2.1 Route Updation by Serving MR

If the new Serving MR finds any relation in ART with old MR as a parent of Old MR, it broadcasts a Handoff message to all siblings regarding the association of new MC. After receiving it, the sibling MR checks its routing table for the entry of the corresponding MC, if any entry found, it updates the next hop IP of the MC is the New serving MR. Else it will send a NACK handoff information. All the sibling MRs follow the same steps. If the new serving MR receives NACK from all the sibling MRs, it simply forwards the RSM to the neighboring parent MRs for that corresponding MC. If the New Serving MR is the Relationship with old MR as a child, it will not forward the RSM to the Gateway. If the new serving MR is the Sibling with the relationship with old MR, it simply forwards the RSM to the neighboring parent MRs for that corresponding MC.

3.2.2 Route Updation by Gateway

The RSM passes through the different MRs. This process goes till Gateway receives the RSM. After receiving RSM, The Gateway first looks into the database for finding the entry of the communicating MC in the Routing Table. If there exists an entry of the corresponding MC, then it checks whether the sequence number of old RSM is the same or less than new RSM and accordingly update its next hop IP is the IP of new serving MR and sequence number will be the RSM's sequence. If there is no entry of the Communicating MC, then Gateway creates a new entry for that MC and fills the next hop IP and sequence number with the new Serving MR of the corresponding MC and the sequence number with RSM. Finally, the Gateways sends back a RAC acknowledgment message to the communicating MC through intermediate MR.

3.2.3 Route Updation by Intermediate MR

When an Intermediate MR receives a Message, it first checks its type whether the message is a MRs or a RAC message. According to the nature of the message, the intermediate MRs Follow the belonging steps: If the intermediate MRs receives a RSM then it looks for the existence of the communicating MC in the ART. If no entry exists then it will create an entry and store the corresponding next hop IP and sequence no.. If there is an entry with less sequence no., then remove the old entry and update the table with the sequence no. of new RSM and next hop IP. After storing the information, the intermediate MRs Forward the RSM to its Parent. If the intermediate MRs receives any RAC message dedicated to an MC, it first looks into the entry for the communicating MC in the routing table and accordingly forwards the message to all other MRs which are the next hop MR in the routing table. Every Intermediate MRs follow these steps. Finally, the RAC message reaches to the Destination serving MR and RSM reaches to the Gateway. The Handoff Process completes with receiving with RAC message to the destination communicating MC.

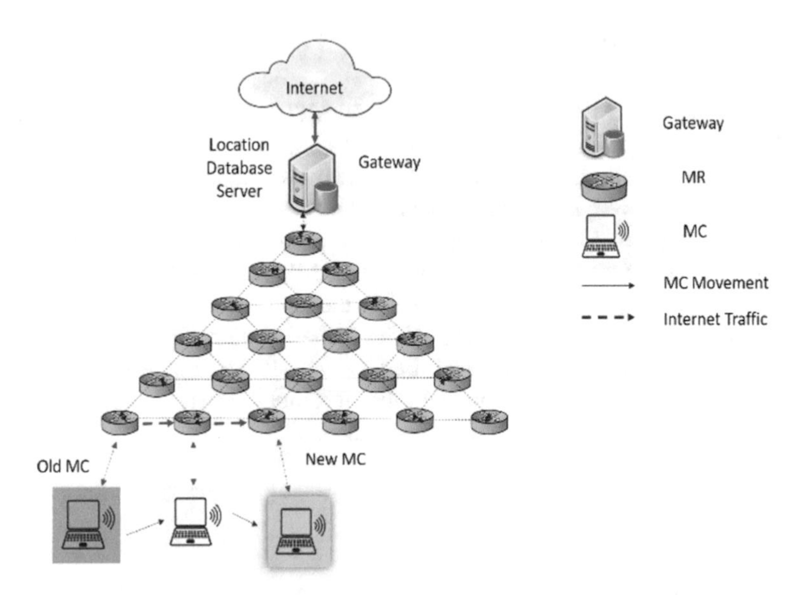

Fig. 2 Pyramidal mesh topology

4 System Model and Assumptions

In this phase, System Model and Assumptions have been described for analyzing different parameters. Let the total no. of MRs present in WMN is M_{Total} and also it could be considered that MRs will follow a Pyramid Structure based Network, which is shown in Fig. 2. Average Distance between two MRs is $D_{\text{MR to MR}}$. Average distance from MR to gateway is $D_{\text{MR to GW}}$. It can be assumed that MCs moves independently in the WMN with equal probability. N_{avg}, P_{avg} and C_{avg} shows the average no. of neighbor parent and client in the network. The residence mobility time and session arrival time follows exponential distribution and normal distribution along with parameter T_t and T_p respectively. Detailed default parameter values are explained in Table 2 separately.

Table 2 Default parameter values

Sr. No	Symbol	Values	Sr. No.	Symbol	Values
1	M_{total}	500	6	C_{avg}	8
2	$D_{\text{MR to GW}}$	5	7	$D_{\text{MR to MR}}$	4
3	N_{avg}	14	8	P_{NACK}	0.85
4	P_{avg}	2	9	$T_{p\,\text{delay}}$	0.001
5	$P_{\text{FRR Message}}$	0.9	10	T_t	10, 20, 30, 40, 50

5 Numerical Analysis

In this section, we analyze the effectiveness of the proposed FPBR scheme. For minimizing the total communication cost. We use a forward pointer which reduces the frequent location update cost. Along with the performance of the Handoff Cost, Signaling Cost, Packet Delivery cost, and the Total communication cost of FPBR scheme, we also introduce a comparison of handoff cost and signaling cost with the existing two schemes: iMesh and MEMO.

5.1 *Handoff Cost*

In iMesh, when the MC moves in the wireless network, it broadcasts a message to all the nodes in the network for updating the routing, in case of MEMO Serving MR of the corresponding MC does not any broadcast a message, rather it sends a spatial Route reply message which contains the IP of the communicating MC then the Serving MR forwards the first Data Packet. In FPBR, MR simply adds a Forward Pointer with the New MR, which reduces the Handoff Cost. Let the Handoff cost for iMesh, MEMO and FPBR be $\text{iMesh}_{\text{handoff Cost}}$, $\text{MEMO}_{\text{handoff Cost}}$ and $\text{FPBR}_{\text{handoff Cost}}$.

The Handoff Cost/Unit time can be calculated as

$$\text{iMesh}_{\text{handoff cost}} = D_{\text{MR to GW}}.T_{p\text{ delay}} \tag{1}$$

$$\text{MEMO}_{\text{handoff cost}} = D_{\text{MR to GW}}.T_{p\text{ delay}} \tag{2}$$

$$\begin{aligned}\text{FPBR}_{\text{handoff cost}} = 2.T_{p\text{ delay}}.\{D_{\text{MR to GW}}.\left(N_{\text{avg}} - P_{\text{avg}} - C_{\text{avg}}\right) \\ + P_{\text{avg}}.(1 - P_{\text{NACK}} - D_{\text{MR to GW}}.P_{\text{NACK}})\}/N_{\text{avg}}\end{aligned} \tag{3}$$

5.2 *Signaling Cost*

In iMesh, whenever an MC changes its location and comes into the vicinity of a new MR, it broadcast an RSM by using OLSR Routing Protocol. In MEMO instead of broadcasting an RSM, MRs only sends RSM to the gateway for updating the entry of the communicating MC. In case of FPBR, it sends an RSM only when the new MR is the parent or the Sibling of the Old serving MR. Along with RSM, FPBR uses a special RAC message to establishment and cancelation of route through the old serving MR. Let the signaling cost for iMesh, MEMO, and FPBR be $\text{iMesh}_{\text{Signalling Cost}}$, $\text{MEMO}_{\text{Signalling Cost}}$, and $\text{FPBR}_{\text{Signalling Cost}}$.

The Signaling Cost/Unit time can be calculated as

$$\text{iMesh}_{\text{Signalling Cost}} = T_t.M_{\text{total}} \tag{4}$$

$$\text{MEMO}_{\text{Signalling Cost}} = T_t.P_{\text{avg}}^{D_{\text{MR to GW}}}.P_{\text{FRR Message}} \tag{5}$$

$$\begin{aligned} \text{FPBR}_{\text{Signalling Cost}} = T_t.\{&N_{\text{avg}} - C_{\text{avg}} - P_{\text{avg}} + P_{\text{NACK}}.P_{\text{avg}}) \\ &.(2.P_{\text{avg}}^{D_{\text{MR to GW}}}.P_{\text{FRR Message}} + D_{\text{MR to MR}}) \\ &+ (N_{\text{avg}} - C_{\text{avg}} - P_{\text{avg}}).P_{\text{avg}}\}/N_{\text{avg}} \end{aligned} \tag{6}$$

5.3 *Packet Delivery Cost*

Unlike other schemes, FPBR scheme uses pointer forwarding technique so that the flight messages can reach to the destination MC. Packet from Gateway to the MC follows the same path until the Handoff process is complete.

Let the Packet Delivery cost/unit for FPBR be $\text{FPBR}_{\text{Packet Delivary Cost}}$ and is calculated as

$$\text{FPBR}_{\text{Packet Delivary Cost}} = T_t.T_p.\text{FPBR}_{\text{handoff cost}} \tag{7}$$

5.4 *Total Communication Cost*

The Total Communication Cost for FPBR can be denoted as $\text{FPBR}_{\text{Total Communication Cost}}$. The total Communication Cost can be calculated with Handoff cost/unit time, Packet Delivery Cost/unit time, and Signaling Cost/unit time. So, the Total Communication Cost/unit can be calculated as

$$\begin{aligned} \text{FPBR}_{\text{Total Communication Cost}} = \; &2.T_{p\,\text{delay}}.\{D_{\text{MR to GW}}.(N_{\text{avg}} - P_{\text{avg}} - C_{\text{avg}}) \\ &+ P_{\text{avg}}.(D_{\text{MR to MR}} - P_{\text{NACK}} - D_{\text{MR to GW}}.P_{\text{NACK}})\}/N_{\text{avg}} \\ &+ T_t.T_p.\text{FPBR}_{\text{handoff cost}} \\ &+ T_t.\{N_{\text{avg}} - C_{\text{avg}} - P_{\text{avg}} + P_{\text{NACK}} * P_{\text{avg}}) \\ &.(2.P_{\text{avg}}^{D_{\text{MR to GW}}}.P_{\text{FRR Message}} + D_{\text{MR to MR}}) \\ &+ (N_{\text{avg}} - C_{\text{avg}} - P_{\text{avg}}).P_{\text{avg}}\}/N_{\text{avg}} \end{aligned} \tag{8}$$

6 Performance Analysis

In this section, the proposed FPBR scheme has been analyzed. Along with this, we try to examine the comparison between iMesh, MEMO, and FPBR for Handoff Cost and Signaling Cost.

Figure 3 shows the Comparison of Handoff delay/unit time of the proposed scheme with iMesh and MEMO. With the proposed scheme, when an MC changes its location and comes into the vicinity of a New MR from Old MR, it adds a forward pointer rather than updating its location frequently. In iMesh, each time MC sends RSM to the GW whenever MC moves and comes into the locality of any new MR. In MEMO, whenever MC moves from the proximity of one MR to another MR, MC does not send location update frequently, rather it will send a spatial RSM. Figure 3 shows the proposed scheme handoff cost/unit time has been decreased.

Figure 4 shows the comparison of the signaling delay/unit time of the proposed scheme with iMesh and MEMO. In the case of FPBR, RAC message along with RSM is also used to limit the forward chain. In the proposed scheme, MC sends RSM to the limited number of neighbors. Figure 4 shows a sharp difference between the signaling costs among the schemes. The Signaling cost/unit time in FPBR is least among these three schemes.

Figure 5 shows the Packet Delivery Cost/unit time for the proposed FPBR scheme with different values of mobility rate and session activity rate. The effect of

Fig. 3 Comparison of handoff cost for iMesh, MEMO, and FPBR

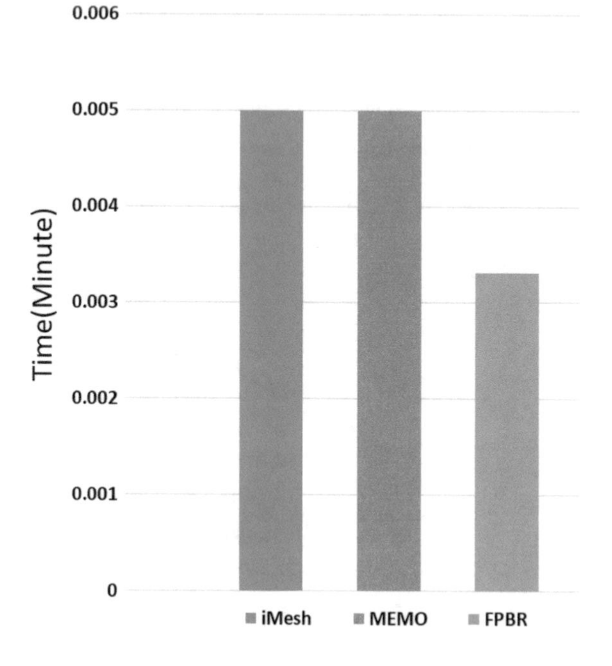

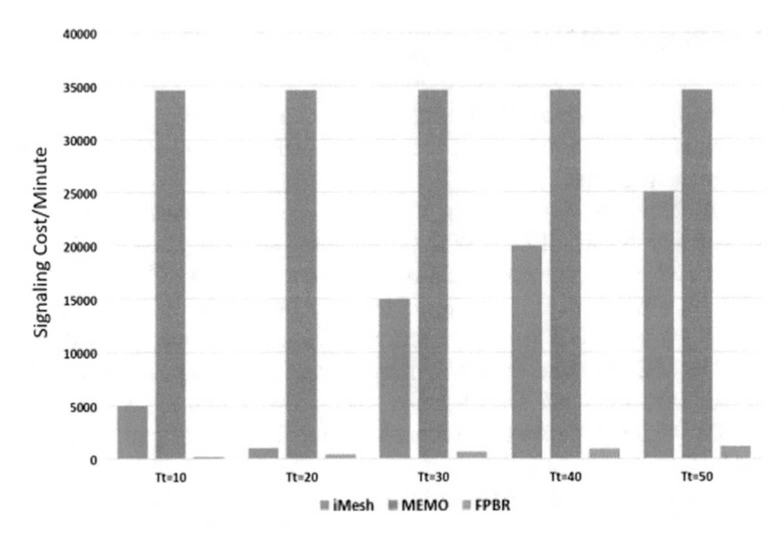

Fig. 4 Comparison of signaling cost for iMesh, MEMO, and FPBR

changing rate with different values of mobility rate and session activity rate for forwarding packets has been demonstrated in Fig. 5.

Figure 6 shows the Total Communication Cost/unit time for the FPBR scheme with different values of Mobility Rate and Session Activity Rate. As the proposed scheme follows the pointer forwarding technique for packet delivery, gradually total communication cost is also going down. The Packet Delivery and Total Communication Cost becomes more with the increase of mobility and large session activity.

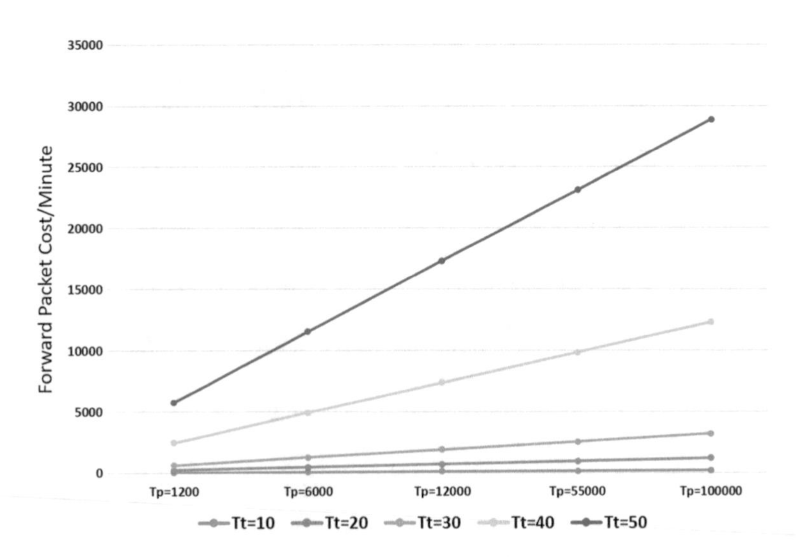

Fig. 5 Comparison of packet delivery cost of FPBR for different values of T_t and T_p

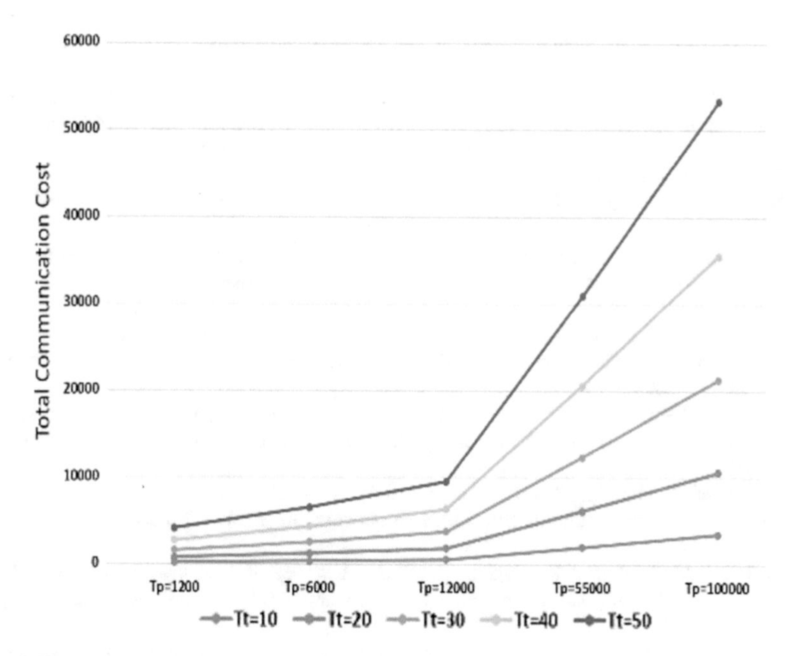

Fig. 6 Total communication cost of FPBR for different values of T_t and T_p

7 Conclusion

In this paper, we proposed an FPBR mechanism for improving the performance of wireless network. In this scheme, whenever any MC moves from the vicinity of one MR to another MR it tries to reduce the frequent location update messages to or from the gateway, thus location update cost reduces drastically. Because of its dynamic pathfinding nature, this scheme is highly efficient and resilient in case of link failure. Moreover, this scheme uses RAC message as an ACK of RSM. Since the mobility of an MC is dynamic, the use of pointer forwarding is beneficial. This scheme reduces the per-hop communication cost compared to iMesh and MIMO.

References

1. Akyildiz, I.F., Wang, X.: Wireless mesh networks: a survey. Comput. Netw. Int. J. Comput. Telecommun. Netw. **47**(4), 445–487 (2005)
2. Akyildiz, I.F., Wang, X.: A survey on wireless mesh networks. IEEE Commun. Mag. **43**(9), S23–S30 (Sept 2005)
3. Long. L.: Multipath routing for wireless mesh network. In: IEEE Global Telecommunication Conference, pp. 1–6 (2011)
4. Kowsar, M.M.S., Karim, M.R.: A survey of mobility management in wireless mesh network. In: IEEE ICCIT, pp. 316–320 (2012)

5. Akyildiz, I.F., et al.: Mobility management for next generation wireless system. In: Proc. IEEE **87**(8), 1347–1384 (Aug 1999). ISSN: 1558–2256
6. Majumder, A., Roy, S.: A performance comparison of mobility management schemes in handling internet traffic of wireless mesh network. In: IEEE Conference-ICCCI, Jan 2014
7. Guo-jun, S., Shu-Qun, S.: Smooth handoff in infrastructure wireless mesh network. In: IET 2nd International Conference, pp. 171–174 (2008)
8. Navda, V., Kashyap, A., Das, R.S.: Design and evaluation of iMesh: an infrastructure-mode wireless mesh network. In: WoWMoM, IEEE Conference, pp. 164–170 (2005)
9. Ren, M., Liu, C., Zhao, H., Zhao, T., Yan, W.: MEMO: an applied wireless mesh network with client support and mobility management. In: IEEE GLOBECOM'07, pp. 164–170 (Nov 2007)
10. Pack, S., Jung, H., Kwon, T., Choia, Y. SNC: a selective neighbour catching scheme for fast handoff in 802.11 wireless networks. AMC SIGMOBILE Mob. Comput. Commun. Rev. **9**(4), 39–49 (Oct 2005). USA
11. Jiang, X., Wang, X. (University of North Caroline at Charlotte, Teranovi Technologies, Inc.): A survey of mobility management in hybrid wireless mesh network, IEEE, pp. 0890–8044 (2008)
12. Majumder, A., Roy, S., Dhar, K.K.: Design and analysis of an adaptive mobility management scheme for internet traffic in wireless mesh network. In: IEEE ICMiCR'13, pp. 1–6, June 2013
13. Sheikh, S.M., Wolhuter, R., Engelbrecht, H.A.: A survey of cross-layer protocols for IEEE 802.11 wireless multi-hop mesh network. Int. J. Commun. Syst. **30**(6) (April 2016). https://doi.org/10.1002/dac.3129
14. Majumder, A., Roy, S., Arab, J.: Implementation of forward pointer-based routing scheme for wireless mesh network. Arab. J. Sci. Eng. **41**(3), 1109–1127 (March 2016)

No-Reference Image Corrosion Detection of Printed Circuit Board

Anandita Bhardwaj and Pragya Varshney

Abstract Image processing techniques are highly recommended for industrial purposes. Its advantages include fast processing speed, high-quality inspection, and great efficiency. Also, the process can be repeated as many times as required without affecting the efficiency. The aim of this project is the detection of corrosion on PCB by the use of *k*-means clustering for colour-based segmentation and image processing theories. This project requires a test image and no-reference image. The algorithm can be applied directly to the coloured image. The corrosion is detected on the basis of the colour information by applying a clustering algorithm, viz. *k*-means. This is an iterative clustering technique. As compared to other hierarchical clustering techniques, *k*-means is the simplest. Its implementation is easy and massive data is reorganized into more formal groups. Moreover, this method produces tighter clusters than other hierarchical clustering. It is computationally faster provided that the number of clusters chosen is small.

Keywords *k*-means clustering · Image processing · Corrosion detection Colour-based segmentation

1 Introduction

Visual inspection performed by a human operator can be erroneous, time consuming, and exhaustive. The speed of human operator will be very slow and might result in poor quality of inspection. Whereas visual inspection through image processing is very fast and repetitive tasks can be performed as many times as required without compromising the efficiency. PCB inspection has two major processes:

A. Bhardwaj (✉) · P. Varshney
Division of ICE, NSIT, New Delhi, India
e-mail: ananditab.ic@nsit.net.in

P. Varshney
e-mail: pragya.varshney1@gmail.com

© Springer Nature Singapore Pte Ltd. 2019

H. Malik et al. (eds.), *Applications of Artificial Intelligence Techniques in Engineering*, Advances in Intelligent Systems and Computing 697,
https://doi.org/10.1007/978-981-13-1822-1_15

(i) Detection of defects
(ii) Classification of defects

Defects can be functional or cosmetic. Functional defects hinder the performance of PCB whereas cosmetic defects have an impact on how the PCB is visible and in some cases, the functioning also gets affected. Defect detection of PCB can be carried out in two ways, using: (i) Contact methods and (ii) Non-contact methods. The connectivity of a circuit is tested by contact method but it cannot detect cosmetic defects. [1, 2]. The use of image processing along with X-ray imaging, thermal imaging, and optical inspection are few of the non-contact methods [3–6]. In Khalid's work [7], the first step of the methodology is image difference operation. In this operation, the test image (image of the PCB under inspection) is compared pixel-by-pixel with the reference image (image of the ideal PCB) using XOR operation. The stated operation cannot be carried out in absence of a reference image.

The work of Putera [8] combines the research of Heriansyah [9] and Khalid [7]. For inspecting one PCB image (using a reference image), 20 images are formed and the detection is carried out by comparison of the image under inspection and template image.

k-means clustering is an iterative refinement technique. It is easy to implement and its execution is very fast. It is used for grouping a large number of data points into clusters. This technique is widely used in the area of medical science, computer graphics, etc. [10–14].

In this paper, corrosion detection on a PCB is performed using k-means clustering for colour-based segmentation. This work is an improvement over the works of Putera [8] and Khalid [7], as in this method, the need of a reference image is eliminated. Also, the test image can be fed or processed into the algorithm in RGB (Red–Green–Blue) format. Usually, detection of defects in PCB is performed using mathematical morphology and image comparison. In this paper, the authors have applied a different method of corrosion detection in PCB using the *k*-means clustering technique. The proposed method has been compared with other methods existing in the literature. It has been observed that given the limitation of non-availability of the reference image, the proposed method yields best results. Following is the layout of the paper: Sect. 2 gives literature review and discusses the research methodology. Section 3 covers the details of *k*-means clustering. In Sect. 4, the proposed procedure for corrosion detection is presented and the results are discussed. Conclusions are summarized in Sect. 5.

2 Literature Review and Research Methodology

The main advantage of the work presented in this paper is that corrosion detection is performed without the use of any reference image. In [7, 8], the authors have been able to perform error detection by comparison of test and reference images. Also in

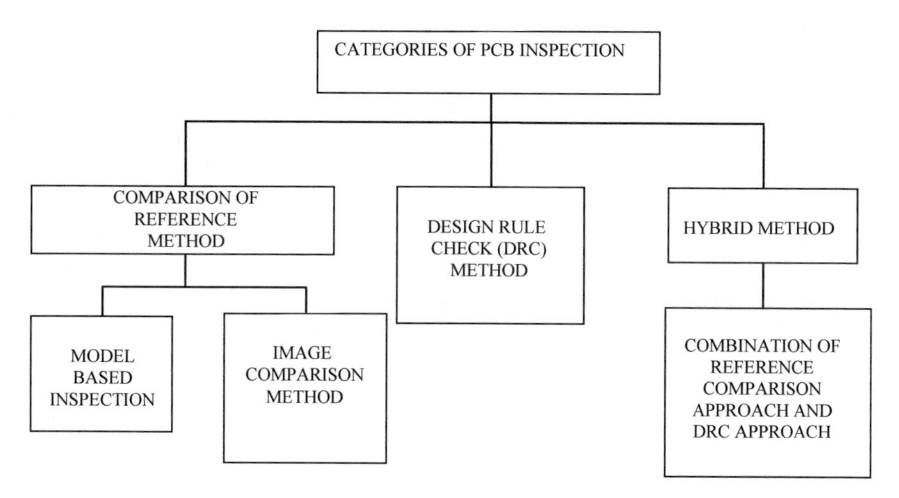

Fig. 1 Image under consideration (Test image)

[7, 8], the hybrid algorithm used utilizes greyscale images whereas in the proposed work, colour (RGB) images are also acceptable, thereby reducing the computational time. In Putera's work [8], twenty intermediate images were created during the execution of the hybrid algorithm. In the proposed work, only four intermediate images are created. For defect detection, it is imperative that the alignment of the template and the test image should be precise [1]. Otherwise, the results obtained will be erroneous. In the method suggested in this paper, alignment is not required. The image in consideration is fed into the algorithm directly in coloured format, making the technique easily implementable. Under the reference comparison approach, model-based inspection is used for pattern matching of PCB under inspection with predefined models as shown in the flowchart of Fig. 1. Pattern matching is performed on the basis on topological, structural and geometric properties of image. It is very complex and time consuming [1].

3 *k*-Means Clustering

This is a clustering method in which the process of vector quantization is involved and it is generally used in cluster analysis. It is an iterative method also used for detecting brain tumours. Vector quantization is a technique used for modelling of density functions, probability etc., i.e. it is used for signal processing. Clustering is a technique of analysis of statistical data. Cluster analysis is a task of grouping a set of objects in such a manner that the objects in one cluster have more similarity than those in other groups. The objective of *k*-means clustering involves grouping the data into '*k*' clusters each having '*n*' data points, in such a manner that the data points are included in that group which has the nearest mean. The mean is a characteristic feature of the cluster.

In this work, *k* clusters are made on the basis of the colour information of each pixel. In a coloured image, each pixel has a certain RGB value. For a PCB, there are three major colours—Copper (or yellow) colour, greyish brown (colour of the corrosion) and green colour (background colour of the PCB). The number of pixels is treated as observations or data points. As a result, three clusters are created and each cluster will have a mean value. The mean is calculated on the basis of the information of the chromaticity layer of the image. The criterion of segregation of pixels into these three clusters will be on the basis of their corresponding colour information.

4 Work Done

In this paper, the corrosion detection is done by *k*-means clustering method. The procedure is described in this section. The image given in Fig. 2 is the test image (image under inspection). This RGB image is converted to Ll * a * b * colour space. In Ll * a * b*, colour space:

(i) Ll* is the luminosity layer.
(ii) a* is the chromaticity layer. It contains red–green axis.
(iii) b* is the chromaticity layer. It contains blue-yellow axis.

a*, b* being the chromaticity layers, contain the colour information of an image.

Applying the *k*-means clustering technique to the test image yields three sub-images as follows:

(i) Image containing copper slots only (Fig. 3)
(ii) Image containing green background and corrosion (Fig. 4), and
(iii) Image containing a few sections of background with slightly different shade of green colour (figure not shown).

Fig. 2 Image under consideration (Test image)

Fig. 3 "Copper-content only" image of the test image

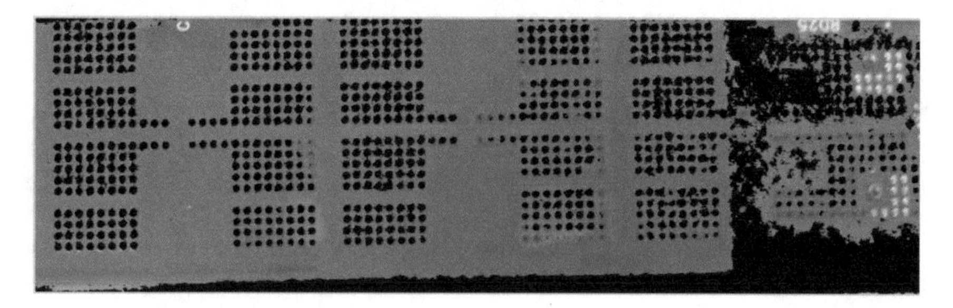

Fig. 4 "Corrosion and background" image

Again, *k*-means clustering is applied to the image containing green background and corrosion (Fig. 4). The second segmentation of the sub-image results in further sub-images:

(i) Some sections of green background only and
(ii) Image containing "corrosion-content only" (Fig. 5).

Thus, the image with the "corrosion-content only" is obtained which is further analyzed for finding the percentage of corrosion.

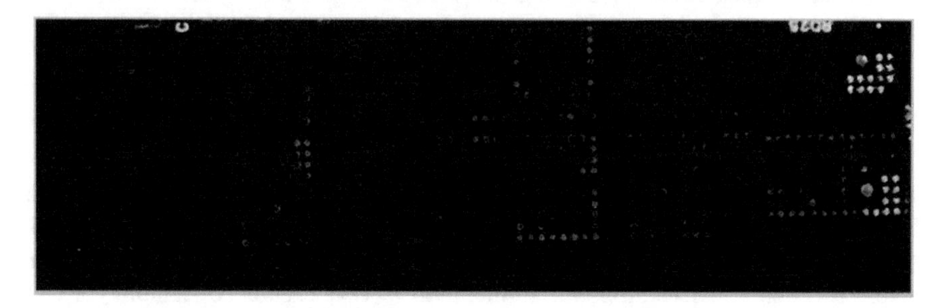

Fig. 5 "Corrosion-content only" image

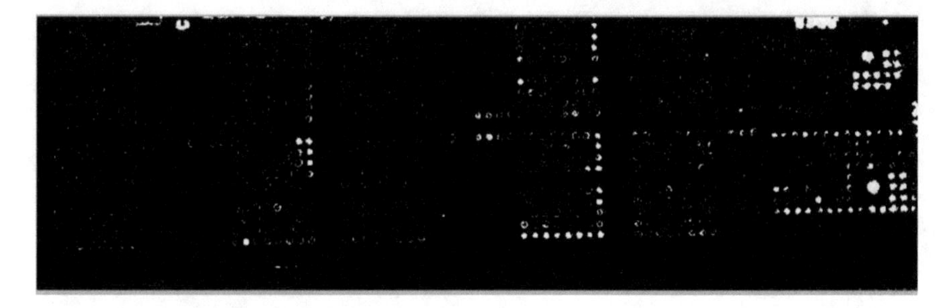

Fig. 6 “Binary corrosion-content only” image

4.1 Binarization of the Image with Corrosion-Content Only

The image containing only the corrosion-content (Fig. 5) is further processed by converting it to a binary image. In the “binary corrosion-content only” image (Fig. 6), the white pixels represent the corrosion content and the rest of the (non-corroded) pixels are black.

The total number of white pixels and black pixels is determined and the percentage of the corroded chip area is obtained as

$$\text{corrosion percentage} = \frac{\text{number of white pixels}}{\text{number of white pixels} + \text{number of black pixels}} * 100\% \tag{1}$$

5 Simulation and Results

Simulations were performed for 20 corroded images using the procedure mentioned in Sect. 4. Table 1 lists the result of five such images. It is observed that manual inspection gives lower results as compared to the results obtained using this approach.

Percentage corrosion of the image under inspection = 2.1055%.

6 Conclusion

The proposed technique successfully detects corrosion percentage of a PCB from the sample experiment. The results of five test images are listed in Table 1. This work improves the shortcomings of the previous works by eliminating the need for a reference image. The proposed formulation is also capable of handling RGB (colour) scale image without any prior conversion which is an improvement on

Table 1 Result for five test images

S. No.	Test image	Percentage corrosion (%)
1		2.1055
2		4.5317
3		0.8944
4		9.4837
5		20.9746

Khalid's work. In addition to this, a lesser number of images were produced during the execution of the proposed algorithm.

The limitation of the proposed method is that it has been tested only for corrosion detection on the PCB surface. The project under discussion can be improved by utilizing it to form a closed-loop system by the use of image capturing system.

References

1. Moganti, M., Ercal, F., Dagli, C.H., Tsunekawa, S.: Automatic PCB inspection algorithms: a survey. Comput. Vis. Image Underst. Elsevier **63**(2), 287–313 (1996)
2. Taniguchi, T., Kacprzak, D., Yamada, S., Iwahara, M., Miyogashi, T.: Defect Detection on Printed Circuit Board by using Eddy Current Technique and Image Processing, vol. 37, pp. 330–335. 101 Press (2000)
3. Wu, W.Y., Wang, M.J.J., Liu, C.M.: Automated inspection of printed circuit board through machine vision. Comput. Ind. **28**, I03–III (1996)
4. Ibrahim, Z., AI-Attas, S.A.R., Aspar, Z.: Analysis of the wavelet based image difference algorithm for PCB inspection. In: 41st SICE Annual Conference, Japan, vol. 4, pp. 2108–2113 (2002)
5. Ibrahim, Z., AI-Attas, S.A.R., Aspar, Z., Mokji, M.M.: Performance evaluation of wavelet-based PCB defect detection and localization algorithm. In: ICIT, Thailand, vol. 1, pp. 226–231 (2002)
6. Ibrahim, Z., AI-Attas, S.A.R., Aspar, Z.: Coarse resolution defect localization algorithm for an automated visual PCB inspection. Jurnal Teknologi **37**(D) Dis., 2629–2634 (2002)
7. Khalid, N.K.: An Image Processing Approach Towards Classification of Defects on Printed Circuit Board. Universiti Teknologi, Malaysia, Projek Sarjana Muda (2007)

8. Indera Putera, S.H., Ibrahim, Z.: Printed circuit board defect detection using mathematical morphology and MATLAB image processing tools. In: 2nd International Conference on Education Technology and Computer (ICETC), pp. V5–359–363 (2010)
9. Heriansyah, R., AI-Attas, S.A.R., Ahmad Zabidi, M.M.: Segmentation of PCB Images into Simple Generic Patterns using Mathematical Morphology and Windowing Technique, pp. 233–238. CoGRAMM Melaka, Malaysia (2002)
10. Saadi, M., Ahmad, T., Zhao, Y., Wuttisttikulkij, L.: An LED based indoor localization system using k-means clustering. In: 15th IEEE International Conference on Machine Learning and Applications (ICMLA), pp. 246–252 (2016). https://doi.org/10.1109/icmla.2016.0048
11. Bou Assi, E., Sawan, M., Nguyen, D.K., Rihana, S.: A 2D clustering approach investigating inter-hemispheric seizure flow by means of a directed transfer function. In: 3rd Middle East Conference on Biomedical Engineering (MECBME), pp. 68–71 (2016). https://doi.org/10.1109/mecbme.2016.7745410
12. Ren, S., Lu, L., Zhao, L., Duan, H.: Circuit board defect detection based on image processing. In: 8th International Congress on Image and Signal Processing (CISP), Shenyang, pp. 899–903 (2015)
13. Guo, M., Wang, R.: The introduction of AOI in PCB defect detection based on linear array camera. In: International Forum on Management, Education and Information Technology Application (IFMEITA 2016), Atlantis Press, pp. 767–770 (2016)
14. Li, Y., Li, S.: Defect detection of bare printed circuit boards based on gradient direction information entropy and uniform local binary patterns. Circuit World. https://doi.org/10.1108/CW-06-2017-0028

Extraction of Equivalent Circuit Parameters of Metamaterial-Based Bandstop Filter

Priyanka Garg and Priyanka Jain

Abstract This article presents the design and analysis of a metamaterial-inspired Bandstop Filter (BSF) providing suppression of frequency at 3 GHz. The overall size of the proposed BSF is 20 mm × 20 mm × 1.6 mm. This paper presents the extraction of lumped parameters of the designed BSF using simulated results and validation of the results using equivalent circuit simulation has also been presented. The analysis is performed using transmission coefficient, reflection coefficient, and impedance curve.

Keywords Bandstop filter · Transmission line · Metamaterial · Open slot split-ring resonator (OSSRR)

1 Introduction

Bandstop filters are used in various microwave and RF communication systems to reject a particular band of frequencies. Various bandstop filters are available in the state-of-the-art that are based on planar technology. Most of them are developed by etching slots either on microstrip line or on the ground plane.

Verelago's left-handed metamaterials [1] have added new dimensions to the planar technology by providing exceptional properties. The presence of negative permittivity and permeability for such metamaterials at certain band of frequencies can be helpful to provide bandstop behavior with sharp rejection level. Split-ring resonators (SRRs) proposed by Pendry et al. [2] gives rise to an effective magnetic response without the need for magnetic materials. A cut-band filter was presented by Carver et al. [3] that were realized by combing SRR alongside of microstrip line.

P. Garg (✉) · P. Jain
Department of Electronics and Communication Engineering,
Delhi Technological University, New Delhi, India
e-mail: garg.priyanka16@yahoo.com

P. Jain
e-mail: priyajain2000@rediffmail.com

© Springer Nature Singapore Pte Ltd. 2019

H. Malik et al. (eds.), *Applications of Artificial Intelligence Techniques in Engineering*, Advances in Intelligent Systems and Computing 697,
https://doi.org/10.1007/978-981-13-1822-1_16

Complementary split-ring resonators (CSRRs) were also used to design BSF, giving wideband performance [4] as compared to SRR. Another CSRR-based filter is presented by Li et al. [5].

This paper presents parameter extraction of the circuit using equivalent circuit model of Open Slot Split-Ring Resonator (OSSRR). OSSRR was designed by Karthikeyan et al. [6] and it was observed that OSSRR offers lower resonant frequency compared to the CSRR of similar dimensions. The work proposed in this paper provides an extension by utilizing OSSRR to operate at 3 GHz, further by extraction of equivalent circuit model. The simulated transmission coefficient and impedance curve are used to determine the lumped equivalent circuit and a mathematical model is described in detail for the extraction of circuit parameters. The design is further validated using circuit simulation results. The equivalent circuit extraction is important to study the electrical behavior of the planar design in order to provide ease of integration with any external electrical circuit. All the design and circuit simulations are carried out using Computer simulation Technology (CST) Microwave studio (MWS) [7] and CST design studio [8], respectively.

2 Design and Circuit Simulation

The proposed bandstop filter is designed using a 1.6 mm thick FR-4 substrate ($\varepsilon_r = 4.4$) with dimensions 20×20 mm^2. The substrate consists of a 50 Ω microstrip line on the top and an OSSRR on the bottom. Figure 1 shows the design of the proposed bandstop filter along with the port assignments. The bandstop filter is designed and simulated in CST microwave studio and results are obtained.

Figure 2 shows the simulated transmission coefficient. Figure 2 shows that bandstop filter shows resonance at 3.024 GHz, thus performing complete suppression of frequency at 3.024 GHz. From the impedance curve as shown in Fig. 3, it was observed that the BSF shows inductive effect below the resonant frequency and capacitive effect above it. Thus, the equivalent circuit must contain a parallel resonant circuit. Since, the insertion loss is negative as shown in Fig. 2, we conclude that the equivalent circuit of proposed BSF is a series connected parallel resonator.

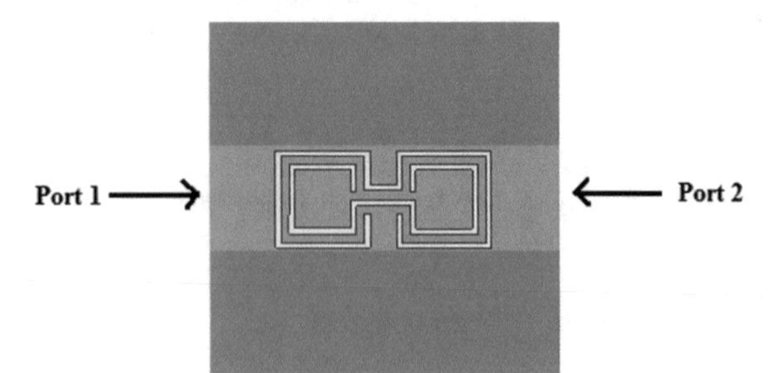

Fig. 1 Design of the proposed bandstop filter (light gray color shows 50 Ω microstrip line on top and dark gray color shows OSSRR on bottom of substrate)

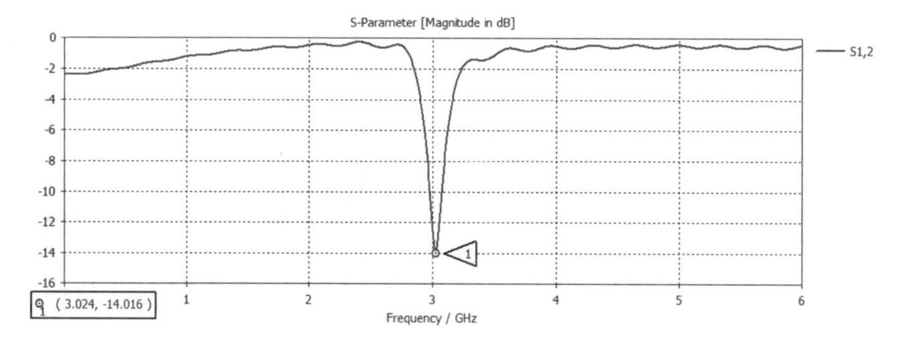

Fig. 2 Simulated transmission coefficient of the proposed bandstop filter design

Fig. 3 Impedance curve of the proposed design

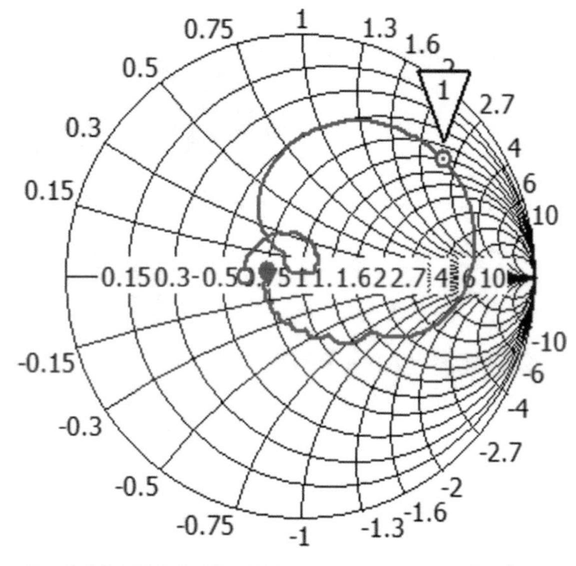

The resultant series connect parallel resonant circuit arrangement, assuming no losses, is shown in Fig. 4. The values of L (inductor) and C (capacitor) is obtained using the following procedures:

1. At the resonant frequency, say $f_2, |S_{21}| = 0$.
2. At the intersection of $|S_{11}|$ and $|S_{21}|$, say at $f_1, |S_{21}| = 1/\sqrt{2}$ (since, $|S_{11}|^2 + |S_{21}|^2 = 1$).

For a series-connected parallel resonator, S-parameter matrix is given as

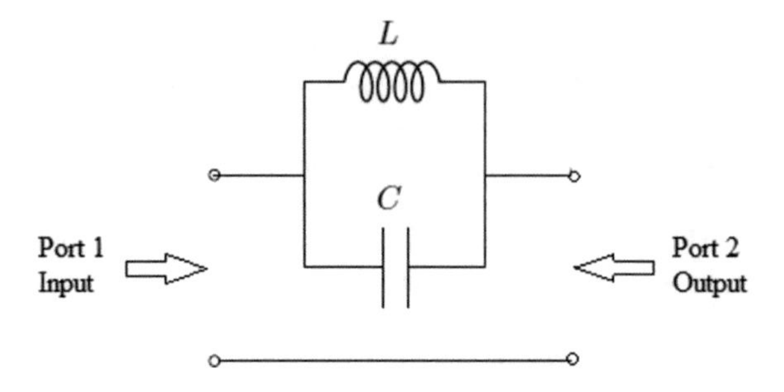

Fig. 4 Equivalent circuit of the proposed bandstop filter

$$[S] = \begin{bmatrix} \frac{Y_0}{Y_0 + 2Y} & \frac{2Y}{Y_0 + 2Y} \\ \frac{2Y}{Y_0 + 2Y} & \frac{Y_0}{Y_0 + 2Y} \end{bmatrix}$$

Using condition 1, $|S_{21}| = 0$

$$S_{21} = \frac{2Y}{Y_0 + 2Y} = \frac{2Z_0}{Z + 2Z_0} = 0 \tag{1}$$

From the circuit as shown in Fig. 4, we obtain

$$Z = \frac{j\omega_2 L}{1 - \omega_2^2 LC}$$

Putting in Eq. (1), we get

$$\begin{aligned} C &= \frac{1}{\omega_2^2 L} \\ &= \frac{1}{4\pi^2 f_2^2 L} \end{aligned} \tag{2}$$

Now, using condition 2, $|S_{21}| = 1/\sqrt{2}$

$$\frac{2Z_0}{\sqrt{\frac{\omega_1^2 L^2}{\left(1 - \omega_1^2 LC\right)^2} + 4Z_0^2}} = \frac{1}{\sqrt{2}}$$

On solving the above equation using Eq. (2), we obtain

$$L = \frac{Z_0}{\pi f_1}\left[1 - \frac{f_1^2}{f_2^2}\right] \tag{3}$$

3 Results and Discussion

After the design simulations performed on CST Microwave Studio [6], it was obtained that the intersection of $|S_{11}|$ and $|S_{21}|$ occurs at frequency f_1 = 2.88 GHz and resonance at f_2 = 3.024 GHz is observed. Using Eqs. (2) and (3), we obtain the value of L and C as 0.5 nH and 5.549 pF, respectively. The equivalent circuit is designed and simulated in CST design studio [8]. The transmission and reflection coefficient obtained after design and circuit simulations are shown in Figs. 5 and 6, respectively. Design and circuit simulated impedance curves are also shown in Fig. 7 simultaneously. Figures 5 and 6 shows similar results. Little deviation in the impedance may be due to substrate losses. Thus, we conclude that the equivalent circuit is a close approximation of the proposed OSSRR based BSF.

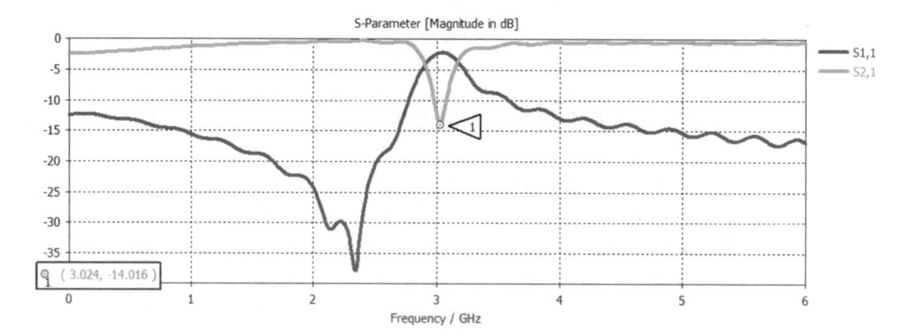

Fig. 5 Reflection and transmission coefficient of simulated design

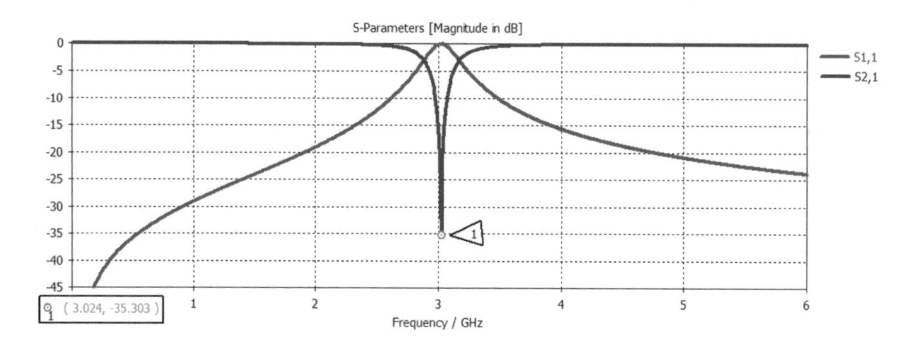

Fig. 6 Reflection and transmission coefficient of simulated equivalent circuit

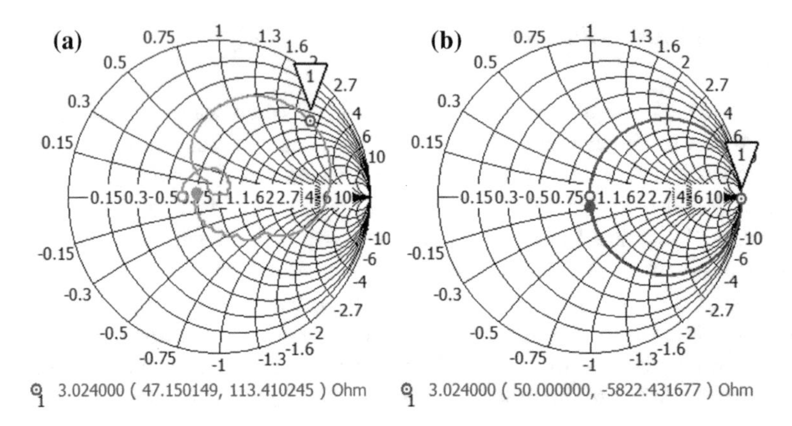

Fig. 7 Impedance Curve **a** Design; **b** Equivalent circuit

4 Conclusion

OSSRR-based bandstop filter has been designed and analyzed using equivalent circuit model. The BSF was designed to operate at 3.024 GHz. A methodology based on the simulated design transmission coefficient and impedance curve to select an appropriate circuit model for the BSF was also described. It was found that the proposed BSF could be represented as series-connected parallel resonator. The *L*, *C* lumped parameters were obtained using suitable mathematical model. Further, the circuit equivalent was simulated and results obtained were comparable to the simulated design results. Thus, the obtained circuit could closely approximate the design.

References

1. Veselago, V.G.: The electrodynamics of substances with simultaneously negative values of ε and μ. Soviet Phy. Uspekhi **10**, 509–514 (1968). https://doi.org/10.1070/PU1968v010n04ABEH003699
2. Pendry, J.B., Holden, A.J., Robbins, D.J., Stewart, W.J.: Magnetism from conductors and enhanced non linear phenomena. IEEE Trans. Microwave Theory Tech. **47**(11), 2075–2084 (1999). https://doi.org/10.1109/22.798002
3. Carver, J., Reignault, V., Gadot, F.: Engineering of the metamaterial-based cut-band filter. Appl. Phys. A **117**, 513–516 (2014). https://doi.org/10.1007/s00339-014-8694-7
4. Oznazı, V., Erturk, V.B.: A comparative investigation of SRR–and CSRR-based bandreject filters: simulations, experiments, and discussions. Microwave Opt. Tech. Lett. **50**(2), 519–523 (2008). https://doi.org/10.1002/mop.23119
5. Li, C., Liu, K.Y., Li, F.: A microstrip highpass filter with complementary split ring resonators. PIERS Online (2007). https://doi.org/10.1049/el:20072945

6. Karthikeyan, S.S., Kshetrimayum, R.S.: Composite right/left handed transmission line based on open slot split ring resonator. Microwave Opt. Tech. Lett. **52**(8), 1729–1731 (2010). https://doi.org/10.1002/mop.25330
7. Computer simulation technology microwave studio (CST MWS). Available at https://www.cst.com/
8. Computer simulation technology design studio (CST DS). Available at https://www.cst.com/products/csts2

Design of a Narrow-Band Pass Asymmetric Microstrip Coupled-Line Filter with Distributed Amplifiers at 5.5 GHz for WLAN Applications

Karteek Viswanadha and N. S. Raghava

Abstract The design of an asymmetric microstrip coupled-line filter with distributed amplifiers is presented in this paper. The proposed filter along with the distributed amplifier decreases attenuation with the increase in the selectivity. These amplifiers further reduce the fringing fields which occur in the preceded filter sections. The proposed distributed amplifier is designed at 5.5 GHz using Agilent Advanced Design System-2009 simulator. An irregular spacing between the coupled lines is introduced in order to improve insertion loss and reduce reflection loss. This filter further possesses high skirt rate along with the high selectivity. The insertion loss of −0.006 dB and reflective loss of −28.772 dB are achieved with a bandwidth of 400 MHz. These specifications make the proposed filter suitable for Wireless Local Area Network (WLAN) applications.

Keywords Microstrip · Coupled line · Insertion loss · Distributed amplifiers ADS-2009 · Reflection loss · WLAN

1 Introduction

Filters play an important role in rejecting the unwanted frequencies during the microstrip filters that often suffer from leaky waves due to discontinuities at the patch edges, dispersion losses, and interference of higher order modes. In order to reduce the surfaces wave, the ratio of patch area to the ground area has to be increased [1–19]. In [2], a novel microstrip filter is designed using Photonic Bandgap Structures or Electromagnetic Bandgap Structures (EBGS). The periodic feature of these structures enhances the bandwidth of the filter. Defective Ground Structures (DGSs) can be introduced in both microstrip line [3] as well as the

K. Viswanadha (✉) · N. S. Raghava
Delhi Technological University, New Delhi, India
e-mail: karteekviswanath@gmail.com

© Springer Nature Singapore Pte Ltd. 2019

H. Malik et al. (eds.), *Applications of Artificial Intelligence Techniques in Engineering*, Advances in Intelligent Systems and Computing 697,
https://doi.org/10.1007/978-981-13-1822-1_17

ground layer in order to achieve high selectivity and suppress harmonics. Instead of introducing periodicity in the ground layer, changing the width and length of a microstrip line and periodically cascading the microstrip lines will yield stepped impedance filters [4–7]. These filters are used for the wideband applications. Stepped impedance filters achieve good harmonics [8, 9] suppression with compactness. These filters cannot achieve the sharp cutoff frequencies, as a result, their selectivity is poor. Many applications sought flat pass-band characteristics. Butterworth filters are widely used in the wireless communications where the pass band should be flat so as to pass the band of interest. Generally, the performance of a filter depends on the insertion loss [10]. If the applications are limited to narrow-band type then the order of the Butterworth filter should be increased in order to fulfill the system specifications. Increase in the order of a filter increases roll-off [11]. Increase in the roll-off decreases the stopband range due to which interference of intermodulation products occurs. Moreover, increase in the order of the Butterworth filter increases the size of the filter. Chebyshev filters [12, 13] which act as an alternate solution to the Butterworth filters and have equi-ripples in pass-band stopbands. These filters have a wide stopband and constant group delay in the pass band [14]. Microstrip lines usually suffer from fringing fields and high attenuation. Therefore, achieving low ideal insertion loss becomes difficult, if the successive sections of the filter are designed with microstrip lines. Coupled lines are prone to high attenuation while operating in a narrowband frequency range. Though the proposed filter is highly selective, the successive sections may be prone to high losses. In order to overcome all the above problems, the proposed filter is connected to distributed amplifiers which will incorporate the transmission line equivalent to conventional amplifier design. The use of distributed amplifiers not only improves insertion loss but also reduces reflective losses with reasonable impedance matching at the output ports.

2 Design of the Proposed Filter

Fourth-order coupled-line filters are designed by varying the dimensions of the coupled lines using the ADS-2009 schematic tuner. Lengths (l_1, l_2, l_3, l_4, l_5 & l_6), widths (w_1, w_2, w_3, w_4, w_5 & w_6), and spacings (s_1, s_2, s_3 & s_4) of the microstrip lines are deviated from dimensions obtained numerically from the above formulae. Iterations are carried by varying the lengths of all microstrip lines independently until the filter works at 5.5 GHz. Figure 1 shows the schematic diagram of the fourth order asymmetric coupled-line filters. The lengths of the microstrip lines are tuned to 2, 4.7841, 4.39, 6.1202, 6.78, and 2 mm respectively. The spacing between the lines is chosen to be 0.1, 0.475, 0.4815, and 0.35925 mm, respectively.

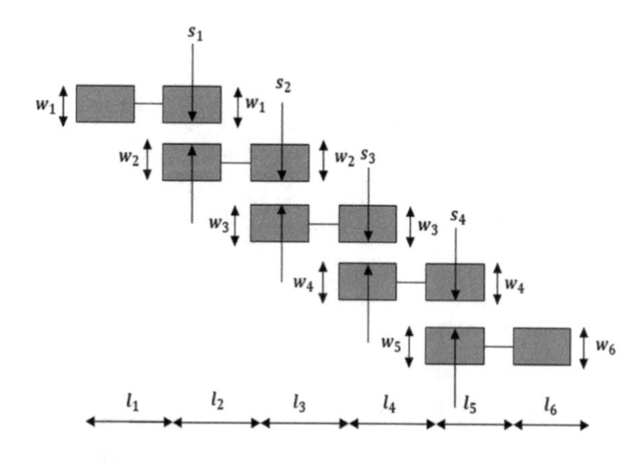

Fig. 1 Schematic of the proposed coupled-line filter

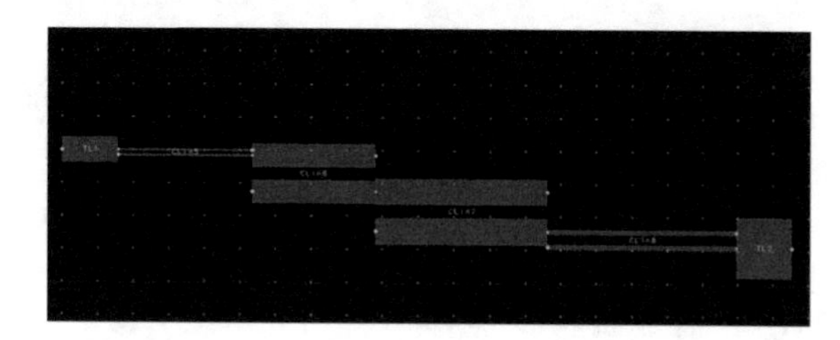

Fig. 2 The layout of the coupled-line proposed filter

Widths of the lines are chosen to be 1, 0.1, 0.914, 1.05, 0.211, and 2.34 mm, respectively. The proposed filter is mounted on a RT-Rogers/Duroid 5880 ($\varepsilon_r = 2.2$) with the substrate thickness of 2.5 mm

The layout of the filter is shown in Fig. 2.

3 Results and Discussions

A consolidated plot of insertion loss (S_{21}) and reflection coefficient (S_{11}) of the proposed asymmetric coupled-line filter are shown in Fig. 3. Tuning the parameters of the filter and carrying the iteration process on the parameters of the filter will give the desired insertion loss and reflection coefficient. The proposed device possesses narrow bandwidth at 5.5 GHz. The insertion loss of the parallel coupled lines depends on the spacing between them. Proper choice of the width of the lines improves reflection coefficient. The bandwidth of the proposed filter is observed to be 200 MHz.

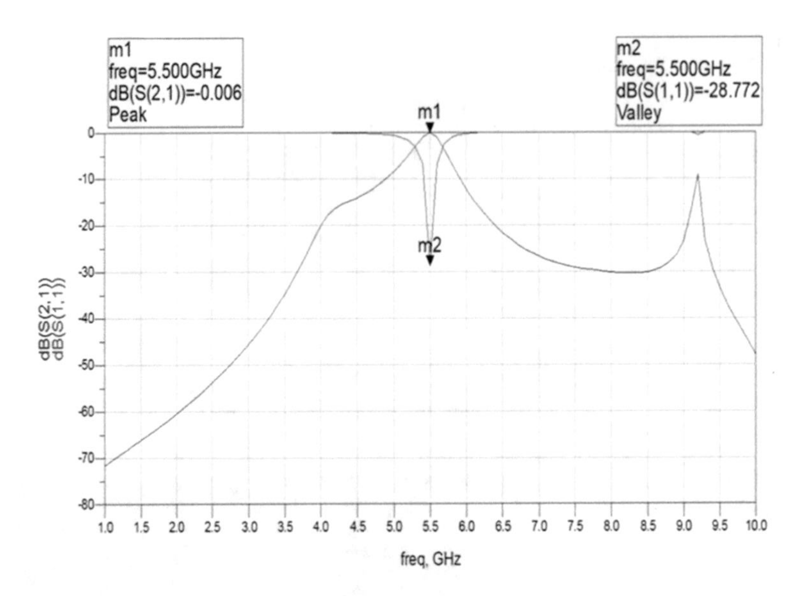

Fig. 3 A consolidated plot of insertion loss and reflection coefficient

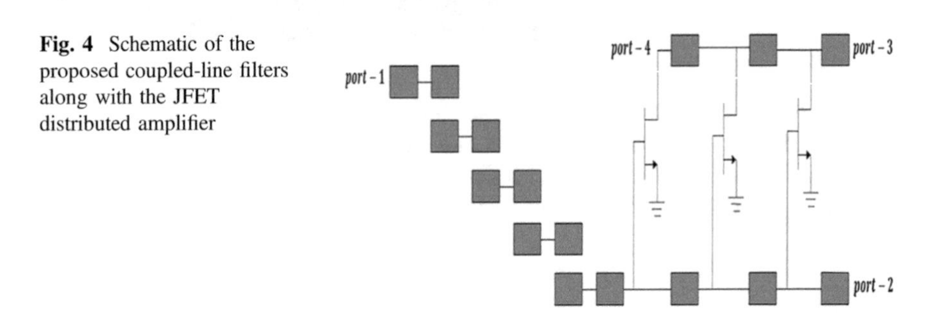

Fig. 4 Schematic of the proposed coupled-line filters along with the JFET distributed amplifier

4 Application in Distributed Amplifiers

The distributed amplifiers are designed using the JFETS with equal gate widths (0.635 mm) and channel lengths (100 mm). Figure 4 shows the schematic of the proposed coupled-line filter applied to the input of the JFET distributed amplifier.

Figure 5 shows the plot of insertion loss of the device shown in the Fig. 4.

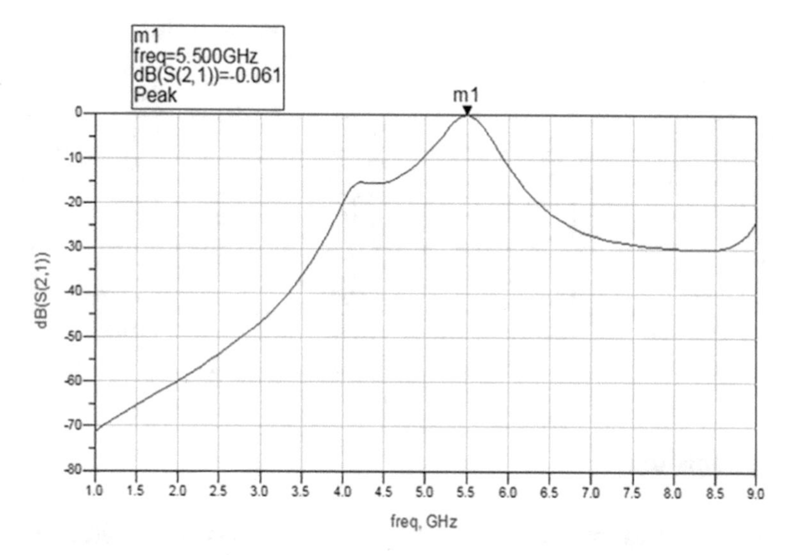

Fig. 5 Insertion loss graph of the coupled-line filter applied to the input of the JFET distributed amplifier

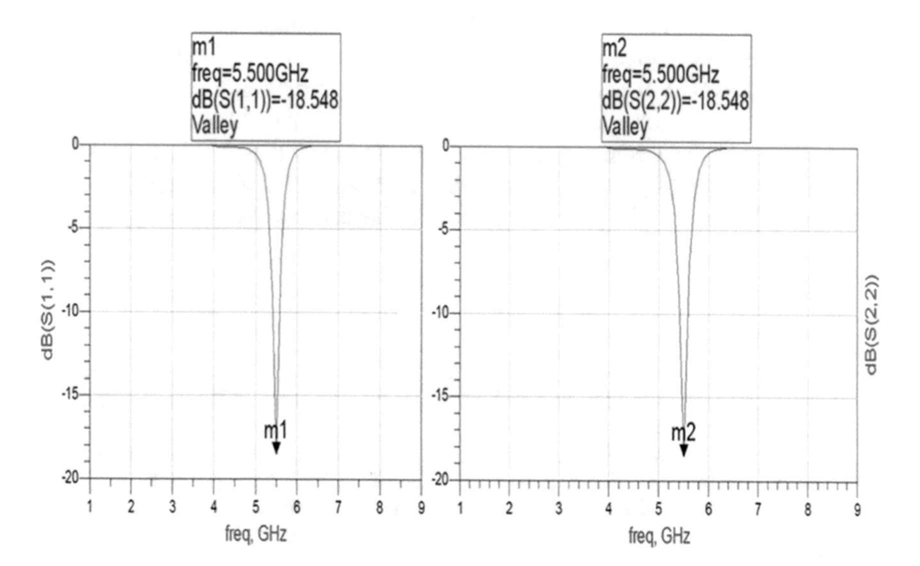

Fig. 6 A consolidated plot of reflection coefficients at port-1 and port-2

Figure 6 shows the reflection coefficients at the port-1 and port-2 of the device. Ports-3 and -4 are matched perfectly for the device shown in Fig. 4.

As shown in Table 1, the proposed filter with and without distributed amplifiers has low insertion and reflective losses.

Table 1 Comparisons of the proposed filters with published band-pass filters

No. of references	Frequency (GHz)	Reflection coefficient S_{11} (dB)	Insertion loss S_{21} (dB)
Ref. [15]	5.5	−25	−0.62
Ref. [16]	1.75	−30	−0.032
Ref. [17]	5.75	−14.5	−4.1
Ref. [18]	5.5	−12.8	−0.61
Ref. [19]	5.5	−15.2	−2.46
Proposed filter	5.5	−28.772	−0.006
Proposed filter with distributed amplifiers	5.5	−18.548	−0.061

5 Conclusions

The asymmetric narrow-band pass coupled-line filter possesses insertion loss of −0.006 dB at 5.5 GHz and the reflection coefficient of −28.772 dB at 5.5 GHz in Fig. 4. Further, varying the lengths of microstrip lines connected the front and end of the filters shifts the frequency from 5.5 GHz and proportionately insertion loss decreases and reflection coefficient increases. The proposed filter along with the distributed amplifiers gives the insertion loss of −0.061 dB and reflection coefficients of −18.548 dB at ports-1 and -2, respectively. Insertion loss can be further improved by using the asymmetric coupled narrow-band-pass filter of higher orders along with the hybrid configurations like Bipolar junction Field-effect Transistor (Bi-FET) or Bipolar Junction Metal-Oxide Semiconductor Transistor (Bi-MOS) distributed amplifiers. The proposed filter is under the process of fabrication.

References

1. Hong, J.S.G., Lancaster, M.J.: Microstrip filters for RF microwave applications. Wiley, New York, NY (2001)
2. Ahmad, B.H., Mazlan, M.H., Husain, M.N., Zakaria, Z., Shairi, N.A.: Microstrip filter design techniques: an overview. ARPN J. Eng. Appl. Sci. **10**(1819–6608), 1–5 (2015)
3. Prajapati, P.R.: Application of defected ground structure to suppress out-of-band harmonics for WLAN microstrip antenna. Int. J. Microwave Sci. Tech. **2015**, 1–9 (2015). https://doi.org/10.1155/2015/210608
4. Patel, N., Parikh, N., Katare, P., Kathal, K., Chaitanya, G.: Stepped impedance low pass microstrip line filter. Int. J. Innovative Res. Electr. Electron. Instrum. Control Eng. **3**(5), 1–2 (2015)
5. Jacob, W. H.: Design and implementation of RF and microwave filters using transmission lines. Journal of Bobylon University /Engineering Science. **2**(22), 1–13 (2014)
6. Liu, S., Xu, J., Xu, Z.: Compact lowpass filter with wide stopband using stepped impedance hairpin units. IEEE Electron. Lett. **51**(1), 67–69 (2015). https://doi.org/10.1049/el.2014.3673
7. Patankar, S., Kumar, D.: Object, analysis of stepped-impedance microstrip low pass filter for l-band applications. Int. J. Eng. Res. Tech. **5**(2), 1–57 (2013)

8. Nath, M.: Review of filter techniques. Int. J. Eng. Trends Tech. **3**(3), 1–6 (2013)
9. Rajasekaran, K., Jayalakshmi, J., Jayasankar, T.: Design and analysis of stepped impedance microstrip low pass filter using ADS simulation tool for wireless applications. Int. J. Sci. Res. Publ. **3**(08), 1–3 (2013)
10. Hong, J.S.: Microstrip filter design. In: Practical Aspects of Microwave Filter Design and Realization, IEEE-MTTS, IMS Workshop-WMB, pp. 1–40 (2005). https://doi.org/10.1109/mwsym.2005.1516781
11. Chernobrovkin, R.E., Ivanchenko, I.V., Korolev, A.M., Popenko, N.A., Sirenko, KYu.: The novel microwave stop-band filter. Act. Passive Electron. Compon. 1–5 (2008). https://doi.org/10.1155/2008/745368
12. Tsigkas, Dimitrios E., Alysandratou, Diamantina S., Karagianni, Evangelia A.: Butterworth filter design at RF and X-band using lumped and step impedance techniques. Int. J. Adv. Res. Electron. Commun. Eng. (IJARECE) **3**(4), 403–409 (2014)
13. Mishra, S., Mishra, A.: Chebyshev filters for microwave frequency applications a literature review. IJSRD—Int. J. Sci. Res. Dev. **3**(4), 1587–1591 (2015)
14. Singh, K., Ramasubramanian, R., Pal, S.: Coupled microstrip filters: simple methodologies for improved characteristics, pp. 1–7 (2006)
15. Xinlin, X., Liu, Y., Lin, H., Yang, T., Jin, H.: Novel UWB BPF with a controllable notched band using hybrid structure. IEICE Electronics Express **14**(6), 1–6 (2017). https://doi.org/10.1587/elex.14.20170083
16. Taher, Hany: A New artificial intellegent technique for designing of microstrip stepped impedance low pass filter. Microw. Opt. Tech. Lett. **52**(9), 1946–1949 (2010). https://doi.org/10.1002/mop.25401
17. Ahmad, A., Othaman, A.R.: Narrow dual bandpass filter using microstrip coupled line with bell shaped resonator. In: International Conference on Advances in Electrical, Electronic and System Engineering, pp. 550–554, Malaysia (2016). https://doi.org/10.1109/icaees.2016.7888106
18. Ouf, E.G., Mohra, A.S., Abdallah, E.A., Elhennawy, H.: Ultra-wideband bandpass filter with sharp tuned notched band rejection based on CRLH transmission-line unit cell. Prog. Electromagn. Res. Lett. **69**, 9–14 (2017)
19. Darwis, F., Setiawan, A., Daud, P.: Performance of narrow hairpin bandpass filter square resonator with folded coupled line. In: International Seminar on Intelligent Technology and its Applications, pp. 291–294 (2016)

Cooperative Spectrum Sensing Using Eigenvalue-Based Double-Threshold Detection Scheme for Cognitive Radio Networks

Chhagan Charan and Rajoo Pandey

Abstract Sensing spectrum in a reliable and efficient manner is a fundamental problem in cognitive radio networks. The energy-based detection methods are highly noise uncertainty conditions. The eigenvalue-based detection schemes address the noise uncertainty problem, however, performance degrades under low SNR. The double-threshold-based eigenvalue detection scheme performs well under low SNR, but it is not reliable under fading environment and hidden terminal problem. Thus, ensuring cooperation among spectrum sensing devices is an appropriate method when cognitive radio network is under deep shadowing and in a fading environment. The artificial intelligent techniques can be used to find the spectrum hole. In this paper, a cooperative spectrum sensing scheme using double-threshold-based detection scheme is proposed. The threshold is obtained from ratios of eigenvalues of the sample covariance matrix. The random matrix theory is employed to quantify the threshold. The simulation results show that the proposed cooperative spectrum sensing scheme performs better than the existing eigenvalue-based detection scheme and energy-based detection method.

Keywords Cognitive radio · Eigenvalue-based double-threshold detection (EVDD) · Sensing failure · Spectrum sensing · Cooperative spectrum sensing

1 Introduction

As the everlasting growth of wireless communication technologies and high demand of capacity for wireless services, the wireless frequency spectrum has become a scarce resource in the past years. The analysis by the FCC have concede that large portion of spectrum band is underutilized, due to existing fixed spectrum band allocation policies [1]. Thus, to increase the utilization of unused spectrum in

C. Charan (✉) · R. Pandey
ECE Department, NIT Kurukshetra, Kurukshetra, India
e-mail: chhagan.charan@nitkkr.ac.in

© Springer Nature Singapore Pte Ltd. 2019

H. Malik et al. (eds.), *Applications of Artificial Intelligence Techniques in Engineering*, Advances in Intelligent Systems and Computing 697,
https://doi.org/10.1007/978-981-13-1822-1_18

'Cognitive radio' (CR), unlicensed secondary user (SU) exploits the licensed spectrum without interfering with the licensed primary user (PU) [2–4].

To provide sufficient protection to PU, spectrum sensing (SS) becomes a vital stage in CR networks [5–12]. In the literature various spectrum detection techniques have been discussed to find the status (active/idle) of PU, the frequency spectrum band, such as energy detection (ED) [5, 6, 11], feature detection techniques [7], eigenvalue-based detection method [9, 10] and covariance-based method [8]. In these spectrum detection techniques, each scheme has some advantages and some limitations. The ED-based scheme is blind detection technique as it does not require any PUs information and is easy to implement. However, the performance of these schemes degrades at low SNR and under noise uncertainty. Feature detection-based spectrum sensing methods perform well under low SNR but require knowledge of PUs signal characteristics and more processing time is needed. Covariance and eigenvalue-based detection schemes employs the sample covariance matrix of signal and noise, that are generally different. This characteristic is used to decide the presence/absence of PU. The eigenvalue-based detection scheme is implemented by obtaining the thresholds based on the ratio of maximum and minimum eigenvalue of the sample covariance matrix. However, the performance of these schemes degrades under low SNR, which is addressed by the presented eigenvalue-based double-threshold detection scheme.

One of the major challenges of implementing spectrum sensing is hidden terminal problem, in deep multipath fading or inside building with high penetration loss, while a licensed user is active in the vicinity [12]. Due to hidden terminal problem, an SU may fail to detect the presence of PU and will access the licensed spectrum band, which will cause the interference to PU. This problem can be addressed as multiple SUs can cooperate to sense the spectrum band. In [12], an optimal voting rule is suggested for cooperative spectrum sensing.

In this paper, the cooperative spectrum sensing with optimal voting rule is used to address the hidden terminal problem. Also, as in the case of ED-based scheme, the proposed technique also does not need prior PUs information, synchronization and channel.

The rest of the paper is organized as follows. First, the existing spectrum detection schemes are briefed in Sect. 2. The proposed cooperative eigenvalue-based double-threshold sensing method is described in Sect. 3. Section 4 presents simulation results and a comparison with existing schemes. Finally, Sect. 5 concludes the paper.

2 Conventional Spectrum Detection Schemes

This section describes the commonly used sensing model and existing eigenvalue-based spectrum sensing scheme.

2.1 Spectrum Sensing Model

The received signal at SU can be expressed in the following two hypotheses as follows [8, 10]:

$$H_0 : x(n) = \eta(n) \tag{1}$$

$$H_1 : x(n) = s(n) + \eta(n) \tag{2}$$

where

$x(n)$ sample of signal received by SU.
$s(n)$ sample of PUs transmitted signal with mean 0 and variance σ_s^2.
$\eta(n)$ additive white Gaussian noise (AWGN) with zero mean and variance σ_η^2.

The hypothesis H_0 represents that the absence of PU, whereas hypothesis H_1 indicates the presence of PU. Following probabilities are of interest in spectrum sensing: probability of detection P_d, probability of detecting the presence of PU under hypothesis H_1 and probability of miss detection P_m, which means a busy spectrum band is detected to be idle. Probability of false alarm P_f, detection of PU under hypothesis H_0. For reliable spectrum sensing scheme, the P_m should be low while P_fa should be as small as possible [12].

2.2 Eigenvalue-Based Spectrum Sensing Scheme

The method in eigenvalue-based detection scheme is calculated in terms of ratio maximum to minimum eigenvalues (MME) of the covariance matrix of the received signal defined in [8, 9] as

$$R_x(N_\mathrm{s}) = \begin{bmatrix} f(0) & f(1) & \ldots & f(L-1) \\ f^*(0) & f(1) & \ldots & f(L-1) \\ \cdot & \cdot & \cdot & \cdot \\ \cdot & \cdot & \cdot & \cdot \\ \cdot & \cdot & \cdot & \cdot \\ f^*(L-1) & f^*(L-2) & \ldots & f(0) \end{bmatrix} \tag{3}$$

where

$$f(l) = \frac{1}{N_\mathrm{s}} \sum_{m=0}^{N_\mathrm{s}-1} x(m)x(m-l)^*$$

with $l = 0, 1, \ldots, L-1$;

N_s denotes the number of samples and L denotes the smoothing factor. The threshold is computed in [8] as

$$\gamma = \frac{\left(\sqrt{N_s} + \sqrt{L}\right)^2}{\left(\sqrt{N_s} - \sqrt{L}\right)^2}\left(1 + \frac{\left(\sqrt{N_s} + \sqrt{L}\right)^{-2/3}}{(N_s L)^{1/6}} F_1^{-1}(1 - P_f)\right) \tag{4}$$

where F_1 is Tracy–Widom distribution of order 1 [13, 14]

3 Proposed Eigenvalue-Based Double-Threshold Sensing Method

The proposed scheme can be summarized as in (5). If test statistic $T(N_s)$ is greater than upper threshold γ_2, then the SU decides in favour of hypothesis H_1. The spectrum band considered as vacant if test statistic is less than γ_1. The SU will re-sense the spectrum if the test statistic is between two thresholds. Thus, the decision can be described as

$$\text{Decision} = \begin{cases} \text{Spectrum is vacant} & T(N_s) < \gamma_1 \\ \text{resense the spectrum} & \lambda_1 < T(N_s) < \gamma_2 \\ \text{Spectrum is occupied} & T(N_s) > \gamma_2 \end{cases} \tag{5}$$

3.1 Analytical Analysis Using RMT Theory

The sample covariance matrices of the received signal at SU can be written as

$$\boldsymbol{R}_x(N_s) = \boldsymbol{R}_s(N_s) + \boldsymbol{R}_\eta(N_s) \tag{6}$$

where $\boldsymbol{R}_x(N_s), \boldsymbol{R}_s(N_s), \boldsymbol{R}_\eta(N_s)$ represents the sample covariance matrices of received signal at SU, signal transmitted by PU and noise, respectively.

As noise is AWGN and IID, then Eq. (6) can also be expressed as

$$\boldsymbol{R}_x(N_s) = \boldsymbol{R}_s(N_s) + \sigma_\eta^2 \boldsymbol{I}_L \tag{7}$$

Let the eigenvalues of $\boldsymbol{R}_x(N_s)$ and $\boldsymbol{R}_s(N_s)$ be $\lambda_1 \geq \lambda_2 \geq \cdots \geq \lambda_L$ and $\rho_1 \geq \rho_2 \geq \cdots \geq \rho_L$, respectively.

It is obvious $\lambda_n = \rho_n + \sigma_\eta^2$.

Therefore, in the absence of PU signal, we have

$$\lambda_1 = \lambda_2 = \cdots \lambda_L = \sigma_\eta^2 \quad \text{hence}\, \lambda_1/\lambda_L = 1 \tag{8}$$

When the primary user is active, then it can be written as

$\rho_1 > \rho_L$ and $\lambda_1/\lambda_L > 1$, because in practice PU signals are correlated signals. Thus, the ratio $T(N_s) = \lambda_1/\lambda_L$ can be chosen as the test statistic for the detection of primary user, as shown in [9].

When the PU is absent, then $\boldsymbol{R}_x(N_s)$ will convert to $\boldsymbol{R}_\eta(N_s)$. As $\boldsymbol{R}_\eta(N_s)$ is a special Wishart random matrix. Then, the distribution of largest eigenvalue is described as follows:

$$\text{Let}\, \boldsymbol{A}(N_s) = \frac{N_s}{\sigma_\eta^2}\boldsymbol{R}_\eta(N_s) \tag{9}$$

$$\text{With mean}\, \mu = \left(\sqrt{(N_s - 1}\right) + \sqrt{L})^2 \tag{10}$$

$$\text{Variance}\, v = \left(\sqrt{(N_s - 1}\right) + \sqrt{L})\left(\frac{1}{\sqrt{(N_s - 1}} + \frac{1}{L}\right)^{1/3} \tag{11}$$

Assume $L \ll N_s$, then $\frac{\lambda_{\max}(\boldsymbol{A}(N_s)) - \mu}{v}$ converges to the TW of order 1.

The numerical computation table is provided. The value of F_1 for some points that can be used to calculate $F_1^{-1}(y)$, is provided in Table 1.

Also, as $N_s \to \infty$ the minimum eigenvalue ($\lambda_{\min}$) of sample covariance matrix $\boldsymbol{R}_\eta(N_s)$ approaches the following value.

$$\lambda_{\min} \approx \frac{\sigma_\eta^2}{N_s}\left(\sqrt{N_s} - \sqrt{L}\right)^2 \tag{12}$$

3.2 Computation of Upper Threshold

For reliable detection, the value of P_f should be low. Upper threshold (γ_2) is calculated for pre-specified value of probability of false alarm as

Table 1 Numerical table for Tracy–Widom distribution of order 1

$F_1(t)$	0.01	0.10	0.30	0.50	0.70	0.90	0.95	0.99
t	−3.90	−2.78	−1.91	−1.27	0.59	−0.45	0.98	2.02

$$\begin{aligned} P_{\rm f} &= P(\lambda_{\max} > \gamma_2 \lambda_{\min}) \\ &= P\left(\frac{\sigma_\eta^2}{N_{\rm s}} \lambda_{\max}(\boldsymbol{A}(N_{\rm s})) > \gamma_1 \lambda_{\min}\right) \\ &\approx P\left(\lambda_{\max}(\boldsymbol{A}(N_{\rm s})) < \gamma_1 \left(\sqrt{N_{\rm s}} - \sqrt{L}\right)^2\right) \end{aligned}$$

Using RMT theory, it can written as

$$\gamma_2 = \frac{\left(\sqrt{N_{\rm s}} + \sqrt{L}\right)^2}{\left(\sqrt{N_{\rm s}} - \sqrt{L}\right)^2} \left(1 + \frac{\left(\sqrt{N_{\rm s}} + \sqrt{L}\right)^{-2/3}}{(N_{\rm s} L)^{1/6}} F_1^{-1}(1 - P_{\rm f})\right) \tag{13}$$

3.3 Computation of Lower Threshold

To provide sufficient protection to the PU, $P_{\rm m}$ should be below the pre-specified value. In the presented algorithm, lower threshold γ_2 is calculated on the basis of specified value of $P_{\rm m}$ as shown below. In the presence of PU, the sample covariance matrix $\boldsymbol{R}_x(N_{\rm s})$ is no longer a Wishart matrix. Till now, the distribution of its eigenvalues is unknown. Therefore, it is very difficult (mathematically intractable) to obtain a precise closed-form expression for probability of detection. However, by approximation, it can be driven by some empirical formulae.

$$\boldsymbol{R}_x(N_{\rm s}) = \boldsymbol{R}_{\rm s}(N_{\rm s}) + \boldsymbol{R}_\eta(N_{\rm s}) \tag{14}$$

where $\boldsymbol{R}_\eta(N_{\rm s})$ can be approximated to $\sigma_\eta^2 \mathrm{I}_L$. So, it can be written as

$$\lambda_{\max}(\boldsymbol{R_x}(N_{\rm s})) = \rho_1 + \lambda_{\max}\left(\boldsymbol{R_\eta}(N_{\rm s})\right) \tag{15}$$

$$\lambda_{\min}(\boldsymbol{R_x}(N_{\rm s})) \approx \rho_L + \sigma_\eta^2 \tag{16}$$

Here, using some approximation, the formula for probability of miss detection is obtained

$$P_{\rm m} = P(\lambda_{\max}(\boldsymbol{R_x}(N_{\rm s})) < \gamma_2 \lambda_{\min}(\boldsymbol{R_x}(N_{\rm s}))) \tag{17}$$

using Eqs. (24) and (25), we get as

$$\begin{aligned}&\approx P\left(\lambda_{\max}\left(\boldsymbol{R}_{\boldsymbol{\eta}}(N_{\mathrm{s}})\right)<\gamma_2\left(\rho_L+\lambda_{\min}\left(\boldsymbol{R}_{\boldsymbol{\eta}}(N_{\mathrm{s}})\right)\right)-\rho_1\right)\\&=P\left(\lambda_{\max}\left(\boldsymbol{R}_{\boldsymbol{\eta}}(N_{\mathrm{s}})\right)<\gamma_2\lambda_{\min}\left(\boldsymbol{R}_{\boldsymbol{\eta}}(N_{\mathrm{s}})\right)+\gamma_2\rho_L-\rho_1\right)\\&=P\left(\lambda_{\max}(\boldsymbol{A}(N_{\mathrm{s}}))<\frac{N_{\mathrm{s}}}{\sigma_\eta^2}\left(\gamma_2\lambda_{\min}\left(\boldsymbol{R}_{\boldsymbol{\eta}}(N_{\mathrm{s}})\right)+\gamma_2\rho_L-\rho_1\right)\right)\end{aligned}$$

After normalization, it is given as

$$P_{\mathrm{m}}=P\left(\frac{\lambda_{\max}(\boldsymbol{A}(N_{\mathrm{s}}))-\mu}{v}<\frac{\frac{N_{\mathrm{s}}}{\sigma_\eta^2}\left(\gamma_2\lambda_{\min}\left(\boldsymbol{R}_{\boldsymbol{\eta}}(N_{\mathrm{s}})\right)+\gamma_2\rho_L-\rho_1\right)-\mu}{v}\right)$$

$$\frac{\frac{N_{\mathrm{s}}}{\sigma_\eta^2}\left(\gamma_2\lambda_{\min}\left(\boldsymbol{R}_{\boldsymbol{\eta}}(N_{\mathrm{s}})\right)+\gamma_2\rho_L-\rho_1\right)-\mu}{v}=F_1^{-1}(P_{\mathrm{m}}) \tag{18}$$

$$P_{\mathrm{m}}=F_1\left(\frac{\frac{N_{\mathrm{s}}}{\sigma_\eta^2}\left(\gamma_2\lambda_{\min}\left(\boldsymbol{R}_{\boldsymbol{\eta}}(N_{\mathrm{s}})\right)+\gamma_2\rho_L-\rho_1\right)-\mu}{v}\right)$$

Let $B=\rho_1-\gamma_2\rho_L$; as PUs signal are correlated so B will be a positive quantity. Let us consider two cases:
Case 1.

$$P_{\mathrm{m1}}=F_1\left(\frac{\frac{N_{\mathrm{s}}}{\sigma_\eta^2}\left(\gamma_2\lambda_{\min}\left(\boldsymbol{R}_{\boldsymbol{\eta}}(N_{\mathrm{s}})\right)-B\right)-\mu}{v}\right) \tag{19}$$

Case 2.

$$P_{\mathrm{m2}}=F_1\left(\frac{\frac{N_{\mathrm{s}}}{\sigma_\eta^2}\left(\gamma_2\lambda_{\min}\left(\boldsymbol{R}_{\boldsymbol{\eta}}(N_{\mathrm{s}})\right)\right)-\mu}{v}\right) \tag{20}$$

Therefore, it can be seen that $P_{\mathrm{m1}}<P_{\mathrm{m2}}$. For calculating the lower threshold (γ_1), Eq. (20) is used with consideration of a certain fixed probability of miss detection (P_{m2}), because Eq. (20) is not depending on eigenvalues of signal covariance matrix. It can be verified that $P_{\mathrm{m1}}<P_{\mathrm{m2}}$ using the obtained lower threshold value (γ_1). So, in practical scenario, it will provide more protection to PU, as probability of miss detection will be less.

Therefore, lower threshold (γ_1) can be calculated as

$$\gamma_2=\frac{vF_1^{-1}(P_{\mathrm{m}})+\mu}{\frac{N_{\mathrm{s}}}{\sigma_\eta^2}\lambda_{\min}\left(\boldsymbol{R}_{\boldsymbol{\eta}}(N_{\mathrm{s}})\right)}=\frac{vF_1^{-1}(P_{\mathrm{m}})+\mu}{\left(\sqrt{N_{\mathrm{s}}}-\sqrt{L}\right)^2} \tag{21}$$

3.4 Cooperative Spectrum Sensing

In cooperative spectrum sensing, each SU node makes a binary decision based on its local observation and forwards a 1-bit decision (D_i) to the fusion centre, where 1 stands for presence of PU and 0 for absence of PU. Therefore, according to optimal voting rule [12], which minimizes the probability of error, the probability of false alarm (Q_{f}) and probability of miss detection (Q_{m}) can be given as

$$Q_{\mathrm{f}} = \mathrm{Prof}\{H_1|H_0\} = \sum_{l=n_{\mathrm{opt}}}^{K} \binom{K}{l} P_{\mathrm{f}}^{l}(1-P_{\mathrm{f}})^{K-l} \tag{22}$$

$$Q_{\mathrm{m}} = \mathrm{Prof}\{H_0|H_1\} = \sum_{l=n_{\mathrm{opt}}}^{K} \binom{K}{l} P_{\mathrm{d}}^{l}(1-P_{\mathrm{d}})^{K-l} \tag{23}$$

where K is number of SU nodes in cooperative CR network;

$$n_{\mathrm{opt}} = \min\left(K, \left\lceil \frac{K}{1+\alpha} \right\rceil\right) \tag{24}$$

n_{opt} is optimal number of SUs for optimal 'n-out-K' rule with $\alpha = \frac{\ln\frac{P_{\mathrm{f}}}{1-P_{\mathrm{m}}}}{\ln\frac{P_{\mathrm{m}}}{1-P_{\mathrm{f}}}}$.
Therefore, probability of error in optimal voting rule can be given as

$$P_{\mathrm{e}} = P(H_0)Q_{\mathrm{f}} + P(H_1)Q_{\mathrm{m}} \tag{25}$$

where $P(H_0), P(H_1)$ are probability of hypothesis H_0 and H_1, respectively.

4 Simulation Results and Discussion

This section presents simulation results of Eigenvalues-based double-threshold detection (EVDD) scheme under optimal voting rule cooperative spectrum sensing. Also, performance is compared with some existing spectrum detection techniques such as ED-based and Maximum–Minimum eigenvalues detection (MME).

The number of SU nodes (K) in cognitive radio network used for cooperative spectrum sensing is 5 and N_{s} for each SU node is 10,000. The L is taken as 5 and the thresholds are calculated considering the pre-specified value as $P_{\mathrm{f}} = 0.1$ and $P_{\mathrm{m}} = 0.1$. The optimal number of SU nodes for cooperative optimal voting spectrum sensing is $n_{\mathrm{opt}} = 3$. The probability of PU to be active in spectrum band $P(H_1)$ and to be inactive $P(H_0)$ both are taken as 0.5 in the simulation (Fig. 1).

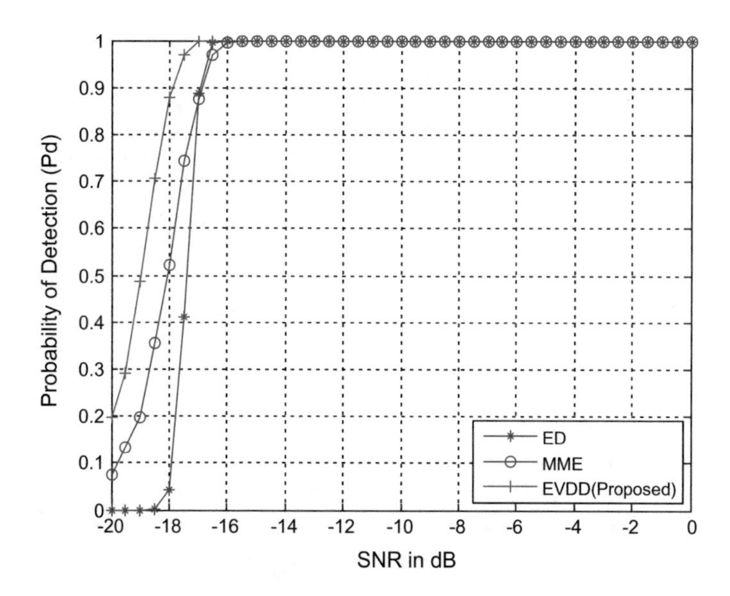

Fig. 1 P_d versus SNR, with cooperative spectrum sensing of optimal voting rule under noise uncertainty of 0 dB

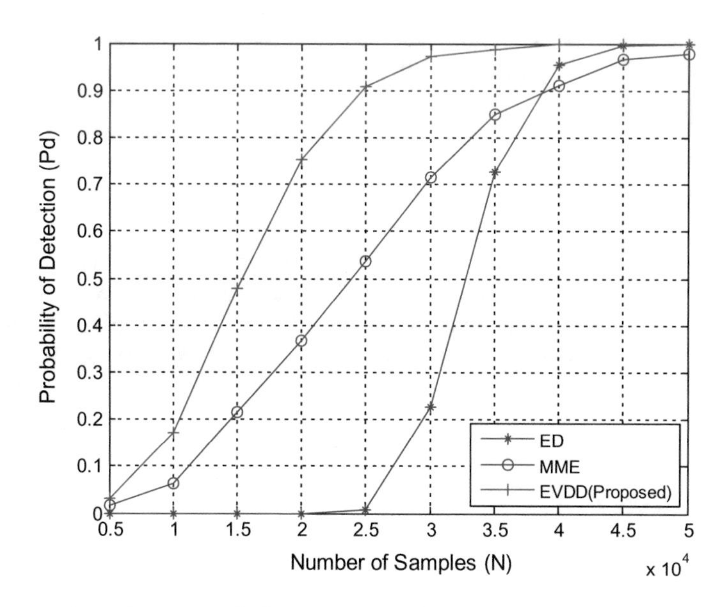

Fig. 2 Probability of detection (P_d) versus number of samples under cooperative spectrum sensing of optimal voting rule with noise uncertainty of 0 dB

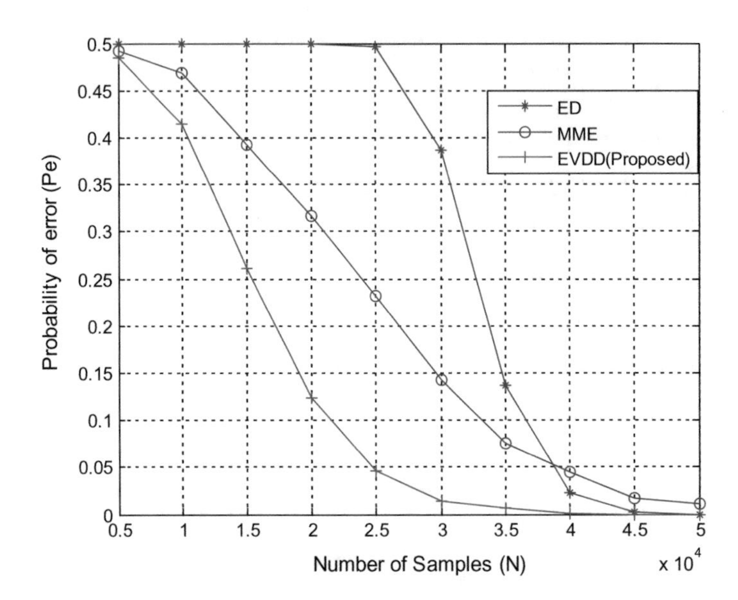

Fig. 3 Probability of error (P_e) versus number of samples under cooperative spectrum sensing of optimal voting rule with noise uncertainty of 0 dB

Figure 2 shows the variation of P_d with SNR. This figure verifies the better performance of proposed scheme over the other two methods. The EVDD scheme achieves P_d of 0.88 at SNR = −18 dB, same in MME and ED have as 0.521 and 0.042, respectively.

The variation of P_d. with number of samples is shown in Fig. 2 at SNR = −20 dB. From Fig. 2, it is observed that EVDD scheme works better than ED and MME method, when the number of samples is $N_s = 30,000$, the P_d is 716 and 0.227 for MME method and ED method, respectively.

Figure 3 shows the variation of probability of error (P_e) with number of samples at SNR = −20 dB. This figure verifies the less P_e of EVDD scheme than ED and MME method, when the number of samples $N_s = 25,000$, the P_e of proposed method is 0.046, whereas P_e is 0.2315 and 0.496 for MME scheme and ED method, respectively.

5 Conclusions

This paper presents a cooperative spectrum sensing scheme using eigenvalue-based double-threshold detection technique. The thresholds are obtained from the RMT theory. Also, it addresses the hidden terminal problem as well using optimal voting cooperative spectrum sensing. This scheme is called as blind spectrum detection method, as it does not require any knowledge of the signal characteristic of PU and the noise power. The better performance of presented scheme is verified through extensive simulation results, however, the computational complexity in comparison to ED is higher.

References

1. Federal Communications Commission: Spectrum policy task force report. pp. 02–155 (2002)
2. Mitola, J., Maguire, G.Q.: Cognitive radio: making software radios more personal. IEEE Pers. Commun. **6**, 13–18 (1999)
3. Haykin, S.: Cognitive radio: brain-empowered wireless communications. IEEE J. Select. Areas Commun. **23**, 201–220 (2005)
4. Ghaznavi, M., Jamshidi, A.: A reliable spectrum sensing method in the presence of malicious sensors in distributed cognitive radio network. IEEE Sens. J. **15**, 1810–1816 (2015)
5. Bogale, T.E., Vandendorpe, L.: Max-min SNR signal energy based spectrum sensing algorithms for cognitive radio networks with noise variance uncertainty. IEEE Trans. Wirel. Commun. **13**, 280–290 (2014)
6. Atapattu, S., Tellambura, C., Jiang, H., Rajatheva, N.: Unified analysis of low-SNR energy detection and threshold selection. IEEE Trans. Veh. Technol. **64**(11), 5006–5019 (2015)
7. Oner, M., Jondral, F.: Cyclostationary-based methods for the extraction of the channel allocation information in a spectrum pooling system. In: Proceedings IEEE Radio and Wireless Conference, Atlanta, GA, pp. 279–282 (2004)
8. Zeng, Y., Liang, Y.-C.: Covariance based signal detections for cognitive radio. In: Proceedings IEEE International Symposium on and New Frontiers in Dynamic Spectrum Access Networks, Dublin, Ireland, pp. 202–207 (2007)
9. Zeng, Y., Liang, Y.-C.: Maximum-minimum eigenvalue detection for cognitive radio. In: Personal, Indoor and Mobile Radio Communications, 2007 IEEE 18th International Symposium, pp. 1–5 (2007)
10. Yucek, T., Arslan, H.: A survey of spectrum sensing algorithms for cognitive radio applications. IEEE Commun. Surv. Tutor. **11**(1) (2009)
11. Liang, Y.-C., Zeng, Y., Peh, E., Hoang, A.T.: Sensing-throughput tradeoff for cognitive radio networks. IEEE Trans. Wirel. Commun. **5**(7), 1326–1337 (2008)
12. Zhang, W., Mallik, R.K., Letaief, K.: Optimization of cooperative spectrum sensing with energy detection in cognitive radio networks. IEEE Trans. Wirel. Commun. **12**, 5761–5766 (2009)
13. Tulino, A.M., Verdu, S.: Random Matrix Theory and Wireless Communication. Now Publisher Inc., Hanover, USA (2004)
14. Johnstone, I.M.: On the distribution of largest eigenvalue in principle components analysis. Comm. Math. Phys. **29**(2), 295–327 (2001)

Impact of Mathematical Optimization on OSPF Routing in WDM Optical Networks

Himanshi Saini and Amit Kumar Garg

Abstract High-speed communication systems require robust routing protocol. The selection and optimization of the routing protocol in Wavelength Division Multiplexed (WDM) optical networks decide utilization of bandwidth capacity offered by these networks. When Open Shortest Path First (OSPF) protocol is used for routing, link weight selection is one of the main concerns in order to optimize the routing. In this paper, OSPF link weight selection is performed through Ant Colony Optimization (ACO) and Evolutionary Algorithm (EA). The novelty of the work lies in the implementation of optimization techniques in offline network mode on networks of variable densities with an objective is to reduce utilization of bottleneck link, i.e., to reduce congestion. As a result, knowledge of optimization specific to type of network can be retained in online networking mode. ACO and EA are implemented on random network which is composed of 4 nodes and 5 links (4n5e), standard National Science Foundation NETwork (NSFNET), and standard COST 239 networks in order to test the optimization on networks of different densities. Maximum End-to-End (E2E) latency and bottleneck Link Utilization (LU) obtained after application of optimization techniques is compared for networks considered. It is observed that EA optimization has a better optimized denser network (COST 239) and ACO has optimized NSF and 4n5e networks better than EA optimization.

Keywords OSPF · ACO · EA · Optimization · Link utilization
Latency

H. Saini (✉) · A. K. Garg
ECE Department, D.C.R.U.S.T, Murthal, Sonepat 131039, Haryana, India
e-mail: himanshi.4887@gmail.com

© Springer Nature Singapore Pte Ltd. 2019

H. Malik et al. (eds.), *Applications of Artificial Intelligence Techniques in Engineering*, Advances in Intelligent Systems and Computing 697,
https://doi.org/10.1007/978-981-13-1822-1_19

1 Introduction

Ever increasing speed and capacity of optical networks demand promising approaches for network design, routing and optimization. OSPF is one of the link state routing protocols. Its dynamic behavior makes it capable to perform under all network conditions. Each node holds complete network information and calculates all required shortest path to all other nodes in a network. Criterion of link weight is used for calculation of shortest paths in OSPF protocol. It is proportional to inverse of the bandwidth of the link [1]. Some network administrators set link weights proportional to physical distance of the link or priority of link [2]. For optimized network functioning, link weights should match expected traffic on the links. Link weights decide network performance parameters like E2E delay, utilization ratio, and congestion. Network can be controlled by the operator by modifying link weights according to traffic demands [3]. In this paper, ACO and EA optimization techniques are used for selection of link weights to optimize OSPF routing in networks with different sizes. ACO copies the foraging behavior of ants in finding their path from food to nest facing and tackling the obstacles on the path [4]. EA mimics selection criteria of species followed by nature, i.e., evolution and survival of fittest. Steps in each ACO iteration are summarized in Fig. 1 and each EA run is briefly discussed in Fig. 2.

A number of previously performed analysis have contributed various approaches for OSPF link weight assignment. Buriol et al. [6] have worked on determination of link weight and capacity for OSPF routed network under any link or node failure. They have proposed a genetic algorithm for addressing the issue of link weight and capacity assignment. They have shown that the proposed algorithm avoids overloading in the network in condition of network failure. Squalli et al. [7] implemented tabu search iterative heuristic to select link weights for a network under normal network functions and under single link failure scenario. It has been proved that link failures do not influence network performance if load on the network is low. Lin et al. [8] proposed a shortest path routing approach by using priority-based GA. They have shown that GA efficiently meets QoS requirements and implements OSPF weight setting. Buriol et al. [9] have dealt with OSPF weight problem with local improvement in genetic algorithm. The proposed procedure implements

In each iteration of ACO loop following steps are executed till optimal solution is obtained

$\boldsymbol{A}$, number of ants are created
Each ant gives solution of routing problem by greedy randomized scheme based on pheromone
Set of solution components $\boldsymbol{C}$ contain solution by all the ants
Each solution component, $\boldsymbol{c} \in \boldsymbol{C}$ is assigned $\boldsymbol{P_C}$ of pheromone
Update the obtained solution
Increase $\boldsymbol{P_C}$ of traversed solution components (Reinforcement)
Reduce $\boldsymbol{P_C}$ of all solution components (Evaporation)

Fig. 1 Steps in each ACO algorithm iteration [5]

In each iteration of EA loop following steps are executed till optimal solution is obtained
Initial population of problem solutions, P is created using pure or greedy randomized procedure
Couples of solutions ***(x, y)*** *from* ***P*** *are chosen (Parents)*
Offspring is created from each ***(x, y)*** *pair inheriting characteristics of both* **x** *and* **y** *(Crossover)*
Offspring are randomly changed (Mutation)
Apply solution selection process as $\boldsymbol{P} = \boldsymbol{P} \cup \boldsymbol{Offspring}$

Fig. 2 Steps in each EA algorithm iteration [5]

dynamic shortest path algorithm after any modification in link weights in order to reestablish shortest paths. It has been shown that the proposed procedure produces near-optimal solutions. Garg et al. [10] optimized OSPF algorithm using improved moth flame optimization. They have validated the performance of optimized OSPF algorithm in case of dynamic traffic by comparing the parameters like end-to-end delay, throughput for traditional, modified, optimized, and adaptively optimized OSPF algorithm. Valadarsky et al. [11] have introduced the concept of intelligently adapting the routing by foreseeing network demands, i.e., the concept of machine learning in deciding favorable routing configurations. They have introduced methods to learn probable traffic demands and routing via reinforcement learning for small networks. Sousa et al. [12] proposed an evolutionary algorithm-based optimization framework with multiple objectives like reducing network congestion, meeting end-to-end delay constraints. The framework offers various routing configurations which are adaptive to traffic demands and resilient to network link failures. This framework can be utilized by network administrators in order to select routing configurations fulfilling specific network requirements. Magnani et al. [13] proposed an optimization technique for OSPF weights which is traffic adaptive. They have not reduced network congestion for an average population but have proposed an optimization which can work according to individual user perspective.

In this paper, the impact of optimizations such as ACO and EA algorithms on performance of OSPF routing protocol in networks: NSFNET, COST239, and 4n5e are analyzed. Simulation environment is discussed in Sect. 2. This section covers the networks and their characteristics considered in the present analysis. Results are presented in Sect. 3 with the help of tabular representation for each network considered and plots for bottleneck LU and Maximum E2E Latency. Conclusion is framed in Sect. 4.

2 Simulation Environment

Net2Plan tool is used to perform this analysis. It is an **open-source** JAVA-based tool in which networks can be planned and optimized [14]. The three-test networks simulated in Net2Plan are shown in Fig. 3. Fig. 3a shows COST 239 network with

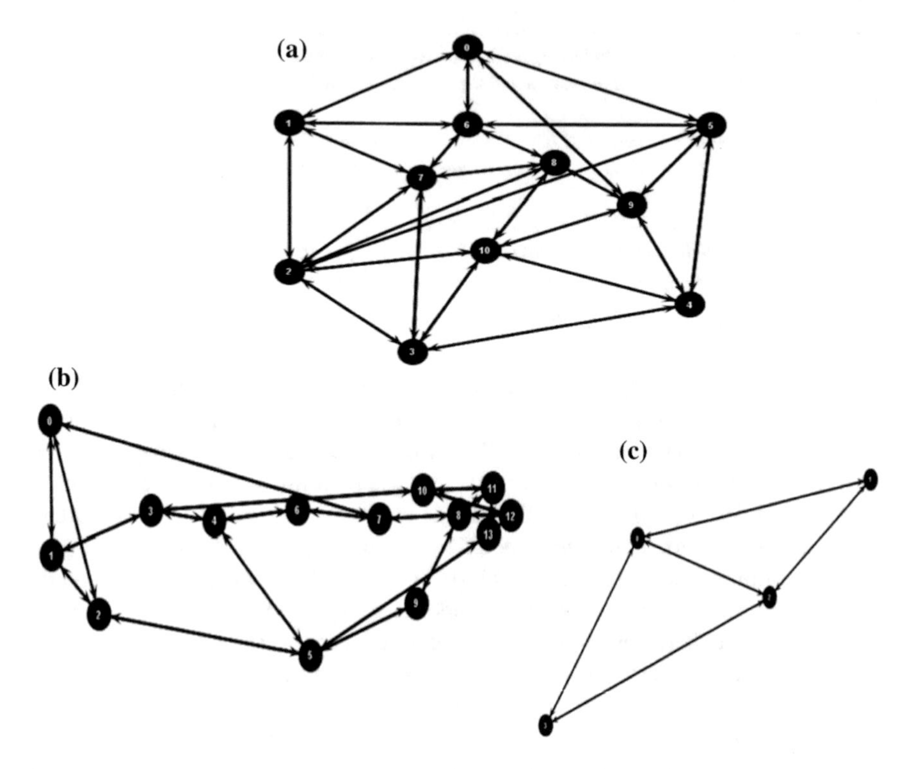

Fig. 3 Test networks: **a** COST239 network. **b** NSFNET network. **c** 4n5e network

11 nodes and 26 links. NSFNET with 14 nodes and 21 links is shown in Fig. 3b, and 4 nodes and 5 links (4n5e) network is shown in Fig. 3c. The networks are implemented in offline network design mode. ACO and EA heuristic algorithms are implemented on all the networks. These algorithms assume that network nodes are implementing OSPF protocol. The aim of these heuristics is to find network link weights in order to optimize the routing in terms of minimizing network congestion. Both the algorithms, ACO and EA are executed for 400 s. Optimized solution is provided in less than this time limit which is observed by repeated simulations. EA is executed for maximum number of 100 iterations, 200 numbers of children in offspring, and 1000 number of elements in population. ACO is executed for maximum number of 100 iterations and 10 numbers of ants. OSPF link weights are constrained between 1 and 16 for both ACO and EA. The objective is to minimize network congestion, and this is computed as function of maximum and average link utilization.

Few parameters of networks under consideration are shown in Table 1. COST 239 has highest connectivity among all test networks. The performance of various heuristics is observed to depend on network connectivity and hence different heuristics are required to optimize networks of different densities. Traffic demands indicate route from each node to all other nodes in network. It is highest for

Table 1 Network parameters

Parameter	Test networks		
	NSFNET	COST239	4n5e
Traffic demands	182	110	12
Nodes	14	11	4
Links	21	26	5
Maximum degree	4	6	3
Minimum degree	2	4	2

NSFNET as it has more number of nodes as compared to other test networks. Performance of small networks with less connectivity is not influenced by optimization technique.

The objective of ACO and EA considered for present analysis is minimizing the congestion. Bottleneck link utilization (link utilization of most utilized link) in a network clearly indicates the level of congestion. It is calculated as ratio of load on most utilized link to link capacity of most utilized link. The target of OSPF weight selection should result into reduction of bottleneck link utilization.

3 Results and Discussions

ACO and EA are applied to networks shown in Fig. 3. Table 2 indicates impact of ACO and EA on OSPF protocol in NSFNET, COST239, and 4n5e network. EA takes more time than ACO to reach at an optimal solution, i.e., finding link weights targeted toward reduction in network congestion. EA offers slightly low bottleneck LU as compared to ACO. Maximum E2E latency for ACO is lower than EA. For ACO, maximum E2E latency is for traffic demand from node 5 to node 7, routed as 5-2-0-7 network path. Traffic demand from node 9 to node 7 is maximum time-consuming demand for EA routed as 9-8-7 and 9-5-2-0-7 network paths. In COST239 network, bottleneck LU has reduced to a much greater extent by EA as compared to ACO and overall reduction in LU is observed as compared to LU in NSFNET. Maximum E2E latency for ACO calculated for demand from node 8 to node 1 routed as 8-9-0-1 network path is higher than EA calculated for demand

Table 2 OSPF performance with ACO and EA algorithms

Performance parameter	NSFNET		COST239		4n5e	
	ACO	EA	ACO	EA	ACO	EA
Optimization time	37.8 s	374 s	20 s	204 s	0.889 s	6.6 s
Bottleneck link utilization/network congestion	0.662	0.591	0.369	0.253	0.053	0.049
Max. E2E latency	32 ms	38 ms	13.9 ms	11 ms	4.22 ms	4.22 ms

from node 4 to node 1, routed as 4-5-2-1 network path. EA heuristic has optimized OSPF routing in COST 239 network better than ACO. In 4n5e network, EA offers slightly lower bottleneck LU as compared to ACO. Maximum E2E latency for both the heuristics is same and calculated for demand from node 1 to node 3, routed as 1-2-3 network path.

Network congestion measured in terms of bottleneck LU is plotted in Fig. 4 for NSFNET, COST239, and 4n5e networks. OSPF link weights are calculated by ACO and EA with an objective to minimize network congestion. LU for all the networks, optimized by ACO or EA is compared. It is observed that EA results in lower bottleneck LU as compared to ACO in all the networks. Maximum latency calculated from maximum time-consuming demand is plotted in Fig. 5 for all three networks. EA results in higher E2E latency as compared to ACO in NSFNET network, whereas ACO gives higher value in COST 239 network. COST239 network is denser as compared to NSFNET and 4n5e networks. The optimization behavior indicates that denser networks are better optimized by EA heuristic.

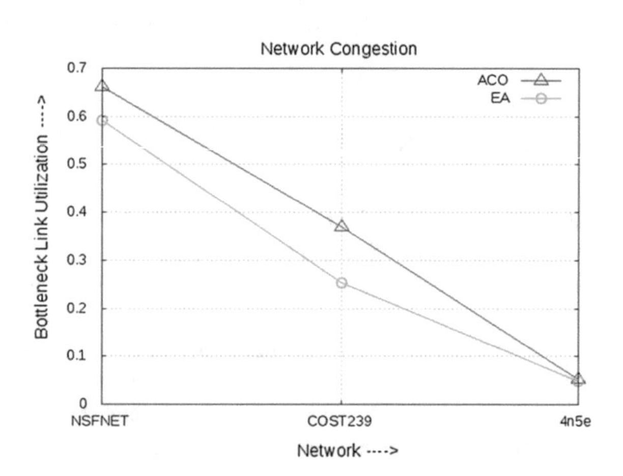

Fig. 4 Bottleneck link utilization for all networks with EA and ACO algorithms

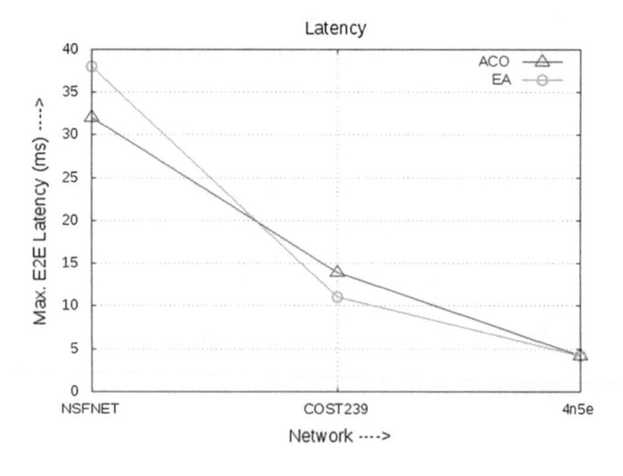

Fig. 5 Maximum E2E latency for all networks with EA and ACO algorithms

4 Conclusion and Future Scope

In order to exploit merits of WDM optical networks, it is imperative to implement optimized routing technique in these networks. In the presented work, problem of optimum weight assignment in OSPF routing protocol is addressed. Three test networks NSFNET, COST239, and 4n5e are considered to observe the impact of ACO and EA heuristics on OSPF routing. OSPF weight assignment problem is targeted towards an objective of minimizing network congestion. It is observed that in 4n5e network, OSPF protocol performance is almost identical for ACO and EA. Performance of small size network is almost the same for any mathematical optimization technique. Bottleneck LU obtained by EA algorithm is less than ACO for all networks under consideration as compared to ACO algorithm. Reduction in LU of COST239 network by EA algorithm is more as compared to the reduction in NSFNET and 4n5e. Reduction in maximum E2E latency is also obtained by optimizing OSPF by EA algorithm in COST239, whereas in NSFNET network, lesser maximum E2E delay value is obtained by ACO algorithm. The analysis performed shows that EA can better optimize dense network as compared to ACO algorithm. The present analysis can be extended with more network topologies. The work can be extended with analyzing the impact of optimization techniques on network resilience under single and multiple link failures in networks of different densities, thereby implementing resilient WDM optical network.

References

1. Hosny, W., Mahmoud, M., Aly, M.H., El-Badawy, E.S.A.: New optical parameters dependent metric for OSPF protocol for OBS networks. In: 2009 IFIP International Conference on Wireless and Optical Communications Networks, Cairo, pp. 1–4 (2009). https://doi.org/10.1109/wocn.2009.5010577
2. Pereira, T.B., Ling, L.L.: Network performance analysis of an adaptive OSPF routing strategy —effective bandwidth estimation. In: International Telecommunication Symposium—ITS, Natal, Brazil (2002)
3. Srivastava, S., Agrawal, G., Pioro, M., Medhi, D.: Determining link weight system under various objectives for OSPF networks using a Lagrangian relaxation-based approach. IEEE Trans. Netw. Serv. Manage. **2**(1), 9–18 (2005). https://doi.org/10.1109/TNSM.2005.4798297
4. Pavani, G.S., Waldman, H.: Routing and wavelength assignment with crankback re-routing extensions by means of ant colony optimization. IEEE J. Sel. Areas Commun. **28**(4), 532–541 (2010). https://doi.org/10.1109/JSAC.2010.100503
5. Mariño, P.P.: Optimization of Computer Networks—Modeling and Algorithms: A Hands-On Approach. Wiley, Chichester (2016)
6. Buriol, L.S., Franca, P.M., Resende, M.G.C., Thorup, M.: Network design for OSPF routing. In: Proceedings of Mathematical Programming in Rio, Búzios, Rio de Janeiro, Brazil, pp. 40–44 (2003)
7. Sqalli, M.H., Sait, S.M., Asadullah, S.: OSPF weight setting optimization for single link failures. Int. J. Comput. Netw. Commun. (IJCNC) **3**(1), 168–183 (2011)

8. Lin, L., Gen, M.: Priority-based genetic algorithm for shortest path routing problem in OSPF. In: Gen, M., et al. (eds.) Intelligent and Evolutionary Systems. Studies in Computational Intelligence, vol. 187. Springer, Berlin (2009)
9. Buriol, L.S., Resende, M.G.C., Ribeiro, C.C., Thoup, M.: A hybrid genetic algorithm for the weight setting problem in OSPF/IS-IS routing. Networks **46**, 36–56 (2005)
10. Garg, P., Gupta, A.: Adaptive optimized open shortest path first algorithm using enhanced moth flame algorithm. Indian J. Sci. Technol. **10**(23), 1–7 (2017). https://doi.org/10.17485/ijst/2017/v10i23/112150
11. Valadarsky, A., Schapira, M., Shahaf, D., Tamar, A.: A machine learning approach to routing (2017). arXiv:1708.03074v2 [cs.NI]
12. Sousa, P., Pereira, V., Cortez, P., Rio, M., Rocha, M.: A framework for improving routing configurations using multi-objective optimization mechanisms. J. Commun. Softw. Syst. **12**(3), 145–156 (2016)
13. Magnani, D.B., Carvalho, I.A., Noronha, T.F.: Robust optimization for OSPF routing. IFAC-PapersOnLine **49–12**, 461–466 (2016)
14. Network Simulator: Net2Plan. http://www.net2plan.com/

Optimized Power Allocation in Selective Decode-and-Forward Cooperative Communication

E. Bindu and B. V. R. Reddy

Abstract Cooperative relay transmission is an emerging concept in wireless communication, in which single-antenna devices are allowed to take benefit of spatial diversity to improve reliability and error performance. In this work, Selective Decode-and-Forward (SDF) cooperative relaying scheme is analyzed, in which decoding and subsequent retransmission are undertaken by the relay node only when the received signal meets certain threshold constraints at that node. This paper describes power optimization in relay nodes for a dual-hop half-duplex SDF-based cooperative relay communication system, with all nodes using single antenna. For the system, average end-to-end probability of error is derived, leading to evaluation of asymptotically tight upper bound under high signal-to-noise ratio (SNR). Optimal power allocation to minimize the bit error rate (BER) is formulated and a closed-form solution for this convex optimization problem is derived and verified through simulation. The BER performance is compared with that of direct source to destination link which indicated performance enhancement, owing to virtual MIMO effect. A variety of numerical results reveal that cooperate scheme with optimal power allocation has an improved performance, compared to its direct link counterpart.

Keywords Cooperative communication · Selective decode-and-forward relay Virtual MIMO · Convex optimization

---

E. Bindu (✉)
Department of ECE, Amity School of Engineering & Technology,
GGS Indraprastha University, New Delhi, India
e-mail: bindue25@gmail.com

E. Bindu · B. V. R. Reddy
USICT, GGS Indraprastha University, New Delhi, India

© Springer Nature Singapore Pte Ltd. 2019

H. Malik et al. (eds.), *Applications of Artificial Intelligence Techniques in Engineering*, Advances in Intelligent Systems and Computing 697,
https://doi.org/10.1007/978-981-13-1822-1_20

1 Introduction

Relay-assisted communication is a promising strategy that uses the spatial diversity available among a collection of distributed single-antenna terminals, for both centralized and decentralized wireless networks [1]. Wireless signal transmission is subjected to many attenuation factors like path loss, channel fading, and shadowing. Multiple-input multiple-output (MIMO) techniques can improve the communication reliability by sending multiple copies of same signal through different paths, hence allowing better reception at the destination. Cooperative relay scheme having single-antenna relay stations is called virtual MIMO system as it emulates a MIMO channel with two transmit antennas, for the destination node. Virtual MIMO system helps the single-antenna devices to attain some benefits of spatial diversity without the need of actual antenna arrays [2]. This is beneficial in scenarios where there are physical limitations to implement multiple antennas on the device due to constraints on cost, size, or hardware implementation [3]. There could be a single relay or multiple relays. At the destination, the signals received from multiple sources are compared/combined to select/obtain the optimum performance [4]. In most of the relaying networks, a two-stage strategy is used. In the first stage, the source broadcasts and all relays listen. In the second stage, the relays forward the source symbols to the destination. There are two types of relay protocols, amplify-and-forward (AF) and decode-and-forward (DF). DF relay protocol receives information from source and retransmits the signal to destination after decoding and re-encoding the message. A sub-variant of DF strategy is selective DF (SDF) in which the relay decides whether or not to forward the message to destination, based on decodability of the received signal, by applying signal-to-noise ratio (SNR) threshold criterion at the relay [5]. The decision can be taken based on many parameters and one such is SNR of signal at the relay. Fixed DF transmission does not offer diversity gains for large SNR since the relay has to fully decode the information received from the source node. SDF performs better under such conditions. Moreover, this strategy reduces the chance of error propagation [6, 7].

Various researchers have been working to investigate application of SDF in different communication scenarios. For a dual-hop relay system using SDF, the energy efficiency optimization is proposed in [8]. Performance analysis of SDF-based single and multiple-hop cooperative MIMO relay under fading scenario is done in [9] and applied to spatial modulation (SM) is carried out in [10]. A new variant of SDF, named dynamic SDF(D-SDF) is proposed in [11] under slow fading channel conditions. Energy consumption issues in wireless sensor networks (WSNs) using SDF protocol is studied in [12].

Our focus is on dual-hop SDF-based cooperative relay system. Since multiple antenna relay nodes can increase the system cost, we are assuming single antenna at participating nodes. Simulations show diversity gain due to virtual MIMO effect. Analysis of average bit error rate (BER) is carried out and closed-form expression for asymptotically tight upper bound is formulated. Power optimization to obtain minimal end-to-end BER is carried out by formulating and solving the convex objective function. The theoretical results were substantiated by numerical results obtained through simulations.

The rest of the paper is arranged as follows: Sect. 2 explains system model and analysis. System model for source-destination (S-D) direct link, followed by source-relay (S-R) and relay-destination (R-D) are analyzed. Then SNR and probability of error for each link is evaluated and symbol error rate (SER) of S-D and S-R-D link is calculated. This is followed by formulation of power optimization problem. Section 3 contains simulation results which validate the proposed solutions. Section 4 concludes the paper.

2 System Model

We consider a wireless relay-based cooperative communication network consisting of source (S), a single relay node (R), and destination (D), as shown in Fig. 1. It is assumed that each node in the network is having single antenna and operate in half-duplex mode, in which relay cannot transmit and receive simultaneously. The proposed cooperative relay scheme emulates a virtual MIMO system [13]. The relaying protocol assumed is selective decode and forward (SDF). This protocol is divided into two phases. In the first phase, the information is broadcast by source. The received message at the relay is analyzed. In the second phase, the relay re-encodes and transmits the message to destination, only if the signal received in the first phase is error-free and decodable.

Frequency-selective, independent, and identically distributed (iid) Rayleigh fading channel is assumed with perfect channel state information (CSI) knowledge at the receiver. The perfect CSI means the relay knows the channel condition in S-R link and the destination knows the channel perfectly for S-D, S-R, and R-D links, so as to perform Maximum Likelihood (ML) detection at the destination [14].

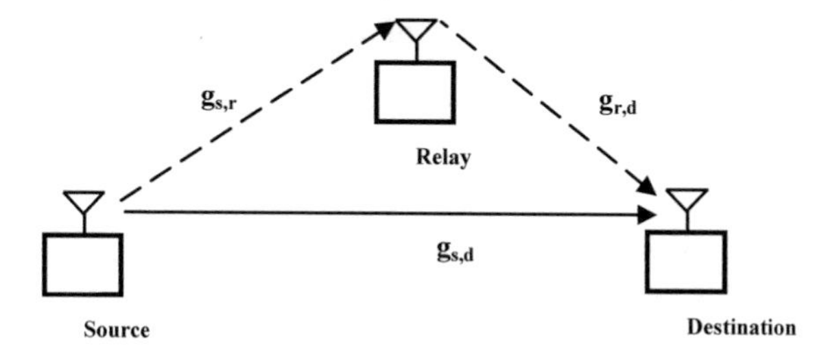

Fig. 1 System model

2.1 Selective DF Transmission

Half-duplex communication is assumed, between S-R and S-D, in which message transfer happens in two phases.

Phase1: Source broadcasts signal to both destination and relay nodes

$$r_{\mathrm{s,d}} = \sqrt{P_{\mathrm{S}}} g_{\mathrm{s,d}} s + n_{\mathrm{s,d}} \tag{1a}$$

$$r_{\mathrm{s,r}} = \sqrt{P_{\mathrm{S}}} g_{\mathrm{s,r}} s + n_{\mathrm{s,r}} \tag{1b}$$

$g_{\mathrm{s,d}}$, $g_{\mathrm{s,r}}$ are frequency-selective, independent, and identically distributed (iid) Rayleigh fading channel coefficients for S-D and S-R links. P_{S} is source transmit power. $n_{\mathrm{s,d}}$, $n_{\mathrm{s,r}}$ are complex circularly symmetric additive white Gaussian noise of power $\eta 0/2$. Assume transmit symbol (s) has average power $E\{|s|^2\} = 1$.

Average channel gains corresponding to the individual links are given as

$$E\left\{\left|g_{\mathrm{s,d}}\right|^2\right\} = \xi_{\mathrm{s,d}}^2 \text{ and } E\left\{\left|g_{\mathrm{s,r}}\right|^2\right\} = \xi_{\mathrm{s,r}}^2 \tag{2}$$

Phase2: In selective DF relay transmission, relay forwards decoded symbols to destination if relay decodes the symbol correctly

$$r_{\mathrm{r,d}} = \sqrt{P_{\mathrm{R}}} g_{\mathrm{r,d}} s + n_{\mathrm{r,d}} \tag{3}$$

$g_{\mathrm{r,d}}$ is channel coefficient for R-D link. P_{R} is transmit power employed at the relay and is given as

$$P_{\mathrm{R}} = \begin{cases} P_{\mathrm{R}} & \text{if decoded correctly} \\ 0 & \text{if decoded incorrectly} \end{cases} \tag{4}$$

Decoding at the relay is done based on SNR threshold criterion. At R, if the received signal meets certain SNR threshold, then only the signal is decoded, re-encoded, and transmitted to D node. If SNR criteria are not met, then R-D link will not be established in Phase 2. In that case, only the direct link exist between S and D, and corresponding SNR at destination is

$$\mathrm{SNR_d} = P_{\mathrm{S}} \frac{\left|g_{\mathrm{s,d}}\right|^2}{\sigma^2} \tag{5}$$

If no error, then there are two paths existing between S and D, i.e., S-D and S-R-D, and received signals can be represented as

$$\begin{bmatrix} r_{\mathrm{s,d}} \\ r_{\mathrm{r,d}} \end{bmatrix} = \begin{bmatrix} \sqrt{P_{\mathrm{S}}} g_{\mathrm{s,d}} \\ \sqrt{P_{\mathrm{R}}} g_{\mathrm{r,d}} \end{bmatrix} s + \begin{bmatrix} n_{\mathrm{s,d}} \\ n_{\mathrm{r,d}} \end{bmatrix} \tag{6}$$

This equation reminds us of a virtual MIMO system with two transmit antennas. So, a diversity gain of 2 can be expected from this system. At the destination, optimal receiver performs MRC with the weighing coefficients is calculated as $w = \frac{g}{\|g\|}$

$$\text{SNR at destination is } \text{SNR}_\text{d} = \frac{\|g\|^2 E\left\{|s|^2\right\}}{\sigma^2} = \frac{P_\text{S}|g_{\text{s,d}}|^2 + P_\text{R}|g_{\text{r,d}}|^2}{\sigma^2} \tag{7}$$

2.2 SER Analysis

Using Craig's formula to obtain SER of BPSK system [15], we have

$$Q\left(\sqrt{\frac{P}{\sigma^2}}\right) = \frac{1}{\pi}\int_0^{\frac{\pi}{2}} \exp\left(-\frac{\frac{P}{\sigma^2}}{2\sin^2\theta}\right)\mathrm{d}\theta = \frac{1}{\pi}\int_0^{\frac{\pi}{2}} \exp\left(-\frac{\rho}{2\sin^2\theta}\right)\mathrm{d}\theta \tag{8}$$

Let Ψ denote the event of error at the relay. Then, the error probability is

$$P_\text{e}(\Psi) = Q\left(\sqrt{\frac{P_\text{S}|g_{\text{s,r}}|^2}{\sigma^2}}\right) = \frac{1}{\pi}\int_0^{\frac{\pi}{2}} \exp\left(-\frac{\rho_\text{S}\Omega_{\text{s,r}}^2}{2\sin^2\theta}\right)\mathrm{d}\theta \tag{9}$$

where $\Omega_{\text{s,r}}$ is Rayleigh fading channel coefficient with probability distribution function (PDF)

$$f_{\text{A}_{\text{s,r}}}(\Omega_{\text{s,r}}) = \frac{2\Omega_{\text{s,r}}}{\xi_{\text{s,r}}^2}\exp\left(\frac{\Omega_{\text{s,r}}^2}{\xi_{\text{s,r}}^2}\right) \tag{10}$$

Therefore, the average probability of error at the relay is evaluated as

$$\overline{P_\text{e}}(\Psi) = \frac{1}{\pi}\int_0^{\infty}\int_0^{\frac{\pi}{2}} \exp\left(-\frac{\rho_\text{S}\Omega_{\text{s,r}}^2}{2\sin^2\theta}\right) f_{\text{A}_{\text{s,r}}}(\Omega_{\text{s,r}})\,\mathrm{d}\theta\,\mathrm{d}\Omega_{\text{s,r}} = \frac{1}{\pi}\int_0^{\frac{\pi}{2}} \left(\frac{1}{1+\frac{\rho_\text{S}\xi_{\text{s,r}}^2}{2\sin^2\theta}}\right)\mathrm{d}\theta \tag{11}$$

The probability of error $\boldsymbol{e}$ at destination, knowing Ψ is

$$P_\text{e}(e|\Psi) = Q\left(\sqrt{\frac{P_\text{S}|g_{\text{s,d}}|^2}{\sigma^2}}\right) = \frac{1}{\pi}\int_0^{\frac{\pi}{2}} \exp\left(-\frac{\rho_\text{S}\Omega_{\text{s,d}}^2}{2\sin^2\theta}\right)\mathrm{d}\theta \tag{12}$$

where $\Omega_{s,d}$ is Rayleigh fading channel coefficient with PDF

$$f_{A_{s,d}}\left(\Omega_{s,d}\right) = \frac{2\Omega_{s,d}}{\xi_{s,d}^2}\exp\left(\frac{\Omega_{s,d}^2}{\xi_{s,d}^2}\right) \tag{13}$$

Therefore, the average probability of error at the destination is evaluated as

$$\overline{P_e}(e|\Psi) = \int_0^{\infty} P_e(e|\Psi) f_{A_{s,d}}\left(\Omega_{s,d}\right) d\Omega_{s,d} = \frac{1}{\pi}\int_0^{\frac{\pi}{2}} \left(\frac{1}{1+\frac{\rho_S \xi_{s,d}^2}{2\sin^2\theta}}\right) d\theta \tag{14}$$

Ψ is the event of error at the relay. Then, $\overline{\Psi}$ is the event of correct decoding. Hence in this event, the relay retransmits the message to the destination and destination employs MRC of source and relay signals. The SNR at destination, given $\overline{\Psi}$ is

$$\mathrm{SNR_d} = \frac{P_S\left|g_{s,d}\right|^2 + P_R\left|g_{r,d}\right|^2}{\sigma^2} = \rho_S\left|g_{s,d}\right|^2 + \rho_R\left|g_{r,d}\right|^2 \tag{15}$$

Probability of error $\boldsymbol{e}$ at destination, given $\overline{\Psi}$ $(P_e(e|\overline{\Psi}))$, and corresponding average probability of error $(\overline{P_e}(e|\overline{\Psi}))$ are

$$P_e(e|\overline{\Psi}) = Q\left(\sqrt{\frac{P_S\left|g_{s,d}\right|^2 + P_R\left|g_{r,d}\right|^2}{\sigma^2}}\right) = \frac{1}{\pi}\int_0^{\frac{\pi}{2}} \exp\left(-\frac{\rho_S\Omega_{s,d}^2 + \rho_R\Omega_{r,d}^2}{2\sin^2\theta}\right) d\theta \tag{16}$$

$$\overline{P_e}(e|\overline{\Psi}) = \int_0^{\infty}\int_0^{\infty} P_e(e|\overline{\Psi}) f_{A_{s,d},A_{r,d}}\left(\Omega_{s,d},\Omega_{r,d}\right) d\Omega_{s,d} d\Omega_{r,d} \tag{17}$$

Since Rayleigh channel fading coefficients are assumed independent, the joint PDF can be written as product of individual PDFs.

$$f_{A_{s,d},A_{r,d}}\left(\Omega_{s,d},\Omega_{r,d}\right) = f_{,A_{s,d}}\left(\Omega_{s,d}\right) \times f_{,A_{r,d}}\left(\Omega_{r,d}\right) \tag{18}$$

Substituting and solving

$$\overline{P_e}(e|\overline{\Psi}) = \frac{1}{\pi}\int_0^{\frac{\pi}{2}} \left(\frac{1}{1+\frac{\rho_S\xi_{s,d}^2}{2\sin^2\theta}} \times \frac{1}{1+\frac{\rho_R\xi_{r,d}^2}{2\sin^2\theta}}\right) d\theta \tag{19}$$

In the above equation, assuming large value for ρ, an approximate solution can be obtained. The average end-to-end error probability of error $\boldsymbol{e}$ at destination is

$$P_e(e) = P_e(e|\Psi) \times P_e(\Psi) + P_e(e|\overline{\Psi}) \times P_e(\overline{\Psi}) \quad (20)$$

$$E\{P_e(e)\} = \overline{P_e}(e|\Psi) \times \overline{P_e}(\Psi) + \overline{P_e}(e|\overline{\Psi}) \times (1 - \overline{P_e}(\Psi))$$

Combining Eqs. (11), (14), and (19), average end-to-end error probability of error $\boldsymbol{e}$ at destination can be evaluated. Assuming high SNR, asymptotically tight upper bound for end-to-end SER is obtained from this result as

$$P_e(e) = \frac{\sigma^4}{4P_S^2\xi_{s,d}^2\xi_{s,r}^2} + \frac{3\sigma^4}{4P_S P_R \xi_{s,d}^2\xi_{r,d}^2} \quad (21)$$

Let total power be constrained as $P_S + P_R = P$ and let $P_S = \gamma P$, then $P_R = (1-\gamma)\,P$, where γ is the cost function for fractional power allocation.

$$P_e(e) = \frac{1}{\mathrm{SNR}^2}\left(\frac{1}{4\gamma^2\xi_{s,d}^2\xi_{s,r}^2} + \frac{3}{4\gamma(1-\gamma)\xi_{s,d}^2\xi_{r,d}^2}\right) \quad (22)$$

Diversity order can be seen to be 2. Thus, cooperative diversity improves end-to-end BER of the system.

2.3 *Optimal Power Allocation*

We need to find optimal γ which minimizes $P_e(e)$. The objective function is formulated as

$$P1.\min.P_e(e) \equiv \min.\left(\frac{1}{\gamma^2\xi_{s,r}^2} + \frac{3}{\gamma(1-\gamma)\xi_{r,d}^2}\right) \quad (23)$$

$$\mathrm{C1.\,s.t}\,P_S + P_R = P$$

This is a quadratic problem and can be solved through convex optimization [16] to obtain the optimal cost function.

Applying Karush–Kuhn–Tucker (KKT) conditions to find optimal γ, viable root for this quadratic equation is

$$\gamma = \left(\xi_{s,r} + \sqrt{\xi_{s,r}^2 + \frac{8}{3\xi_{r,d}^2}}\right) \Big/ \left(3\xi_{s,r} + \sqrt{\xi_{s,r}^2 + \frac{8}{3\xi_{r,d}^2}}\right) \quad (24)$$

Therefore, the optimal power allocation is

$$P_S = \gamma P \text{ and } P_R = (1 - \gamma)P \tag{25}$$

If all average gains are equal to unity, then optimal γ yields 0.6 and we get $P_S = 0.6P$ and $P_R = 0.4P$.

3 Results and Discussion

Simulation results are now presented to demonstrate the performance of the proposed source power optimization in wireless SDF relay network. Dual-hop half-duplex communication is assumed for the cooperative relay channel. Source, relay, and destination are having single antenna, and the system has perfect CSI at the receiver. Channel model is frequency-selective, *iid* Rayleigh fading channel with channel gains taken as unity. Modulation scheme is BPSK with 10,000 symbols per block. Monte Carlo simulation is executed in Matlab with 1000 iterations.

Figure 2 shows error rate performance of cooperative communication system against full transmit power, without optimized power allocation. The BER curve for cooperative links follow the same slope as that of direct link, for lower transmit powers, but drops faster at high SNR. For 20 dB, the BER for direct link is 3.7×10^{-3}, and using cooperative relay communication, the value comes down to 3.114×10^{-5}. Use of dual-hop relay system with single antenna at all three nodes, brings in a huge reduction of BER of the order of 10^{-2} in cooperative link. This is due to the virtual MIMO effect in which the destination receives multiple copies of the same message and hence attaining the diversity gain. But in a worst-case scenario in which the relay link breaks, there is a graceful degradation in performance with an approximate roll off from BER of 0.213 at 0 dB to 1.3×10^{-3} at 25 dB.

Our objective is to optimize the average BER performance through optimal power allocation to the source. Solving the objective functions manually as in Eq. (23), it is obtained that, for an optimal power of 60% allotted to the source, the cooperative system is giving minimum BER. Convex optimization technique is used to simulate the optimization problem and to get the optimal value.

Table 1 contains simulated values of the power allocation cost function and corresponding values of BER for direct and cooperative links. Values for 50 and 60% power allocation to source are presented. Minimum BER is obtained for 60% power allocation to source. Hence the simulation results are in agreement with theoretical calculations.

Figure 3 is graphical representation of this data. It is observed that the rate of decline is more steep as we increase total transmit power from 0 to 25 dB. This is because the SNR improves as power increases and BER is nonlinearly related to SNR, as given by Eqs. (12) and (16).

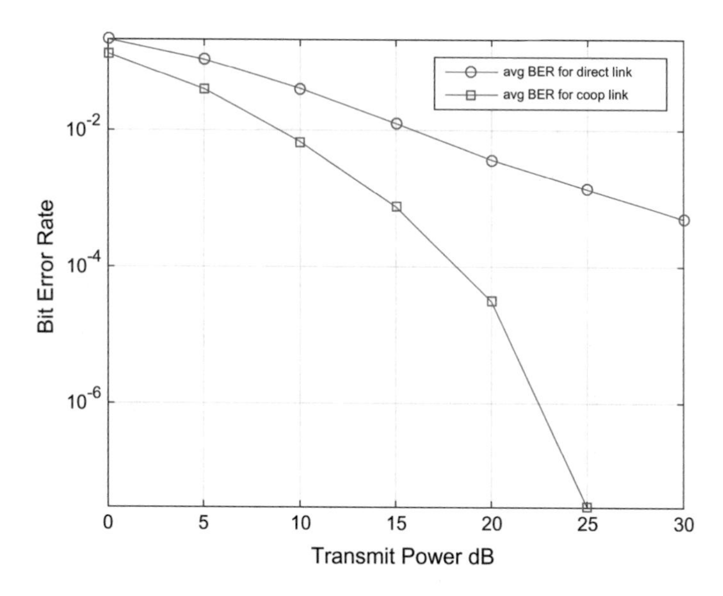

Fig. 2 BER versus transmit power for a wireless cooperative SDF relay channel, without power optimization

Table 1 Transmit power (P) versus BER for direct and cooperative links with optimal power allocation

P (dB)	BER for direct link (S-D)		BER for cooperative link (S-R-D)	
	$\gamma = 0.5$	$\gamma = 0.6$ (optimal)	$\gamma = 0.5$	$\gamma = 0.6$ (optimal)
0	0.2776	0.2632	0.2038	0.2068
5	0.1736	0.1516	0.0921	0.085
10	0.0814	0.0653	0.0217	0.0209
15	0.0277	0.0233	0.0032	0.0028
20	0.0090	0.0078	1.822×10^{-4}	2.745×10^{-4}
25	0.0030	0.0021	9.5×10^{-6}	2.4×10^{-6}
30	0.0012	5.36×10^{-4}	9×10^{-7}	0

Figure 4 displays BER variation with respect to power allocation factor γ for a constant transmit power of 0 dB. The simulation results confirm the correctness of analytical expressions. The objective function formulated in Eq. (23) is convex in nature, which is proven by the BER curve of cooperative relay link, with the optimal (minimal) error rate of 3.99×10^{-4} achieved at $\gamma = 0.6$

Figures 5 and 6 indicates BER for direct and cooperative communication for different transmit powers ranging from 0 to 25 dB. As total transmit power increases, the BER for direct as well as cooperative links improve, which is very intuitive. But the rate of variation is different for high transmit power. This is due to the nonlinear relation between SNR and average probability of error. For an optimal power allocation of 60%, BER of direct and cooperative links are 6.7×10^{-3} and 3.2×10^{-4} for 20 dB transmit power. As total transmit power increase to 25 dB, the BER values become 2.7×10^{-3} and 1×10^{-5}.

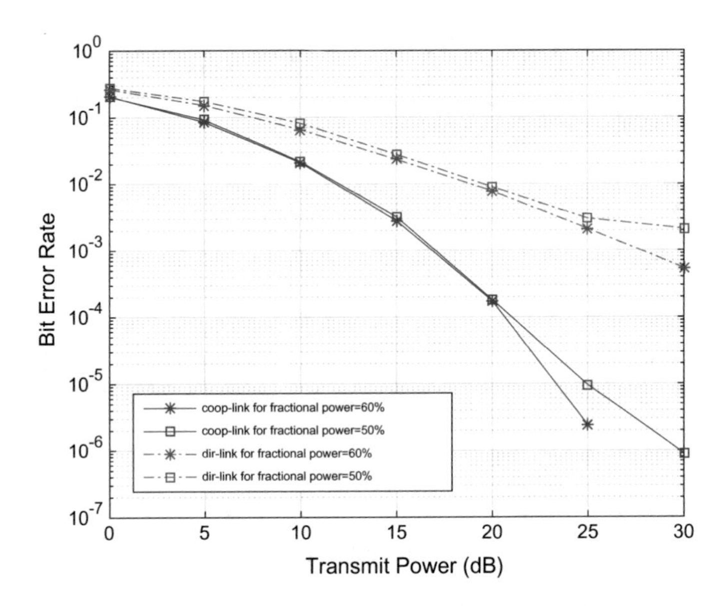

Fig. 3 BER versus transmit power for a wireless cooperative SDF relay channel, with power optimization in cooperative link for fractional powers 50 and 60% (optimal)

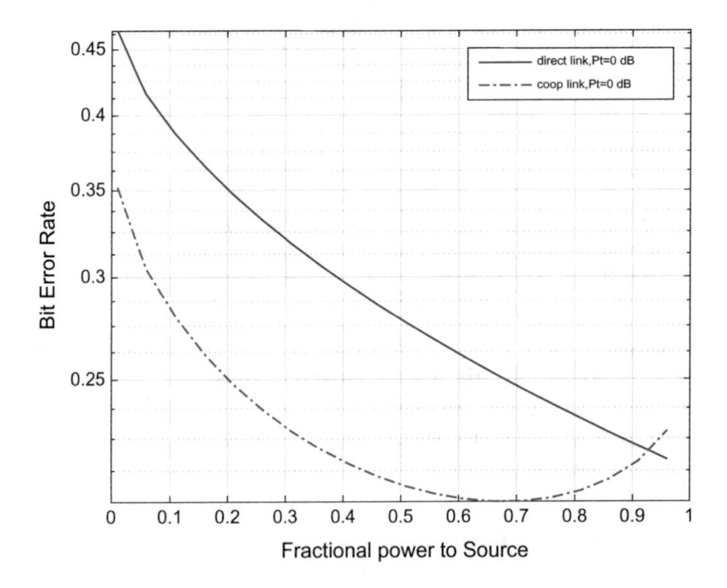

Fig. 4 Comparison of direct and cooperative SDF relay link for power allocation factor ranging from 1 to 99% for transmit power = 1 W

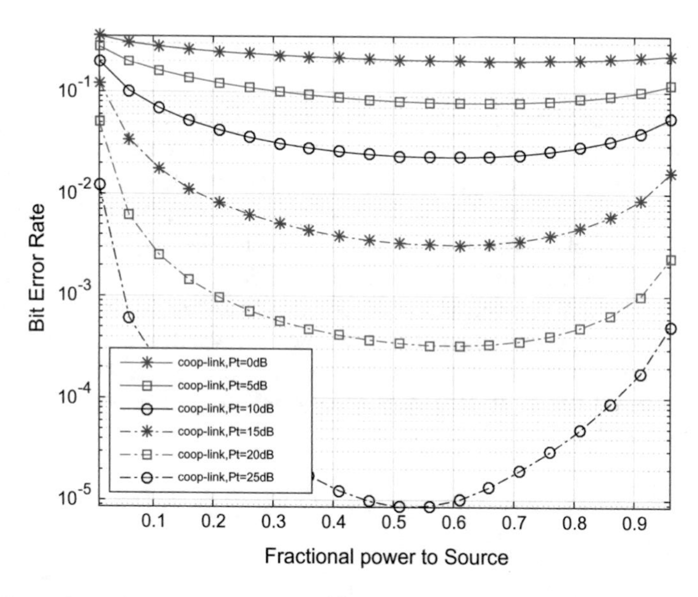

Fig. 5 Comparison of error performance for SDF cooperative communication link (S-R-D) with optimal power allocation for different transmit powers

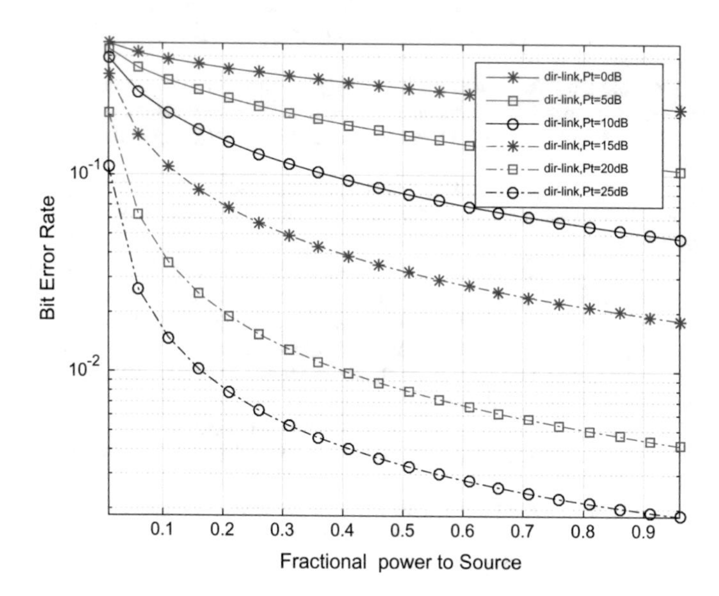

Fig. 6 Comparison of error performance for direct communication link (S-D) with optimal power allocation for different transmit powers

4 Conclusion

This paper discussed average end-to-end error performance of cooperative relay using selective decode-and-forward scheme with single-antenna nodes under optimized source power allocation. System simulations indicate improved system performance in terms of BER, with diversity gain due to virtual MIMO effect. The formulated power allocation optimization problem for the relay system is observed to be convex in nature with a constrained power. Closed-form analytical expression is derived for optimal power allocation. Monte Carlo simulation of the proposed solution is carried out for different source powers to validate the analytical result. The simulation results are in perfect agreement with the analytical solution and indicate improvement in BER performance at the optimal power allocation cost function at $\gamma = 0.6$.

References

1. Zhang, C., Yin, H., Wang, W., Wei, G.: Selective partial decode-and-forward schemes for distributed space-time coded relaying networks. In: 69th Vehicular Technology Conference, pp. 1–5. IEEE, Barcelona, Spain (2009). https://doi.org/10.1109/vetecs.2009.5073844
2. Moço, A., Teodoro, S., Silva, A., Gameiro, A.: Performance evaluation of virtual MIMO schemes for the UL OFDMA based systems. In: 4th International Conference on Wireless and Mobile Communications, pp. 71–76. IEEE, Athens, Greece (2008). https://doi.org/10.1109/icwmc.2008.61
3. Nahas, M., Saadani, A., Hachem, W.: Outage probability and power optimization for asynchronous cooperative networks. In: 17th International Conference on Telecommunications, pp. 153–159. IEEE, Doha, Qatar (2010). https://doi.org/10.1109/ictel.2010.5478646
4. Zhang, J., Zhang, T.Y., Huang, J.X., Yuan, R.P.: Selective decode-and-forward cooperation over Nakagami-m fading channels. Electron. Lett. **45**(15), 12–13 (2009). https://doi.org/10.1049/el.2009.1543
5. Atay, F., Fan, Y., Yanikomeroglu, H., Poor, H.V.: Threshold-based relay selection for detect-and-forward relaying in cooperative wireless networks. EURASIP J. Wireless Commun. Networking, 1–9 (2010). https://doi.org/10.1155/2010/721492
6. Li, Y., Wang, W., Kong, J., Peng, M.: Subcarrier pairing for amplify-and-forward and decode-and-forward OFDM relay links. IEEE Commun. Lett. **13**(4), 209–211 (2009). https://doi.org/10.1109/LCOMM.2009.080864
7. Kim, S.-I., Kim, S., Kim, J.T., Heo, J.: Opportunistic decode-and-forward relaying with interferences at relays. Wireless Personal Commun. **68**(2), 247–264 (2013). https://doi.org/10.1007/s11277-011-0449-6
8. Farhat, J., Brante, G., Souza, R.D., Rebelatto, J.L.: Secure energy efficiency of selective decode and forward with distributed power allocation. In: International Symposium on Wireless Communication Systems, pp. 70–75. IEEE, Brussels, Belgium (2015). https://doi.org/10.1109/iswcs.2015.7454439
9. Varshney, N., Krishna, A.V., Jagannatham, A.K.: Selective DF protocol for MIMO STBC based single/multiple relay cooperative communication: end-to-end performance and optimal power allocation. IEEE Trans. Commun. **63**(7), 2458–2473 (2015). https://doi.org/10.1109/tcomm.2015.2436912

10. Varshney, N., Goel, A., Jagannatham, A.K.: Cooperative communication in spatially modulated MIMO systems. In: Wireless Conference and Networking Conference, pp. 1–6. IEEE, Doha, Qatar (2016). https://doi.org/10.1109/wcnc.2016.7564938
11. Mohamad, A., Visoz, R., Berthety, A.O.: Dynamic selective decode and forward in wireless relay networks. In: 7th International Congress on Ultra-Modern Telecommunications and Control Systems and Workshops, pp. 189–95. IEEE, Brne, Czech Republic (2015). https://doi.org/10.1109/icumt.2015.7382426
12. Grira, L., Bouallegue, R.: Energy consumption analysis of SDF and non-cooperative schemes over Nakagami-m channel and under outage probability constraint. In: 13th International Wireless Communications and Mobile Computing Conference, pp. 1834–39. IEEE, Valencia, Spain (2017). https://doi.org/10.1109/iwcmc.2017.7986563
13. Yang, W., Yang, W., Cai, Y.: Outage performance of OFDM-based selective decode-and-forward cooperative networks over Nakagami-m fading channels. Wirel. Personal Commun. **56**(3), 503–515 (2011). https://doi.org/10.1007/s11277-010-9986-7
14. Zhou, G., Wang, T., Wu, Y., Zheng, G., Yang, G.: Energy-efficient power allocation for decode-and-forward OFDM relay links. Mob. Wireless Technol. J. 13–24 (2016). https://doi.org/10.1007/978-981-10-1409-3_2 (Springer)
15. Tellambura, C., Annamalai, A.: Derivation of Craig's formula for Gaussian probability function. Electron. Lett. **35**(17), 1424–1425 (1999). https://doi.org/10.1049/el:19990949
16. Boyd, S., Vandenberghe, L.: Convex Optimization. Cambridge University Press, Cambridge (2004)

Design of Flip-Flops Using Reversible DR Gate

Anurag Govind Rao and Anil Kumar Dhar Dwivedi

Abstract Nowadays, due to transistors miniaturization and advancement in CMOS VLSI Design Technology, Engineers and Scientist are adding more and more functionality to a device by increasing component density in a very small chip area. Functions are executed at a very high-speed transistor switching, resulting in increase of power dissipation and consumption. To overcome such problems, new technologies have to be developed and Reversible Logic Technology (RLT) is one such new paradigm, which has ideally zero power dissipation. In this paper, reversible logic-based flip-flops design methodology has been developed to remove earlier problems such as the use of various types of Reversible Gates for different logical operations, generation of complimented and un-complimented flip-flop output, and fanout. In this paper, Reversible Logic-based SR Latch, Clocked SR, D, T, and JK Flip-Flop (FF) have been designed using only 3 $\times$ 3 Dwivedi-Rao Gate 4 (DRG4) as AND, NAND, and conventional NOT gate, which is the only gate from a conventional family qualifying for the Reversible Logic. The novelty of the proposed design methodology is that, first, design of SR latch is done and then clocked SR flip-flop was implemented on this, followed by D, T, and JK flip-flop and that too using DRG4 and NOT gate only. The results obtained are verified using VHDL simulation and mathematical analysis, which proves the proposed design methodology.

Keywords Reversible logic · Quantum computing · Flip-flops
DRG4

A. G. Rao (✉) · A. K. D. Dwivedi
National Institute of Electronics & Information Technology, Gorakhpur, India
e-mail: anurag_govind@yahoo.co.in

A. K. D. Dwivedi
e-mail: Indiadwivedi_anil1@yahoo.com

© Springer Nature Singapore Pte Ltd. 2019

H. Malik et al. (eds.), *Applications of Artificial Intelligence Techniques in Engineering*, Advances in Intelligent Systems and Computing 697,
https://doi.org/10.1007/978-981-13-1822-1_21

1 Introduction

According to Moore's law [1, 2], numbers of transistor integrated on a VLSI chip will double every 18 months, resulting in periodic increase in computing power due to transistor miniaturization. This is also increasing component density. Due to increase in component density and reduction in transistor size, power dissipation and consumption has also been increased [3–5] and thus challenging the further growth and development of Moore's law using conventional standard Silicon technologies like CMOS [3–5]. Thus, designing of chips, i.e., Computer in 3D (three-dimensional), Molecular Computers, and Quantum Computers in nanotechnology domain are the biggest challenge to engineers and scientist throughout the world due to power dissipation. Reversible Logic Technology (RLT) is one such paradigm for design where power dissipation is ideally zero and thus is best solution for design of chips for computer, Molecular, and Quantum computers in nanotechnology era. In this paper, Reversible Logic-based SR latch, clocked SR, D, T, and JK flip-flop has been designed using 3-input 3-output (3×3) DRG4 and inverter which is the only conventional gate which qualifies for Reversible logic operations.

2 Reversible Logic Technology

Moore's law [1, 2], stated that numbers of transistor integrated on a VLSI chip will double every 18 months resulting in increase in power dissipation and consumption [3–5]. This problem is challenging the further growth and development of Moore's law using conventional CMOS VLSI Design Technology [3–5]. Thus, designing of chips, i.e., Computer in 3D (three-dimensional), Molecular Computers, and Quantum Computers in nanotechnology domain are the biggest challenge to engineers and scientist throughout the world due to power dissipation.

RLT is one such paradigm for design where power dissipation is ideally zero and thus is the best solution for design of Chips for Computer, Molecular, and Quantum computers in nanotechnology era. This paper section includes Introduction to RLT, Reversible Logic Gate Design and Constraints, Description of earlier proposed Dwivedi-Rao Gate (DRG), Dwivedi-Rao Gate (DRG1), Dwivedi-Rao Gate (DRG2), Dwivedi-Rao Gate (DRG3), and Dwivedi-Rao Gate (DRG4), Design of SR Latch, Clocked SR Latch, Clcoked D, Clcoked T, and Clocked JK Flip-Flop, and followed by Results Discussion and Conclusion.

3 Reversible Logic Gate Design and Constraints

Conventional Logical operation erases bits during operation, for example, AND operation, i.e., 2 input and 1 output, resulting in erasing of 1 bit. According to Landauer [6] this is the main reason of power dissipation $P_{\mathrm{Dissipation}}$ and is given by

$$P_{\mathrm{Dissipation}} = nKT \ln 2 \tag{1}$$

where "n" is the No. of bits erased, "K" is the Boltzmann's constant $(1.3807 \times 10^{-23}\,\mathrm{JK}^{-1})$ and "T" is the operating temperature (300 K). For single-bit erase, $KT \ln 2$ approximately 2.8×10^{-23} J. However, it is a small magnitude, but it cannot be ignored considering component density and transistor miniaturization. According to Bennett [7], power dissipation produced can ideally be reduced to zero if Reversible Logic Gates. Circuit designed using Reversible gates is called as Reversible Circuits. Applications of Reversible Logic are in Nanotechnology, Quantum Computing, and Cryptography [8–14]. For example, Reversible Fredkin gate [9] is shown in Fig. 1 and its truth table in Table 1.

4 DRG, DRG1, DRG2, DRG3, and DRG4

Series of Reversible logic-based DR gates has been designed, i.e., 6 × 6 DR [11] gate for ALU design, N × N DRG1 [12] for Binary-to-Gray code conversion, N × N DRG2 [12] for Gray-to-Binary code conversion, 5 × 5 DRG3 [13] for

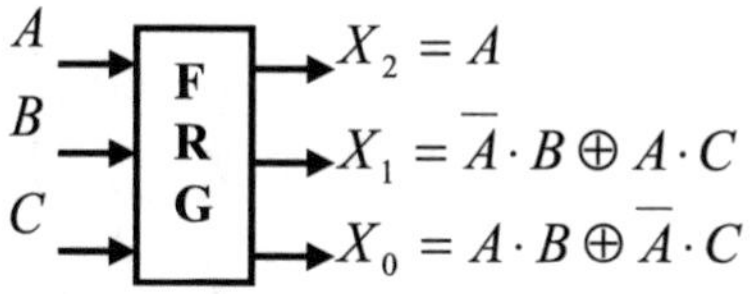

A, B, C → FRG → $X_2 = A$; $X_1 = \overline{A} \cdot B \oplus A \cdot C$; $X_0 = A \cdot B \oplus \overline{A} \cdot C$

Fig. 1 Fredkin gate (FRG)

Table 1 Fredkin gate truth table

Input			Output		
A	*B*	*C*	X_2	X_1	X_0
0	0	0	0	0	0
0	0	1	0	0	1
0	1	0	0	1	0
0	1	1	0	1	1
1	0	0	1	0	0
1	0	1	1	1	0
1	1	0	1	0	1
1	1	1	1	1	1

Fig. 2 DRG4 gate

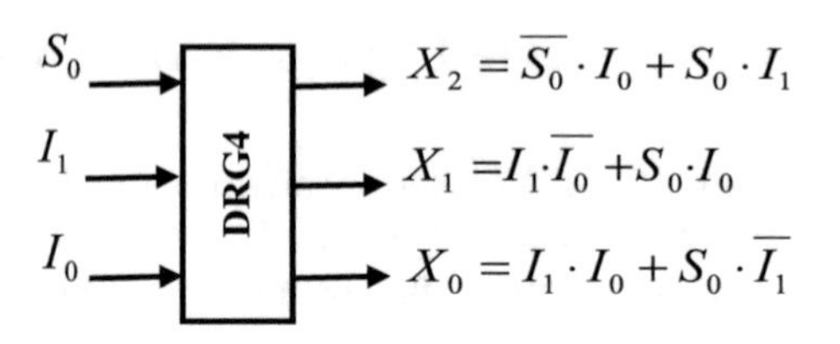

multifunctional operations like BCD to EX-3, Ex-3 to BCD, Full Adder, and Full Subtraction, and 3 × 3 DRG4 [13] gate which performs all conventional logical operations and 2:1 MUX. Based on DRG4 and DRG2, Exclusive Sum of Product-Reversible Programmable Logic Array (ESOP-RPLA) is proposed in [14]. Due to universal property of DRG4 gate has been selected for implementation of

Table 2 DRG4 gate truth table

Input			Output		
S_0	I_1	I_0	X_2	X_1	X_0
0	0	0	0	0	0
0	0	1	1	0	0
0	1	0	0	1	0
0	1	1	1	0	1
1	0	0	0	0	1
1	0	1	0	1	1
1	1	0	1	1	0
1	1	1	1	1	1

Fig. 3 DRG4 gate as AND operation

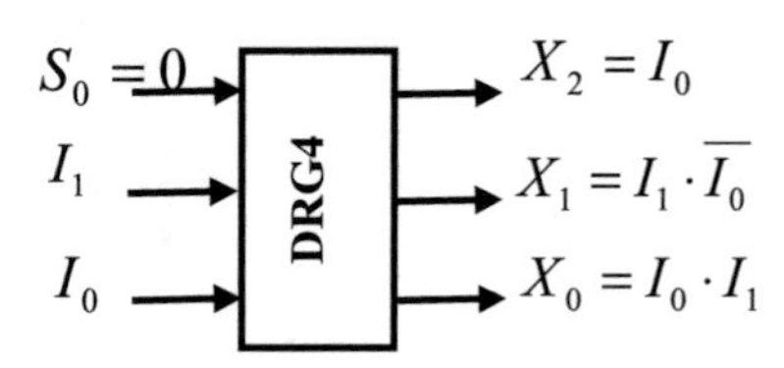

Fig. 4 DRG4 gate as NAND operation

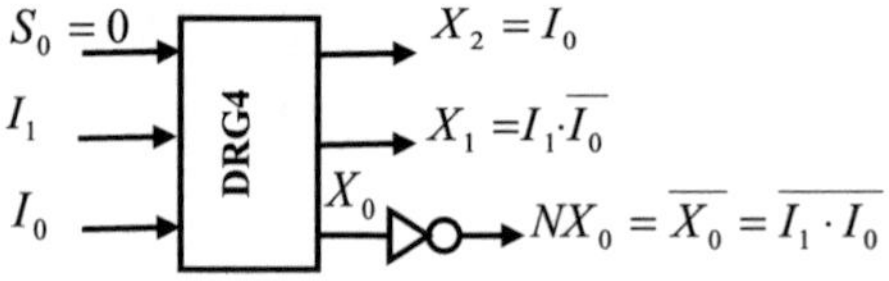

Fig. 5 DRG4 gate as Copier operation

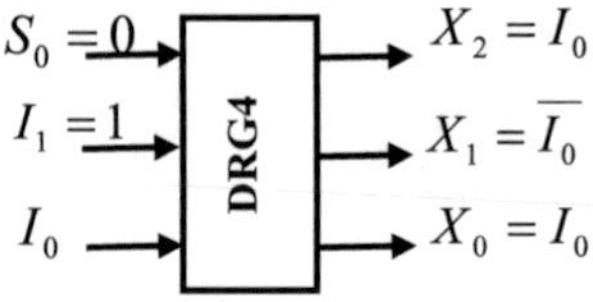

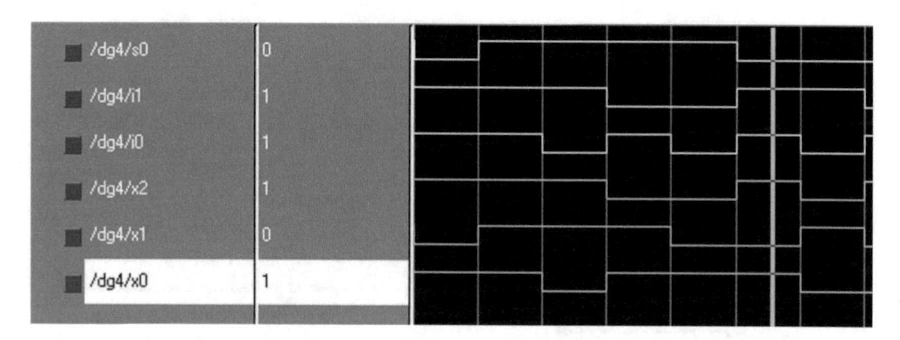

Fig. 6 VHDL simulation of DRG4

flip-flops. Figure 2 shows DRG4 gates and its truth table is shown in Table 2. DRG4 gate as AND, NAND and Copier are shown in Figs. 3, 4, and 5, respectively. VHDL simulation is shown Fig. 6.

5 SR Latch Design Using DRG4

For designing of any type of flip–flop, SR latch is the gateway. Figure 7 shows the SR Latch using NAND gate and its truth table is shown in Table 3. Its implementation using DRG4 is shown in Fig. 8 and VHDL simulation is shown in Fig. 9. Table 4 shows the evaluation of the Reversible logic-based SR Latch in terms of Total Numbers of Gates, Quantum Cost, Total Clock Cycle, and Total Logical Operations. This verifies the operations of SR latch using DRG4 gates only.

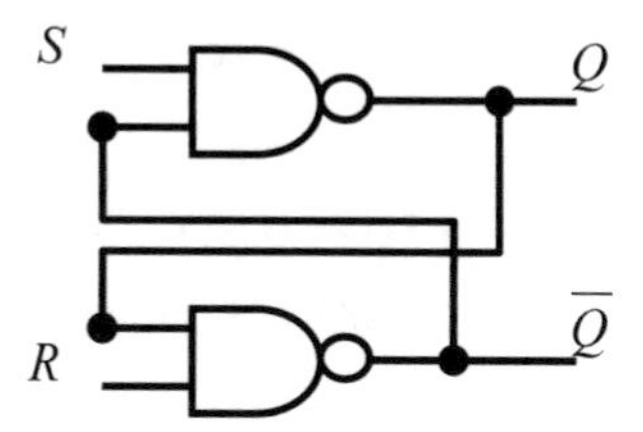

Fig. 7 SR latch using NAND gate

Table 3 Truth table of SR latch

S	R	Q	$\overline{Q}$
0	0	Not used	
0	1	1	0
1	0	0	1
1	1	Memory	

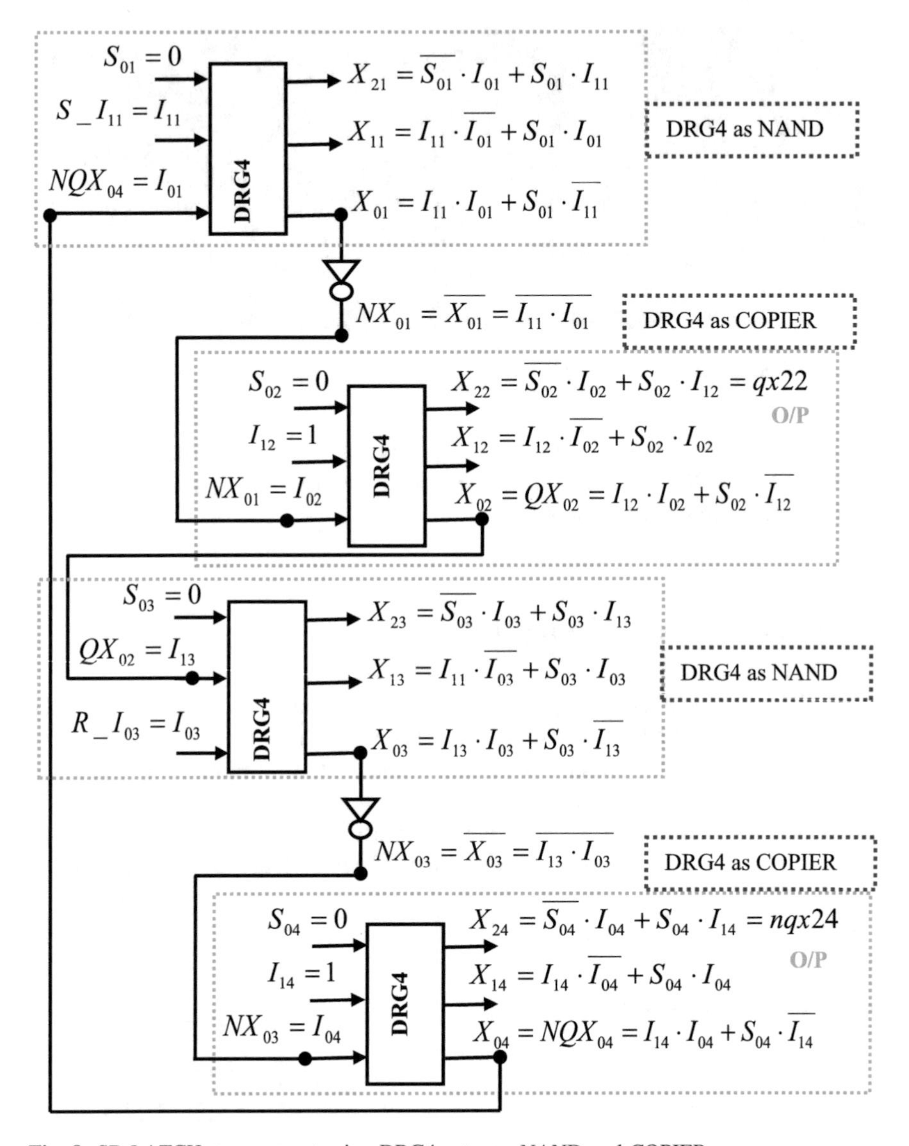

Fig. 8 SR LATCH arrangement using DRG4 gates as NAND and COPIER

6 Clocked SR Flip-Flop Design Using DRG4

Figure 10 shows clocked SR flip-flop using NAND gates and its equivalent using DRG4 is shown in Fig. 11. Operation of clocked SR flip-flop is shown in Table 5. VHDL simulation is shown in Fig. 12. Evaluation Table is shown in Table 6. This verifies the operations of clocked SR flip-flop using DRG4 gates only.

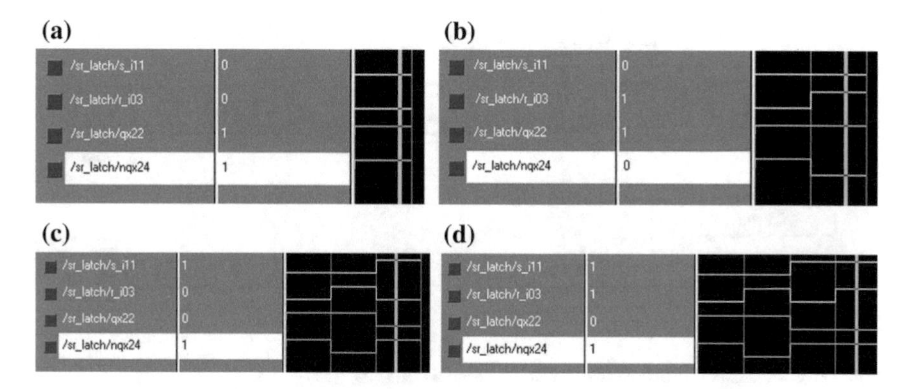

Fig. 9 **a** VHDL simulation of SR latch when $S(i11) = R(i03) = 0$, (not allowed). **b** VHDL simulation of SR latch when $S(i11) = 0, R(i03) = 1$, Set Operation. **c** VHDL simulation of SR latch when $S(i11) = 1, R(i03) = 0$, Reset Operation. **d** VHDL simulation of SR latch when $S(i11) = R(i03) = 1$, Memory Operation

Table 4 Evaluation table of SR latch using DRG4 gate

Parameters	Remarks	Total
Total no. of gates	06 DRG4 gate 04 NOT gate	10 gates
Quantum cost	$6 \times 42 = 252$ for DRG4 $4 \times 1 = 2$ for NOT	256
Total clock cycle	$10q$	$10q$
Total logical operations	$6 \times (6x + 3y) + 4 \times (3z)$ for DRG4	$36x + 18y + 12z$

Where q = Clock Cycle − Unit
x = AND − 2 Input Calculation
y = Inverter Calculation
z = OR − 2 Input Calculation

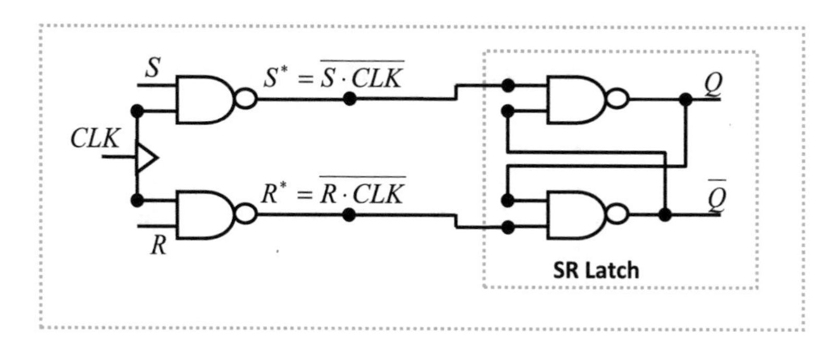

Fig. 10 Clocked SR flip-flop

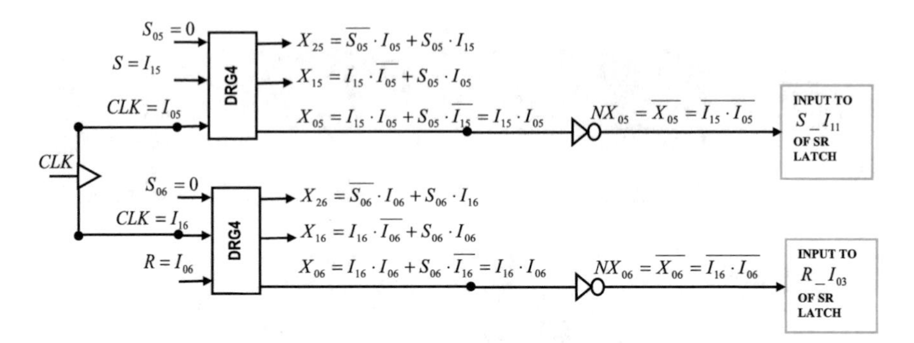

Fig. 11 Clocked SR flip-flop using DRG4 gates as NAND

Table 5 Operation of clocked SR-FF

CLK	*S*	*R*	*Q*	$\overline{Q}$
0	×	×	Memory	
1	0	0	Memory	
1	0	1	0	1
1	1	0	1	0
1	1	1	Not allowed	

7 Clocked D Flip-Flop Design Using DRG4

Symbol of D-FF using clocked SR-FF is shown in Fig. 13 and truth table is shown in Table 7. DRG4 gates implementation of clocked D-FF is shown in Fig. 14. Table 8 shows the evaluation of the Reversible logic-based clocked D-FF. VHDL simulation is shown in Fig. 15, which verifies the operations of clocked D flip-flop using DRG4 gates only.

8 Clocked T Flip-Flop Design Using DRG4

Figure 16 shows the symbolic representation of clocked T flip-flop. Its truth table is shown in Table 9. DRG4 implementation of T flip-flop is shown in Fig. 17. Evaluation table is shown in Table 10 in terms of Total Numbers of Gates, Quantum Cost, Total Clock Cycle, Total Clock Cycle, and Total Logical Operations.

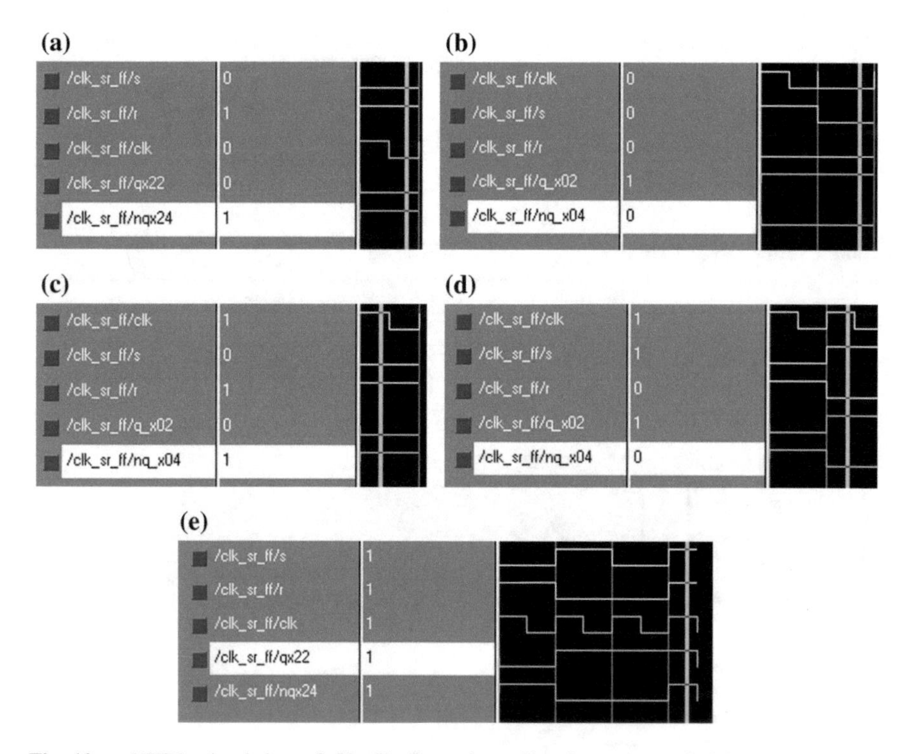

Fig. 12 **a** VHDL simulation of SR flip-flop when $clk = 0, s = 0, r = 1$, Memory operation. **b** VHDL simulation of SR flip-flop when $clk = 1, s = 0, r = 0$, Memory operation. **c** VHDL simulation of SR flip-flop when $clk = 1, s = 0, r = 1$, Reset Operation **d** VHDL simulation of SR flip-flop when $clk = 1, s = 1, r = 0$, Set Operation **e** VHDL simulation of SR flip-flop when $clk = 1, s = 1, r = 1$, Operation Not Allowed

Table 6 Evaluation table of clocked SR-FF using DRG4 gate

Parameters	Remarks	Total
Total no of gates	06 DRG4 gate 04 NOT gate	10 gates
Quantum cost	$6 \times 42 = 252$ for DRG4 $4 \times 1 = 2$ for NOT	256
Total clock cycle (q)	$10q$	$10q$
Total logical operations	$6 \times (6x + 3y) + 4\times (3z)$ for DRG4	$36x + 18y + 12z$

Where q = Clock Cycle − Unit
x = AND − 2 Input Calculation
y = Inverter Calculation
z = OR − 2 Input Calculation

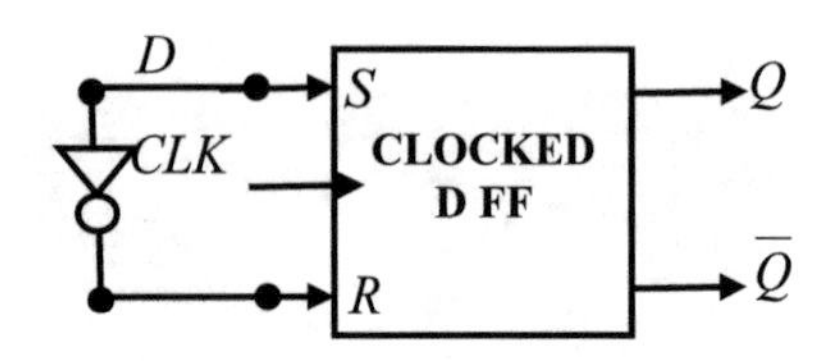

Fig. 13 Symbol of Clocked D-FF

Table 7 Truth table of D-FF

CLK	*D*	*Q*	$\overline{Q}$
0	×	Memory	
1	0	0	1
1	1	1	0

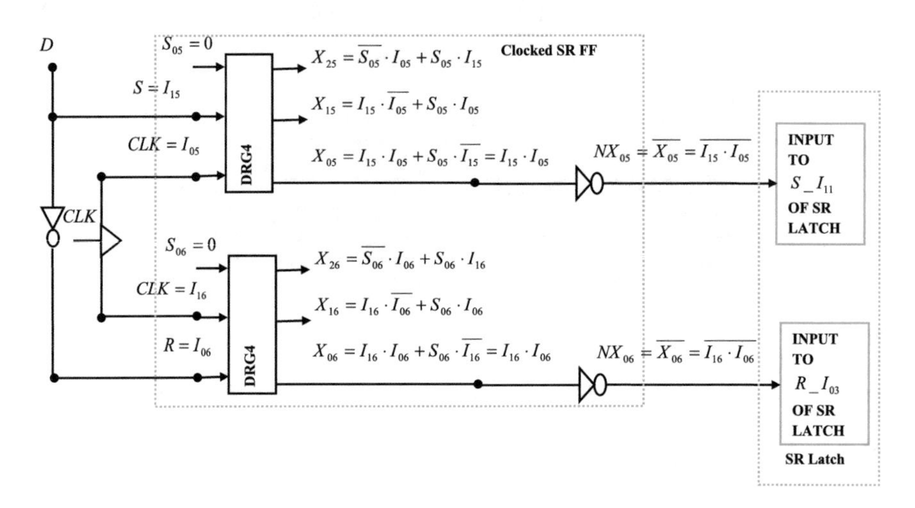

Fig. 14 Clocked D flip-flop using DRG4 gates as NAND

Table 8 Evaluation table of clocked D-FF using DRG4 gate

Parameters	Remarks	Total
Total no of gates	06 DRG4 gate 05 NOT gate	11 gates
Quantum cost	6 × 42 = 252 for DRG4 5 × 1 = 5 for NOT	257
Total clock cycle (q)	11q	11q
Total logical operations	6 × (6x + 3y) + 5 × (3z) for DRG4	36x + 18y + 15z

Where q = Clock Cycle − Unit
x = AND − 2 Input Calculation
y = Inverter Calculation
z = OR − 2 Input Calculation

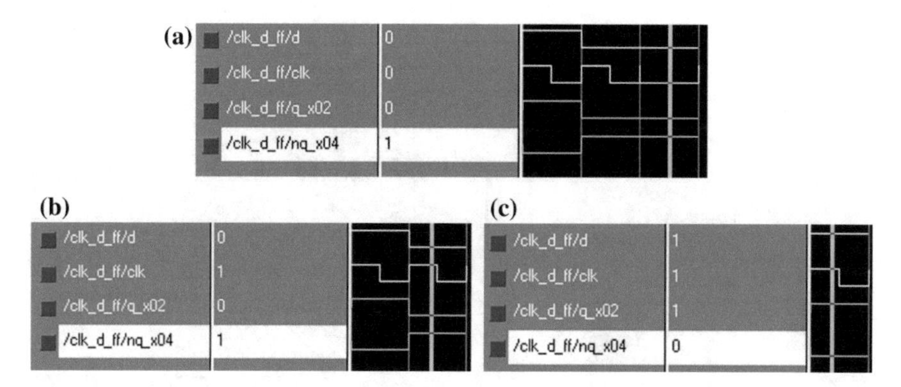

Fig. 15 **a** VHDL simulation of D-FF when $clk = 0, d = 0$, Memory Operation. **b** VHDL simulation of D-FF when $clk = 1, d = 0$. **c** VHDL simulation of D-FF when $clk = 1, d = 1$

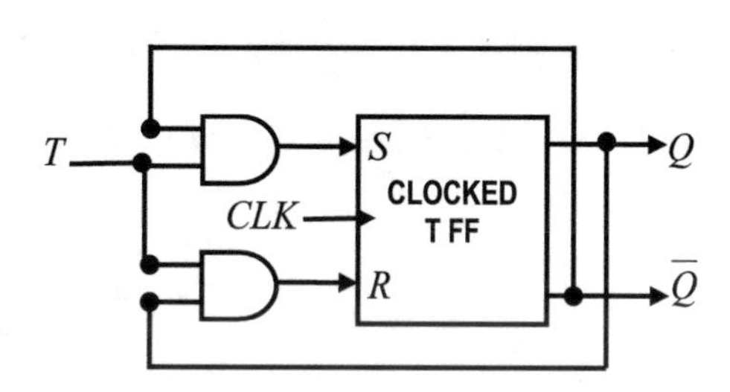

Fig. 16 Symbol of clocked T-FF

Table 9 Truth table of T-FF

CLK	*T*	*Q*	$\overline{Q}$
0	×	Memory	
1	0	Memory	
1	1	$\overline{Memory}$	

9 Clocked JK Flip-Flop Design Using DRG4

Figure 18 shows the symbolic representation of clocked JK flip-flop. Its truth table is shown in Table 11. DRG4 implementation of JK flip-flop is shown in Fig. 19. Evaluation table is shown in Table 12 in terms of Total Numbers of Gates, Quantum Cost, Total Clock Cycle, Total Clock Cycle, and Total Logical Operations.

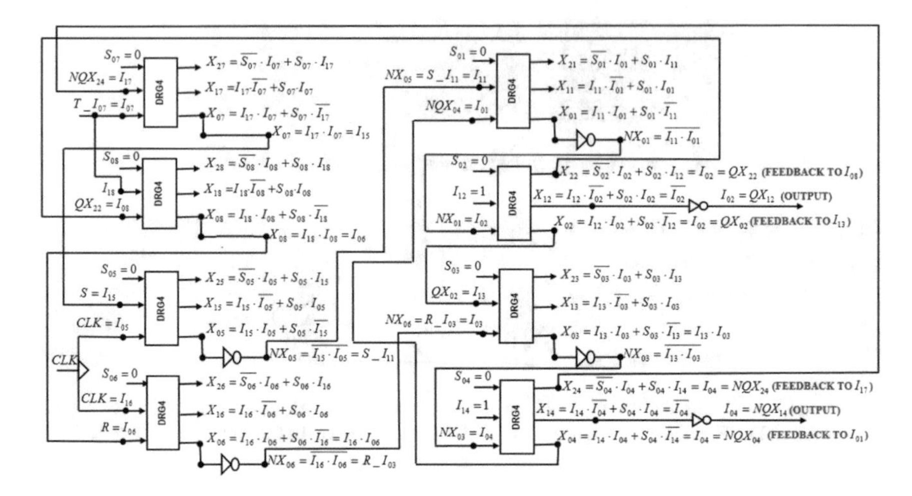

Fig. 17 Clocked T flip-flop using DRG4 gate as AND–NAND from SR clocked flip-flop

Table 10 Evaluation table of clocked T-FF using DRG4 gate

Parameters	Remarks	Total
Total no of gates	08 DRG4 gate 04 NOT gate	12 gates
Quantum cost	8 × 42 = 336 for DRG4 4 × 1 = 4 for NOT	257
Total clock cycle (q)	12q	11q
Total logical operations	8 × (6x + 3y) + 4 × (3z) for DRG4	48x + 24y + 12z

Where q = Clock Cycle − Unit
x = AND − 2 Input Calculation
y = Inverter Calculation
z = OR − 2 Input Calculation
w = NOT − 2 Input Calculation

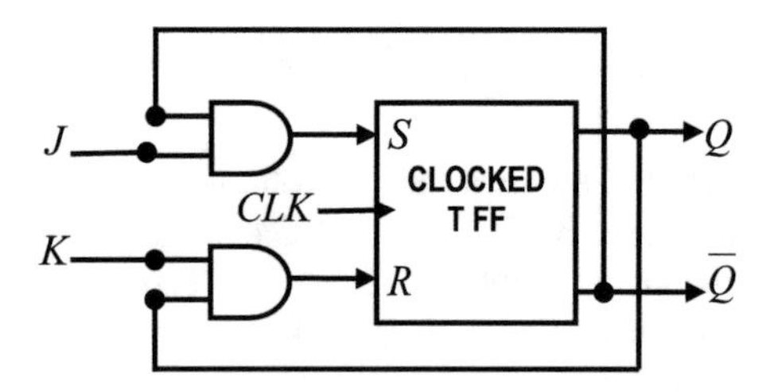

Fig. 18 Symbol of clocked JK-FF

10 Results and Discussion

During literature survey [15–18], it was found lacking of novel and systemic approach for FF design as emphasis was given more on reduction of reversible logic design constraints factors as discussed above. Thus, no clear and acceptable

Table 11 Truth table of clocked JK-FF

CLK	J	K	Q	$\overline{Q}$
0	×	×	Memory	
1	0	0	Memory	
1	0	1	0	1
1	1	0	1	0
1	1	1	Not allowed	

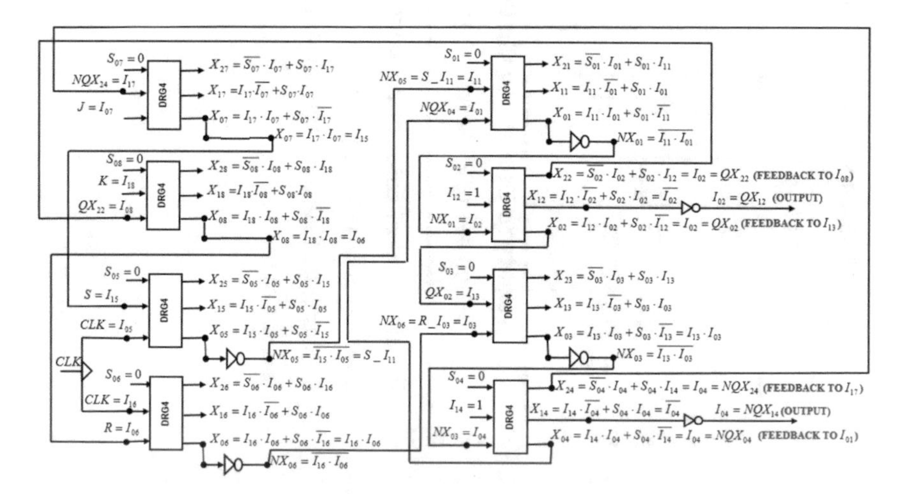

Fig. 19 Clocked JK flip-flop using DRG4 gate as AND–NAND from SR clocked flip-flop

Table 12 Evaluation table of clocked JK-FF using DRG4 gate

Parameters	Remarks	Total
Total no of gates	08 DRG4 gate 04 NOT gate	12 gates
Quantum cost	8 × 42 = 336 for DRG4 4 × 1 = 4 for NOT	257
Total clock cycle (q)	12q	11q
Total logical operations	8 × (6x + 3y) + 4 × (3z) for DRG4	48x + 24y + 12z

Where q = Clock Cycle − Unit
x = AND − 2 Input Calculation
y = Inverter Calculation,
z = OR − 2 Input Calculation
w = NOT − 2 Input Calculation

sequential circuit design methodology is developed which matches with the conventional digital circuit design. Therefore, this paper focuses more on design methodology instead of reducing the design constraints factors. Comparison has been done with existing [15–18] and is shown in Table 13. Results of Evaluation

Table 13 Comparison of design methodology comparison with [15–18]

Proposed in this paper
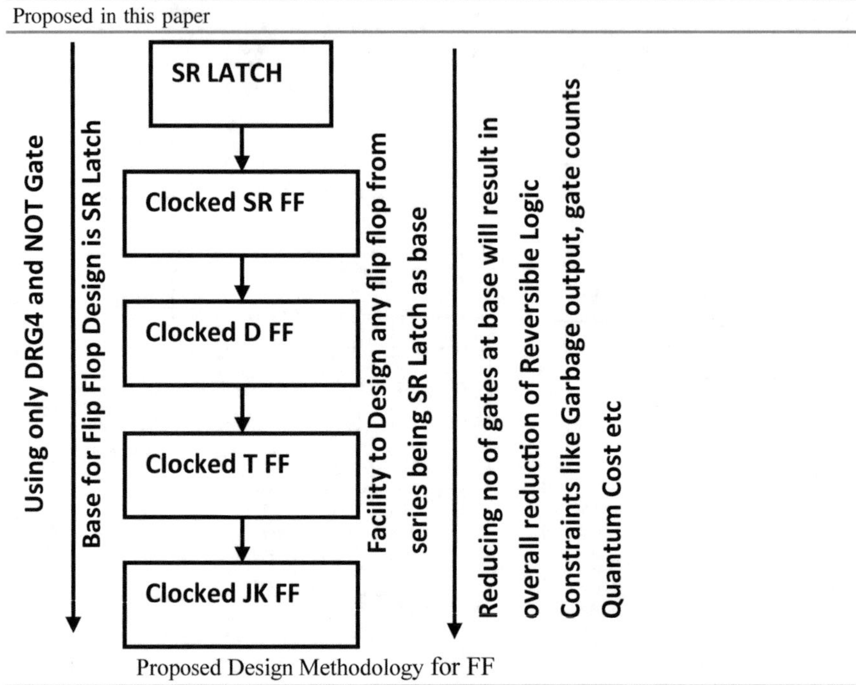 Proposed Design Methodology for FF
Proposed in [15]: design of only Master Slave T-FF using 01 Feynman and 02 SG gate
Proposed in [16]: clocked SR-FF using 04 Fredkin and 02 CNOT gate proposed using AND–NOR combination. Clocked D-FF using 04 Fredkin and 03 CNOT gate in NAND combination
Proposed in [17]: proposed 4 × 4 RM gate clocked T-FF with 01 RM gate having only Q output. Another 01 RM and 01 Feynman gate needed for generation of $\overline{Q}$
Proposed in [18]: proposed 3 × 3 AG gate and MPG. SR-FF design using 01 MPG, 01 AG and 01 F2G, JK-FF design using 02 AG and 01 F2G, D-FF design using 01 AG and 01 F2G, T-FF design using 01 AG, 01 F2G, Master Slave D-FF design using 01 AG, 01 F2G, Master Slave JK-FF design using 05 AG

Tables 4, 6, 8, 10, 12, and comparison Table 13 of SR Latch, SR-FF, D-FF, T, and JK–FF, respectively, and these flip-flop architectures designed using DRG4 gate and its VHDL simulation verifies all the results.

11 Conclusion

The main aim of this paper is the development of novel design approach for flip-flop. This paper proves that reduction in reversible design constraints can be achieved if the reduction is done at base level. On the other hand, in the future, improvement is needed in Total Numbers of Gates, Quantum Cost, Total Clock Cycle, Total Clock Cycle, and Total Logical Operations.

References

1. Moore, G.E.: Cramming more components onto integrated circuits. In: Proceedings of the IEEE, vol. 86, no. 1, pp. 82–85 (1998)
2. Moore, G.E.: Progress in digital integrated electronics. In: Proceedings of Technical Digest International Electron Devices Meeting, vol. 21, pp. 11–13 (1975)
3. Bishop, D.: Nanotechnology and the end of Moore's Law. Bell Labs Tech. J. **10**(3), 23–28 (2005)
4. Stanley Williams, R.: What next? [The end of Moore's Law]. Comput. Sci. Eng. IEEE J. Mag. **19**(2), 7–13 (2017)
5. Theis, T.N., Wong, H.S.P.: A new beginning for information technology. Comput. Sci. Eng. IEEE J. Mag. **19**(2), 41–50 (2017)
6. Laundauer, R.: Irreversibility and heat generation in the computational process. IBM J. Res. Dev. **5**(6), 183–191 (1961)
7. Bennett, C.H.: Logical reversibility of computation. IBM J. Res. Dev. **17**(6), 525–532 (1973)
8. Thapliyal, H., Ranganathan, N.: Reversible logic: fundamentals and applications in ultra-low power. In: 25th International Conference on VLSI Design Fault Testing and Emerging Nanotechnologies and Challenges in Future. IEEE Computer Society, pp 13–15 (2012)
9. Fredkin, E., Toffoli, T.: Conservative logic. Int. J. Theoret. Phys. **21**(3/4), 219–253 (1982)
10. Peres, A.: Reversible logic and quantum computers. Phys. Rev. A **32**, 3266–3276 (1985)
11. Rao, A.G., Dwivedi, A.K.D.: Design of multifunctional DR Gate and its application in ALU design. In: International Conference on Information Technology. IEEE Computer Society, pp 339–344 (2014)
12. Rao, A.G., Dwivedi, A.K.D.: New multi-functional DR Gates and its application in code conversion. In: 6th IEEE Power India International Conference. IEEE, pp. 1–5 (2014)
13. Rao, A.G., Dwivedi, A.K.D.: New versatile 3 × 3 reversible universal DRG4 gate design and its application in logical and arithmetical operations. In: IEEE Sponsored International Conference on Computing, Communication and Control Technology (IC4T) (2016). ISBN:978-93-5258-814-5
14. Rao, A.G., Dwivedi, A.K.D.: Design of ESOP-RPLA array using DRG2 and DRG4 gates based on reversible logic technology. In: IEEE International Symposium on Nano-electronics and Information System IEEE Computer Society, pp. 218–223 (2016)
15. Sahu, I., Joshi, P.: A review paper on design of an asynchronous counter using novel reversible SG Gate. In: International Conference on Innovative Mechanisms for Industry Applications. IEEE Conference Publication, pp. 617–621 (2017)
16. Rohini, H., Rajashekar, S., Kumar, P.: Design of basic sequential circuits using reversible logic. In: International Conference on Electrical, Electronics, and Optimization Techniques IEEE Conference Publication, pp. 2110–2115 (2016)
17. Singh, R., Pandey, M.K.: Design and optimization of sequential counters using a novel reversible gate. In: International Conference on Computing, Communication and Automation (ICCCA). IEEE Conference Publication, pp. 1393–1398 (2016)
18. Abir, M.A.I., Akhter, A., et al.: Design of parity preserving reversible sequential circuits. In: 5th International Conference on Informatics, Electronics and Vision. IEEE Conference Publication, pp. 1022–1027 (2016)

Utilizing CMOS Low-Noise Amplifier for Bluetooth Low Energy Applications

Malti Bansal and Jyoti

Abstract In this paper, we have surveyed the design procedure of CMOS LNA for Bluetooth Low Energy (BLE) and Bluetooth Technology (BT) applications. The design specifications of LNA such as noise figure, power dissipation, supply voltage, linearity, etc., have been analyzed. The design of LNA has been targeted for BLE applications range, i.e., 2.4–2.46 GHz. Bluetooth low energy is a new version of Bluetooth Technology brand, as well as it borrows plenty of practical applications from its source. Due to this, BLE should be contemplated as a brand new technology which addresses various design goals with distinct market segments. Generally, Bluetooth Low Energy (BLE) is a wireless personal area network (WPAN) technology which was marketed as well as designed by Bluetooth SIG for unique applications in the field of security, entertainment industries as well as health care. Finally, a comparison of different LNA topologies and their parameters for BT and BLE application range has been presented. The purpose of Bluetooth low energy is to provide minimum power consumption with low supply voltage and is cost-effective, with the same communication range, as Bluetooth.

Keywords Low-noise amplifier (LNA) · Complementary metal-oxide semiconductor (CMOS) · Bluetooth low energy (BLE) · Bluetooth technology (BT) Common gate (CG) · Common source (CS) · Noise figure (NF)

1 Introduction

In earlier times, Bluetooth Low Energy was known as Bluetooth smart technology. Basically, BLE is a Wireless Personal Area Network technology (WPAN) marketed by Bluetooth Special Interest Group (SIG), aimed at new fangled high-technology applications [1, 2]. The remarkable progress in the mobile, portable applications

---

M. Bansal (✉) · Jyoti
Department of Electronics and Communication Engineering,
Delhi Technological University, New Delhi 110042, India
e-mail: maltibansal@gmail.com

© Springer Nature Singapore Pte Ltd. 2019

H. Malik et al. (eds.), *Applications of Artificial Intelligence Techniques in Engineering*, Advances in Intelligent Systems and Computing 697,
https://doi.org/10.1007/978-981-13-1822-1_22

domain is approaching for minimum power consumption solutions with cost-effectiveness. With this tremendous growth of wireless applications in the market, one of the ultimate fascinating fields emerging in wireless personal area communication development is the Bluetooth technology. Short-range communication, i.e., 2.4 GHz ISM radio band is a wireless standard for associating several communication devices like printers, laptops, mobile phones, wireless headsets, and PC as well as notebook computers. The most prominent design issue for BT application is to reduce power consumption of the radio-frequency-integrated circuit [3].

The 2.4 GHz ISM radio band has prodigious advantages and disadvantages. The characteristics of this band are poor propagation because the spectrum of radio energy is promptly absorbed by every little thing, preeminently by water. So, that is why ISM band is considered as an atrocious place to use and design a portable wireless communication technology. The advantages of this band of radio spectrum are no license requirements and are easily available worldwide. Obviously, this free license sign meant that various wireless applications or technologies are using this band such as wireless LAN, Bluetooth technology, and Wi-Fi networks. Basically, ISM band has rules which are primarily associated to specify the output capacity of the various devices that are using this spectrum. Furthermore, licensed spectrums are paying heavily due to this ISM band. As a result, we need to select Industrial Scientific and Medical radio band which is cost-effective. Bluetooth Low Energy has been optimized for low power consumption [4].

The purpose of Bluetooth technology is to consolidate different worlds of communications and computing, connecting various devices such as mobile phones to mobile phones, cell phones to notebook(s), etc. The applications of Bluetooth technology include audio and video connection between laptops and mobile phones. This brand new technology grew and enormous number of applications was added such as file transfer, wireless printing, notebook or phone book, and stereo music streaming loads from the cell phones to car, etc. Another data rate, i.e., Enhanced Data Rate (EDR) was further added in Bluetooth version 2.0 to accumulate the Physical Layer in which the value of data rates up to 3 Mbps [4, 5].

Generally, a low-noise amplifier (LNA) is the main element in Radio Frequency (RF) transceiver. The primitive constraint of the LNA is the amplification of weak radio frequency signal while simultaneously introducing limited noise as much as possible. Generally, its performance parameters adversely affect the overall transceiver performance. LNA should provide high linearity. In case of wireless communication systems which are ruined by power batteries, so that the design consideration of low-noise amplifier like power consumption should be very small to get the device longer battery life or run on single battery. The metrics of good LNA are high linearity, minimum power consumption, higher gain, minimum noise figure, and better input and output impedance matching but normally there are trade-offs in these design considerations of LNA [6]. This paper is organized as follows: BLE and IoT applications are given in Sect. 2, Low-Noise amplifier is given in Sect. 3; CMOS LNA for BLE is applications given in Sect. 4; lastly, the conclusion is given in Sect. 5.

2 BLE and IoT Applications

The terminology Internet of things (IoT) has been growing tremendously in recent years. In the last few years, it has acquired more consideration due to the development of wireless technology. The fundamental principle behind the IoT is the variety of protocols and devices, including Radio-frequency Identification (RFID), sensors, actuators, Near-Field Communication (NFC), and mobile phones, etc. Linking of these devices is done by acquiring a different protocol address. The term IoT employs various generous facts to detect, listen, think, and perform various activities by talking to each other, also transfer the data, facts, and information [7, 8] (Fig. 1).

However, Bluetooth Low Energy (BLE) is the brand new operational mode. BLE became live in fourth release of the Bluetooth wireless technology standard. Furthermore, it has its applications for movable devices which are powered by coin-cell batteries, among sensors and wearable biomedical devices. Under BLE standard, reduced power consumption is required for the long-term functionality that these devices are working on [9]. Bluetooth technology was designed to permit wireless for connecting an extensive array of devices by short-range communication, ad hoc networks such as WANET or MANET known as piconets. Piconets are designed dynamically in which two or more Bluetooth enabled devices are surrounded by radio range for abrupt connectivity and exchange of information. For that reason, several devices with Bluetooth-enabled radio can design a piconet to instantly transfer information wirelessly to a linking device [10].

Bluetooth-enabled devices design piconets for point-to-point (P2P) as well as point-to-multipoint peer (P2MP) connecting devices. To avoid data transfer collisions within the piconet, a master and slave relationship are established. In the case of Bluetooth, a piconet can establish a maximum of one master and seven slave devices. However, each device itself can also be an element of various other piconets concurrently, called as a scatternet. Scatternets are piconets which share the similar coverage area. Therefore, to reduce the surplus interference between

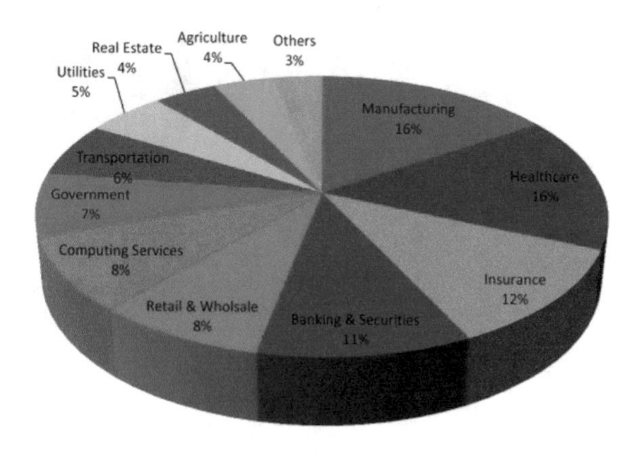

Fig. 1 Projected market share of dominant IoT applications by 2020 [29]

scatternets, Bluetooth operates on adaptive frequency-hopping (AFH) spread spectrum scheme to minimize the probability of two transmitters occupying the similar transmitting channel (Fig. 2).

(a) **Modulation:**
The Gaussian frequency shift keying modulation scheme with modulation index between 0.45 and 0.55 is implemented in Bluetooth Low Energy [11]. GFSK is a variation of Continuous Phase Frequency Shift Keying (CPFSK), where digital information is intimidated into different discrete frequency changes of a sinusoidal carrier wave, usually by changing the voltage applied to a voltage controlled oscillator (VCO). This implementation minimizes the superior emissions which are generated from phase discontinuities from fast-changing bit sequences when two independent oscillators are operated. The option of pulse applied to the VCO is also critical to minimize surplus emissions and can be enhanced by investigating the frequency-domain of the conventional rectangular pulse [12].

(b) **Modes of BLE:**
Bluetooth low energy generally consists of different modes. There are two types of devices, namely dual-mode and single-mode devices. First, a single-mode device is a Bluetooth technology device that only supports for Bluetooth low energy. On the other hand, dual-mode device is a Bluetooth technology device that has sustained for classic Bluetooth as well as Bluetooth Low Energy. In addition to this, a third type of device mode which is Bluetooth classic-only device. Therefore, it sustains both classic Bluetooth and Bluetooth Low Energy in which a dual-mode device can connect with millions of presented Bluetooth devices [4].

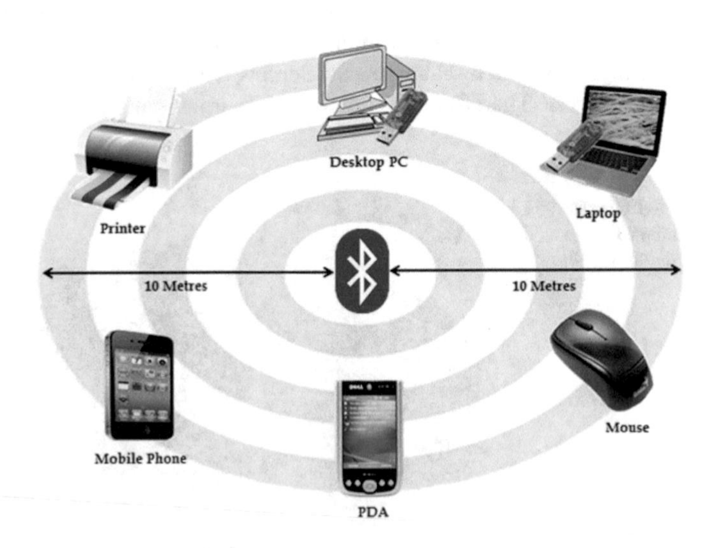

Fig. 2 Bluetooth network [30]

Table 1 Types of bluetooth technology [4]

Type	Technology supported
Classic mode	Bluetooth classic
Single mode	Bluetooth low energy
Dual mode	Bluetooth classic + Bluetooth low energy

(i) **Dual-mode**
In Bluetooth wireless technology, dual-mode devices are the latest entrants. The necessary requirement of dual-mode devices is the new hardware as well as software in the host, and firmware in controller. The parts for existing Bluetooth classic controllers are easily replaced by dual-mode controllers. This empowers designers of personal computers, cellular phones, and as well as other devices to substitute their current classic Bluetooth controllers among dual-mode controllers very efficiently.

(ii) **Single Mode**
In single-mode devices, Bluetooth Low Energy does not connect with the existing Bluetooth devices. In other respects, Bluetooth Low Energy can connect to other dual-mode devices as well as single-mode devices. Furthermore, these brand new single-mode devices extremely improve the power consumption factor. They have been designed for the components which are powered up using coin-cell batteries. In case of Bluetooth classic, single-mode devices cannot be used because the single-mode device in Bluetooth Low Energy cannot maintain stereo music as well as audio for headsets and higher value of data rates for transferring the files from one place to another.

Table 1 shows which types of device can connect with another, types of devices, and also what type of Bluetooth technology could be used for connecting the various devices to each other. In case of single-mode devices would connect with other single device modes and it would also connect among dual-mode devices. In another case of dual-mode devices would connect with another dual-mode devices as well as classic devices using bit rate and enhanced data rate. On the other hand, single-mode device cannot connect a classic Bluetooth device [4].

3 Low-Noise Amplifier

With the advancement of technology, the requirement for immensely integrated complementary metal-oxide semiconductor radio frequency elements among low noise has been increasing continuously. Consequently, the applications of low-noise amplifier became significant as they are required to analyze area of low noise and power CMOS RFIC design [13]. LNA is basically an electronic amplifier which is used to amplify weak RF signals as received by the antenna. It is generally placed very near to the antenna or the detection device to minimize the feeder and

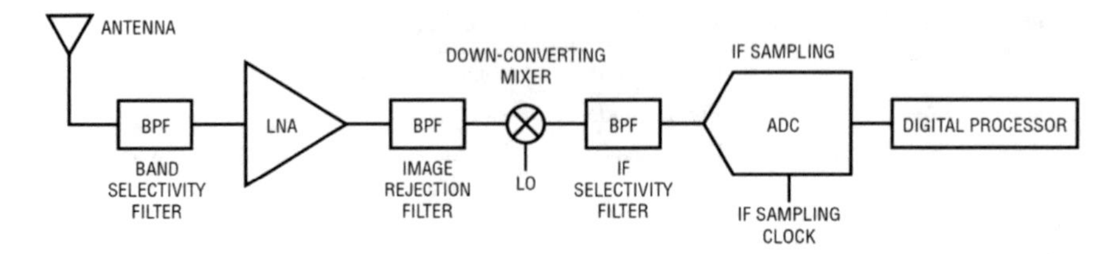

Fig. 3 RF receiver [31]

branching losses. LNA is the main element located at the foremost block of a wireless transceiver circuit. In a low-noise amplifier, the various design considerations such as effect of noise from consecutive stages of the transceiver which can be dominated by the gain of the amplifier [14, 15]. Low-noise amplifier should ideally have a low-noise figure, high gain, better output, and input impedance matching and stability without oscillation over entire frequency range [16]. There are various kinds of low-noise amplifiers and are classified on the basis of high gain bandwidth, highest input offset voltage, simple slew rate, packaging type as well as number of channels [14, 17] (Fig. 3).

3.1 LNA Design Considerations

There are various design specifications of low-noise amplifier such as noise figure, S parameters, 1 dB compression point, Dynamic range, etc. First, noise figure is explained as the deterioration of the SNR (signal-to-noise ratio) as the signal proceeds through the device [18]. 1 dB compression point is defined as considering the input signal as a sinusoid, i.e., $x(t) = A \cos wt$, then the output of the nonlinear system would be

$$Y(t) = \alpha_1 A \cos wt + \alpha_2 A \cos^2 wt + \alpha_3 A^3 \cos^3 wt$$

where A is the amplitude of the signal and α_1, α_2, and α_3 coefficients [19]. Gain compression is explained as the level of input signal which causes the gain to drop by 1 dB. Dynamic Range is usually represented as the maximal input level that a circuit can endure to the minimal input level that a circuit can tolerate so that the circuit provides a reasonable signal quality. This definition of the Dynamic Range can be used differently for different applications [20]. Spurious free dynamic range [21] can be defined as follows:

$$\mathrm{SFDR} = \mathrm{P_{in,max}} - \mathrm{P_{in,min}} = \frac{2}{3}(\mathrm{IIP3} - \mathrm{F}) - \mathrm{SNR_{min}}$$

S Parameters: At very high RF and microwave frequencies, it is not possible to measure *Y*-, *Z*-, or *H*- parameters directly due to the unavailability of the equipment

(s) to measure RF/MW total current and difficulty of attaining perfect shorts/opens. Also, the active devices could be unstable under short/open conditions [22]. There are four S parameters for a two-port devices, namely $S_{11,}$ $S_{12,}$ $S_{21,}$ S_{22} where S_{11} is forward reflection coefficient, S_{12} is reverse reflection coefficients, S_{21} is forward gain, and S_{22} is reverse gains with source and load impedance having value of 50 Ω [23, 24].

4 CMOS LNA for BLE Application

In the following section, some of the popular LNA topologies for Bluetooth and BLE applications are reviewed briefly.

4.1 Circuit Topology for Bluetooth

(1) **Cascode Stage Topology**
Shaikh K. Alam represented a completely differential gain LNA at 2.4 GHz in a 0.18 μm CMOS technology process [25]. The Variable Gain Low-Noise Amplifier (VGLNA) acquires a maximum small signal gain of 18.37 dB, −11.35 dBm 1-dB compression points and a 6.4 dB minimum small gain of 6.4 dB. In high gain mode, the LNAs high gain mode is 2.47 dBm. The current consumption of LNA is only 4.3 mA from a 1.5 V power supply. This cascode technology is the most promising topology for LNA design as shown in Fig. 4. The configuration of cascode amplifier basically consists of a common source amplifier with inductively degenerated followed by transconductors, $M_1 - M_2$, and gate as well as source inductors, L_g and L_s ,respectively. Generally, using the inductive-source degeneration through L_s has an advantage of concurrently obtaining input matching with noise matching. For optimizing the noise or NF calculation, the transistor is selected such that the noise is reduced. Figure 4 represents the VGLNA without a bias circuit. The advantage of using such a circuit is that it is less susceptible to common mode injected noise such as substrate noise as compared to single-ended LNA.

(2) **Common Source Topology with Inductive-Source Degeneration**
Deepak Balodi et al. presented a design procedure of Low-Noise Amplifier. Finally, common source topology has been designed for an optimum performance with 0.13 μm CMOS process [25]. In this paper, designing of LNA has been targeted on Bluetooth application range (2.4–2.6 GHz), because of its extensive usage in addition to popularity with supply range in 1.5/2.5 V. The author describes three basic topologies such as Common source topology using inductive-source degeneration technique, Cascade-LNA, and Cascode-LNA with CS stage. These topologies are also compared on the basis of various

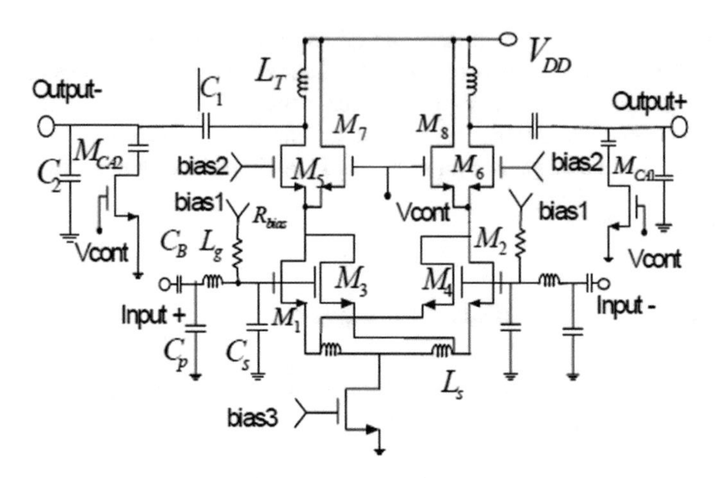

Fig. 4 Differential gain LNA [25]

parameters like linearity, gain, noise, and power. The comparison of all three LNA topologies provide with the noise figure of 1.25 dB along with a forward gain of 33 dB plus a very low power consumption with a two-stage cascaded common source LNA Topology. The software used for designing and simulation is Advanced Design System (ADS) with UMC 0.13 μm RF CMOS process [26].

4.2 Circuit Topology for BLE Applications

(1) **Cascode Topology**
Chin-To Hsiao proposed a circuit design for a 2.4 GHz CMOS LNA for BLE specific applications using 45 nm technology file [27]. The author(s) has considered three dominant LNA topologies, i.e., Common Gate (CG), Cascode, and Common Source (CS). The main benefit of cascode amplifier is that it provides the highest gain, broad bandwidth but slightly compensating in NF parameter and design complexity. The cascode methodology is the most protrusive topology for LNA. Generally, the systematic schematic of this amplifier consists of a CS stage and CG stage. First, Common source stage has an advantage of providing a better stability that has surpassing temperature variation immunity, sensitivity to process, component variations and power supply. So, that is why this topology is used to design the LNA. The architecture of this topology LNA with distinct four blocks is shown in Fig. 5.
The author has presented a schematic of LNA in which four different blocks such as input matching network, biasing circuit, source degeneration as well as output matching network. There are four factors which measure the

Fig. 5 Structure of cascode with four different blocks [27]

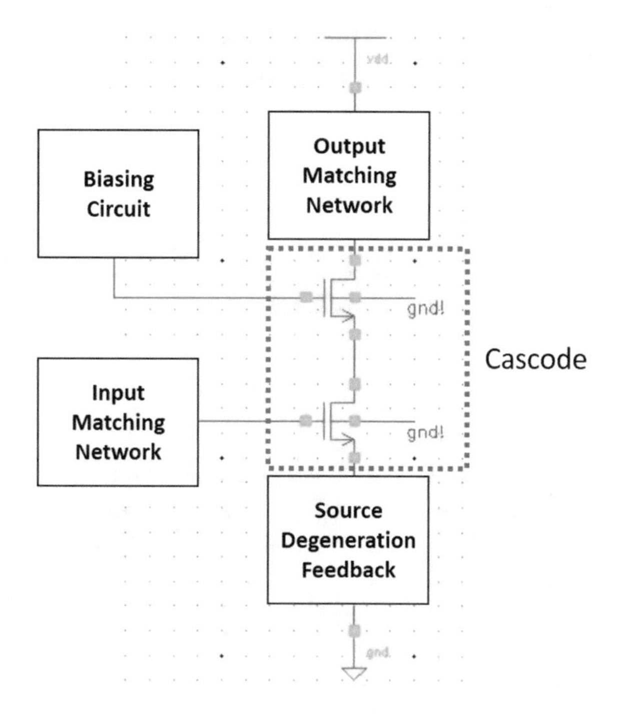

performance of LNA: gain, power consumption, linearity and noise figure. The specifications of Low-Noise Amplifier for BLE front-end receiver noise figure lies close to 0.98 dB, power consumption of 1.01 mW, and gain of 14.53 dB. In this work, LNA has been designed for LNA minimizing the power consumption up to 1.01 mW, which maintains more power to wireless devices. A major aim of reducing power was achieved [27].

(2) **Common Gate Topology**

The author Anith Selvakumar et al. designed a low-noise amplifier which accomplishes the quadrature on RF signal path. The architecture of receiver front end in which single RC network behaves, simultaneously, as a LPF (Low-Pass Filter) for the Common Gate Stage (CG), is realized with M0I, as well as a HPF (High-Pass Filter) for the Common Gate Stage (CG), implemented by M0Q. It generates a 90° swing among the two output currents. By choosing C_0 larger than gate-source capacitances M0I/M0Q, a wideband quadrature is guaranteed, although amplitude matching is attained approximately around the filter cutoff frequency $1/R_0C_0$. Despite that, since the BLE standard range of frequencies is only between the 83.5 MHz around 2.44175 GHz, in which they have obtained a decent amplitude-matching plus the image-rejection over the complete range of frequencies. The topology used in this paper is a single RC network. The above implementation can be done by using the triode region in series with the resistor [28].

4.3 Comparison of Different Parameters of LNA for BT and BLE Applications

See Table 2.

4.4 Comparison of Different LNA Topologies for BT and BLE Applications

See Table 3.

Table 2 Comparison of different parameters of LNA for BT and BLE applications

References	[32]	[27]	[33]	[34]	[35]	[36]	[37]
Frequency (GHz)	2.4 GHz	2.44 GHz	2.4–2.483 GHz	2.4 GHz	2.4–2.5 GHz	2.4 GHz	2.45 GHz
NF (dB)	1.1 dB	0.98	3.2 dB	1.47 dB	1.98 dB	2.27 dB	3 dB
S_{21} (dB)	27 dB	–	–	–	–	–	7 dB/−17 dB
IIP3	−30 dBm	−10.67	–	−8.1 dBm	5 dBm		
Supply voltage	0.65 V	1 V	1.2 V	1.8 V	1.2 V	0.36	1.8 V
Power	4.6mW	1.01 mW	1.2 mW	11.1 mW	7.33 mW	2.57 dB	7.66 mW
Technology	0.18 μm	45 nm	130 nm	0.18 μm	130 nm	180 nm	180 nm
Applications	Bluetooth	BLE	BLE	Bluetooth	Bluetooth	Bluetooth	Bluetooth

Table 3 Comparison of different LNA topologies for BT and BLE applications

LNA topologies	Features	Specifications
(1) Common source stage with inductive feedback as source degeneration [38]	Simplest design and area efficient • High degree of linearity • Power efficient • Insufficient gain • Poor isolation	Gain: 13.71 dB [20] Isolation: −16.07 dB NF: 0.324 dB (ideal inductors) 1 dB compression point: 10.64 dBm (input), 14.07 dBm (output) Power dissipation: 0.825 mW
(2) Common gate topology [38–40]	Sufficient linearity Low input impedance Superior stability High noise figure	Voltage gain: 23 dB NF: 2.98 dB Frequency range: 2.44 GHz
(3) Cascode topology with inductive-source degeneration [38]	High voltage gain Improve voltage gain at high frequency Suppressed miller effect Higher gain Higher input and output impedances	Noise figure: 0.98 dB Power consumption: 1.01 m Gain: 14.53 dB

(continued)

Table 3 (continued)

LNA topologies	Features	Specifications
(4) Common source with cascode	High gain Moderate linearity Low-noise figure Limited voltage swing	Gain: 17.83 dB [20] Isolation: −48.92 dB NF: 1.256 dB 1 dB compression point: −2.059 dBm (input), 4.414 dBm (output) Power dissipation: 1.375 mW
(5) Two stage LNA	Area insufficient Low-noise figure High gain Nonlinearity	Gain: 33.54 dB [20] Isolation: −48.9 dB NF: 1.256 dB 1 dB compression point: −16.52 dBm (input) 7.414 dBm (output) Power dissipation: 3.875 mW

5 Conclusion

The design procedure of CMOS LNA for BLE and BT applications has been contemplated. The design specifications of LNA are low-noise figure, low power dissipation, nominal supply voltage, high linearity, etc. The design of LNA has been targeted for BLE applications range, i.e., 2.4–2.46 GHz. Bluetooth Low Energy is an advanced technology of Bluetooth and it derives lot of technology from its parent. BLE is a WPAN (Wireless Personal Area Network) technology which was advertised by Bluetooth SIG, aiming at unique applications in the security, entertainment industries as well as health care. Finally, a comparison of different LNA topologies and their parameters for BT and BLE application range. BLE is intended for providing reduction in power consumption with low supply voltage and at the same time being cost-effective with same communication range as of the Bluetooth technology.

Acknowledgements One of the authors (Jyoti) acknowledges the fellowship support she is receiving from Delhi Technological University (DTU), for carrying out this work, as a part of her Ph.D. thesis work in the domain of RF Microelectronics. She also acknowledges the guidance support from her thesis supervisor, Dr. Malti Bansal, Assistant Professor, Department of Electronics and Communication Engineering, DTU, for carrying out this research work.

References

1. https://en.wikipedia.org/wiki/Bluetooth_Low_Energy
2. https://www.Bluetooth.com/what-is-Bluetooth-technology/where-to-find-it/retail-location-based-services
3. Khan, Z., Wang, Y.: Comparison of different CMOS low-noise amplifiers topologies for Bluetooth applications. In: 2005 IEEE Conference of Wireless and Microwave Technology, p. 15 (2005)

4. Heydon, R.: Bluetooth Low Energy: The Developer's Handbook. Prentice Hall, Upper Saddle River, NJ (2013)
5. Bansal, M., Jyoti: CMOS LNA for BLE applications. In: 2017 4th International Conference on Science, Technology and Management, pp. 654–661 (2017)
6. Razavi, B.: RF microelectronics. Prentice Hall (1998)
7. Azori, L. Iera, A., Morabito, G.: The internet of things: a survey. Comput. Netw. **54**. doi.: https://doi.org/10.1016/comnet.2010.05.010
8. Shah, S., Yaqoob, I.: A survey: internet of things (IOT) technologies, applications and challenges. In: 2016 IEEE Smart Energy Grid Engineering (SEGE), pp. 381–385 (2016)
9. Selvakumar, A. Sub-mw receiver front-end for Bluetooth low energy in 130 nm technology. Master thesis (2015)
10. Bluetooth Specification Version 4.1. Bluetooth (2013) [Online]. Available: http://www.Bluetooth.com
11. Chi, A., Wong, W., Dawkins, M., Devita, G., Kasparidis, N., Katsiamis, A., King, O., Lauria, F., Schiff, J., Burdett, A.J., Member, S.: A 1 V 5 mA multimode IEEE 802.15.6/Bluetooth low-energy WBAN transceiver for biotelemetry applications. J. Solid State Circuits **48**(1), 186–198 (2013)
12. Darabi, H., Ibrahim, B., Rofougaran, A.: An analog GFSK modulator in 0. 35 um CMOS. J. Solid State Circuits **39**(12), 2292–2296 (2004)
13. Kumar, S.: CMOS low noise RF amplifier design and parameters using ANN. Master thesis. Thapar University (2012)
14. http://www.futureelectronics.com/en/amplifiers/low-noise.aspx
15. Bansal, M., Jyoti: A review of low noise amplifier for 2.4 GHz frequency Band. In: 2017 International Conference on Innovations in Control, Communication and Information System (ICICCI), Greater Noida, Uttar Pradesh, pp. 63–69, 12–13 Aug 2017
16. Patial, V.: Design a low noise amplifier for WCDMA reception range. Master thesis. Thapar University (2012)
17. Bansal, M., Jyoti: A review of various of applications of low noise amplifiers. Presented at 2017 International Conference on Innovations in Control, Communication and Information System (ICICCI), Greater Noida, Uttar Pradesh, pp. 142–147, 12–13 Aug 2017
18. http://www.docente.unicas.it/useruploads/000725/files/spectrum_analysis_basics_ii.pdf
19. Nga, T.T.T.: Ultra low-power low-noise amplifier designs for 2.4 GHz ISM band applications. Doctor Philosophy of Engineering (2012)
20. Chang, J.: An integrated 900 MHz spread-spectrum wireless receiver in 1-μm CMOS and a suspended inductor technique. University of California (1998)
21. Zhu, L.: RF engineering techniques course notes. Master thesis, Nanyang Technological University (2008)
22. Dores, J.M.H.M.: LNA for a 2.4 GHz ISM receiver. Master thesis, July 2010
23. Lee, T.H.: The design of CMOS radio-frequency integrated circuits. 1st edn., pp. 134–140. Cambridge University Press, New York (1998) (Chapter 6)
24. Orfanidis, S.: Scattering parameters. Electromagn. Waves
25. Alam, S.K., DeGroat, J.: A 1.5-V 2.4 GHz differential CMOS low noise amplifier for bluetooth and wireless LAN applications. In: 2006 IEEE North East Workshop on Circuits and Systems, pp. 13–16 (2006)
26. Balodi, D., Verma, A., Govidacharyulu, P.A.: A high gain low noise amplifier design & comparative analysis with other MOS-topologies for Bluetooth applications at 130 nm CMOS, pp. 378–383 (2016)
27. Hsiao, C.-T.: Design of a 2.4 GHz CMOS LNA for Bluetooth low energy application using 45 nm technology. Master's theses, p. 4802 (2017)
28. Selvakumar, A., Zargham, M., Liscidini, A.: Sub-mW current re-use receiver front-end for wireless sensor network applications. IEEE J. Solid-State Circuits **50**(12) (2015)
29. http://monacotrades.com/2015/06/
30. https://www.ictlounge.com/html/Bluetooth_wi-fi.htm

31. http://www.electronicspecifier.com/mixed-signal-analog/ltc6957-ltc2153–14-linear-es-design-magazine-finding-a-differential-solution
32. Khan, M.Z., Wang, Y.: Comparison of different CMOS low-noise amplifiers topologies for Bluetooth applications. In: 2005 Wireless and Microwave Technology, p. 15 (2005)
33. Liao, L., Kaehlert, S., Wang, Y., Atac, A., Zhang, Y., Schleyer, M., Wunderlich, R., Heinen, S.: A low power LNA for Bluetooth low energy application with consideration of process and mismatch. In: Proceedings of APMC 2012 (2012)
34. Yang, L., Yan, Y., Zhao, Y., Ma, J., Qin, G.: A high gain fully integrated CMOS LNA for WLAN and Bluetooth application. In: 2013 IEEE International Conference of Electron Devices and Solid-State Circuits, EDSSC 2013, vol. 2012, pp. 3–4 (2013)
35. Nadia, A., Belgacem, H., Aymen, F.: A low power low noise CMOS amplifier for Bluetooth applications. In: 2013 International Conference of Applied Electronics, p. 14 (2013)
36. Karimi, G.: Designing and modeling of ultra low voltage and ultra low power LNA using ANN and ANFIS for Bluetooth applications. Neurocomputing **120**, 505–508 (2013)
37. Beffa, F., Bachtold, W.: A switched-LNA in 0.18 pm CMOS for Bluetooth applications. In: 2003 Topical Meeting on Silicon Monolithic IC in RF Systems, pp 80–83 (2003)
38. Gyamlani, S., Zafar, S., Sureja, J., Chaudhari, J.: Comparative study of various LNA topologies used for CMOS LNA design. Int. J. Comput. Sci. Emerg. Technol. **3** (2012)
39. Devi, A., Kumar, R., Singh, L., Talukdar, F.A.: A review on the low noise amplifier for wireless application. Int. J. Comput. Appl. (2015)
40. Allstot, D.J., Li, X., Shekhar, S.: Design considerations for CMOS low-noise amplifiers. In: IEEE Radio Frequency Integrated Circuits Symposium, pp. 97–100 (2004)

Steady-State Analysis of Permanent Magnet Synchronous Generator with Uncertain Wind Speed

Vikas Kumar Sharma and Lata Gidwani

Abstract This paper represents a platform for steady-state analysis of permanent Magnet Synchronous Generator (PMSG) based on uncertain wind speed. To evaluate the system performance such as torque, speed, current, voltage, and power of PMSG with considering the uncertain wind speed. This system consists of a wind speed model, wind turbine model, and PMSG model. The proposed wind speed model is put forward that could reflect the natural wind speed characteristics. The wind speed for wind turbine is used as an input, so that it captures the optimal power of wind and generates mechanical torque for PMSG. Mathematical analysis is used to demonstrate the efficacy of the model in d*q*-synchronous rotating reference frame of the generator. The steady-state analysis of the proposed PMSG uncertain wind speed model is evaluated with MATLAB/Simulink software.

Keywords Steady-state analysis · Permanent magnet synchronous generator (PMSG) · Uncertain wind speed · Wind turbine

1 Introduction

In the end of the twentieth century, it was nautical that the environment is getting polluted due to excess production of CO_2 and other similar elements. Advancement in the field of generation technology, countries all over the world are looking for green and clean generation sources. Renewable energy sources, especially wind generation is considered as the possible solution for this requirement [1]. Wind energy is the easily available source of nonrenewable source of energy. It is an indirect form of solar energy. The kinetic energy produced by the wind turbine is used to generate electrical energy. As by the use of wind energy, the carbon discharge also reduces. Among the entire energy source, wind energy is having maximum share because of low cost and less space requirement for the equipment installation [2].

V. K. Sharma (✉) · L. Gidwani
University College of Engineering, RTU, Kota, India
e-mail: vikasvs1985@gmail.com

© Springer Nature Singapore Pte Ltd. 2019

H. Malik et al. (eds.), *Applications of Artificial Intelligence Techniques in Engineering*, Advances in Intelligent Systems and Computing 697,
https://doi.org/10.1007/978-981-13-1822-1_23

At present, generators for wind turbines is PMSG and induction generators which includes squirrel cage and wound rotor. Wind turbines with low power rating are of mainly PMSG and squirrel cage induction-type generators. These are often used because these are found suitable in terms of cost and reliability induction generators [3].

The main function of wind energy conversion system is that it extracts energy from the blade of the hub and transfer to the rotor of generator via a gearbox. We generally use PMSG because of its efficiency and in this, we have no need of gearbox. Elimination of gearbox is the main advantage of PMSG which are interfaced with voltage source converter [4].

Synchronous generator excitation is provided through permanent magnet with high energy. Because of relatively large number of poles, the permanent magnet construction would enable reduction in black iron and stator yoke. Nowadays, development of low speed and innovative synchronous machine, especially with permanent magnet excitation has received attention all over the world [5]. Permanent magnets could be considered as a potential solution for the excitation of synchronous generators. This would replace the excitation winding of synchronous generators. Hence, it reduces magnet price while improving the magnetic material characteristic [6]. Nowadays, variable speed turbine system is interfaced with electronics equipment and by doing this, we extracted maximum power with the control system. PMSG-based wind turbines are suitable in wind energy conversion system because of self-excitation property. These could operate at acceptable power factor and efficiency while offering economical solutions [7]. Steady-state model is formed considering PMSG, wind speed model ,and wind turbine. In wind turbine, rotor blades catch the wind energy and then, this is transferred to synchronous generator. PMSG converts mechanical energy into electrical energy. In steady-state analysis, wind generator is connected to load (see Fig. 1).

In this regard, this paper puts forward in detail modeling and steady-state analysis of PMSG-based wind turbine system and their control. The rest of the paper is structured as the detail mathematical modeling of wind speed, wind turbine, and PMSG are presented in Sect. 2. In Sect. 3, PMSG based with uncertain wind speed model simulation is discussed and conclusions are described in Sect. 4.

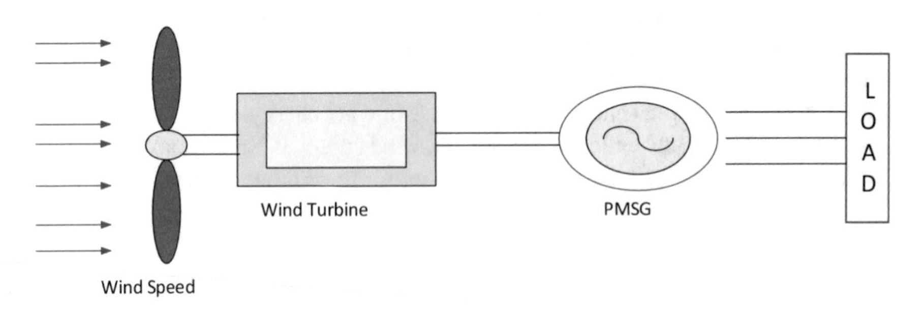

Fig. 1 Steady-state model of PMSG based wind power system

2 Mathematical Model of System

2.1 Mathematical Modeling of Uncertain Wind

The simulation of the uncertain wind in the MATLAB without wind can be done by mathematical model of wind speed. Here, this wind speed signal has four components, i.e., the basic speed of wind (V_{B}), a fluctuating wind speed (V_{N}) that describes a steady increase in wind speed, gust component (V_{G}), and a gradient wind component (V_{R}). The uncertain wind speed is sum of these components.

$$V_{\mathrm{w}}(t) = V_{\mathrm{B}}(t) + V_{\mathrm{N}}(t) + V_{\mathrm{G}}(t) + V_{\mathrm{R}}(t) \quad (1)$$

2.1.1 Basic Wind Speed

It reflects the variation of mean wind speed during the whole process of wind turbines and also determines the size of rated power to the system. The basic systems find the change in mean speed of wind and determine size of rated wind turbine.

$$V_{\mathrm{B}} = 0.1 + 0.824 N_{W}^{1.505} \quad (2)$$

where V_{B} is the basic wind speed of n-tire wind in m/s, N_{W} is the series of wind.

2.1.2 Fluctuating Wind Speed

To describe the random behaviors of wind and change of the speed at different heights and at different attitudes, the simulation of fluctuating wind is done. When the atmosphere is quiet or it is stable, the fluctuating wind can be considered as a stationary random process with sample at one point for a long time observation.

$$V_{\mathrm{N}} = \frac{1}{\Delta t}\int\limits_{t1}^{t2} V_{\mathrm{W}}\mathrm{d}t = 0 \quad (3)$$

2.1.3 Gust of Wind

It tells us about characteristics of unexpected change in wind speed, i.e., wind speed with cosine characteristics with time. It is used to access the dynamic characteristics in the case of large wind speed change. It is the ratio of gust wind speed and average wind speed. It also related to turbulence intensity. It is given as

$$g(t) = 1 + 0.42\varepsilon u \ln \frac{3600}{t} \quad (4)$$

2.1.4 Gradient Wind

The gradient change characteristics of wind speed are a function of wind speed. It is used to describe the increase and decrease in the magnitude of wind speed.

$$V_{\mathrm{R}} = \begin{matrix} 0 & 0 < T < t_{1r} \\ & V_{\mathrm{ramp}} t_{1r} \leq T < r \\ & R_{\max} t_{2r} < T \leq t_{2r} + t_r \end{matrix} \quad (5)$$

2.2 *Mathematical Modeling of Wind Turbine*

The mechanical power output of the wind turbine is given as

$$P_{\mathrm{wind}} = \frac{1}{2} C_{\mathrm{p}}(\lambda, \beta) \pi r^2 \rho_{\mathrm{air}} V_{\mathrm{W}}^3 \quad (6)$$

where ρ is the air density (kg/m^3), rotor radius of blades (m) is r, and V_{W} is the uncertain wind speed (m/s) C_{p} is the performance coefficient of the wind turbine. Which λ is the tip speed ratio, and the performance coefficient C_{p} is calculated by [8].

$$C_{\mathrm{p}} = C_1 \left(C_2 \frac{1}{\alpha} - C_3 \beta - C_4 \beta^x - C_5 \right) \mathrm{e}^{-C_6 \frac{1}{\alpha}} \quad (7)$$

β is function of pitch angle of rotor blades (in degree) when β is equal to zero, The tip speed ratio is defined as

$$\lambda = \frac{\omega_{\mathrm{a}} r}{V_{\mathrm{w}}} \quad (8)$$

where w_{a} is the rotor angular velocity (rad/sec) of the wind turbine generator. The output of wind turbine mechanical torque is defined as

$$T_{\mathrm{mech}} = \frac{0.5 C_{\mathrm{p}} \pi r^2 \rho_{\mathrm{air}} V_{\mathrm{w}}^3}{\omega_{\mathrm{a}}} \quad (9)$$

2.3 *Mathematical Modeling of PMSG*

The dynamic model of PMSG is developed in dq rotating reference frame. The direct axis and quadrature axis voltage equation of generator is defined as

$$V_{\mathrm{d}} = -R_{\mathrm{s}}i_{\mathrm{d}} - L_{\mathrm{d}}\frac{\mathrm{d}i_{\mathrm{d}}}{\mathrm{d}t} + \omega_{\mathrm{e}}L_{\mathrm{q}}i_{\mathrm{q}} \tag{10}$$

$$V_{\mathrm{q}} = -R_{\mathrm{s}}i_{\mathrm{q}} - L_{\mathrm{q}}\frac{\mathrm{d}i_{\mathrm{q}}}{\mathrm{d}t} - \omega_{\mathrm{e}}L_{\mathrm{d}}i_{\mathrm{d}} + \omega_{\mathrm{e}}\mu_{\mathrm{g}} \tag{11}$$

where μ_{g} the permanent flux of generator is, ω_{e} is the electrical speed rotor in rad/s of the generator. The electromagnetic torque of generator is defined as

$$T_{\mathrm{elect}} = \frac{3}{4}P\left[\mu_{\mathrm{g}} + (L_{\mathrm{d}} - L_{\mathrm{q}})i_{\mathrm{d}}\right]i_{\mathrm{q}} \tag{12}$$

3 Simulation Result and Discussion

The simulation is carried out to demonstrate the effectiveness of modeling of wind turbine and PMSG. Figure 2 shows simulation model implemented in Sim Power System of the Mat Lab to analyze the performance of the PMSG based wind power system operation in steady-state mode. It is operated with load to determine the dynamic performance of the model. Table 1 provides simulation parameter of wind turbine and generator. The steady-state analysis of system has been carried out with uncertain wind speed. When system reaches at steady state, mechanical torque is equal to electrical torque and rotor angular speed becomes stable.

Figure 3 shows the uncertain wind output generated by the wind speed model. In this output the wind speed varies from 7 to 12 m/s in the duration of 0–1 s.

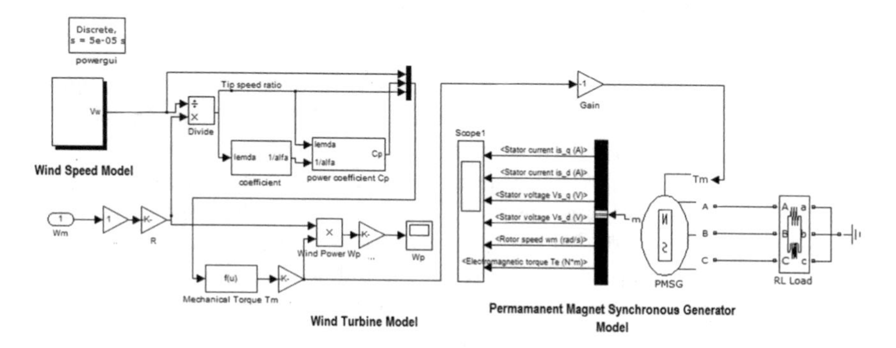

Fig. 2 Simulink modal of PMSG based wind power system

Table 1 PMSG parameter and wind turbine specification

Parameter	Value
Rated generated power (P)	2 MW
Stator resistance (R_s)	0.7305 m Ω
Stator d-axis inductance (L_d)	1.21 mH
Stator q-axis inductance (L_q)	2.31 mH
Permanent magnet flux (μ_g)	6.61 Wb
Equivalent inertia (J_{eq})	10,000 kg m^2
Number of pole pairs	30
Rotor radius of blades (r)	38 m

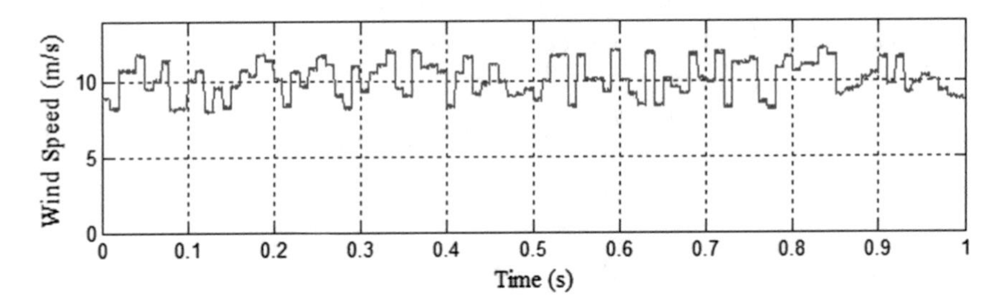

Fig. 3 Uncertain wind speed

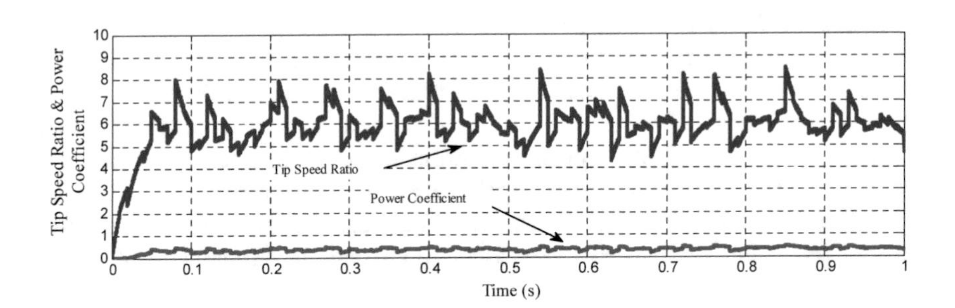

Fig. 4 Coefficient of power and TSR of wind turbine

It provides the uncertain wind input to wind turbine model. In Fig. 4, power coefficient of the wind turbine is maintained at 0.3–0.45 and in this, duration tip speed ratio is maintained 5–8 as per wind speed. Figure 5 show gives the steady-state conditions of wind turbine, so that mechanical torque equal to electrical torque. Figure 6 shows rotor speed of the generator. Figure 7 shows generated electrical power by the PMSG are gradually changed. Figures 8 and 9 shows the three-phase output voltage of PMSG is the line to line voltage and current. In the steady state, the entire performance variables include current, voltage, and power which are kept at their rated value.

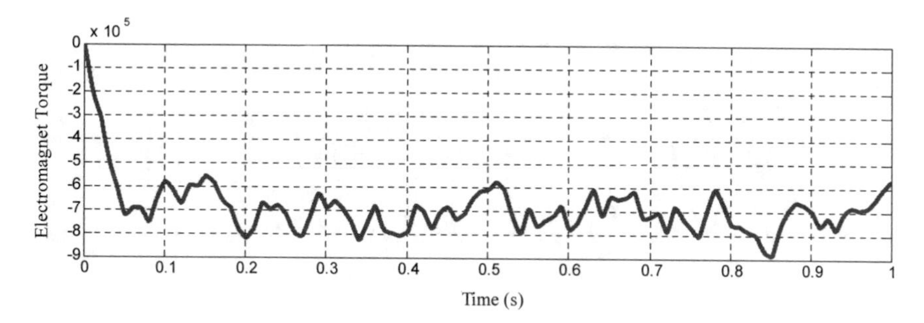

Fig. 5 Electromagnet torque developed by PMSG

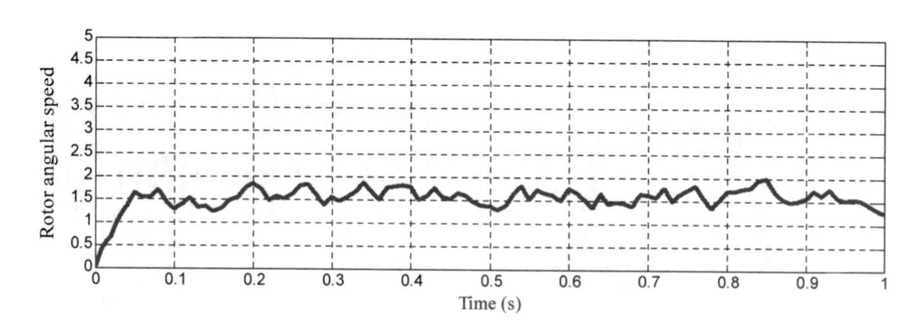

Fig. 6 Rotor angular speed of PMSG

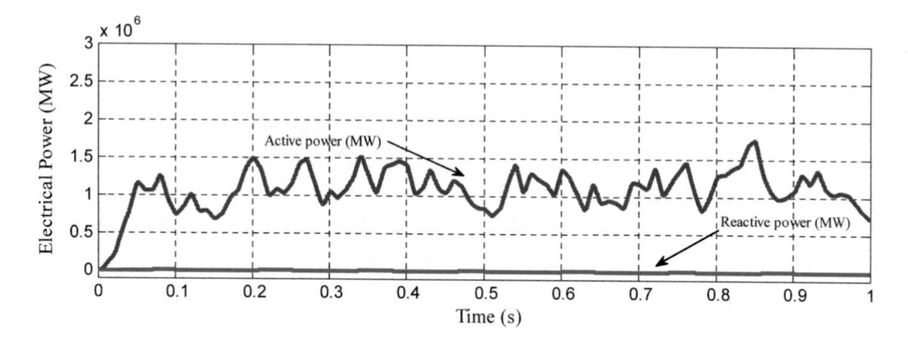

Fig. 7 Electrical power generated by PMSG

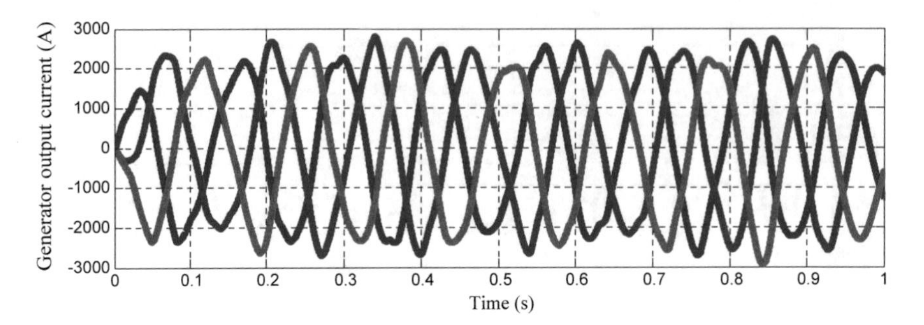

Fig. 8 Three phase output current of PMSG

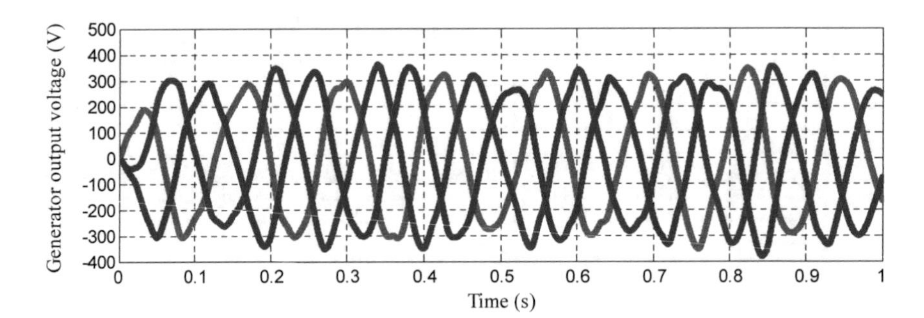

Fig. 9 Three phase output voltage of PMSG

4 Conclusion

This paper presents detail modeling and steady-state analysis of PMSG with uncertain wind speed system. The proposed model includes PMSG model, wind speed model, and wind turbine model. The proposed model is designed and implemented the MATLAB/Simulink platform. The efficacy of the proposed designed model is validated by building the dynamic model of the PMSG with uncertain system. Wind speed model quantifies the uncertain and randomness of wind speed. PMSG model is implemented with dq synchronous rotating reference frame and analysis is carried out for the steady-state mode. Simulation results verify the robustness and legitimacy of the implemented model in terms of quantification of uncertain wind speed while offering the parameters within the required limits. This model has the potential to provide a foundation for efficient quantification of uncertain wind speed.

References

1. Stiebler, M.: Wind Energy Systems for Electric Power Generation. Green Energy and Technology. Springer, New York, NY, USA (2008)
2. Mellah, H., Hemsas, K.E.: Simulations analysis with comparative study of a PMSG performances for small WT application by FEM. Int. J. Energy Eng. **3**(2), 55–64 (2013)
3. Bang, D.J., Polinder, H., Shrestha, G., Ferreira, J.A.: Review of generator systems for direct-drive wind turbines. In: Presented at the European Wind Energy Conference Exhibition Belgium (2008)
4. Rolan, A., Luna, A., Vazquez, G., Aguilar, D., Azevedo, G.: Modeling of a variable speed wind turbine with a permanent magnet synchronous generator. In: IEEE International Symposium on Industrial Electronics, 5–8 July Seoul, Korea, pp. 734–739 (2009)
5. Bhende, C.N., Mishra, S., Malla, S.G.: Permanent magnet synchronous generator-based standalone wind energy supply system. IEEE Trans. Sustain. Energy **2**(4), 361–373 (2011)
6. Babu, N.R., Arulmozhivarman, P.: Wind energy conversion systems-a technical review. J. Eng. Sci. Tech. **8**, 493–507 (2013)
7. MarufHossain, M., Hasan, M.: Future research directions for the wind turbine generator system. Renew. Sustain. Energy Rev. **49**, 481–489 (2015)
8. Carrillo, C., Diaz-Dorado, E., Silva-Ucha, M., Perez-Sabín, F.: Effects of WECS settings and PMSG parameters in the performance of a small wind energy generator. In: International Symposium on Power Electronics Electrical Drives Automation and Motion (SPEEDAM), Pisa, pp. 766–771(2010)

Short-Term Wind Speed Forecasting for Power Generation in Hamirpur, Himachal Pradesh, India, Using Artificial Neural Networks

Amit Kumar Yadav and Hasmat Malik

Abstract In this paper, wind speed (WS) forecasting in the mountainous region of Hamirpur in Himachal Pradesh, India is presented. The time series utilized are 10 min averaged WS data are utilized. In order to do WS forecasting, ANN models are developed to forecast WS 10, 20, 30 min, and 1 h ahead. Statistical error measures such as the mean square error (MSE), mean absolute error (MAE), root mean square error (RMSE), and mean error (ME) were calculated to compare the ANN models at 10, 20, 30 min, and 1 h ahead forecasting. It is found that statically error of 10 min ahead forecasting error is least. This study is useful for online monitoring of wind power.

Keywords Forecasting · Wind speed · Artificial neural network

1 Introduction

Due to the growing awareness of limited fossil fuels, interest in renewable energy particularly has increased in recent years. It is estimated that by 2020, about 12% of the entire world electricity demands should be fulfilled from wind energy resources [1]. Wind energy must address mostly by providing accurate short-term prediction of wind power which is dependent wind speed (WS). Long-term WS prediction is used for management of energy distribution and short-term WS is a key anticipatory control of wind turbines and online monitoring. The published literatures using Artificial Neural Network (ANN) for WS prediction at different time scales have

---

A. K. Yadav (✉) · H. Malik
Electrical and Electronics Engineering Department, NIT Sikkim,
Ravangla, Barfung Block, South Sikkim 737139, India
e-mail: amit1986.529@rediffmail.com

H. Malik
Electrical Engineering Department, IIT Delhi,
Hauz Khas 110016, New Delhi, India
e-mail: hmalik.iitd@gmail.com

© Springer Nature Singapore Pte Ltd. 2019

H. Malik et al. (eds.), *Applications of Artificial Intelligence Techniques in Engineering*, Advances in Intelligent Systems and Computing 697,
https://doi.org/10.1007/978-981-13-1822-1_24

been presented in [2–9]. Studies on short-time (measured in minutes) forecasting are deficient. This study fills this gap by focusing on developing short-time forecasting at 10, 20, 30 min, and 1 h ahead.

In the recent study, Ramasamy et al. [10] used MLP with LM algorithm for prediction of daily value WS for different cities in Himachal Pradesh, India. ANN model uses inputs as altitude, air pressure, temperature, and solar radiation. MAPE and *R* value are 4.55 and 0.98%, respectively. In this study, ANN is developed to forecast WS at 10, 20, 30 min, and 1 h ahead which is different from the previous study.

2 Methodology

2.1 Data Measurement

To develop a network, time series data of WS for CEEE NIT-H (lat: 31.68°N, long: 78.52°E, altitude of 775 m above mean sea level) are used (Fig. 1). The daily variation of WS for the year 2012 is shown in Fig. 2 and its normalized value are shown in Fig. 3. The measuring instruments and sensors specification are shown in Table 1. Out of 45,000, normalized value of WS 44,634 value are used for training and 366 values are used for testing the model.

2.2 Forecast Error

The forecast errors are evaluated by mean square error (MSE), mean absolute error (MAE), root mean square error (RMSE), and mean error (ME) are as follows:

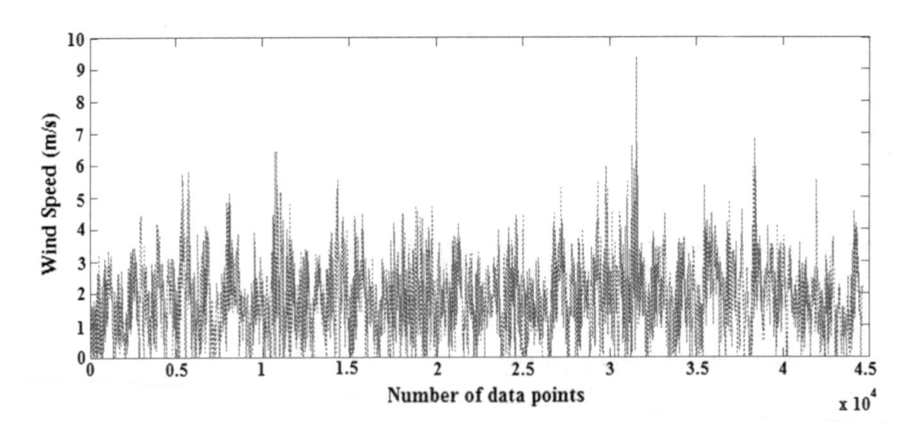

Fig. 1 Time series value of wind speed

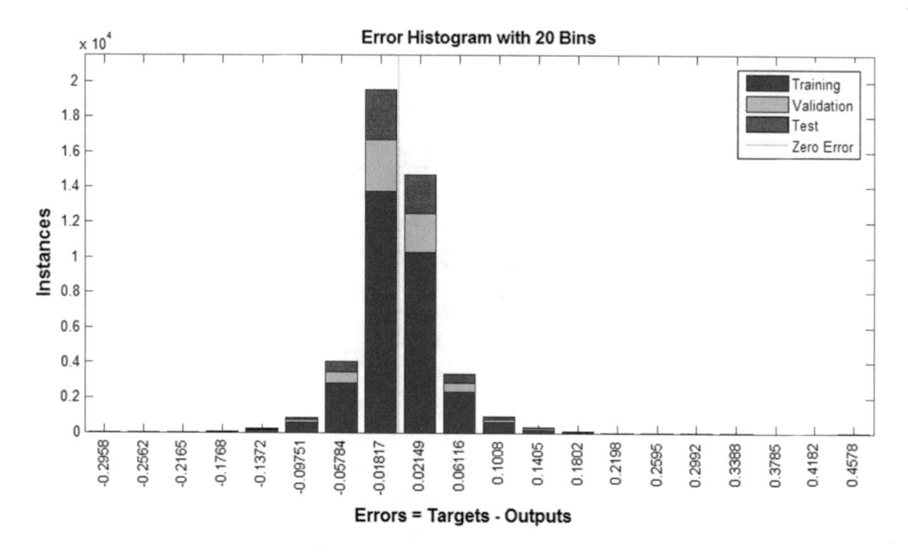

Fig. 2 EH plot of ANN-1 model for forecasting WS at 10-min ahead

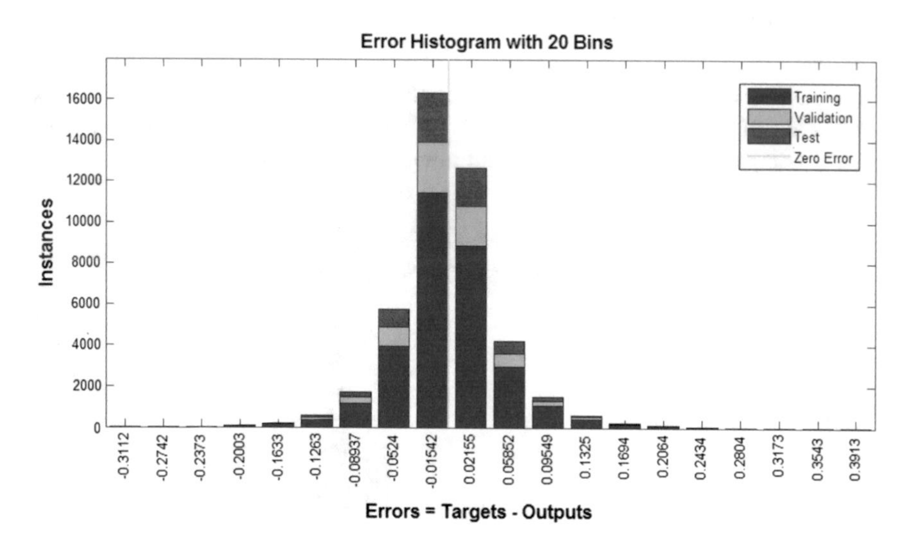

Fig. 3 EH plot of ANN-2 model for forecasting WS at 20-min ahead

Table 1 Forecasting accuracy evaluations of models

Models	Time step forecasting	ME	RMSE	MAE	MSE
ANN-1	10 min	0.0012	0.0267	0.0180	7.1288×10^{-4}
ANN-2	20 min	0.00042	0.0313	0.0219	9.8048×10^{-4}
ANN-3	30 min	0.0012	0.0328	0.0241	0.0011
ANN-4	1 h	−0.0032	0.0486	0.0343	0.0024

$$E_t = \mathrm{WS}_{i,\mathrm{measured}} - \mathrm{WS}_{i,\mathrm{forecast}} \tag{1}$$

$$\mathrm{MSE} = \frac{1}{N}\sum_{i=1}^{N} E_t^2 \tag{2}$$

$$\mathrm{MAE} = \frac{1}{N}\sum_{i=1}^{N} |E_t| \tag{3}$$

$$\mathrm{RMSE} = \sqrt{\frac{1}{N}\sum_{i=1}^{N} (E_t)^2} \tag{4}$$

$$\mathrm{ME} = \frac{1}{N}\sum_{i=1}^{N} E_t \tag{5}$$

3 Results and Discussion

Four ANN models (ANN-1, ANN-2, ANN-3, and ANN-4) are developed for forecasting of WS at 10, 20, 30 min, and 1 h ahead, respectively. The error histogram (EH) plot, regression plots, and comparison between measured and forecasted WS for ANN-1, ANN-2, ANN-3, and ANN-4 are shown in Figs. 2, 3, 4, 5,

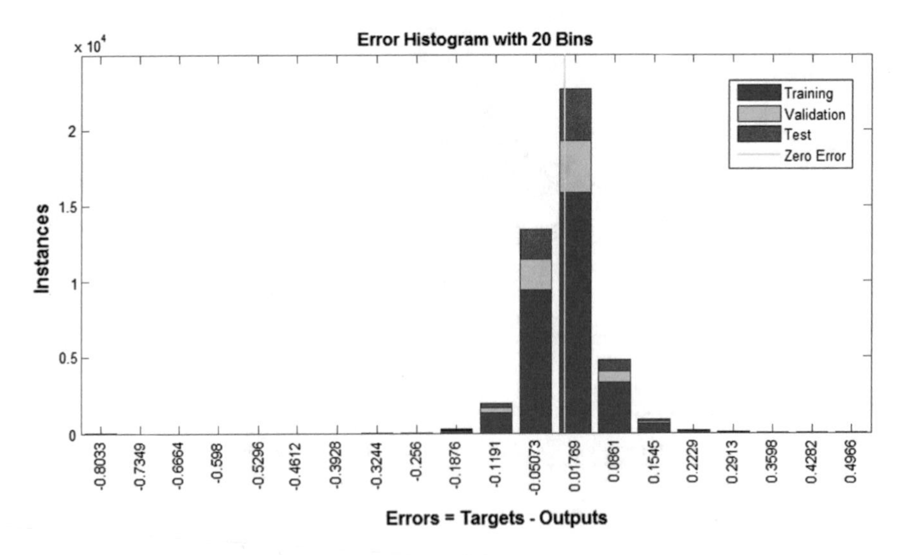

Fig. 4 EH plot of ANN-3 model for forecasting WS at 30-min ahead

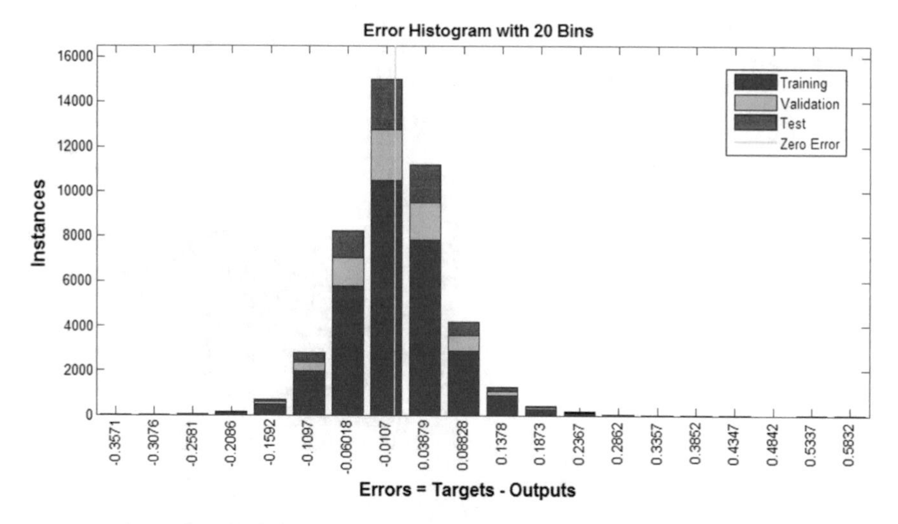

Fig. 5 EH plot of ANN-4 model for forecasting WS at 1-h ahead

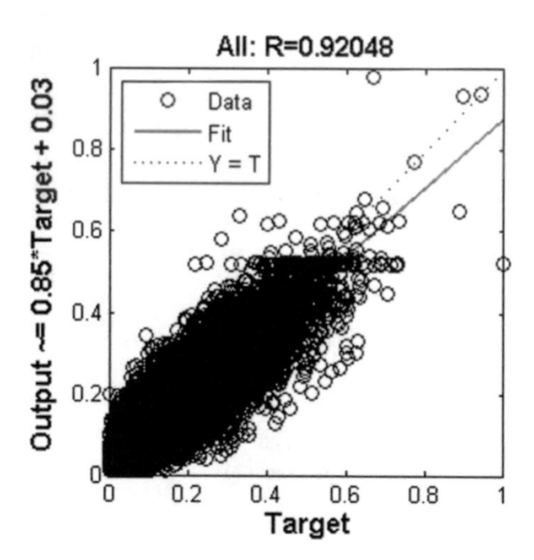

Fig. 6 Regression plot of ANN-1 model for forecasting WS at 10-min ahead

6, 7, 8, 9, 10, 11, 12, and 13, respectively. The regression *R* value for forecasting of WS at 10, 20, 30 min, and 1 h ahead are found to be 92.04, 87.10, 83.83, and 76.25%, respectively. For detailed analysis and implementation of ANN, references [11–19] can be referred by the reader.

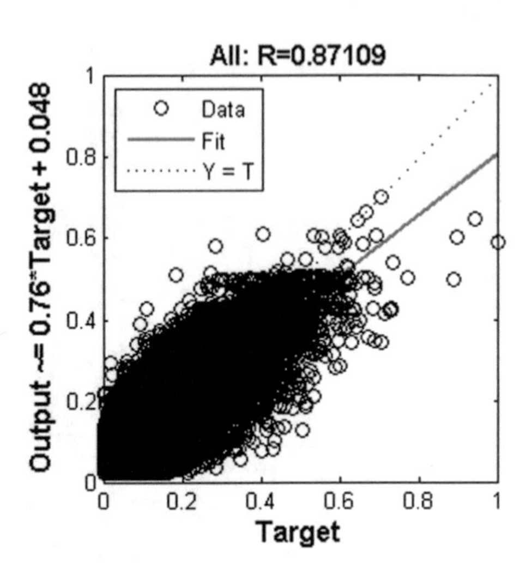

Fig. 7 Regression plot of ANN-2 model for forecasting WS at 20-min ahead

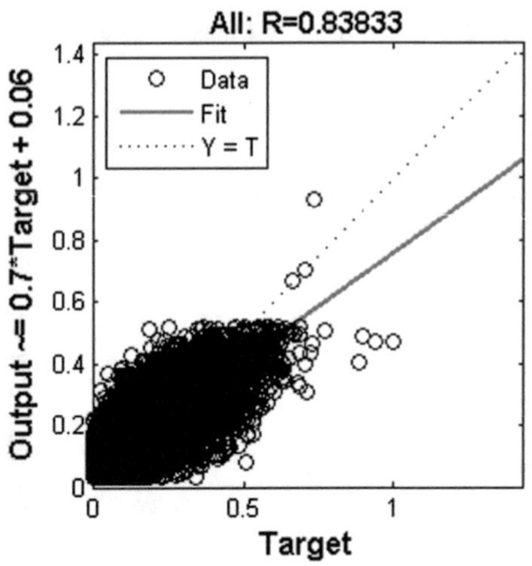

Fig. 8 Regression plot of ANN-3 model for forecasting WS at 30-min ahead

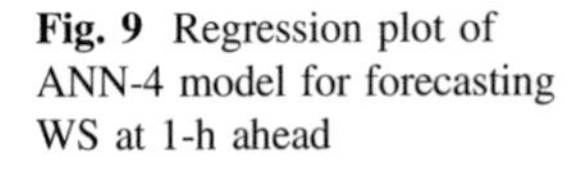

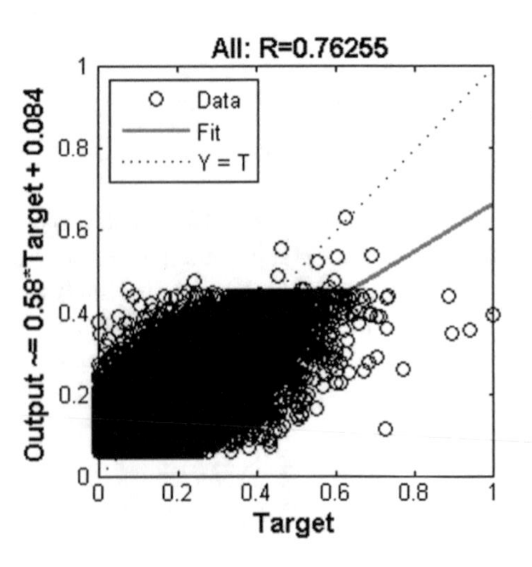

Fig. 9 Regression plot of ANN-4 model for forecasting WS at 1-h ahead

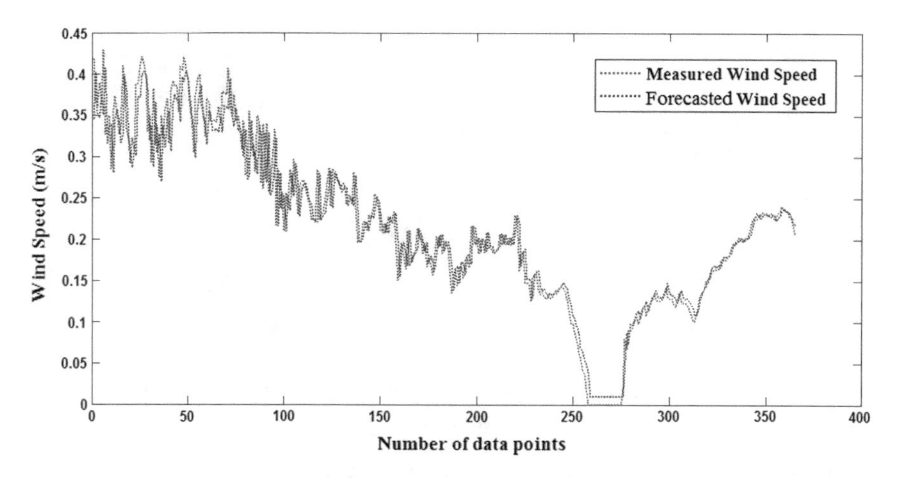

Fig. 10 Comparison between measured and forecasted WS ANN-1 model at 10-min ahead

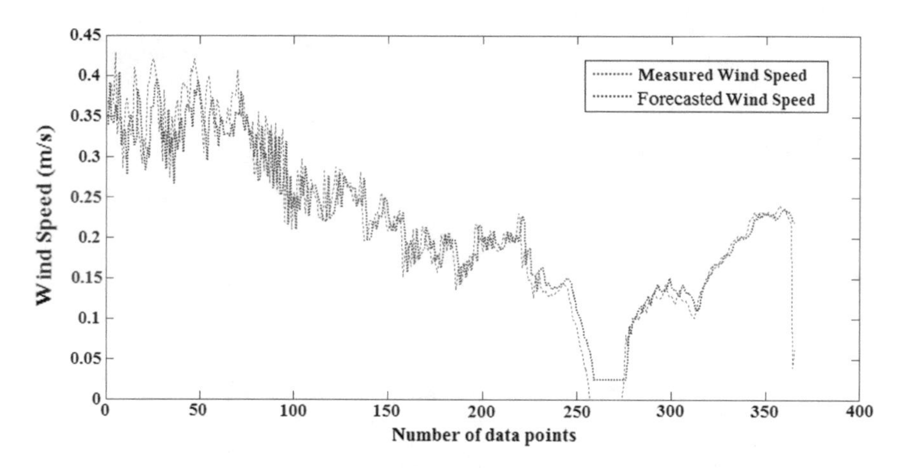

Fig. 11 Comparison between measured and forecasted WS ANN-2 model at 20-min ahead

4 Conclusion

ANN models have been proposed and developed for the WS forecasting at different time steps in mountainous region of Hamirpur Himachal Pradesh, India. The regression R value for forecasting of WS at 10, 20, 30 min, and 1 h ahead are found to be 92.04, 87.10, 83.83, and 76.25%, respectively which shows the suitability of the proposed approach in WS forecasting.

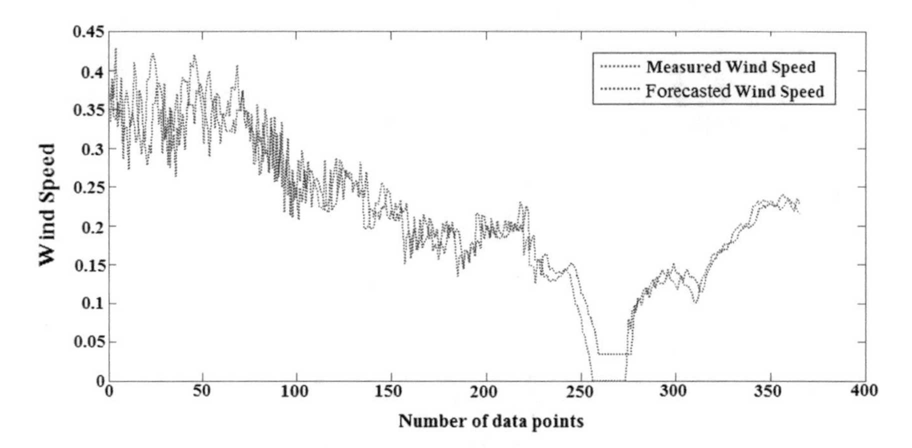

Fig. 12 Comparison between measured and forecasted WS ANN-3 model at 30-min ahead

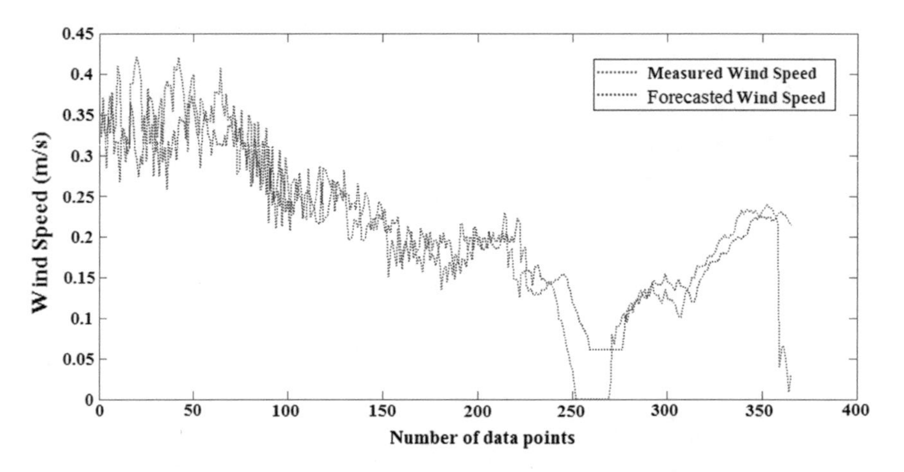

Fig. 13 Comparison between measured and forecasted WS ANN-4 model at 1-h ahead The forecasting accuracy of ANN models at 10, 20, 30 min, and 1 h ahead are evaluated by ME, RMSE, MAE, and MSE are shown in Table 1. The RMSE, MAE, and MSE for ANN-1 model is least

References

1. European Wind Energy Association: Wind force 12. Available: http://www.ewea.org/doc/WindForce12.pdf (2002)
2. İzgi, E., Öztopal, A., Yerli, B., Kaymak, M.K., Şahin, A.D.: Determination of the representatives time horizons for short-term wind power prediction by using artificial neural networks. Energy Sources Part A Recovery, Utilization Environ. Effects **36**, 1800–1809 (2014)

3. Xiaojuan, H., Xiyun, Y., Juncheng, L.: Short-time wind speed prediction for wind farm based on improved neural network. In: 8th World Congress on Intelligent Control and Automation, 5186–5190 (2010)
4. Sanz, S.S., Perez-Bellido, A., Ortiz-Garcia, E., Portilla-Figueras, A., Prieto, L., Paredes, D., Correoso, F.: Short-term wind speed prediction by hybridizing global and mesoscale forecasting models with artificial neural networks. In: International Conference on Hybrid Intelligent Systems, 608–612 (2008)
5. Zhang, W., Wang, J., Wang, J., Zhao, Z., Tian, M.: Short-term wind speed forecasting based on a hybrid model. Appl. Soft Comput. **13**, 3225–3233 (2013)
6. Daraeepour, A., Echeverri, D.P.: Day-ahead wind speed prediction by a neural network-based model. In: Innovative Smart Grid Technologies Conference, pp. 1–5 (2014)
7. Kusiak, Andrew, Li, Wenyan: Estimation of wind speed: a data-driven approach. J. Wind Eng. Ind. Aerodyn. **98**, 559–567 (2010)
8. Bilgili, M., Sahin, B., Yasar, A.: Application of artificial neural networks for the wind speed prediction of target station using reference stations data. Renew. Energy **32**, 2350–2360 (2007)
9. Mohandes, M.A., Rehman, S., Halawani, T.O.: A neural networks approach for wind speed prediction. Renew. Energy **13,** 345–354 (1998)
10. Ramasamy, P., Chandel, S.S., Yadav, A.K.: Wind speed prediction in the mountainous region of India using an artificial neural network model. Renew. Energy **80**, 338–347 (2015)
11. Malik, H., Mishra, S.: Artificial neural network and empirical mode decomposition based imbalance fault diagnosis of wind turbine using TurbSim, FAST and Simulink. IET Renew. Power Gener. **11**(6), 889–902 (2017). https://doi.org/10.1049/iet-rpg.2015.0382
12. Yadav, A.K., Malik, H., Chandel, S.S.: Application of rapid miner in ANN based prediction of solar radiation for assessment of solar energy resource potential of 76 sites in Northwestern India. Renew. Sustain. Energy Rev. **52**, 1093–1106 (2015). https://doi.org/10.1016/j.rser.2015.07.156
13. Azeem, A., Kumar, G., Malik, H.: Artificial neural network based intelligent model for wind power assessment in India. In: Proceedings IEEE PIICON-2016, pp. 1–6, 25–27 Nov. 2016. https://doi.org/10.1109/poweri.2016.8077305
14. Yadav, A.K., Malik, H., Chandel, S.S.: Selection of most relevant input parameters using WEKA for artificial neural network based solar radiation prediction models. Renew. Sustain. Energy Rev. **31**, 509–519 (2014). https://doi.org/10.1016/j.rser.2013.12.008
15. Saad, S., Malik, H.: Selection of most relevant input parameters using WEKA for artificial neural network based concrete compressive strength prediction model. In: Proceedings IEEE PIICON-2016, pp. 1–6, 25–27 Nov 2016. https://doi.org/10.1109/poweri.2016.8077368
16. Yadav, A.K., Sharma, V., Malik, H., Chandel, S.S.: Daily array yield prediction of grid-interactive photovoltaic plant using relief attribute evaluator based radial basis function neural network. Renew. Sustain. Energy Rev. **81**(2), 2115–2127 (2018). https://doi.org/10.1016/j.rser.2017.06.023
17. Malik, H., Yadav, A.K., Mishra, S., Mehto, T.: Application of neuro-fuzzy scheme to investigate the winding insulation paper deterioration in oil-immersed power transformer. Electr. Power Energy Syst. **53**, 256–271 (2013). http://dx.doi.org/10.1016/j.ijepes.2013.04.023
18. Malik, H., Savita, M.: Application of artificial neural network for long term wind speed prediction. In: Proceedings IEEE CASP-2016, pp. 217–222, 9–11 June 2016. https://doi.org/10.1109/casp.2016.7746168
19. Azeem, A., Kumar, G., Malik, H.: Application of Waikato environment for knowledge analysis based artificial neural network models for wind speed forecasting. In: Proceedings IEEE PIICON-2016, pp. 1–6, 25–27 Nov 2016. https://doi.org/10.1109/poweri.2016.8077352

Cost-Effective Power Management of Photovoltaic-Fuel Cell Hybrid Power System

Tushar Sharma, Gaurav Kumar and Nidhi Singh Pal

Abstract Solar energy is the most promising but yet an unreliable source of renewable energy. To overcome continuity and reliability issues in photovoltaic (PV) generation system, solar PV system is integrated with the other forms of energy sources, either conventional or non-conventional. In this paper, a photovoltaic (PV) system is combined with an unconventional source like fuel cell (FC) and a battery system to provide continuous power flow to the utility grid. The system usually focused on the power management of the overall system taking cost factor under consideration. The power management strategies took the high operating cost of the fuel cell under consideration. So, the fuel cell is required when power from PV array and battery is not sufficient for maintaining the constant power flow to the utility. The control operation is carried in such a way that provides effective, reliable, and quality power generation and also minimizes the overall operating cost of the system.

Keywords Photovoltaic (PV) system · Fuel cell (FC) · Utility grid
Power management · Battery

Nomenclature

SOFC	Solid oxide fuel cell
PV	Photovoltaic
SO_x	Sulphur oxides
NO_x	Nitrogen oxides

T. Sharma (✉) · G. Kumar · N. S. Pal
Electrical Engineering, Gautam Buddha University, Greater Noida, India
e-mail: tushar95sharma@gmail.com

G. Kumar
e-mail: realgauravkumar904@gmail.com

N. S. Pal
e-mail: nidhi@gbu.ac.in

© Springer Nature Singapore Pte Ltd. 2019

H. Malik et al. (eds.), *Applications of Artificial Intelligence Techniques in Engineering*, Advances in Intelligent Systems and Computing 697,
https://doi.org/10.1007/978-981-13-1822-1_25

DG	Distributed generation
DC, AC	Direct current, Alternating current
H_2, O_2	Hydrogen, Oxygen
IGBT	Induced gate bipolar transistor
SOC	State of charge

1 Introduction

The challenges in power generation are expected to increase more and more in the near future due to the increasing load demands worldwide. Fossil fuels are widely used as a primary source for power generation but at the cost of polluting the environment. They produce highly toxic SO_x and NO_x gases, which results in depleting air quality. Renewable and clean energy sources give rise to the awareness for environmental protection and reduce the desire for fossil fuels [1].

Renewable energy sources like solar, wind, hydrogen fuel cells, etc., have the potential to become the primary source for power generation and these are the sources for clean energy [2]. Photovoltaic (PV) system is not very reliable due to the varying solar irradiance and temperature other sources such as fuel cells are more reliable but they have economic issues due to their dependency on hydrogen gas and hydrocarbons. Merging two or more energy sources either non-conventional or conventional provides reliable operation and intelligent operation also gives the efficient and cost-effective generation [3]. Fuel cell raises up to be an effective power source for hybrid generation this is due to the high efficiency, power matching, fast response and remote application. The hybrid operation of solid oxide fuel cell (SOFC) and microturbines are well suited for the efficiency, performance-related issues, and reliability [4–8].

The hybrid operation of a fuel cell with PV is considered reliable in an interconnected utility mode as well as for standalone application. The enhancement in performance is achieved by adding a simple storage device like battery, capacitors, etc. Storage devices are simply needed for the times like peak loading or employed when the primary source is unavailable due to the varying environmental conditions. Batteries are used mainly as the storage device during hybrid operation. Devices like supercapacitors can also be implemented due to their faster response but the initial cost of installation is marginally large as the storage is required in large capacity for any DG operation [9].

The three hybrid power systems, i.e., PV/battery system, PV/FC system, and PV/FC/battery systems are analysed, optimized, and compared. Results favours the PV/FC/battery system, and the system is equipped with higher efficiency, lower cost, and less capacity of PV array are required [10]. The hybrid PV-FC and battery system have the same advantage as PV-battery and diesel hybrid system but the previous is a clean energy plant and gives quality power generation without polluting the environment.

The hybrid PV/FC generation system along with battery as a storage device is used to provide reliable and cost-effective power generation and maintains the power flow to the grid side. Solid oxide fuel cell (SOFC) is more suitable in areas like distributed generation (DG) system due to its high efficiency, better fuel flexibility, and high-temperature operation [11]. All the sources are linked to a single DC bus through various DC-DC converters. The system is grid-connected and works as a DG plant. A fixed/reference power is set by the inverter controller. Power management strategies are implemented in supplying the reference power to the utility grid while making sure about the cost-effective power generation.

2 The Hybrid Power Generation System

The overall system comprises of a PV array, a SOFC stack, and lithium-ion battery. All the energy sources are connected to a single 500 V DC bus through befitting DC-DC converters. PV array comprises of an MPPT boost converter and the battery is equipped with bidirectional converter for the charging and discharging purposes.

Figure 1 decodes the typical simulation model into a block diagram. As Fig. 1 suggests that the DC bus is connected to a three-level bridge converter. The bridge converter is controlled by the voltage source controller. The bridge converter converts 500 V DC into 260 V AC. The overall system passes through a transformer and then fed to the 25 kV utility grid.

A fixed/reference power is delivered to the utility grid with the help of inverter control. Almost, all the reference power is fulfilled by power from PV array, and the remaining power is provided by the battery system. Thus, the battery acts as the second source for power generation. The fuel cell is not used in place of the battery due to the economic issues related to the operation of SOFC while SOFC delivers

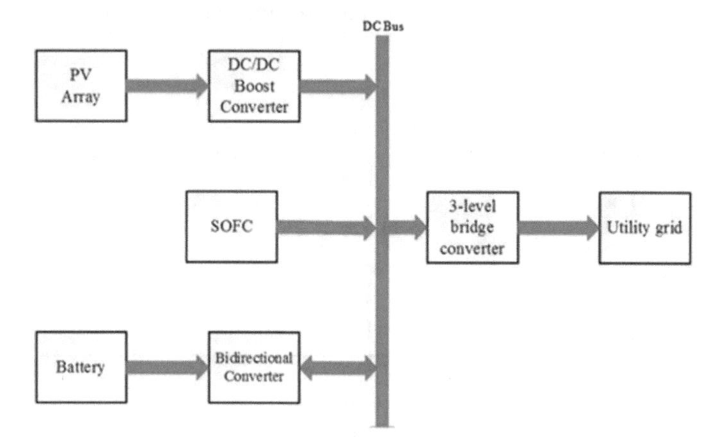

Fig. 1 Basic block diagram of the proposed model

the power to the grid when power from PV array and battery is not sufficient to reach the reference power.

The aim is to maintain the system stability and power flow along with the minimization of operating cost. The PV array and battery bank does not require any cost in their operation. There is only the requirement of one-time installation capital while the operating cost is only affected by SOFC usage due to the high price and high consumption of hydrogen.

3 System Modelling

3.1 Modelling of Photovoltaic Cell

The power from a PV array varies according to the solar irradiance level and temperature of the array. Figure 2 shows the equivalent circuit of PV cell for one-diode model. The cell photocurrent is represented by current source Iph. Rs, and Rsh are the intrinsic series and shunt resistance of the cell, respectively.

Equation of different currents of solar cell is given as I_{ph}—photocurrent.

$$I_{ph} = [I_{sc} + K_i(T - 298)] * I_r/1000 \quad (1)$$

It directly depends upon solar irradiance and ambient temperature of the cell. I_{sc} is the short-circuit current (A) and K_i is the short-circuit current of the cell at ideal temperature 250 °C.

I_{rsat}—Modules reverse saturation current

$$I_{rsat} = I_{sc} / [\exp(q.V_{oc} / N_s nKT) - 1] \quad (2)$$

where electron charge, $q = 1.6 \times 10^{-19}$ °C, V_{oc} is the open-circuit voltage, n is ideality factor of diode D, and K is the Boltzmann's constant, $K = 1.3805 \times 10^{-23}$ J/K

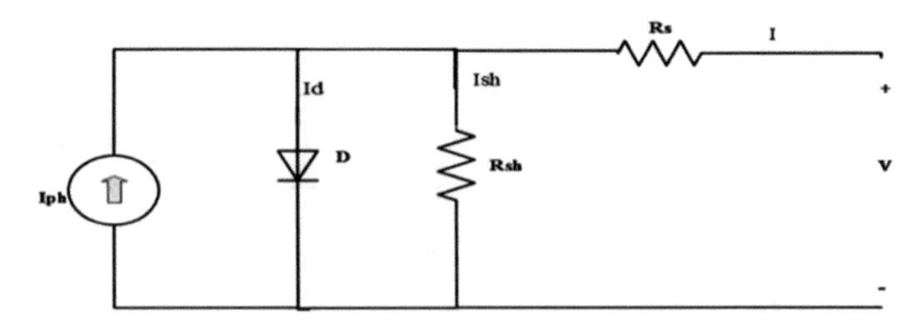

Fig. 2 Circuit diagram for one-diode model of solar cell [12]

Table 1 Parameters of the PV module

Symbol	Parameter	Value
T_n	Nominal temperature	298 K
R_{sh}	Shunt resistance	269.59 Ω
R_s	Series resistance	0.3715 Ω
V_{oc}	Open-circuit voltage	64.2 V
I_{sc}	Short-circuit current	5.96 A
N_s	Number of cells connected in series	96
N_p	Number of cells connected in parallel	1
n	Diode ideality factor	0.94504

I_0—The module saturation current

$$I_0 = I_{rsat}\,[T/T_n]3\exp\left[q * E_{go}\left[1/T_n - 1/T\right]/nK\right] \tag{3}$$

where E_{go} is the band gap energy of the semiconductor,

E_{go} = 1.1 eV, T_n is the nominal temperature
T_n = 298 K I—PV module output current

$$\begin{aligned} &I = N_p * I_{ph} - N_p * I_0 * \left[\exp\left(\left(V/N_s + I * R_s/N_p\right)/n * V_t\right) - 1\right] - I_{sh} \\ &V_t = K * T/q \\ &\text{And } I_{sh} = V * N_p/N_s + IR_s/R_{sh} \end{aligned} \tag{4}$$

where

N_p = number of modules connected in parallel
R_{sh} = shunt resistance (Ω)
R_s = series resistance (Ω)
V_t = diode thermal voltage (V)

Table 1 gives the values of various parameters used in simulation of PV module. Combining PV modules in various ways of series and parallel connection gives the different voltage and current from the array.

3.2 *Maximum Power Point Tracking (MPPT)*

Incremental conductance algorithm (IC) MPPT is used to retrieve the maximum power from PV array. IC works effectively on vast irradiance changing environment and improves the tracking time. The maximum power point (MPP) can be determined using d*I*/d*V* and −*I*/*V* relation. The algorithm IC works on the equations (Fig. 3).

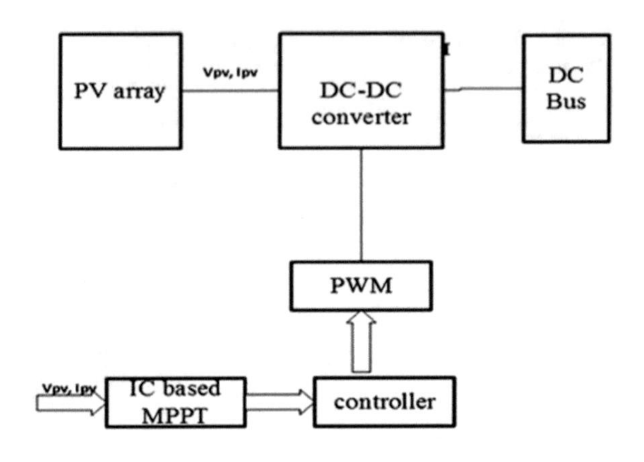

Fig. 3 Block diagram of MPPT converter

$$\begin{aligned} \mathrm{d}P/\mathrm{d}V &= \mathrm{d}(V.I)/\mathrm{d}V = I.\mathrm{d}V/\mathrm{d}V + V.\mathrm{d}I/\mathrm{d}V \\ &= I + V.\mathrm{d}I/\mathrm{d}V \end{aligned} \tag{5}$$

Maximum power point is reached when

$$\begin{aligned} &\mathrm{d}P/\mathrm{d}V = 0, \\ &\text{so}, \mathrm{d}I/\mathrm{d}V = -I/V \end{aligned} \tag{6}$$

3.3 Modelling of SOFC

The dynamic model for the fuel cell for performance in normal operation is simulated. Based on dynamic model, fuel cell control strategies, power section, and fuel processors response function is added to SOFC model [11, 13]. SOFC is ideal for DG applications where power is generated at the load side itself. SOFC is fed with H_2 and O_2 and the gases are ideal. Fuel utilization factor is necessary in calculation for efficiency of the stack. Practically, 80–90% fuel utilization is used. The fuel cell demand current is restricted for input hydrogen flow in the range.

$$0.8q^{\mathrm{in}}\mathrm{H}_2/2K_r \leq 0.9q^{\mathrm{in}}\mathrm{H}_2/2K_r \tag{7}$$

The overall fuel cell reaction is

$$\mathrm{H}_2 + 1/2\mathrm{O}_2 \rightarrow \mathrm{H}_2\mathrm{O} \tag{8}$$

Stoichiometric ratio of oxygen to hydrogen is 1–2. The electrochemical reactions occurring within the cell are:

Anode

$$1/2O_2 + 2e^- \rightarrow O \tag{9}$$

Cathode

$$H_2 + 1/2O_2 \rightarrow H_2O + 2e^- \tag{10}$$

CO and hydrocarbons such as CH_4 can also be used as fuels in SOFC. Due to high temperatures within the cell, it is feasible for the water gas shift reaction (Fig. 4).

$$CO + H_2O \rightarrow H_2 + CO_2 \tag{11}$$

And the steam reforming reaction

$$CH_4 + H_2O \rightarrow 3H_2 + CO \tag{12}$$

Nernst equation relates the standard electrode potential of an electrochemical reaction to a reduction potential. Nernst equation (Table 2)

$$E = E^o + RT/2F * \ln\left[PH2.\,P^{1/2}O_2/PH_2O\right] \tag{13}$$

where E = electromotive force or open-circuit voltage E^o = emf at standard pressure, and R = universal gas constant = 8.314 JK^{-1} mol^{-1}.

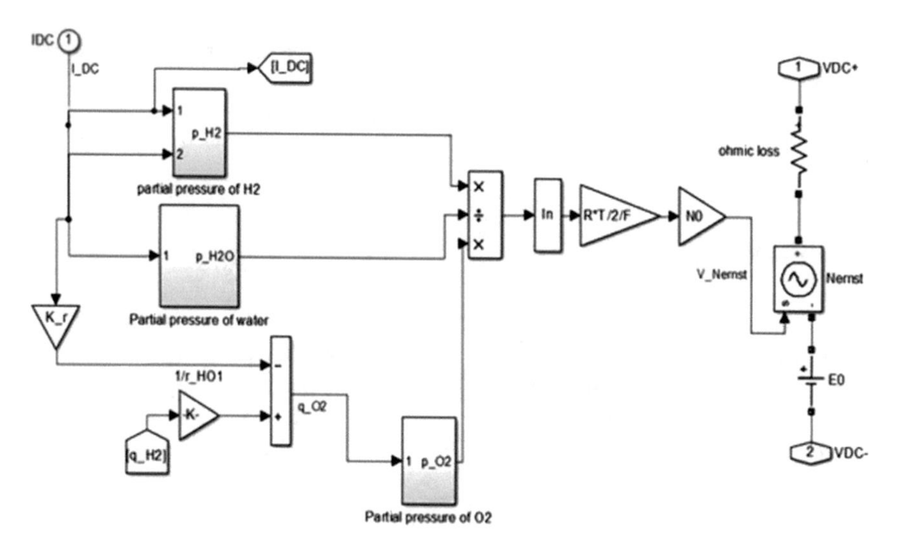

Fig. 4 SOFC model in Matlabsimulink with Nernst equation and calculations of various partial pressures

Table 2 Gives the values of SOFC used while modelling the SOFC

Symbol	Parameter	Value
P_{rate}	Rated power	50 kW
T	Absolute temperature	1273 K
F	Faraday's constant	96,487 C/mol
N_0	Number of cells in series in stack	500
r	Ohmic losses	3.28e−4 Ω
R	Universal gas constant	8314 J/(k mol K)
Eo	Ideal standard potential	1.18 V
K_r	Constant, K_r = No/4F	0.996e−6 k mol/(sA)
RH_O	Ratio of hydrogen to oxygen	1.145
PF	Power factor	1

Table 3 parameters for battery modelling

Symbol	Parameter	Value
Ah	Rated capacity	6.5 Ah
SOC	Initial state of charge	85%
I	Nominal discharge current	2.8261 A
Ω	Internal resistance	0.76923 Ω
V	Nominal voltage	500 V
	Fully charged voltage	581.9 V
Ω	Internal resistance	0.76923 Ω

3.4 Battery Model

The battery used in the modelling is lithium–ion battery. Battery works as a secondary source and fulfills the power requirement when solar irradiance is low. Batteries are usually rated in ampere hours (Ah) and Ah is usually the size of battery. Table 3 gives of battery system.

4 Simulation Results

To study the proposed system, a Matlab simulation model of hybrid PV/FC/battery-based DG system is prepared. A 260 V, 100 kW PV array, a 500 V, 50 kW SOFC, and a 500 V, 6.5 Ah battery system is connected to a single DC bus of 500 V. The model is simulated for 10 s. Figure 5 is the simulated model of the hybrid system. IC algorithm-based MPPT is implemented for determining the maximum power point (MPP). A DC–DC boost converter is used to step up the PV array voltage from 260 to 500 V. Battery is connected to the DC bus using a bidirectional converter for charging and discharging of battery. The gate pulse for bidirectional converter is provided by the battery controller. It maintains the battery voltage at 500 V and switches in when power from PV array is not sufficient

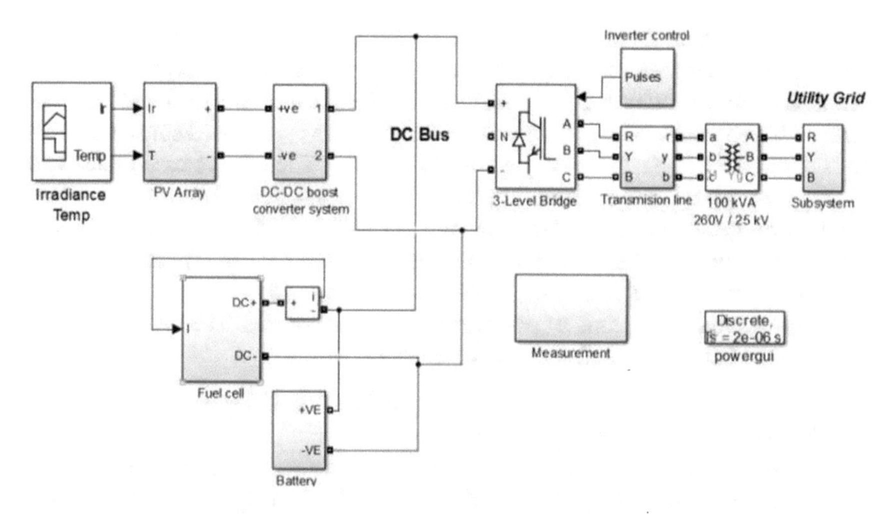

Fig. 5 Matlab Simulink model of PV/FC/battery connected to the grid

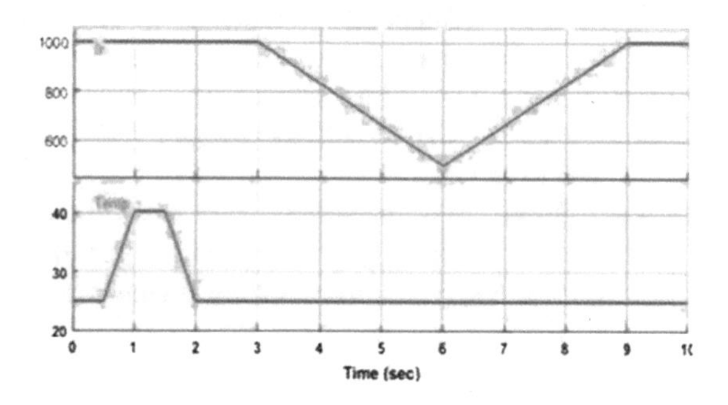

Fig. 6 Signal builder to create input pulses for PV array. Input pulses are temperature and irradiance

enough to fulfil power demand. The 500 V DC bus is connected to a three-level bridge converter made up of IGBT/diodes. It converts 500 V DC to 260 Vph-ph AC. The inverter controller sets the reference power delivered to the grid. The AC side of three-level bridge converter is connected to a small transmission line containing only inductance and resistance. The transmission line is then connected to a 3-ϕ star-delta transformer of 100 KVA that converts 260 V to 25 kV. Transformer supplies the power to the utility grid of 25 kV.

Figure 6 shows the pulses by signal builder, it helps in manually setting the magnitude of the input pulses of PV array. Irradiance and temperature are the pulses from signal builder. Both the pulses are varied manually to show effect of temperature and irradiance on PV power. Initially the irradiance is taken 1000 W/m^2 and temperature is kept at 25 °C. Magnitude of irradiance is kept 1000 and temperature is kept at 25 °C from 0 to 0.50 s. Temperature is increased constantly up to 40 °C from 0.5 to 1 s and it is kept at 40 °C till 1.5 s. Then, it decreased linearly to 25 °C in period of 1.5–2 s. From time 0 to 3 s, the irradiance is kept at 1000 W/m^2. It is linearly decreased up to 500 W/m^2 in time 3–6 s and linearly brought back to 1000 W/m^2 from 6 to 9 s. From 9 to 10 s, the irradiance level is kept at 1000 W/m^2 and from 2 to 10 s, the temperature is kept constant at 25 °C.

Figure 7 depicts the graph of power from different sources. The reference power (P_{ref}) is kept constant at 103 kW using inverter control.

a. At instance of 0.5 s, P_{ref} is 103 kW. PV provides 100 kW power and remaining 3 kW power is provided by battery.
b. From 0.5 s, the power from PV array is linearly decreased to show the effect of temperature on PV array. Due to the marginally slow dynamics of battery, it starts providing power from 0.58 s. PV power keeps on decreasing from 100 to 91.2 kW up to time 1 s and again brought back to 100 kW from time 1.5 to 2 s. To fulfil the reference power demand, battery power is increased from 3 to 9.8 kW and brought back to 3 kW at time 2 s. SOFC have no role in providing power to the grid in that interval.
c. From 2 to 3 s, the PV power is kept at 100 kW and the remaining supply is fulfilled by battery.
d. Again PV power is decreased to 50 kW from time 3 to 6 s due to the decreasing irradiance level. Battery gives power at higher discharge rate from 3 to 3.2 s and from 3.2 to 6 s, power from battery is slowly increased to 22 kW. SOFC power is linearly increased to 38 kW from 3.2 to 6 s.

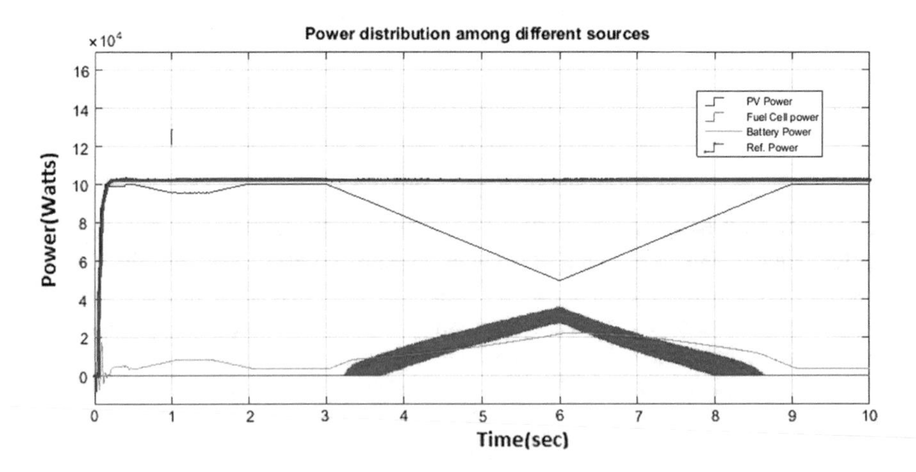

Fig. 7 Power distribution graph for the different sources

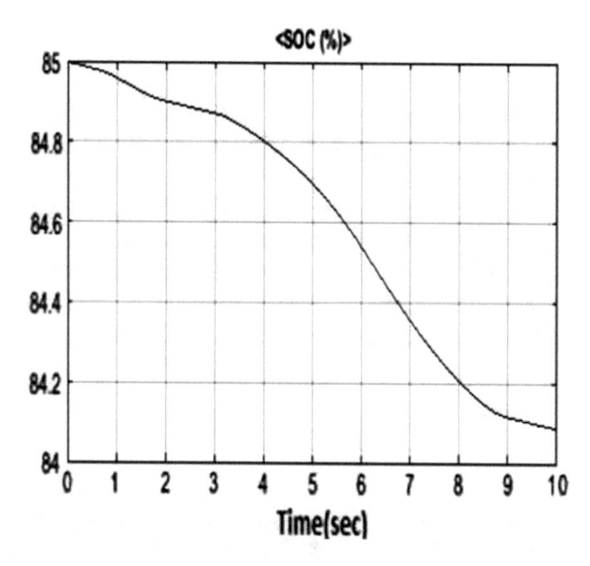

Fig. 8 State of charge of battery during the operation from 0 to 10 s

Table 4 Cost estimation of SOFC operation

Symbol	Provided energy (kWh)	Cost (operating for 10 s) INR	Cost (operating for 1 h)
SOFC	0.057	0.9405 $ = 60.85 INR	21,906 INR
Difference	0.005	0.0825 $ = 5.33 INR	1921 INR
SOFC used as secondary source	0.062	1.023 $ = 66.188 INR	23,828 INR

e. PV power is linearly increased from 50 to 100 kW from time 6 to 9 s. So, the SOFC power is decreased from 38 to 0 kW from time 6 to 8.7 s. SOFC is inactive from 8.7 to 10 s. Battery power is decreased to 3 kW till 9 s and kept constant up to 10 s.

Figure 8 shows the state of charge of battery while operating it to 10 s. SOC gives the amount of battery charged in percentile. It may be varied from 0 to 1.

Table 4 gives the cost estimation while operating SOFC. Cost of operation of fuel cell is 16.5 $ per kWh. This is calculated by number of litres of hydrogen required by fuel cell and the current price of hydrogen barrels. The energy provided by SOFC in the model that runs for time interval of 10 s is 0.057 kWh. Cost of operation of fuel cell is 0.9405 $ and this is the overall cost while supplying 103 kW to the grid (Table 4).

If SOFC is used as a secondary source in place of battery and battery is used in place of SOFC. From Fig. 7, the graph of battery is taken as the graph of SOFC and vice versa then, the energy by SOFC would be 0.062 kWh and the cost of operation becomes 1.023 $. The difference while operating as secondary source is 0.0825 $. If such operations would carry on for 1 h then this difference becomes 29.7 $ or 1921 INR.

5 Conclusion

This paper gives the effective power management strategy for the operation of PV/FC/battery as a DG system. The power flow from three-level bridge inverter to the grid is maintained by switching intelligently between all the sources. The cost-effectiveness of the model is determined by finding the energy usage of fuel cell as secondary as well as tertiary source of energy. Fuel cells have high installation and operational cost as well and operational cost of the whole system is only affected by fuel cell operation. Result shows the continuous power flow to the grid and also proves the authenticity as a tertiary source for operation.

References

1. Shuo-Ju, C., Chang, K.T., Yen, C.Y.: Residential photovoltaic energy storage system. IEEE Trans. Ind. Electron. **45.3**, 385–394 (1998). https://doi.org/10.1109/41.678996
2. Erdinc, O., Uzunoglu, M.: Optimum design of hybrid renewable energy systems: overview of different approaches. Renew. Sustain. Energy Rev. **16**(3), 1412–1425 (2012)
3. Thounthong, P., et al.: Intelligent model-based control of a standalone photovoltaic/fuel cell power plant with supercapacitor energy storage. IEEE Trans. Sustain. Energy **4.1**, 240–249 (2013). https://doi.org/10.1109/tste.2012.2214794
4. Nayak, S.K., Gaonkar, D.N.: Fuel cell based hybrid distributed generation systems "a review". In: 2013 8th IEEE International Conference on Industrial and Information Systems (ICIIS). IEEE (2013). https://doi.org/10.1109/iciinfs.2013.6732039
5. Li, W., Zhu, X., Cao, G.: Modeling and control of a small solar fuel cell hybrid energy system. J. Zhejiang Univ. Sci. A **8**(5), 734–740 (2007)
6. Jiang, Z.: Power management of hybrid photovoltaic-fuel cell power systems. Power Engineering Society General Meeting, 2006. IEEE. IEEE (2006). https://doi.org/10.1109/pes.2006.1709000
7. Hossain, S.M., Uddin, M.N., Palash, K.A.: Feasibility of photovoltaic–fuel cell hybrid system to meet present energy demand. Am. Sci. Res. J. Eng. Technol. Sci. (ASRJETS) **26.1**, 204–212 (2016)
8. Kisacikoglu, M.C., Uzunoglu, M., Alam, M.S.: Load sharing using fuzzy logic control in a fuel cell/ultracapacitor hybrid vehicle. Int. J. Hydrogen Energy **34**(3), 1497–1507 (2009)
9. Hwang, J.J., et al.: Dynamic modeling of a photovoltaic hydrogen fuel cell hybrid system. Int. J. Hydrogen Energy **34**(23), 9531–9542 (2009)
10. Thounthong, P., et al.: Energy management of fuel cell/solar cell/supercapacitor hybrid power source. J. Power Sources **196.1**, 313–324 (2011)
11. Zhu, Y., Tomsovic, K.: Development of models for analyzing the load-following performance of microturbines and fuel cells. Electr. Power Syst. Res. **62**(1), 1–11 (2002)
12. Nguyen, X.H., Nguyen, M.P.: Mathematical modeling of photovoltaic cell/module/arrays with tags in Matlab/Simulink. Environ. Syst. Res. **4.1** (2015)
13. Cingoz, F., Elrayyah, A., Sozer, Y.: Optimized resource management for PV–fuel-cell-based microgrids using load characterizations. IEEE Trans. Ind. Appl. **52**(2) (2016). https://doi.org/10.1109/tia.2015.2499287

Long-Term Solar Irradiance Forecast Using Artificial Neural Network: Application for Performance Prediction of Indian Cities

Hasmat Malik and Siddharth Garg

Abstract Solar radiation data is extremely useful for utilizing solar energy in applications like solar power plants and solar heating. As the fossil fuel resources are degrading, renewable sources of energy like solar energy can reduce our dependence on fossil fuels. Solar energy is also a clean form of energy. In this paper, we have made use of Artificial Neural Network (ANN) for Solar Radiation Prediction (SRP). The ANN network used is Feed Forward with Backpropagation (FFBP), Backpropagation being the learning algorithm. A three-layer network has been used with one hidden layer. The data used was from 67 cities in India, which was further divided in two sets, namely—training and testing. The testing data was not used to train the network. There were 19 input parameters for the network with one output parameter solar radiation.

Keywords ANN (Artificial neural network) · MLP (Multi-layer perceptron) FFBP (Feedforward backpropagation)

1 Introduction

Solar radiation is an important constraint for solar energy study but it is not readily available for sites because of unavailability of solar radiation measuring equipment (pyranometer is a solar radiation measurement instrument) at the meteorological station. So, it becomes necessary to predict the solar radiation by making use of different climatic variables. These different climatic variables can be obtained from weather stations and also from satellite data.

H. Malik (✉) · S. Garg
Electrical Engineering Department, IIT Delhi, Hauz Khas, New Delhi 110016, India
e-mail: hmalik.iitd@gmail.com

S. Garg
ICE Division, NSIT Delhi, Dwarka, New Delhi 110078, India
e-mail: gdsidd123@gmail.com

© Springer Nature Singapore Pte Ltd. 2019

H. Malik et al. (eds.), *Applications of Artificial Intelligence Techniques in Engineering*, Advances in Intelligent Systems and Computing 697,
https://doi.org/10.1007/978-981-13-1822-1_26

In order to harness solar energy, an accurate SRP is required in the proposed locations. Solar energy is measured with the use of solar measuring equipment, but due to unavailability of these devices in some remote and rural locations which especially have high potential for solar plant installation, making use of the prediction tools to estimate the solar potential is a very good option.

Artificial neural network as an estimation tool is very efficient. It has has been proved that in predicting different parameters using other parameters that their relationship with each other is not specific. Two important parameters in indicating the amount of solar radiation in a given region are climatological and meteorological. Prediction of solar radiation and the effects of meteorological can be very efficiently found out using artificial neural networks.

In the previous many studies, ANN has been applied to predict solar radiation. Azadeh [1] et al. have used MLP network, Backpropagation learning in order to SRP for six cities in Iran.

Eleuch et al. [2] have used an adaptive α-model for the Jeddah site that can predict the hour-by-hour solar irradiance with respectable accuracy results.

Hasni et al. [3] have used month, day, hour, temperature and relative humidity data for Bechar city as training data set for training an FF-ANN using Backpropagation algorithm.

Yadav et al. [4] have prepared a review of solar radiation prediction methods making use of ANN techniques. The aim of their study was to recognize the apt methods already in literature and also discover the research gaps. They found that the ANN model with MLP architecture was the most apt method for SRP.

The regression value (R^2) or the R-squared value is the statistical measure of how close the predicted output is from the actual output.

2 ANN Approach

ANN has the ability to correctly make out a non-linear mapping between inputs and outputs and store it in the form of weights and biases. The final aim is to construct a model that fittingly maps out input to output. The historical data is fed as input to the model with our target data as the output. The model can then be subsequently used to obtain the output by feeding the input parameters to the model. ANN model consists of:

- input layer to which we feed the input parameters;
- output layer which sends information to the computer;
- one or more hidden layers which are placed in between input layer and output layer.

The model representing multi-layered neural network is shown in Fig. 1.

2.1 The input signals are represented as inputs $X_{k,1}$, $X_{k,2}$ …., $X_{k,\mathrm{p}}$ for a k layer network.

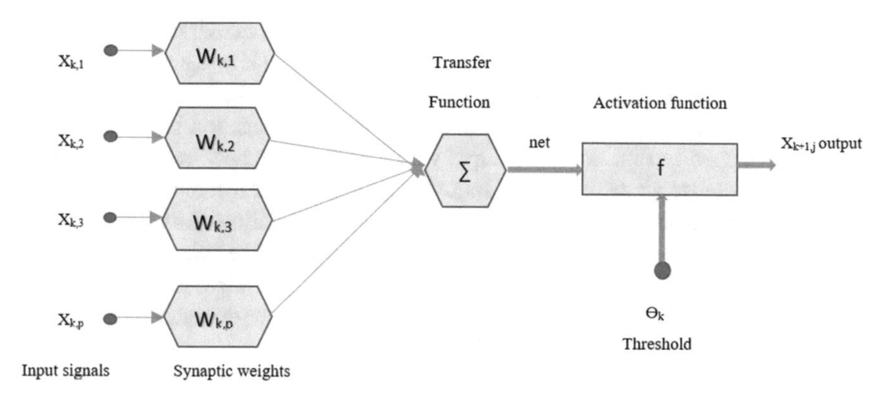

Fig. 1 Architecture of a multi-layered neural network

2.2 Signal O for the output to the next layer is given by

$$O = f(\text{net}) = f(w^T x_k) = f\left(\sum_{j=1}^{p} w_{k,j}, x_{k,j}\right) \tag{1}$$

$W_{k,j}$ represents synaptic weights (where k represents layer index and j represents neuron index) and f(net) represents Transfer Function or an Activation Function.

2.3 The net is scalar product of the input vectors and weight which is then given as input to the Activation Function

2.4 The Activation Function (f) then transforms the input accordingly and gives the output accordingly

2.5 The multiple steps for ANN implementation consist of selecting:

- ANN size number of input layers. Output layers and hidden layers;
- ANN structure;
- Learning Algorithms;
- Training set and test set;
- Input data.

For more information of ANN implementation and mathematical modelling for different prediction and forecasting problems, Refs. [6–21] can be referred.

3 ANN Implementation

The steps followed for implementation of ANN are:

3.1 **Data Collection**—The data that has been collected for this study is mixed, i.e. from both Atmospheric Science Data Centre (ASDC) at NASA Langley Research Centre [5] and recorded data from CWET, India. Data from 67 cities were used and cities location is represented by a chart in Fig. 2. The 19 input variables were, namely–relative humidity, altitude, air temperature, latitude, longitude, earth temperature, heating degree days, cooling degree days, elevation, heating design temperature, atmospheric pressure, cooling design temperature, standard deviation, earth temperature amplitude, frost days at site, air density power law index (PLI), mean monthly wind speed, energy pattern factor and monthly wind power density.

3.2 **Data Processing**—We have divided the 12 months data into 8 months for training and 4 months for testing the results. We have managed the data in a sequence in such a way that, we have taken January to August data as train data and September to December as test data for first city. Then, we have taken May to December data as train data and January to April data as test data for second city. Then, we have taken May to August data as test data and January to April and September to

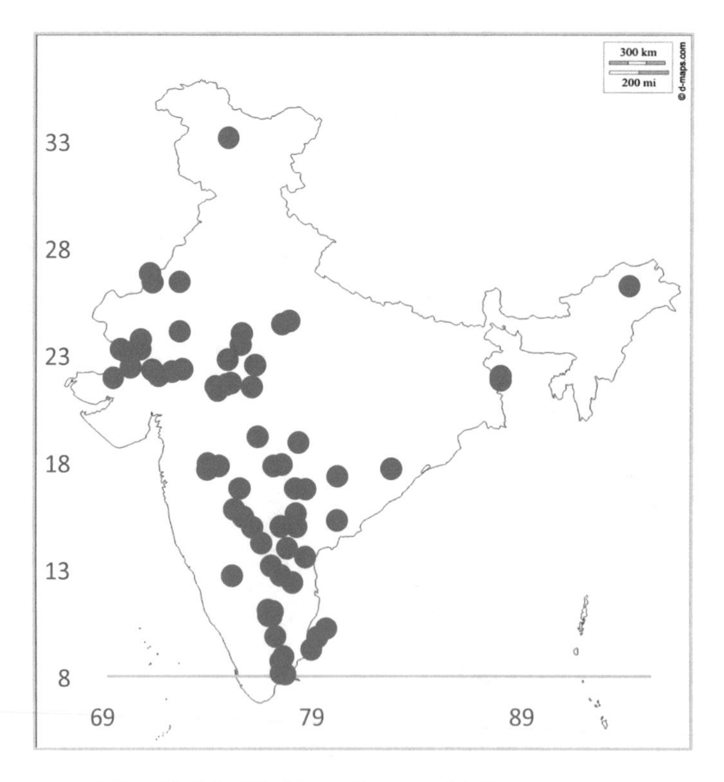

Fig. 2 A representation of all the 67 cities on the map of India

December data as train data for third city. Similar pattern is followed for rest 67 cities.

3.3 **Structure**—Three-layer MLP using feed-forward backpropagation. The number of internal neurons in the hidden layer is calculated using the following formula.

$$\text{Neurons} = \left[\left(\frac{\text{Input} + \text{Output}}{2} \right) + \sqrt{\text{Sample}} \right] \pm 10\% \tag{2}$$

where Input = 19 and Output = 1, Samples = 536, we get number of hidden neurons = 37.

3.4 **Training**—Using normalization process for the network and data was set as—70% for Training, 15% for Validation and 15% for Testing by randomizing. A total of 40 iterations for each neuron is achieved using Nftool in Matlab.

3.5 **Testing**—The target is to achieve the Mean Square Error minimization. Determination of regression value between output and target can also be achieved.

4 Results

The maximum regression value we get is 0.9465 (94.65%) for 18 hidden layer neurons. The minimum regression value we get is 0.7826 (78.86%) for 30 hidden layer neurons. The training and test regression values are depicted in Fig. 3 (test results) and Fig. 4 (training results). The graphical validation for training and testing phases of the proposed approach is presented below. The regression graph obtained after training and testing the network with 18 hidden layer neurons is shown in Figs. 5 and 6 respectively.

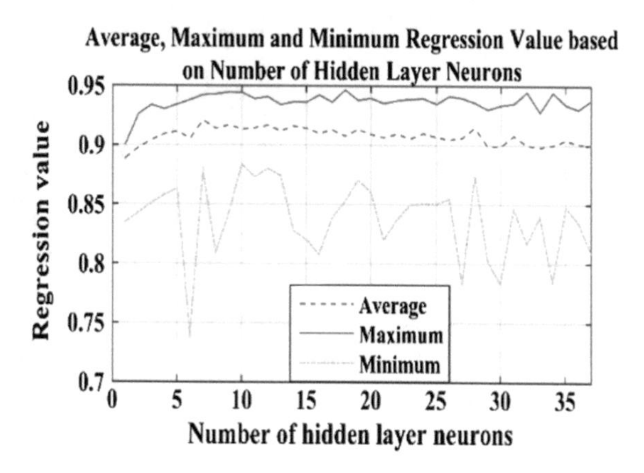

Fig. 3 Test result for 37 hidden layer neurons

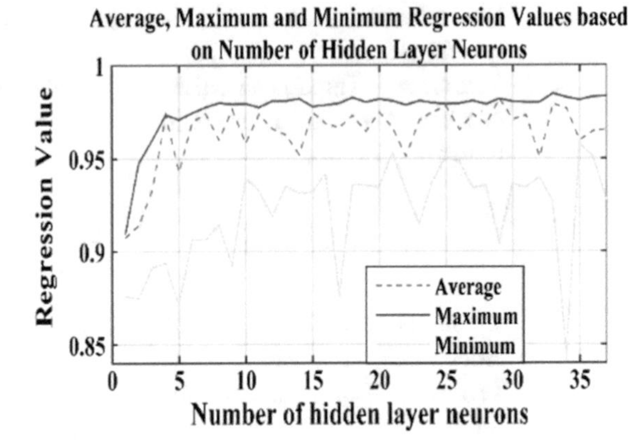

Fig. 4 Training result for 37 hidden layer neurons

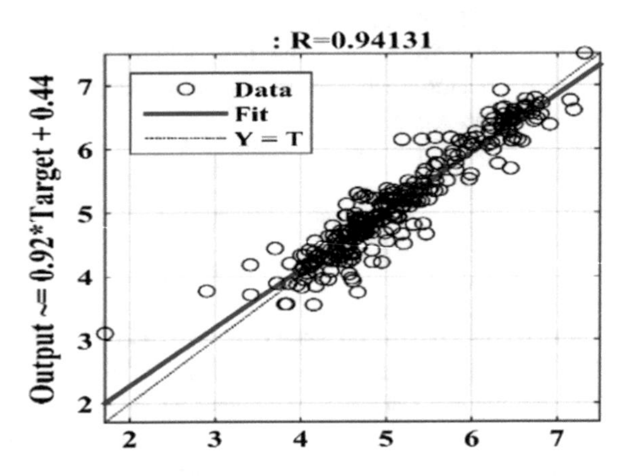

Fig. 5 Regression graph after testing with 18 hidden layer neurons

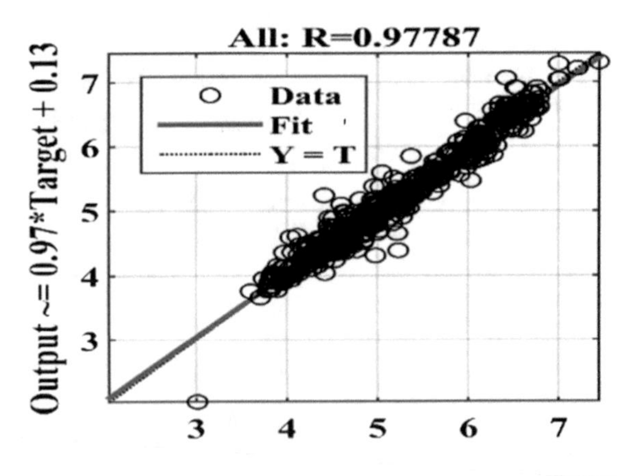

Fig. 6 Regression graph for training network with 18 hidden layer neurons

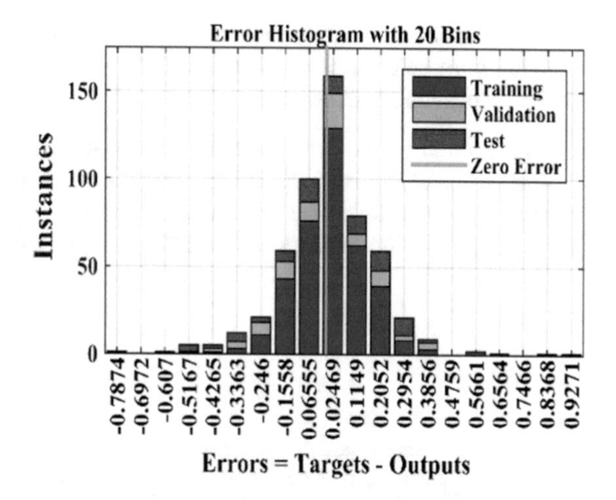

Fig. 7 Histogram after training network with 18 hidden layer neurons

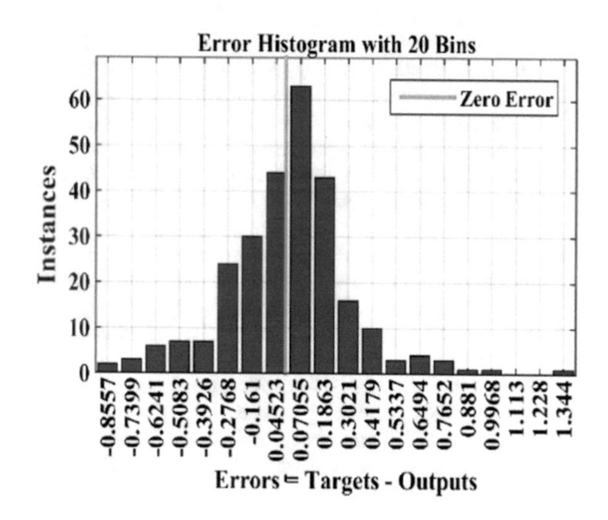

Fig. 8 Histogram after testing network with 18 hidden layer neurons

The error histogram for training results is shown in Fig. 7 and the error histogram for test results are shown in Fig. 8.

5 Conclusion

Understanding of radiation potential of different locations is important for the efficient usage of the solar energy. An important prediction tool nowadays is the artificial neural network. For the purpose of this article, ANN model was used for prediction of solar radiation in India. This model involved entering different climatological parameters as input to the ANN network and solar radiation was predicted. It can be used for prediction of solar radiation in remote areas where the availability of solar measurement is scarce.

Scope of future interest are:

1. Comparison of the present study with Generalized regression neural network (GRNN) and Radial basis function neural network (RBFNN).
2. Relevant Input Selection out of the 19 inputs.
3. Use optimization techniques like PSO to train ANN.

References

1. Azadeh, A., Maghsoudi, A., Sohrabkhani, S.: Using an integrated artificial neural networks model for predicting global radiation: the case study of Iran. Energy Convers. Manag. **50**(6), 1497–1505 (2009)
2. Mellit, A., et al.: An adaptive model for predicting of global, direct and diffuse hourly solar irradiance. Energy Convers. Manag. **51**, 771–782 (2010)
3. Hasni, A., et al.: Estimating global solar radiation using artificial neural network and climate data in the south-western region of Algeria. Energy Procedia **18**, 531–537 (2012)
4. Yadav, A.K., Malik, H., Chandel, S.S.: Application of rapid miner in ANN based prediction of solar radiation for assessment of solar energy resource potential of 76 sites in Northwestern India. Renew. Sustain. Energy Rev. **52**, 1093–1106 (2015)
5. https://eosweb.larc.nasa.gov–Atmospheric Science Data Center (ASDC) at NASA Langley Research Center
6. Malik, H., Mishra, S.: Artificial neural network and empirical mode decomposition based imbalance fault diagnosis of wind turbine using TurbSim, FAST and simulink. IET Renew. Power Gener. **11**(6), 889–902 (2017). https://doi.org/10.1049/iet-rpg.2015.0382
7. Yadav, A.K., Malik, H., Chandel, S.S.: Selection of most relevant input parameters using WEKA for artificial neural network based solar radiation prediction models. Renew. Sustain. Energy Rev. **31**, 509–519 (2014). https://doi.org/10.1016/j.rser.2013.12.008
8. Yadav, A.K., Sharma, V., Malik, H., Chandel, S.S.: Daily array yield prediction of grid-interactive photovoltaic plant using relief attribute evaluator based radial basis function neural network. Renew. Sustain. Energy Rev. **81**(2), 2115–2127 (2018). https://doi.org/10.1016/j.rser.2017.06.023
9. Malik, H., Sharma, R.: EMD and ANN based intelligent fault diagnosis model for transmission line. J. Intell. Fuzzy Sys. **32**(4), 3043–3050 (2017). https://doi.org/10.3233/JIFS-169247
10. Malik, H., Yadav, A.K., Mishra, S., Mehto, T.: Application of neuro-fuzzy scheme to investigate the winding insulation paper deterioration in oil-immersed power transformer. Electr. Power Energy Sys. **53**, 256–271 (2013). https://doi.org/10.1016/j.ijepes.2013.04.023
11. Arora, P., Malik, H., Sharma, R.: Wind speed forecasting model for Northern-Western region of India using decision tree and multi layer perceptron neural network approach. Interdis. Environ. Rev. **19**(1), 13–30 (2018). https://doi.org/10.1504/IER.2018.089766
12. Yadav, A.K., Malik, H., Mittal, A.P.: Artificial neural network fitting tool based prediction of solar radiation for identifying solar power potential. J. Electr. Eng. **15**(2), 25–29 (2015)
13. Yadav, A.K., Singh, A., Malik, H., Azeem, A.: Cost analysis of transformer's main material weight with artificial neural network (ANN). In: Proceedings IEEE International Conference on Communication System's Network Technologies, pp. 184–187 (2011). https://doi.org/10.1109/csnt.2011.46

14. Yadav, A.k., Singh, A., Malik, H., Azeem, A., Rahi, O.P.: Application research based on artificial neural network (ANN) to predict no load loss for transformer design. In: Proceedings IEEE International Conference on Communication System's Network Technologies, pp. 180–183 (2011). https://doi.org/10.1109/csnt.2011.45
15. Rahi, O.P., Yadav, A.K., Malik, H., Azeem, A., Bhupesh, K.: Power system voltage stability assessment through artificial neural network. Elsevier Procedia Eng. **30**, 53–60 (2012). https://doi.org/10.1016/j.proeng.2012.01.833
16. Yadav, A.K., Malik, H.: Comparison of different artificial neural network techniques in prediction of solar radiation for power generation using different combinations of meterological variables. In: Proceeding IEEE International Conference on Power Electronics, Drives and Energy Systems (PEDES), pp. 1–5 (2014). https://doi.org/10.1109/pedes.2014.7042063
17. Yadav, A.K., Malik, H., Chandel, S.S.: ANN based prediction of daily global solar radiation for photovoltaics applications. In: Proceeding IEEE India Annual Conference (INDICON), pp. 1–5 (2015). https://doi.org/10.1109/indicon.2015.7443186
18. Malik, H.: Application of Artificial Neural Network for Long Term Wind Speed Prediction. In: Proceeding IEEE CASP-2016, pp. 217–222, 9–11 June 2016. https://doi.org/10.1109/casp.2016.7746168
19. Azeem, A., Kumar, G., Malik, H.: Artificial neural network based intelligent model for wind power assessment in India. In: Proceedings IEEE PIICON-2016, pp. 1–6, 25–27 Nov 2016. https://doi.org/10.1109/poweri.2016.8077305
20. Saad, S., Malik, H.: Selection of most relevant input parameters using WEKA for artificial neural network based concrete compressive strength prediction model. In: IEEE PIICON-2016, pp. 1–6, 25–27 Nov 2016. https://doi.org/10.1109/poweri.2016.8077368
21. Azeem, A., Kumar, G., Malik, H.: Application of Waikato environment for knowledge analysis based artificial neural network models for wind speed forecasting. In: Proceedings IEEE PIICON-2016, pp. 1–6, 25–27 Nov 2016. https://doi.org/10.1109/poweri.2016.8077352

Operation of DC Drive by Solar Panel Using Maximum Power Point Tracking Technique

Alok Jain, Sarvesh Kumar, Suman Bhullar and Nandan Kumar Navin

Abstract Solar panel systems are extensively increasing due to significant, procure, depletable, and widely accessible resource as a succeeding energy supply. In this paper, a DC drive is associated to a solar panel system. A DC drive is energized by the arrangement of boost converter overwhelming solar power via maximum power point tracking (MPPT) regulator device. MPPT technique improves the efficiency of the solar cell. The DC drive is served by the maximal power assimilated from the MPPT of solar system. Its main objective is to track the maximal power, so that the maximum possible power can be extracted from the solar cell. So, here the operation of DC drive-by solar panel in terms of its speed, armature current, field current, and electrical torque with respect to time using MPPT technique have been shown. Both the boost converter and solar cell are demonstrated via MATLAB/Simulink.

Keywords Boost converter · DC drive · Insulated-gate bipolar thyristor (IGBT) MPPT · PV cell

1 Introduction

Today, electricity has become one of the necessities for humans. There are many types of energy sources but in this paper, we are utilizing solar energy as it is renewable energy. So, the electricity produced by solar energy conversion is not

A. Jain (✉) · S. Kumar
Department of Electrical Engineering, Indian Institute of Technology (BHU), Varanasi, Uttar Pradesh, India
e-mail: alok.rs.eee14@itbhu.ac.in

S. Bhullar
Department of Electrical and Instrumentation Engineering, Thapar University, Patiala, Punjab, India

N. K. Navin
Department of Instrumentation and Control, NSIT, New Delhi, India

© Springer Nature Singapore Pte Ltd. 2019

H. Malik et al. (eds.), *Applications of Artificial Intelligence Techniques in Engineering*, Advances in Intelligent Systems and Computing 697,
https://doi.org/10.1007/978-981-13-1822-1_27

only environment-friendly but also improving generations by reducing our dependency on conventional sources like coal, etc. A MPPT is utilized for extracting the maximum power from the solar panel and transfer of that power to the load. A boost converter is serving the aim of reassigning maximal power from the solar panel to the load. A boost converter behaves interfacially amongst load and the solar panel. Since solar panels are comparatively valuable, more research work has been focused to amend the employ of solar energy. Physically, the power furnished by the panels reckons on umpteen integral component such as isolation (incident solar radiation) degree, temperature, and load condition. A MPPT technique is utilized to find maximal power output of a solar system. The MPPT is used in association with boost converter so that power output from solar panel, which is independent of the temperature, radiation conditions, and the load electrical characteristics for solar systems. So, it can be operated in the field of exploitation of inexhaustible energy reservoir to create the dc drive with the support of the dc machine.

Clean, reliable, and domestically secure production of electricity is done using photovoltaic energy production and these factors are promoting the increasing technology usage. The power conversion of PV array from DC to AC is done using inverter technology and it is utilized along some man-to-man end-user or concentrated grid levels. Generally, inverters are used to efficiently transfer maximum amount of power. For achieving optimized transfer of power, MPPTDC-DC converter is one of the specific means. MPPT is utilized to guarantee that PV panel or array is creating power which is always close to the knee of its I–V curve possibly. It will allow us to get the maximal quantity of power at any nominative time. A system consisting of two-phased trailing that contours a photovoltaic power-increment-aided incremental-conductance (PI-INC) MPPT to enhance the trailing demeanor of the formal INC MPPT has been presented [1]. The trial for PV inverter MPPT efficiency has been examined carefully and reported to rating index of PV inverter MPPT efficiency [2].

An artificial neural network, MPPT restrainer has been presented from the rule of perturbation and observation (P&O) method [3]. A new hybrid MPPT technique has been presented and discussed having advanced MPPT efficiency and advanced speed of trailing held through way of an accelerated forecast of the maximum power voltages of the inverter and PV modules [4]. For both centralized and diffused MPPT systems, a PSCAD model has been developed by randomly varying the solar irradiation for the reflexion of constancy and caliber of turnout voltage for all systems [5]. An extreme low power MPPT circuit having single sample and hold and cold-start transcription has been discoursed that has enablement of MPPT all over the reach of light strengths [6]. A subpanel MPPT converter that links to each PV cell chain have been analyzed and discussed with a purpose to cut down the monetary value and simplify the apportioned MPPT system [7]. An investigation has been done on MPPT controller by using P&O and hill climb search algorithm by implementing it in hybrid system consisting of both the renewable resources [8]. A solar PV-powered BLDC fans have been formulated without utilizing of repositing and affected for their suitableness in daylight and used in offices, schools, colleges [9]. A MPPT algorithm based on artificial fish swarm algorithm have been proposed for controlling purpose of a singular-stage photovoltaic grid-connected system [10]. For photovoltaic (PV) systems an MPPT method has been acquainted that meliorates the employed customary P&O method in dynamic ecological variations by exploiting the Fractional Short-Circuit Current (FSCC) [11]. A complete critique of MPPT technique used in grid bound PV systems with elaborated presentation study of the INC method on d (delta) control has been demonstrated [12]. A reconciling MPPT technique for PV array has been presented which can still execute in heterogeneous irradiance status [13]. In PV generation system, space vector pulse width modulation (SVPWM) controlling strategy has been used for three phase PWM inverter [14]. A single-phase grid bound PV system supported on MPPT technique has been executed by using particle swarm optimization method [15]. The MPPT techniques w.r.t the quantity of energy pulled out of the photovoltaic (PV) panel related to the accessible power, wavelet of PV voltage, driving response, and sensors expend have been evaluated [16]. MPPT which is based on algorithm has been suggested for getting maximum possible power from PV module based upon P&O method [17]. A fresh MPPT hybrid technique, i.e., coordinated sequence of two elemental techniques P&O and Fractional Open-Circuit Voltage (FOCV) technique have been proposed to surmount the genetic insufficiencies established in P&O technique [18]. A load current supported MPPT digitalized controller along an accommodative step-size and adaptive perturbation frequency algorithm has been presented [19]. A reconciling govern algorithm have been suggested, by judging the short-circuit current, where current disruption is advised to meliorate the trailing speed [20]. A PI-based MPPT algorithm have been proposed, where it is applied to a buck–boost converter to increase the, bring forth power from photovoltaic panels [21]. An initial voltage tracking function have been acquainted to allot recent first voltage for trailing the implicit MPP once the PV array is at mirky situations [22].

In this paper, a DC drive system utilizing photovoltaic cell has been used. This DC drives runs effortlessly permitting to the power engendered by the PV system. This PV system is embraces of a single-diode model. The P&O algorithm is utilized for the MPPT system. A generalized dc drive system model which represents the PV cell, boost converter and MPPT has been utilized with MATLAB/ Simulink. Irradiance of sunlight and temperature of cell are taken as an input parameters in the proposed model and the output V–I and P–V characteristics under assorted conditions and which are then given to the dc motor. So, here the operation of DC drive-by solar panel in terms of its speed, armature current, field current, and eslectrical torque with respect to time using MPPT technique have been shown. Both the boost converter and solar cell are demonstrated via MATLAB/Simulink.

The paper has been divided into eight sections. Section 1 introduces some of the renewable energy resources, their possible causes, and adverse effect of these problems to power system networks and human beings. Section 1 also sets the objective of the paper. The characteristics of solar PV cell and its modelling are presented in Sect. 2. Section 3 presents the working of the boost controller, Sect. 4 presents the Simulink models of the proposed system, Sect. 5 presents the results of the operation of DC drive by solar panel in terms of its speed, armature current, field current, and electrical torque with respect to time using MPPT technique and finally, Sect. 6 presents conclusions of the paper.

2 Modelling and Characteristics of Solar Photovoltaic (PV) Cell

A solar cell is the basic block of a solar panel. A solar panel consists of legion of solar cells in series/parallel connection. Regarding entirely a single solar cell, it might be explained by exploiting a current source, i.e., a diode and two resistors. A p–n junction fancied in a thin wafer or layer of semiconductor is basically constituting a solar cell. Figure 1 shows a basic structure of PV cell. The electromagnetic radiation of solar energy can be straightly regenerated to produce electricity through photovoltaic upshot using single diode model of solar system. Electrical feature, namely current, voltage, or resistances alter when uncovered to light similarly as in a case of photoelectric cell. Solar cell is delineated as being photovoltaic disregarding of whether the reservoir is sunlight or an artificial light. For sleuthing light or other electromagnetic radiation near the seeable ambit or quantifying light intensity, they are used as a photodetector.

The basic attributes needed for the operation of a photovoltaic cell are:

- Either electron–hole pair or excitations generated by the absorption of light.
- Opposite type charge carrier separation.
- To an external circuit the separate extraction of those carriers.
- The disunite origin of those bearer to an outer circuit.

For the aim of either target heating or mediate power generation from heat, a solar thermal collector provides heat by gripping sunlight.

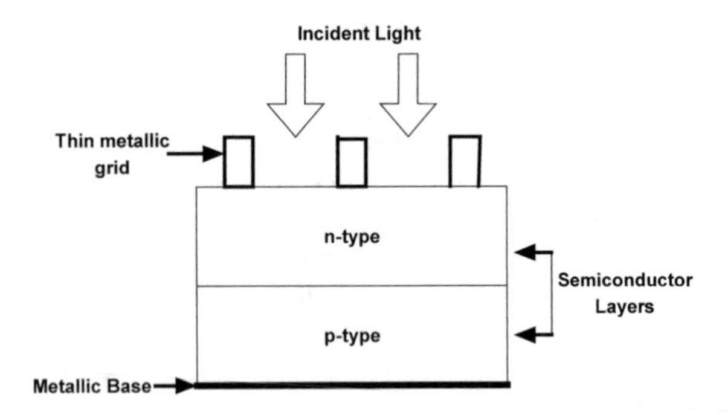

Fig. 1 Basic structure of a PV cell

2.1 Model of Solar Cell

The general circuit of the model consists of a diode, photocurrent, a diode, a resistance in series which shows an interior resistance to the path of current flow in the circuit and a parallel resistor showing leakage current. Figure 2 shows single diode model of solar cell. From Fig. 2, it is shown that

By appling KCL at diode's terminal

$$I_{ph} = I_{\mathrm{D}} + I_{\mathrm{RS}} + I \tag{1}$$

where I_{ph} = Light-generated Current or Photocurrent, i.e., bring forth straight away by utilizing incident of sunlight on the PV cell. This current alters linearly on sun irradiation and reckons upon temperature which is acknowledged by equation

$$I_{ph} = [I_{\mathrm{SC}} + K_i(T_C - 25)] * \lambda \tag{2}$$

where

I_{SC} = Reverse Saturation Current;
K_i = Current temperature coefficient;
T_C = Temperature; λ = Sun irradiation

In Eq. (1), I_{D} = Diode Current is given by

$$I_{\mathrm{D}} = I_{\mathrm{S}} * \left[\mathrm{e}^{\frac{(V + I * R_{\mathrm{S}})}{V_{\mathrm{T}}}} - 1 \right] \tag{3}$$

where,

V = Cell Voltage;
I = Cell Current;
K = Boltzmann constant = 1.38×10^{-23} J/K;
V_{T} = Thermal Voltage = KT/Q;

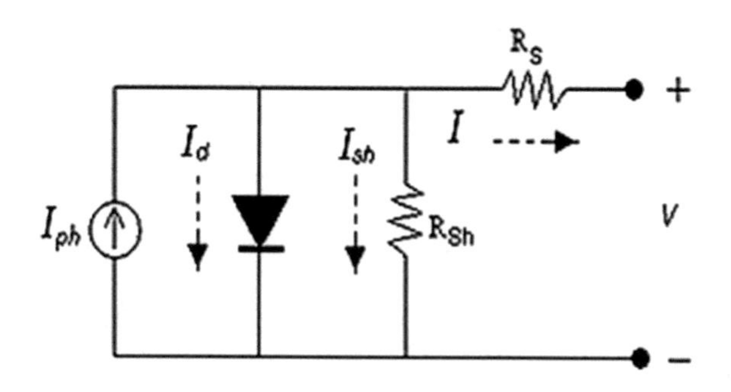

Fig. 2 Single diode model of solar cell

Q = Charge of an electron = 1.607×10^{-19} C;
R_S = Series Resistance;
I_S = Saturation current=

$$I_S = I_{RS}\left(\frac{T_C}{25}\right)^2 * e^{[qE_g(\frac{1}{25}-\frac{1}{T_C})/kA]} \tag{4}$$

where

I_{RS} = Reverse saturation current;
T_C = Temperature;
A = Ideal factor;

$$I = I_{ph} - I_S * \left[e^{\frac{(V+I*R_S)}{V_T}} - 1\right] - \left[\frac{V + I * R_S}{R_P}\right] \tag{5}$$

2.2 *Characteristics of Solar Cell*

2.2.1 V–I Output Characteristics with Different T_C

The output V–I characteristics at different temperatures T_C is shown in Fig. 3.

2.2.2 P–V Output Characteristics with Different T_C

The output P–V characteristics at different temperature is shown in Fig. 4.

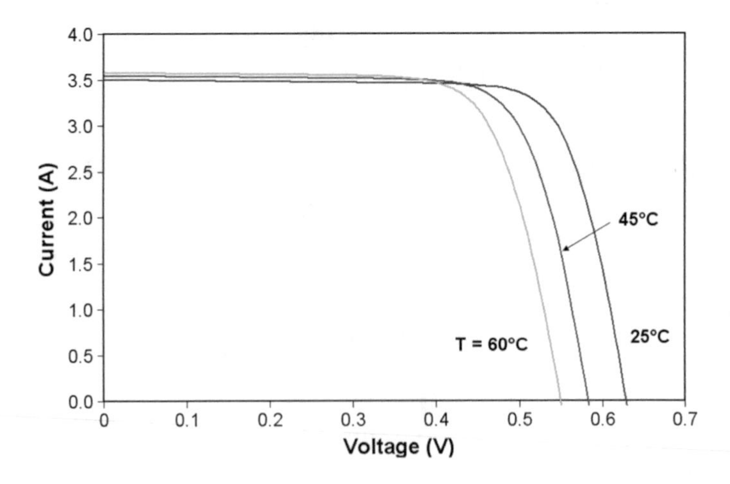

Fig. 3 V–I output characteristics with different *Tc*

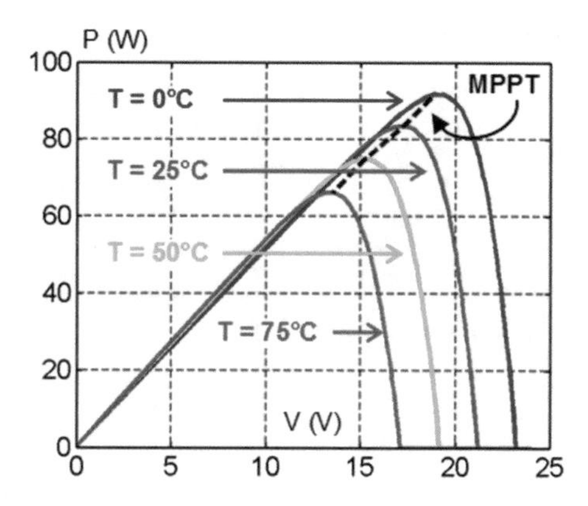

Fig. 4 P–V output characteristics with different T_C

3 Boost Converter

Boost converter is a step-up converter/dc-dc converter whose input voltage is less than its output voltage. Boost converter is a type of switched mode power supply (SMPS) and has leastwise two semiconductors (a diode and a transistor), at least one energy depot device, a capacitor, inductor, or the combination of two as shown in Fig. 5. To cut back output voltage ripples, it has filters consisting of capacitor, could be with inductors, ordinarily summated to the output of the converter. The basic principle to drive the boost converter is the property of an inductor that resists any modification done in current by generating and demolishing a magnetic field.

MOSFET, IGBT, or BJT can be used as a switch. Batteries, solar panels, rectifiers, or DC generators may be used to supply power for boost converter. Due to conservation of power (P = VI), the output current is lower than the source current in a boost converter.

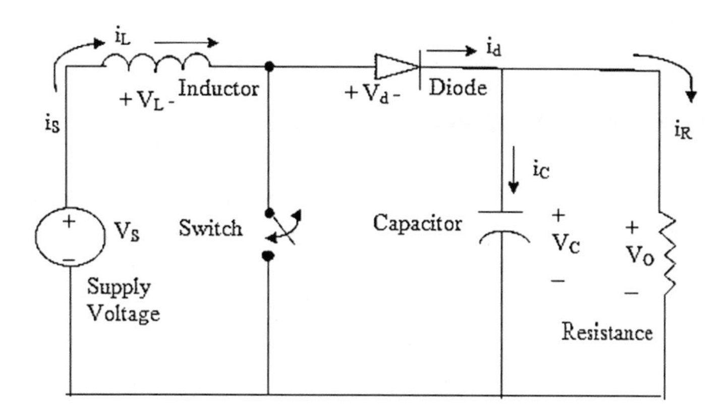

Fig. 5 Basic diagram of a boost converter

Fig. 6 Mode of charging

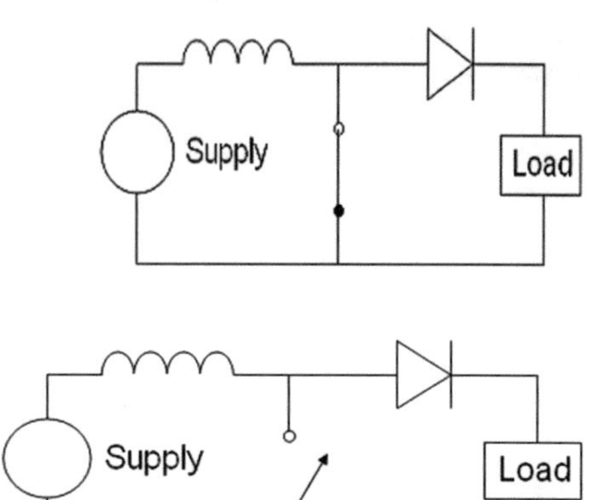

Fig. 7 Mode of discharging

3.1 Modes of Charging

With switch in closed position, the current flows through inductor in clockwise direction and the inductor stores some energy by creating a magnetic field. The polarity of the left side of inductor is positive in this case. Mode of charging is shown in Fig. 6.

3.2 Modes of Discharging

With switch in open position, current will be reduced as the impedance is higher. To maintain the current flow to load, the magnetic field created earlier will be destroyed. Thus, the polarity will be reversed (showing that left side of the inductor will become negative). As a result, two sources will be in series. Mode of discharging is shown in Fig. 7.

4 System Simulink Model

MATLAB is a synergistic, matrix-based bundle for technological and engineering GUI, signal processing, and numerous others. The number of optimizations, statistics, and neural networks of different kinds of toolboxes increase with newer adaptation of MATLAB calculation and visualization.

4.1 Simulink Model of DC Drive

Modeling of drive DC by solar cell with boost converter utilizing MPPT technique is divided mainly into three parts:

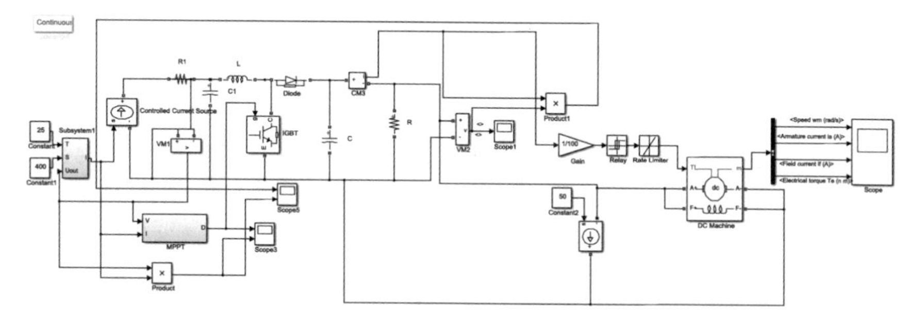

Fig. 8 Simulink diagram of proposed system

(1) Behavioral PV modeling;
(2) Electrical power limited driver and;
(3) Load.

The Simulink model of DC drive is shown in Fig. 8.

Boost converter is built by diode, resistance, IGBT, inductor, and resistor. In this algorithm, the proposed MPPT technique has various scopes connected at several stages to display the results.

4.2 Simulink Model of MPPT

MPPT technique is meant for tracking of maximum power point. Through this technique, solar battery charges and similarly other devices are used to get the maximum achievable power from one or more solar cells that have difficult kinship among solar irradiance (W/m^2) temperature and full opposition which gives a nonlinear output efficiency which can be studied established on the V–I curve. Simulink model of MPPT is shown in Fig. 9. The aim of the MPPT technique is to have maximum power output from the system of the PV cells and by applying a proper resistance maximum power from the system may be obtained at any given environmental conditions.

4.3 Simulink Model of Control Technique

MPPT devices are usually used in an electrical power converter system, which renders voltage or current conversion percolating and regulating operating different loads such as power grids, batteries, or motors, etc. MPPT is an electronic method of capturing the maximum power from solar cell. The cell operating voltage to battery voltage can be converted in this and the output current in the process can be raised. Figure 10 shows the Simulink diagram of the control technique.

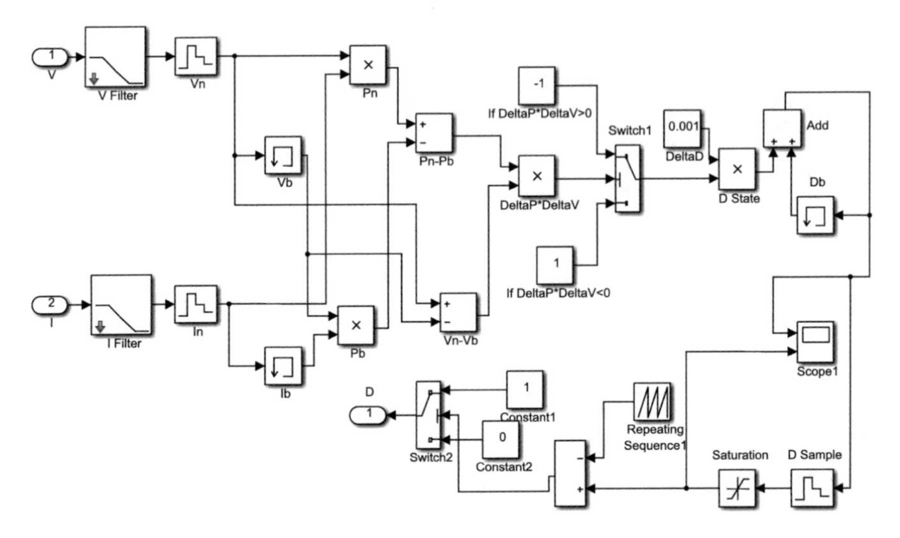

Fig. 9 Simulink diagram of MPPT technique

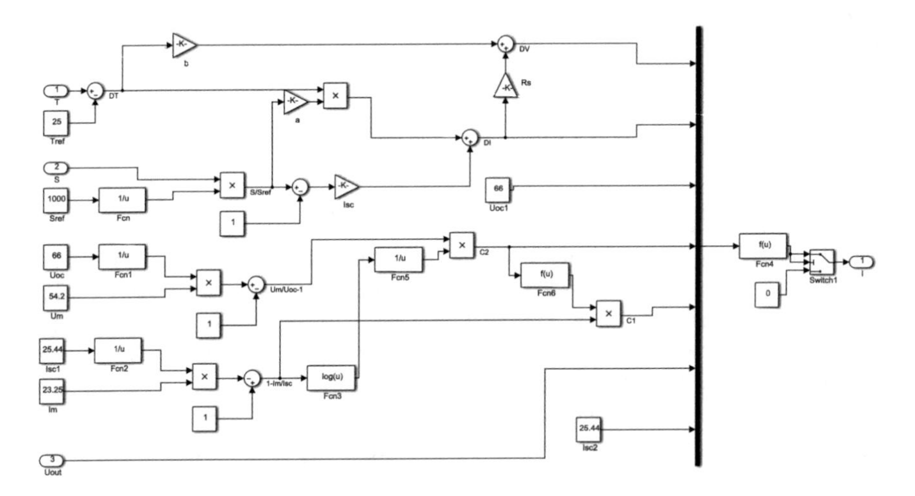

Fig. 10 Simulink diagram of control technique using MPPT methodology

The word "tracking" has nothing to do with moving the PV modules mechanically to track the sun.

4.4 *Perturb and Observe (P&O) Method*

In this method, a small amount of voltage adjustment is made from the solar cell and power is measured by the MPPT controller and if the measured power increases

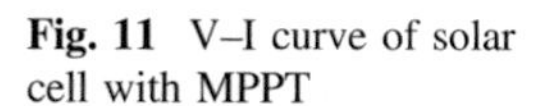

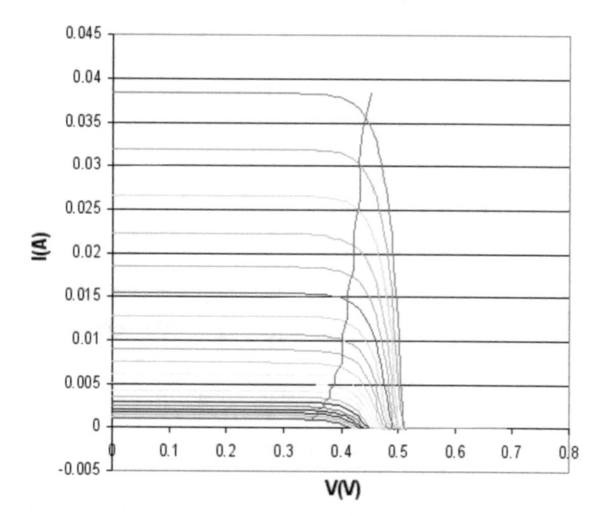

Fig. 11 V–I curve of solar cell with MPPT

in the particular direction more adjustment will be made until it stops increasing. It is mostly used, though oscillations of power output may be caused by this method.

The solar cell V–I curve which crosses the knee of the curve and the maximum power transfer point can be set is shown in Fig. 11. The same thing is done by all non-MPPT charge controllers: the solar cell and the battery are directly connected. The solar cell operating voltage is reduced as the battery acts as a load. To reach the set points of the charge controller the non-MPPT charge controller will keep pushing the battery voltage up. This leads to the development of a charge controllers.

5 Results

Modeling of DC Drive-by solar cell with boost converter using MPPT technique has few results versus time. Figure 12a–d shows speed, armature current, field current and electrical torque versus time graph.

6 Conclusions

A DC drive system utilizing photovoltaic cell has been used in power system network. This DC drives runs effortlessly permitting to the power engendered by the PV system. This PV system is embraces of a single diode model. The P&O algorithm is utilized for the MPPT system. A generalized dc drive system model which represents PV cells, boost converter, and MPPT has been utilized with MATLAB/SIMULINK. Irradiance of sunlight and cell temperature are taken as an

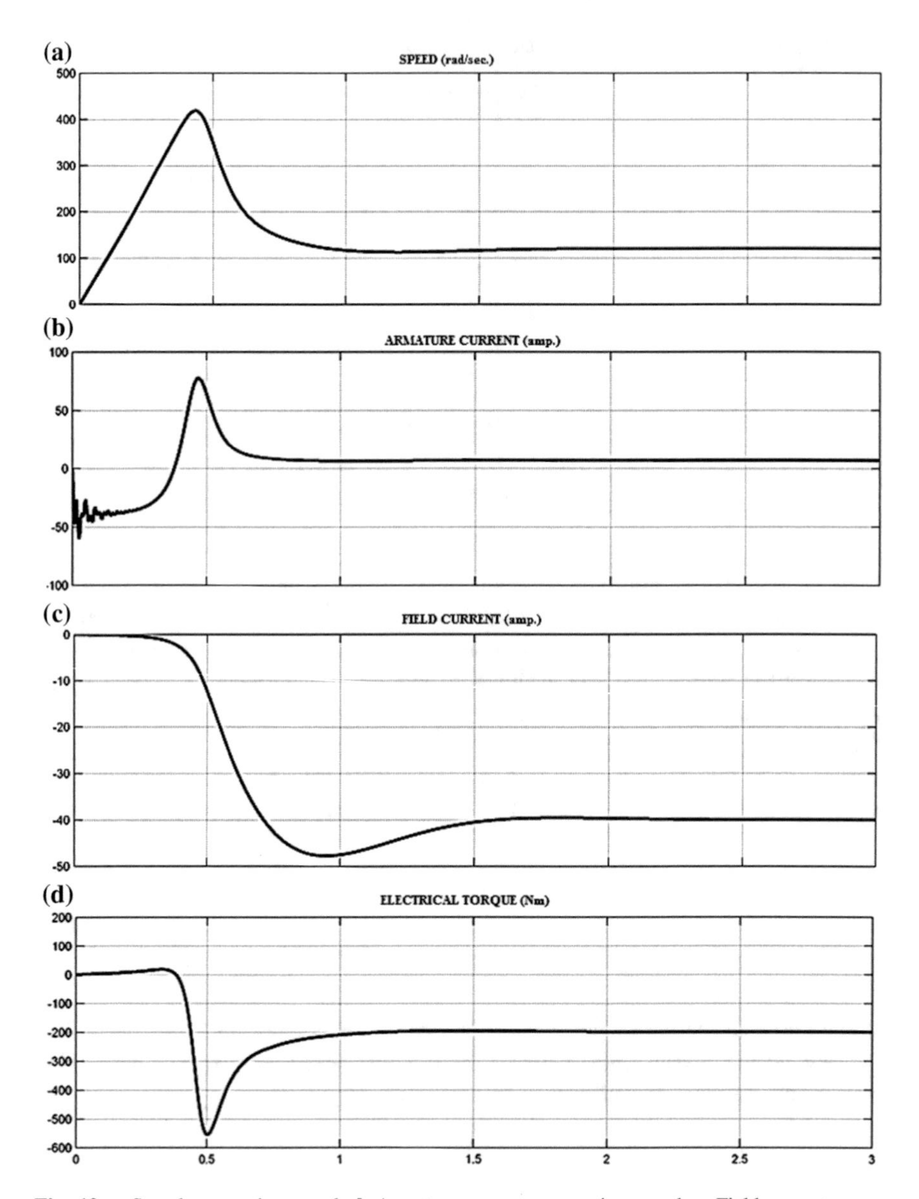

Fig. 12 **a** Speed versus time graph. **b** Armature current versus time graph. **c** Field current versus time graph. **d** Electrical torque versus time graph

input parameters in the proposed model and the output V–I and P–V characteristics under assorted conditions and which are then given to the dc motor.

It can be used in every field like agriculture (water pumping, centrifugal pump, reciprocating pump, water treatment, and hybrid vehicles. Also in industries in lathes machine, drills machine boring mill, etc. It may also be used in vehicles like trains, metro, cranes, elevators and so many automotive devices and most of the home appliances as sewing machine, hair drier, washing, vacuum cleaner, and street lightin, etc. This technique may be used effectively in communication satellite, radio tower, and hospitals.

References

1. Hsieh, G.-C., Taiwan, C.-L., Hsieh, H.-I., Tsai, C.-Y., Wang, C.-H.: Photovoltaic power-increment-aided incremental-conductance MPPT with two-phased tracking. IEEE Trans. Power Electr. **28**(6), 2895–2911 (2013)
2. Dong, Y., Ding, J., Huang, J., Xu, L.: Investigation of PV inverter MPPT efficiency test platform. In: IEEE International Conference on Renewable Power Generation, Beijing, China, October 2015, pp. 1–4
3. Messalti, S., Harrag, A.G., Loukriz, A.E.: A new neural networks MPPT controller for PV systems. In: IEEE International Conference on Renewable Energy Congress, Sousse, March 2015, pp. 1–6
4. Balato, M., Vitelli, M.: A hybrid MPPT technique based on the fast estimate of the maximum power voltages in PV applications. In: IEEE International Conference and Exhibition on Ecological Vehicles and Renewable Energies, Monte Carlo, March 2013, pp. 1–7
5. Barchowsky, A., Parvin, J.P., Reed, G.F., Korytowski, M.J.: A comparative study of MPPT methods for distributed photovoltaic generation. In: IEEE PES Conference on Innovative Smart Grid Technologies, Washington, DC, January 2012, pp. 1–7
6. Weddell, Alex S., Merrett, Geoff V., Al-Hashimi, Bashir M.: Photovoltaic sample-and-hold circuit enabling MPPT indoors for low-power systems. IEEE Trans. Circuits Syst. I **59**(6), 1196–1204 (2011)
7. Wang, Feng, Xinke, Wu, Lee, Fred C., Wang, Zijian: Analysis of unified output MPPT control in subpanel PV converter system. IEEE Trans. Power Electron. **29**(3), 1275–1284 (2013)
8. Pavan Kumar, A.V., Parimi, A.M, Uma Rao, K.: Implementation of MPPT control using fuzzy logic in solar-wind hybrid power system. In: International Conference on Signal Processing, Informatics, Communication and Energy Systems (SPICES), Kozhikode, February 2015, pp. 1–5
9. Veeraraghavan, S., Kumaravel, M., Vasudevan, K., Jhunjhunwala, A.: Experimental studies and performance evaluation of solar PV powered BLDC motor drive with an integrated MPPT in fan applications. In: IEEE Photo-Voltaic Specialist Conference (PVSC), Denver, CO, June 2014, pp. 3713–3718
10. Li, S., Zhang, B., Xu, T., Yang, J.: A new MPPT control method of photovoltaic grid-connected inverter system. In: IEEE Chinese Control and Decision Conference, Changsha, May 2014, pp. 2753–2757
11. Sher, H.R., Murtaza, A.F., Noman, A., Addoweesh, K.E.: A new sensorless hybrid MPPT algorithm based on fractional short-circuit current measurement and P&O MPPT. IEEE Trans. Sustain. Energy **6**(4), 1426–1434 (2015)
12. Lenin Prakash, S., Arutchelvi, M., Stephy Sharon, S.: Simulation and performance analysis of MPPT for single stage PV grid connected system. In: IEEE International Conference on Intelligent Systems and Control (ISCO), Coimbatore, January 2015, pp. 1–6
13. Choudhury, S., Rout, P.K.: Comparative study of M-FIS FLC and modified P&O MPPT techniques under partial shading and variable load conditions. In: IEEE India Conference (INDICON), New Delhi, December 2015, pp. 1–6

14. Li, J., Wang, H.: A novel stand-alone PV generation system based on variable step size INC MPPT and SVPWM control. In: IEEE International Conference on Power Electronics and Motion Control, Wuhan, May 2009, pp. 2155–2160
15. de Oliveira, F.M., Oliveira da Silva, S.A., Durand, F.R., Sampaio, L.P.: Grid-tied photovoltaic system based on PSO MPPT technique with active power line conditioning. IET J. Power Electr. **9**(6), 1180–1191 (2016)
16. de Brito, M.A.G., Galotto, L., Sampaio, L.P., e Melo, G.D.A.: Evaluation of the main MPPT techniques for photovoltaic applications. IEEE Trans. Ind. Electr. **60**(3), 1156–1167 (2012)
17. Sharma, D.K., Purohit, G.: Advanced perturbation and observation (P&O) based maximum power point tracking (MPPT) of a solar photo-voltaic system. In: IEEE International Conference on Power Electronics, Delhi, December 2012, pp. 1–5
18. Murtaza, A.F., Sher, H.A., Chiaberge, V., Boero, D.: A novel hybrid MPPT technique for solar PV applications using perturb & observe and fractional open circuit voltage techniques. International Symposium on Mechatronika, Prague, December 2012, pp. 1–8
19. Jiang, Y., Abu Qahouq, J.A., Haskew, T.A.: Adaptive step size with adaptive-perturbation-frequency digital MPPT controller for a single-sensor photovoltaic solar system. IEEE Trans. Power Electr. **28**(7), 3195–3205 (2012)
20. Kollimalla, S.K., Mishra, M.K.: Novel adaptive P&O MPPT algorithm for photovoltaic system considering sudden changes in weather condition. In: IEEE International Conference on Clean Electrical Power (ICCEP), Alghero, June 2013, pp. 653–658
21. Kabalci, E., Gokkus, G., Gorgun, A.: Design and implementation of a PI-MPPT based Buck-Boost converter. In: International Conference on Electronics, Computers and Artificial Intelligence, Bucharest, June 2015, pp. SG-23–SG-28
22. Chin, C.S., Tan, M.K., Neelakantan, P., Chua, B.L.: Optimization of partially shaded PV array using fuzzy MPPT. In: IEEE Colloquium on Humanities, Science and Engineering, Penang, December 2011, pp. 481–486

Feature Extraction Using EMD and Classifier Through Artificial Neural Networks for Gearbox Fault Diagnosis

Hasmat Malik, Yogesh Pandya, Aakash Parashar and Rajneesh Sharma

Abstract Fault diagnosis has been extensively used to identify the faults of gearbox, which is beneficial to minimize the retaining and operation cost and improve the reliability and feasibility of the wind gearbox. Although there is a considerable number of papers related to gearbox, fault diagnosis have been found in the literature survey take up an impeccable and economical technique. In this paper, recorded vibration signals are utilized to diagnose the fault of gearbox. For this purpose, the features are extracted by using EMD and classify through ANN-based diagnostic model, which is helpful for real-life gearbox diagnosis.

Keywords Gearbox · Wind turbine · Condition monitoring

1 Introduction

For reducing the greenhouse effect, wind energy is the most viable renewable resource. Wind turbine is the main source of producing wind energy. When the wind turbine is placed in severe operating condition and strident environmental conditions, fault arises in component and assemblies of wind turbine. Key

H. Malik
Electrical Engineering Department, IIT Delhi, New Delhi, India
e-mail: hmalik.iitd@gmail.com

Y. Pandya
School of Mechatronics and Automobile Engineering,
Symbiosis University of Applied Sciences, Bada Bangadda, Super Corridor,
Near Airport, Indore 453112, India
e-mail: yogeshpandyaa@gmail.com

A. Parashar (✉) · R. Sharma
Division of ICE, NSIT Delhi, New Delhi, India
e-mail: paaksh2011@gmail.com

R. Sharma
e-mail: rajneesh496@gmail.com

© Springer Nature Singapore Pte Ltd. 2019

H. Malik et al. (eds.), *Applications of Artificial Intelligence Techniques in Engineering*, Advances in Intelligent Systems and Computing 697,
https://doi.org/10.1007/978-981-13-1822-1_28

components of wind turbine are gearbox, generator, rotor blades, tower, and break. Whenever there is failure of any one component, downtime occurs in the wind turbine and operation and maintenance cost increases.

Studies reveal that failure rate of gearbox is lesser than the other components of wind turbine but downtime in wind turbine due to gearbox is higher than the other components. Average time for maintenance of gearbox is 256 h. Majority of the wind turbine failure (59%) are caused by the gearbox, 76% of the wind turbine gearbox failure is attributed to the bearings, and 17.1% of the wind turbine failure is attributed to the gears. Breakage during transportation, misalignments of gearbox component, wrong material, and underestimated design loads, and torque overloads error in manufacture, these are the faults encountered in the wind turbine gearbox. Therefore, condition monitoring and fault diagnosis are essential. After the fault diagnosis of gearbox downtime in wind turbine decreases and replacement and repair costs are reduced. There are several methods evolve in the past decades. These methods are online pressure analysis, online temperature monitoring, oil condition monitoring, performance analysis, and vibration analysis. After analysis of literature survey, fault diagnosis techniques are divided into five categories, i.e., lubrication-based fault diagnosis system, vibration-based fault diagnosis system, temperature-based fault diagnosis system, frequency-based fault diagnosis system, and torque- and speed-based fault diagnosis. Vibration signal analysis is the commonly used approach for examining the faults of wind turbine gearbox, which is the nonstationary.

2 Types of Faults Encountered in Gearbox

Fault is termed as a defect, something that minimizes from perfect and for completing the function of an object by the capability of terminate. One of the most significant issues that raise complaints about the wind turbine system is that the predicted/promised energy gain over the planned lifetime cannot be realized. One of the main reasons for this is that the gearboxes which have a designated lifetime of 20 years fail in a shorter period of time. Most wind turbines which were originally designed to run for 20 years is necessary for important repairs or complete modernization in a range of 5–7 years. The major operating cost for wind turbines is the high cost of repairs. Gearboxes are one of the primary reasons for increasing the cost to generate energy from these free renewable resources. Most of the downtime of wind turbines is due to gearbox issues. The approximate cost of a gearbox replacement is $250,000 (for a turbine of 1.5 MW) which includes labor, crane rental and lost profits. Gearbox failures have been a problem from the beginning of the design of wind turbines. Design flaws account for most of the problems with gearboxes. During failure, innermost part of wind turbine for example, in gearbox temperature of oil is high; the control units of gearbox records the failure precisely or registers the repercussion of the fault, and acknowledge referring to the malfunction of different types. There is absolutely a gap between measuring the indication to come up to a characterization of fault type and its sincerity. Operating parameters of gearbox are deeply interconnected.

There are six different types of faults occur in the gearbox as (1) Teeth breakage, (2) Wear, (3) Plastic flow, (4) Spalling, (5) Surface Fatigue, Scoring (Scuffing), and (6) Deterioration of The Rollers of Bearing.

3 Dataset/Experimental Setup

Gear Fault diagnosis data for CM of gearbox has been created using Spectra Quest's Gear Fault Diagnosis Simulator which is freely available at open EI (https://openei.org/datasets/dataset/gearbox-fault-diagnosis-data) and at data world (https://data.world/gearbox/gear-box-fault-diagnosis-data-set) [1].

For prognostics investigation and diagnostics, Spectra Quest's Gearbox Fault Diagnostics Simulator (GFDS) has been intended to reproduce industrial gearbox. Dataset has been created using Spectra Quest's Gearbox Fault Diagnostics Simulator (GFDS) for two cases, i.e., healthy condition and with a broken tooth condition under variation of different loading conditions from zero load to 90% loading condition. Four sensors are placed in four different directions on the body of gearbox recorded vibration signals are utilized for further study.

4 Signal Processing and Feature Extraction

For the purpose of future extraction, EMD method is applied to decompose the raw signals into intrinsic mode functions (IMFs). Obtained IMFs are utilized for the gearbox diagnosis purpose as an input vector to the ANN-based classifier. The detail explanation of EMD and ANN-based classifier is represented in subsequence sections.

4.1 Empirical Mode Decomposition (EMD) [2–5]

For utilizing to preprocess and decompose any complex experiment dataset into small and limited number of segments approach used for this known as EMD. These segments are known as IMFs. It is basically a process of shifting for converted nonlinear signals and stationary signals into corresponding mono segment and symmetric segment.

Two conditions which are mentioned below are satisfied by intrinsic mode functions, i.e., (1) In the total experimental dataset, between zero crossings, there must be one and only one extreme and (2) mean value must be zero at any point. The detail analysis of IMFs extraction from EMD has been given in Refs. [2–5].

Extracted IMFs depends only on signal. When *pn* becomes monotonic process will be terminated. The algorithm described in a step-wise manner mentioned below:

Algorithms

Step 0: Initialize: p0= y(t), and j =1
Step 1: Extract the jth IMF di

(a) Initialize: cj(m-1) = pj-1, n =m
(b) select the local minima and maxima of cj(m-1)
(c) Interpolate the local maxima and the minima by cubic spline lines to form upper and lower envelops of cj(m-1)
(d) Compute the mean mi(k-1) of the lower and upper envelops of cj(m-1)
(e) Let cjm= cj(m-1) - ni(m-1)
(f) If cjm is an IMF then set di = hik, else move to step (b) with k =k+1

Step 2: assign the remainder pj +1=pj - di
Step 3: If pj+1 still has minimum 2 extreme then move to step (1) with j=j+1 else the disintegration process is completed and pj+1 is the residue of the signal.

5 Classification Method

Classification model is utilized to determine in which aggregate each experimental data instance is interrelated in a given experimental dataset. According to a few constrains, using classification model, different classes can be classified. Numerous kinds of algorithms based on classification models including ANN, k-nearest neighbor classifier, ID3, Naive Bayes, SVM, and C4.5 are mainly useful for classification purpose. In this study, ANN-based MLP classifier has been implemented to diagnose the gearbox.

5.1 MLP Neural Network [6–13]

Multi-layer Perceptron was introduced by Werbos in 1974 and revised by Rumelhart, McClelland, and Hinton in 1986 which is also called feed-forward networks. It is a model of feed-forward artificial neural network in which a set of input data are mapped on to the relevant sets of output data.

The two fundamental activation functions utilized as part of modern applications are both sigmoid, and are illustrated by: (a) Hyperbolic tangent: $y(\mathrm{l}) = \tanh(\mathrm{l})$ (range: −1 to 1) and (b) Logistic sigmoid: $y(\mathrm{vi}) = 1/(1+e^{-x})$, (range: 0 to 1).

A three-layer MLP neural network has been implemented with one input layer, one hidden layer, and one output layer. The number of neurons at input layer is equal to the number of input variables, while neurons at hidden layer may be selected by hidden and trial method. In this study, hidden layer neurons are calculated as per Eq. (1). The neurons at output layer are two.

6 Results and Discussion

In this study, four different case studies have been presented to analyze the gearbox faults conditions. Case study#1 includes the input data as a raw data as an input variable, case study#2 has presented the diagnosis results by using IMFs as input variables. In the case study#3 and case study#4, calculated 29 features from raw data and IMFs, respectively, are utilized as an input variables.

The MLP neural network architecture used in this study is *i-n-o*, where *i*, *n*, and *o* are the number of neurons at input layer, hidden layer, and output layer, respectively. Two conditions are identified in this study, so output layer neurons are two, whereas hidden layer neurons are evaluated as per Eq. 1 for each case.

$$\text{Hidden Layer Neurons} = \left(\frac{\text{Number of Input} + \text{Number of Output}}{2} + \sqrt{\text{Number of Training Samples}}\right) + 10\% \tag{1}$$

6.1 CASE 1: Gearbox Fault Diagnosis Using Raw Data as Input

For this case, Training and Testing dataset are of the order 4 × 40,000 containing 40,000 samples and 4 × 40,000 containing 40,000 samples, respectively, are used as input to neural network. Number of hidden neurons for the neural network is varied according to the Eq. (1) up to 217 and obtained results have been presented in Fig. 1.

6.2 CASE 2: Gearbox Fault Diagnosis Using IMFs as Input

For this case, Training and Testing dataset are of the order 16 × 40,000 containing 40,000 samples and 16 × 40,000 containing 40,000 samples, respectively, are used as input to neural network. Number of hidden neurons for the neural network is varied according to the Eq. (1) up to 223, and we set a target of order 2 × 40,000. The output results for this case study have been presented in Fig. 2.

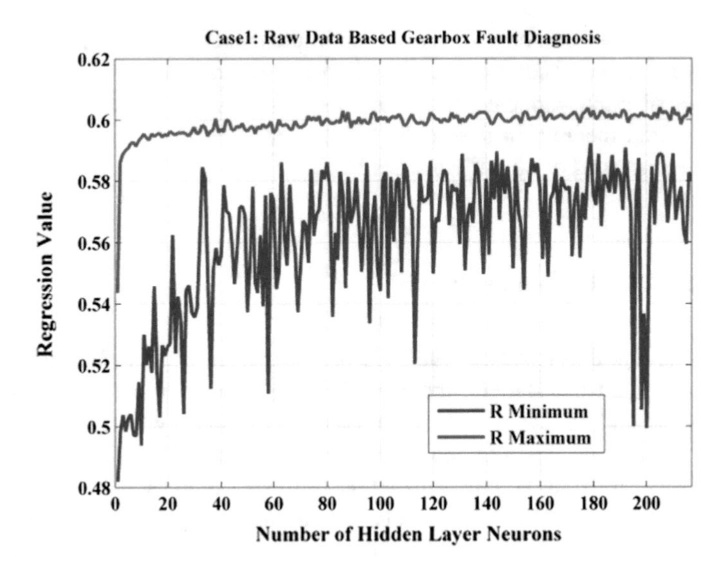

Fig. 1 Regression plot for case 1

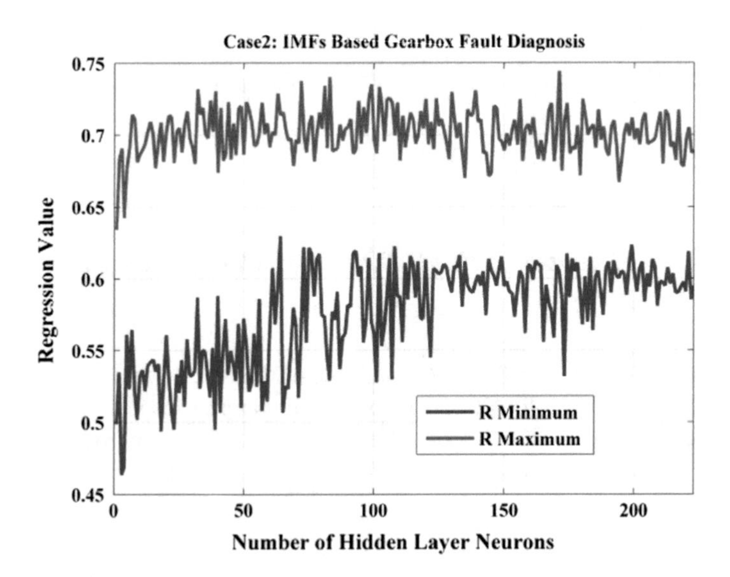

Fig. 2 Regression plot for case 2

6.3 *CASE 3: Gearbox Fault Diagnosis Using Calculated Features from Raw Data as Input*

For this case, Training and Testing dataset are of the order 29 × 40 containing 40 samples and 29 × 40 containing 40 samples, respectively, are used as input to neural network. Number of hidden neurons for the neural network is varied according to the equation and we set a target 2 × 40. The obtained results for case study#3 have been shown in Fig. 3.

6.4 *CASE 4: Gearbox Fault Diagnosis Using Calculated Features from IMFs as Input*

For this case, Training and Testing dataset is of the order 29 × 1280 containing 1280 samples and 29 × 160, 29 × 640, 29 × 64 containing 160, 640, and 64 samples, respectively, are used as input to neural network. Number of hidden neurons for the neural network is varied according to the equation and we set a target 2 × 64, 2 × 640, 2 × 160 and obtained results for case study#4 have been represented in Fig. 4.

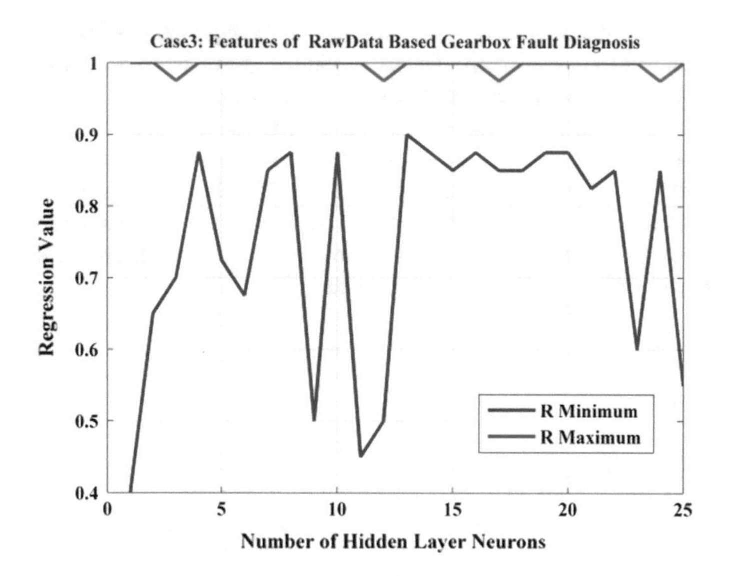

Fig. 3 Regression plot for case 3

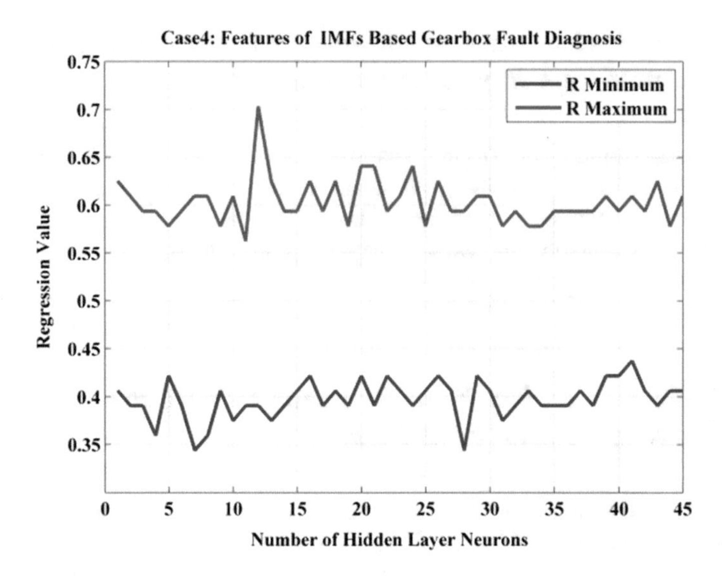

Fig. 4 Regression plot for case 4

7 Conclusions

In this study, it is shown that how the training and testing matrices for each case has been created and applied to the classifier. Signals have been decomposed using EMD technique to reduce the nonlinearity of the recorded signals. Four different case studies have been carried out with four different input variables. Case study#3 (Gearbox Fault Diagnosis Using Calculated Features from Raw Data as Input vector) will give better results when compared with other case studies.

In the future, some work can be done in this field for the complete protection of the gearbox. Other internal effects on gearbox can be studied and other type of fault's impact on the gearbox can be studied.

References

1. Gear Box Fault Diagnosis Data Set. https://data.world/gearbox/gear-box-fault-diagnosis-data-set%23. Accessed Jan 2018
2. Malik, H., Mishra, S.: Artificial neural network and empirical mode decomposition based imbalance fault diagnosis of wind turbine using Turbsim, FAST and Simulink. IET Renew. Power Gener. **11**(6), 889–902 (2017a). https://doi.org/10.1049/iet-rpg.2015.0382
3. Malik, H., Mishra, S.: Application of GEP to investigate the imbalance faults in direct-drive wind turbine using generator current signals. IET Renew. Power Gener. **11**(6), 889–902 (2017b). https://doi.org/10.1049/iet-rpg.2016.0689
4. Malik, H., Sharma, R.: Transmission line fault classification using modified fuzzy Q learning. IET Gener. Transm. Distrib. **11**(16), 4041–4050 (2017a). https://doi.org/10.1049/iet-gtd.2017.0331

5. Malik, H., Sharma, R.: EMD and ANN based intelligent fault diagnosis model for transmission line. J. Intell. Fuzzy Syst. **32**(4), 3043–3050 (2017b). https://doi.org/10.3233/JIFS-169247
6. Yadav, A.K., Malik, H., Chandel, S.S.: Application of rapid miner in ANN based prediction of solar radiation for assessment of solar energy resource potential of 76 sites in Northwestern India. Renew. Sustain. Energy Rev. **52**, 1093–1106 (2015). https://doi.org/10.1016/j.rser.2015.07.156
7. Yadav, A.K., Malik, H., Chandel, S.S.: Selection of most relevant input parameters using WEKA for artificial neural network based solar radiation prediction models. Renew. Sustain. Energy Rev. **31**, 509–519 (2014). https://doi.org/10.1016/j.rser.2013.12.008
8. Yadav, A.K., Sharma, V., Malik, H., Chandel, S.S.: Daily array yield prediction of grid-interactive photovoltaic plant using relief attribute evaluator based radial basis function neural network. Renew. Sustain. Energy Rev. **81**(Part 2), 2115–2127 (2018). https://doi.org/10.1016/j.rser.2017.06.023
9. Malik, H., Yadav, A.K., Mishra, S., Mehto, T.: Application of neuro-fuzzy scheme to investigate the winding insulation paper deterioration in oil-immersed power transformer. Electr. Power Energy Syst. **53**, 256–271 (2013). http://dx.doi.org/10.1016/j.ijepes.2013.04.023
10. Malik, H., Savita: Application of artificial neural network for long term wind speed prediction. In: Proceedings of IEEE CASP-2016, 9–11 June 2016, pp. 217–222. https://doi.org/10.1109/casp.2016.7746168
11. Azeem, A., Kumar, G., Malik, H.: Artificial neural network based intelligent model for wind power assessment in India. In: Proceedings of IEEE PIICON-2016, 25–27 Nov 2016, pp. 1–6. https://doi.org/10.1109/poweri.2016.8077305
12. Saad, S., Malik, H.: Selection of most relevant input parameters using WEKA for artificial neural network based concrete compressive strength prediction model. In: Proceedings of IEEE PIICON-2016, 25–27 Nov 2016, pp. 1–6. https://doi.org/10.1109/poweri.2016.8077368
13. Azeem, A., Kumar, G., Malik, H.: Application of Waikato environment for knowledge analysis based artificial neural network models for wind speed forecasting. In: Proceedings of IEEE PIICON-2016, 25–27 Nov 2016, pp. 1–6. https://doi.org/10.1109/poweri.2016.8077352

PSO-NN-Based Hybrid Model for Long-Term Wind Speed Prediction: A Study on 67 Cities of India

Hasmat Malik, Vinoop Padmanabhan and R. Sharma

Abstract Wind speed forecasting for longer time horizons is important for efficient unit commitment and maintenance scheduling in wind power industry. Highly stochastic nature of the wind calls for accurate prediction techniques. This paper analyses Artificial Neural Network-based techniques for long-term wind speed prediction using average monthly weather data from 65 Indian cities. The data collected is divided into train and test data by 2:1 ratio and train data is used to train the neural network. The trained system is then tested with the remaining data. A total of 19 inputs like location coordinates, temperature, pressure, relative humidity, solar radiation, etc., are given to the neural network. Neural network with different number of neurons in hidden layer is used and each model is trained for considerable number of times to confirm the dependability.

Keywords ANN · Wind speed forecasting · Long-term prediction
PSO

1 Introduction

India stands fifth in total installed capacity of wind power in the world with more than 32 GW of installed capacity and 46,011 GWh of energy units generated annually. With an estimated installable potential at 80 m of 102,788 MW [1], the future of wind energy production in India is highly encouraging.

H. Malik
Electrical Engineering Department, IIT Delhi, New Delhi, India
e-mail: hmalik.iitd@gmail.com

V. Padmanabhan (✉) · R. Sharma
Division of ICE, NSIT Delhi, New Delhi, India
e-mail: vinoopp@gmail.com

R. Sharma
e-mail: rajneesh496@gmail.com

© Springer Nature Singapore Pte Ltd. 2019

H. Malik et al. (eds.), *Applications of Artificial Intelligence Techniques in Engineering*, Advances in Intelligent Systems and Computing 697,
https://doi.org/10.1007/978-981-13-1822-1_29

The wind by its very nature is highly intermittent. Years of recorded data of wind speed shows that it does not follow any particular pattern [2]. These make it very difficult to predict and develop models for estimating power generation capacity of a particular location, which is essential for efficient integration to power systems, plan scheduled maintenances and most importantly in calculations of unit commitment for a wind turbine.

Many statistical models are employed in wind prediction and out of which machine learning models are found to be more accurate and dependable. In [3] Brka et al. examine two types of neural networks Radial Basis Function (RBF) NN and Feed Forward (FF) NN in estimating wind speeds for different time horizons using metrological data of various frequencies. RBF NN was found to give superior results for shorter time horizons [4]. Comparing four different intelligent techniques out of which two are hybrid. Hybrid systems gave more accurate results. In [5] another hybrid model, WT-FA is compared with BPNN, RBFNN, ANFIS, NNPSO, FA, WT+BPNN, WT-ANFIS and WT+NNPSO for different time horizons. The proposed model found to give superior results [6] and investigated long-term wind speed and power prediction using GRNN. Monthly average wind speed is first predicted using weather and location information and is then translated to wind power using turbine power curve. In [7], Cao et al. explains how multivariate wind data improves prediction accuracy over univariate wind data. They also found out that RNN gave improved results than ARIMA model for very short-term wind forecasting. In [8], Liu et al. compare two hybrid models for short-term wind prediction. [9] discusses three recurrent neural networks to predict wind speed and power for 72-h time horizon. Two novel training algorithms, namely GRPE and DRPE are used and all the models gave more accurate results than static MLP-BP method. A hybrid model based on the model of error forecast correction (WSP-MFEC) is proposed by Mao et al. in [10]. Here, an error prediction model is formed to forecast errors in the traditional models and is then superimposed onto the traditional model output to reduce the error. These models reduced the prediction error significantly.

This paper compares ANN-BP model with a hybrid model where PSO is used to optimise initial weights of the neural network rather than choosing arbitrary values. The objective is to predict average monthly wind speed for a location from its weather data.

2 Data Set Collection

Metrological data of 67 cities are procured from Centre for Wind Energy Technology, Chennai. C-WET has dedicated met-masts of varying heights installed in various parts of the country. Data loggers from M/s NRG Systems, USA and M/s Second Wind Inc., USA are used to collect sensor data at every 2 s and are then averaged to 10 min and stored.

Average monthly wind speed and other weather data like air temperature, relative humidity, daily solar radiation, atmospheric pressure, earth temperature, heating degree days, cooling degree days, latitude, longitude, elevation, Heating design temperature, cooling design temperature, Earth temperature amplitude, Frost days at site, Frost days at site, Power Law Index (PLI), Energy Pattern Factor, Air Density and Standard Deviation are used in this study.

For each city, 12 set of data are available corresponding to each month from January to December. This total data set is divided into two parts one each for setting up the prediction model called training set and testing the set model called testing set. Data is divided in such a way that 8 months of each city are used for training and the remaining 4 months for testing.

3 Methodology

3.1 Artificial Neural Network (ANN)

ANN mimics human brain by storing experiential knowledge as weights of its interconnections [11]. A typical neural network may consist of many layers of interconnected neurons. By presenting a set of inputs and their corresponding outputs, a NN can be tuned to develop an input–output relation, which can then be used to predict new outputs corresponding to new set of inputs. Since this method lacks a model-based system, it is found to be very useful in cases where input–output relations are highly nonlinear or there is no significant relationship between inputs and outputs exist at all. Figure 1 shows the structure of a single neuron.

The output is given by the following equation:

$$y = f\left(\sum_{i=1}^{n} W_i X_i\right) \tag{1}$$

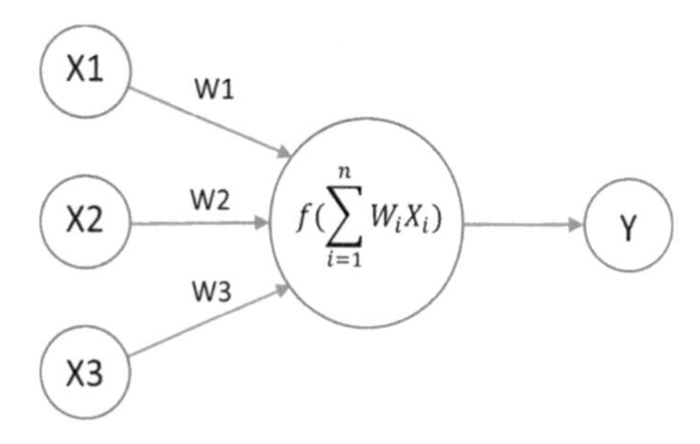

Fig. 1 Structure of a single neuron

where X_i are inputs of the network, W_i are weights associated with each input and f is activation function. For studying of ANN implementation in detail, references [12–20] can be referred in which all steps have been explained and implementation of ANN for distinct application have been presented.

3.1.1 Training

Backpropagation is the most widely used training method for NN. In this method, weights and biases of the NN are initialised arbitrarily and the output generated by a set of inputs is compared with its corresponding target to generate the error signal. This error signal is then propagated backward to update the weights and biases of each neuron in each layer so as to reduce the error. By exposing the NN to significant number of training data sets, a reliable input–output relationship can be obtained.

3.2 *PSO-NN*

One of the main disadvantages of the ANN training process is that it could be trapped to local minimum. This can be avoided by optimising the initial weights and biases of the network beforehand rather than taking arbitrary values. Particle Swarm optimisation is a stochastic optimisation technic inspired from behaviour of bird flocks in search of food. Each individual in the flock changes its velocity and position in accordance with its own previous best position and the previous best position of its neighbourhood. In this way, the flock move towards its target.

3.2.1 Implementation

A swarm is initialised with certain N number of particles with positions as uniformly distributed between upper and lower limits. Initial velocities are taken to be zero. Velocity and position of each particle are updated by the following equation until the performance criteria is reached.

$$v_{i+1} = v_i + c1 * \text{rand}() * (\text{Pbest}_i - P_i) + c2 * \text{rand}() * (\text{Gbest}_i - P_i) \quad (2)$$

$$P_{i+1} = P_i + v_{i+1} \quad (3)$$

v_{i+1} is the new velocity, v_i is current velocity, P_i is the current position, P_{i+1} is the new position, Pbest_i is the current individual best position and Gbest_i is the current best position of the neighbourhood.

Using the above equations, weights and biases of the neural network is updated until the performance criteria, in this case, mean square error, is reached. These optimised values are then loaded to the neural network and is further trained using back propagation method.

4 Results and Discussion

In the first method where only ANN is used for prediction, the training data set is fed directly to the ANN. The network subsequently optimises its weights to generate a prediction model. Testing data set is then introduced to this model and the resulting outputs are compared with the actual measured wind data. In the second method, initial weights of the ANN are first optimised with PSO and are then updated to the ANN. This network is further trained using training data set. The network is then tested with testing data and outputs are recorded. Mean square of the differences between predicted and measured wind data is used to validate the model.

$$\mathrm{MSE} = \frac{1}{n}\sum_{i=1}^{n}\left(\mathrm{MWS}_m - \mathrm{MWS}_p\right)^2 \quad (4)$$

Here, n is the no of samples, MWS_m is the measure mean monthly wind speed and MWS_p is the predicted mean monthly wind speed.

Figure 2 shows the regression curves of NN (a, b) and PSO-NN (c, d) models for training and testing phases. The PSO-NN model is found to give better regression value than the NN one. In Fig. 3, the test outputs of PSO-NN (Orange) and NN (Blue) models are plotted with actual measured monthly average wind speed (Red) for 268 samples from 67 cities. Both the models satisfactorily follow the measured data.

Figure 4 shows the comparison chart of both the models under consideration in terms of the mean squared error between measured and predicted wind speed of the testing samples and are averaged over 40 iterations, plotted against different number of hidden neurons varying from 1 through 37. It can be clearly seen that except for a few instances, the PSO-NN model outperformed the ANN only model. The best result of 0.1005 was obtained with three neurons in the hidden layer.

In Fig. 5, regression values obtained with testing signals, averaged over 40 iterations are plotted against corresponding number of hidden neurons in the network. Here, also PSO-based NN model provided superior results over NN-only model. Regression values as high as 0.9806 was obtained with three neurons in the hidden layer for PSO-NN model.

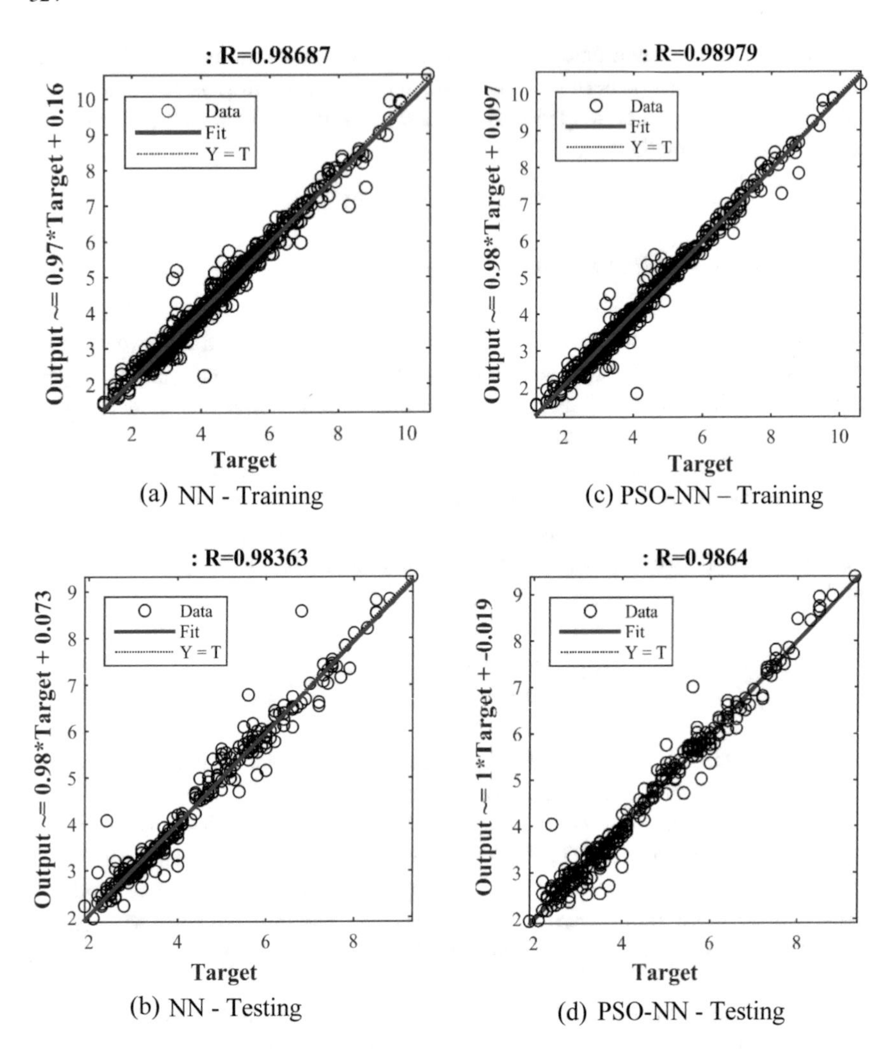

Fig. 2 Regression values during training—testing phases

From Figs. 3, 4 and 5, it is found that PSO-based ANN hybrid model yields better and consistent results than using ANN alone for long-term wind speed prediction.

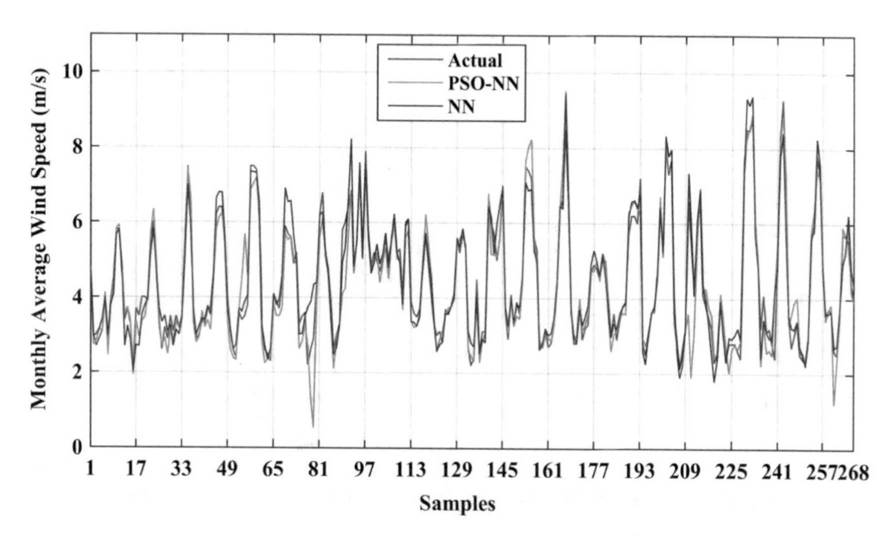

Fig. 3 Plot of measured wind speed with corresponding outputs of test models

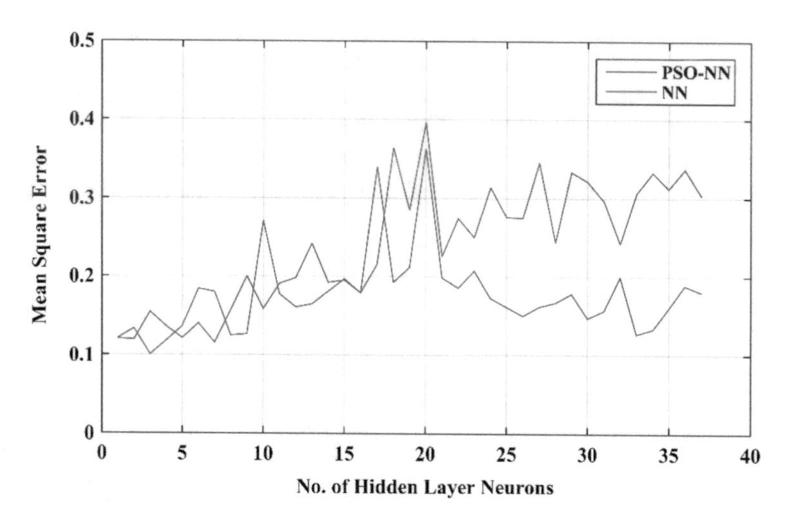

Fig. 4 Averaged MSE of 40 iterations for different number of hidden neurons

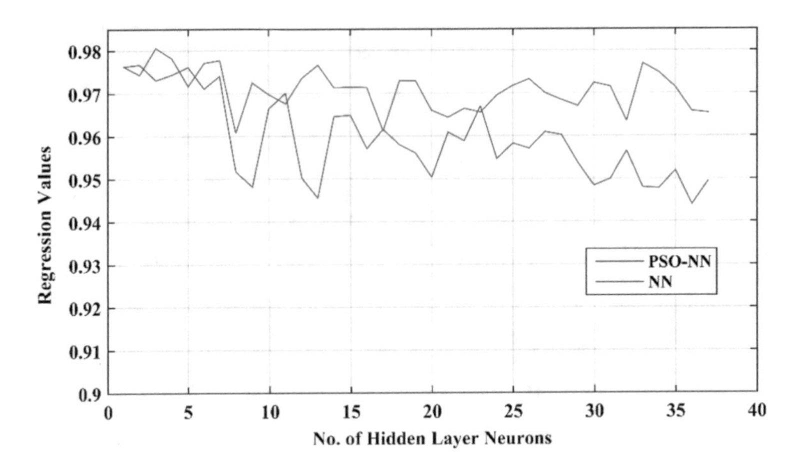

Fig. 5 Averaged regression values over 40 iterations for different number of hidden neurons

5 Conclusion

This paper compared two artificial intelligent methods for their accuracy in predicting monthly average wind speed from weather and location data. For the purpose, data from 67 Indian cities were used. Data from each city is divided into two parts, one for training and setting up the network and other for testing the accuracy of trained network in predicting with new data. Both ANN and PSO-based ANN methods returned satisfactory results and in comparison of the two, it was found that PSO-based ANN method performed more consistently than the one using only ANN.

References

1. Indian Wind Energy Association, http://www.inwea.org/. Accessed 10 Nov 2017
2. Azad, H.B., Mekhilef, S., Ganapathy, V.G.: Long-term wind speed forecasting and general pattern recognition using neural networks. IEEE Trans. Sustain. Energy **5**(2), 546–553 (2014)
3. Brkaa, A., Al-Abdeli, Y.M., Kothapalli, G.: Influence of neural network training parameters on short-term wind forecasting. Int. J. Sustain. Energy (2014)
4. Huang, C.-M., Kuo, C.-J., Huang, Y.-C.: Short-term wind power forecasting and uncertainty analysis using a hybrid intelligent method. IET Renew. Power Gener. **11**(5), 678–687 (2017)
5. Haque, A.U., Mandal, P., Meng, J., Negnevitsky, M.: Wind speed forecast model for wind farm based on a hybrid machine learning algorithm. Int. J. Sustain. Energy, **34**(1), 38–51 (2015)
6. Savita, Ansari, M.A., Pal, N.S., Malik, H.: Wind speed and power prediction of prominent wind power potential states in India using GRNN. In: 1st IEEE International Conference on Power Electronics, Intelligent Control and Energy Systems (ICPEICES-2016)
7. Cao, Q., Ewing, B.T., Thompson, M.A.: Forecasting wind speed with recurrent neural networks. Eur. J. Oper. Res. **221**, 148–154 (2012)

8. Liu, H., Tiana, H., Chen, C., Li, Y.: An experimental investigation of two Wavelet-MLP hybrid frameworks for wind speed prediction using GA and PSO optimization. Int. J. Electr. Power Energy Syst. **52**, 161–173 (2013)
9. Barbounis, T.G., Theocharis, J.B., Alexiadis, M.C., Dokopoulos, P.S.: Long-termwind speed and power forecasting using local recurrent neural network models. IEEE Trans. Energy Convers. **21**(1), 273–284 (2006)
10. Mao, M., Ling, J., Chang, L., Hatziargyriou, N.D., Zhang, J., Ding, Y.: A novel short-term wind speed prediction based on MFEC. IEEE J. Emerg. Sel. Top. Power Electr. **4**(4) (2016)
11. Haykin, S.: Neural Network A Comprehensive Foundation. Pearson Education (2005)
12. Malik, H., Mishra, S.: Artificial neural network and empirical mode decomposition based imbalance fault diagnosis of wind turbine using TurbSim, FAST and Simulink. IET Renew. Power Gener. **11**(6), 889–902 (2017). https://doi.org/10.1049/iet-rpg.2015.0382
13. Yadav, A.K., Malik, H., Chandel, S.S.: Application of rapid miner in ANN based prediction of solar radiation for assessment of solar energy resource potential of 76 sites in Northwestern India. Renew. Sustain. Energy Rev. **52**, 1093–1106 (2015). https://doi.org/10.1016/j.rser.2015.07.156
14. Yadav, A.K., Malik, H., Chandel, S.S.: Selection of most relevant input parameters using WEKA for artificial neural network based solar radiation prediction models. Renew. Sustain. Energy Rev. **31**, 509–519 (2014). https://doi.org/10.1016/j.rser.2013.12.008
15. Yadav, A.K., Sharma, V., Malik, H., Chandel, S.S.: Daily array yield prediction of grid-interactive photovoltaic plant using relief attribute evaluator based radial basis function neural network. Renew. Sustain. Energy Rev. **81**(Part 2), 2115–2127 (2018). https://doi.org/10.1016/j.rser.2017.06.023
16. Malik, H., Yadav, A.K., Mishra, S., Mehto, T.: Application of neuro-fuzzy scheme to investigate the winding insulation paper deterioration in oil-immersed power transformer. Electr. Power Energy Syst. **53**, 256–271 (2013). http://dx.doi.org/10.1016/j.ijepes.2013.04.023
17. Malik, H., Savita: Application of artificial neural network for long term wind speed prediction. In: Proceedings of IEEE CASP-2016, 9–11 June 2016, pp. 217–222. https://doi.org/10.1109/casp.2016.7746168
18. Azeem, A., Kumar, G., Malik, H.: Artificial neural network based intelligent model for wind power assessment in India. In: Proceedings of IEEE PIICON-2016, 25–27 Nov 2016, pp. 1–6. https://doi.org/10.1109/poweri.2016.8077305
19. Saad, S., Malik, H., Ishtiyaque, M.: Selection of most relevant input parameters using WEKA for artificial neural network based concrete compressive strength prediction model. In: Proceedings of IEEE PIICON-2016, 25–27 Nov 2016, pp. 1–6. https://doi.org/10.1109/poweri.2016.8077368
20. Azeem, A., Kumar, G., Malik, H.: Application of Waikato environment for knowledge analysis based artificial neural network models for wind speed forecasting. In: Proceedings of IEEE PIICON-2016, 25–27 Nov 2016, pp. 1–6. https://doi.org/10.1109/poweri.2016.8077352

Intelligent Traffic Light Scheduling Using Linear Regression

Ajinkya Khadilkar, Kunal Sunil Kasodekar, Priyanshu Sharma and J. Priyadarshini

Abstract A standard traffic light scheduler applies fixed time intervals for the green lights. This might sometimes result in traffic congestion and blocking of roads. To overcome this problem, this paper obtains a solution for scheduling the traffic light using a multiple linear regression model. This paper presents a model of how to achieve a traffic light schedule which is more efficient in terms of the number of vehicles stopping at the red light in each direction, and more effective to reduce traffic congestion as compared to traditional scheduling methods. Green time is the amount of time a directional light will have a green signal. This work presents an algorithm and a model to analyze the traffic density at different times and predict the green time for each directional traffic light such that the probability of traffic congestion is reduced and orderly flow of vehicles on the road is achieved. After applying this model, the number of vehicles stopping at the red light at a given time decrease by around 11%.

Keywords Intelligent traffic scheduling · Scheduling · Optimization
Linear regression · Machine learning · Green ratio

A. Khadilkar (✉) · K. S. Kasodekar · P. Sharma · J. Priyadarshini
Vellore Institute of Technology, Chennai, India
e-mail: ajinkya.khadilkar2015@vit.ac.in

K. S. Kasodekar
e-mail: kasodekarkunal.sunil2015@vit.ac.in

P. Sharma
e-mail: priyanshu.sharma2015@vit.ac.in

J. Priyadarshini
e-mail: priyadarshini.j@vit.ac.in

© Springer Nature Singapore Pte Ltd. 2019

H. Malik et al. (eds.), *Applications of Artificial Intelligence Techniques in Engineering*, Advances in Intelligent Systems and Computing 697,
https://doi.org/10.1007/978-981-13-1822-1_30

1 Introduction

With the increase in population and number of vehicles per capita in cities around the world, road traffic control has continued to be a big problem for city planners. It is very difficult for the cities to expand their road infrastructure. Even if it is possible to increase the number of roads, it is almost impossible to apply it in the central areas of a city. Thus, they have to rely on various traffic control mechanisms. Traditional traffic light consists of red and green light, where green light means it is safe to cross the road and flashing red light means do not cross the road. Traditionally a standard traffic light applies fixed green signal time to each and every direction irrespective of the traffic density in a particular lane. This paper makes the use of traffic density details from different directions in order to calculate the optimum green ratio.

Instead of using a fixed green signal time for each direction, this paper presents an algorithm which dynamically calculates the green signal time for each direction by using multiple linear regression model to predict optimal green ratios. It is important to have a proper and consistent data in order to optimize the traffic scheduling. For this to happen, it is important to capture the number of vehicles on the particular lane and for this purpose, one can make the use of a highly efficient camera such as AIS-IV Camera [1]. An inductive loop can also be used in order to detect the presence of vehicles. The inductive loop makes use of a moving magnet or an alternating current. The green ratio of a direction is defined as the ratio of green signal time of that direction to the total cycle time of that signals. This model has two applications, one when adequate resources are available to count the current number of vehicles on an intersection where the vehicle density and time can be given as an input to the model, and output will be the green ratio. Another application is when these resources are not available, the values of directional traffic can be taken from previous data and only current time will be required to obtain the green ratios.

2 Related Works

The standard traffic light scheduling method is efficient for handling daily traffic. But this model fails in places with unusual traffic patterns. This paper deals with change in traffic pattern every 15 min. A sudden change in traffic can be easily handled with this model.

There are few works done in this field. In [2], wireless sensor networks are used to monitor real-time traffic. Their work only determines the number of vehicles but does not consider time as a factor. Though collecting real-time data for scheduling traffic light might be accurate but predicting the traffic density at different times will be more efficient. The paper [3] focuses on vehicular ad hoc technology, which requires that every vehicle in the influential radius of the traffic light must have a

GPS (Global Positioning System) and should be able to communicate with the system. The position, speed, and vehicle density are calculated from each vehicle in the radius. This requires an existing infrastructure in all the vehicles. Whereas, the Intelligent Traffic Light Scheduling using Linear Regression requires cameras and image recognition to calculate the number of vehicles. The authors in [4] have calculated the average arrival rate of vehicles using an algorithm. The arrival rate is based on the various factors including the current number of vehicles in a queue. This model also does not consider the time as a parameter. The throughput can be increased, if the expected number of vehicles at a given time is predicted beforehand.

3 Proposed Work

In this paper, the traffic is analyzed and a traffic light schedule is developed for a four-way intersection. First, a green ratio is calculated for each direction for each cycle of the traffic signal with the mentioned algorithm. An optimum green ratio is the distribution of cycle time of the traffic signal in such a way that the number of vehicles stopping at the red light is minimum. A cycle time of a traffic light is the total time for the traffic light to complete its cycle of green lights for each direction. The obtained green ratio along with traffic densities and time are fit in a multiple linear regression model. Then, the green ratios are predicted based on the input test data.

3.1 Analysis of Traffic Patterns

The current traffic patterns of an intersection are analyzed. The average number of vehicles crossing the intersection in a day is plotted with respect to time.

From Fig. 1, it is clear that the maximum traffic flow is around 8 a.m. in the morning and 6 p.m. in the evening. A similar pattern occurs in all the other directions. The probability of traffic congestion is high during these peak hours. Efficient traffic scheduling is required the most around these times.

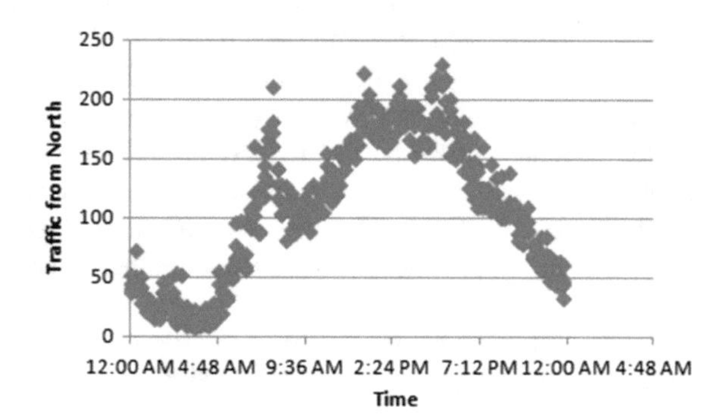

Fig. 1 Time versus north traffic

3.2 Architecture

The green ratios calculated from the multiple linear regression model are passed to the dynamic traffic scheduler. This scheduler calculates the actual green signal timing for each direction and passes it further to the traffic signal. The regression model is trained from the ratios calculated by the mentioned algorithm. To predict the green ratio values, the model must be given the time and number of vehicles in each direction as input. The number of vehicles at a given time can be calculated using the methods mentioned above, i.e., cameras and inductive loops. With the multiple linear regression model, green ratios at a given time can be predicted. The green ratio is calculated for each direction for each cycle of the traffic signal. Direction with the maximum traffic density will have the maximum green ratio as compared to other directions. In this way, we can predict the green ratios of all the directions and hence we can effectively switch between the red light and green light of different directions.

3.3 Algorithm to Calculate Green Ratio

A simulation program was developed in Python to practically simulate and apply the changes of signal timings and its effect on the traffic on each road. Assume the frequencies at which the vehicles arrive at the intersection from all four directions be nfreq, efreq, wfreq, and sfreq, respectively. The abovementioned simulator will increment the respective variables based on these frequencies. Assume the total cycle time as "T", number of lanes as "N", the time required for a vehicle to cross the stop line as "t". Here, it is assumed that the number of vehicles is equally distributed on the number of lanes for a given direction. Let nratio, eratio, sratio, and wratio be the green ratios of north, east, west, and south direction traffic lights, respectively. Initially, all four values are set to 0.25, indicating that all the directions will have an equal share of green light where 0.25 represents the ratio of green time of a particular direction to the total cycle time. This algorithm will change this ratio to the optimal green ratio.

There are four lanes from north, south, east, and west directions which meet at an intersection. Initially, all the four directions are assumed to have equal traffic density. Hence, equal green ratios are set for all the directions for the first cycle. Only for this first cycle, the green ratio is 0.25. After the first cycle, algorithm will work with these initial green ratios in order to calculate the optimal green ratios for all the directions based on the traffic density. This algorithm works on the basis of greedy paradigms hence it is necessary to give equal green ratios to all the four lanes (1/4 lanes = 0.25). The given scenario is simulated for one traffic light cycle. After which, the direction with the most number of vehicles is obtained. The green ratio of this direction is raised slightly by a value of "d". To maintain the sum of ratios as 1, a value of $d/3$ has to be subtracted from all other ratios, thus slightly decreasing the green time for that direction.

For example, after the first cycle if west direction has the most traffic, its green ratio will be wratio = 0.25 + d, nratio = 0.25 − (d/3), eratio = 0.25−(d/3), and sratio = 0.25−(d/3).

But if the traffic from one direction is extremely high, the ratio for that direction may reach 1, disabling the green light for all other directions. To overcome this problem, some constraints are added while changing the ratio. For the program, we used the values of variables which are as follows T = 120 s [4], N = 3, the roads have three lanes, d = 0.03, t = 2.5 s [5], where "t" is the time required for a vehicle to the cross the stop line. The more iteration the above program runs on, the more accurate are the ratios. Once the green ratios are obtained for each direction, they can be applied in the Multiple Linear Regression Model.

From the dataset, the traffic density of the particular direction can be easily obtained. The algorithm which is based on multiple linear regression model is used to calculate the optimal green ratio of the particular direction based on its traffic density.

The green ratio of a particular direction is solely based upon the traffic density. Suppose if the north direction has maximum traffic density then the green ratio of the north direction will be maximum. The algorithm sets a maximum and minimum limiting values for the green ratios such that each and every direction could get a fair and optimal green ratio time. Hence in this way, the optimal solution for the traffic congestion management is obtained.

3.4 Model—Linear Regression

Linear Regression is a statistical method that finds the relationship between at least two attributes. From all these attributes, one is dependent and all the others are independent. With the help of this statistical model, we can predict the dependent variable using the other independent variables.

The representation of a multiple linear regression is a linear equation that combines a specific set of input values (x) the solution to which is the predicted output for that set of input values (y). For a simple regression (a single x and a single y), the equation would be

$$y = B0 + B1^{*} x$$

This equation represents the equation of a line. The multiple linear equation is given by

$$y = B0 + B1^{*} x1 + B2^{*} x2 + \ldots Bn^{*} xn$$

Here, B0, B1, B2 ... Bn are regression coefficients.

For multiple linear regression, the equation represents a plane or a hyperplane. For this paper, a multiple regression model is applied. The green ratios obtained from the algorithm are fitted to a multiple linear regression model along with the time and traffic from each direction at that time. The green ratio is the target attribute. After plotting the regression model from the training data, one can obtain the green ratios by entering the time and number of vehicles.

3.5 Comparative Analysis

Comparative analysis is done to compare the traditional traffic light scheduling with this model. For the comparison between the two, the traffic from all the direction is monitored at regular intervals of 5 s. This comparison is done for the data values of 7:30 a.m (Fig. 2).

We obtained the traffic data of 45th St. and Main Ave., Fargo, North Dakota, US. We used the website [6] to obtain the datasets which contained monthly average values of traffic from each direction of the intersection at a 15-min interval.

There are no particular ratios in traditional scheduling. The ratios vary from intersection to intersection based on the type of the road and general traffic flow. But, all these ratios are constant with respect to time. For the purpose of comparative analysis, the green ratios for each direction for traditional scheduling were calculated based on the traffic density ratio in each direction. The average values for both the graphs are:

- For Traditional Ratio Distribution = 18.37 vehicles.
- For Predicted Ratio = 16.81 vehicles.

This is around 11% decrease in the number of vehicles stopping at the red light. It can be concluded that this model is more efficient than the traditional model with equal distribution.

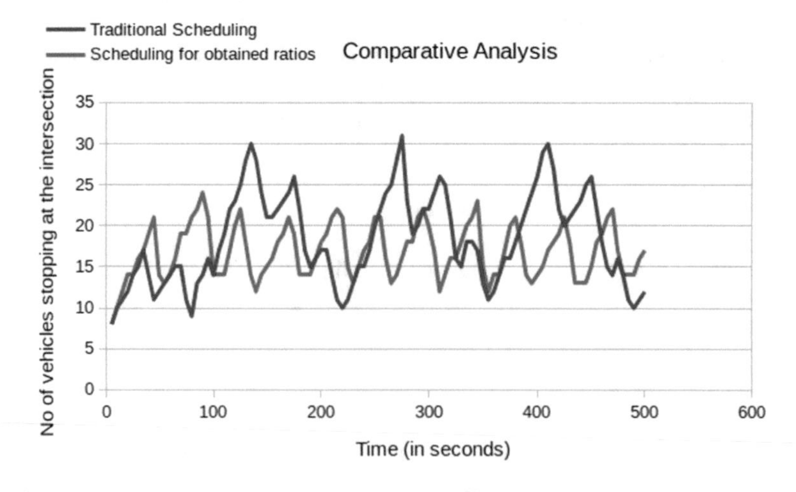

Fig. 2 Comparison for north direction at 7.30 a.m. traffic between traditional and new model

4 Conclusion

This paper has highlighted the need for a new traffic scheduling method. When the obtained linear regression model was applied to a test data, the accuracy of prediction of green ratio was in the range of 55–60%. It was found that this model is better than the traditional scheduling model on the basis of avoiding traffic congestion. This model decreases the number of vehicles stopping at the red light at a given time by around 11%.

References

1. Nabaasa, E., Bulega, T.E.: Scheduling and optimization of traffic lights on a T-junction using a wireless sensor network. Int. J. Sci. Knowl. **3**(3) (July 2013)
2. Younes, M.B., Boukerche, A.: An intelligent traffic light scheduling algorithm through VANETs. In: 10th IEEE International Workshop on Performance and Management of Wireless and Mobile Networks, P2MNET 2014, Edmonton, Canada (2014)
3. Abas, Z.A., Ee-Theng, L., Rahman, A.F.N.A., Abidin, Z.Z., Shibghatullah, A.S.: Enhanced scheduling traffic light model using discrete event simulation for improved signal timing analysis. ARPN J. Eng. Appl. Sci. **10**(18) (Oct 2015)
4. Signal timing. http://cce.oregonstate.edu/sites/cce.oregonstate.edu/files/pw-sigtime.pdf
5. Signal timing design. http://www.webpages.uidaho.edu/niatt-labmanual/Chapters/signaltiming design/professionalpractice/
6. Zhang, G., Avery, R.P., Wang, Y.: A video-based vehicle detection and classification system for real-time traffic data collection using uncalibrated video cameras. Transp. Res. Rec.: J. Transp. Res. Board pp. 138–147. https://doi.org/10.3141/1993-19
7. NDSU The Upper Great Plains Transportation Institute (UGPTI). http://www.atacenter.org/programs/ops/trafficvol/

A Novel p-Norm-like-Set Membership Affine Projection Algorithm in Sparse System Identification

Rajni and C. S. Rai

Abstract This paper presents a p-norm-like-set membership affine projection (p-norm-like-SMAP) algorithm. This algorithm finds an optimum value of norm constraint on each system coefficient by comparing each coefficient with all other system coefficients. This leads to imposing a zero-attraction constraint or not any sparsity constraint on each coefficient according to its comparative value to other coefficients. The simulation results are carried out in MATLAB both for colored input and Gaussian white input. The estimation performance confirms the better result than other existing algorithms.

Keywords p-norm · Set membership · Colored signal · Convergence rate
Sparsity · Gaussian

1 Introduction

Adaptive filtering finds application in a time-varying environment such as system identification, channel equalization, echo cancellation, and inverse system modeling [1–3]. In most of the substantial structures, the channel response is sparse in nature or we can say there are some large coefficients interlarded between many insignificant ones. The most known instance for this kind of system is echo paths [4, 5], in which the time spanned by the active region of echo path is only 8–12 ms as compared to the total time of network echo path of 64–128 ms. Some other instances are terrestrial transmission channel of high-definition digital TV (HDTV) [6], broadband wireless communication channel [7], and underwater acoustic channel [8]. These channels have only a few number of nonzero significant taps.

Rajni (✉) · C. S. Rai
University School of Information, Communication & Technology,
GGSIPU, New Delhi, Delhi, India
e-mail: radhagulati1986@gmail.com

© Springer Nature Singapore Pte Ltd. 2019

H. Malik et al. (eds.), *Applications of Artificial Intelligence Techniques in Engineering*, Advances in Intelligent Systems and Computing 697,
https://doi.org/10.1007/978-981-13-1822-1_31

Many researchers have considered the standard LMS and NLMS as the constructive methods for system identification problem because of low complexity and higher convergence rate [1–3].

However, in standard algorithms, the characteristics of system are not considered prior. Hence, the performance of sparse system identification can be further effectively improved by providing the system, prior information. Recent research in the field of compressive sensing promotes the various adaptive filtering algorithms by incorporating a penalty term acquainted with information regarding the sparseness of system into the objective function of the standard adaptive algorithms [7–16]. The most commonly proposed sparsity constraints in the above sparsity-aware adaptive algorithms are the ℓ_1-norm and reweighted ℓ_1-norm, which generate a zero-attractor in the iterations of filtering algorithms [7, 8, 16]. Research developed an ℓ_p-norm-based sparsity constraint, which shows better results than the ℓ_1-norm-based and reweighted ℓ_1-norm-based adaptive algorithm [13, 15].

Further, set membership adaptive filters have been proposed to reduce the computational complexity [17–20]. In this paper, we have added an ℓ_p-norm-based penalty term in set membership affine projection (SMAP) algorithm for sparse system identification. An ℓ_p-norm-like penalty adds a zero attraction on system coefficients in compliance with the comparative value of each system coefficient among all other values of coefficients. The main advantage of an ℓ_p-norm-like penalty is that all coefficients were not shrunk uniformly. In this penalty, zero attraction of the coefficient depends upon its relative value among other coefficients.

This paper is divided as follows: In Sect. 2, the SM-theory is demonstrated and, in Sect. 3, SMAP algorithm is reviewed; in Sect. 4, the proposed algorithm is derived. In Sect. 5, the simulation outputs of the proposed algorithm in MATLAB software are discussed. Finally, conclusions are drawn.

2 Set Membership Theory

Set membership algorithm imposes a bound on estimation of tap weights vector $\hat{\mathbf{w}} \in R^L$ by putting a constraint on magnitude of estimation error $\xi(k) = d(k) - \mathbf{u}^{\mathrm{T}}(k)\hat{\mathbf{w}}(k)$ over a model space of interest, where $\mathbf{u}(k) = [u(k)\ u(k-1)\ u(k-2)\ \ldots\ u(k-L+1)]^{\mathrm{T}}$ is the system input vector of dimension $\boldsymbol{L} \times \mathbf{1}$ and $d(k)$ is the desired output of the unknown system.

The model space φ consists of input signal vector–desired output signal pairs over which a bound will be specified on estimation error. This algorithm imposes a bound on the estimation of tap weights vector depending upon a parameter γ for all values in φ [14].

The possible solution set χ also called the feasibility set of $\hat{\mathbf{w}}(k)$ with estimation errors upper bounded in magnitude γ, whenever $(\mathbf{u}(k), d(k)) \in \boldsymbol{\varphi}$ can be defined as

$$\chi = \bigcap_{(u(k),d(k))\in\varphi} \left\{ \hat{\mathbf{w}}(k) \in R^L : \left| d(k) - \mathbf{u}^{\mathrm{T}}(k)\hat{\mathbf{w}}(k+1) \right|^1 \leq \gamma \right\}. \tag{1}$$

We have considered the input signal and desired sequences ($u(m)$, $d(m)$) for m = 0,1, … , k are available for training of adaptive filter. At time instant k, the constraint set $\mathbf{S}_k$ is defined as

$$\mathbf{S}_k = \left\{ \hat{\mathbf{w}}_k \in R^L : \left| d(k)) - \mathbf{u}^{\mathrm{T}}(k)\,\hat{\mathrm{w}}(k) \right|^1 \leq \gamma \right\}. \tag{2}$$

Now, we can define the membership set $\Theta(k)$, which is a subset of feasibility set

$$\chi\Theta(k) = \bigcap_{m=1}^{k} \mathrm{S}_m. \tag{3}$$

Thus, the main goal of SMF is to find adaptively an estimate that belongs to the constraint set.

3 Set Membership Affine Projection Algorithm

The theory of set membership affine projection (SMAP) algorithm is based on past N constraint set used in updating. Now, we can define $\Theta(k)$ as

$$\Theta(k) = \Theta^{k-N}(k) \cap \Theta^{N}(k), \tag{4}$$

where Θ^N is the intersection of the last N constraint sets, and Θ^{k-N} is the intersection of the first $k - N$ constraint sets. Based on the above definition of set membership algorithm theory, the objective of SMAP algorithm is to minimize $\|\hat{\mathbf{w}}(k+1) - \hat{\mathbf{w}}(k)\|^2$ according to the constraint $\hat{\mathbf{w}}(k) \in \Theta^N$. But for those $\hat{\mathbf{w}}(k) \notin \Theta^N$, the objective function of SMAP becomes

$$\textbf{Minimize}\ \|\hat{\mathbf{w}}(k+1) - \hat{\mathbf{w}}(k)\|^2 \text{ s.t. } \mathbf{d}(k) - \mathbf{U}^{\mathrm{T}}(k)\hat{\mathbf{w}}(k+1) = \boldsymbol{\gamma}(k), \tag{5}$$

where $\mathbf{U}(k) = [\mathbf{u}(k), \mathbf{u}(k-1), \ldots, \mathbf{u}(k-N+1)]$ is the input signal matrix of dimension $(L) \times (N)$, and $\mathbf{d}(k) = [d(k), d(k-1), \ldots, d(k-N+1)]^{\mathrm{T}}$ is the desired signal vector, and where $\boldsymbol{\gamma}(\mathrm{k}) = [\gamma_1(k)\ \gamma_2(k), \ldots, \gamma_N(k)]$ is the error-bound vector that specifies the point in Θ^N. Here, we have selected all the values of $\gamma_i(\mathrm{k})$ such that $|\gamma_i(\mathrm{k})| \leq \bar{\gamma}$. Let $\boldsymbol{\xi}(\mathrm{k})$ be a posteriori error signal given as

$$\boldsymbol{\xi}(\mathrm{k}) = [\xi_1(k), \xi_2(k), \ldots, \xi_N(k)]^{\mathrm{T}}, \tag{6}$$

where $\xi(\mathrm{k}) = \mathrm{d}(\mathrm{k}-\mathrm{i}+1) - \mathbf{u}^T(\mathrm{k}-\mathrm{i}+1)\hat{\mathbf{w}}(\mathrm{k}+1)$, for i = 1, …, N

By the method of Lagrange multiplier, the weight update equation of SMAP algorithm becomes

$$\hat{\mathbf{w}}(k+1) = \begin{cases} \hat{\mathbf{w}}(k) + \mathbf{U}(k)\left[\mathbf{U}(k)\mathbf{U}^{\mathrm{T}}(k)\right]^{-1}[\boldsymbol{\xi}(k) - \boldsymbol{\gamma}(k)] \text{ if } |\xi_1(k)| > \gamma \\ \hat{\mathbf{w}}(k) \text{ otherwise} \end{cases}. \tag{7}$$

Recently S. Werner and Paulo S. R. Diniz have proposed a very simple choice of $\gamma(\mathrm{k})$ in SMAP by providing $\gamma_i(k) = \xi_i(k)$ for $i \neq 1$ and $\gamma_1(\mathrm{k}) = \mathrm{sign}(\xi_1(\mathrm{k}))$ for i = 1 [19, 20].

4 Proposed *p*-Norm-like-SMAP-Algorithm

The proposed algorithm exploits the sparsity of system by incorporating p-norm-like penalty into the objective function of standard SMAP algorithm. The objective function of p-norm-like-SMAP becomes

$$\begin{aligned} G_{\ell p}(k) = \|(\hat{\boldsymbol{w}}(k+1) - \hat{\boldsymbol{w}}(k)\|^2 + \Lambda_{\ell p}^{\mathrm{T}}(k)\,(\mathbf{d}(k) - \boldsymbol{U}^{\mathrm{T}}(k)\hat{\mathbf{w}}(k+1) \\ - \boldsymbol{\gamma}(k)) + q\|\hat{\boldsymbol{w}}(k+1)\|_p^p, \end{aligned} \tag{8}$$

where $\Lambda_{\ell p}$ is a vector of Langrage multiplier of dimension $N \times 1$ and $\gamma(k)$ is similarly selected as above in SMAP algorithm and q is an adjustable factor to balance the convergence rate and estimation error and $||\hat{\boldsymbol{w}}(k+1)||_p^p$ is p-norm-like, whose definition is given as

$$||\hat{\boldsymbol{w}}(k)||_p^p = \sum_{j=1}^{L} \left|\hat{w}_j(n)\right|^p, \quad \text{where } p \in (0,1). \tag{9}$$

From the above definition, it can be seen that the ℓ_0- and ℓ_1-norm can be derived from the above p-norm-like as

$$\lim_{p\to 0}||\hat{\boldsymbol{w}}(k)||_p^p = \|\hat{\boldsymbol{w}}(k)\|_0 \text{ is known as } \ell_0\text{−norm} \tag{10}$$

$$\lim_{p\to 1}||\hat{\boldsymbol{w}}(k)||_p^p = \|\hat{\boldsymbol{w}}(k)\|_1 \text{ is known as } \ell_1\text{−norm.} \tag{11}$$

By Lagrange multiplier, the weight update equation of the proposed p-norm-like-SMAP algorithm becomes

$$\hat{\boldsymbol{w}}(k+1) = \hat{\boldsymbol{w}}(k) + \frac{\mathbf{U}(k)}{\mathbf{U}^{\mathrm{T}}(k)\,\mathbf{U}(k)}(\boldsymbol{\xi}(k) - \boldsymbol{\gamma}(k)) + \left(\frac{\mathbf{U}(k)\mathbf{U}^{\mathrm{T}}(k)}{\mathbf{U}^{\mathrm{T}}(k)\mathbf{U}(k)} - I\right) q \frac{\mathrm{psgn}(\hat{\mathbf{w}}(k+1))}{2|\hat{\mathbf{w}}(k+1)|^{1-p}}. \tag{12}$$

Let

$$H(k) = \frac{\mathbf{U}(k)\mathbf{U}^{\mathrm{T}}(k)}{\mathbf{U}^{\mathrm{T}}(k)\,\mathbf{U}(k)} - I \tag{13}$$

$$\hat{w}(k+1) = \hat{w}(k) + \frac{\mathbf{U}(k)}{\mathbf{U}^{\mathrm{T}}(k)\,\mathbf{U}(k)}(\xi(k) - \gamma(k)) + H(k)\,q\frac{\mathrm{psgn}(\hat{\mathbf{w}}(k+1))}{2|\hat{\mathbf{w}}(k+1)|^{1-p}}. \tag{14}$$

To prevent the overflow caused by dividing by zero, we add a small coefficient σ in denominator,

$$\hat{w}(k+1) = \hat{w}(k) + \frac{\mathbf{U}(k)}{\mathbf{U}^{\mathrm{T}}(k)\,\mathbf{U}(k)}(\xi(k) - \gamma(k)) + H(k)\,q\frac{\mathrm{psgn}(\hat{\mathbf{w}}(k+1))}{\sigma + 2|\hat{\mathbf{w}}(k+1)|^{1-p}}. \tag{15}$$

However, parameter p affects biases during estimation and sparsity measure independently. So, parameter p cannot be taken usually. To overcome this problem, different values of p should be used for different values of $\hat{\mathbf{w}}(k)$. The variable p-norm can be defined as

$$\lim_{p\to 1}||\hat{w}(k)||_p^p = \sum_{j=1}^{L}|\hat{w}_j(n)|^{p_j} \text{ where } p \in (0,1). \tag{16}$$

Using the above analysis, the weight update equation becomes

$$\hat{w}_j(k+1) = \hat{w}_j(k) + \frac{\mathbf{U}_j(k)}{\mathbf{U}_J^{\mathrm{T}}(k)\mathbf{U}_j(k)}(\xi(k) - \gamma(k)) + H_j(k)\,q\frac{\mathrm{psgn}\big(\hat{w}_j(k+1)\big)}{\sigma + 2\big|\hat{w}_j(k+1)\big|^{1-p_j}}. \tag{17}$$

Hence, the incorporation of nonuniform value of p-norm as $p = [p_0, p_1, p_2 \ldots p_L]$ will introduce different values of p to different weight taps vector, and this will make a good balance between sparsity level measure and bias during the estimation of weight taps.

So, we have to optimize p to minimize the estimation bias when $\hat{w}_j(k)$ is large. To address this problem, we should introduce a metric for the magnitude of estimated tap weights. Based on the value of tap weights, the metric differentiates the weights into large and small classes. We have considered the expectation of magnitude of estimated tap weights as the metric, which can be written as

$$\hat{s}_j(k) = E\{|\hat{w}_j(k)|\} \quad \forall\, 0 \le j \le L. \tag{18}$$

From the above metric, it is clear that the effect caused by p-norm-like sparsity constraint should be zero for large class of tap weights, which can be written as

$$\text{Minimize } \frac{\text{p}_j}{\left|\hat{w}_j(k+1)\right|^{1-p_j}} \quad \forall\, \hat{s}_j(k) \geq \hat{w}_j(k) \tag{19}$$

So, the above inclusive optimization theory used in calculation of p-norm-like constraint will make the variable p to take the value either 1 or 0 for smaller and larger classes of tap weights.

Hence, the weight update equation becomes

$$\hat{w}_j(k+1) = \hat{w}_j(k) + \frac{\mathbf{U}_j(k)}{\mathbf{U}_j^{\mathrm{T}}(k)\mathbf{U}_j(k)}(\xi(k) - \gamma(k)) + H_j(k)\, q\, \operatorname{sgn}\left(\hat{w}_j(k+1)\right) t_j, \tag{20}$$

where

$$t_j = \frac{\operatorname{sgn}\left(\hat{s}_j(\mathrm{k}) - \left|\hat{w}_j(k)\right|\right) + 1}{2} \quad 0 \leq j \leq L. \tag{21}$$

From Eq. (20), it is clear that the sparsity constraint imposed by irregular value of p-norm is completely different from ℓ_0- and ℓ_1-norm. This sparsity constraint depends on the value of tap weights' individual value. This will accelerate the convergence of small weights and remove the bias of large weights during estimation.

The calculation of $H(k)$ increases the computational complexity. So, to reduce complexity, optimization problem using above definition of p-norm-like constraint can be written as

$$G_{\ell p}(k) = \|(\hat{\boldsymbol{w}}(k+1) - \hat{\boldsymbol{w}}(k)\|^2 + \Lambda_{\ell p}(k)\|(\xi(k) - \gamma(k)\|_{D(k)}^2 + q\|\hat{w}(k+1)\|_p^p, \tag{22}$$

where $D(k) = \frac{1}{\mathbf{U}^{\mathrm{T}}(k)\mathbf{U}(k)}$ and

$$\|(\xi(k) - \gamma(k))\|_{D(k)}^2 = (\xi(k) - \gamma(k))^{\mathrm{T}} D(k)\, (\xi(k) - \gamma(k)).$$

By the method of Langrage multiplier, the weight update equation becomes

$$\hat{\boldsymbol{w}}(k+1) = \hat{\boldsymbol{w}}(k) + \frac{\mathbf{U}(k)}{\mathbf{U}^{\mathrm{T}}(k)\,\mathbf{U}(k)}(\xi(k) - \gamma(k)) - q\frac{\operatorname{sgn}(\hat{\boldsymbol{w}}(k))}{1 + \varepsilon|\hat{\boldsymbol{w}}(k)|}\mathrm{T}, \tag{23}$$

where T is the diagonal matrix such that

$$T = \begin{bmatrix} t_0 & 0 & 0 & \ldots \\ 0 & t_1 & 0\ldots0 & \\ 0 & \cdots & & 0 \\ \vdots & \ddots & & \vdots \\ 0 & \cdots & & t_{L-1} \end{bmatrix}_{L\times L} \tag{24}$$

5 Simulation Results

In this section, we evaluate the performance of the proposed p-norm-like-SMAP algorithm in sparse system identification. For this, we have considered an unknown system with 128 taps with different sparsity levels $K = \{4, 8, 16\}$. The number of taps for unidentified system and adaptive filter is assumed to be the same. We have taken the average mean square deviation (MSD) as a criterion to assess the performance of sparse system identification. The nonzero coefficients of unknown system are Gaussian with zero mean and unit variance, and locations of these taps are randomly generated.

The parameter q affects the performance of the proposed algorithm, and hence it should be optimized. Figure 1 shows the performance for different values of q. The performance of system is going better when q goes below from 8×10^{-3} to 8×10^{-4}, but it starts diminishing when q goes below 5×10^{-4}. So, we have taken $q = 8 \times 10^{-4}$ in all simulations to compare the performance with other affine projection algorithms.

Now, we will evaluate the performance of system for colored input, which is produced by passing white Gaussian input through second-order AR filter with $H(z) = [1 - 0.95z^{-1}]$. We have taken SNR = 30 dB,$\gamma = \sqrt{5\sigma_z^2}$ for set membership normalized least mean square (SMNLMS), set membership affine projection (SMAP), and the proposed p-norm-like-SMAP algorithm, and projection order, $N = 6$ for affine projection algorithm (APA), zero-attracting APA (ZA-APA) [16], and residual zero-attracting APA (RZA-APA) [16].

From Figs. 2, 3, and 4, it is clear that the proposed algorithm performs better than other algorithms even when the system is getting denser. The performance of other algorithms starts deteriorating when the system becomes less sparser.

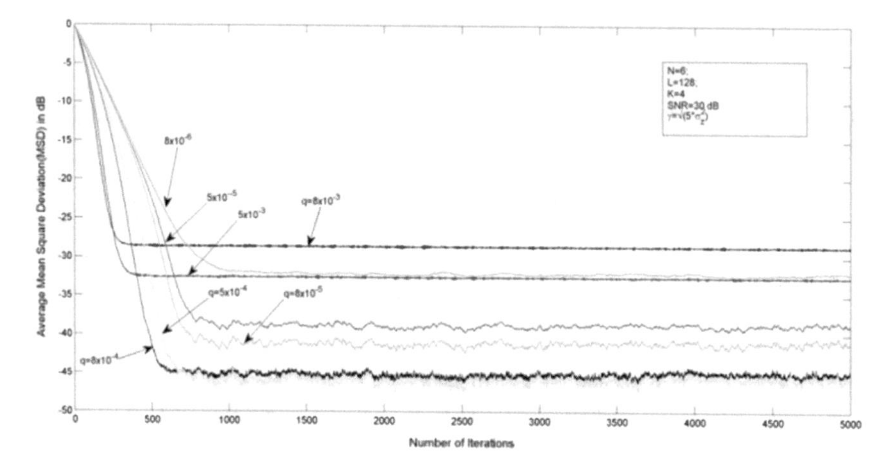

Fig. 1 Performance of the proposed algorithm for different values of q

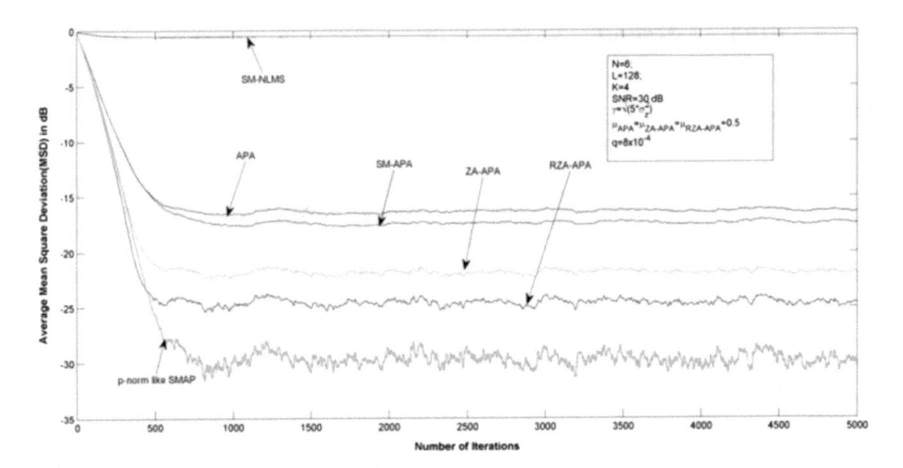

Fig. 2 Comparison of average MSD of the proposed algorithm with existing affine projection algorithms for system sparsity level, $K = 4$ with (colored input)

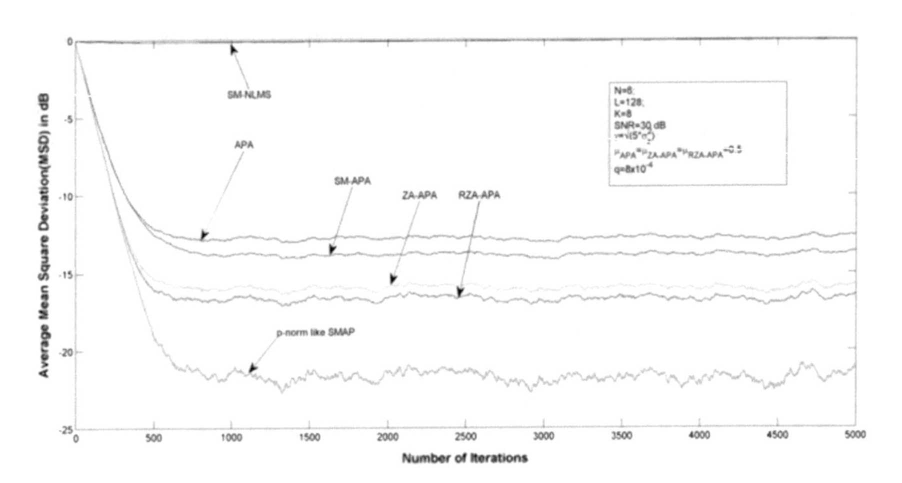

Fig. 3 Comparison of average MSD of the proposed algorithm with existing affine projection algorithms for system sparsity level, $K = 8$ with (colored input)

Second, we consider the input signal as white Gaussian input with zero mean and unity variance. We have taken SNR equal to 30 dB, $\gamma = \sqrt{5\sigma_z^2}$, and $N = 6$ in all simulations. The simulation results are averaged to 50 Monte Carlo iterations. In Figs. 2, 3, and 4 we compare the proposed algorithm with the standard existing algorithm for colored input. Figures 5, 6, and 7 represent the performance of the proposed algorithm for white Gaussian input.

From the above simulation results, it is clear that p-norm-like-SMAP performs better than the other existing algorithms in all cases of sparsity level for both colored input and white Gaussian input. Thus, the proposed algorithm is not

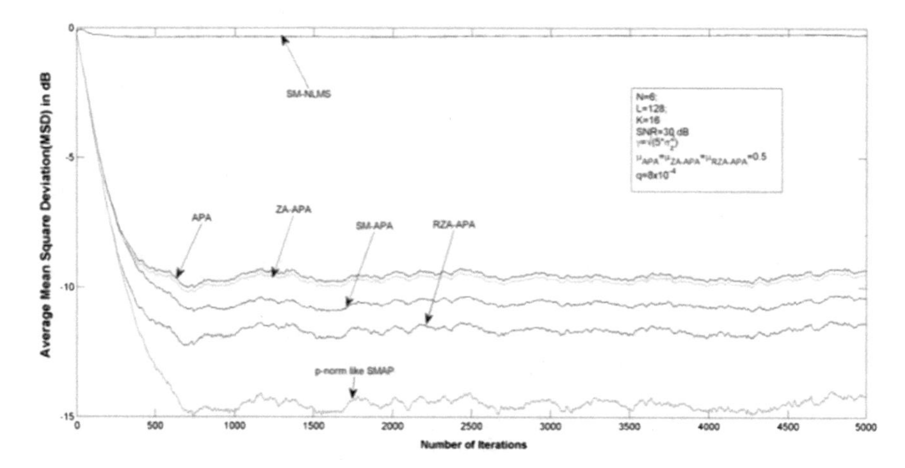

Fig. 4 Comparison of average MSD of the proposed algorithm with existing affine projection algorithms for system sparsity level, K = 16 with (colored input)

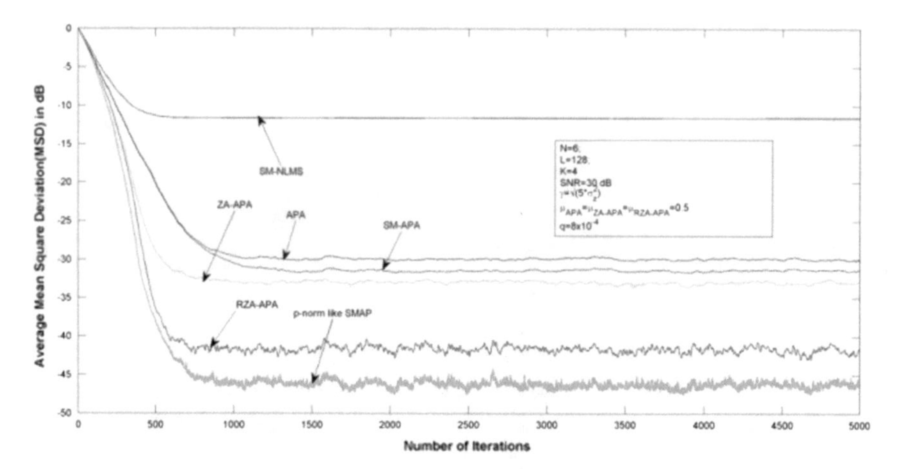

Fig. 5 Comparison of average MSD of the proposed algorithm with existing affine projection algorithms for system sparsity level, K = 4 with (white input)

affected by sparsity level as well as kind of input. Hence, the proposed algorithm performs better in terms of lower average mean square deviation and higher convergence rate than the other existing algorithms.

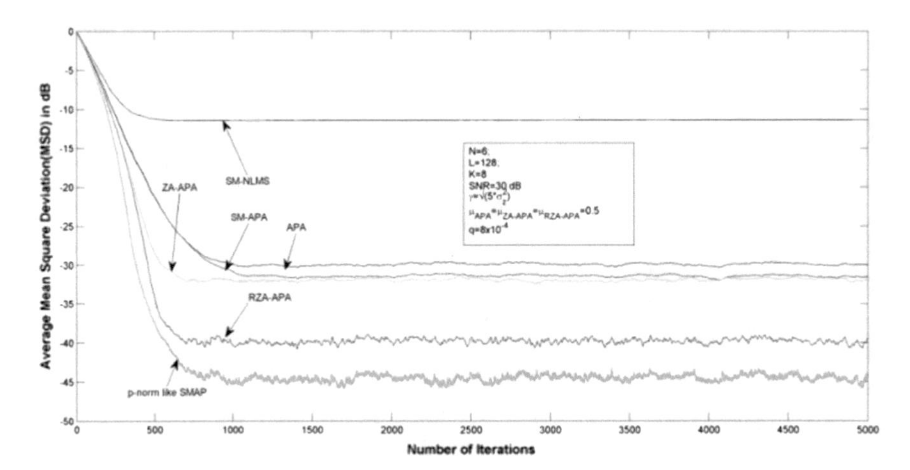

Fig. 6 Comparison of average MSD of the proposed algorithm with existing affine projection algorithms for system sparsity level, $K = 8$ with (white input)

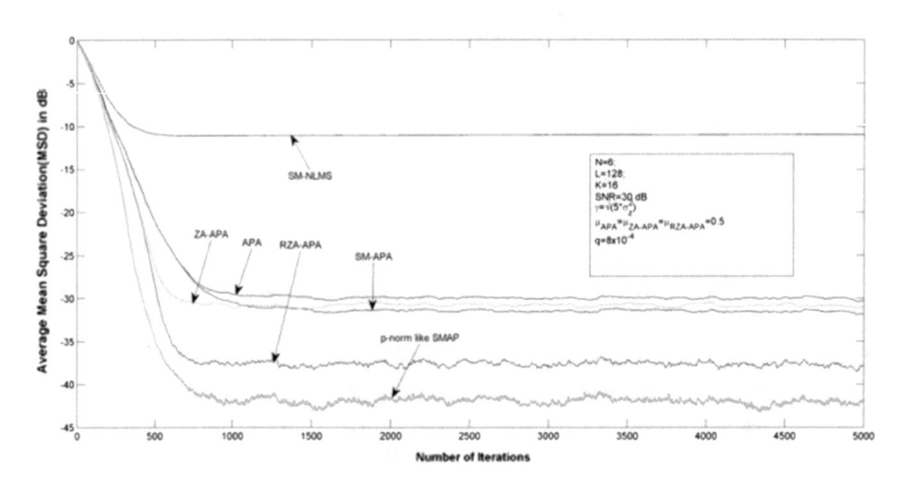

Fig. 7 Comparison of average MSD of the proposed algorithm with existing affine projection algorithms for system sparsity level, $K = 16$ with (white input)

6 Conclusion

This paper presents a novel p-norm-like-SMAP algorithm, which incorporates an irregular p-norm penalty into the cost function of SMAP algorithm. This penalty is based on the comparative value of each system coefficient among all coefficients. Finally, the simulation results are executed to demonstrate the better performance of the proposed algorithm in terms of lower MSD and faster convergence rate for both colored input and white input.

References

1. Sayed, A.H.: Fundamental of Adaptive Filtering. Wiley, NY (2003)
2. Haykin, S.: Adaptive Filter Theory, 4th edn. Pearson Education India (2008)
3. Diniz, P.S.R.: Adaptive Filtering: Algorithms and Practical Implementation. Kluwer Academic Publishers, Norwell, MA, USA (2002)
4. Paleologu, C., Benesty, J. Ciochina, S.: Sparse Adaptive Filters for Echo Cancellation, 1st edn, p. 124. Morgan & Claypool (2011)
5. Radecki, J., Zilic, Z., Radecka, K.: Echo cancellation in IP networks. In: 45th Midwest Symposium on Circuits and Systems, MWSCAS, vol. 2, pp. 219–222 (2002)
6. Fevrier, J., Gelfand, S.B., Fitz, M.P.: Reduced complexity decision feedback equalization for multipath channels with large delay spreads. IEEE Trans. Commun. **47**(6), 927–937 (1999)
7. Chen, Y., Gu, Y., Hero, A.: Sparse LMS for system identification. In: Proceedings of IEEE International Conference on Acoustic Speech and Signal Processing, pp. 3125–3128 (April 2009)
8. Wang, Y., Li, Y., Jin, Z.: An improved reweighted zero-attracting NLMS algorithm for broadband sparse channel estimation. In: IEEE International Conference on Electronic Information and Communication Technology (ICEICT), Harbin, pp. 208–213 (2016)
9. Chen, P., Rong, Y., Nordholm, S., Duncan, A.J., He, Z.: Compressed sensing based channel estimation and impulsive noise cancelation in underwater acoustic OFDM systems. In: 2016 IEEE Region 10 Conference (TENCON), Singapore, pp. 2539–2542 (2016)
10. Gu, Y., Jin, J., Mei, S.: l0 norm constraint LMS algorithm for sparse system identification. IEEE Signal Process. Lett. **16**(9), 774–777 (2009)
11. Gui, G., Peng, W., Xu, L., Liu, B., Adachi, F.: Variable-step-size based sparse adaptive filtering algorithm for channel estimation in broadband wireless communication systems. EURASIP J. Wirel. Commun. Netw. (2014)
12. Yoo, J.W., Shin, J.W., Park, P.G.: An improved NLMS algorithm in sparse systems against noisy input signals. IEEE Trans. Circuits Syst.—ii: express briefs **62**(3) (March 2015)
13. Weruaga, L., Jima, S.: Exact NLMS algorithm with l_p-norm constraint. IEEE Signal Process. Lett. **22**, 366–370 (Mar 2015)
14. Jahromi, M.N.S., Salman, M.S., Hocanin, A., Kukre, O.: Convergence analysis of the zero-attracting variable step-size LMS algorithm for sparse system identification. Springer SIViP. https://doi.org/10.1007/s11760-013-0580-9 (2013)
15. Taheri, O., Vorobyov, S.A.: Sparse channel estimation with lp-norm and reweighted l1 -norm penalized least mean squares. In: IEEE International Conference on Acoustic, Speech and Signal Processing (ICASSP), Prague, Czech Republic, pp. 2864–2867, 22–27 May 2011
16. Meng, R., de Lamare, R.C., Nascimento, V.H.: Sparsity-aware affine projection adaptive algorithms for system identification. In: Sensor Signal Processing for Defence (SSPD 2011), London, pp. 1–5 (2011). https://doi.org/10.1049/ic.2011.0144
17. Deller, J.R.: Set-membership identification in digital signal processing, IEEE Acoust. Speech Signal Process. Mag. **6**, 4–22 (Jan 1989)
18. Zhang, S., Zhang, J.: Set-membership NLMS algorithm with robust error bound. IEEE Trans. Circuits Syst. II Express Briefs **61**(7), 536–540 (2014)
19. Werner, S., Diniz, P.S.R.: Set-membership affine projection algorithm. IEEE Signal Process. Lett. **8**(8), 231–235 (2001)
20. Yazdanpanah, H., Diniz, P.S.R., Lima, M.V.S.: A simple set-membership affine projection algorithm for sparse system modeling. In: 24th European Signal Processing Conference (EUSIPCO), Budapest, pp. 1798–1802 (2016)

Space Shuttle Landing Control Using Supervised Machine Learning

Sunandini Singla and Niyati Baliyan

Abstract Supervised machine learning as a data mining tool has seen a huge growth in interest in the recent past, and is being widely used in real-world problems of diverse domains such as informatics, mechanical engineering, automation, and geology, among others. Space shuttle auto landing control using supervised machine learning has been attempted by various researchers previously. It has been seen that the performance of the control system in terms of prediction time and accuracy was determined by the underlying classification algorithm. A few works are available on binary classification of landing control, i.e., manual or automatic. However, the application of support vector machine based classifier has not been researched thus far. Particularly, the support vector machine based classifiers are prevalent in their use as they work really well with distinct segregation margin in greater dimension region. Moreover, owing to the usage of a subset of training instances in the classifier, support vector machine approach is also memory efficient. In this paper, we have less number of tuples and hence reduced training time of the support vector classifier. The primary goal of our experiment is to train the classifier for finding out the situation in which an auto landing would be preferred to manual landing of the space shuttle.

Keywords Shuttle landing · Control · Supervised machine learning
Classification · Support vector machine

1 Introduction

Machine learning is the technique employed to make the computer system learn as a human does, with the intention of making sense of real-world data and facilitate decision-making. Machine learning is employed in diverse applications in which explicit programming is either infeasible or inefficient [1].

S. Singla · N. Baliyan (✉)
Department of Computer Science and Engineering, Thapar University,
Bhadson Road, Patiala 147004, Punjab, India
e-mail: niyati.baliyan@gmail.com

© Springer Nature Singapore Pte Ltd. 2019

H. Malik et al. (eds.), *Applications of Artificial Intelligence Techniques in Engineering*, Advances in Intelligent Systems and Computing 697,
https://doi.org/10.1007/978-981-13-1822-1_32

1.1 Supervised Machine Learning

Further, supervised machine learning makes use of training tuples to predict a possible pattern in a new data tuple. It is like the system is being provided with sample inputs and their corresponding output, by a supervisor and the objective is to learn a pattern or generic rule that will generate output from inputs.

1.2 Classification

Classification is a type of supervised machine learning technique where a learner must obtain a model that assigns labels or classes to unseen inputs. The training data is composed of input attributes and output labels or predicted classes. The goodness in terms of accuracy, of a supervised machine learning technique is evaluated based on its run on the test tuples [2]. In case, there are two labels to choose from, the task is known as binary classification problem [3].

1.3 Support Vector Machine

The classifier used in this research paper is Support Vector Machine (SVM). The fundamental idea in SVM is the specification of decision or classification boundary using decision planes. It is aimed to achieve maximum interclass separation in the case of SVM classifier [4]. A classifier that is able to divide the data points into classes with the help of a linear function or line is known as linear classifier. While a classifier that is able to divide the data points into their respective classes with the help of a curve is known as nonlinear classifier [5].

SVM is a supervised learning approach which is majorly used in classification problems. It is one of the most accurate approaches used for classifications. In SVM, each of the data items is plotted as a point in n-dimensional space (where n is number of attributes/features that we have in our dataset) with the value of each attribute/feature being the value of a specific coordinate as shown in Fig. 1 [6]. Support vectors are simply the coordinates of individual instances. SVM is a frontier which best segregates the two classes (hyperplane/line).
Various advantages of using SVM for space shuttle landing control are:

- Effective in high dimensional spaces.
- Effective when the number of dimensions is larger than the number of samples.
- Has less memory requirements.

However, the abovementioned advantages come with a trade-off as follows:

- The training time required is higher.

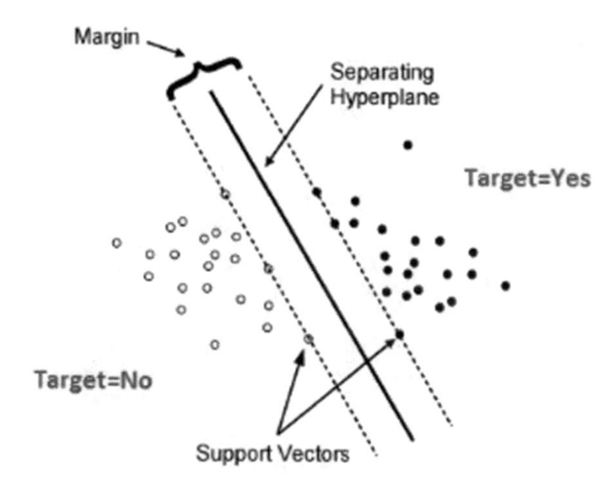

Fig. 1 Support vector machine

- It performs poorly when the data set has more noise, i.e., target classes are overlapping.
- SVM does not directly provide probability estimates, these are computed using cross validation.

2 Dataset Description

For the research work in this paper, we have obtained the space shuttle landing control data from UCIs repository which is a conglomeration of standard datasets [7]. This has been worked upon previously as well [8, 9], however, not much work has been done for exploring SVM-based classification on the shuttle landing control dataset.

The dataset used for space shuttle landing control [7] is a rather tiny database, with 15 instances (tuples or rows), and 7 attributes (or columns) including the class attribute. Table 1 describes the attributes and their data types in the dataset under consideration.

The numeric values assigned to each attribute range from 1 to the number of possible values, for example, stability ranges from stab (1) to xstab (2), and magnitude ranges from low (1) to out of range (4). However, the output attribute, i.e., class is assigned either of the two values: no auto (0) and auto (1).
The class distribution is as follows:

- Class 1: uses no auto control: six instances
- Class 2: uses automatic control: nine instances.

Table 1 Attribute description

SNo.	Attribute	Type	Possible values	Description
1	Class	Binary	noauto; auto	Helps to take decision whether or not autolanding is preferred to manual control
2	Stability	Nominal	Stab; xstab	Based on stability analysis, xstab implies spacecraft faces unstable conditions
3	Size of error	Ordinal	XL, LX, MM, SS	Keeps track of size of errors faced in a tuple reading
4	Sign	Binary	pp, nn	NA
5	Wind	Binary	Head, tail	Determines if direction of spacecraft is same as that of wind (tail) or not (head)
6	Magnitude	Ordinal	Low, medium, strong, out of range	NA
7	Visibility	Binary	Yes, no	Based on a threshold of visibility range that determines if the view ahead is good

3 Data Preprocessing

There are no missing values in the dataset, however, there are various "don't care" values marked with an asterisk (*). Table 2 gives attribute number and the corresponding count of their "don't care" values.

The * values were replaced with every value that existed for a particular variable. Tables 3 and 4 show the before and after preprocessing scenario owing to "don't care" terms, respectively.

The data is preprocessed by replacing the "don't care" terms by any plausible random values, as in Table 4.

Table 2 Don't care values of attributes

Attribute no.	Number of don't care values
2	2
3	3
4	8
5	8
6	5
7	0

Table 3 Dataset before preprocessing

Class	Stability	Error	Sign	Wind	Magnitude	Visibility
2	*	*	*	*	*	2
1	2	*	*	*	*	1
1	1	2	*	*	*	1
1	1	1	*	*	*	1
1	1	3	2	2	*	1
1	*	*	*	*	4	1
2	1	4	*	*	1	1
2	1	4	*	*	2	1
2	1	4	*	*	3	1
2	1	3	1	1	1	1
2	1	3	1	1	2	1
2	1	3	1	2	1	1
2	1	3	1	2	2	1
1	1	3	1	1	3	1
2	1	3	1	2	3	1

Table 4 Dataset after preprocessing

Class	Stability	Error	Sign	Wind	Magnitude	Visibility
2	1	3	1	2	1	2
1	2	2	2	3	2	1
1	1	2	1	4	1	1
1	1	1	2	1	3	1
1	1	3	2	2	3	1
1	2	1	1	1	4	1
2	1	4	1	3	1	1
2	1	4	2	4	2	1
2	1	4	2	2	3	1
2	1	3	1	1	1	1
2	1	3	1	1	2	1
2	1	3	1	2	1	1
2	1	3	1	2	2	1
1	1	3	1	1	3	1
2	1	3	1	2	3	1

4 Research Methodology

The SVM-based classifiers are dominant in their use as they perform well with different distinct boundaries for higher dimensions or larger number of features. Additionally, due to the application of a part of training tuples in the classifier, the SVM approach does not consume much storage. Our work, therefore, decreases the training time of the SVM classifier. We intend to finding out the situation in which an auto landing would be preferred to manual landing of the space shuttle after training the classifier.

The overall workflow of our prediction model is illustrated in Fig. 2.

In this paper, for the space shuttle landing scenario, one needs to predict whether manual or automated landing would be preferred, based on various parameters that act as a determinant of this crucial decision. This is a manifestation of binary (manual versus automated landing) classification problem. The pseudocode for the working of proposed method can be summarized as below:

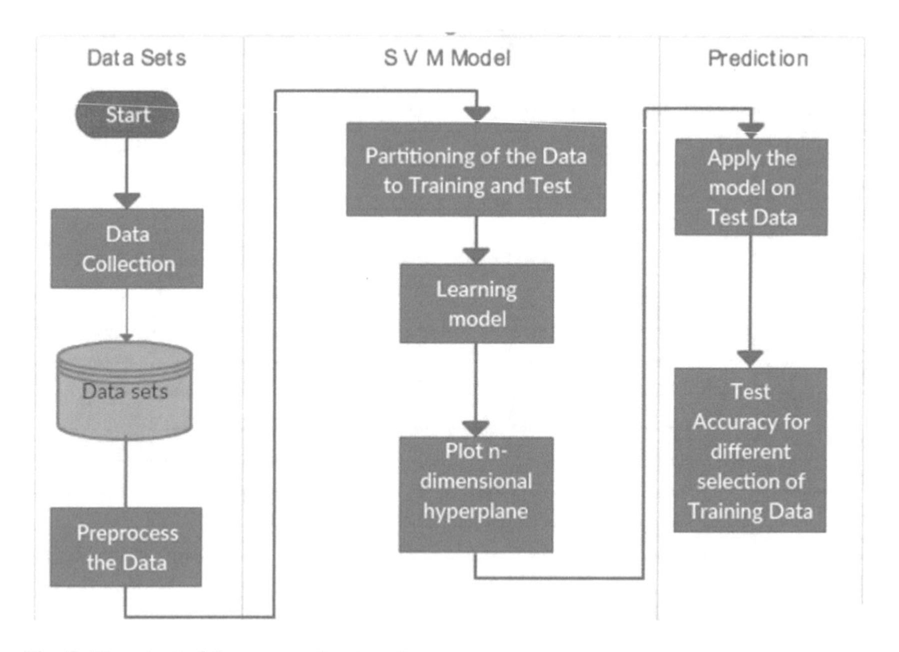

Fig. 2 Flowchart of the proposed approach

Input:dataset with 6 attributes and 2 classes
Tuple (stability, error, sign, wind, magnitude, visibility)

Initialization:
Shuffle the dataset

For each distinct dataset partition into trainDataSet
  For each shuffle of trainDataSet
    Take n initial tuples as training set and rest as test set
    Transform dependent variables to factor variables
    Implement an n-dimensional hyperplane using attributes
Test obtained model on testData
Validate obtained (predicted) values with actual values, to get accuracy

5 Results and Discussion

The model is implemented in R Studio [10] and accuracy value is nearly 84% in arithmetic mean of multiple runs, which is a highly satisfactory value. The accuracy was obtained with multiple shuffles within the dataset itself. The shuffling selects random samples which are to be trained in the dataset. Additionally, a variation on the percentage of tuples taken into training data set from the original one is carried out to validate its accuracy and is plotted on the graph as shown in Fig. 3.

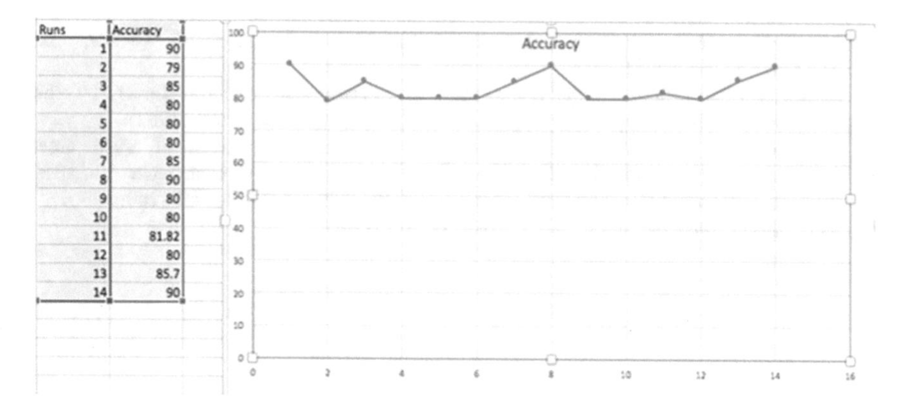

Runs	Accuracy
1	90
2	79
3	85
4	80
5	80
6	80
7	85
8	90
9	80
10	80
11	81.82
12	80
13	85.7
14	90

Fig. 3 Accuracy of SVM on space shuttle landing control prediction

6 Conclusion

Space shuttle safety is a crucial research domain and landing is one of the most significant aspects that need to be taken into account in the context of space shuttle safety. There is not adequate advancement in this research area although some progress has been made on the landing control issues. This paper proposes a model on prediction of landing control using support vector machine (SVM) method. First, the factors influencing manual versus automated landing safety were explored and the relevant shuttle data were procured. Next, shuttle data were preprocessed and SVM model was employed to train the space shuttle data sample. According to the results, the optimal SVM model was demonstrated effectively. In the future, ensemble machine learning instead of SVM can be used to further improve the landing outcome prediction accuracy.

References

1. Samuel, A.: Some studies in machine learning using the game of checkers. IBM J. Res. Dev. **3** (3) (1959). https://doi.org/10.1147/rd.33.0210
2. Han, J., Jian, P., Kamber, M.: Data Mining: Concepts and Techniques. Elsevier (2011)
3. Witten, I.H., Frank, E.: Data Mining: Practical Machine Learning Tools and Techniques. Morgan Kaufmann, p. 664 (2011). ISBN 978-0-12-374856-0
4. Hill, T., Lewicki, P.: Statistics: Methods and Application: A Comprehensive Reference for Science, Industry, and Data Mining. StatSoft, Tulsa, OK (2006)
5. Kotsiantis, S.B., Zaharakis, I., Pintelas, P.: Supervised machine learning: a review of classification techniques. **3**(24) (2007)
6. http://dni-institute.in/blogs/building-predictive-model-using-svm-and-r/. Available online, last accessed on 2 June 2017
7. https://archive.ics.uci.edu/ml/datasets/Shuttle+Landing+Control. Available online, last accessed on 1 Mar 2017
8. Michie, D.: The fifth generation's unbridged gap. In Herken, R. (ed.) The Universal Turing Machine: A Half-Century Survey, pp. 466–489. Oxford University Press (1988)
9. Chaudhari, N. S., Tiwari, A., Thomas, J.: Performance evaluation of SVM based semi-supervised classification algorithm. In: 10th International Conference on IEEE Control, Automation, Robotics and Vision, ICARCV 2008, pp. 1942–1947 (2008)
10. rstudio.com. Available online, last accessed on 3 July 2017

Realization of Recursive Algorithm for One-Dimensional Discrete Sine Transform and Its Inverse

Pragati Dahiya and Priyanka Jain

Abstract This paper proposes algorithms for computation of Discrete Sine Transform (DST) and Inverse Discrete Sine Transform (IDST) using the recursive technique. The presented algorithms are realized using Infinite Impulse Response (IIR) filter structures. The proposed algorithms are efficient in terms of hardware and computational complexity. Hardware complexities of presented algorithms are determined by the total count of multipliers, adders, and latches required in realized structures. Computational complexities of suggested algorithms are measured in terms of number of multiplications and addition operations required to compute a coefficient. The comparison of realized structures is done with previously existing structures in terms of these complexities and it was observed that the proposed structures are better than them in one or more aspects. The realized structures are simple, modular, provides reliable results. Therefore, it is suitable for parallel VLSI implementation.

Keywords Discrete sine transform (DST) · Discrete cosine transform (DCT) IIR (Infinite impulse response) filter · VLSI · Recursive algorithm

1 Introduction

The Discrete trigonometric transforms—Discrete Cosine Transform (DCT) and Discrete Sine Transform (DST) are widely used in the field of signal processing and image processing for various applications. DCT performs equivalent to optimal Karhunen–Loeve Transform (KLT). The DCT and DST have fixed sinusoidal

P. Dahiya (✉) · P. Jain
Department of Electronic and Communication Engineering,
Delhi Technological University, Bawana Road, New Delhi 110042,
Delhi, India
e-mail: pragatidahiya@gmail.com

P. Jain
e-mail: priyajain2000@rediffmail.com

© Springer Nature Singapore Pte Ltd. 2019

H. Malik et al. (eds.), *Applications of Artificial Intelligence Techniques in Engineering*, Advances in Intelligent Systems and Computing 697,
https://doi.org/10.1007/978-981-13-1822-1_33

kernel which is independent of the input statics, whereas the KLT depends on the properties of input (correlation matrix). Based on the value of correlation coefficient of the input, DCT and DST can be used in place of KLT. If correlation coefficient of signal to be transformed is tending to unity ($\rho = [-0.5, 0.5]$) that is for low correlated signal, DST shows better performance than DCT [1]. Other applications include noise estimation [2] closed-form design of FIR frequency selective filter (FSF) [3], image resizing [4], image coding [5], designing of fixed Riesz fractional order differentiator (RFOD) [6], image mirroring and rotation [7], and solving of tolpetiz equation [8]. DST formulation is computationally intensive but due to its numerous applications, various approaches to simplify its computational complexity have become an active area of research. Perera et al. [9] derived fast, recursive, and numerically stable radix-2 algorithms for DST. Algorithms for DCT and DST are developed using: (i) Indirect approach in which already existing fast algorithm like FFT (fast algorithm for Discrete Fourier Transform) are used to generate transform coefficient with requirement of some additional mathematical operations [10], (ii) sparse matrix factorization in which transform matrix is decomposed into product of sparse matrices [8, 9, 11], (iii) recursive technique in which lower order transform coefficients are used to generate higher order coefficients [12–16], and (iv) odd prime length DST can be computed using cyclic convolution or cyclic correlation [17]. The recursive technique of algorithm design avoids global butterfly cross connections of fast algorithms. In this technique, IIR filter-based structures are realized based on derived recursive relation. These structures are simple, modular, parallel and regular such that they can be easily implemented in VLSI. The proposed algorithms involve folding of input sequence which results in reduction of number of computational cycles, arithmetic operations as well as computation time required to compute DST/IDST. Also, the reduction in count of computational cycles reduces the truncation error.

2 Derivation of Proposed Recursive Algorithm for DST

For an input sequence $x(n) : n = 1, 2, \ldots, N$, where N has even value, DST-II is given by the following equation.

$$X(k) = \sqrt{\frac{2}{N}} A_k \sum_{n=1}^{N} x(n) \sin\left[\frac{k(2n-1)\pi}{2N}\right], \quad k = 1, 2, \ldots, N \tag{1a}$$

where

$$A_k = \begin{cases} \frac{1}{\sqrt{2}}, & k = N \\ 1, & \text{otherwise} \end{cases} \tag{1b}$$

Omitting the scalar multiplier $A_k\sqrt{\frac{2}{N}}$ from Eq. (1a) as it is constant so it is not included in derivation procedure.

$$X(k) = \sum_{n=1}^{N} x(n)\sin\left[\frac{(2n-1)}{2}\alpha_k\right], \quad k = 1, 2, \ldots, N \tag{2a}$$

where

$$\alpha_k = \frac{k\pi}{N} \tag{2b}$$

Rewriting Eq. (1), we get

$$X(k) = \sum_{n=1}^{N/2} w_k(n)\sin\left[\frac{(2n-1)\alpha_k}{2}\right], \quad k = 1, 2, \ldots, N \tag{3a}$$

where

$$w_k(n) = \left\{x(n) + (-1)^{k-1}x(N-n+1)\right\} \tag{3b}$$

Let

$$c_i \sin\alpha_k = \sum_{n=i+1}^{N/2} w_k(n)\sin[(n-i)\alpha_k], \quad k = 1, 2, \ldots, N, \; i = 0, 1, \ldots, \left(\frac{N}{2} - 1\right) \tag{4}$$

Taking the first term out of summation in Eq. (4)

$$c_i \sin\alpha_k = w_k(i+1)\sin\alpha_k + \sum_{n=i+2}^{N/2} w_k(n)\sin[(n-i)\alpha_k], \quad k = 1, 2, \ldots, N \tag{5}$$

Equation (5) is further solved to get the following formula.

$$\begin{aligned} c_i \sin\alpha_k = {} & w_k(i+1)\sin\alpha_k + 2\sum_{n=i+2}^{N/2} w_k(n)\{\sin[(n-i-1)\alpha_k]\cos\alpha_k \\ & - \sum_{n=i+2}^{N/2} w_k(n)\{\sin[(n-i-2)\alpha_k], \quad k = 1, 2, \ldots, N \end{aligned} \tag{6}$$

$$c_i \sin\alpha_k = w_k(i+1)\sin\alpha_k + 2c_{i+1}\sin\alpha_k\cos\alpha_k - c_{i+2}\sin\alpha_k, \quad \mathrm{k} = 1, 2, \ldots, N \tag{7}$$

Finally, we have obtained the following recursive expression:

$$c_i = w_k(i+1) + 2c_{i+1}\cos\alpha_k - c_{i+2}, \quad i = 0, 1, \ldots, \left(\frac{N}{2} - 1\right) \text{ and } c_i = 0 \text{ if } i \geq \frac{N}{2} \tag{8}$$

Next equation is obtained by multiplying Eq. (3a) by $\sin\alpha_k$

$$X(k)\sin\alpha_k = \sum_{n=1}^{N/2} w_k(n)\sin\left[\frac{(2n-1)}{2}\alpha_k\right]\sin\alpha_k, \quad k = 1, 2, \ldots, N \tag{9}$$

$$X(k)\sin\alpha_k = w_k(1)\sin\left(\frac{\alpha_k}{2}\right)\sin\alpha_k + \sum_{n=2}^{N/2} w_k(n)\sin\left[\frac{(2n-1)}{2}\alpha_k\right]\sin\alpha_k, \quad k = 1, 2, \ldots, N \tag{10}$$

Equation (10) is further expanded to get

$$X(k)\sin\alpha_k = \sum_{n=1}^{N/2} w_k(n)\left\{\sin(n\alpha_k)\,\sin\left(\frac{\alpha_k}{2}\right)\right\} + \sum_{n=2}^{N/2} w_k(n)\left\{\sin[(n-1)\alpha_k]\,\sin\left(\frac{\alpha_k}{2}\right)\right\}, \quad k = 1, 2, \ldots, N \tag{11}$$

$$X(k)\sin\alpha = c_0\sin\alpha_k\sin\left(\frac{\alpha_k}{2}\right) + c_1\sin\alpha_k\sin\left(\frac{\alpha_k}{2}\right), \quad k = 1, 2, \ldots, N \tag{12}$$

$$X(k) = \sqrt{\frac{2}{N}}A_k(c_0 + c_1)\sin\left(\frac{\alpha_k}{2}\right), \quad k = 1, 2, \ldots, N \tag{13}$$

Based on the recursive expression mentioned in Eq. (8), DST structure is realized as shown in Fig. 1.

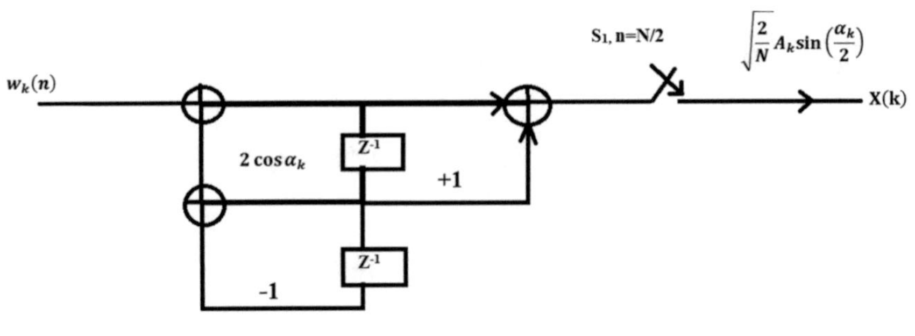

Fig. 1 IIR filter structure for N-point DST

3 Recursive Output Folding Algorithms (OFR) for IDST-II

IDST-II for sequence $X(k) : k = 1, 2, \ldots, N$ where N has even value is given by the following expression:

$$x(n) = \sqrt{\frac{2}{N}} A_k \sum_{k=1}^{N} X(k) \sin\left[\frac{k(2n-1)\pi}{2N}\right], \quad n = 1, 2, \ldots, N \tag{14}$$

We get Eq. (15) by reversing the order of $x(n)$

$$x(N+1-n) = \sqrt{\frac{2}{N}} A_k \sum_{k=1}^{N} X(k) \sin\left[\frac{(2N-2n+1)k\pi}{2N}\right], \quad n = 1, 2, \ldots, N \tag{15}$$

Equations (16) and (17) are obtained from Eqs. (14) and (15) by folding the sequence $x(n)$.

$$x(n) + x(N+1-n) = \sqrt{\frac{2}{N}} A_k \sum_{k=1}^{N} X(n)(1 - \cos k\pi) \sin\left[\frac{k(2n-1)\pi}{2N}\right], \quad n = 1, 2, \ldots, \frac{N}{2} \tag{16}$$

$$x(n) - x(N+1-n) = \sqrt{\frac{2}{N}} A_k \sum_{k=1}^{N} X(n)(1 + \cos k\pi) \sin\left[\frac{k(2n-1)\pi}{2N}\right], \quad n = 1, 2, \ldots, \frac{N}{2} \tag{17}$$

3.1 OFR Kernel for Even Index Coefficients of DST

For even values of k, Eq. (17) can be written as

$$x(n) - x(N+1-n) = 2 \sum_{k=1}^{N/2} X(2k) \sin(2\beta_n k), \quad n = 1, 2, \ldots, \frac{N}{2} \tag{18a}$$

where

$$\beta_n = \frac{(2n-1)\pi}{2N}, \quad n = 1, 2 \ldots \frac{N}{2} \tag{18b}$$

Let

$$r_i \sin\beta_n = \sum_{k=i+1}^{N/2} X(2k)\sin[2(k-i)\beta_n], \quad n = 1,2,\ldots,\frac{N}{2}, \quad i = 0,1,\ldots,\frac{N}{2}-1 \tag{19}$$

Equation (19) can be further solved to get

$$\begin{aligned} r_i \sin\beta_n = {} & X(2(i+1))\sin 2\beta_n + 2\sum_{k=i+2}^{N/2} X(2k)\sin[2(k-i-1)\beta_n]\cos 2\beta_n \\ & - \sum_{k=i+2}^{N/2} X(2k)\sin[2(k-i-2)\beta], \quad n = 1,2,\ldots,\frac{N}{2} \end{aligned} \tag{20}$$

$$\begin{aligned} & r_i \sin\beta_n = X(2(i+1))\sin 2\beta_n + 2r_{i+1}\sin\beta_n \cos 2\beta_n - r_{i+2}\sin\beta_n, \\ & n = 1,2,\ldots,\frac{N}{2} \end{aligned} \tag{21}$$

$$\begin{aligned} & r_i = 2X(2(i+1))\cos\beta_n + 2r_{i+1}\cos 2\beta_n - r_{i+2} \\ & \text{for } i = 0,1,\ldots,\frac{N}{2}-1 \text{ and } r_i = 0 \text{ if } i > \frac{N}{2}-1 \end{aligned} \tag{22}$$

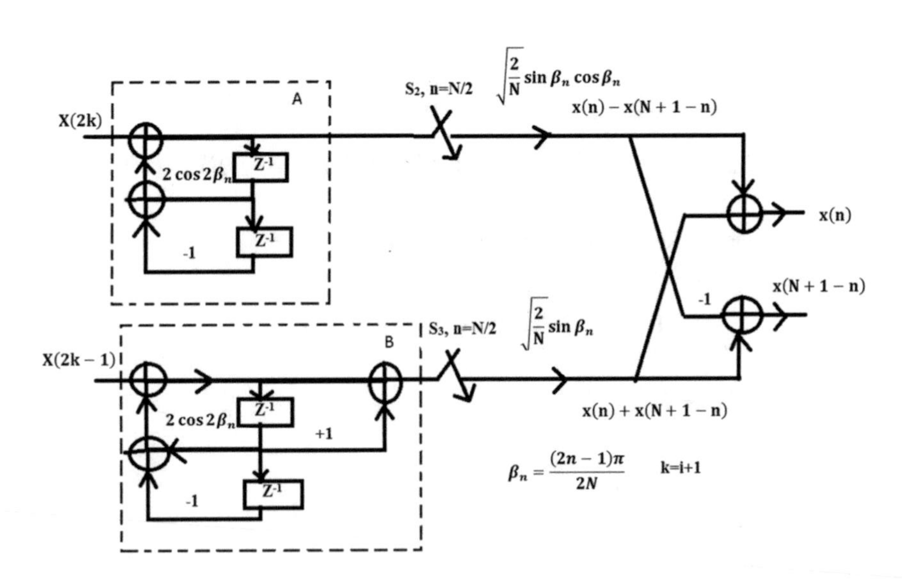

Fig. 2 IIR filter structure for N-point IDST

Using Eqs. (19) and (18a), we get the following relation:

$$x(n) - x(N+1-n) = 2\sqrt{\frac{2}{N}}A_k r_0 \sin\beta_n, \quad n = 1, 2, \ldots, \frac{N}{2} \tag{23}$$

IIR filter structure is realized by utilizing Eq. (22) which is shown in Fig. 2. Thus, thereafter we can determine the value of $x(n) - x(N+1-n)$ using even-indexed $X(k)$.

3.2 OFR Kernel for the Odd-Index Coefficients of DST

For odd values of k, Eq. (16) can be rewritten as

$$x(n) + x(N+1-n) = 2\sum_{k=1}^{N/2} X(2k-1)\sin[(2k-1)\beta_n], \quad n = 1, 2, \ldots, \frac{N}{2} \tag{24}$$

Let

$$p_i \sin\beta_n = \sum_{k=i+1}^{N/2} X(2k-1)\sin[(2k-2i-1)\beta_n], \quad n = 1, 2, \ldots, \frac{N}{2} \tag{25}$$

Using the similar procedures as for even-indexed sequence, we can obtain following recursive relation.

$$\begin{aligned} p_i &= X(2i+1) + X(2i+3) + 2p_{i+1}\cos 2\beta_n - p_{i+2} \\ &\text{for } i = 0, 1, \ldots, \frac{N}{2} - 1 \text{ and } p_i = 0 \text{ if } i > \frac{N}{2} - 1 \end{aligned} \tag{26}$$

Using Eqs. (24) and (25), we get following relation.

$$x(n) + x(N+1-n) = 2\sqrt{\frac{2}{N}}A_k p_0 \sin\beta_n, \quad n = 1, 2, \ldots, \frac{N}{2} \tag{27}$$

IIR filter structure is realized using Eq. (26) as shown in Fig. 2. Thus, thereafter we can determine the value of $x(n) - x(N+1-n)$ using odd-indexed $X(k)$. Equations (28) and (29) are obtained by adding and subtracting Eqs. (23) and (27) respectively. Clearly, having computed the values of $x(n) - x(N+1-n)$ and $x(n) - x(N+1-n)$ for $n = 1, 2, \ldots, \frac{N}{2}$, we can get all the values of $x(n) n = 1, 2, \ldots, N$.

Table 1 Hardware complexity of DST/IDST algorithms

	Performance factor	[12]	[13]	[14]	[15]	Proposed
DST/ IDST	Number of multipliers	4/5	2/2	2/2	6/9	2/4
	Number of adders	6/8	3/3	3/2	10/16	3/7
	Number of latches	4/4	3/3	2/2	6/8	2/4
	Number of computation cycles	*N*/2/*N*/2	*N*/*N*	*N*/*N*	*N*/4/*N*/4	*N*/2/*N*/2
	DTPT	2/2	1/1	1/1	2/4	1/2

$$x(n) = \sqrt{\frac{2}{N}} A_k \sin\left[\frac{(2n-1)k\pi}{2N}\right](p_0 + r_0), \quad n = 1, 2, \ldots, \frac{N}{2} \tag{28}$$

$$x(N+1-n) = \sqrt{\frac{2}{N}} A_k \sin\left[\frac{(2n-1)\pi}{2N}\right](p_0 - r_0), \quad n = 1, 2, \ldots, \frac{N}{2} \tag{29}$$

4 Discussion

The proposed algorithms for one-dimensional DST and IDST are compared with other existing algorithms in terms of number of arithmetic operations (multiplication, addition) and hardware components (multiplier, adder and latches) required to realize the presented structures. The hardware complexity comparison is given in Table 1 from which the following points are concluded—Even though DTPT (Data Throughput per Transformation) for DST computation in [12] is twice that of proposed, hardware requirement in the suggested algorithm is exactly 50% less than that of [12]. For IDST, presented algorithm and [12] has same DTPT and computational cycles but former requires one less multiplier and adder, respectively. In comparison to [13], the proposed DST algorithm has same DTPT and an equal number of multipliers and adders as that of [13] but suggested algorithm requires one less latch and half computational cycle than [13]. Even though hardware requirement is more in the proposed IDST compared to [13], former requires half computational cycle and has twice DTPT compared to [13]. The proposed DST algorithm has same hardware requirement and DTPT as in [14] with half of the computational cycle requirement. Proposed IDST require half computational cycle and has twice DTPT compared to [14]. Compared to [15] the proposed DST and IDST algorithm has twice the computational cycles and half DTPT but the proposed DST and IDST requires three times and two times less hardware than [15], respectively. It can be observed from Table 2 that the proposed algorithm will compute DST with 50% reduction in number of multiplication operations as compared to [12–15]. Also, reduction in number of addition operations is an added advantage to the overall computational complexity of the algorithm. Also, it can be

Table 2 Comparison of number of multiplications/additions for DST and [**IDST**] *algorithms*

N	[12] $(N)/(2N-4)$ [(**2N − 1**)/(**3N − 7**)]	[13] $(N)/(3N-4)$ [(**2N**)/(**3N − 4**)]	[14] $(N)/(2N-2)$ [(**N**)/(**2N − 3**)]	[15]		Proposed $(N/2)/(3N/2-2)$ [(**N**)/(**5N/2 − 6**)]
				Even-indexed values $(N/2)/(3N/2-4)$	Odd-indexed values $(N-1)/(2N-4)$	
4	4/4 [**7/15**]	4/8 [**8/8**]	4/6 [**4/5**]	2/2	3/4	2/4 [**4/4**]
8	8/12 [**17/17**]	8/20 [**16/20**]	8/14 [**8/13**]	4/8	7/12	4/10 [**8/14**]
16	16/28 [**31/41**]	16/44 [**32/44**]	16/30 [**16/29**]	8/20	15/28	8/22 [**16/34**]
32	32/60 [**63/91**]	32/92 [**64/92**]	32/62 [**32/61**]	16/44	31/60	16/46[**32/74**]
64	64/124 [**127/185**]	64/188 [**128/188**]	64/126 [**64/125**]	32/95	63/124	32/94 [**64/84**]

seen from Table 2, IDST algorithm requires less number of multiplication operations, i.e., $\left(\frac{3}{2}N\right)$ and addition operation, i.e., $(2N - 2)$ as compared to [12–14]. The proposed IDST algorithm has twice the data throughput along with half computational cycle as compared to [12, 13].

5 Conclusions

A recursive algorithm for discrete sine transforms (DST) and its inverse has been proposed in this paper which is realized through IIR filter structures. The algorithms are efficient in terms of hardware and computation complexity. Realized structures are suitable for parallel VLSI implementation as they are stable, parallel, regular, modular, and simple with local connectivity which results in a reduction of space complexity and simple interconnection in designing process. With suitable input, these structures can compute desired transform coefficient independent of the computation of other coefficients. The proposed structure can be fully exploited for computation of DCT and DHT as it is well-known DCT can yield DHT [18]. Hence, the proposed structure is potential for computation of DCT, DST as well as DHT.

Acknowledgements The Authors wish to acknowledge reviewers for their valuable comments and research fellowship from University Grants of commission, Government of India.

References

1. Jain, P., Kumar, B., Jain, S. B.: An efficient approach for realization of discrete sine transform and its inverse. IETE J. Res **54**(4), 285–296 (2008)
2. Dhamija, S., Jain, P.: Comparative analysis for discrete sine transform as a suitable method for noise estimation. IJCSI Int. J. Comput. Sci. Issues **8**(5) (2011)
3. Tseng, C.C., Lee, S.L.: Closed-form design of FIR frequency selective filter using discrete sine transform. In: Circuits and Systems (APCCAS) IEEE Asia Pacific Conference, pp. 591–594 (2016). https://doi.org/10.1109/APCCAS.2016.7804039
4. Park, Y.S., Park, H.W.: Arbitrary–ratio image resizing using fast DCT of composite length for DCT-based transcoder. IEEE Trans. Image Process. **15**(2), 494–500 (2006). https://doi.org/10.1109/TIP.2005.863117
5. Lim, S.C., Kim, D.Y., Lee, Y.L.: Alternative transform based on the correlation of the residual signal. In Image and Signal Processing, 2008. CISP'08 (1), pp. 389–394. https://doi.org/10.1109/CISP.2008.226
6. Kumar, M., et al.: Design of Riesz fractional order differentiator using discrete sine transform. In: 3rd International Conference on Signal Processing and Integrated Networks (SPIN). IEEE (2016), pp. 702–706. https://doi.org/10.1109/SPIN.2016.7566788
7. Kim, D.N., Rao, K.R.: Two-dimensional discrete sine transform scheme for image mirroring and rotation. J. Electron. Imaging **17**(1) (2008). https://doi.org/10.1117/1.2885257
8. Gupta, A.., Rao, K.R.: A fast recursive algorithm for the discrete sine transform. IEEE Trans. Acoust. Speech Sign. Proces.**38**(3), 553–5557 (1990). https://doi.org/10.1109/29.106875

9. Perera, S.M., Olshevsky, V.: Fast and stable algorithms for discrete sine transformations having orthogonal factors. In: Interdisciplinary Topics in Applied Mathematics, Modeling and Computational Science, pp. 347–354 (2015). https://doi.org/10.1007/978-3-319-12307-3_50
10. Wang, Z.D.: A fast algorithm for the discrete sine transform implemented by the fast cosine transform. IEEE Trans. Acoust. Speech Signal Process. **30**(5), 814–815 (1982). https://doi.org/10.1109/TASSP.1982.1163963
11. Perera, S.M.: Signal flow graph approach to efficient and forward stable DST algorithms. Linear Algebra Appl., 1–31 (2017). https://doi.org/10.1016/j.laa.2017.05.050
12. Murthy, M.N.: Radix-2 algorithms for implementation of type-II discrete cosine transform and discrete sine transform. Int. J. Eng. Res. Appl. **3**, 602–608 (2013)
13. Jain, P., Kumar, B., Jain, S.B.: Discrete sine transform and its inverse—realization through recursive algorithms. Int. J. Circuit Theory Appl. **36**(4), 441–449 (2007). https://doi.org/10.1002/cta.447
14. Jain, P., Jain, A.: Regressive structures for computation of DST-II and its inverse. Int. Sch. Res. Netw. ISRN Electron., 1–4 (2012). https://doi.org/10.5402/2012/537469
15. Murthy, M.N.: Recursive algorithms for realization of one-dimensional discrete sine transform and inverse discrete sine transform. Int. J. Recent Res. Appl. Stud. IJRRAS **14**(2), 340–347 (2013)
16. Dahiya, P., Jain, P.: Realization of first-order structure for the recursive algorithm of discrete sine transform. In: 8th IEEE International Conference on Computing, Communication and Networking Technologies (ICCCNT), pp. 1–5 (2017). https://doi.org/10.1109/ICCCNT.2017.8204014
17. Meher, P.K., Swamy, M.N.: New systolic algorithm and array architecture for prime-length discrete sine transform. IEEE Trans. Circuits Syst. II: Express Briefs **54**(3), 262–266 (2007). https://doi.org/10.1109/TCSII.2006.889453
18. Jain, P., Kumar, B., Jain, S. B.: Fast computation of the discrete Hartley transform. Int. J. Circuit Theory Appl. **38**(4), 409–417 (2008)

Identification of Dynamic Systems Using Intelligent Water Drop Algorithm

Anuli Dass, Smriti Srivastava and Monika Gupta

Abstract Fuzzy systems are widely known to be universal approximators. They are very widely used for identification of unknown systems and plants. In the field of control engineering, it is a general practice to control complex plants and systems. But this requires the knowledge about the exact structure of the plant or the system to be controlled or analyzed. But in most of the practical cases, this is not available. Identification is a process to determine the structure of unknown plants and systems. Fuzzy systems, when used as approximators, mimic the plant and learn how to behave exactly like the plant. This learning process requires updation of the fuzzy parameters, so that ultimately the fuzzy system starts behaving exactly like the unknown plant. This paper discusses a very recent and popular scheme of optimization called 'Intelligent Water Drop (IWD) Algorithm'. This algorithm is inspired by the nature and is based on intelligence of the drops of water present in nature. The paper demonstrates with examples the actual implementation procedure of the IWD algorithm. Also, the results clearly show that the fuzzy system optimized using IWD algorithm successfully identifies unknown complex plants and systems.

Keywords Intelligent water drop algorithm · Optimization · Fuzzy systems
Swarm intelligence · Identification

1 Introduction

With an objective to develop better intelligent systems and algorithms, scientists are constantly seeking inspiration from the nature. Various optimization algorithms based on the intelligence of a swarm have been proposed and implemented. These optimization algorithms are basically used to improve or optimize the intelligence of intelligent systems such as fuzzy, neural networks, etc [1–12]. One of the major

---

A. Dass (✉) · S. Srivastava · M. Gupta
ICE Division, NSIT, New Delhi, India
e-mail: anulidass@gmail.com

© Springer Nature Singapore Pte Ltd. 2019

H. Malik et al. (eds.), *Applications of Artificial Intelligence Techniques in Engineering*, Advances in Intelligent Systems and Computing 697,
https://doi.org/10.1007/978-981-13-1822-1_34

application areas of this concept in the field of engineering is identification and control of complex linear as well as nonlinear systems [1, 2]. Reference [3] proposes a recurrent fuzzy neural network (RFNN) structure for identifying and controlling nonlinear dynamic systems. Reference [4] presents a TSK-type recurrent fuzzy network (TRFN) structure. The proposed TRFN structure can be designed by either neural network or genetic algorithms depending on the learning environment. Reference [5] presents the basic designing of intelligent systems based on fuzzy and neural networks and also throws some light on nature-inspired optimization algorithms. Reference [6] discusses feed-forward and recurrent fuzzy systems which are rule based and present a unique learning method for them. In order to achieve minimum complexity and the smallest rule base for the fuzzy logic based systems, it uses genetic algorithm. A current and very popular nature-inspired swarm-based algorithms is the 'Intelligent water drop algorithm' [7]. The principle is based on intelligence of the water drops in nature which finds their way to rivers and lakes. In doing so, these water drops always choose the most optimum path available. A straight line path would, in general, be the shortest and most optimum but in reality, due to the presence of obstacles and hurdles in the path, the option of straight line is automatically ruled out. Reference [8] investigates the effectiveness of the selection method in the solution construction phase of the IWD algorithm. Reference [9] provides an improvement in the IWD algorithm to solve node selection problem in the field of wireless sensor networks (WSNs). It does so by assuming that every drop of water acts as a proxy whose task is to find the minimum number of sensor nodes with maximum possible efficiency. Reference [10] uses the IWD algorithm in creating an automated approach to help in software project planning which would, in turn, reduce the cost and total duration of the project.

Section 2 presents the fundamentals of the Intelligent Water Drop-based algorithm. Section 3 presents the results and simulations of the identification process using Fuzzy system optimized by the IWD algorithm. It gives the details about the parameters to be updated using the optimization algorithm. The paper is concluded in Sect. 4.

2 Intelligent Water Drop Algorithm

There are many nature-inspired algorithms being used in the field of engineering and control. The intelligence of the swarm has always inspired scientists to create new algorithms based on them. Some of the nature-inspired swarm-based algorithms are particle swarm optimization (PSO), fish swarm optimization, ant-colony optimization (ACO), etc. Rigorous study and research on these algorithms have been done and are still in progress. These algorithms are basically used in order to optimize the solutions generated by intelligent systems such as fuzzy and neural networks. This section explains the concept of a very recently developed nature-inspired swarm-based optimization algorithm called 'Intelligent water drop algorithm.' The algorithm was proposed by Hamed Shah-Hosseini. The intelligence of the drops of water present in the nature is used in the development of the algorithm. These water drops always choose the most optimum path while travelling from the rivers to reach seas and oceans. The intelligence of these natural water drops is used in the IWD algorithm. The basic principle behind the algorithm is the fact that a water drop always chooses the path with minimum soil content. This is because greater the soil content higher will be the friction offered by the path to the water drop. To physically implement the problem, we design a graph (N, E) where N is the number of nodes and E the number of edges. This is shown in Fig. 1. The water drops are made to travel through each and every node. The edges of the graph serve as the path through which the water drops travel. The nodes shaded orange are the starting and terminating nodes. The water drop requires travelling all the nodes with fixed starting and termination points in the most optimized way possible such that no node is travelled twice.

The algorithm involves certain static parameters which once initialized remain same or constant throughout the process. Then, there are dynamic parameters which need to be initialized at the start of the algorithm but there value changes after every iteration. It is also essential to know that each water drop is characterized by two properties:

1. Soil content in the drop of water: This is the amount of soil in which the drop of water carries with itself and is denoted by *soil* (*IWD*). As the soil moves from one point or node to the other, it washes away with itself certain quantity of soil from the path which adds to the soil content of the soil.
2. Velocity of the water drop: It is denoted by *vel* (*IWD*). Larger the velocity of the water drop, higher will be the soil content being washed away from the path by the water drop.

The algorithm starts as follows:

Identify the parameters to be optimized. Choose a range of possible suitable values of each. Each individual value in the set represents the nodes of the graph. Thus, we can say that against each parameter to be optimized, there are a certain number of possible values of that parameter itself which serves as the nodes for the water drops. Choose a certain number of water drops (say 10). Initialize each water drop with an initial velocity, InitVel = 4. Soil(i, j) represents the soil between the ith and the jth node which serves as path for the water drops. Initialize soil(i,j) by InitSoil = 1000.

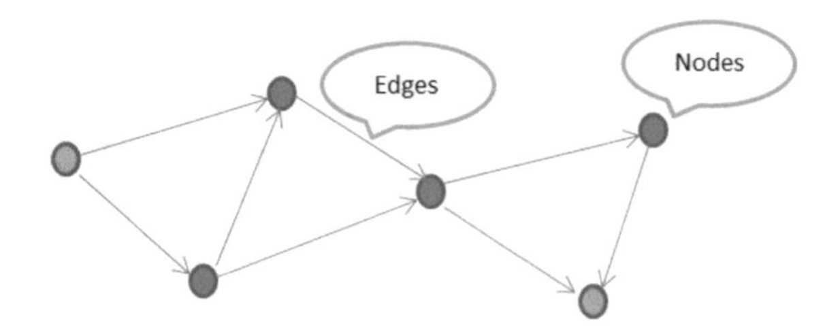

Fig. 1 Graphical representation of the problem

Now, each water drop is made to generate an output. After every iteration ends, solutions of all the water drops are compared and an iteration-best solution is selected. This iteration-best solution is then compared to the global-best solution. If the iteration-best solution is superior to the global-best solution, then the global-best solution is changed accordingly. At the end of all the iterations, the global-best solution serves as the final outcome of the algorithm.

The most peculiar step is the generation of solution by each water drop. This is carried out as follows: consider the first water drop. Out of all the nodes for the first parameter, a random one is selected. From that node, the probability of moving to all the other nodes of the next parameter is calculated.

$$p(i,j;\mathrm{IWD}) = \frac{f(\mathrm{soil}(i,j))}{\sum_{k \nexists vc(\mathrm{IWD})} f(\mathrm{soil}(i,j))} \tag{1}$$

where

$$(\mathrm{soil}(i,j)) = \frac{1}{\epsilon_s + g(\mathrm{soil}(i,j))}.$$

The function $g(\mathrm{soil}(i,j))$ is calculated as

$$g(\mathrm{soil}(i,j)) = \left\{ \begin{array}{c|c} \mathrm{soil}(i,j) & \text{if } \min(\mathrm{soil}(i,l)) \geq 0; l \notin vc(\mathrm{IWD}) \\ \mathrm{soil}(i,j) - \min(\mathrm{soil}(i,l)) & \text{else} \end{array} \right\} \tag{2}$$

ϵ_s is chosen to be a small positive number. Equation (1) calculates the probability of travelling from *i*th to the *j*th node. Water drop chooses the path with maximum probability. This process is repeated till all the parameters are covered. Also, as the water drop moves from node *i* to node *j*, its velocity and soil content changes. Also, the path soil, i.e. soil (*i*, *j*) changes. These changes are calculated as below

$$\Delta \mathrm{vel}^{\mathrm{IWD}}(t) = \frac{a_v}{b_v + c_v.\mathrm{soil}^{2\alpha}(i,j)} : a_v, b_v, c_v, 2\alpha \tag{3}$$

are static parameters which have positive values selected by the user.

$$\Delta \mathrm{soil}(i,j) = \frac{a_s}{b_s + c_s.\mathrm{time}^{2\theta}\left(i,j;\mathrm{vel}^{\mathrm{IWD}}\right)} : a_s, b_s, c_s, 2\theta \tag{4}$$

are static parameters which have positive values selected by the user.

$$\mathrm{time}\left(i,j;\mathrm{vel}^{\mathrm{IWD}}\right) = \frac{\mathrm{HUD}(i,j)}{\mathrm{vel}^{\mathrm{IWD}}} \tag{5}$$

Equation (4) is the time taken by the water drop to move from node i to node j with velocity vel^{IWD}.

$\text{HUD}(i,j)$ is a heuristic function which is chosen depending upon the nature of the problem. The heuristic function should be such that the excellence of the output generated is evaluated using the quality of the heuristic function.

$$\text{soil}(i,j) = \rho_0.\text{soil}(i,j) - \rho_n.\Delta\text{soil}(i,j) \tag{6}$$

$$\text{soil}^{\text{IWD}} = \text{soil}^{\text{IWD}} + \Delta\text{soil}(i,j) \tag{7}$$

Equation (6) gives the updating equation for the update of soil between nodes i and j. Similarly, Eq. (7) gives the update equation for the drop of water's soil content.

Thus, at the end, once all the drops of water have generated the solutions and the upper limit of the iteration count is achieved, the value of the global-best solution is treated as final outcome generated by the algorithm.

3 Results and Simulations

The section confers the simulation results of the identification process using fuzzy systems being optimized by the IWD algorithm. The number of water drops chosen for each example is 10. In this paper, the inputs are fuzzified using Gaussian Membership Function. A Gaussian membership function is given as

$$\mu_{ij} = \exp^{\left(\frac{x_i - c_{ij}}{d_{ij}^2}\right)^2} \tag{8}$$

where x_i is the ith input, c_{ij} *is the centre and* d_{ij} is the width of the Gaussian membership function associated with the jth fuzzy set of the ith input.

Every fuzzy system functions and generates output based on a rule base. This rule base consist if-then-else rules formed on the basis of expert's knowledge. A general TSK-fuzzy rule is of the form:

'If x_1 is P_1 and x_2 is P_2 then $y = b_{10} + b_{11}x_1 + b_{12}x_2$'.

The fuzzy parameters optimized using IWD algorithm are the membership function (Gaussian) parameters, i.e. center of the Gaussian membership function (c_{ij}) and its associated width (d_{ij}) and the coefficients associated with each fuzzy rule output $(b_{i0}, b_{i1}, b_{i2}$ etc.)

Example 1 Consider the plant given by the following differential equation:

$$y_p(k+1) = \frac{y_p(k).y_p(k-1).\left(y_p(k)+2.5\right)}{1+y_p^2(k)+y_p^2(k-1)} + u(k) \tag{9}$$

where $u(k) = \sin(2\pi k/250)$ is the input to the plant.

As evident from the above differential equation, the necessary inputs to the *fuzzy system* for the process of identification would be $y_p(k)$ and $y_p(k-1)$ and $u(k)$.

Figure 2 shows the simulation results for the desired identification using IWD algorithm. As it is clearly seen when a fuzzy model is optimized using IWD algorithm, it identifies the unknown plant very quickly and efficiently. The efficiency is evident from the mean square error shown in Fig. 3.

Example 2 Consider the plant given by the following differential equation:

$$y_p(k) = \frac{y_p(k-1).y_p(k-2).y_p(k-3).y_p(k-4).\left(y_p(k-3)-1\right).u(k-2)+u(k-1)}{2+y_p^2(k-2)+y_p^2(k-3)+y_p^2(k-4)} \tag{10}$$

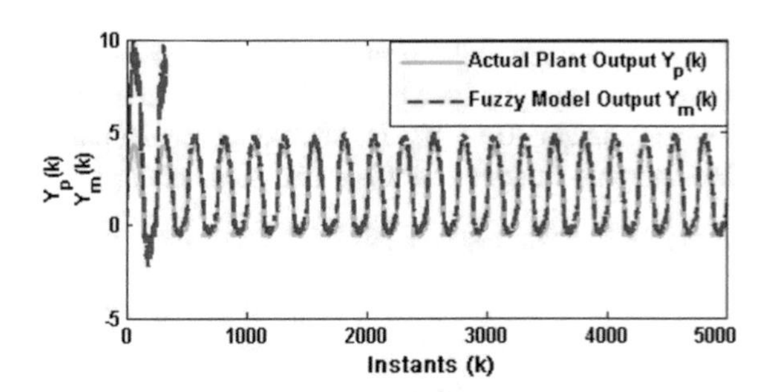

Fig. 2 Identification of example 1 using fuzzy system optimized by IWD algorithm

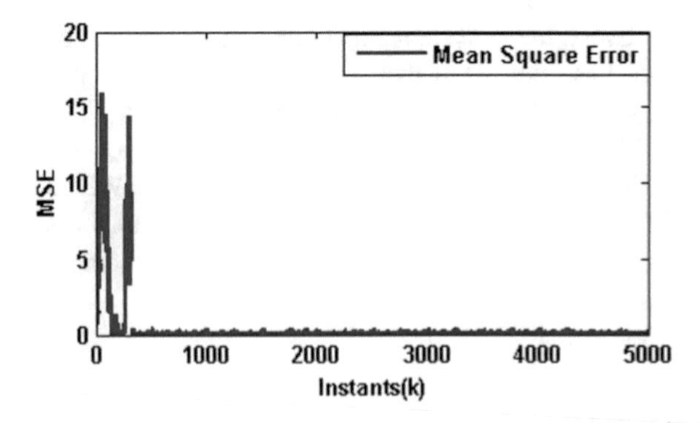

Fig. 3 Mean square error obtained using IWD algorithm during identification

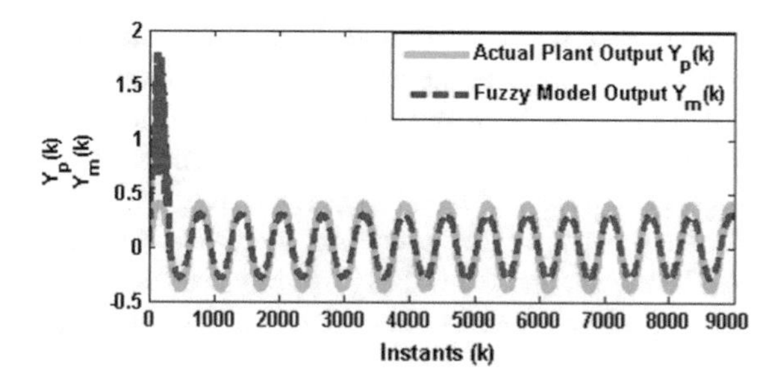

Fig. 4 Identification of example 2 using fuzzy system optimized by IWD algorithm

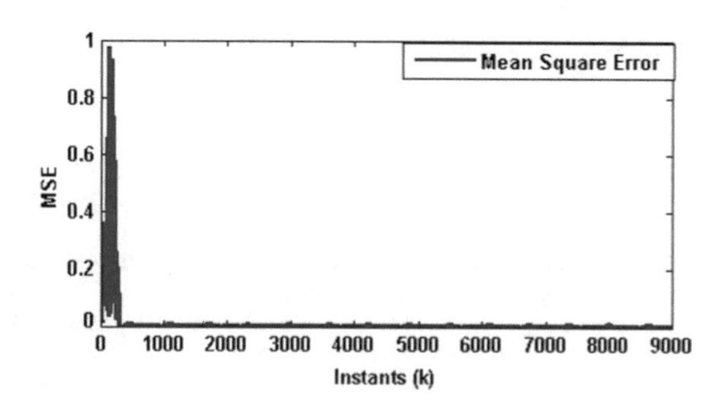

Fig. 5 Mean square error obtained using IWD algorithm during identification

where

$$u(k) = \left\{ \begin{array}{c|c} 0.5\sin(0.01k) & k < 2500 \\ -1 & 2500 \le k < 5000 \\ 1 & 5000 \le k < 7000 \\ 0.4\sin(0.01k) + 0.6\sin(0.01k) & 7000 \le k < 9000 \end{array} \right\}$$

This plant is different from the previous example in the sense that the input signal $u(k)$ is an integral part of the internal structure of the plant. The result after simulation has been shown in Fig. 4 depicts the successful identification of unknown plant using fuzzy model being optimized by the IWD algorithm. The mean square error is shown in Fig. 5.

4 Conclusion

Nature has always been a source of inspiration to the scientists with an objective to improve their quality of work. No matter how far we reach in achieving perfection in artificial intelligence but can never beat the intelligence of the nature. This led to the development of many nature-inspired algorithms with an objective to optimize the solution generated by intelligent systems. The process of optimization using nature-inspired swarm based algorithms namely ant-colony optimization, PSO, cuckoo search, etc. have already been investigated and researched a lot. A very recent and popular optimization algorithm named 'Intelligent Water Drop Algorithm' has been discussed in depth in this paper. Identification of unknown structures using fuzzy system has been shown in this paper. It also demonstrates implementation of the IWD algorithm to optimize the fuzzy system when employed for the purpose of identification. The results very clearly show that IWD algorithm can be used in the field of engineering as a learning algorithm to optimize the result of an intelligent fuzzy system. This fuzzy system after being optimized by the IWD algorithm generates satisfactory results for the process of identification. The paper very vividly discusses the concept of the IWD algorithm and also presents the approach of implementing the algorithm on real-world engineering problems.

References

1. Narendra, K. S., Parthasarathy, K.: Identification and control of dynamical systems using neural networks. IEEE Trans. Neural Netw. **1**(1), 4–27 (1990)
2. Takagi, T., Sugeno, M.: Fuzzy identification of systems and its applications to modeling and control. IEEE Trans. Syst. Man Cybern. **1**, 116–132 (1985)
3. Lee, C.-H., Teng, C.-C.: Identification and control of dynamic systems using recurrent fuzzy neural networks. IEEE Trans. Fuzzy Syst. **8**(4), 349–366 (2000)
4. Juang, C.-F.: A TSK-type recurrent fuzzy network for dynamic systems processing by neural network and genetic algorithms. IEEE Trans. Fuzzy Syst. **10**(2), 155–170 (2002)
5. Martinez, D., Valdez, F.: An improved intelligent water drop algorithm to solve optimization problems. In: Melin, P., Castillo, O., Kacprzyk, J. (eds.) Design of Intelligent Systems Based on Fuzzy Logic, Neural Networks and Nature-Inspired Optimization. Studies in Computational Intelligence, vol. 601. Springer, Cham (2015)
6. Hartmut, S., Maniadakis, M.: Learning feed-forward and recurrent fuzzy systems: a genetic approach. J. Syst. Archit. **47**(7), 649–662 (2001)
7. Shah-Hosseini, H.: The intelligent water drops algorithm: a nature-inspired swarm-based optimization algorithm. Int. J. Bio-Inspired Comput. **1**(1–2), 71–79 (2009)
8. Alijla, B.O., Wong, L.P., Lim, C.P., Khader, A.T., Al-Betar, M.A.: A modified intelligent water drops algorithm and its application to optimization problems. Expert Syst. Appl. **41** (15), 6555–6569 (2014)
9. Vajdi, A., Zhang, G., Wang, Y., Zhao, Y., Liu, D., Wang, T.: A new approach based on intelligent water drops algorithm for node selection in service-oriented wireless sensor networks. In: 2014 IEEE Fourth International Conference on Big Data and Cloud Computing (BdCloud). IEEE, pp. 33–40 (2014)
10. Crawford, B., Soto, R., Astorga, G., Olguín, E.: Intelligent water drop algorithm (IWD) to solve software project scheduling problem. In: 2016 11th Iberian Conference on Information Systems and Technologies (CISTI). IEEE, pp. 1–4 (2016)
11. Gorrini, V., Bersini, H.: Recurrent fuzzy systems. In: Proceedings of the Third IEEE Conference on Fuzzy Systems, 1994. IEEE World Congress on Computational Intelligence, pp. 193–198. IEEE (1994)
12. Lin, C.-T., George Lee, C.S.: Neural Fuzzy Systems. PTR Prentice Hall, Upper Saddle River (1996)

Weighted Percentile-Based Context-Aware Recommender System

Veer Sain Dixit and Parul Jain

Abstract Context-Aware Recommender Systems (CARS) focus on the improvement of accuracy and user's satisfaction by incorporating context features while making recommendations. Using too many context features aggravate the data sparsity problem and may impair predictive performance while few context features fail to capture the contextual effects. Though genre-based ratings called implicit ratings play an important role while making recommendations, despite that most of the studies have focused on utilizing explicit ratings. Addressing these issues, we propose a novel framework that demonstrates multiple rating prediction algorithms based on user neighborhood and item neighborhood approaches exploiting explicit and implicit ratings. The algorithms incorporate context communities to alleviate the data sparsity problem. We have also used weighted percentile method to increase the precision. Furthermore, we extended our research to Group Recommendations to see the effectiveness of the proposed algorithms. Finally, the results using two datasets indicate that the proposed context-aware weighted percentile algorithms are superior than the baseline approaches. The item neighborhood-based approaches are more accurate than user neighborhood-based approaches and the performance of explicit and implicit ratings are dataset dependent. The results obtained also prove the effectiveness of the algorithms for Group Recommendations.

Keywords User neighborhood · Item neighborhood · Weighted percentile
Implicit rating · Group recommendations

1 Introduction

Context-aware recommender systems (CARS) [1, 2] emerged as a novel type of Recommender systems (RSs) which utilize contextual information in making recommendations for different state of affairs. In the real world, the final decisions

V. S. Dixit · P. Jain (✉)
Atma Ram Sanatan Dharama College, Department of Computer Science,
Delhi University, New Delhi, India
e-mail: paruljainpj@rediffmail.com

© Springer Nature Singapore Pte Ltd. 2019

H. Malik et al. (eds.), *Applications of Artificial Intelligence Techniques in Engineering*, Advances in Intelligent Systems and Computing 697,
https://doi.org/10.1007/978-981-13-1822-1_35

of the user are also affected by the situation, which can be characterized by the contextual information [1, 2]. For example, one would plan a different movie if he/she is going to see it with friends rather than kids or will prefer to listen to different music if emotional condition is happy rather than sad. Therefore, to improve the recommendation precision and user's amusement, CARS are focused on generating techniques which include contextual information into traditional RS [1, 2, 11, 12, 14].

Limitations of existing methods. User-based and Item-based collaborative filtering (CF) do have a primary focus on explicit ratings which make these algorithms lightweight and flexible. In some practical scenarios, however, explicit ratings are unavailable because there are no incentives for users to rate items. (a) Existing approaches in CARS do not pay much attention to genre-based preferences called as implicit ratings [9] assigned under different contextual conditions.

However, there are several challenges open in CARS. Data sparsity is one such concern which becomes more severe when user preferences are diluted with contextual conditions [1, 2, 11, 12, 14, 16]. Hence, we attempt to explore this issue by including context communities and implicit ratings.

Recently, the demand of recommendations to groups of users in the social network has increased tremendously that led to the evolution of Group Recommender Systems [4, 6]. We hardly find research work on group recommendations using CARS.

Our approach. The main contribution of this work is presented as follows:

- Here, a novel framework is presented to do qualitative analysis of five variations of rating prediction algorithms based on user and item model which utilizes the explicit ratings in an implicit way without prompting users to give their preferences to different genres in context-aware scenario. It also resolves the sparsity problem of CARS using context communities.
- We propose to add weighted percentile approach to our algorithms that improve the performance without incurring extra cost in terms of coverage and diversity, the well-known metrics.
- We further propose to expand research area to Group Recommendation to analyze the efficacy of the proposed algorithms for group of users where two group recommendation techniques are compared on different types and sizes of groups and finally, the proposed framework is experimentally verified.

The remaining of the paper is organized as follows. Section 2 mentions some related works. The model framework and detailed constructions of our recommender system are introduced in Sect. 3, whereas Sect. 4 presents the results and analysis of the experiments. The conclusions are as follows in Sect. 5 and finally, we give some perspective of the future research.

2 Related Work

Several CARS algorithms have been developed to assist item recommendations in contexts including differential context relaxation (DCR) [10], context-aware matrix factorization [2, 15], reduction-based approach [1, 2], and contextual sparse linear method [14]. Following [1–3], CF based on neighborhood [9, 11] are well known and efficient methods because of their simplicity and ability to produce accurate and personalized recommendations in many real-world systems. In order to handle data sparsity problem, [11] describes a context feature selection process and [12] composed an approach based on differential context modeling. Instead, we aim to match some of the selected context features based on previous work [8, 11, 13, 16] to resolve the data sparsity problem. Furthermore, to serve group of users [4, 6] describes various group recommendations techniques. However, the challenges in CARS are pretty much open.

3 The Proposal

In this section, we will specify our framework depicted in Fig. 1 that provides techniques to generate recommendations within domains where items tend to be explored by individuals as well as group of users. In the framework, the Context-Aware

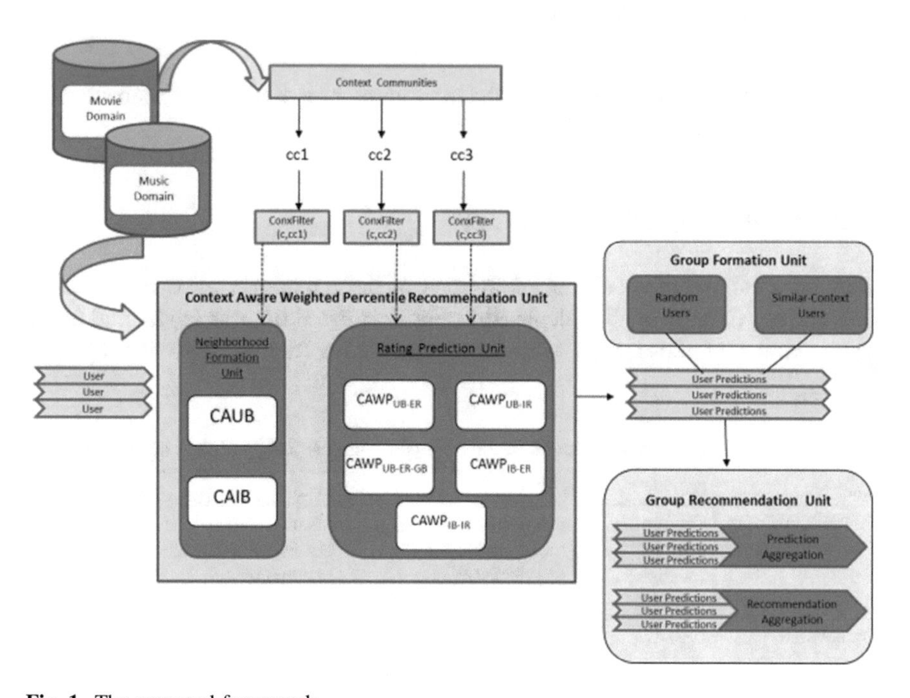

Fig. 1 The proposed framework

Weighted Percentile Recommendation Unit (CAWPRU) uses five variations of context-aware prediction algorithms based on user and item model utilizing weighted percentile of ratings (CAWP) instead of weighted average (CAWA).

3.1 Context Communities

In order to alleviate the sparsity issue and produce contextual effects, we formed three different communities of context factors for both datasets based upon the previous work [5, 8, 11, 13, 16].

Instead of considering all users to form neighborhood, $ConxFilter(\mathrm{c}, \mathrm{cc1})$ filters those users whose all context features belong to context community cc1 match exactly to that of the active user. Consequently, $ConxFilter(\mathrm{c}, \mathrm{cc2})$ in Fig. 1 filters those users whose any of the features of context community $cc2$ match exactly to that of the active user, hence utilized in rating prediction as stated in Eqs. 5–9. $ConxFilter(\mathrm{c}, \mathrm{cc3})$ in Fig. 1 filters those users where any of the features belong to context community cc3 match exactly to that of the active user to find the average rating as described in Eq. 6 (see Table 1).

3.2 Context-Aware Weighted Percentile Recommendation Unit

In this section, we elaborate on all the algorithms proposed in our framework in detail. The notations used in Eqs. 1–9 are described in Table 2.

3.2.1 Neighborhood Formation Unit

This unit is aimed to acquire neighborhood of the active user u_{a}. Context-Aware User-Based (CAUB) method depicted in Fig. 1 is based on user model that forms neighborhood of those users who have rated active item i_{a} after filtering with

Table 1 Description of context factors in different communities (cc1, cc2, and cc3) used

Dataset	cc1	cc2	cc3	References
LDOS-CoMoDa	Daytype, social, location, time	domEmo, endEmo, mood, interaction, physical decision	Movie genre, movie year, movie language	[8, 11, 13, 16]
IncarMusic	NULL	Two context having highest relevance value	Music genre	[5]

Table 2 Notations and their descriptions

Notation	Notation description
$U = \{u_{xo} : xo = 1.2..., m\}$	A set of m users
$I = \{i_{xo} : xo = 1.2..., n\}$	A set of n items
$G = \{g_{xo} : xo = 1, 2, ..., p\}$	A set of p genres
$R = \{r_{xo} : xo : 1, 2, ..., q \text{ where } q \leq m * n\}$	A set of q explicit ratings
$R' = \{r'_{xo} : xo = 1, 2, ..., t \wedge t \leq m * p\}$	A set of t implicit ratings
r_{u_{xo}, i_t}	Rating of u_{xo} on item i_t
$\overline{r_{u_{xo}}}$	Average rating of user u_{xo}
$r'_{u_{xo} g_{xo}}$	Implicit rating of u_{xo} to g_{xo}
$\bar{r}_{u_{a,ConxFilter(c,c3)}}$	Average rating of u_a after filtering the preferences with $ConxFilter(c, c3)$

$ConxFilter(c, cc1)$ and Context-Aware Item-Based (CAIB) method (see Fig. 1) is based on item model that forms neighborhood of those items which are rated by active user u_a after filtering with $ConxFilter(c, cc1)$.

3.2.2 Rating Prediction Unit

This unit provides techniques to predict rating utilizing context community cc2 and cc3 which further generates recommendations to individuals and group of users. The approaches $CAWP_{UB\text{-}ER}$, $CAWP_{UB\text{-}IR}$, $CAWP_{UB\text{-}ER\text{-}GB}$, $CAWP_{IB\text{-}ER}$,and $CAWP_{IB\text{-}IR}$ depicted in Fig. 1 predicts the rating as described below.

Similarity Measures.

In this section, a set of user-based and item-based similarity measures are presented and the detailed description about them is given in [9].

User-based similarity.

Based on explicit ratings.

$$\alpha_{xo,yo} = \text{sim}(u_{xo}, u_{yo}) = \frac{\sum_{t=1}^{n'} \left(r_{u_{xo}, i_t} - \overline{r_{u_{xo}}}\right) - \left(r_{u_{yo}, i_t} - \overline{r_{u_{yo}}}\right)}{\sqrt{\sum_{t=1}^{n'} \left(r_{u_{xo}, i_t} - \overline{r_{u_{xo}}}\right)^2} \sqrt{\sum_{t=1}^{n'} \left(r_{u_{yo}, i_t} - \overline{r_{u_{yo}}}\right)^2}} \quad (1)$$

Based on implicit ratings.

$$\gamma_{xo,yo} = \text{sim}(u_{xo}, u_{yo}) = \frac{\sum_{t=1}^{p} \left(r'_{u_{xo}, g_t} - \overline{r_{u_{xo}}}\right) - \left(r'_{u_{yo}, g_t} - \overline{r_{u_{yo}}}\right)}{\sqrt{\sum_{t=1}^{p} \left(r'_{u_{xo}, g_t} - \overline{r_{u_{xo}}}\right)^2} \sqrt{\sum_{t=1}^{p} \left(r'_{u_{yo}, g_t} - \overline{r_{u_{yo}}}\right)^2}} \quad (2)$$

Item-based similarity.
Based on explicit ratings.

$$\beta_{xo,yo} = \text{sim}\left(i_{xo}, i_{yo}\right) = \frac{\sum_{t=1}^{m'} \left(r_{u_t,i_{xo}} - \overline{r_{i_{xo}}}\right) - \left(r_{u_t,i_{yo}} - \overline{r_{i_{yo}}}\right)}{\sqrt{\sum_{t=1}^{m'} \left(r_{u_t,i_{xo}} - \overline{r_{i_{xo}}}\right)^2} \sqrt{\sum_{t=1}^{m'} \left(r_{u_t,i_{yo}} - \overline{r_{i_{y0}}}\right)^2}} \quad (3)$$

Based on item-genre matrix.

$$\mu_{xo,yo} = \text{sim}\left(i_{xo}, i_{yo}\right) = \frac{\sum_{t=1}^{p} \left(v_{g_t,i_{xo}} - \overline{v_{i_{xo}}}\right) - \left(v_{g_t,i_{yo}} - \overline{v_{i_{yo}}}\right)}{\sqrt{\sum_{t=1}^{p} \left(v_{t,i_{xo}} - \overline{v_{i_{xo}}}\right)^2} \sqrt{\sum_{t=1}^{p} \left(v_{g_t,i_{yo}} - \overline{v_{i_{yo}}}\right)^2}} \quad (4)$$

Weighted Percentile.

The key intuition behind using this weighted percentiles notion, rather than weighted averages, is that high percentiles (such as in the 70–90% range) makes more real estimates of preference of a candidate item by a target user based on the experiences of his/her neighbors. It is to be noted that since both low and high ratings devote to the estimation, hence ranks are affected and thus, the percentile quantity of interest.

Illustrative Example.

Table 3 represents a matrix where $N_i(u)$ is the neighborhood of user u of size 4, $w_{N_i(u)} = \left(w_{u,v} \text{ where } v \in N_i(u) \text{ and } i = 1, 2, \ldots, n\right)$ are the similarity values, $r_{N(u),xo}$ and $r_{N(u),yo}$ represents ratings given by $N(u)$ to item xo and $N(u)$ to item yo.

Table 4 presents the results using Eq. 5 which clearly shows that item xo would be recommended if weighted average is used. However, item yo would be recommended if weighted 80th percentile of the variable $r_{N(u),i}$ is used since the specific percentile amounts to a higher rating for item yo than item xo. In other words, the probability of assigning a rating greater or equal to 4 is more for item yo than xo by user u.

Table 3 Scenario of use for the method

$w_{N_i(u)}$	0.2	0.4	0.3	0.1
$r_{N(u),xo}$	2	3	3	4
$r_{N(u),yo}$	2	2	4	4

Table 4 Predicted ratings for the items by neighborhood of user u in Table 3

Average weighted	$\hat{r}_{N(u),xo}$	$\frac{0.2*2+0.4*3+0.3*3+0.1*4}{0.2+0.4+0.3+0.1} = 2.9$
	$\hat{r}_{N(u),yo}$	$\frac{0.2*2+0.4*2+0.3*4+0.1*4}{0.2+0.4+0.3+0.1} = 2.8$
Weighted 80th percentile	$\hat{r}_{N(u),xo}$	$\frac{0.2*0.2*2+0.4*0.2*3+0.3*0.2*3+0.1*0.8*4}{0.2*0.2+0.4*0.2+0.3*0.2+0.1*0.8} = 3.15$
	$\hat{r}_{N(u),yo}$	$\frac{0.2*0.2*2+0.4*0.2*2+0.3*0.8*4+0.1*0.8*4}{0.2*0.2+0.4*0.2+0.3*0.8+0.1*0.8} = 3.45$

User-Based Context-Aware Weighted Percentile with Explicit Ratings (**CAWP$_{\mathbf{UB-ER}}$**).

$$P_{u_a,i_a,c} = \bar{r}_{u_{a,ConxFilter(c,c3)}} + \frac{\sum_{t=1}^{m'} \text{perc}*\alpha_{a,t}\left(r_{u_t,i_a,ConxFilter(c,cc2)} - \bar{r}_{u_{t,ConxFilter(c,cc2)}}\right)}{\sum_{t=1}^{m'} \text{perc} * \alpha_{a,\text{t}}} \quad (5)$$

User-Based Context-Aware Weighted Percentile with Explicit Ratings and Genre Boosted (**CAWP$_{\mathbf{UB-ER-GB}}$**).

$$P_{u_a,i_a,c} = \bar{r}_{u_{a,ConxFilter(c,c3)}} + \frac{\sum_{t=1}^{m'} \text{perc}*\alpha_{a,\text{t}}\left(r_{u_t,i_a,ConxFilter(c,cc2)} - \bar{r}_{u_{t,ConxFilter(c,cc2)}}\right)}{\sum_{t=1}^{m'} \text{perc} * \alpha_{a,\text{t}}} \quad (6)$$

User-based Context-Aware Weighted Percentile with Implicit Ratings (**CAWP$_{\mathbf{UB-IR}}$**).

$$P_{u_a,i_a,c} = \bar{r}_{u_{a,ConxFilter(c,c3)}} + \frac{\sum_{t=1}^{m'} \text{perc}*\gamma_{a,\text{t}}\left(r_{u_t,i_a,ConxFilter(c,cc2)} - \bar{r}_{u_{t,ConxFilter(c,cc2)}}\right)}{\sum_{t=1}^{m'} \text{perc} * \gamma_{a,t}} \quad (7)$$

Item-based Context-Aware Weighted Percentile with Explicit Ratings (**CAWP$_{\mathbf{IB-ER}}$**).

$$P_{u_a,i_a,c} = \bar{r}_{i_{a,ConxFilter(c,c3)}} + \frac{\sum_{t=1}^{n'} \text{perc}*\beta_{a,\text{t}}\left(r_{u_t,i_a,ConxFilter(c,cc2)} - \bar{r}_{u_{t,ConxFilter(c,cc2)}}\right)}{\sum_{t=1}^{m'} \text{perc} * \beta_{a,t}} \quad (8)$$

Item-based Context-Aware Weighted Percentile with Implicit Ratings (CAWP$_{\text{IB-IR}}$).

$$P_{u_a,i_a,c} = \bar{r}_{i_{a,ConxFilter(c,c3)}} + \frac{\sum_{t=1}^{n'} \text{perc}*\mu_{a,t}\left(r_{u_t,i_a,ConxFilter(c,cc2)} - \bar{r}_{u_{t,ConxFilter(c,cc2)}}\right)}{\sum_{t=1}^{m'} \text{perc} * \mu_{a,\text{t}}} \quad (9)$$

3.3 Group Formation Unit

The main goal of this unit is to describe two different types of groups which we have been used to evaluate all techniques. One is Random Group (RG) which represents users without any social relationship such as people watching a TV channel or listening a radio station. Another is Similar-Context Groups (SCG) which represents people with some social relation, such as groups of friends going to see a movie or colleagues going to dine together or family members travelling in a car. These are very natural cases in real life picture.

3.4 Group Recommendation Unit

This unit provides two different types of group techniques to obtain a small number of recommendation solutions to a group of people. Several works [4, 6] have pointed out that the Recommendation Aggregation (RA) and Prediction aggregation (PA) approaches obtain the best results.

4 Experiments

In consequence of the experiments, we aimed to address the following questions:

- How do different context-aware algorithms utilizing explicit ratings or implicit ratings affect performance of prediction and recommendation?
- How does the proposed approach perform compared with the baseline approaches?
- How do weighted percentile method influence the performance?
- Are these proposed algorithms effective for group recommendations?

4.1 Experimental Setups

We evaluate our approach on two real-world datasets specially designed for context-aware personalization research. The summarized statistics of these datasets are presented in Table 5.

Each dataset is partitioned into threefolds and the experiments are repeated five times using onefold in the test set while the rest twofolds in the training set. We further used the popular measures such as mean absolute error (MAE) and F1-score (F1) for top ten items as metrics to perform the task of item recommendations. We consider an item to be relevant (a hit) only if the active user has rated it greater than or equal to 4 (among 5) in both the InCarMusic dataset and LDOS-CoMoDa dataset. As the final results, we recorded the average of five runs for all measures. In order to test weighted percentiles, two percentiles perc $\in$ {70, 90} are used after careful analysis. Regarding the type and size of group, each approach is evaluated

Table 5 The statistics of datasets

Datasets	# of users	# of items	# of ratings	# of contexts factors	# of user attributes	# of item attributes	Rating scale
InCarMusic	42	139	4012	8	1	8	1–5
LDOS-CoMoDa	121	1232	2296	12	4	11	1–5

using Small Group (SG) (size 3) and Large Group (LG) (size 6) with five Random Groups (RGs) and five SCG.

4.2 Baseline Approaches

To elaborate the effectiveness of our framework, we choose the following baseline approaches.

PF (all context). As suggested by [1, 2], we use exact Pre-filtering (PF) where preferences are first filtered using target context and then user neighborhood-based CF approach is used to make predictions.

DCR. Differential Context Relaxation (DCR) in [10] is demonstrated a novel context-aware prediction algorithm based on user model. We kept{exact(social), exact(daytype)}, {any(domEmo)} and {any(endEmo)}, respectively, in first, second and third components of the algorithm using LDOS-CoMoDa dataset. While using InCarMusic dataset, {NULL}, {any(relevant factor)} and {exact(music genre)}, respectively, are placed in first, second, and third components of the algorithm.

CF utilizing explicit rating and implicit ratings methods. The $CF_{UB\text{-}ER}$, $CF_{UB\text{-}ER\text{-}GB}$, and $CF_{UB\text{-}IR}$ based on user model and $CF_{IB\text{-}ER}$ and $CF_{IB\text{-}IR}$ based on item model have been reported as novel and efficient recommendation algorithms [9] in non-context-aware scenario.

Location-based CF method. $CF_{LOCATION\text{-}BASED}$ is a CF approach that considers active user's location in providing recommendations as location is a key factor [7]. We set constant value $\alpha = 0.1$ after the parameter tuning is carefully performed.

4.3 Results and Analysis

This section presents the experimental results for LDOS-CoMoDa and InCarMusic datasets.

Performance comparisons of algorithms for Individual Recommendations. In general, Fig. 2 clearly shows that the performance of weighted average context-aware prediction algorithms is much superior over baselines in all the cases. The algorithms based on item model are more effective than the algorithms based on user model. Efficacy of implicit rating-based algorithms over explicit rating ones is dataset dependent. $CAWP_{UB\text{-}ER\text{-}GB}$ enhances the precision in terms of F1-score than $CAWP_{UB\text{-}ER}$ but it is a minor increment w.r.t. the extra computations made to utilize genre information, so it is less significant. In contrast to other algorithms except $CAWP_{IB\text{-}ER}$, $CAWA_{UB\text{-}ER}$ performs much better using LDOS-CoMoDa dataset than InCarMusic dataset. F1-score is slightly improved when compared with

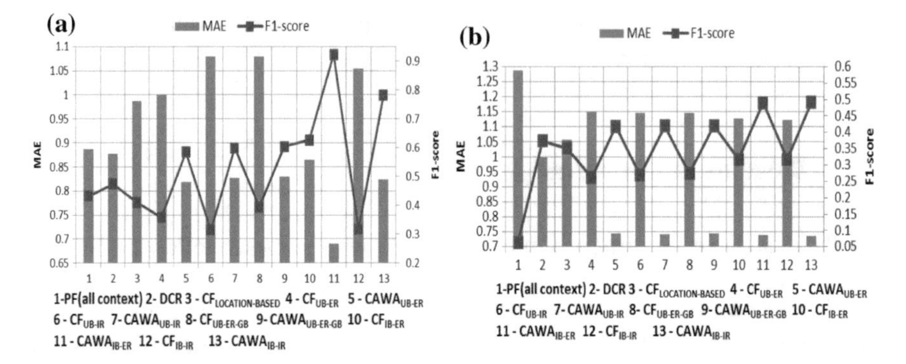

Fig. 2 Comparison of different algorithms using MAE and F1-score on two datasets. **a** LDOS-CoMoDa dataset. **b** InCarMusic dataset

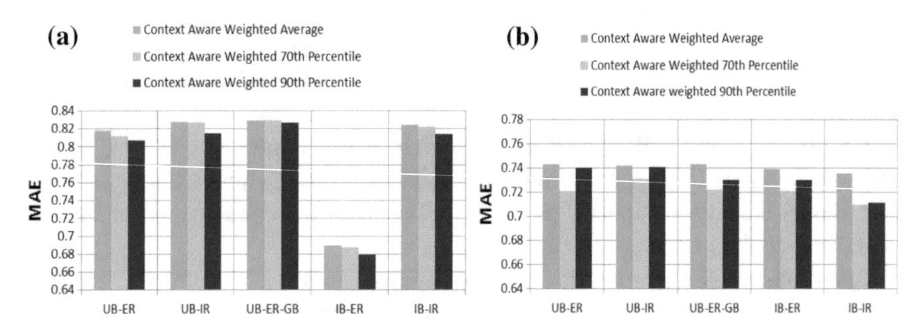

Fig. 3 Effect of weighted average and different weighted percentiles on different techniques in terms of MAE on two datasets. **a** LDOS-CoMoDa dataset. **b** InCarMusic dataset

explicit rating in $CAWA_{UB\text{-}IR}$ using both datasets. Using InCarMusic dataset, $CAWA_{IB\text{-}IR}$ is slightly better than $CAWA_{IB\text{-}ER}$.

Effectiveness of Weighted Percentile approach.

Figure 3 demonstrates that the proposed method outperforms. The weighted percentile method increases the accuracy over weighted average of ratings using both datasets. We have observed that the maximum F1-score (0.9298) for the LDOS-CoMoDa dataset is achieved using the 90th percentile whereas for InCarMusic, the maximum F1-score (0.4901) is obtained using the 70th percentile. Therefore, this suggests that amount of percentile to be considered as weight is dataset dependent. Moreover, this performance enhancement is not attained at the cost of diversity and coverage.

Effectiveness of algorithms for Group Recommendations.

In general, Fig. 4 clearly says that SCG get more accurate recommendations, even better than individual recommendations. RA technique performs uniformly worse than PA method. It is to be noted that the improvement in MAE score reveals that the aggregation of the ranked lists of the group members is able to fix errors

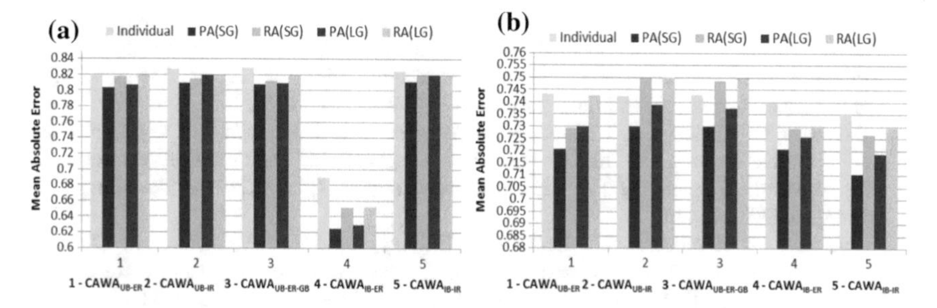

Fig. 4 Comparison of PA and RA grouping techniques for Similar-Context Group (SCG) on SG and LG using context-aware weighted percentiles techniques in terms of MAE on two datasets. **a** LDOS-CoMoDa dataset. **b** InCarMusic dataset

which otherwise are produced by the individual predictions. Similar trend is seen in RG but due to lack of space the results are not presented here. Hence, the proposed algorithms are effective for group recommendations also.

5 Conclusions and Future Work

This paper presents five weighted percentile-based context-aware rating prediction algorithms for individual and group recommendations. These algorithms utilize similarity measures based on either explicit or implicit ratings. Experimental analysis showed that the performance of item-based context-aware prediction algorithms is predominant over user-based context-aware prediction algorithms. The performance of implicit ratings-based algorithms is dataset dependent while Genre boosted technique is comparatively better when combined with explicit ratings. Furthermore, the proposed algorithms are proved to be very effective for group recommendations, especially for SCG.

As a future work, we are planning to expand the framework functionality by improving the solution using particle swarm optimization. In addition, there is still need to explore the proposed framework on large datasets to make it more feasible in the big age of data.

References

1. Adomavicius, G., Sankaranarayanan, R., Sen, S., Tuzhilin, A.: Incorporating contextual information in recommender systems using a multidimensional approach. ACM Trans. Inf. Syst. (TOIS) **23**(1), 103–145 (2005). https://doi.org/10.1145/1055709.1055714
2. Adomavicius, G., Tuzhilin, A.: Context-Aware Recommender Systems: Recommender Systems Handbook. Springer, Boston, pp. 217–253 (2011). https://doi.org/10.1007/978-0-387-85820-3_7

3. Adamopoulos, P., Tuzhilin, A.: Recommendation opportunities: improving item prediction using weighted percentile methods in collaborative filtering systems. In: RecSys' 13 Proceedings of the 7th ACM Conference on Recommender Systems, pp. 351–354. ACM, Hong Kong, China (2013). https://doi.org/10.1145/2507157.2507229
4. Baltrunas, L., Makcinskas, T., Ricci, F.: Group recommendation with rank aggregation and collaborative filtering. In: RecSys '10 Proceedings of the Fourth ACM conference on Recommender Systems, pp. 119–126. ACM, New York, NY, USA (2010). https://doi.org/10.1145/1864708.1864733
5. Baltrunas, L., Kaminskas, M., Ludwig, B., Moling, O., Ricci, F., Aydin, A., Luke, K., Schwaiger, R.: InCarMusic: context-aware music recommendations in a car. In: Huemer, C., Setzer, T. (eds.) EC-Web, LNBIP 85, pp. 89–100. Springer, Heidelberg (2011). https://doi.org/10.1007/978-3-642-23014-1_8
6. Christensen, I.A., Schiaffino, S.: Entertainment recommender systems for group of users. Expert Syst. Appl. **38**, 14127–14135 (2011). https://doi.org/10.1016/j.eswa.2011.04.221
7. Madadipouya, K.: A location-based movie recommender system using collaborative filtering. Int. J. Found. Comput. Sci. Technol. **5**, 13–19 (2015). https://doi.org/10.5121/ijfcst/2015.5402
8. Odic, A., Tkalcic, M., Tasic, J.F., Kosir, A.: Relevant context in a movie recommender system: users opinion vs. statistical detection. In: Proceedings of the 4th International Workshop on Context-Aware Recommender Systems. Dublin, Ireland (2012)
9. Papagelis, M., Plexousakis, D.: Qualitative analysis of user-based and item-based prediction algorithms for recommendations agents. Eng. Appl. Artif. Intell. **18**, 781–789 (2005). https://doi.org/10.1016/j.engappai.2005.06.010
10. Zheng, Y., Burke, R., Mobasher, B.: Differential context relaxation for context-aware travel recommendation. In: 13th International Conference on Electronic Commerce and Web Technologies EC-Web, LNBIP 85, pp. 88–99. Springer, Berlin (2012). https://doi.org/10.1007/978-3-642-32273-0_8
11. Zheng, Y., Burke, R., Mobasher, B.: Optimal feature selection for context-aware recommendation using differential relaxation. In: Conference Proceedings of the 4th International Workshop on Context-Aware Recommender Systems, Dublin, Ireland: ACM RecSys (2012).https://doi.org/10.13140/2.1.3708.7525
12. Zheng, Y., Burke, R., Mobasher, B.: Recommendations with differential context weighting. In: UMAP, pp. 152–164. Springer (2013) https://doi.org/10.1007/978-3-642-38844-6_13
13. Zheng, Y., Burke, R., Mobasher, B.: The role of emotions in context aware recommendation. In: Decisions@RecSys Workshop in Conjunction with the 7th ACM Conference on Recommender Systems, pp. 21–28. ACM, Hong Kong, China (2013)
14. Zheng, Y., Mobasher, B., Burke, R.: Integrating context similarity with sparse linear recommendation model. In: Ricci, F., Bontcheva, K., Conlan, O., Lawless, S. (eds.) User Modeling, Adaptation and Personalization. UMAP 2015. Lecture Notes in Computer Science, vol. 9146. Springer, Cham (2015). https://doi.org/10.1007/978-3-319-20267-9_33
15. Zheng, Y., Mobasher, B., Burke, R.: Similarity-based context-aware recommendation. In: Wang, J., et al. (eds.) Web Information Systems Engineering—WISE 2015. Lecture Notes in Computer Science, vol. 9418. Springer, Cham (2015). https://doi.org/10.1007/978-3-319-26190-4_29
16. Zheng, Y.: A revisit to the identification of contexts in recommender systems. In: 20th International Conference on Intelligent Users Interfaces, ACM IUI, pp. 109–115. Atlanta, GA, USA (2015)

Classification of PASCAL Dataset Using Local Features and Neural Network

Ritu Rani, Amit Prakash Singh and Ravinder Kumar

Abstract In this paper, the various binary patterns especially, the texture operators, namely Local Binary Pattern, Compound Local Binary Pattern, and Local Quinary Pattern (LQP) are implemented on the PASCAL 2012 dataset. The binary patterns are very useful for a lot of computer vision applications. Therefore, through this paper, an attempt is made to explore the suitability of backpropagation algorithm for classification of object recognition dataset using these texture operators. The extracted features through these operators are then fed to the input layer of the backpropagation neural network (BPNN). The training of the network is initially done using the 20 different classes of the dataset on known features and then, this trained network is used to classify the entire image. A comparative evaluation of these techniques is done using the dataset on the basis of the parameters such as execution time, number of features extracted, and the classification accuracy to show their effectiveness for the matching and object category classification purpose. The results show that how LQP is best in terms of classification accuracy and speed.

Keywords Binary descriptors · Local binary pattern · Compounded local binary pattern · Local quinary pattern · Neural networks

Ritu Rani (✉) · Amit Prakash Singh
USICT, GGSIPU, Dwarka, India
e-mail: ritujangra00@gmail.com

Amit Prakash Singh
e-mail: aps.ipu@gmail.com

Ravinder Kumar
HMRITM, GGSIPU, Dwarka, India
e-mail: ravinder_y@yahoo.com

© Springer Nature Singapore Pte Ltd. 2019

H. Malik et al. (eds.), *Applications of Artificial Intelligence Techniques in Engineering*, Advances in Intelligent Systems and Computing 697,
https://doi.org/10.1007/978-981-13-1822-1_36

1 Introduction

Recently, researchers have shown a keen interest and enthusiasm in Local Binary Patterns (LBPs) and its variants and their applicability and effectiveness in numerous areas of computer vision and image processing. LBP which is basically a nonparametric descriptor summarizes the local structures of images. Some interesting properties of LBP are their robustness towards illumination changes less computational complexity. As a result, they are useful for various applications like analyses of face image, video and image retrieval, motion analysis, remote sensing, visual inspection, biomedical, and aerial image analysis. The main aim of the ongoing research is to improve the robustness of the systems against various factors and transformations. In this paper, a sincere attempt has been made to empirically evaluate and compare the texture operators LBP, an extended version of LBP, i.e., Compound Local Binary Pattern (CLBP) and Local Quinary Pattern (LQP). Also, an attempt has been made to show the suitability of using the Backpropagation Neural Network (BPNN) for the classification of the object category dataset.

The flow of the paper is as follows. Section 2 gives the overview of the various texture operators and their extensions and variations with their applicability in various areas. Section 3 shows the proposed flowchart design and Sects. 4 and 5 illustrate the results and conclusion, respectively.

2 Related Work

Texture analysis is one of the emerging research trends with vast literature. Texture is a commonly used term in computer vision which basically involves the interaction, combination, and intertwinement of elements into a complex whole. The texture can be recognized visually when the author sees it, however, its concept is rather subjective and imprecise and defining it formally is still quite a task. It is basically a property of an area which is dependant on the scale of an image and involves variation in appearance and that is seemed as the combination of some basic patterns. According to Davies [1], texture is basically a pattern that involves both randomness and regularity, whereas according to Petrou and García-Sevilla [2], texture is the variation of data at scales smaller than the scale of interest. Haralick [3] divided texture descriptors into two classes

- *Statistical*
- *Structural*

Wu and Chen [4] further divided statistical methods into the following five subclasses:

- *Spatial gray-level dependence methods*
- *Stochastic model-based features*
- *Spatial frequency-based features*

- *Heuristic approaches*
- *Filtering methods*

Tuceryan and Jain [5] proposed a classification into four categories, i.e.,

- *Statistical*
- *Model-based*
- *Geometrical*
- *Signal processing methods*

LBP- and LBP-related methods were based on image patches. However, according to Ahonen et al. [6], it can be perceived as a filter operator which is basically based on local derivative filters and a vector quantization function.

2.1 Local Binary Patterns

LBP [7] proposed in 1990 have been received and attracted the researchers for the pattern recognition community and is used for classification in computer vision. It has been widely used in image retrieval, object recognition, and leads to good results in face recognition. Since it has an effective discriminative power and simple computation technique, it is one of the popularly used approaches in numerous applications. Also, it is computationally simple which helps to analyze images in difficult real-time applications. Silva et al. [8] gave the comparisons of numerous improvements of the original LBP in terms of background subtraction in 2015. The LBP operator as introduced by Ojala et al. is one of the effective operators, wherein the labeling of the pixels of an image is done by thresholding the neighborhood of each pixel and the result is considered as a binary number. In this operator, the thresholding of the 3 × 3 neighborhood of each pixel with the center value is done to form the labels and the result is considered as a binary number. The histogram of these $2^8 = 256$ different labels can then be used as a texture descriptor. The LBP operator was then further extended to use neighborhoods of various different sizes [7].

$$f_{\mathrm{LBP}}(x) = \sum_{j=0}^{7} b(I_j - I_c)2^j \tag{1}$$

The basic algorithm for the LBP feature vector goes in the following manner:

1. First, divide the window to be examined into cells (say 16 × 16 pixels for each cell).
2. Now, start comparing the center pixel with each of the 8 neighboring pixel along the circle either in clockwise or counterclockwise manner.
3. If the value of the center pixel is greater than the neighbor value, then write "0" else consider it as "1" and thus form an 8-bit binary number.

4. Next step is to compute the histogram of the frequency of each "number" occurring. This histogram can be perceived as a 256-dimensional feature vector.
5. Now normalize the histogram.
6. Lastly, concatenate the histograms of all cells to get the feature vector for the entire window.

Similar to LBP, there is another operator called Improved local binary patterns (ILBP) which is based on the same basic idea, however, the only difference is that the thresholding of the entire 3×3 neighborhood is done by its average gray-scale value [8] and gives $(2^9 - 1)$ possible binary patterns. The kernel function used in ILBP operator is

$$f_{\mathrm{ILBP}}(x) = b(I_c - T_{\mathrm{mean}})2^8 + \sum_{j=0}^{7} b\left(I_j - T_{\mathrm{mean}}\right)2^j - 1 \tag{2}$$

where T_{mean} is the average gray-scale value.

Another operator is the Median binary patterns (MBPs) [9] which are quite similar to ILBP, but for thresholding, the gray-scale values of the 3×3 neighborhood uses their median value, instead of their average value. The kernel function used for MBP operator can be expressed as below

$$f_{\mathrm{MBP}}(x) = b(I_c - T_{\mathrm{median}})2^8 + \sum_{j=0}^{7} b\left(I_j - T_{\mathrm{median}}\right)2^j - 1 \tag{3}$$

Center-symmetric local binary patterns (CS-LBPs) [8] use a quite different method to compare the pixels in the neighborhood. CS-LBP considers the following center-symmetric couples of pixel values (I_0, I_4), $(I_1, I_5), (I_2, I_6), (I_3, I_7)$ and thresholding of the gray-level differences is done with a parameter T which further generates a set of 24 possible patterns. It does not consider the center pixel.

The kernel function of CS-LBP operator can be expressed as

$$f_{\mathrm{CS-LBP}}(x, T) = \sum_{j=0}^{3} b\left(I_j - I_{j+4} - T - 1\right)2^j \tag{4}$$

However, Centralized binary patterns (CBPs) [10] considers the central pixel along with the center-symmetric couples of pixels and the comparison is based on the absolute difference of gray-scale values thresholded at a predefined small positive value T. The kernel function of CBP can be expressed as follows:

$$f_{\mathrm{CBP}}(x, T) = b(|I_c - T_{\mathrm{mean}}| - T)2^4 + \sum_{j=0}^{3} b\left(\left|I_j - I_{j+4}\right| - T\right)2^j \tag{5}$$

where T_{mean} is defined.

In Gradient-based local binary patterns (GLBPs) [11], the absolute difference between the central pixel and each surrounding pixel is thresholded at the mean absolute difference between (I_0, I_4) and (I_2, I_6).

In formulas

$$f_{\text{GLBP}}(x) = \sum_{j=0}^{7} b(I_{+} - |I_j - I_c|)2^j \tag{6}$$

where

$$I_{+} = \frac{1}{2}(|I_0 - I_4| + |I_2 - I_6| \tag{7}$$

2.2 Compounded Local Binary Patterns

CLBP [12] is basically an extension of LBP operator that assigns a 2*P*-bit code to the center pixel based on the gray values of a local neighborhood comprising *P* neighbors. Unlike LBP, it also takes into consideration the magnitude information of the difference between the center and the neighbor gray values along with the sign information and uses two bits for each neighbor for representation, the first bit represents the sign of the difference between the center and the corresponding neighbor gray values and the other bit encodes the magnitude of the difference with respect to a threshold value. If the value of the difference between the center and the corresponding neighbor is greater than the threshold M_{avg}, then set this bit to 1 otherwise set it to 0. This method helps to produce consistent codes

$$s(i_{\text{p}}, i_{\text{c}}) = \begin{Bmatrix} 00 & i_{\text{p}} - i_{\text{c}} < 0, & |i_{\text{p}} - i_{\text{c}}| \le M_{\text{avg}} \\ 01 & i_{\text{p}} - i_{\text{c}} < 0, & |i_{\text{p}} - i_{\text{c}}| > M_{\text{avg}} \\ 10 & i_{\text{p}} - i_{\text{c}} \ge 0, & \\ 11 & \text{otherwise} & |i_{\text{p}} - i_{\text{c}}| \le M_{\text{avg}} \end{Bmatrix} \tag{8}$$

Here, i_{c} is the gray value of the center pixel, i_{p} is the gray value of a neighbor *p*, and M_{avg} is the average magnitude of the difference between i_{p} and i_{c} in the local neighborhood.

2.3 Local Quinary Patterns

In LQPs [13], two thresholds *T* and T' are used to compute the five discrete levels (i.e., −2, −1, 0, 1, and 2) which basically encodes the gray-level difference between the central pixel and the surrounding pixels. It is quite similar to LTP except that in LQP, the encoding levels are five, whereas in LTP it is three. The quinary pattern is split into four binary patterns through the following method:

$$b_{j,i} = \left\{ \begin{array}{ll} 1, & \text{if } q_{j=i} \\ 0, & \text{otherwise} \end{array} \right\} \tag{9}$$

where $b_{j,i}$ and q_j are the binary and quinary values corresponding to pixel j and level i; $i \in \{-2, -1, 1, 2\}$. The kernel function used for each level is

$$f_{\mathrm{LQP},i}(x, T, T') = \sum_{j=0}^{7} \delta\left[q\left(I_c - I_j, T, T'\right) - i\right] 2^j \tag{10}$$

The feature vector is obtained as follows:

$$h_{\mathrm{LQP}} = h_{\mathrm{LQP},-2} \| h_{\mathrm{LQP},-1} \| h_{\mathrm{LQP},1} \| h_{\mathrm{LQP},2} \| \tag{11}$$

2.4 Classification Using the Backpropagation Neural Network

The Backpropagation (BP) algorithm was originally introduced in the 1970s. It has multilayer feed-forward neural network consisting of input layer, hidden layer, and output layer. Each unit from these layers is associated with other units through the weighted connections. The BPNN was trained on 20 classes of the PASCAL 2012 dataset on known features and then, this trained network was used to classify the entire image. The BP algorithm involves providing the data into the input layer, calculating the error and then, backpropagating the error and accordingly adjusting the weights. This is a continuous process and is repeated a number of times (Fig. 1).

3 Proposed Work

In the given framework, the various object categories of the PASCAL 2012 dataset are classified using the texture operators LBP, CLBP, and TQP and passing the features extracted through the BPNN. The total dataset is divided into testing and training set. To see the robustness of these operators, the dataset is divided into different sets of ratios. Used dataset in this study has been collected from http://host.robots.ox.ac.uk/pascal/VOC/voc2012/ (Fig. 2).

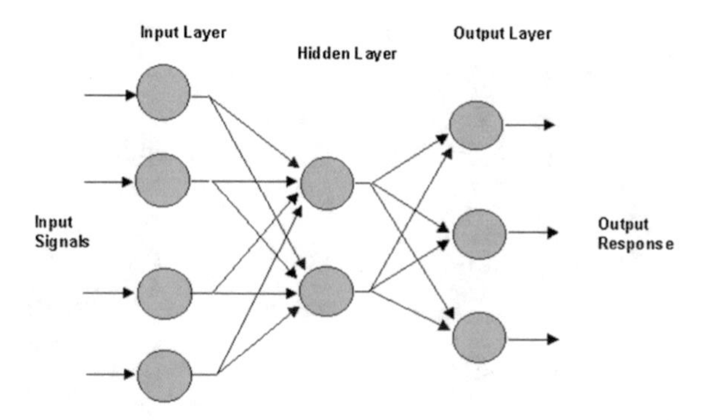

Fig. 1 Architecture of backpropagation neural network

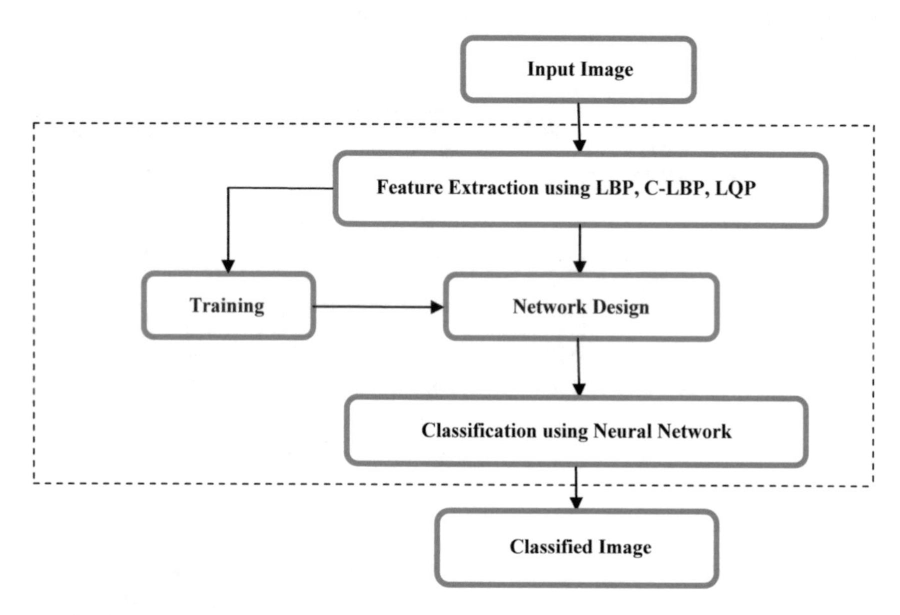

Fig. 2 The flowchart showing the network design

4 Results and Observation

The experiments are performed on MATLAB 2013b on the computer having Intel i7 processor with 8 GB RAM and 1 TB hard disk using the dataset PASCAL 2012. The database chosen for the analysis is the PASCAL 2012 database. Total images

Fig. 3 Sample images from the PASCAL VOC2012 database

Table 1 The recognition rates of the techniques with different ratios of testing and training set of dataset

Technique	Training set	Testing set	Recognition rate (%)
LBP	7000	3000	83.610
LBP	5000	5000	82.456
LBP	3000	7000	81.812
LQP	7000	3000	88.567
LQP	5000	5000	87.083
LQP	3000	7000	86.672
CLBP	7000	3000	87.353
CLBP	5000	5000	86.759
CLBP	3000	7000	84.362

taken into consideration are 10,000 of 20 classes, wherein each class is having 500 images. These classes contain the images like aeroplanes, bicycles, birds, bottles, buses, etc. (Fig. 3).

The data is split into training set and the testing set in 70:30, 50:50, and 30:70 ratios, respectively. The recognition rates of all techniques, i.e., LBP, LQP and CLBP are given in Table 1.

Figure 4 gives the recognition rate for PASCAL 2012 dataset when the training set and the testing set are in the ratio 70:30, 50:50, and 30:70 (Figs. 5 and 6; Table 2).

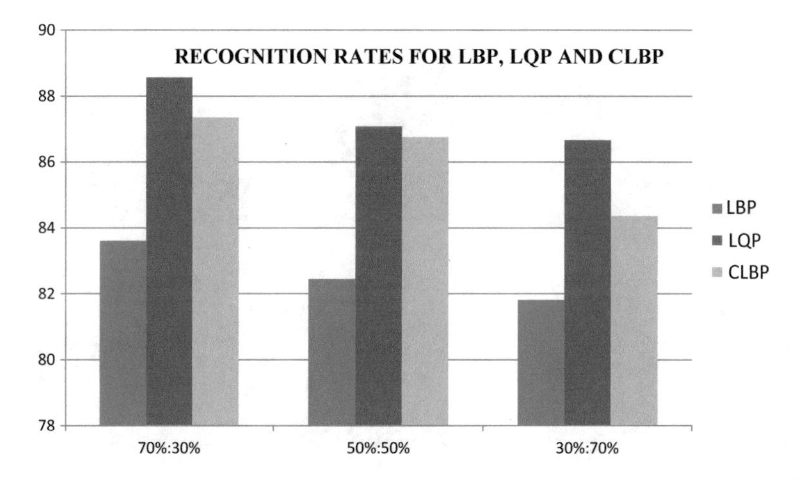

Fig. 4 Recognition rates for LBP, LQP, and CLBP for different ratios of training and testing dataset

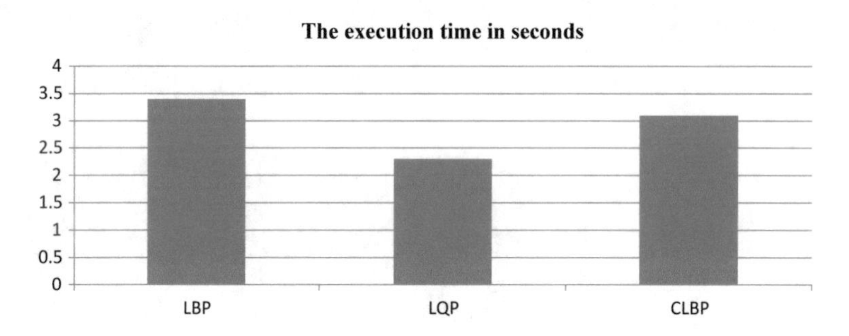

Fig. 5 The execution time for the techniques LBP, LQP, and CLBP

5 Conclusion and Discussion

From the experiments, it can be seen that the empirical evaluation of the texture operators LBP, CLBP, and LQP are done for the object category classification. The computation is done using the PASCAL 2012 dataset on MATLAB. The whole dataset is being divided into the training and testing set and the training and testing of the features for classification is done using the BPNN. The results show that the CLBP is much more robust and provides higher classification rates as compared to the LBP. The reason for this robustness is that it takes into consideration the magnitude information of the difference between the center and the neighbor gray values along with the sign information and uses two bits for each neighbor for representation. The testing and training sets are used in various ratios, one in equal ratio, one where training set is large, and one where training set is small and testing set is large. But for all these ratios, the techniques are quite robust and give

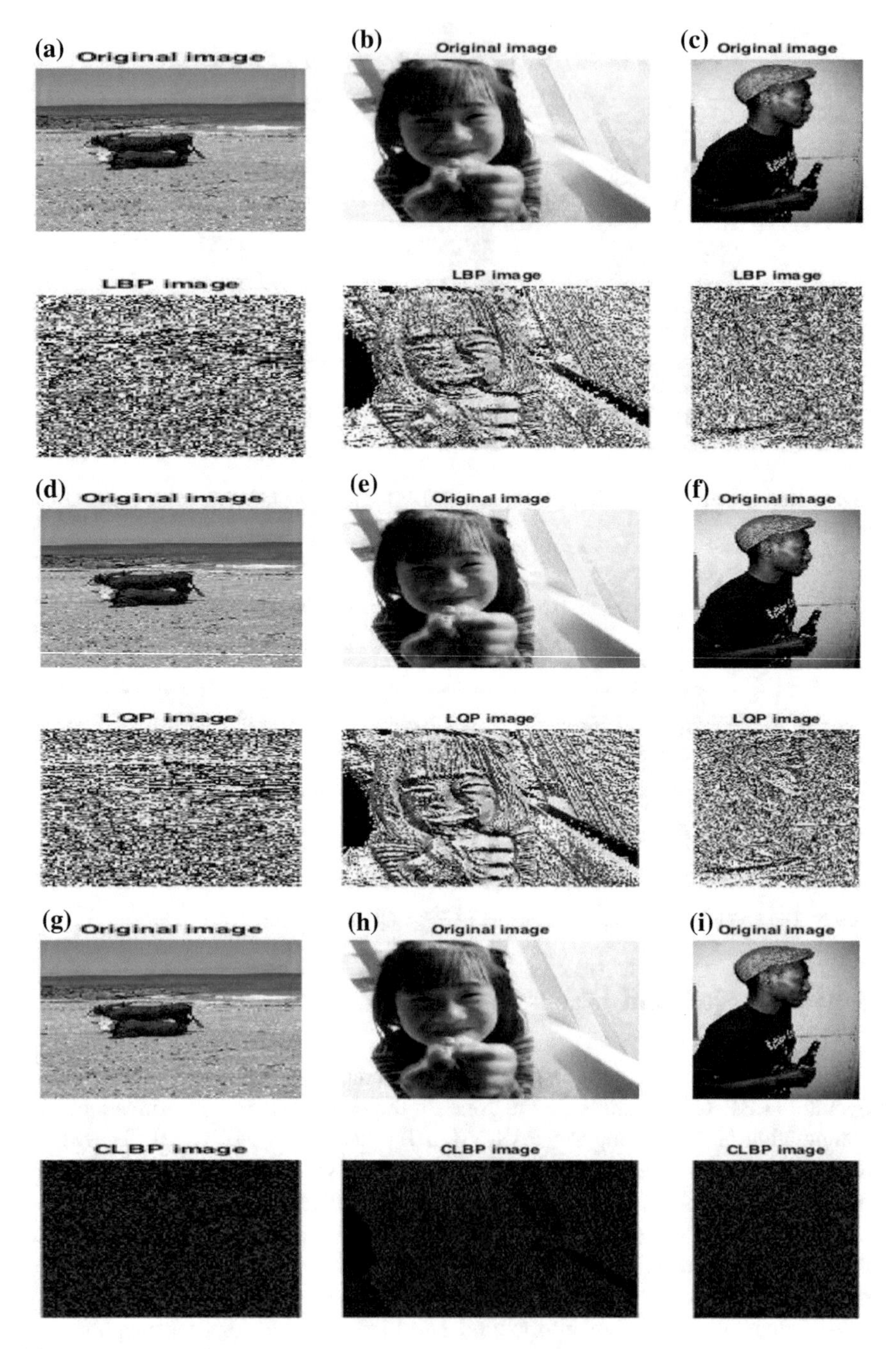

Fig. 6 The original images with their binary pattern images

Table 2 The execution time of the techniques

Technique	Execution time (s)
LBP	3.4
LQP	2.3
CLBP	3.1

remarkable and good classification rates. The LQP gives the best classification rates and is fastest also in terms of the execution time.

Acknowledgements The author is very thankful to http://host.robots.ox.ac.uk/pascal/VOC/voc2012/ to provide the necessary data for the study and also thankful to the reviewers to provide the necessary comments.

References

1. Davies, E.R.: Introduction to texture analysis. In: Mirmehdi, M., Xie, X., Suri, J. (eds.) Handbook of Texture Analysis, pp. 1–31. Imperial College Press, London (2008)
2. Petrou, M., Sevilla, P.G.: Image Processing. Dealing with Texture. Wiley Interscience, Chichester (2006)
3. Haralick, R.M.: Statistical and structural approaches to texture. Proc. IEEE **67**(5), 786–804 (1979)
4. Wu, C.-M., Chen, Y.-C.: Statistical feature matrix for texture analysis. CVGIP: Graph. Models Image Process. **54**(5), 407–419 (1992)
5. Tuceryan, M., Jain, A.K.: Texture analysis. In: Chen, C.H., Pau, L.F., Wang, P.S.P. (eds.) Handbook of Pattern Recognition and Computer Vision, 2nd edn, pp. 207–248. World Scientific Publishing, Singapore (1998)
6. Ahonen, T., Matas, J., He, C., Pietikäinen, M.: Rotation invariant image description with local binary pattern histogram Fourier features. In: Proceedings of the 16th Scandinavian Conference (SCIA 2009), Lecture Notes in Computer Science, vol. 5575, pp. 61–70. Springer (2009)
7. Ojala, T., Pietikäinen, M., Harwood, D.: A comparative study of texture measures with classification based on feature distributions. Pattern Recogn. **29**(1), 51–59 (1996)
8. Silva, C., Bouwmans, T., Frélicot, C.: An eXtended center-symmetric local binary pattern for background modeling and subtraction in videos. In: International Joint Conference on Computer Vision, Imaging and Computer Graphics Theory and Applications, VISAPP 2015 (2015)
9. Hafiane, A., Seetharaman, G., Zavidovique, B.: Median binary pattern for textures classification. In: Proceedings of the 4th International Conference on Image Analysis and Recognition (ICIAR 2007), Lecture Notes in Computer Science, vol. 4633, pp. 387–398. Montreal (2007)
10. Fu, X., Wei, W.: Centralized binary patterns embedded with image euclidean distance for facial expression recognition. In: Proceedings of the Fourth International Conference on Natural Computation (ICNC'08), vol. 4, pp. 115–119 (2008)
11. He, Y., Sang, N.: Robust illumination invariant texture classification using gradient local binary patterns. In: Proceedings of 2011 International Workshop on Multi-Platform/Multi-Sensor Remote Sensing and Mapping, pp. 1–6. Xiamen (2011)

12. Ahmed, F., Hossain, E., Bari, A., Shihavuddin, A.: Compound local binary pattern for robust facial expression recognition. In: Proceedings of the 12th IEEE International Symposium on Computational Intelligence and Informatics, Hungary, pp. 391–395 (2011)
13. Nanni, L., Lumini, A., Brahnam, S.: Local binary patterns variants as texture descriptors for medical image analysis. Artif. Intell. Med. **49**(2), 117–125 (2010)

Comparative Study of Random Forest and Neural Network for Prediction in Direct Marketing

Arushi Gupta and Garima Gupta

Abstract This paper deals with prediction in direct marketing for banks. Nowadays, direct marketing methods are used to reach potential customers as they involve reduced promotional cost. For getting maximum business from the huge customer data sets, accurate prediction of response becomes crucial. The data considered here is of a Portuguese banking institution. Input is a 20-attribute data set about bank clients while the response is to predict if the client will subscribe to term deposit or not. Comparative analysis is done between random forest and neural networks in R platform for the stated prediction problem.

Keywords Direct marketing · Prediction · Random forest · Neural networks Term deposit

1 Introduction

Direct marketing is a method to reach customers without the use of an advertising middleman. Direct marketing can be done in various ways like telephone calls, emails, brochures, newsletters, text messages, emails and many others. Direct marketing aims in reducing campaigning cost and maximizing return rate. For direct marketing, it is important to identify target audience. In a particular area, the said audience is identified as per our needs and their ability.

Once potential customers are identified, it is always easy to identify the customers need and its expectation from the product. The client can be explained on

A. Gupta (✉)
Department of Computer Science Engineering, Kalinga Institute of Industrial Technology, Bhubaneswar, Odisha, India
e-mail: 1505104@kiit.ac.in

G. Gupta
Department of Computer Science Engineering, Maharaja Agrasen Institute of Technology, New Delhi, India
e-mail: garimagupta@mait.ac.in

© Springer Nature Singapore Pte Ltd. 2019

H. Malik et al. (eds.), *Applications of Artificial Intelligence Techniques in Engineering*, Advances in Intelligent Systems and Computing 697,
https://doi.org/10.1007/978-981-13-1822-1_37

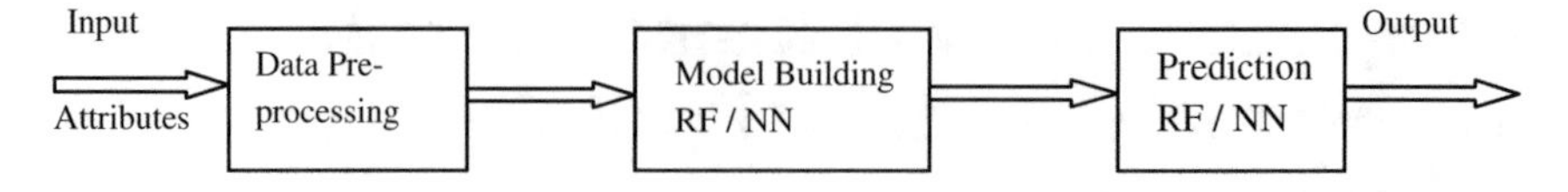

Fig. 1 Block diagram for prediction

the benefits of the product by direct interacting with him either meeting him directly or giving him utmost attention over phone. By this, the mood of the customer can be easily judged and all his misconception or preconceived notions can be corrected immediately. With direct marketing, the company representative develops a personal touch with the customer and can expect the results faster. In today's scenario relationship, between the customer and company representative, plays a very important role in success or failure of the product. One can expect results faster and more. By direct marketing, the company can make the feel of the product to the customer, which in turn clears any doubt for the same. Thus, it can be stated that profitability of direct marketing largely depends upon accurate prediction of the potential customer.

Data considered here for classification and prediction of potential customers is of a Portuguese banking institution. Phone calls were the basis of the data collected. Inputs are twenty different attributes about the client mentioned in detail in Sect. 4 and output is if client has opted for term deposit (yes or no). Contact to the client was required more than once for the assessment of whether he/she would take the term deposit or not.

We have done a prediction for the mentioned data set using random forest (RF) and neural network (NN). Leo Breiman [1] developed random forest. Literature shows that since then RF has been used for classification and prediction for various problems like finance, biomedical analysis and genetic analysis [2–9]. Neural network has also successfully used in prediction problems [10, 12–14]. A general block diagram for data training and prediction is shown in Fig. 1. Detail of techniques used is explained in the succeeding sections.

Sections 2 and 3 brief about theory of random forest and neural networks, respectively. Section 4 explains the data of Portuguese bank. Section 5 gives the results and finally Sect. 6 concludes the paper.

2 Random Forest

Random forest (RF) is an effective algorithm for prediction problems and is widely used for data regression and classification. The advantage of using random forest is that accuracy is maintained even with inconsistent data; is easy to use, fast and robust. RF saves data preprocessing and data preparation time for numerical data. Drawback is that if there are many wrong trees then prediction may go wrong and it

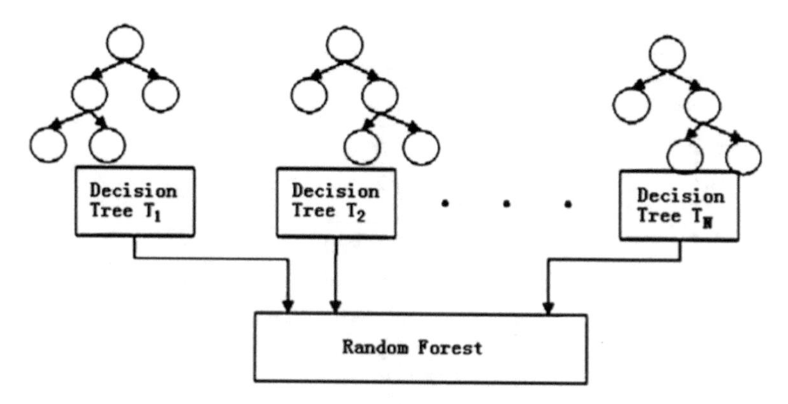

Fig. 2 Random forest—supervised classification machine learning algorithm

is difficult to interpret. RF is widely implemented in R or Python. In our work, we are working on R platform. R is a open source software.

In RF algorithm, various decision trees are made from random subsets of data. Or put in other words, RF is based on bagging approach to get a combination of decision trees with same distribution for all the trees in the forest. This is pictorially depicted in Fig. 2.

Stage 1: Boosting: In this stage, random data is bagged together to create 'n' random trees. Collection of these trees makes a random forest.
Stage 2: Bagging: In this stage, outcome of each tree for same variable is combined to get the final predicted output. Final prediction is either based on result of each decision tree or that appears maximum times in decision trees.

Figure 3 shows steps used in our study for data preprocessing, training and prediction using random forest.

RF algorithm can be divided in two stages.

3 Neural Network

Neural network has applications in various problems like pattern recognition, classification, image processing, system identification and control, signal processing and many more. Here, neural network is used as a predictor. The first model of neural network is built for the Portuguese banking institution data; by training the weights of neural network. Once training of weights is done, values of weights are freezed and this trained model of neural network is used for prediction [10, 12–14]. Basic element of neural network is a neuron and is represented in (1).

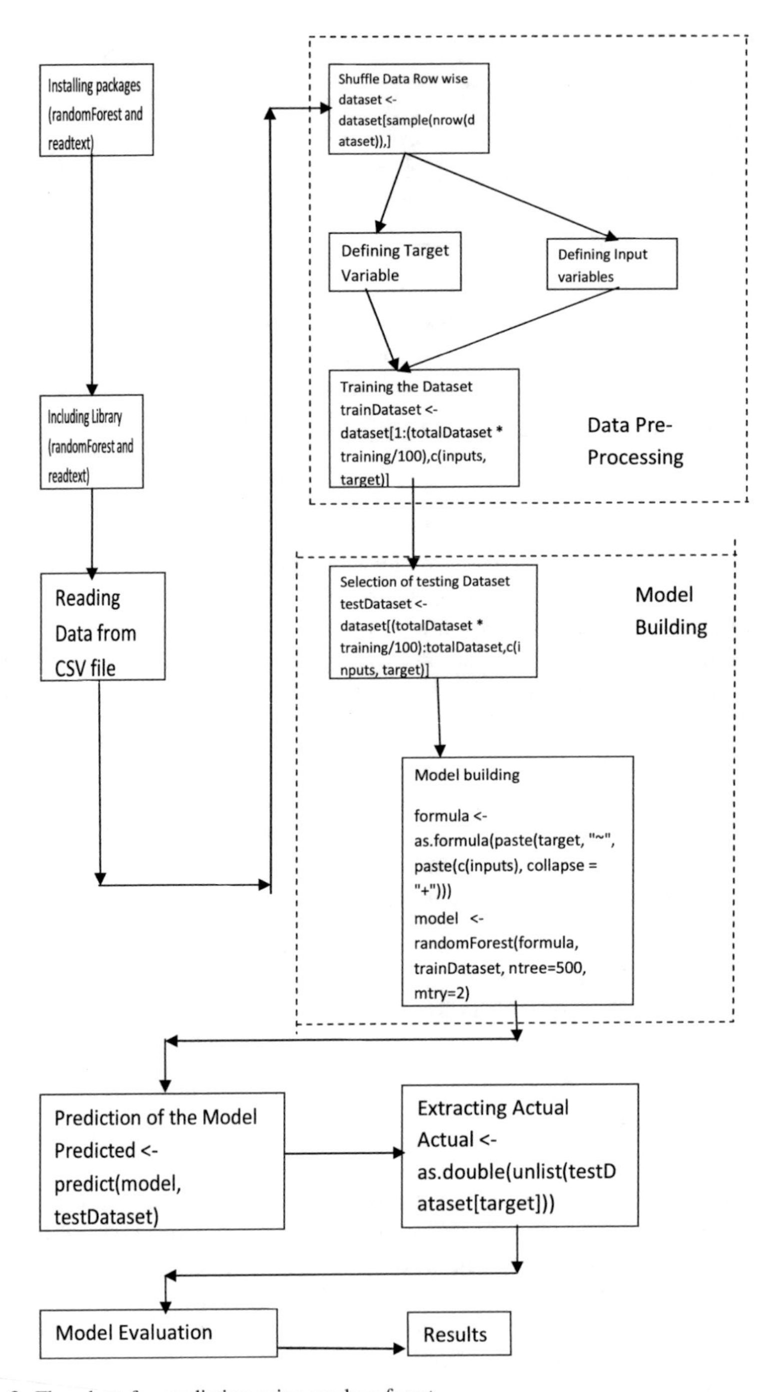

Fig. 3 Flowchart for prediction using random forest

$X(1, 2 \ldots n)$ = Input samples f_{pn} (k) = Output sample of the neural network

$$\left.\begin{aligned} v(k) = \sum_{k=0}^{n} w_k u_k + B \\ f_{\text{pn}}(k) = y[v(k)] \end{aligned}\right\} \quad (1)$$

$\psi(\bullet)$ nonlinear activation function
w weights of neural network
u input to neural network
b threshold
n number of input samples
f_{pn} output of neural network.

The power of neural network increases when number of neurons is combined together in different layers. Some types of neural network are multi-layer feed-forward network, recurrent neural networks and Hopfield neural network. We have used multi-layer feed-forward neural network (MLFFNN) in our problem. Figure 4 shows basic diagram for MLFFNN.

4 The Data set Under Study [11]

Data under study is of a Portuguese banking institution for direct marketing of term deposits. Phone calls are the basis of the data. Twenty different attributes which include information about potential customer like age, occupation, etc., were inputs of the model and output is if the client would take term deposit or not (yes or no). The data set is a.csv file with 41,188 examples (number of clients), from May 2008 to November 2010, 20 inputs were taken, with date wise ordering. Details of various attributes are given in Table 1. First 20 attributes are input and last attribute is the target output.

5 Results

Training and prediction was successfully done for the above-mentioned Portuguese banking institution data using random forest and neural network.

5.1 Prediction Using Random Forest

Prediction results were analysed in terms of out-of-bag (OOB) error, classification error (CE) and root mean square error (RMSE) in random forest with different

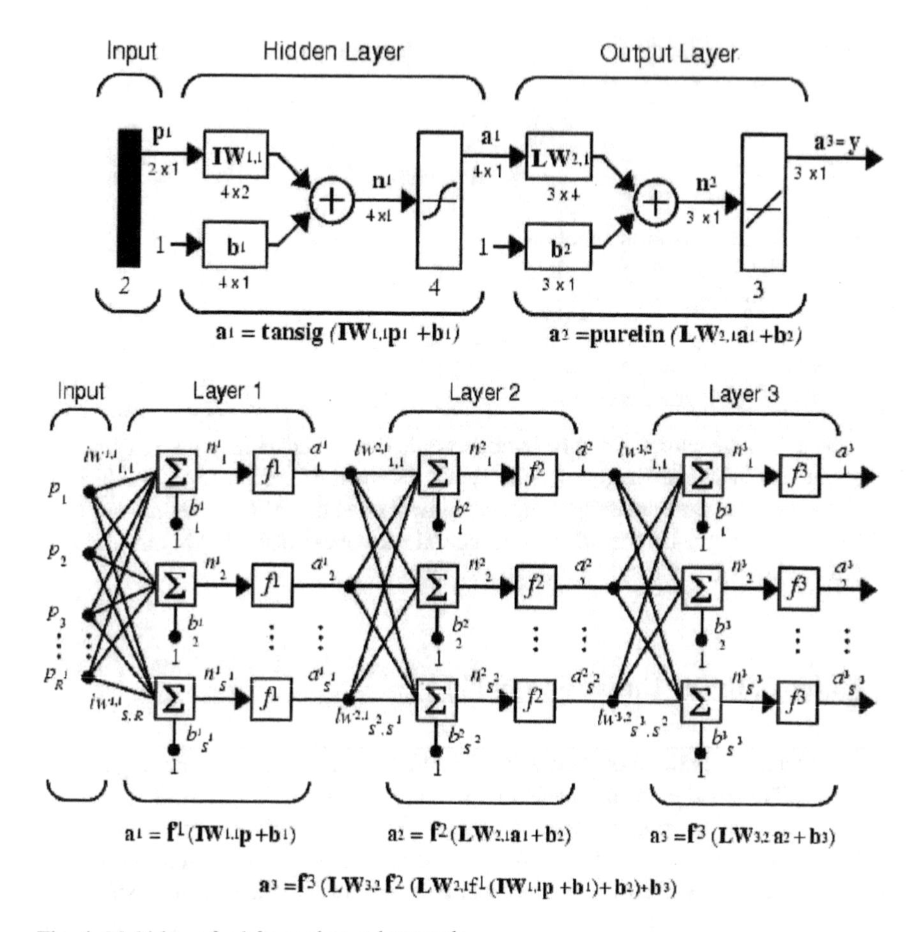

Fig. 4 Multi-layer feed-forward neural network

number of trees. Out-of-bag is the error rate of the out-of-bag classifier on the training set. Classification error (CE) is computed as in Eq. 2.

$$CE = \frac{g}{a} \times 100 \quad (2)$$

where

g number of incorrectly classified samples;
a total number of samples.

Table 2 shows performance of random forest for prediction of Portuguese Banking Institution with different number of trees (n = 50, 100, 200, 300, 400 and 500). This variation is directly related to number of randomly selected variables at each node for cluster building (tree). Literature shows that performance of RF is enhanced with increasing number of trees [10]. Our results for prediction also verify

Table 1 Attribute information

S. no.	Feature	Description	Type
1	Age	Age of client in years	Numeric
2	Balance	Client's average annual balance in euros (€)	Numeric
3	Day	Client's last contact day	Numeric
4	Duration	How long it takes to contact the client in seconds	Numeric
5	Campaign	Number of contacts performed for client during the current campaign	Numeric
6	Pdays	Number of days elapsed since last contact from previous campaign	Numeric
7	Previous	Number of contacts performed for the client before current campaign	Numeric
8	Employment	Variation rate—quarterly indicator	Numeric
9	Consumer price index	Monthly indicator	Numeric
10	Consumer confidence index	Monthly indicator	Numeric
11	Number of employees	Quarterly indicator	Numeric
12	Job	Type of job held by client	Categorical-'admin.', 'blue-collar', 'entrepreneur', 'housemaid', 'management', 'retired', 'self-employed', 'services', 'student', 'technician', 'unemployed', 'unknown'
13	Marital	Client's marital status	Categorical-'divorced', 'married', 'single', 'unknown'; note: 'divorced' means divorced or widowed
14	Education	Client's highest educational qualification	Categorical-'basic.4y', 'basic.6y', 'basic.9y', 'high.school', 'illiterate', 'professional.course', 'university. degree', 'unknown'
15	Default	Is the client in default of credit facility?	Categorical-'no', 'yes', 'unknown'
16	Housing	Does the client have housing loan?	Categorical-'no', 'yes', 'unknown'
17	Loan	Does the client have personal loan?	Categorical-'no', 'yes', 'unknown

(continued)

Table 1 (continued)

S. no.	Feature	Description	Type
18	Contact	Client's contact communication type	Categorical-'cellular', 'telephone'
19	Month	Client's last contact month	Categorical-'jan', 'feb', 'mar', …, 'nov', 'dec'
20	Poutcome	Outcome of the previous marketing campaign	Categorical-: 'mon','tue','wed','thu','fri'
21	Output variable y or n (desired target)	Has the client subscribed a term deposit?	binary: 'yes', 'no'

Table 2 Performance of random forest

Model (RF)	Number of trees (n)			
	OOB (%)	CE (%)	RMSE (%)	Execution time (s)
n = 50	9.25	4.56	9.45	5.12
n = 100	8.8	4.44	9.27	6.28
n = 200	8.87	4.40	8.95	10.58
n = 300	8.92	4.52	8.93	16.8
n = 400	8.82	4.43	8.45	19.69
n = 500	8.78	4.42	8.04	23.09

the same. But there are always exceptions [10]; as in our case, when number of trees was increased from 200 to 300, OOB decreases by 5%.

5.2 *Prediction Using Neural Network*

Prediction and results of the given data using NN were analysed in terms of sum of squared errors (SSE), root mean square error (RMSE). In both cases, error is between actual output and predicted output. Neural network's performance is sensitive towards the number of layers in the network. As number of layers is increased, the performance of neural network is improved at the cost of its complexity. In our study, we have varied number of layers of NN in the range of 3–10. For NN program execution, categorical attributes were changed to numeric values.

Table 3 summarizes the various errors under study (SSE, RMSE) and execution time for different number of layers in NN (3, 5, 7 and 10).

Table 3 Performance of neural network

Model (NN)	SSE (%)	RMSE (%)	Execution time (s)
Neural network layers (l = 3)	5.56	10.2	4.50
Neural network layers (l = 5)	4.89	9.56	5.68
Neural network layers (l = 7)	4.12	9.15	9.96
Neural network layers (l = 10)	3.18	8.25	16.18

Table 4 Comparative analysis of RF and NN

Model	RMSE (%)	Execution time
RF (n = 500)	8.04	23.09
NN (l = 10)	8.25	16.18

5.3 Comparative Analysis of RF and NN

RMSE and execution time were computed in the prediction problem of Portuguese banking institution using RF and NN. In Table 4, comparative analysis of RF and NN is done on this basis. From Table 2, the least value of RMSE using RF is compared with the least Value of RMSE using NN from Table 3.

From Table 4, it is seen that value of RMSE is slightly lesser in RF as compared with NN.

6 Conclusion

This paper explored random forest and neural network for prediction in direct marketing for a Portuguese banking Institution. We cannot choose best machine learning among the various techniques available in literature. Different techniques need to be validated for a particular problem. This study compares well-known neural network and random forest for the stated problem. In RF prediction, analysis was done by taking different number of trees and in neural network, different number of layers was considered. For both, the techniques prediction results were good; but in case of RF value of RMSE is slightly lesser than with NN. So, we can conclude that for our particular Portuguese banking institution problem RF is giving better results than NN for prediction.

References

1. Olaya-Marín, E.J., Martínez-Cape, F., Vezza, P.: A comparison of artificial neural networks and random forests to predict native fish species richness. Mediterranean river. Knowl. Manage. Aquat. Ecosyst. **409**, 07–26 (2013)
2. Elsalamony, H.A.: Bank direct marketing analysis of data mining techniques. Int. J. Comput. Appl. **85**(7), 12–22 (2014). (0975–8887)
3. Bharathidason, S., Venkataeswaran, C.J.: Improving classification accuracy based on random forest model with uncorrelated high performing trees. Int. J. Comput. Appl. **101**(13), 26–30 (2014). (0975–8887)
4. Ahmad, M.W., Mourshed, M., Rezgui, Y.: Trees vs neurons: comparison between random forest and ANN for high-resolution prediction of building energy consumption. Energy Build. **147**, 77–89 (2017)
5. Cui, G., Wong, M.L.: Modeling direct marketing response: Bayesian networks with evolutionary programming. Lingnan University (2015)
6. Ling, C.X., Li, C.: Data mining for direct marketing problems and solutions. In: Plenary Presentation. KDD-98
7. Olson, D.L., Chae, B.K.: Direct marketing decision support through predictive customer response modelling. Decis. Support Syst. **54**, 443–451 (2012)
8. Apampa, O., Tilburg University: Evaluation of classification and ensemble algorithms for bank customer marketing response prediction. J. Int. Technol. Inf. Manage. **25**(4), 6 (2016)
9. Penpece, D., Elma, O.E.: Predicting sales revenue by using artificial neural network in grocery retailing industry: a case study in Turkey. Int. J. Trade Econ. Finan. **5**(5), 435 (2014)
10. Duran, E.A., Pamukcu, A., Bozkurt, H.: Comparison of data mining techniques for direct marketing campaings. J. Eng. Nat. Sci. **32**, 142–152 (2014)
11. Moro, S., Cortez, P., Rita, P.: A data-driven approach to predict the success of bank telemarketing. Decis. Support Syst. **62**, 22–31 (2014)
12. Kalogirou, S.A., Bojic, M.: Artificial neural networks for the prediction of the energy consumption of a passive solar building. Energy **25**(5), 479–491 (2000)
13. Azadeh, A., Ghaderi, S., Sohrabkhani, S.: Annual electricity consumption forecasting by neural network in high energy consuming industrial sectors. Energy Convers. Manag. **49**(8), 2272–2278 (2008)
14. Thirunavukarasu Anbalagan, T., Uma Maheswari, S.: Classification and prediction of stock market index based on fuzzy metagraph. Proc. Comput. Sci. **47**, 214–221 (2015)

Analyzing Risk in Dynamic Supply Chains Using Hybrid Fuzzy AHP Model

Umang Soni and Vaishnavi

Abstract In today's dynamic era, analysis of risk is a critical process in supply chain management. With the proliferation in the magnitude of vulnerabilities, the esteem given to risk examination has increased manifold. This study illustrates how to minimize risk taking into account the two major dimensions of risk management, i.e., qualitative and quantitative. To set up a value-added supply chain integrated, fuzzy AHP model has been used for weighing various factors such as market characteristics, political/economic stability, natural/man-made disaster, management-related issues, legal issues, financial issues, technical resources, and market strength, which are responsible for overall risk assessment. The analysis will be useful for manufacturing and service industries.

Keywords Supply chain management · Decision-making · Fuzzy logic Analytic hierarchy process (AHP)

1 Introduction

Potentially as a result of the long history, the expression "risk" is unclear and imprecise. Even though in day-to-day conversation, the word is often used and comfortably understood [1], the hidden notion is hard to define.

Over time, we have got various definitions of risk. For example, "fear or adventure", "risk indicate the fear that economic activities lead to a downfall in the performance of the business", "risk is fear of losing a venture", "risk is probability of an incident that results in diminution". The last definition has led to the development of probability theory [2].

U. Soni (✉)
Netaji Subhas Institute of Technology, Sector-3 Dwarka,
New Delhi 110078, Delhi, India
e-mail: umangsoni.1@gmail.com

Vaishnavi
Banasthali University, Vanasthali, Rajasthan, India

© Springer Nature Singapore Pte Ltd. 2019

H. Malik et al. (eds.), *Applications of Artificial Intelligence Techniques in Engineering*, Advances in Intelligent Systems and Computing 697,
https://doi.org/10.1007/978-981-13-1822-1_38

As the establishment of supply chains has become increasingly uncertain due to the unknown and unexpected changes, the authors have shifted risk concept based on probability to the risk management in supply chains.

Supply chain risk management talks about the recognition of triggering event and the evaluation of its probability of happening. March and Shapira [3] were the first authors to set up a supply chain risk definition. The risk present in the supply chain was defined as "the variation in the distribution of possible supply chain outcomes, their likelihood and their subjective values". In a collective paper, Juttner, Peck and Christopher [4] define supply chain risk as "anything that disrupts or impedes the information, material or product flow from original suppliers to the delivery of the final product to the ultimate end-user". Zsidisin [5] explains supply chain risk and connects the incident of an event to the uncertainty of the supply chain to fulfill the customer's demand. Supply chain risks are also enhanced because of globalization [6]. It is also interesting to note that more complex a network is more vulnerable the supply chain will be which means exposure to risk will be more likely [7]. Thus by the literature cited above, it can be concluded that the concept of supply chain risk is equally important for industry and academia. But the definition or meanings of supply network hazard are unclear, obscure, and elude evaluation.

The structure of the paper is as follows: Sect. 2 audits contribution of the existing literature in supply chain risk domain. Section 3 depicts the methodology applied. Finally, Sect. 4 depicts a case illustration.

2 Contribution to the Existing Literature

Distinct techniques have been used by numerous researchers which have broadly been divided into five categories which include exploratory longitudinal, exploratory cross-sectional, empirical, descriptive, and conceptual. This classification was given by Malhotra and Grover [8]. Descriptive methodology has been applied to almost 50 percent of the papers, whereas qualitative and descriptive methodologies made majority of almost 80 percent.

Tang categorized different qualitative and quantitative approaches upon the supply chain unit, which deals with risk [9]. Singhal et al. gave an analysis that categorizes literature by the supply chain risk management key issues and research approaches [10]. In order to minimize the firm risk, Cucchiella and Gastaldi gave a structure of uncertainty management [1–18]. In the same year, AHP model was applied to minimize the inbound risk present in supply chain by Wu et al. [11] and for supplier selection using fuzzy AHP [12]. Mixed-integer programming approach has been previously used in majority of the papers. Nonlinear approaches are rarely chosen.

This paper is different from previous approaches as it applies fuzzy AHP model to minimize the risk in the supply chain by finding out the weights of the factors considered. Both the qualitative and quantitative aspects have been considered while choosing the parameters (factors).

3 Applied Methodology

For multi-criteria decision-making, fuzzy analytic hierarchy process (FAHP) is a very useful technique. Fuzzy AHP is a powerful tool for risk prediction as risks are subjective in nature [13]. In dealing with the fuzziness and vagueness of the undetermined environment fuzzy set theory is very useful. All things considered, fuzzy set theory expels the vulnerability in basic leadership by evaluating subjective esteem [14].

Considering its simplicity in calculations and its advantage over other methods, extent analysis model was used by Chang. This approach is more robust in nature and deals with the fuzziness convoluted in decision-making. For pair-wise comparison of different factors, triangular fuzzy number is used.

A triangular-shaped fuzzy number H can be symbolized by (p', q', r') with its enrollment work as indicated Fig. 1 [4, 6, 14].

$$\mu_h(x) = \begin{cases} \frac{x-p'}{q'-p'} (p' \leq x \leq q') \\ \frac{r'-x}{r'-q''} (q' \leq x \leq r') \\ 0, \text{ otherwise} \end{cases}$$

With $-\infty < p' \leq q' \leq r' \leq \infty$

Here, p' and r' are the lower and upper limits and participation of the parameter q' is most noteworthy, $f_M(q') = 1$. In Fig. 2, $H_1(h_1^-, h_1, h_1^+)$ and $H_2(h_2^-, h_2, h_2^+)$ are the two triangular fluffy numbers.

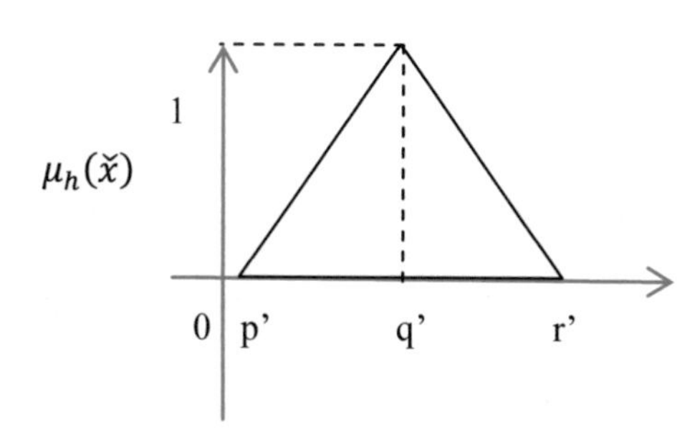

Fig. 1 Triangular fuzzy number

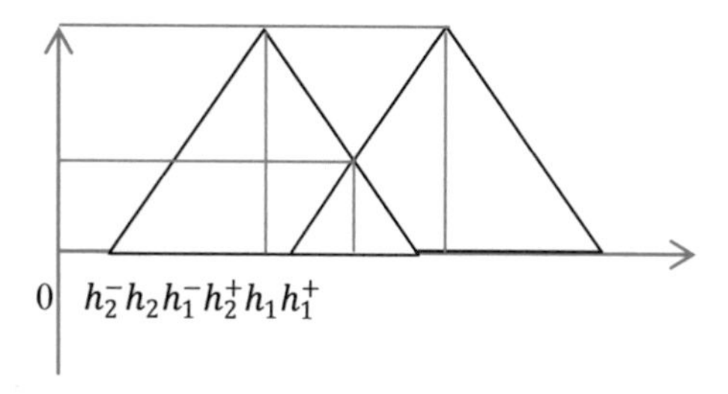

Fig. 2 H_1 and H_2 are the two triangular fuzzy numbers [4]

Now, when $h_1^- \geq h_2^-, h_1 \geq h_2, h_1^+ \geq h_2^+$
Then, the equation which represents the degree of possibility is

$$V(H_1 \geq H_2) = 1$$

Otherwise, the most astounding crossing point is figured as [4, 15]

$$\begin{aligned} V(H_2 \geq H_1) &= \text{hgt}(H_1 \cap H_2) = \mu(d') \\ &= \frac{h_1^- - h_2^+}{\left(h_2 - h_2^+\right) - \left(h_1 - h_1^-\right)} \end{aligned}$$

Fuzzy synthetic extent value can be calculated by these equations [4, 15]

$$F_i = \sum_{j=i}^{m} H_{gi}^j \oplus \left(\sum_{i=1}^{n}\sum_{j=1}^{m} H_{gi}^j\right)^{-1}, \quad i = 1, 2, \ldots, n$$

$$\sum_{j=i}^{m} H_{gi} = \left(\sum_{j=i}^{m} H_{ij}^-, \sum_{j=i}^{m} H_{ij}, \sum_{j=i}^{m} H_{ij}^+\right), \quad i = 1, 2, \ldots, n$$

$$\left(\sum_{i=1}^{n}\sum_{j=1}^{m} H_{gi}^j\right)^{-1} = \left[\frac{1}{\sum_{i=1}^{n}\sum_{j=1}^{m} H_{ij}^+}, \frac{1}{\sum_{j=1}^{n}\sum_{j=1}^{m} H_{ij}}, \frac{1}{\sum_{i=1}^{n}\sum_{j=1}^{m} H_{ij}^-}\right]$$

Convex number is defined as,

$$\begin{aligned} V\left(F^3, F_1, \ldots, F_K\right) &= \text{Minimum } V\left(F^3 F_i\right), \quad i = 1, 2, \ldots, K \\ d(F_i) &= \text{Minimum } V(F \geq F_K) = W_i, K \\ &= 1, 2, \ldots, n \text{ and } K \neq i \end{aligned}$$

Weights $W_i^{'}$, of the factors, based on the above factors are

$$W' = \left(W_1', W_2', \ldots, W_n'\right)^{\mathrm{T}}$$

The priority weights, after normalization is as follows:

$$W' = (W_1, W_2, \ldots, W_n)^{\mathrm{T}}$$

3.1 Process of Computation

To decide the relative weights of the variables considered, fuzzy AHP has been utilized. The strategy of model computation is as follows:

Step 1: The risk factors are identified.
Step 2: Sample is set up for pair-wise correlation by specialists.
Step 3: Reactions are utilized to figure fuzzy significance weights. It is consolidated to acquire a triangular fuzzy number.

$$\check{T} = (l^{-}, l, l^{+})$$

where

$$l^{-} = \left(\prod_{t=1}^{s} L_t\right)^{\frac{1}{s}}, \quad t = 1, 2, \ldots, s$$

$$l = \left(\prod_{t=1}^{s} N_t\right)^{\frac{1}{s}}, \quad t = 1, 2, \ldots, s$$

$$l^{+} = \left(\prod_{t=1}^{s} N_t\right)^{\frac{1}{s}}, \quad t = 1, 2, \ldots, s$$

Furthermore, (L_t, M_t, N_t) are lower, center and max points of confinement of the fuzzy reaction gathered from the tth expert.

4 Case Illustration

A case study of an automobile company has been used to fuse risk management criteria in its supply chain processes. The organization administration welcomed specialists from promoting, production, quality, and research divisions. A meeting to generate new ideas was sorted out to recognize and stipulate risk management criteria that could be measured using brainstorming sessions. The criteria which were finalized are discussed here. They are market characteristics, political/economic stability, natural/man-made disaster, management-related issues, legal issues, financial issues, technical resources, and market strength.

The fuzzy pair-wise association amongst the measures is given in the accompanying Table 1.

Table 1 Fuzzy pair-wise correlation among the criteria

	Market characteristics	Political/ economic stability	Natural/man-made disaster	Management issues	Financial issues	Legal issues	Technical resources	Market strength
Market characteristics	1.00 1.00 1.00	1.51 1.82 2.94	0.48 0.66 0.93	1.59 2.08 3.17	1.59 2.08 2.83	1.62 1.91 3.05	1.59 2.08 3.17	1.35 1.82 2.62
Political/ economic stability	0.34 0.55 0.66	1.00 1.00 1.00	1.00 1.44 2.00	1.59 2.08 3.17	1.59 2.08 3.17	1.51 1.82 2.94	1.26 1.44 2.24	2.14 2.85 3.96
Natural/ man-made disaster	1.07 1.52 2.08	0.50 0.69 1.00	1.00 1.00 1.00	1.41 1.73 2.52	1.51 2.18 2.94	1.19 2.62 3.70	1.59 2.08 3.17	1.51 2.04 3.14
Management issues	0.31 0.48 0.63	0.31 0.48 0.63	0.40 0.58 0.71	1.00 1.00 1.00	1.00 1.12 2.14	0.46 0.56 0.87	0.92 1.07 1.73	0.41 0.45 0.78
Financial issues	0.35 0.48 0.63	0.31 0.48 0.63	0.34 0.46 0.66	0.47 0.89 1.00	1.00 1.00 1.00	1.26 1.44 2.52	1.41 1.73 2.83	1.00 1.00 1.41
Legal issues	0.33 0.52 0.62	0.34 0.55 0.66	0.27 0.38 0.52	1.15 1.78 2.15	0.40 0.69 0.79	1.00 1.00 1.00	1.26 1.44 2.52	1.12 1.20 1.59
Technical resources	0.31 0.48 0.63	0.45 0.69 0.79	0.31 0.48 0.63	0.58 0.93 1.09	0.35 0.58 0.71	0.40 0.69 0.79	1.00 1.00 1.00	0.89 0.89 1.78
Market strength	0.38 0.55 0.74	0.25 0.35 0.47	0.32 0.49 0.66	1.29 2.24 2.42	0.71 1.00 1.00	0.63 0.83 0.89	0.56 1.12 1.12	1.00 1.00 1.00

$$\sum_{i=1}^{n}\sum_{j=1}^{m} H_{gi}^{j} = (1,1,1) + (1.51,1.82,2.94) + \ldots(1,1,1)$$
$$= (57.94, 74.71, 103.15)$$

$$F_1 = \sum_{j=1}^{m} H_{g1}^{j} \otimes \left[\sum_{i=1}^{n}\sum_{j=1}^{m} H_{gi}^{j}\right]^{-1}$$
$$= (10.72, 13.44, 19.71) \otimes (0.0097, 0.0134, 0.0173)$$
$$= (0.10, 0.18, 0.34)$$

$$F_2 = \sum_{j=1}^{m} H_{g2}^{j} \otimes \left[\sum_{i=1}^{n}\sum_{j=1}^{m} H_{gi}^{j}\right]^{-1}$$
$$= (10.43, 13.27, 19.15) \otimes (0.0097, 0.0134, 0.0173)$$
$$= (0.10, 0.19, 0.34)$$

$$F_3 = \sum_{j=1}^{m} H_{g3}^{j} \otimes \left[\sum_{i=1}^{n}\sum_{j=1}^{m} H_{gi}^{j}\right]^{-1}$$
$$= (10.51, 13.87, 19.55) \otimes (0.0097, 0.0134, 0.0173)$$
$$= (0.10, 0.19, 0.34)$$

$$F_4 = \sum_{j=1}^{m} H_{g4}^{j} \otimes \left[\sum_{i=1}^{n}\sum_{j=1}^{m} H_{gi}^{j}\right]^{-1}$$
$$= (4.82, 5.74, 8.48) \otimes (0.0097, 0.0134, 0.0173)$$
$$= (0.05, 0.08, 0.15)$$

$$F_5 = \sum_{j=1}^{m} H_{g5}^{j} \otimes \left[\sum_{i=1}^{n}\sum_{j=1}^{m} H_{gi}^{j}\right]^{-1}$$
$$= (6.15, 7.48, 10.68) \otimes (0.0097, 0.0134, 0.0173$$
$$= (0.06, 0.10, 0.18)$$

$$F_6 = \sum_{j=1}^{m} H_{g6}^{j} \otimes \left[\sum_{i=1}^{n}\sum_{j=1}^{m} H_{gi}^{j}\right]^{-1}$$
$$= (5.87, 7.57, 9.86) \otimes (0.0097, 0.0134, 0.0173)$$
$$= (0.06, 0.10, 0.17)$$

$$F_7 = \sum_{j=1}^{m} H_{g7}^{j} \otimes \left[\sum_{i=1}^{n}\sum_{j=1}^{m} H_{gi}^{j}\right]^{-1}$$
$$= (4.30, 5.75, 9.86) \otimes (0.0097, 0.0134, 0.0173)$$
$$= (0.04, 0.08, 0.13)$$

$$F_8 = \sum_{j=1}^{m} H_{g8}^{j} \otimes \left[\sum_{i=1}^{n}\sum_{j=1}^{m} H_{gi}^{j}\right]^{-1}$$
$$= (5.14, 7.59, 8.30) \otimes (0.0097, 0.0134, 0.0173)$$
$$= (0.05, 0.10, 0.14)$$

$$V(E_1 \geq E_2) = 1, V(E_1 \geq E_3) = 1, V(E_1 \geq E_4) = 1,$$
$$V(E_1 \geq E_5) = 1, V(E_1 \geq E_6) = 1, V(E_1 \geq E_7) = 1,$$
$$V(E_1 \geq E_8) = 1$$
$$V(E_2 \geq E_1) = 1, V(E_2 \geq E_3) = 1, V(E_2 \geq E_4) = 1,$$
$$V(E_2 \geq E_5) = 1, V(E_2 \geq E_6) = 1, V(E_2 \geq E) = 1,$$
$$V(E_1 \geq E_8) = 1$$
$$V(E_3 \geq E_1) = 1, V(E_3 \geq E_2) = 1, V(E_3 \geq E_4) = 1$$
$$V(E_3 \geq E_5) = 1, V(E_3 \geq E_6) = 1, V(E_3 \geq E_7) = 1,$$
$$V(E_3 \geq E_8) = 1$$
$$V(E_4 \geq E_1) = 0.375, V(E_4 \geq E_2) = 0.444,$$
$$V(E_4 \geq E_3) = 0.444, V(E_4 \geq E_5) = 1,$$

$$V(E_4 \geq E_6) = 1, V(E_4 \geq E_7) = 1,$$
$$V(E_4 \geq E_8) = 0.857$$
$$V(E_5 \geq E_1) = 0.143, V(E_5 \geq E_2) = 0.25,$$
$$V(E_5 \geq E_3) = 0.25, V(E_5 \geq E_4) = 0.80,$$
$$V(E_5 \geq E_3) = 0.25, V(E_5 \geq E_4) = 0.80,$$
$$V(E_5 \geq E_6) = 1, V(E_5 \geq E_7) = 0.833,$$
$$V(E_5 \geq E_8) = 0.667$$
$$V(E_6 \geq E_1) = 0.286, V(E_6 \geq E_2) = 0.375,$$
$$V(E_6 \geq E_3) = 0.375, V(E_6 \geq E_4) = 1,$$
$$V(E_6 \geq E_5) = 1, V(E_6 \geq E_7) = 1,$$
$$V(E_6 \geq E_8) = 0.833$$
$$V(E_7 \geq E_1) = 0.143, V(E_7 \geq E_2) = 0.25,$$
$$V(E_7 \geq E_3) = 0.25, V(E_7 \geq E_4) = 0.80,$$
$$V(E_7 \geq E_5) = 1, V(E_7 \geq E_6) = 0.833,$$
$$V(E_7 \geq E_8) = 0.667$$
$$V(E_8 \geq E_1) = 0.429, V(E_8 \geq E_2) = 0.50,$$
$$V(E_8 \geq E_3) = 0.50, V(E_8 \geq E_4) = 1,$$
$$V(E_8 \geq E_5) = 1, V(E_8 \geq E_6) = 1,$$
$$V(E_8 \geq E_7) = 1$$

The significance (weight) vectors are ascertained:

$$\begin{aligned}
d(e_1) &= \text{Minimum } V(E_1 \geq E_2E_3E_4E_5E_6E_7E_8)\\
&= \text{Minimum}(1,1,1,1,1,1,1)\\
&= 1\\
d(e_2) &= \text{Minimum } V(E_2 \geq E_1E_3E_4E_5E_6E_7E_8)\\
&= \text{Minimum}(1,1,1,1,1,1,1)\\
&= 1\\
d(e_3) &= \text{Minimum } V(E_3 \geq E_1E_2E_4E_5E_6E_7E_8)\\
&= \text{Minimum}(1,1,1,1,1,1,1)\\
&= 1\\
d(e_4) &= \text{Minimum } V(E_4 \geq E_1E_2E_3E_5E_6E_7E_8)\\
&= \text{Minimum}(0.377,0.445,0.445,1,1,1,0.856)\\
&= 0.374\\
d(e_5) &= \text{Minimum } V(E_5 \geq E_1E_2E_3E_4E_6E_7E_8)\\
&= \text{Minimum}(0.144,0.255,0.256,0.80,1,0.832,0.666)\\
&= 0.144\\
d(e_6) &= \text{Minimum } V(E_6 \geq E_1E_2E_3E_4E_5E_7E_8)\\
&= \text{Minimum}(0.286,0.375,0.375,1,1,1,0.833)\\
&= 0.286\\
d(e_7) &= \text{Minimum} V(E_7 \geq E_1E_2E_3E_4E_5E_6E_8)\\
&= \text{Minimum}(0.144,0.25,0.25,0.80,1,0.833,0.667)\\
&= 0.144\\
d(e_8) &= \text{Minimum} V(E_1E_2E_3E_4E_5E_6E_7)\\
&= \text{Minimum in}(0.428,0.50,0.50,1,1,1,1)\\
&= 0.428
\end{aligned}$$

$$\begin{aligned}
W' &= (d(e_1)d(e_2)d(e_3)d(e_4)d(e_5)d(e_6)d(e_7)d(e_8))^{\mathrm{T}}\\
&= (1,1,1,0.374,0.144,0.286,0.144,0.428)^{\mathrm{T}}\\
&= (0.228,0.228,0.228,0.086,0.0333,0.065,0.0333,0.098)
\end{aligned}$$

5 Results and Conclusion

The examination of the results demonstrates that among all variables market characteristics, political/economic stability, and natural/man-made disaster are positioned as the most significant. This demonstrates the significance of these factors in a production (supply) network. The other risk indicators, namely management-related issues, financial issues, legal issues, technical resources, and market strength conveyed the accompanying weights: 0.086, 0.033, 0.065, 0.33, and 0.098. The entirety of these weights was 0.274, which could be seen as very critical and can influence the production network. In this review, all risk variables have been considered to make the risk management model extensive. The insights are worthwhile for supply chain managers from risk perspectives.

References

1. Morgan, M.G., Henrion, M.: Uncertainty in Quantitative Risk and Policy Analysis. Cambridge University Press, Cambridge (1990)
2. Bernstein, Peter L.: Against the gods: the remarkable story of risk. Wiley, New York (1999)
3. March, J.G., Shapira, Z.: Managerial perspectives on risk and risk taking. Manage. Sci. (1987)
4. Lee, A.H.I., Kang, H.Y., Chang, C.T.: Fuzzy multiple goal programming applied to TFT-LCD supplier selection by downstream manufacturers. Expert Syst. Appl. (2009)
5. Zsidisin, G.A.: A grounded definition of supply risk. J. Purchasing Supply Manage. (2003)
6. Cheng, C.-H.: Evaluating weapon systems using ranking fuzzy numbers. Fuzzy Sets Syst. (1999)
7. Peck, H.: Drivers of supply chain vulnerability: an integrated framework. Int. J. Phys. Distrib. Logistics Manage. **35**(4) (2005)
8. Malhotra, M.K., Grover, V.: An assessment of survey research in POM: from constructs to theory. J. Oper. Manage. **16**(4) (1998)
9. Tang, S.: Perspectives in supply chain risk management. Int. J. Prod. Econ. (2006)
10. Singhal, P., Agarwal, G., Mittal, M.L.: Supply chain risk management: review, classification and future research directions. Int. J. Bus. Manage. (2011)
11. Wu, T., Blackhurst, J., Chidambaram, V.: A model for inbound supply risk analysis. Comput. Ind. **57**(4) (2006)
12. Kumar, D., Rahman, Z., Chan, F.T.S.: A fuzzy AHP and fuzzy multi-objective linear programming model for order allocation in a sustainable supply chain: a case study. Int. J. Comput. Integr. Manuf. (2016)
13. Dey, P.K., Ogunlana, S.O.: Selection and application of risk management tools and techniques for build-operate-transfer projects. Ind. Manage. Data Syst. **104**(4) (2004)
14. Lee, A.H.I., Kang, H.Y., Wang, W.P.: Analysis of priority mix planning for the fabrication of semiconductor under uncertainty. Int. J. Adv. Manuf. Technol. (2005)
15. Shaw, K., Shankar, R., Yadav, S.S., Thakur, L.S.: Supplier selection using fuzzy AHP & fuzzy Multi-objective linear programming for developing low carbon supply chain. Expert Syst. Appl. (2012)
16. Berry, J.: Supply chain risk in an uncertain global supply chain environment. Int. J. Phys. Distrib. Logistics Manage. **34**(9) (2004)
17. Cucchiella, F., Gastaldi, M.: Risk management in supply chain: a real option approach. J. Manuf. Technol. Manage. **17**(6) (2006)
18. Peck, H.: Reconciling supply chain vulnerability, risk and supply chain management. Int. J. Logistics Res. Appl. (2006)

Optimal Bidding Strategy in Deregulated Power Market Using Invasive Weed Optimization

A. V. V. Sudhakar, Chandram Karri and A. Jaya Laxmi

Abstract In this paper, Invasive Weed Optimization algorithm has been adopted to address the Optimal Bidding Strategy problem. The utilities take part in the bidding for maximizing benefits. The bidding coefficients are picked strategically by each participant to counter the contenders bidding strategy. The proposed algorithm has been coded in MATLAB and validated on IEEE 30 bus system. The results have been analyzed and compared with the techniques such as Particle Swarm Optimization (PSO), Differential Evolution (DE), and Monte Carlo simulation. The outcomes demonstrate the flexibility of the IWO technique compared with other methods.

Keywords Optimal bidding strategy · Invasive weed optimization Deregulation · Market clearing price · GENCOs

1 Introduction

The electric utilities have been changing their operation from vertically integrated markets to deregulation. The deregulated market provides the choice to market participants and encourages competition [1, 2]. Deregulation creates space for GENCOs and consumers to maximize their own benefits. To realize this, GENCOs and consumers develop optimal bidding strategy. The profit of suppliers and consumers mainly depends on the Market Clearing Price (MCP). Because of changes in the MCP, the market participants try to take part in bidding process to increase their profits/benefits.

The Power Exchange (PX)/ISO develops aggregated demand curve for each hour of the next day, starting from aggregated highest priced demand bid to lowest

A. V. V. Sudhakar (✉) · C. Karri
S R Engineering College, Warangal 506371, India
e-mail: sudheavv@gmail.com

A. J. Laxmi
JNTUH College of Engineering, Hyderabad 500085, India

© Springer Nature Singapore Pte Ltd. 2019

H. Malik et al. (eds.), *Applications of Artificial Intelligence Techniques in Engineering*, Advances in Intelligent Systems and Computing 697,
https://doi.org/10.1007/978-981-13-1822-1_39

bid. Similarly, the ISO develops the aggregated supply curve for each hour of the next day, starting from aggregated lowest-priced supply bid to highest priced bid. The intersection point of these two curves determines the MCP and power to be generated.

The optimal bidding problem has been addressed by researchers previously [3–10].

Even though several techniques have been proposed, it is observed from the literature that the existing algorithms provide suboptimal solution in terms of profit. Hence, there is a scope for maximizing benefits of market participants. The fundamental target of this work is to recommend a method like Invasive Weed Optimization (IWO) for solving optimal bidding strategy. The organization of rest of the paper is as follows: mathematical modeling of optimal bidding strategy in Sect. 2, Invasive Weed Optimization method in Sect. 3, implementation steps of IWO in Sect. 4, simulation results in Sect. 5, and outcome of the paper are mentioned in Sect. 6.

2 Optimal Bidding Strategy Problem

The mathematical formulation of MCP, for suppliers and consumers, is as follows:

$$\mathbf{x_i} + \mathbf{y_i}\mathbf{P_i} = \mathbf{MCP}, \quad \boldsymbol{i} = 1, 2, \ldots, \boldsymbol{n} \tag{1}$$

$$u_j - v_j L_j = \text{MCP}, \quad j = 1, 2, \ldots, m \tag{2}$$

The power equality constraint may be expressed as follows:

$$\sum_{\mathbf{i}=1}^{\mathbf{n}} \mathbf{P_i} = \mathbf{Q}(\mathbf{MCP}) + \sum\nolimits_{\mathbf{j}=1}^{\mathbf{m}} \mathbf{L_j} \tag{3}$$

where, $\mathbf{Q}(\mathbf{MCP})$ is the forecasted cumulative power demand forecasted by PX, made open to all the suppliers/consumers.

The optimal bidding strategy problem is subjected to generator and load inequality constraints.

The generator inequality constraint is expressed as

$$\mathbf{P_{i\,min}} \le \mathbf{P_i} \le \mathbf{P_{i\,max}}, \quad \boldsymbol{i} = 1, 2, \ldots, \boldsymbol{n} \tag{4}$$

The demand inequality constraint is expressed as

$$L_{j\,\min} \le L_j \le L_{j\,\max}, \quad j = 1, 2, \ldots, m \tag{5}$$

The cumulative predicted power demand is expressed (in the linear form) as follows:

$$\mathbf{Q}(\mathbf{MCP}) = \mathbf{Q}_0 - \mathbf{K} * \mathbf{MCP} \tag{6}$$

where $\mathbf{Q}_0$: constant number and K: coefficient.

The solution to Eqs. (1)–(6) are as follows:

$$\mathbf{MCP} = \frac{\mathbf{Q}_0 + \sum_{i=1}^{n} \frac{x_i}{y_i} + \sum_{j=1}^{m} \frac{u_j}{v_j}}{\mathbf{K} + \sum_{i=1}^{n} \frac{1}{y_i} + \sum_{j=1}^{m} \frac{1}{v_j}} \tag{7}$$

Power to be supplied by GENCOs is calculated as follows:

$$\mathbf{P_i} = \frac{\mathbf{MCP} - \mathbf{x_i}}{\mathbf{y_i}}, \quad \boldsymbol{i} = 1, 2, \ldots, \boldsymbol{n} \tag{8}$$

Power allocated to consumer is calculated as follows:

$$\mathbf{L_j} = \frac{\mathbf{u_{j-MCP}}}{\mathbf{v_j}}, \quad \boldsymbol{j} = 1, 2, \ldots, \boldsymbol{m} \tag{9}$$

The objective of optimal bidding strategy is defined as

$$\text{Max.}\, F(\mathbf{x_i}, \mathbf{y_i}) = \mathbf{MCP} * \mathbf{P_i} - \mathbf{C_i}(\mathbf{P_i}) \tag{10}$$

where $\mathbf{C_i}(\mathbf{P_i})$ is fuel cost. In general, it is expressed as follows:

$$C_i(P_i) = f_i P_i^2 + e_i P_i + d_i \tag{11}$$

jth consumer demand (benefit) function is as follows:

$$\mathbf{B_j}(\mathbf{L_j}) = g_j \mathbf{L_j} - h_j L_j^2 \tag{12}$$

where g_j and h_j are demand cost coefficients of jth consumer and $\mathbf{B_j}(\mathbf{L_j})$ is consumer benefit. The aim is to decide $\boldsymbol{x_i}$ and $\boldsymbol{y_i}$ that maximize $\boldsymbol{F}(\boldsymbol{x_i}, \boldsymbol{y_i})$. Similarly, for the jth consumer, the objective is to maximize the benefit. It is defined as follows:

$$\text{Max.}\, H(u_j, v_j) = B_j(L_j) - \text{MCP} * L_j \tag{13}$$

Each GENCO can predict their rivals bidding coefficients using probability distribution function (pdf) [6].

$$\mathrm{pdf}_p(x_i, y_i) = \frac{1}{2\pi\sigma_i^{(x)}\sigma_i^{(y)}\sqrt{1-\rho_i^2}} \times \exp\left\{-\frac{1}{2(1-\rho_i^2)}\left[\left(\frac{x_i-\mu_i^{(x)}}{\sigma_i^{(x)}}\right)^2 - \frac{2\rho_i\left(x_i-\mu_i^{(x)}\right)\left(y_i-\mu_i^{(y)}\right)}{\sigma_i^{(x)}\sigma_i^{(y)}} + \left(\frac{y_i-\mu_i^{(y)}}{\sigma_i^{(y)}}\right)^2\right]\right\} \tag{14}$$

where, $\mu_i^{(x)}$, $\mu_i^{(y)}$, $\sigma_i^{(x)}$ and $\sigma_i^{(y)}$ are the parameters of the joint normal distribution. ρ_i is the correlation between x_i and y_i. The marginal distributions of x_i and y_i are both normal with mean values $\mu_i^{(x)}$, $\mu_i^{(y)}$ and standard deviations $\sigma_i^{(x)}$ and $\sigma_i^{(y)}$, respectively.

3 Invasive Weed Optimization (IWO)

Invasive Weed Optimization (IWO) [11] suggested by Mehrabian and Lucas is a population-based biologically inspired algorithm. It is a derivative-free modern heuristic algorithm that imitates the biological conduct of colonizing weeds. Four stages are involved in this algorithm. They are initialization, reproduction, spatial distribution, and competitive exclusion. The description of these stages is given in [12].

4 Implementation of IWO for Optimal Bidding Strategy

Steps involved in the IWO algorithm for solving bidding strategy are given as follows:

4.1 Step I Input Data

(i) Bidding problem: GENCOs, consumers, fuel cost coefficients, generator output powers, consumer load demands, and demand cost coefficients of consumers.
(ii) Invasive Weed Optimization: Population, iterations, minimum standard deviation, maximum standard deviation, minimum-seeds, and maximum-seeds.

4.2 Step II Initialization

"x" and "u" in Eqs. (1) and (2) are randomly generated and fixed. "y" and "v" values are obtained such that the pdf in Eq. (14) is maximized. The MCP is evaluated for the above values. This process is run for number of times and the best values of "y" and "v" are taken from those runs.

The "y" and "v" are kept constant and "x" and "u" are selected randomly. Profit is the objective function in this case.

4.3 Step III Iterations Start

(i) Determine MCP using Eq. (7).
(ii) Determine powers of GENCOs and consumers using Eqs. (8) and (9).
(iii) Evaluate benefits of both the utilities.

4.4 Step IV Iterations Start for IWO

(i) "x" and "u" are randomly generated. These are population in this problem.
(ii) Profit is fitness function.
(iii) Find minimum and maximum value of fitness function.
(iv) Seeds are generated between minimum and maximum fitness value.
(v) Generated seeds are randomly scattered and Standard Deviation (SD) is calculated.
(vi) Updation of population.
(vii) Determine MCP for updated values of "y" and "v".
(viii) Calculate powers.
(ix) Test the error.
(x) Determine the better result.
(xi) End of iterations for IWO.
(xii) End of iterations.

4.5 Step V Powers and Benefits

The flowcharts of pdf maximization for optimal bidding strategy are shown in Fig. 1.

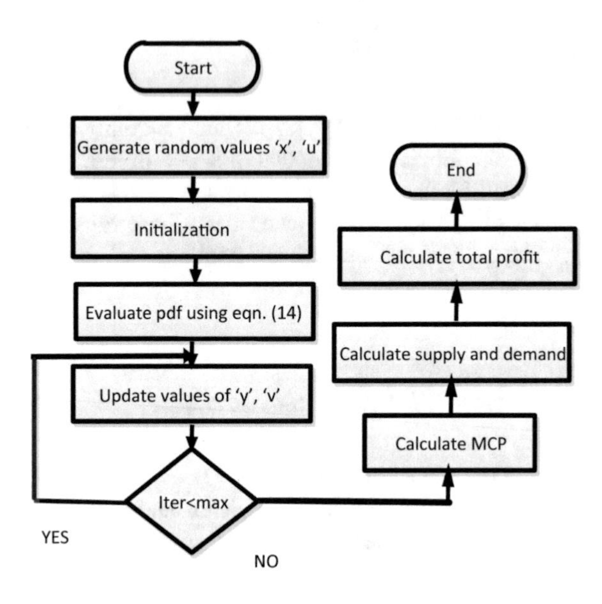

Fig. 1 Flowchart of pdf maximization

5 Simulation Results

The coding of IWO, for optimal bidding problem, is executed in MATLAB. It has been validated on IEEE 30 bus system. The test case consists of six generators and two consumers. The generators and consumers data have been adopted from [13].

While executing the IWO algorithm for optimal bidding problem, different control parameters utilized are shown in Table 1.

The bidding coefficients values, obtained after execution of IWO algorithm for optimal bidding problem, are given in Table 2. The Market Clearing Price obtained by the IWO is 18.6606 $/MWh.

While executing the IWO algorithm, the MCP values obtained for the first 500 iterations are shown in Fig. 2. The best value of the MCP is $18.6606.

The market clearing price plays a vital role in deciding the profit of the bids in optimal bidding strategy. Hence, MCP of various methods is provided in Table 3.

Table 1 Control parameters

S. No.	Control parameter	Value
1	Initial population	100
2	Iterations	100
3	Maximum generations	500
4	Minimum standard deviation	0.05
5	Maximum standard deviation	0.8
6	Minimum number of seeds	4
7	Maximum number of seeds	10

Table 2 Bidding coefficients of suppliers and consumers by the proposed algorithm

Supplier	x	y
1	8.0102	0.0668
2	6.3738	0.1418
3	3.6422	0.2`888
4	11.8371	0.0572
5	10.926	0.1575
6	10.926	0.1575
Consumer	u	v
1	36.4219	0.1080
2	30.3516	0.0810

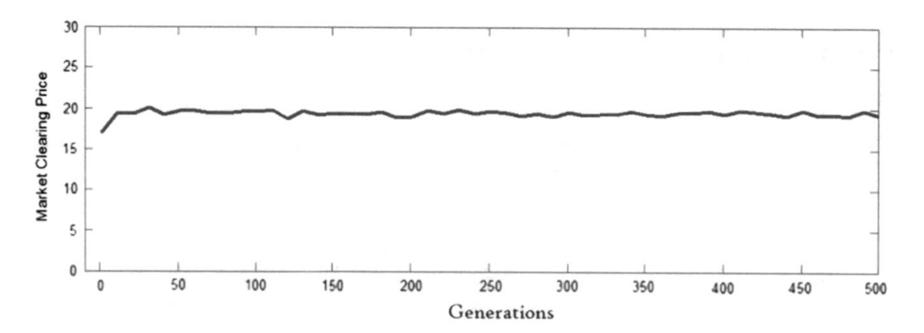

Fig. 2 Variation of MCP with iterations

Table 3 Comparison of market clearing price

	Monte Carlo [10]	GA [10]	PSO [10]	DE [13]	IWO (proposed)
MCP in $/MWh	16.3500	16.3637	16.4729	17.6405	18.6606

The powers with different methods are provided in Table 4.

The profit of generators, consumers, and total profit obtained with IWO algorithm are given in Table 5.

The variation of profit of generators and consumers with iterations is shown in Fig. 3.

The sum of profits attained by the proposed IWO method is higher than the profit gained with Monte Carlo, PSO, GA, and DE algorithms are given in Table 6.

Table 4 The powers (MWs) with various methods

	Powers			
	Monte Carlo [10]	PSO [10]	DE [13]	IWO (proposed)
Generator				
1	160.0000	156.0	160.00	159.4371
2	105.8371	89.37	104.23	86.6488
3	48.5923	45.67	36.66	52.00277
4	120.0000	88.79	120.00	119.2919
5	49.0859	43.09	53.73	49.1085
6	49.0859	43.09	68.49	49.1085
Consumer				
1	170.4639	139.70	180.42	164.4565
2	143.9578	112.06	150.00	144.3333

Table 5 Profit of generators, consumers, and total profit

Profit ($/h)		Total profit ($/h)
Generators	Consumers	
4232.72	1073.02	5305.74

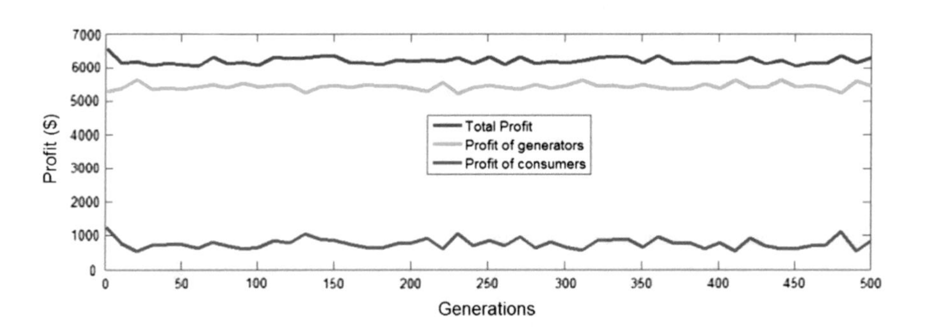

Fig. 3 Variation of profit of generators and consumers with iterations

Table 6 Comparison of profit obtained with different methods

	Monte Carlo [10]	GA [10]	PSO [10]	DE [13]	IWO (proposed)
Profit in $	4723.74	4776.2	4857.41	5076.7	5305.74

6 Conclusion

The optimal bidding strategy problem has been addressed utilizing Invasive Weed Optimization algorithm. The IWO technique is a viable technique and simple to adopt for addressing optimal bidding strategy. From the results, it is evident that the

IWO algorithm is an efficient technique in terms of profit. The outcomes show that the IWO contributes good profit compared to the Particle Swarm Optimization (PSO), Differential Evolution (DE) Genetic Algorithm (GA), and Monte Carlo simulation. The optimal bidding strategy can be developed for GENCOs and consumers.

References

1. Lai, L.L.: Power System Restructuring and Deregulation-Trading, Performance and Information Technology. Wiley, New York (2001)
2. Lamont, J.W., Raman, S.: Strategic bidding in an energy brokerage. IEEE Trans. Power Syst. **12**, 1729–1733 (1997). https://doi.org/10.1109/59.627883
3. Richter Jr., C.W., Shible, G.B., Ashlock, D.: Comprehensive bidding strategies with genetic programming: finite state automata. IEEE Trans. Power Syst. **14**, 1207–1212 (1998). https://doi.org/10.1109/59.801874
4. David, A.K., Wen, F.: Strategic bidding in competitive electricity markets: a literature survey. IEEE Power Eng. Soc. Summer Meet. **4**, 2168–2173 (2000). https://doi.org/10.1109/PESS.2000.866982
5. Song, H., Liu, C.-C., Lawarree, J., Dahlgren, R.W.: Optimal electricity supply bidding by Markov decision process. IEEE Trans. Power Syst. **15**, 618–624 (2000). https://doi.org/10.1109/59.867150
6. Wen, F., David, A.K.: Coordination of bidding strategies in energy and spinning reserve markets for competitive suppliers using genetic algorithm. Power Eng. Soc. Summer Meet., 2174–2179 (2000)
7. Gountis, V.P., Bakirtzis, A.G.: Bidding strategies for electricity producers in a competitive electricity market place. IEEE Trans. Power Syst. **19**, 356–365 (2004). https://doi.org/10.1109/TPWRS.2003.821474
8. Attaviriyanupap, P., Kita, H., Tanaka, E., Hasegawa, J.: New bidding strategy formulation for day-ahead energy and reserve markets based on evolutionary programming. Electr. Power Energy Syst. **27**, 157–167 (2005). https://doi.org/10.1016/j.ijepes.2004.09.005
9. Kumar, J.V., Kumar, D.M.V., Edukondalu, K.: Strategic bidding using fuzzy adaptive gravitational search algorithm in a pool based electricity market. Appl. Soft Comput. **13**, 2445–2455 (2013). https://doi.org/10.1016/j.asoc.2012.12.003
10. Kumar, J.V., Pasha, S.J., Kumar, D.M.V.: Strategic bidding in deregulated market using particle swarm optimization. In: Annual IEEE India Conference, Kolkata, pp. 1–6 (2010). https://doi.org/10.1109/indcon.2010.5712648
11. Mehrabian, A.R., Lucas, C.: A novel numerical optimization algorithm inspired from weed colonization. Ecol. Inf. **1**, 355–366 (2006). https://doi.org/10.1016/j.ecoinf.2006.07.003
12. Sudhakar, A.V.V., Chandram, K., Jaya Laxmi, A.: A hybrid LR—Secant method—invasive weed optimization for profit based unit commitment. Int. J. Power Energy Convers. **9**(1) (2018). https://doi.org/10.1504/ijpec.2018.088256
13. Sudhakar, A.V.V., Chandram, K., Jaya Laxmi, A.: Bidding strategy in deregulated power market using differential evolution algorithm. J. Power Energy Eng. **3**, 37–46 (2015). https://doi.org/10.4236/jpee.2015.311004

Adaptive Fuzzy Logic Controller for the Minimum Power Extraction Under Sensor-Less Control of Doubly Fed Induction Motor (DFIM) Feeding Pump Storage Turbine

Arunesh Kumarsingh and Abhinav Saxena

Abstract This paper presents the minimum power extraction of the Doubly fed Induction machine under motoring mode of operation for feeding the pump storage turbine using sensor-less speed control under different sets of fuzzy rule. The minimum power extraction can be controlled under different fuzzy rules in terms of motor power coefficient after that speed estimated without considering the speed sensor without computing the flux that makes the system more reliable, robust, sophisticated towards error or any unbalancing produced in the system. First, the model of DFIM is developed which includes the grid-side converter (GSC) and rotor-side converter (RSC) which are completely controlled using different fuzzy logic sets under Simulink/MATLAB. Then, motor output feed to pump storage turbine which changes the fluid flow discharge variably with respect to the power flow changes.

Keywords Fuzzy logic · DFIM · GSC · RSC

Symbols

v_s	Voltage of stator
i_s	Current of the stator
v_r	Voltage of the rotor
i_r	Current of the rotor
Φ_s	Stator flux
€	Angle between rotor and stator axis

A. Kumarsingh (✉)
Intelligent Systems Research Lab, University of Saskatchewan, Saskatoon, Canada
e-mail: aru_dei@yahoo.com

A. Saxena
Department of Electrical Engineering, Jamia Millia Islamia, New Delhi, India
e-mail: abhinaviitroorkee@gmail.com

© Springer Nature Singapore Pte Ltd. 2019

H. Malik et al. (eds.), *Applications of Artificial Intelligence Techniques in Engineering*, Advances in Intelligent Systems and Computing 697,
https://doi.org/10.1007/978-981-13-1822-1_40

W_e Angular velocity of rotating flux
W_r Reference angular velocity of rotor
$W_{r(est)}$ Estimated angular velocity of rotor
Ψ_{ds} d Axis stator flux
Ψ_{qs} q Axis stator flux
Ψ_{dr} d Axis rotor flux
Ψ_{qr} q Axis rotor flux
a, b, c Three phases
α, β Reference frame in stationary coordinates
d, q Reference frame in synchronous coordinates

1 Introduction

With the rising demand for energy in the world due to rapid development of nations, the energy crisis is imminent. So, keeping this in mind more number of people are shifting to non-conventional sources of energy or energy based on the renewable sources. As the wind energy is considered to be as one of the major sources to generate electricity cheaply, more and more number of wind farms is placed to generate electricity. With wind source available 24/7 unlike solar power, this makes it more convenient source of energy to generate electricity [1]. The efficiency of these wind farms is a major concern as efficiency is only 33%. And due to the intermittent flow of wind, sometimes it becomes as low as 22%. Keeping all this in mind, today, we are using DFIM (doubly fed induction motor) which is a wound type of rotor of induction motor. This motor provides maximum power at different wind speeds and also has a wide speed control range [2]. To increase the efficiency of this, a better control strategy is to develop and extract the energy. So, we use vector control scheme to increase the efficiency. In this scheme, we separately control the reactive and active power. We convert three-phase to two-DC components and control them easily [3]. The stator of DFIM is connected through the terminals to the grid directly while the rotor terminals are connected via GSC and RSC which depends upon the nature of configuration. The vector control scheme will be applied on the rotor side. The fuzzy logic controller is gently applied to Doubly fed induction motor to obtain the best performance under the different set of rules using different reference frame coordinates like rotor reference frame, synchronous reference frame coordinates. Smoothness over speed control for the long speed range is easily adaptable for the fuzzy logic controller [4]. Basically, the mechanical output of the motor is given to the pump storage motor turbine, which is completely controlled under pitch angle of blades of the turbine using the power coefficient extraction for the extracting the minimum power to generate the maximum mechanical output to maintain the continuous constant flow of the fluid [5].

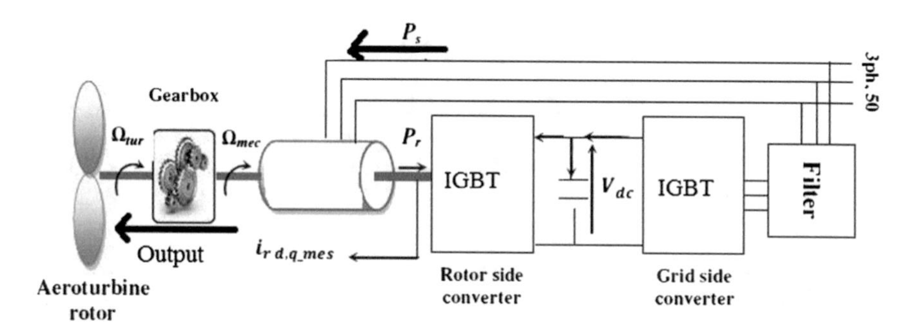

Fig. 1 Layout of doubly fed induction motor

The fuzzy logic controller is used with different set of rules to maintain the constant fluid flow under the unbalance condition or disturbance-like mechanical vibration, grid fluctuations see in Fig. 1 [6–14].

2 Modelling of DFIM

The DFIM is considered to be a wound-type induction motor. To investigate the behaviour of DFIM, we need to consider the reference equations [15]. Therefore, the rotor and stator voltages and current in (*d*-*q*) axis are given as follows:

$$V_{\mathrm{ds}} = R_{\mathrm{s}} i_{\mathrm{ds}} + \frac{\mathrm{d}\varphi_{\mathrm{ds}}}{\mathrm{d}t} - \omega_{\mathrm{s}} \varphi_{\mathrm{qs}} \tag{1}$$

$$V_{\mathrm{qs}} = R_{\mathrm{s}} i_{\mathrm{qs}} + \frac{\mathrm{d}\varphi_{\mathrm{qs}}}{\mathrm{d}t} + \omega_{\mathrm{s}} \varphi_{\mathrm{ds}} \tag{2}$$

$$V_{\mathrm{dr}} = R_{\mathrm{r}} i_{\mathrm{dr}} + \frac{\mathrm{d}\varphi_{\mathrm{dr}}}{\mathrm{d}t} - (\omega_{\mathrm{s}} - \omega_{\mathrm{r}}) \varphi_{\mathrm{qr}} \tag{3}$$

$$V_{\mathrm{qr}} = R_{\mathrm{r}} i_{\mathrm{qr}} + \frac{\mathrm{d}\varphi_{\mathrm{qr}}}{\mathrm{d}t} - (\omega_{\mathrm{s}} - \omega_{\mathrm{r}}) \varphi_{\mathrm{dr}} \tag{4}$$

$$T_{\mathrm{e}} = 1.5p\left(\varphi_{\mathrm{ds}} i_{\mathrm{qs}} - \varphi_{\mathrm{qs}} i_{\mathrm{ds}}\right) \tag{5}$$

$$\varphi_{\mathrm{ds}} = Ls\, i_{\mathrm{ds}} + Lri_{\,\mathrm{dr}} \tag{6}$$

$$\varphi_{\mathrm{qs}} = Ls\, i_{\mathrm{qs}} + Lri_{\,\mathrm{qr}} \tag{7}$$

$$\varphi_{\mathrm{dr}} = Ls\, i_{\mathrm{dr}} + Lr\, i_{\mathrm{ds}} \tag{8}$$

$$\varphi_{qr} = Ls\, i_{qr} + Lri_{qs} \tag{9}$$

$$T_e = T_L + J\, dw_r/dt \tag{10}$$

$$w_r = \int (T_e - T_L)/J\, dt \tag{11}$$

3 Pump Storage Output

Now, we have to see complete aspect of pump storage turbine related with its designing and its mathematical expression which shows the variation of fluid head with power discharge.

3.1 Designing of Pump Storage Turbine

The pump storage turbine is completely driven by DFIM (motoring) which will try to maintain the flow of the fluid variably with respect to the changes the power absorbed irrespective of unbalancing conditions produced using fuzzy set of the controller. Lets defined the torque produced by the motor is:

$$T_m = 0.5^{*}S^{*}\,\rho * C_p(\lambda,\, \beta)^{*}\, V_w^3/w_t \tag{12}$$

Power coefficient C_p performance parameter of the wind turbine [16]. It depends upon the tip speed ratio (λ) and pitch angle (β), swept circular area $S = \pi R^2$

TSR (tip-to-speed ratio) is defined as the ratio of the linear speed of the blades to the wind or fluid speed

$$\mathrm{TSR} = W_{turbine} * R/V_w \tag{13}$$

where R is blades radius, $W_{turbine}$ is linear speed of pump storage motor turbine. V_w is the speed of the motion of the fluid. Power coefficient is given by new strategic relation

$$C_p(\lambda, \beta) = X1\left(\frac{X2}{\lambda i} - X3 * \beta - X4\right)\mathrm{Sin}(X5 * \lambda i) + X6 * \lambda \tag{14}$$

with

$$\frac{1}{\lambda i} = \frac{1}{\lambda + 0.08\beta 2} - \frac{0.035}{1 + \beta^{3.4}} \tag{15}$$

where $X1 = 0.22$, $X2 = 98$, $X3 = 0.28$, $X4 = 6$, $X5 = 29$, $X6 = 0.0058$.

The characteristic between C_p and λ has been plotted for different parametric of the pitch angle (β). For any certain value of the motor turbine, pump turbine output power (P) can be directly controlled by power coefficient (C_p) which directly depends on pitch angle and TSR. Pump-turbine Power (P) generated is fed to the rotor turbine as mechanical input is given as

$$P = 0.5^{*} \mathrm{S}^{*} \rho^{2*} C_{\mathrm{p}}(\lambda, \beta)^{*} V_{\mathrm{w}}^{3} \tag{16}$$

In the above Eq. (14), pump power is directly dependent on the power coefficient (C_p) and rest are constant parameters, if maximum power has to be extracted from the pump, then power coefficient (C_p) has to be maximum under variation of pitch angle and TSR. In the given characteristics ($C_p - \lambda$) as shown in Fig. 2, pitch angle (β) is kept at 0 degree while TSR is 1–14 and it is observed that power coefficient maximization occurs is observed at $\beta = 0$ degree, and it is found that $C_p(\max) = 0.48$ at pitch angle $(\lambda) = 7.64$ for $\beta = 0$.

For the radius of 1 m and suitable value of motor speed and turbine speed with maximum power coefficient is 0.48. The power drawn from the DFIM is 3500 W.

3.2 *Fluid Discharge Variation with Power Absorbed*

We know that the power absorbed by the pump storage turbine is directly proportional to the fluid discharge keeping efficiency of turbine and head constant

Power absorbed, $P = K * Q$. Where, Q is fluid discharge.

The power absorbed is directly dependent on power coefficient as mentioned above accordingly fluid discharge also changing. So we have to choose that power

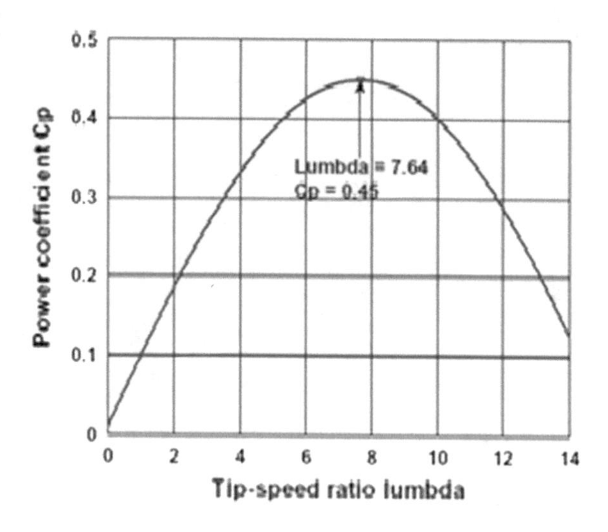

Fig. 2 Variation of the output power coefficient of the pump storage turbine with TSR

Table 1 Variation between power absorbed and fluid discharge with power coefficient

Power coefficient (C_p)	Power absorbed (P) in Watt	Fluid discharge (Q) in m^3
0.36	2800	421
0.40	3250	432
0.48	3500	450
0.52	6800	487
0.56	7220	495

coefficient which has least power coefficient accordingly the minimum power absorption occur see in the Table 1.

4 Fuzzy Logic Controller

We used the complete mechanism of the fuzzy logic controller for different controlling of the DFIM (motoring) see in Fig. 3.

The current in the rotor under three-phase coordinates has been transformed into the two-phase coordinates using the transformation, then the transformed rotor currents applied to the fuzzy logic controller which produces the rotor position angle which can be used for the transformation and variable speed control can be obtained by differentiation the rotor position angle further the rotor speed can be varied using rotor position which is completely controlled by rotor reference frame coordinates under different set of rules of fuzzy logic controller.

Here, the two inputs are given to fuzzy logic controller see in Figs. 4 and 5.

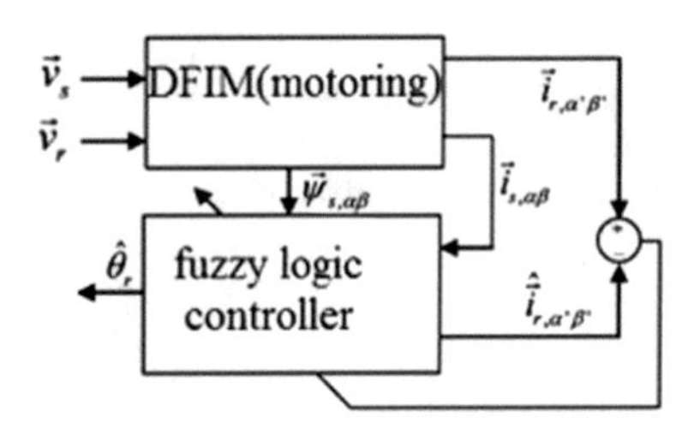

Fig. 3 Layout of DFIM with fuzzy controller

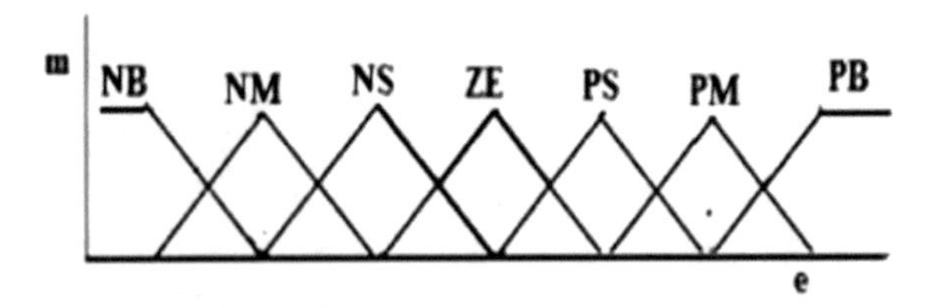

Fig. 4 Membership function varies with error

(a) Error: The produced error is in between the rotor currents and stator currents in the reference frame of the rotor in terms of coordinates which is completely controllable to have good and smooth control on the speed as shown in Fig. 4. Then, error is represented as

$$e = \Delta i_{\mathrm{r}}(\alpha, \beta) - \Delta i_{\mathrm{s}}(\alpha, \beta)$$

(b) Change in error (Δe)

$$\Delta e = \Delta^2 i_{\mathrm{r}}(\alpha, \beta) - \Delta^2 i_{\mathrm{s}}(\alpha, \beta)$$

The above relation shows the change in the error produced between rotor currents and stator currents in the rotor reference frame coordinates as shown in Figs. 5, 6, 7, 8, 9, 10, 11, and 12.

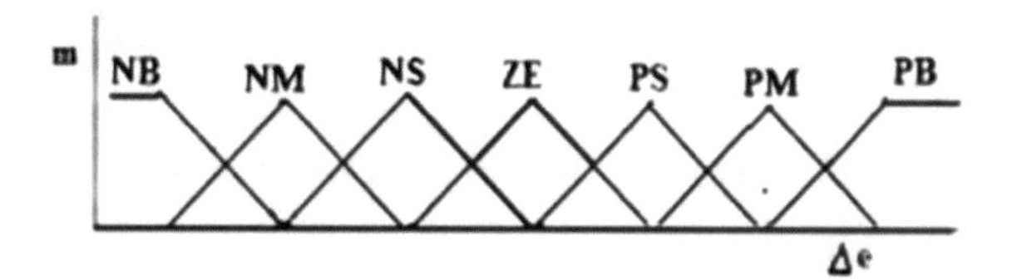

Fig. 5 Membership function varies with change in the error

Fig. 6 Rotor position with membership function. NB = very low, NM = low, NS = medium, ZE = high medium, PS = large, PM = very large, PB = extra large

Δe \ e	NB	NM	NS	ZE	PS	PM	PB
NB	NB	NB	NB	NM	NS	NS	ZE
NM	NB	NM	NM	NM	NS	ZE	PS
NS	NB	NM	NS	NS	ZE	PS	PM
ZE	NB	NM	NS	ZE	PS	PM	PB
PS	NM	NS	ZE	PS	PS	PM	PB
PM	NS	ZE	PS	PM	PM	PM	PB
PB	ZE	PS	PS	PM	PB	PB	PB

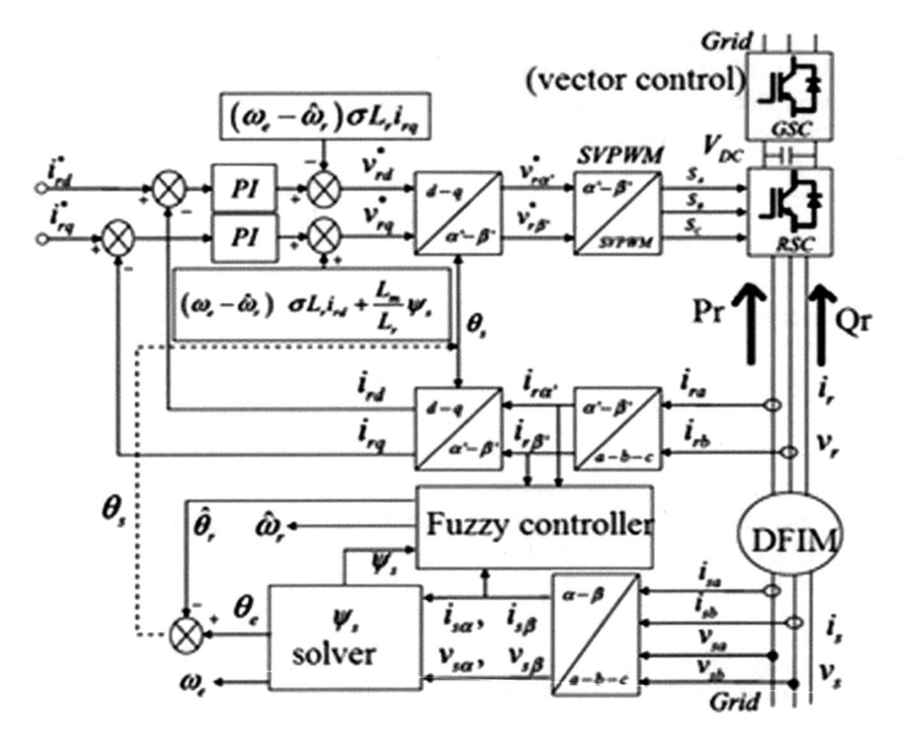

Fig. 7 Fuzzy logic controller for power extraction of the DFIM (motoring) using vector control technique

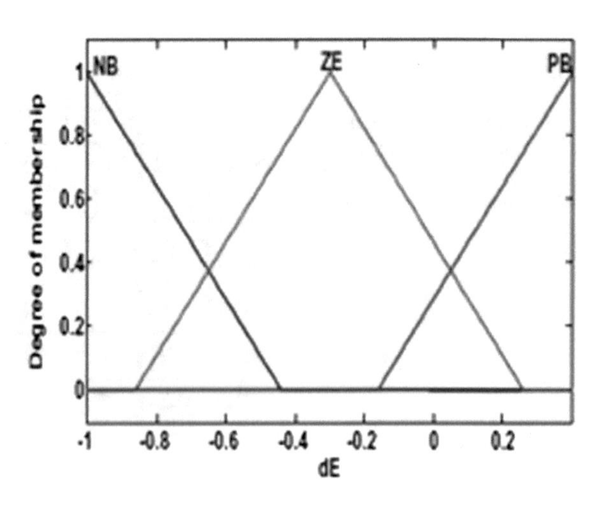

Fig. 8 Graphical representation of change in error

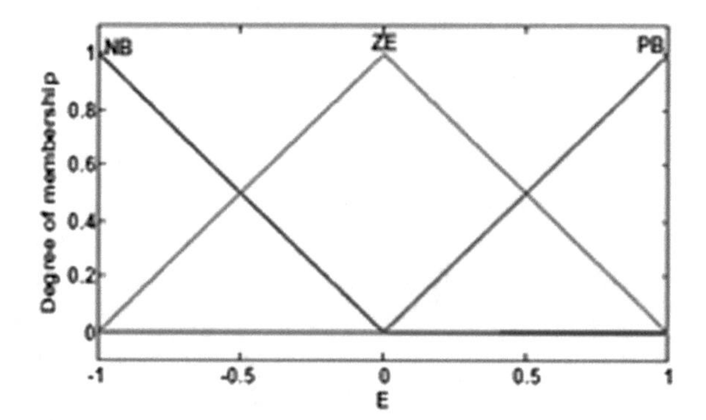

Fig. 9 Graphical representation of error with membership function

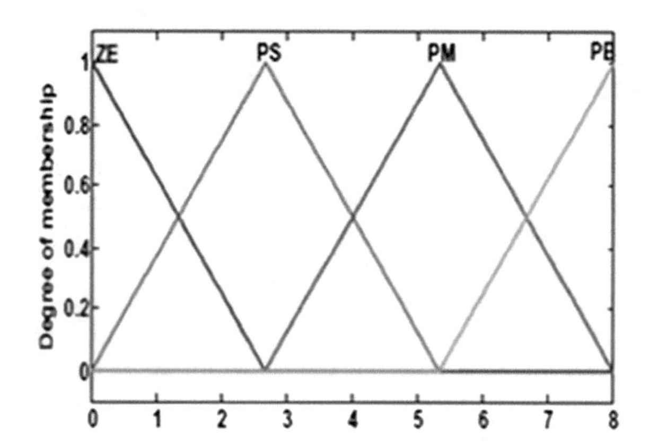

Fig. 10 Graphical representation of rotor position with output

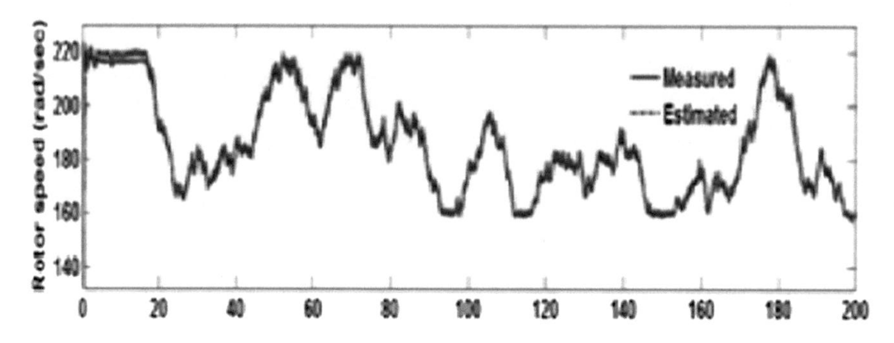

Fig. 11 Speed variation under different membership function

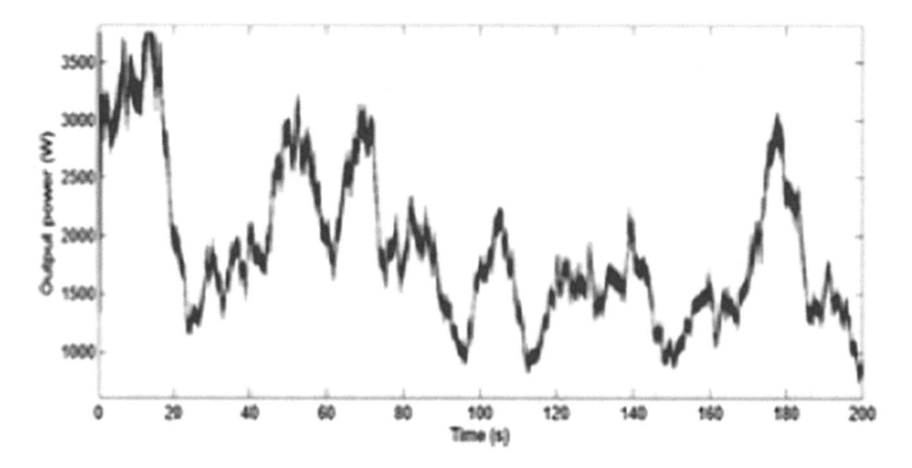

Fig. 12 Power output variation under different membership function

5 Conclusion

The given paper shows real application of fuzzy logic controller to Doubly fed Induction motor for extracting the minimum power to maintain the fluid flow discharge changes with the power absorption of without affecting the vibrations and oscillations produced in the system. The sensor-less operation is observed by taking error between the rotor and stator currents in the synchronous reference frame coordinates and the change in the error is given to fuzzy logic controller and rotor position, and speed will be estimated at the output for the sensor-less mode of the operation applied to the GSC and RSC to make the whole system reliable, economical, and lesser fluctuation in the grid voltage, currents.

6 Future Scope

The above concept of minimum power extraction for doubly fed Induction motor can be applied to other intelligence techniques like ANN, PSO, and genetic algorithm to obtain better performance under sensor-less mode of operation.

Appendix

Doubly fed induction motor:
Three-phase, 400 V, 50 Hz, 5 Hp;
Stator connection: 400 V, Y, 4.8 A;
Rotor Connection: 129 V, Y, 6.6 A.

Stator Rotor
Resistance R_s = 2.13 Ω; Resistance R_r = 2.51 Ω
Inductance L_s = 24.62 mH; Inductance L_r = 25.82 mH
Magnetizing Inductance L_m = 0.2875 H
Magnetizing Inductance L_o = 161.95 mH

References

1. Ghani Varzaneh, S., Abedi, M., Gharehpetian, G.B.: Enhancement of output power smoothing and transient stability of DFIG-based wind energy conversion system using optimized controllers parameters. Electr. Power Syst., 726–738. Taylor Francis (2017)
2. Singh, A.K., Chaturvedi, D., Ibraheem, K., Khatoon, S., Gupta, M.M.: Intelligent controller for eddy current energy absorber of aircraft arrestor barrier system. **25**(1), 1–10 (2017)
3. Ilango, G.S., Nagamani, C., Rani, M.A.A., Reddy, M.J.B.: Sensor-less estimation of rotor position in a doubly fed induction machine. In: Frontier Energy, pp. 171–206. Springer (March 2017)
4. Lin, W.-M., Hong, C.-M., Cheng, F.-S.: Design of intelligent controllers for wind generation system with sensorless maximum wind energy control. Energy Convers. Manage. **52**, 1086–1096. Elsevier (2011)
5. Aurora, C., Ferrara, A.: A sliding mode observer for sensorless induction motorspeed regulation. Int. J. Syst. Sci. **38**(11), 913–929. Taylor & Francis (Nov 2007)
6. Pillai, G.N.: Design and implementation of type-2 fuzzy logic controller for DFIG-based wind energy systems in distribution networks. IEEE Trans. Sustain. Energy **7**(1) (Jan 2016)
7. Rudraraju, V.R.R., Chilakapati, N., Ilango, G.S.: A stator voltage switching strategy for efficient low speed operation of DFIG using fractional rated converters. Renew. Energy **81** (2015)
8. Bedouda, K., Ali-Rachedic, M., Bahid, T., Lakela, R.: Adaptive fuzzy gain scheduling of PI controller for control of the wind energy conversion systems. Energy Proc., 211–215. Elsevier (2015)
9. Belmokhtar, K., Doumbia, M.L., Agbossou, K.: Novel fuzzy logic based sensorless maximum power point tracking strategy for wind turbine systems driven DFIG (doubly-fed induction generator). Energy, 679–693. Elsevier (2014)
10. Kairousa, D., Wamkeue, R.: DFIG-based fuzzy sliding-mode control of WECS with a flywheel energy storage. Electr. Power Syst. Res. **93**, 16–23. Elsevier (2012)
11. Jabr, H.M.: Design and implementation of neuro-fuzzy vector control for wind-driven doubly-fed induction generator. IEEE Trans. Sustain. Energy **2**(4) (Oct 2011)
12. Rüncos, F., Carlson, R., Sadowski, N., Kuo-Peng, P.: Performance analysis of doubly fed twin stator cage induction generator. Front. Energy, 361–373. Springer (2006)
13. Mohammed, O.A., Liu, Z., Liu, S.: A novel sensorless control strategy of doubly fed induction motor and its examination with the physical modeling of machines. IEEE Trans. Magn **41**(5) (May 2005)
14. Xu, L., Cheng, W.: Torque and reactive power control of a doubly fed induction machine by position sensorless scheme. IEEE Trans. Ind. Appl **31**(3). (May, June 1995)
15. Toulabi, M., Bahrami, S.: Modification of DFIG's active power control loop for speed control enhancement and inertial frequency response. IEEE Trans. Sustain. Energy **8**(4) (Oct 2017)
16. Quang, N.P., Dittrich, J.-A.: Nonlinear control structure for wind power plants with DFIM. In: Frontier Energy, pp. 327–343. Springer (May 2015)

Performance Evaluation of Brushless DC Motor Drive Supplied from Hybrid Sources

S. P. Singh, Krishna Kumar Singh and Bhavnesh Kumar

Abstract This paper presents an analysis made on the performance of a Brushless DC motor drive for electric vehicle application powered by two different energy sources. Fuel cells are used to supply the power to the drive. Li-ion battery is also used to supply additional power required in transient states. The drive is tested under motoring as well as regenerative operating conditions. PID controller is used in a speed control loop for better control. Complete model of the drive is modeled in MATLAB/Simulink Software. The hybrid sources are connected through a DC–DC bidirectional converter connected with a VSI.

Keywords PMBLDC motor · PI controller for DC bus · PID speed controller DC–DC bidirectional converter

1 Introduction

Many research work done about sustainable transport like electric vehicle (EV), hybrid electric vehicle (EHV), fuel cell hybrid vehicle (FCHV), and plug-in hybrid vehicle(PHEV) have proven to be effective solution for current energy and environment concerns with continuous improvements in the field of power electronics and energy storage system(ESS) [1, 2]. FCHVs use two energy sources to support their electric power train. In these systems the primary power source is PEMFC and second energy source is energy buffer (battery or super-capacitor). Polymer

S. P. Singh (✉) · K. K. Singh
Rajkiya Engineering College, Ambedkar Nagar, Akbarpur, Uttar Pradesh, India
e-mail: singhsurya12@gmail.com

K. K. Singh
e-mail: krishnakrisingh.90@gmail.com

B. Kumar
Netaji Subhas Institute of Technology, New Delhi, India
e-mail: kumar_bhavnesh@yahoo.co.in

© Springer Nature Singapore Pte Ltd. 2019

H. Malik et al. (eds.), *Applications of Artificial Intelligence Techniques in Engineering*, Advances in Intelligent Systems and Computing 697,
https://doi.org/10.1007/978-981-13-1822-1_41

electrolyte membrane fuel cell (PEMFC) stack is more prominently used in vehicular applications because of its small size, a fast start-up and low operating temperature [3, 4].

There are mainly two type of DC motors based on the mechanism of commutation. One is conventional DC motor where commutation process is implemented through mechanical commutator. Other is brushless DC (BLDC) motor where the permanent magnet produces the necessary flux and commutation is done using electronic commutator. Because of electronics commutation this motor requires less maintenance. BLDC motor essentially is a permanent magnet synchronous motor with back-EMF waveform of trapezoidal nature [5, 6].

Nowadays, BLDC motor drives are gaining popularity for variable speed drive applications in industries and electric vehicle system. BLDC motors are commutator less hence these motor have higher efficiency as compared to commutator motors. The cost of this machine is little high to its counterparts mainly due to the complexity of the control mechanism. BLDC motors have many other advantageous features like wide speed range, fast dynamic response, and noiseless operation. These factors make this motor suitable for the applications where the space available is less and performance required is high. Among its variants, three-phase BLDC motors is most commonly preferred due to high efficiency and low torque ripple [7, 8].

Hybrid electric vehicle is recently getting lot of attention because of low running cost and eco-friendly nature. Despite the high cost, emphasis is on using fuel cell (FC) as a portable, non-conventional energy source in these hybrid electric vehicles but it is costly. Battery is also a portable energy source widely used in electric vehicles. Regenerative braking in these vehicles can improve efficiency by recapturing the energy exhausted during braking and reducing the brake wear [9]. Controller is an integral part of any closed-loop drive. Various research works have been done on the controller designing for BLDC motor drives. These controllers are used to control the speed of the motor. Common types of these controllers are Proportional plus integral (PI), Proportional plus integral plus derivative (PID), GA, fuzzy logic based PID controller. PI controller is the most commonly used speed controller for any drive system. For faster dynamic response in motion control the Artificial Intelligence (AI), Adaptive Neuro-Fuzzy Inference systems are the substitution of the conventional controller like PI, IP, and PID controller [10, 11].

2 System Modeling and Description

Basic block diagram of investigated system using FCs and the battery to supply energy to the motor in an electric vehicle system is shown in Fig. 1. DC–DC converter, Battery, FC, three-phase inverter, BLDC motor, current management system, and the speed controller are important components of this drive system. A boost converter is used to increase the voltage level developed by the FC. Initially the drive is started with battery as FC takes some time for its initialization.

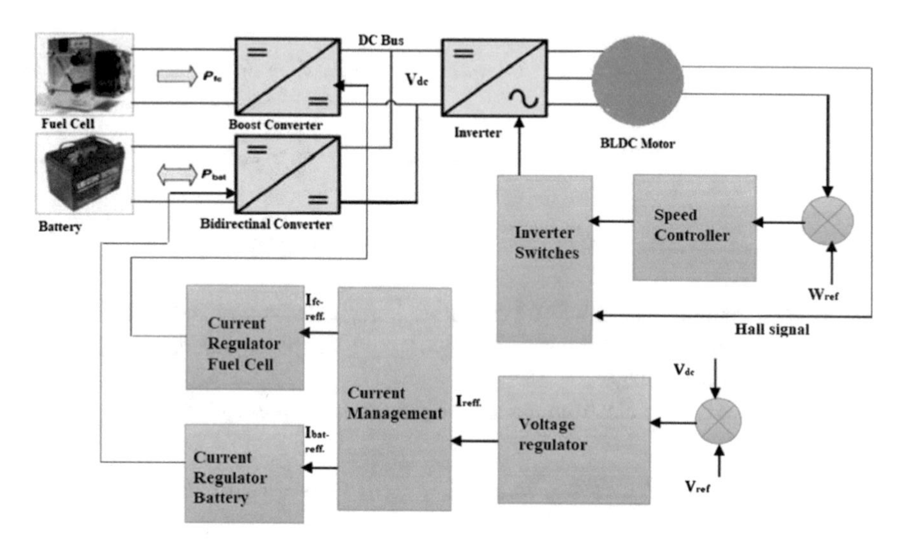

Fig. 1 Schematic of hybrid BLDC motor drive

Basically this time is taken by the FC to achieve the desirable temperature for its efficient operation.

Lithium-ion battery supplies the energy through a bidirectional dc–dc converter. During operation of the drive this bidirectional converter allows discharging and charging of the battery. One of the key challenges in hybrid system is to make the energy balance. FC provides power to the load continuously whereas battery supplies additional power required during the transient operations. In order to maintain the power balance, a current management system during different operating conditions is proposed. As we know that BLDC motors are AC motor because of its sinusoidal property. These motors need high-resolution hall sensors to detect the rotor position at every instant of time to make its operation optimal at every loaded condition and various speeds. With the help of high-resolution hall sensors, we can reduce the operational losses and increase the efficiency of drive system [9, 11]. In case of three-phase BLDC motor, three hall sensors are mounted along with 120° phase differences which are able to detect the rotor position with the help of magnet attached on the rotor shaft. BLDC motor runs at maximum torque when the stator and rotor field lines are perpendicular to each other.

2.1 *Mathematical Model of BLDC Motor Drive*

Mathematical model of BLDC motor with windings in star connection can be expressed using the following equations:

$$v_{ab} = R(i_a - i_b) + L\frac{\mathrm{d}}{\mathrm{d}t}(i_a - i_b) + e_a - e_b \tag{1}$$

$$v_{bc} = R(i_b - i_c) + L\frac{\mathrm{d}}{\mathrm{d}t}(i_b - i_c) + e_b - e_c \tag{2}$$

$$v_{ca} = R(i_c - i_a) + L\frac{\mathrm{d}}{\mathrm{d}t}(i_c - i_b) + e_c - e_a, \tag{3}$$

where: R : Armature resistance, L : Armature inductance, $e_{a,b,c}$: Back-EMF, $i_{a,b,c}$: armature currents flowing through windings, $v_{a,b,c}$: The phase voltages, v_{bc}, v_{ab} and v_{ca}: phase-to-phase voltages

The back-EMF and electromagnetic torque can be expressed as

$$e_a = \frac{k_e}{2}\omega_m F(\theta_e);\ e_b = \frac{k_e}{2}\omega_m F\left(\theta_e - \frac{2\pi}{3}\right);\ e_b = \frac{k_e}{2}\omega_m F\left(\theta_e - \frac{4\pi}{3}\right) \tag{4}$$

$$T_a = \frac{k_t}{2} i_a F(\theta_e);\ T_b = \frac{k_t}{2} i_b F\left(\theta_e - \frac{2\pi}{3}\right);\ T_c = \frac{k_t}{2} i_c F\left(\theta_e - \frac{4\pi}{3}\right), \tag{5}$$

where: k_e is back-EMF constant k_t is electromagnetic torque constant. The function $F(\theta_e)$ is a function of rotor position, which gives the trapezoidal waveform of back-EMF. One period of function can be given as,

$$F(\theta_e) = \begin{cases} 1, & 0 \le \theta_e \le \frac{2\pi}{3} \\ 1 - \frac{6}{\pi}\left(\theta_e - \frac{2\pi}{3}\right), & \frac{2\pi}{3} \le \theta_e \le \pi \\ -1, & \pi \le \theta_e \le \frac{5\pi}{3} \\ 1 + \frac{6}{\pi}\left(\theta_e - \frac{2\pi}{3}\right), & \frac{5\pi}{3} \le \theta_e \le 2\pi \end{cases} \tag{6}$$

2.2 Controllers

A PID speed controller is used to regulate the speed of the drive. While designing the PID controller the gains needs to be properly determined for desired performance. The values obtained for the controller used for speed control of drive are given in Table 1.

Table 1 Speed controller PID values

Controller	k_p	k_i	k_d
PID	18.56	1.456	0.7

In this work for maintaining the DC bus voltage at 400 V another PI controller is used. This voltage is applied to the VSI powering the BLDC motor. The switching of VSI is dependent on the speed and back-EMF of the motor that was detected by the hall sensors. The gain values for this controller are given in Table 2.

2.3 Converter Topology

In this work, two types of converters are used for the fulfillment of power demand of the drive. The first one is voltage source inverter and the second one is DC–DC bidirectional converter. Bidirectional converter is helps to hybridize FC and Li-ion battery. The simulation model of bidirectional converter with hybrid sources is shown in Fig. 2.

Table 2 DC bus voltage PI controller gains

k_p	k_i
4.146	1.13

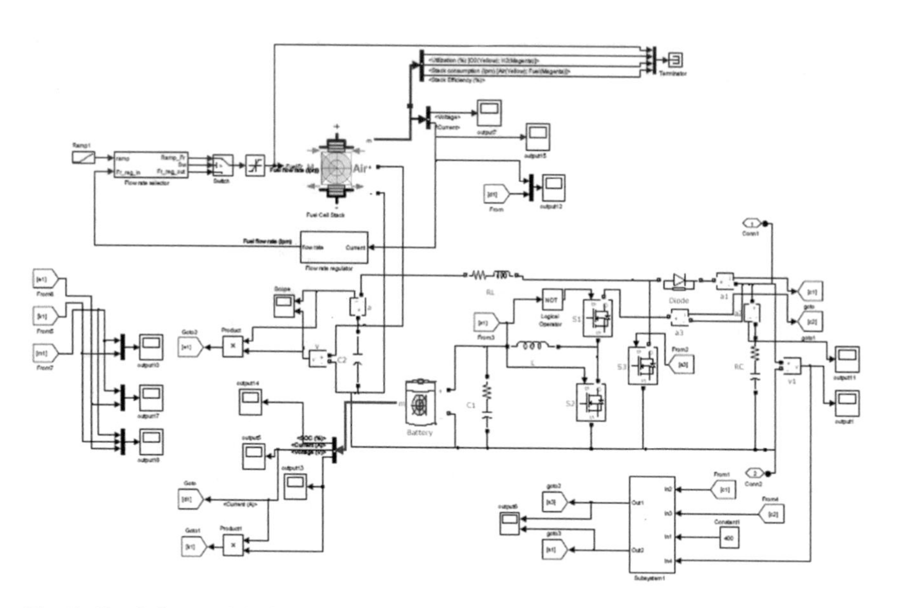

Fig. 2 Simulation model of bidirectional converter with hybrid source

The operation of bidirectional converter can be classified in three modes the first one is discharging mode in which the battery supplies the required power to the BLDC motor and the second one is charging mode in this mode of operation the battery is getting charged. When motor is in regenerative mode the battery gets charged. The third mode of operation is floating mode. In floating mode, the battery as well as the FC supplies the power to the BLDC motor drive system. At initial stage due to the slow chemical reaction of FC it is not capable of driving the motor so, battery is required to supply the power. Between DC–DC bidirectional converter and the voltage source inverter the DC link capacitor is connected to maintain the DC bus voltage constant. The PWM pulses for the bidirectional converter are provided with the help of PI controller. The error signal for PI controller is generated by comparing the reference voltage signal and the actual voltage generated at DC bus.

2.4 PWM Technique for VSI

Generation of PWM pulses for triggering the switches of VSI are generated with the help of position sensor and the PID controller. Triangular wave as a reference signal is compared against the output of controller. Total six pulses are generated for a three-phase VSI.

Table 3 shows the outputs for the back-EMF for clockwise motion and the gate logic to transform EMF to the 6 signal on the gates.

Table 3 Truth table

Hall sensor			EMF			Gate pulses					
A	B	C	A	B	C	Q1	Q2	Q3	Q4	Q5	Q6
0	0	0	0	0	0	0	0	0	0	0	0
0	0	1	0	−1	1	0	0	0	1	1	0
0	1	0	−1	1	0	0	1	1	0	0	0
0	1	1	−1	0	1	0	1	0	0	1	0
1	0	0	1	0	−1	1	0	0	0	0	1
1	0	1	1	−1	0	1	0	0	1	0	0
1	1	0	0	1	−1	0	0	1	0	0	1
1	1	1	0	0	0	0	0	0	0	0	0

Table 4 Parameter of BLDC motor

Parameters	Symbols	Values	Unit
Stator resistance	R	0.846	Ω
Inductance	L	4.28	mH
Number of pole pairs	P	4	poles
Moment of inertia	J	0.00025	(kg m^2)
Flux linkage	ψ_m	21	mwb
Torque constant	K_t	0.168	(N m/A)
Maximum speed	S_{max}	3000	rpm
Friction constant	C	0.0008	(N m s)

3 Results and Discussion

Simulation results for performance investigation of BLDC drive system supplied from hybrid source are discussed under different condition of transient states such as acceleration and de-acceleration condition. The parameters of the BLDC motor are given in Table 4. The important results obtained are as follows.

3.1 Response on Constant Speed of 1000 rpm

In this case, the battery supplies the power up to time of 0.36 s and FC supply power after the time of 0.36 s. State of Charge (SOC) of the battery decrease to level 87.996% when drive speed is settled to 1000 rpm at time 0.2 s. SOC of the battery suddenly decrease up to FC response time, after 0.36 s. the SOC of the battery is increases. So this system works on Hybrid supply at starting time battery supply to the system then FC supply to the system (Figs. 3, 4, 5 and 6).

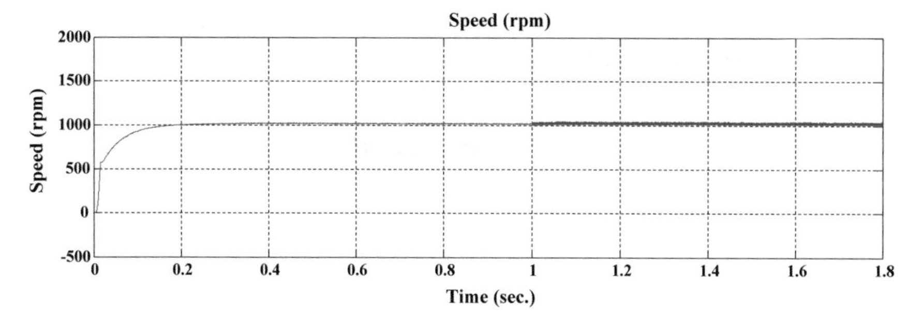

Fig. 3 Speed waveform of BLDC drive for 1000 rpm speed operation

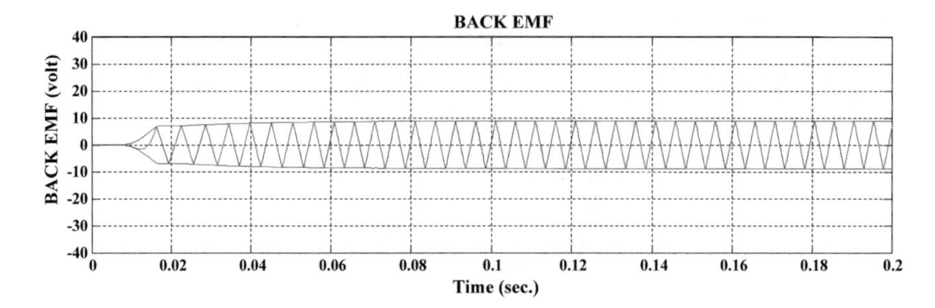

Fig. 4 Three-phase motor back-EMF for 1000 rpm speed operation

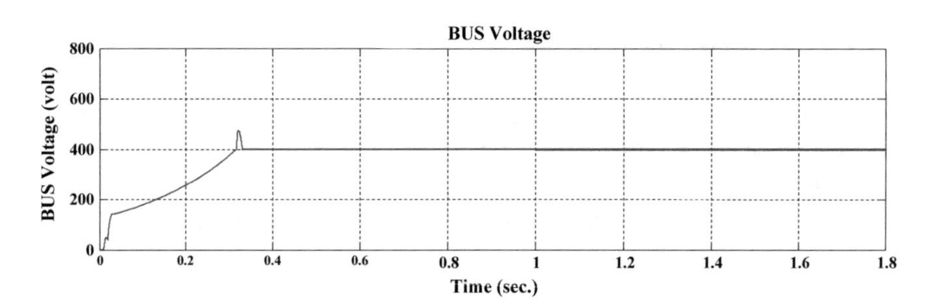

Fig. 5 Bus voltage waveform of drive system for 1000 rpm speed operation

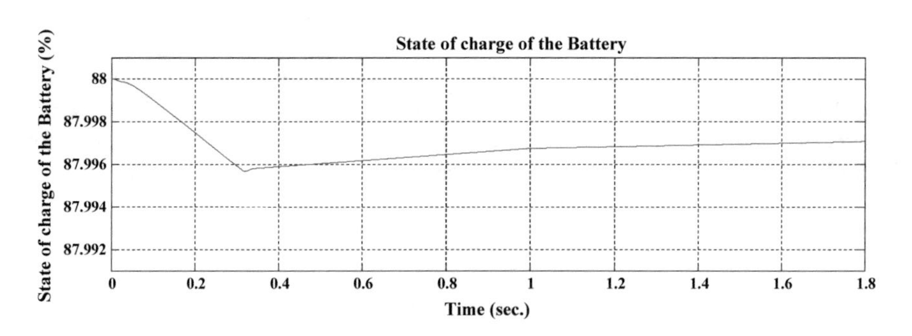

Fig. 6 SOC (%) response of the battery for 1000 rpm speed operation

3.2 Response on Variable Speed Operation

See Figs. 7, 8, 9, 10, 11 and Table 5.

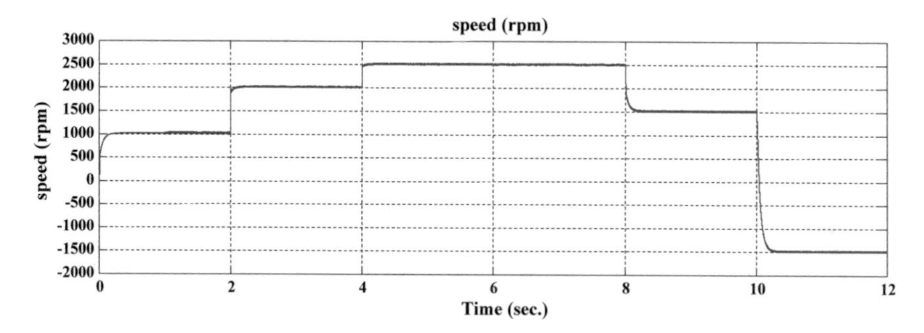

Fig. 7 Speed waveform of BLDC drive on different speed commands

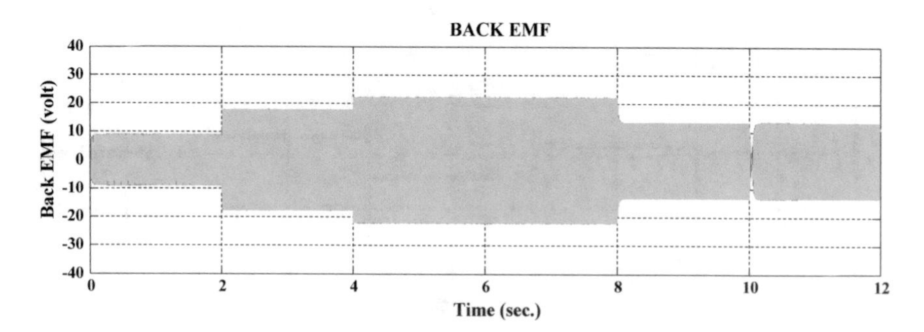

Fig. 8 Back-EMF waveform of BLDC drive with variable speed command

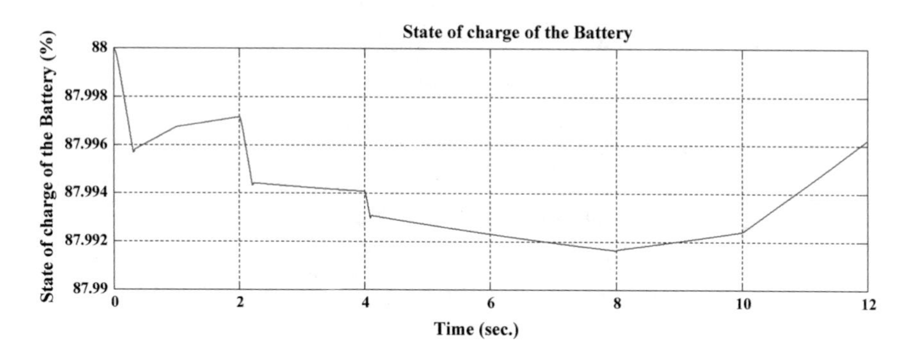

Fig. 9 SOC% the battery with variable speed command

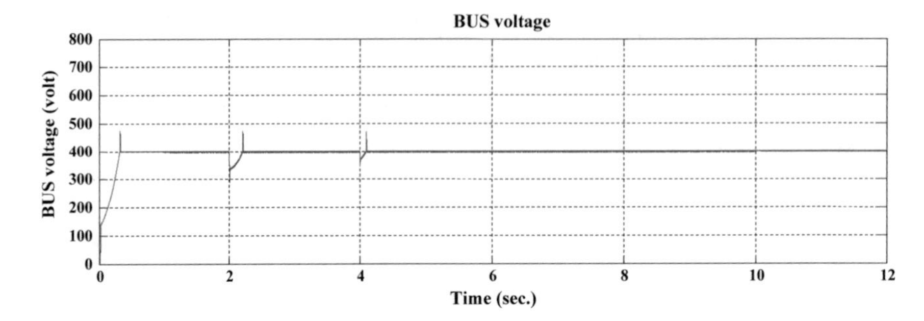

Fig. 10 BUS voltage waveform of drive system with variable speed command

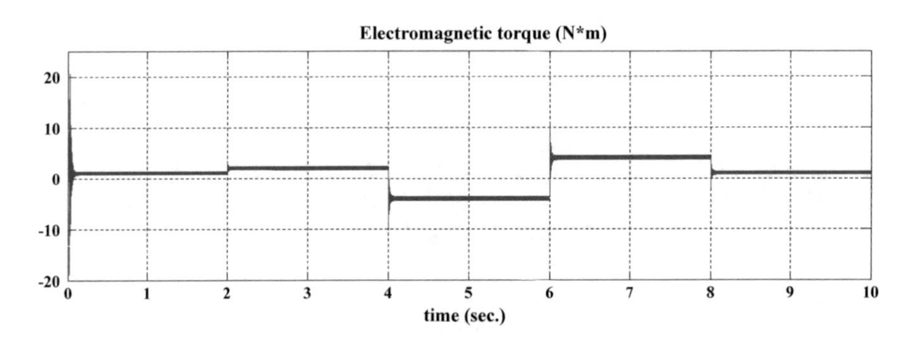

Fig. 11 Electromagnetic torque waveform for variable speed command

Table 5 Performance comparison of hybrid sources at different speed

S. No.	Speed (rpm)	FC current (A)	Battery current (A)	DC bus voltage (V)
1.	1000	16.5 A for 0.36 s then settled at 0.08 A	36.5 A for 0.34 s and settles in 0.36 s	400
2.	2000	16.5 A for 0.2 s then settled at 0.08 A	36.5 A for 0.18 s and settles in 0.215 s	400
3.	2500	16.5 A for 0.08 s then settled at 0.08 A	36.5 A for 0.085 s and settles in 0.125 s	400
4.	1500	0.08 A	−1.7 A settles in 0.02 s	400
5.	−1500	0.08 A	−5 A settles in 0.02 s	400

4 Conclusion

In this paper, performance of a three-phase BLDC motor drive fed from fuel cell and battery is presented. Two different controllers used in drive effectively control the speed and dc link voltage of the drive. The bus voltage is smooth and constant.

Simulation results show that the power flow is managed successfully on various operating conditions. The system is tested for variable speed and load application and also analyzed by dynamically closed-loop operation of BLDC drive.

References

1. Poovizhi, M., Senthil Kumaran, M., Ragul, P., Irene Priyadarshini, L., Lobamba, R.: Investigation of mathematical modelling of brushless dc motor (BLDC) drives by using MATLAB-SIMULINK. In: 2017 International Conference on Power and Embedded Drive Control (ICPEDC), pp. 178–183
2. Bae, J., Jo, Y., Kwak, Y., Lee, D.-H.: A design and control of rail mover with a hall sensor based BLDC motor. In: 2017 IEEE Transportation Electrification Conference and Expo, Asia-Pacific (ITEC Asia-Pacific), pp. 1–6 (2017)
3. Seol, H.-S., Lim, J., Kang, D.-W., Park, J.S., Lee, J.: Optimal design strategy for improved operation of IPM BLDC Motors with low-resolution hall sensors. IEEE Trans. Ind. Electr. **64**(12), 9758–9766 (2017)
4. Seol, H.-S., Kang, D.-W., Jun, H.-W., Lim, J., Lee, J.: Design of winding changeable BLDC motor considering demagnetization in winding change section. IEEE Trans. Magn. **53**(11) (2017)
5. Pongfai, J., Assawinchaichote, W.: Optimal PID parametric auto-adjustment for BLDC motor control systems based on artificial intelligence. In: 2017 International Electrical Engineering Congress (iEECON), pp. 1–4
6. Bhosale, R., Agarwal, V.: Enhanced transient response and voltage stability by controlling ultra-capacitor power in DC micro-grid using fuzzy logic controller. In: 2016 IEEE International Conference on Power Electronics, Drives and Energy Systems (PEDES), pp. 1–6
7. Lu, S., Wang, X.: A new methodology to estimate the rotating phase of a BLDC motor with its application in variable-speed bearing fault diagnosis. IEEE Trans. Power Electr. **99**, 1 (2017)
8. Pany, P., Singh, R.K., Tripathi, R.K.: Performance analysis of fuel cell and battery fed PMSM drive for electric vehicle application. In: 2nd International Conference on Power, Control and Embedded Systems (2012)
9. Renaud, P., Louis, S., Damien, P., Miraoui, A.: Design optimization method of BLDC motors within an industrial context. In: 2017 IEEE International Electric Machines and Drives Conference (IEMDC), pp. 1–7
10. Vinayaka, K.U., Sanjay, S.: Adaptable speed control of bridgeless PFC Buck-Boost converter VSI fed BLDC motor drive. In: 2016 IEEE 1st International Conference on Power Electronics, Intelligent Control and Energy Systems (ICPEICES), pp. 1–5
11. Bist, V., Singh, B.: An adjustable-speed PFC bridgeless buck–boost converter-fed BLDC motor drive. IEEE Trans. Ind. Electr. **61**(6) (2014)

Performance Analysis of Supercapacitor-Integrated PV-Fed Multistage Converter with SMC Controlled VSI for Variable Load Conditions

Shruti Pandey, Bharti Dwivedi and Anurag Tripathi

Abstract This work contains maximum power point tracking fed Photovoltaic (PV) source, in combination with an energy storage device which is supercapacitor. The DC power obtained from PV source is then fed to an inverter controlled by Sliding Mode Controller connected to variable loads. MPPT-controlled boost converter effectively manages the effect of changing solar irradiation. A supercapacitor as a storage device is also deployed to handle intermittent power obtained from the PV source and to meet the stochastic load demands. The implementation of the supercapacitor is facilitated by a bidirectional dc–dc converter with the befitting conversion ratio for interfacing it to the DC link. It bucks and boosts the voltage of supercapacitor and the flow of power is in bidirectional manner for needed charging and discharging of the supercapacitor. The model developed in this paper highlights a seamless action by the storage device for maintaining constant supply under varying irradiation and for varying load conditions. In terms of power quality, the performance of voltage source inverter controlled by sliding mode controller has been found to be superior to PSO-based PI-controlled Voltage Source Inverter. The SMC regulated VSI is found to be robust for any kind of variations at the supply as well as the load ends.

Keywords Photovoltaic (PV) · Particle swarm optimization (PSO) Sliding mode control (SMC) · Voltage source inverter (VSI) · Maximum power point tracking (MPPT) · Supercapacitor · Bidirectional converter

---

S. Pandey (✉) · B. Dwivedi · A. Tripathi
Department of Electrical Engineering, Institute of Engineering and Technology, Lucknow, India
e-mail: shruti.eeee@gmail.com

© Springer Nature Singapore Pte Ltd. 2019

H. Malik et al. (eds.), *Applications of Artificial Intelligence Techniques in Engineering*, Advances in Intelligent Systems and Computing 697,
https://doi.org/10.1007/978-981-13-1822-1_42

1 Introduction

Among all the available non-conventional energy resources the solar energy source is the best suitable resource for almost every power applications. To extract maximum available power under varying irradiation, various algorithms of maximum power point tracking (MPPT) are used. Perturb and Observe (P&O) is chosen over other techniques due to its advantages like low cost and speedy approach to fetch the maximum power.

Inherently PV systems possess low energy conversion characteristics; the MPPT boost converter tracks the maximum power throughout. To get high-quality stabilized output in terms of total harmonic distortion and voltage regulation or in case of transient analysis such as getting stable against variations in supply and load, innovative control approaches are needed to control voltage source inverter(VSI) [1].

The reported controllers are linear and they work reasonably nearby stable point only, to achieve guaranteed total stability of the controlled VSI, the robust nonlinear controllers performs well. Sliding Mode Control (SMC) is nonlinear in nature, which is superior in terms of stability, simplicity, regulation and it shows robust nature under varying operating conditions [2–4]. SMC inherently is variable structure control systems and is best suitable for power converters. To achieve equilibrium point or to get stable and robust response of the system under control, high switching frequency is adopted to make it as close as continuous control.

The supercapacitor has gained a wide attention as complementary source [5, 6] Lower voltage rating and varying terminal voltage issues of supercapacitor are handled satisfactorily with the use of a Bidirectional converter.

Here in the paper, a PV panel is used as a source and is connected to a cascade system with varying loads that are linear and nonlinear in nature. The initial step is to get the maximum power and to boost the voltage of PV source. Variable solar irradiation has been incorporated and an increased DC is obtained as the output voltage of Boost Converter. An energy storage, i.e., supercapacitor is incorporated to complement the PV power under increased loading condition and under a condition when low PV generation is obtained. Thus a desired constant DC-link voltage is obtained in the first stage. In further stage, an inverter controlled by PI or SMC is taken to obtain good class AC power from the DC-link. The inverter is controlled separately by PSO-based PI controller and SMC controller. At load end the linear loads and the nonlinear loads are considered separately.

1.1 Proposed System Configuration

The schematic representation of system considered is presented in the above Fig. 1. The diagram comprises of a PV source providing variation in solar irradiation and its output is given to MPPT interfaced boost converter. Supercapacitor is introduced

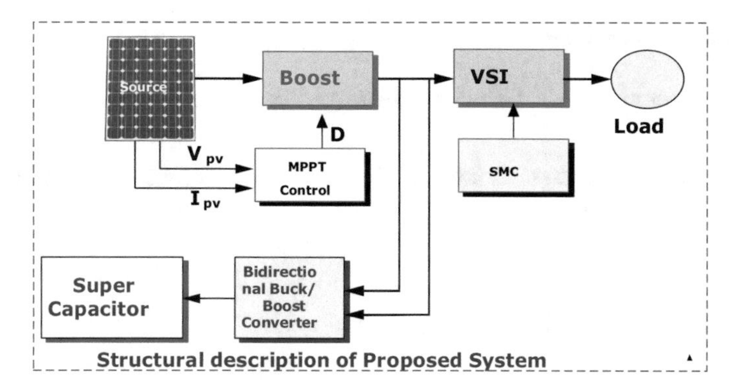

Fig. 1 Block diagram of proposed configuration

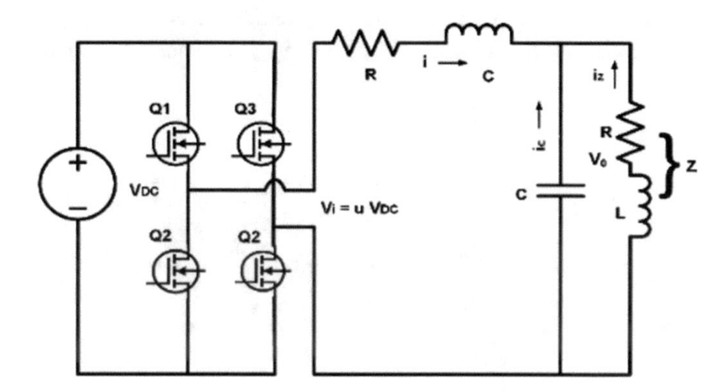

Fig. 2 Inverter

with a suitable battery charger (DC–DC buck-boost converter) to interact with the common bus or DC link connected to the inverter. Further, the PI and SMC fed VSI converts this DC power of the DC bus to a high quality.

Figure 2 shows a single phase VSI for linear load, i.e., RL load, along with an RLC filter. V_o-output of the inverter and u-control variable of the inverter. The state space model of the inverter is used on the sliding surface in Sliding Mode Controller.

Here V_o and i are the state variables. x Consists of state variables.

$$\dot{x} = Fx + Hu\,, \quad F = \begin{bmatrix} -\frac{1}{ZC} & \frac{1}{C} \\ \frac{-1}{L} & 0 \end{bmatrix}, \quad H = \begin{bmatrix} 0 \\ \frac{V_{DC}}{L} \end{bmatrix}$$

The output $V_o C = \begin{bmatrix} 1 & 0 \end{bmatrix}$.

2 Controllers Used

2.1 Supercapacitor Charge Management Controller

Constant current charging is shown in Fig. 3a while constant current discharging is shown in Fig. 3b [7].

2.2 PSO Based PI Controller

This algorithm manages a swarm of particles which represents a candidate function [8]. They follow a behavior in which the velocity is attuned dynamically as per the particle's own as well as its companion's experience. The particles keep track of their coordinates. If that is related to the best solution it is called P_{best} and if the same is related to the global solution it is called G_{best}. The position of particle, '*k*' is

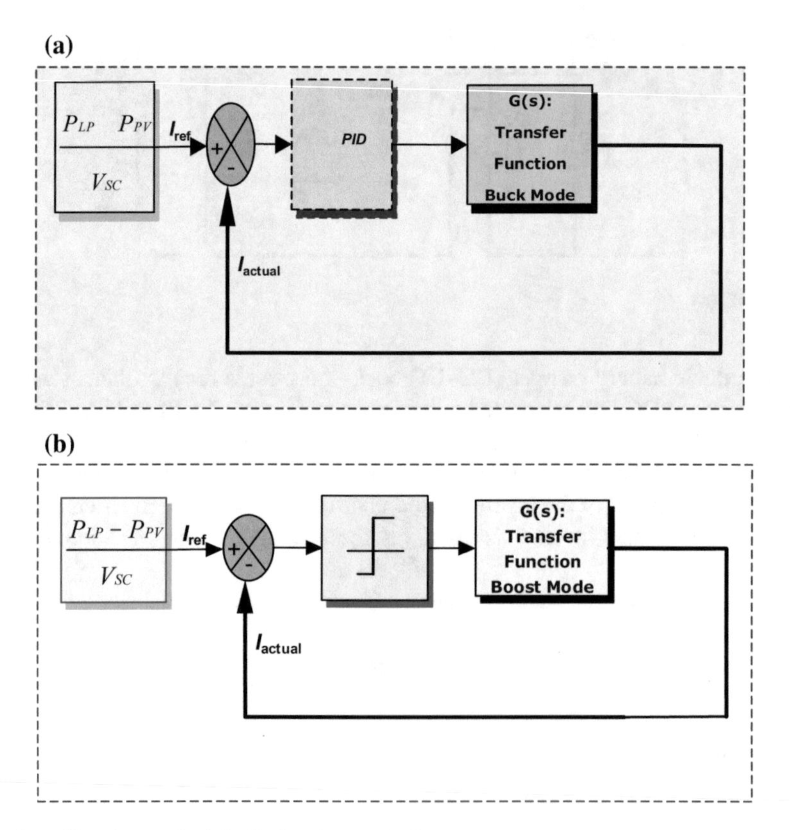

Fig. 3 **a** Charging method, **b** discharging method

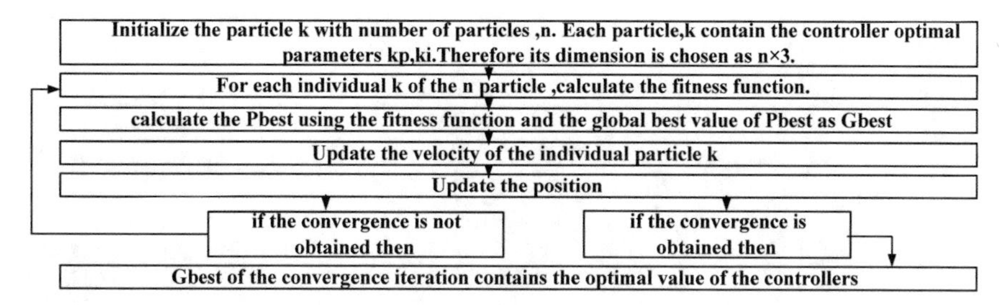

Fig. 4 Algorithm for implementation of PSO based PI controller

adjusted using $k_i^{t+1} = k_i^t + v_i^{t+1}$, where '$v$' is the velocity and can be formulated using $v_i^{t+1} = w.v_i^t + c_i r_i (P_{\text{best}} - k_i^t) + (G_{\text{best}} - k_i^t)$, where 'w' is inertia weight, 'c_1' and 'c_2' are the acceleration coefficients, $r_{1,}\ r_2 \in$ U(0, 1) are random numbers. Above is the algorithm used for the implementation of PSO based PI controller (Fig. 4).

2.3 Proposed Sliding Mode Control

In the proposed SMC (Fig. 5) the current error derived from the error in voltage is used in the inner loop while the voltage error control is used in the outer loop. The Sliding Mode Controller operates in such a way that the voltage across capacitor V_C and the current across inductor i_L tracks closely towards their reference voltage and current.

To get global stability, instead of considering direct voltage and current variables, their respective errors are considered as the state variables for the analysis. The projected surface is sliding in nature and is a linear combination of error state variables

$$\varphi = V_{\text{error}} + mI_{\text{error}} \tag{1}$$

where $V_{\text{error}} = V_{\text{Cref}} - V_C$, $I_{\text{error}} = i_{\text{Lref}} - i_L$ and m is considered as sliding coefficient.

$$i_{\text{Lref}} = \left(k_p + \frac{k_i}{s}\right)[V_{\text{Cref}} - V_C], \tag{2}$$

where K_P and K_i are constants of the PI controller.

The SMC pushes the inverter towards the equilibrium point, and makes $\varphi = 0$

Equating $\varphi = 0$ in Eq. (1)

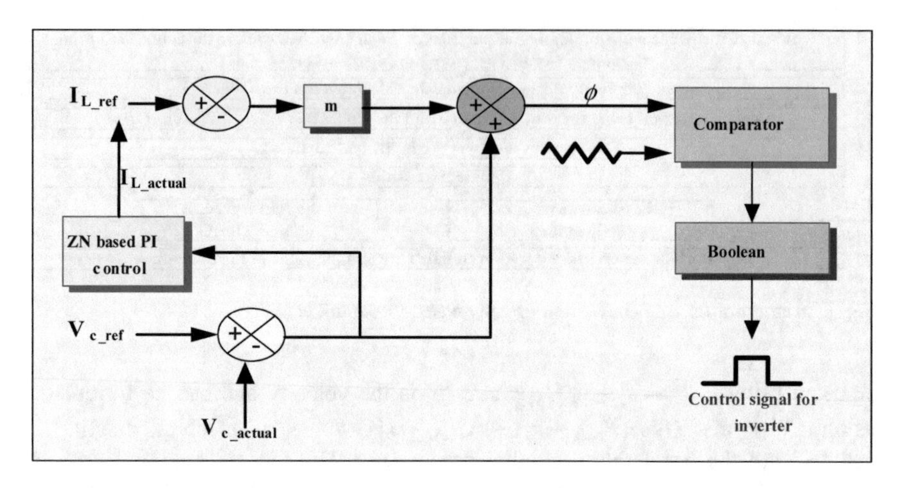

Fig. 5 Proposed control scheme

$$0 = V_{\mathrm{error}} + mI_{\mathrm{error}}$$

Therefore

$$V_{\mathrm{error}} = -mI_{\mathrm{error}} \tag{3}$$

Finding m:
The control effort can be written as

$$u = V_{\mathrm{error}} + mI_{\mathrm{error}} \tag{4}$$

$$u = \begin{bmatrix} 1 & m \end{bmatrix} \begin{bmatrix} V_{\mathrm{error}} \\ I_{\mathrm{error}} \end{bmatrix} \tag{5}$$

As per state variable feedback method

$$u = -kx_{\mathrm{error}} \tag{6}$$

From Eqs. (5) and (6)

$$k = \begin{bmatrix} 1 & m \end{bmatrix}$$

The state space representation with error variables of VSI system is

$$\dot{x}_{\mathrm{error}} = F_{\mathrm{error}} x_{\mathrm{error}} + Hu \tag{7}$$

$$y = Cx_{\mathrm{error}}, \tag{8}$$

where F_{error} is taken from [1] as:

$$F_{\text{error}} = \left[I - H(kH)^{-1}u\right]F \tag{9}$$

Now F_{error} becomes

$$F_{\text{error}} = \begin{pmatrix} -\frac{1}{ZC} & \frac{1}{C} \\ \frac{1}{mZC} & -\frac{1}{mC} \end{pmatrix} \tag{10}$$

Placing (6) in (7)

$$\dot{x}_{\text{error}} = F_{\text{error}} x_{\text{error}} + H(-kx_{\text{error}}) \tag{11}$$

$$\dot{x}_{\text{error}} = [F_{\text{error}} - Hk]x_{\text{error}} \tag{12}$$

$[F_{\text{error}} - Hk]$ can be supposed as F'
m can be calculated with the help of Eigen values

$$\left[\lambda I - F'\right] = 0 \tag{13}$$

$$\lambda^2 + \frac{\lambda}{mC} + \frac{\lambda m V_{DC}}{L} + \frac{\lambda}{ZC} + \frac{mV_{DC}}{LZC} + \frac{V_{DC}}{C} = 0 \tag{14}$$

Now comparing Eq. (14) with the characteristics equation of the system of desired parameters where $T_s = 1 \times 10^{-4}\text{s}, \xi = 0.8$

Then from $T_s = \frac{3}{\xi\omega_n}$, $\omega_n = 3.75 \times 10^4$

Placing above in standard second order characteristics equation $s^2 + 2\xi\omega_n s + \omega_n^2 = 0$

The characteristics equation now comes out to be

$$\lambda^2 + (6 \times 10^4)\lambda + (3.75 \times 10^4)^2 = 0 \tag{15}$$

Substituting V_{dc} = 500 V, impedance calculated from RL load, Z = 528 Ω, L = 1 mH, C = 100 μF in Eq. (14) and from Eqs. (15) and (16) the sliding mode coefficient m = 14.7.

Now the Eigen values are obtained by the following equation:

$$[\lambda I - F_{\text{error}}] = 0 \tag{16}$$

$$\lambda^2 + \frac{\lambda}{mC} + \frac{\lambda}{ZC} = 0 \tag{17}$$

Taking $m = 14.7$, Eigen values calculated is λ_1 = 0, λ_2 = −69.9. These Eigen values will always be located on the left-half side of the s-plane indicating the system stability.

3 Results and Discussion

3.1 Analysis for Linear Load

Figure 6 shows the MATLAB model of the system proposed with SMC and PI controlled VSI for RL Load. The PV source delivers 250 V DC. This voltage obtained is stepped up taking the help of MPPT interfaced boost converter which boosts DC voltage to 500 V. The PV source gives the solar irradiation of 1000 W/m^2 initially. Further the irradiation increases around 1200 W/m^2 and again it is decreased to 500 W/m^2 at the interval of 0.5 and 0.85 s. This varies the voltage of PV source too. Variations are also made in load power to see the effect of load variation on the stability of the system. Load variations are made by placing a parallel RL load in the present load at 0.24 s, due to which the power of load increases from 2500–3000 W, at 0.35 s the parallel load is disconnected due to which power reduces from 3000–2500 W again. The reference value of capacitor voltage is taken as 325 sin ωt. The value of PI controller constants obtained from PSO based technique are $K_p = 0.3$ and $K_i = 42$.

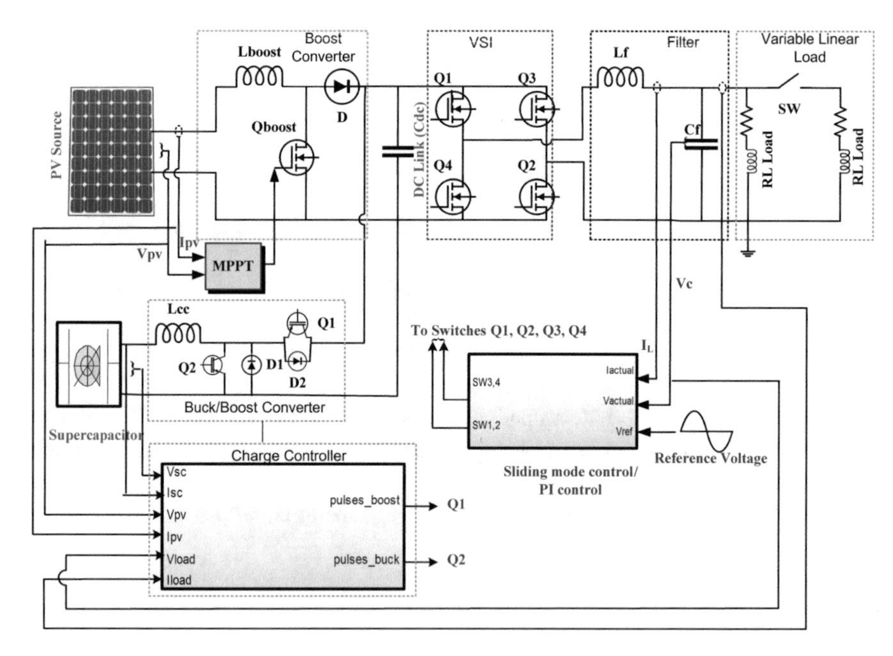

Fig. 6 cascaded converter with SMC and PI controlled VSI

3.1.1 PV Fed Multistage Converter with PSO-Based PI-Controlled VSI for Linear Load

Figure 7a reveals that changes in PV output (Fig. 7a) are efficiently taken care by the supercapacitor as can be seen from Fig. 7a (ii) to meet the load variation in Fig. 7a (i).

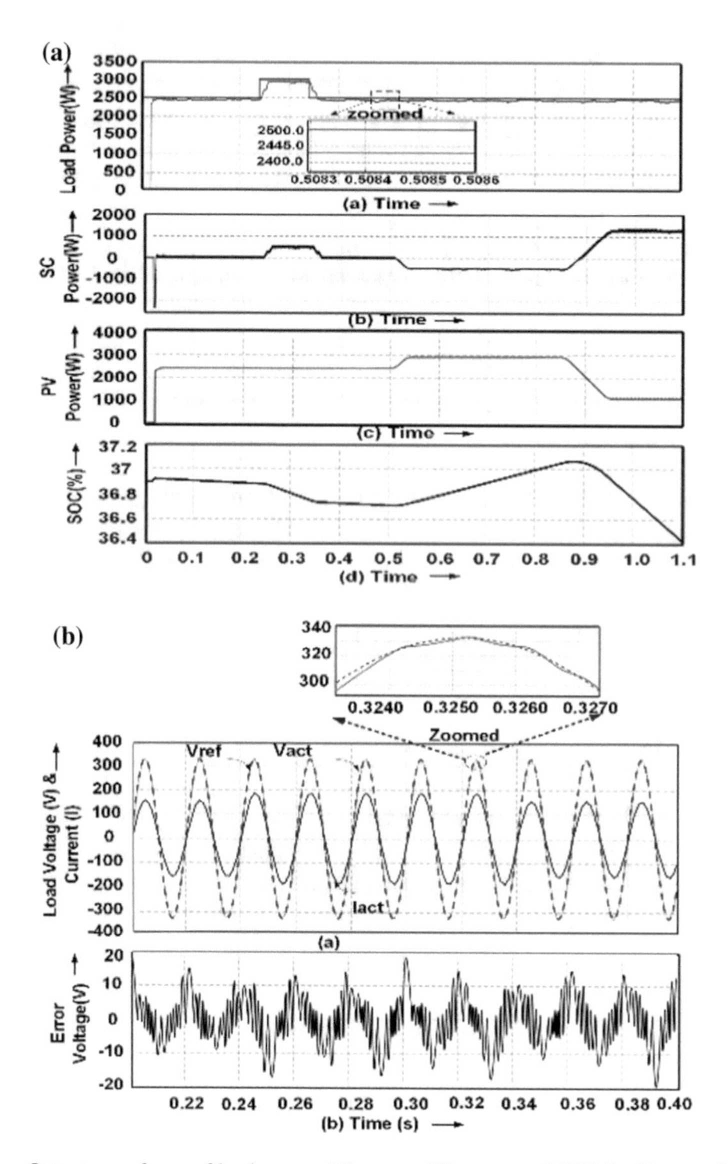

Fig. 7 **a** Output waveforms of load power, SC power, PV power and SOC for PI controlled VSI, **b** PI controlled VSI output-voltage waveform showing Error = 16.52 V

The increment in power of load between 0.24 and 0.34 s as seen in Fig. 7a (i) is met by energy device, supercapacitor which is interfaced with bidirectional converter as shown by Fig. 7a (ii). It can be seen from Fig. 7a (i), (ii), (iii) that the increment of load power is completely met by the supercapacitor. Figure 7a (iv) which is waveform of SCO verifies this fact. Next when there is increment in solar irradiation at 0.5 s from 1000 to 1200 W/m^2 and reduction at 0.85 s from 1200 to 500 W/m^2 Fig 7a (ii), (iii) and (iv) show the charging of that the supercapacitor in the duration from 0.5 to 0.85 s from PV panel maintaining the constant power of 2500 W and when the PV power is reduced between 0.85 and 1.1 s there is discharge in supercapacitor. The dip in the supercapacitor power waveform is due to the initial supercapacitor charging. The zoomed part of power waveform of load in Fig. 7a (i) shows the difference between reference power of load and the actual power of load is 50 W. This difference remains throughout up to 1.1 s.

Figure 7b (i) shows the actual capacitor voltage, reference voltage, and the current waveforms. Figure 7b (ii) waveform shows the error Voltage between ref and actual voltage. The THD in the output voltage of PI controlled VSI is shown is 2.03%.

3.1.2 PV-Fed Multistage Converter with SMC-Controlled VSI for Linear Load

In comparison to that of PI control, the difference between the actual and reference power is found to be 2 W in SMC controlled VSI which is far less than 50 W as obtained in PI. The error obtained is equal to 1.2 V in SMC controlled VSI as against 16.52 V in PI controlled VSI. The THD in the output voltage of SMC controlled VSI is 0.25%.

3.2 Analysis for Nonlinear Load

In the real-world loads use of nonlinear loads such as PCs, TVs, fax machines, computers, refrigerators, printers, PLCs, and electronic lighting ballasts, etc.

The study, therefore, has been carried out for nonlinear loads (Bridge rectifier) also.

3.3 Comprehensive Analysis of Findings

Table 1 shows the comparative analysis of both the controllers PI and SMC, used to control VSI connected to linear as well as nonlinear loads in conjunction with a PV source and supercapacitor.

Table 1 Comparative table

Type of multistage system	SMC fed VSI for linear load	PI fed VSI for linear load	SMC fed VSI for non linear load	PI fed VSI for non linear load
THD (%)	0.25	2.03	0.25	2.17
V_{err} (max) (V)	1.1	16.52	1.3	21.54
Steady-state error in power (W)	2	50	4	100

4 Conclusion

In terms of performance under identical load conditions, PSO-based PI-controlled VSI clearly underperforms SMC-controlled VSI-based PV-fed multistage converter in terms of both THD and output-voltage error. Steady-state error between demand and supply is also found to be high with PI controllers. Thus for a microgrid system, SMC-controlled VSI along with MPPT-controlled PV generation can be a better option. Further, it was interesting to note that the controllers take due care of suitably integrating the supercapacitor as the storage device to neutralize the intermittency of PV source.

Acknowledgements Authors are very thankful to the Department of Electrical Engineering, Institute of Engineering and Technology, Lucknow, India to provide the world level platform for this study.

References

1. Gudey, S.K., Gupta, R.: Second order sliding mode control for a single phase voltage source inverter. IET Power Electronics (2014)
2. Utkin, V.: Sliding Modes in Control Optimization. Springer, Berlin (1992)
3. Pandey, S., Dwivedi, B., Triapthi, A.: Closed loop boost converter control of induction motor drive fed by solar cells. ICETEESES, 11–12 Mar 2016
4. Gudey, S.K., Gupta, R.: Sliding mode control in voltage source inverter based higher order circuits. Int. J. Electron. **102**(4), 668–689 (2015)
5. Sikkabut, S., et. al.: Control strategy of solar/wind energy power plant with supercapacitor energy storage for smart DC microgrid. In: 2013 IEEE 10th International Conference on Power Electronics and Drive Systems (PEDS)
6. Weddell, A.S., Merrett, G.V., Kazmierski, T.J., Al-Hashimi, B.M.: Accurate supercapacitor modeling for energy harvesting wireless sensor nodes. IEEE Trans. Circuits Syst. II Exp. Br. **58**(12), 911–915 (2011)
7. Wang, Z., Li, X., Li, G., Zhou, M., Lo, K.L.: Energy storage control for the photovoltaic generation system in a micro-grid. In: 2010 5th International Conference on Critical Infrastructure (CRIS)
8. Wang, X., Zhou, X., Wang, H.: Optimizing PI controller of the single-phase inverter based on FOA. In: 2017 2nd International Conference on Robotics and Automation Engineering

Optimization of Process Parameters of Abrasive Water Jet Machining Using Variations of Cohort Intelligence (CI)

Vikas Gulia and Aniket Nargundkar

Abstract Abrasive water jet machining is a non-conventional machining process based on sending abrasive material accelerated with high pressure water on to the planes of focused materials with the purpose to cut various engineering materials. Abrasive water jet machining process has various machining process parameters, which in turn will affect the performance parameters. The combination of all the process parameters results in desired output. Hence it is important to find the optimal combination of process parameters. Several optimization techniques have been used to optimize these parameters. Cohort intelligence (CI) is a socio-inspired algorithm based on artificial intelligence conceptions. Further researchers have developed seven variations of cohort intelligence algorithm. The present work investigates the application of four variations of cohort intelligence for the AWJM process parameter optimization. Variations of CI have been applied for the first time in manufacturing optimization. The considered problem involves optimization of commonly used responses Surface Roughness (Ra) and kerf and results are compared with Firefly Algorithm (FA). The performance of cohort intelligence algorithm is found to be much better than firefly algorithm for four variations.

Keywords Abrasive water jet machining · Variations of cohort intelligence
Kerf · Surface roughness (Ra)

V. Gulia (✉) · A. Nargundkar
Symbiosis Institute of Technology, Symbiosis International
(Deemed University), Lavale, Pune 412115, India
e-mail: vikas.gulia@sitpune.edu.in

A. Nargundkar
e-mail: aniket.nargundkar@sitpune.edu.in

© Springer Nature Singapore Pte Ltd. 2019

H. Malik et al. (eds.), *Applications of Artificial Intelligence Techniques in Engineering*, Advances in Intelligent Systems and Computing 697,
https://doi.org/10.1007/978-981-13-1822-1_43

1 Introduction

Abrasive water jet machining (AWJM) is a non-traditional process in which machining is achieved through kinetic energy of water and abrasive particles. It has many advantages over the other machining methods, i.e., intricate shapes can be easily machined with the help of pressurized water by adding abrasive material, and heat-affected zone does not takes place. AWJM was introduced to manufacturing industry a decade ago and has been widely used for machining hard-to-machine and multi-layered materials. Since then it has been used as an alternative tool for milling, turning, drilling [4]. Several materials can be processed with the help of abrasive particles [1]. The major problems of AWJM process while machining the composites are kerf taper, average surface roughness and delamination. The standard of machined components is determined by kerf taper and the average surface roughness [8].

Abrasive water jet machining process has various machining process parameters, i.e., traverse speed, abrasive flow rate, standoff distance, water pressure, and abrasive size. These parameters in turn will affect the performance parameters, like kerf width, surface roughness, taper angle, and material removal rate. It is vital to determine the optimum combination of these process parameters for better process performance. It is observed from past literature that meta-heuristic algorithms like particle swarm optimization (PSO), firefly algorithm (FA), ABC, Simulated annealing (SA), and genetic algorithm (GA) are well-suited for AWJM [7].

The previous work employs a non-traditional optimization technique FA to obtain the optimal solution for electrical discharge machining (EDM) and AWJM process. By using optimum values of objective function, Ra and Kerf has been obtained with corresponding process parameters. The consequences using FA for parameters of EDM and AWJM were proven to be better as compared with the past research [7].

A new optimization approach—Cohort intelligence (CI) based on artificial intelligence has been developed by Kulkarni et al. [3]. In this algorithm, a cohort is formed which is a group of members who interact and compete with one another aimed to attain a predefined goal. Every candidate attempts to enhance its own performance by perceiving the behavior of other candidates in that cohort. As a result, every candidate learns from one another and overall cohort performance gets evolved. The best possible value of goal is said to be attained if objective function converges to optimal value for substantial number of iterations or the performance of few candidates fail to improve significantly [3]. The application of this algorithm includes trusses [9].

The present work investigates application of four variations CI for typical AWJM responses surface roughness and kerf. In Sect. 2, variations of CI are explained. Objective functions and constraints are explained in Sect. 3. Section 4 describes algorithmic parameters, results obtained with variations of CI along with statistical evaluation and computational time. Further results are compared with FA. Convergence plots are also given.

2 Variations of Cohort Intelligence

Variants of CI are proposed by Patankar et al. These are: roulette wheel selection, follow better, follow median, follow best, follow worst, follow itself, and alienate random candidate—and—follow random from remaining. [5] had tested variations of CI for several uni-modal as well as multimodal unconstrained test functions, and investigated the results for performance of variation for a specific category of problem. The results are then matched with other evolutionary algorithms such as PSO, CMAES, Artificial bee colony, CLPSO, and differential evolution algorithm, etc.

2.1 Steps in the Variations of the CI Algorithm

We consider an unconstrained minimization problem of N dimensions where $f(Z)$ is the objective function defined for the variables $Z = [(z_1, \ldots z_i, \ldots, z_N)]$

$$\begin{aligned} &\text{Minimize } f(Z) = f(z_1, \ldots z_i, \ldots, z_N) \\ &\text{Subject to } z_1, \ldots z_i, \ldots, z_N \geq 0 \end{aligned} \tag{1}$$

In the context of CI, elements of Z are considered as the qualities of each candidate in a cohort. In the present work, these variables denote the machining process parameters. Each of these machining parameters have their upper and lower limits based on the experimental study. These are the sampling intervals defined for $Z^C = \left[z_1^c, z_2^c, \ldots, z_N^c\right]$ which results in its own behavior $f(Z^c)$.

Step 1: As per the requirement of CI algorithm, we decide the number of candidates, and reduction factor $r \in [0, 1]$. The algorithm allows the variables to shrink or expand their sampling interval based on this reduction factor.
Step 2: The value of $f^*(Z^c)$ is computed and the behavior selection probability of every candidate c is obtained as follows:

$$P^C = \frac{1/f*(Z^c)}{\sum\limits_{c=1}^{C} 1/f*(Z^c)} (c = 1, \ldots, C) \tag{2}$$

Step 3: Seven different rules for following pattern are proposed by Patankar et al. out of which four variations are considered in the present work.

(a) **Follow best rule**: The candidate c which has the highest probability (i.e., $P = max(P^c), c(c = 1, ..., C))$ is followed in this rule. The corresponding qualities/features of a selected candidate, $X^{c[t]} = [Z_1^{c[t]}, Z_2^{c[t]}, \ldots, Z_N^{c[t]}]$ are followed.

(b) **Follow better rule**: In this rule, each candidate c follows randomly a candidate from the group of candidates having probability of selection better than itself. This implies that each candidate discards candidate with probability lower than itself.
(c) **Roulette wheel selection rule**: In this method, a random number between 0 and 1 is generated. The probabilistic roulette wheel approach is used to decide the following behavior . the roulette wheel selection approach may improve the possibility of any candidate to choose the behavior which is better than itself, as the corresponding probability P^c calculated using Eq. (2) is directly proportional to the value of $f^*(Z)$.
(d) **Alienation-and-random selection rule**: For this technique, every candidate c ignores one candidate at the start of the learning efforts, i.e., the candidate $\boldsymbol{c}$ never follows the alienated candidate in any iteration. The following behavior of the candidates is random and excludes the alienated candidate.

Step 4: Once following pattern is decided, each candidate contracts its sampling interval in the neighborhood of followed candidate. The candidates update the behavior of cohort with candidate C. Steps 1 to 4 are repeated till saturation or convergence is achieved [5].

3 Application of Variations of CI for AWJM Optimization

A case study is considered based on Kechagias et al. [2] work for the present work. Kechagias et al. [2] experimented TRIP steels on AWJM process and used Taguchi L_{18} orthogonal array to perform experiments considering four process parameters (i.e., “thickness”, “nozzle diameter”, “stand-off distance” and “transverse speed”). The response parameters selected for experiments were kerf and Ra. The bounds of different process parameters considered are as used by Kechagias et al. [2] and are given below:

a. Thickness (z_1): (0.9, 1.25 mm),
b. Nozzle diameter (z_2): (0.95, 1.5 mm),
c. Standoff distance (z_3): (20, 96 mm),
d. Transverse speed (z_4): (200, 600 mm/min).

Rajkamal Shukla et al. [6] has considered the same case study of Kechagias et al. [2] and experimented on two of advanced manufacturing processes EDM and AWJM. They have used response surface methodology to find out optimum process parameters using FA. Objective functions are as follows:

$$\begin{aligned} MinRa = & -23.309555 + 16.6968z_1 + 26.9296z_2 + 0.0587z_3 \\ & + 0.0146z_4 - 5.1863z_2^2 - 10.4571z_1z_2 - 0.0534z_1z_3 \\ & - 0.0103z_1z_4 + 0.0113z_2z_3 - 0.0039z_2z_4 \end{aligned} \quad (3)$$

$$\begin{aligned} Min\,Kerf = & -1.15146 + 0.70118z_1 + 2.72749z_2 + 0.00689z_3 - 0.00025z_4 \\ & + 0.00386z_2z_3 - 0.93947z_2^2 - 0.25711z_1z_2 - 0.00314z_1z_3 \\ & - 0.00249z_1z_4 + 0.00196z_2z_4 - 0.00002z_3z_4 - 0.00001z_3^2 \end{aligned} \quad (4)$$

Both these functions are multi-variable unconstrained optimization problems with bounds specified by Kechagias et al. [2]. In the present work of AWJM, the four variations of CI viz. Roulette wheel based approach, Follow better rule, Follow best rule and Alienation-and-random selection rule are applied for the objective functions developed for Kerf and Ra. The same objective functions developed by Rajkumar Shukla et al. [6] are considered for the optimization using variation of cohort intelligence.

4 Results and Discussion

Algorithms for variations of CI for both the objective functions Ra and Kerf was coded in MATLAB(R2016b) and simulations were run on a 2 GB RAM with Windows OS.

The important parameters for designing CI are no. of candidates, reduction factor and no. of iterations. In this work, no. of candidates has been kept constant as 5. Table 1 represents vital parameters for algorithm viz. reduction factor "r" and no. of iterations that has been used in coding. These parameters were decided based on the preliminary trials of the algorithm for solving the problem. Table 2 shows the optimum solution for objective function Ra and Kerf with best value, mean value and standard deviation.

Table 3 compares the solutions obtained from both CI Algorithm and FA in terms of the best solution. FA shows optimum values as for kerf 0.3704 mm and for Ra 4.443 mm respectively. With variations of CI best value for Kerf is 0.3392 mm and for Ra 4.3826 mm respectively. The CI was measured for the computational time and function evaluations (FE). The FE required for CI algorithm is shown in Table 4. The convergenc eplots for the functions Ra and Kerf with "Roulette Wheel" approach is presented in Figs. 1 and 2 respectively. At the start the performance of the individual candidates is notable. As sampling interval decreases

Table 1 CI parameters

Variation	Ra		Kerf	
	Reduction factor r	Iterations	Reduction factor r	Iterations
Roulette wheel	0.90	150	0.96	400
Follow best	0.96	200	0.95	300
Follow better	0.96	200	0.96	300
Alienation	0.90	150	0.95	300

Table 2 Consolidated results for Ra and Kerf using CI variations

Objective function		Roulette	Follow best	Follow better	Alienation
Surface roughness (Ra)	Best	4.386392	4.382609	4.382611	4.382610
	SD	0.029139	0.024858	0.024710	0.022630
	Mean	4.441096	4.414747	4.406331	4.448886
	Run time in sec.	0.035	0.038	0.039	0.044
Kerf	Best	0.339250	0.339253	0.339261	0.339265
	SD	0.003380	0.003236	0.006524	0.010519
	Mean	0.341720	0.340251	0.343383	0.347712
	Run time in sec.	0.045	0.042	0.048	0.059

Table 3 Comparison with FA

Problem	FA	RW	Follow best	Follow better	Alienation
Ra	4.443	4.38639	4.38260	4.38261	4.38261
Kerf	0.3704	0.339250	0.339253	0.339261	0.339265

Table 4 Function evaluations comparisons of different variations

Objective function	Roulette	Follow best	Follow better	Alienation
Surface roughness (Ra)	3001	4001	4001	3000
Kerf	8001	6001	6001	6001

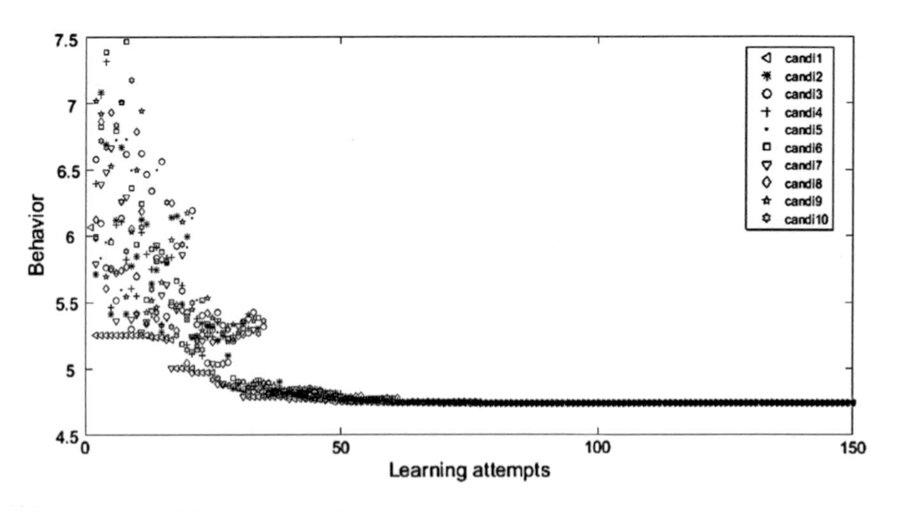

Fig. 1 Plots for objective function Ra

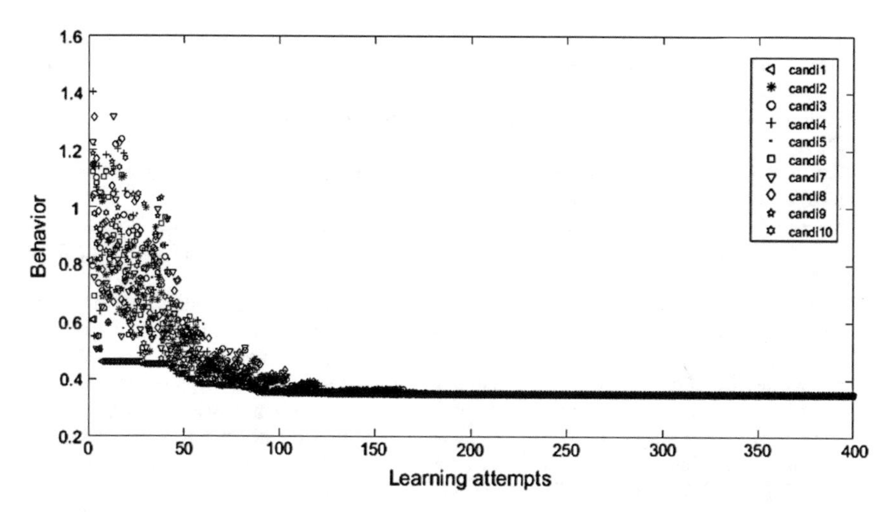

Fig. 2 Plot for objective function Kerf

and the cohort progresses, there is no substantial improvement in the performance of the candidates. At a particular point it is hard to differentiate between the performances of the candidates and convergence is said to have occurred.

5 Conclusion

The present work demonstrates successful application of four variations of Cohort intelligence for optimization of AWJM process. The optimum values of function Kerf and Ra are obtained with corresponding process parameters. For objective function Ra, results with follow best proves to be best while for objective function kerf, using educated roulette wheel philosophy gives best results. The results obtained using variants of CI are found better than original study-Firefly Algorithm. The other two variations of CI did not give promising results. Certain modifications would be required so that these variations are also applicable. In future, we also intend to use variations of Cohort Intelligence for constrained problems.

References

1. Armağan, M., Arici, A.A.: Cutting performance of glass-vinyl ester composite by abrasive water jet. Mater. Manufact. Process. **32**(15), 1–8 (2017). https://doi.org/10.1080/10426914.2016.1269919
2. Kechagias, J., Petropoulos, G., Vaxevanidis, N.: Application of Taguchi design for quality characterization of AWJM of TRIP sheet steels. Int. J. Adv. Manufact. Tech. **62**(5–8), 635–643 (2012). https://doi.org/10.1007/s00170-011-3815-3

3. Kulkarni, A. J., Durugkar, I. P., Kumar, M.: Cohort intelligence: a self-supervised learning behaviour. In: 2013 IEEE International Conference on Systems, Man, and Cybernetics (SMC), pp. 1396–1400 (IEEE 2013, October). https://doi.org/10.1109/smc.2013.241
4. Momber, A.W., Kovacevic, R.: Principles of Abrasive Water Jet Machining. Springer, London (2012)
5. Patankar, N.S., Kulkarni, A.J.: Variations of cohort intelligence. Soft Computing **22**(6), 1–17 (2017). https://doi.org/10.1007/s00500-017-2647-y
6. Shukla, R., Singh, D.: Experimentation investigation of abrasive water jet machining parameters using Taguchi and evolutionary optimization techniques. Swarm Evol. Comput. **32**, 167–183 (2017). https://doi.org/10.1016/j.swevo.2016.07.002
7. Shukla, R., Singh, D.: Selection of parameters for advanced machining processes using firefly algorithm. Eng. Sci. Tech. **20**(1), 212–221 (2017). https://doi.org/10.1016/j.jestch.2016.06.001
8. Shanmugam, D.K., Nguyen, T., Wang, J.: A study of delamination on graphite/epoxy composites in abrasive water jet machining. Composites Part A: Applied Science and Manufacturing **39**(6), 923–929 (2008). https://doi.org/10.1016/j.compositesa.2008.04.001
9. Kale, I. R., Kulkarni, A. J.: Cohort intelligence algorithm for discrete and mixed variable engineering problems. In: International Journal of Parallel, Emergent and Distributed Systems, pp. 1–36 (2017). https://doi.org/10.1080/17445760.2017.1331439

Evaluation of Performance of Electrochemical Honing Process by ANOVA

Md. Mushid, Ravi Kumar and Mohd Musheerahmad

Abstract Electrochemical honing is an electrolytic precision micro-finishing technology based on the hybridization of the electrochemical machining and conventional honing process principles to provide the controlled functional surface generation and fast material removal capabilities in a single action. The present work presents the distinctive findings of comprehensive experimental investigations designed to explore the influence of electrochemical honing process parameters on the work surface micro-geometrical, part-macro-geometrical, and material removal aspects. In the current work, two parameters have been considered—surface rough and cylindricity of job. The current power supply, stick grit size, stick-out pressure and, electrolyte concentration are observed to be the real parameters in the process. The influence of the machining parameters on the surface finish and cylindricity has been investigated and optimized by Taguchi approach and experimental results were analyzed using the analysis of variances (ANOVA).

Keywords Electrochemical honing (ECH) · Work surface characterization Process parameters · Optimization · ANOVA

1 Introduction

Improvement and advancement of manufacturing technologies have somewhat solved the problem but that is not sufficient for the cutting edge finished products to satisfy the requirement of finished products in various applications like in defence,

---

Md. Mushid (✉) · R. Kumar · M. Musheerahmad
Department of Mechanical Engineering, Bharat Institute of Technology,
Meerut 250103, Uttar Pradesh, India
e-mail: murshidrana@gmail.com

R. Kumar
e-mail: er.ravi.49@gmail.com

Md. Mushid · R. Kumar · M. Musheerahmad
Government Polytechnic, Azamgarh, Uttar Pradesh, India

© Springer Nature Singapore Pte Ltd. 2019

H. Malik et al. (eds.), *Applications of Artificial Intelligence Techniques in Engineering*, Advances in Intelligent Systems and Computing 697,
https://doi.org/10.1007/978-981-13-1822-1_44

aerospace, electronics, and many other industries. Therefore these days manufacturing industries focus on the good surface finish, corrosion resistance, fatigue strength, and dimensional accuracy. The quality of surface and shape diversion play an important role in the operative performance of components and also a most ambitious problem in the manufacturing sector for research scholars [1, 2].

Due to their distinguishable capabilities, electrochemical-based hybrid manufacturing technique are obtaining a considerable focuses from the manufacturing sectors to meet the require typical surface quality, tolerances limits, and productivity essential for critical and tribologically importance surface of components made of ultra-hard difficult to machining of materials such as chromium, titanium, nitride, vanadium, cobalt, etc. [3, 4]. Electrochemical honing (ECH) is a hybrid electrolytic high accuracy micro-finishing process characterized by an exclusive coupling of EHC machining and traditional honing process to generate controlled functional surfaces and high material removal capabilities in a single machining operation. ECH is capable of offering a distinct range of benefits to the processed surfaces not gained by either of the conventional mechanical processes when applied independently and increases the service life and efficiency of the components [2].

1.1 Electrochemical Honing (ECH)

Electrochemical Honing (ECH) is one of the most potential hybrid electrochemical–mechanical processes is based on the interface interaction of Electrochemical Machining (ECM) and conventional honing. In this process, most of the material is removed by the electrolytic action of ECM. But during ECM, a thin dull microfilm of metal oxide is deposited on the workpiece surface. This film is insulating in nature and protects the workpiece surface from further being removed. With the help of bonded abrasives, honing acts as a scrubbing agent to remove the thin insulation film from high spots and thus produces fresh metal for further electrolytic action. ECH is a hybrid micro-finishing process combining the high and faster material removal capability of electrochemical machining (ECM) and functional surface generating potentiality of mechanical honing. Hence, it set up the lucrative outcome of the machined surfaces like excellent surface quality, completely stress-free surfaces, generate surfaces with different cross-hatch lay pattern, meet the requirement for oil retention, and surfaces along with compressive residual stresses essential for the components subjected to cyclic loading [5]. Further, to improve ECH process, the optimized power supply and time relation in machining process of the workpiece is required for better results. In this paper, the experimental study was done on the workpiece by ECH process and analyzed using the analysis of variances (ANOVA).The paper is unified as follows: Sect. 2 describes the literature review, Sect. 3 gives the detail of experimental study and parameters and in Sect. 4 result discussion is described.

2 Literature Review

From a decade ago, ECH is developed as a promising cylindrical surface completing method for its miniaturized scale expulsion qualities. ECH is an electrochemical (EC) based cross-breed machining process consolidating the electrochemical machining (ECM) process and conventional honing. ECM gives the faster material expulsion ability and honing gives the controlled useful surface generating capacities. This procedure is five to ten times faster than conventional honing and four times quicker than grinding [6]. In addition, the procedure can give surface finish up to 0.05 μm, which is likewise far superior to other non-traditional cylindrical surfaces finishing processes [7]. The gear finishing by ECH process firstly testified by Capello and Bertoglio [8] in 1970 and the result in term of surface finish were found not acceptable. Chen et al. [9] designed a cathode gear working on ECH principle to develop productivity, accuracy, tool life, and gear finishing of spur gears. ECH process is based on the Faraday's law in which accuracy of any profile can be controlled by duration time and flow of current [10]. Naik et al. [11] performed the ECH to improve the surface roughness and found 80% average and 67% maximum surface roughness respectively. Misra et al. [12] used the mixture of electrolyte in ratio of 3:1 (NaCl and $NaNo_3$) for the better surface finish of helical gear and stated the effects of electrolyte concentration, rotating speed and voltage on the workpiece. After that many researcher has explored ECH process in the gears-finishing and found that it had better finishing and high productivity than other conventional processes [13–17]. As in ECH, a large portion of the material is evacuated by ECM activity, the procedure keeps the workpiece cool, free of heat distortion and produces stress-free surfaces. ECH has no material hardness constraining variable as long as the material is electrically conductive.

3 Experimental Setup

Exhaustive experimental investigations have been carried to establish process input and performance parametric relationships, machined surface characteristics, material removal mechanism, and surface integrity aspects of the electrochemical honing process. Experimentation has also been carried out for process performance optimization including the confirmation experiments conducted for verifying the results of experimental and analytical findings. Experimentation was planned based on the problem objectives ad parametric analysis, keeping in view the best utilization of time and other resources as well as the availability of measuring facilities. A series of experiments were conducted in a phased manner using the developed ECH experimental setup as shown in Fig. 1 and process parameters that were selected for the present experimental investigation are voltage, concentration, the rotational speed of ECH tool, and machining time in Table 1.

Fig. 1 Experimental setup ECH

Table 1 Parameters for ECH process

Sl. No.	Process parameters	Parameter designation	Levels		
			L1	L2	L3
1	Voltage setting	V	20	25	30
2	Electrolyte concentration (g/L)	C (%)	5	7.5	10
3	Rotational speed (RPM)	S	50	65	80
4	Finishing time (min)	T	2	2.5	3

The Ishikawa circumstances and end results graph representing the relationship of different process parameters with ECH execution attributes are represented in Fig. 2.

3.1 Experimental Procedure

A. *Preliminary preparation steps*

i. A general-purpose mild steel cylindrical block of bore 25.4 mm, 40 mm ± 0.5 outer diameter and length 40 ± 0.5 mm was selected as the workpiece specimen in the present study is shown in Fig. 3.
ii. Samples are clean with acetone and weighted before machining process and kept in a protective environment.

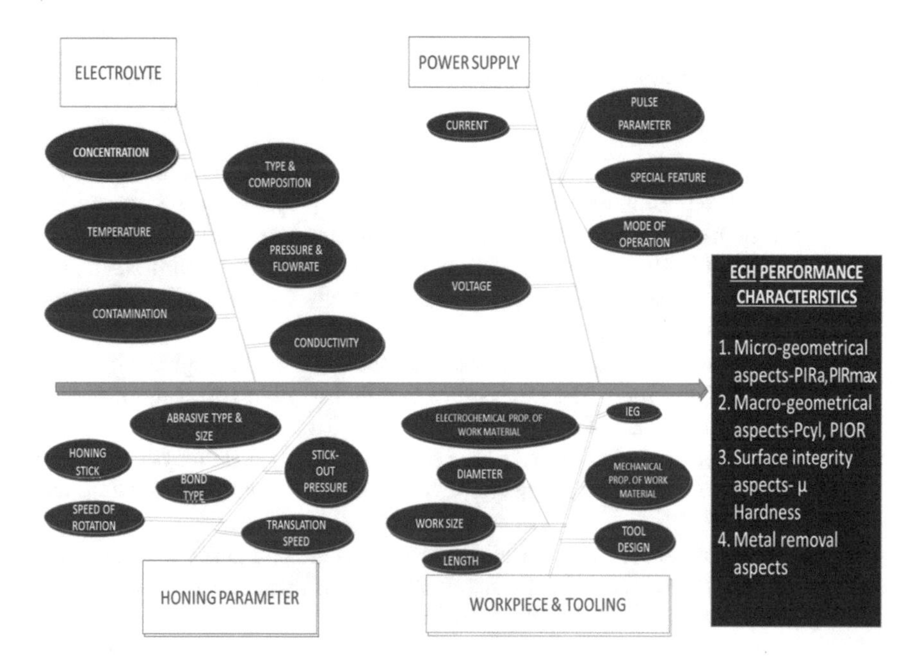

Fig. 2 Ishikawa cause and effect diagram

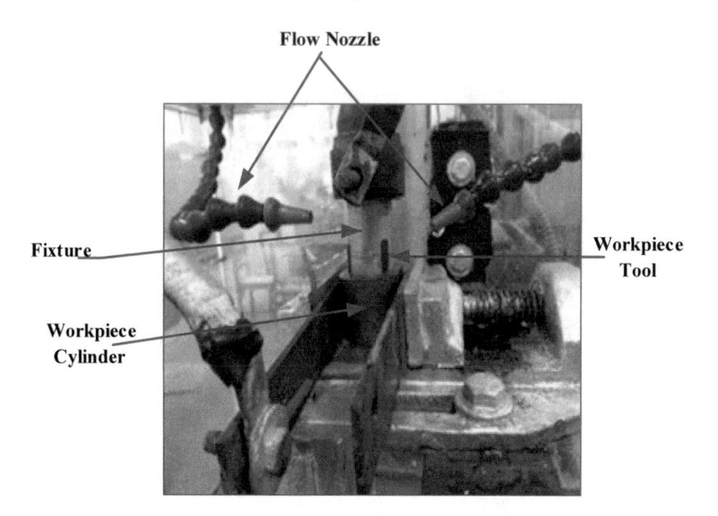

Fig. 3 View of ECH workpiece tool and cylinder

iii. An electrolytic solution (two salts of $NaNO_3$ and NaCl) is of fixed composition in the ratio of 1:3.
iv. Filtration mechanism and electrolyte flow system are to be checked before actual process.

v. Tool rotation rpm is to measure before set as per the levels of rpm.
vi. The power source should be checked for constant voltage mode.

B. *Experimental steps:*

i. The workpiece is mounted on the horizontal parallel plate wise well insulated from the adjoining surfaces via perplex sheet to avoid the formation of the local electrochemical cell and is properly positioned and aligned before clamping. The workpiece was cleaned with acetone to keep impurities and unwanted micro-particles from the machining zone to ensure smooth and error-free machining.
ii. The tool is mounted in a chuck and chuck to a holding spindle of bench drilling machine, and also ensure that tool rotation and reciprocation is not to be hindered by any means.
iii. After putting workpiece in its position, electrolyte flow system is then put into action and to ensure flow behavior to be uniform, smooth, and continuous.
iv. Tool rotation and reciprocation mechanism are then made to run so as to operate these mechanisms in smooth and fair manner, i.e., it is free from any hindrances and vibration if exists.
v. After ensuring everything the power supply is then put on ON mode.
vi. The stopwatch is started as soon as the power supply is made ON and time duration is too observed as the decided levels in every trial.
vii. After the conclusion of machining, the stopwatch is then made to stop and power supply system, electrolyte flow system and tool rotation and reciprocation system are to be stopped.
viii. Workpiece and tool are to be dismantled from their place and washed with acetone and cotton cloth and they are kept in a protective environment so that tool can be used effectively in next trial experiments.
ix. The workpiece is to be weighed again and the difference in weight before and after machining is to be recorded.

The time interval between each trial experiment should be ample (say 5–6 h) so as to ensure electrolyte free from any agitation and impurity. In this experimental work, single set of the electrolyte is to be used in ECH of three samples as shown in Table 1.

4 Result and Discussion

The target of the Taguchi strong plan is to decide the ideal parametric settings while guaranteeing process execution coldhearted to different wellsprings of arbitrary variety. Grid tests utilizing unique networks, called orthogonal exhibits (OA), enable the impacts of a few parameters to be resolved productively. Further, the OAs can be utilized with constant and also discrete factors as opposed to focal composite outlines regularly utilized as a part of established examination plan, particularly in conjunction with the reaction surface strategy for assessing the ebb and flow impacts of the components, which are valuable for just nonstop factors [18]. Analyses were thus arranged utilizing a standard L9 (3^4) OA.

The values of surface roughness and cylindricity of workpiece are measured before and after the execution of the experiment. The observed data table is in tune with the recommendation of Taguchi robust design methodology. Taguchi defines to conduct total nine experiments as we have four performance parameters and three levels for each performance parameter. The L9 orthogonal array is chosen and observed and measured data is represented in Tables 2 and 3 for the percentage improvement in surface roughness (PIRa) and percentage improvement in cylindricity (PIC) respectively.

Table 2 Experimental data for percentage improvement in surface roughness (PIRa)

Experiment no.	Independent variables				Surface roughness (Rmax) μm		PIRmax (%)
	Voltage	Concentration	Rot. speed (RPM)	Processing time	Before ECH	After ECH	
1	V1	C1	S1	T1	34.36	19.21	44.09
2	V1	C2	S2	T2	42.08	22.34	46.91
3	V1	C3	S3	T3	38.82	18.53	52.27
4	V2	C1	S2	T3	37.68	16.49	56.24
5	V2	C2	S3	T1	56.1	24.9	55.61
6	V2	C3	S1	T2	34.91	12.98	62.82
7	V3	C1	S3	T2	36.95	17.63	52.29
8	V3	C2	S1	T3	37.74	10.18	73.03
9	V3	C3	S2	T1	34.12	16.48	51.70

Table 3 Experimental data for percentage improvement in cylindricity (PIC)

Experiment no.	Independent variables				Cylindricity of workpiece (μm)		PICylmax (%)
	Voltage	Concentration	Rot. speed (RPM)	Processing time	Before ECH	After ECH	
1	V1	C1	S1	T1	25.11	14.65	41.66
2	V1	C2	S2	T2	22.5	11.38	49.42
3	V1	C3	S3	T3	38.25	16.5	56.86
4	V2	C1	S2	T3	44.2	24.93	43.60
5	V2	C2	S3	T1	88.1	47.05	46.59
6	V2	C3	S1	T2	44.5	22.68	49.03
7	V3	C1	S3	T2	38.2	15.78	58.69
8	V3	C2	S1	T3	20.4	10.25	49.75
9	V3	C3	S2	T1	30.5	14.32	53.05

4.1 Effect of ECH on Parameters

The effect on the parameters like voltage, electrolyte concentration, tool rotation speed, and machining time, which is considered for the experimental study are shown in Figs. 4 and 5. The X-axis values of this graph have been taken on the upper side of figure that is V, C, S, and time T. The mean percentage variation in parameters has been taken as Y-axis. A detectable conduct of ECH is the expanded PIRa with expanding current powers for voltage. An expansion in current force expands the present thickness in the inter-electrode gap (IEG), which prompts

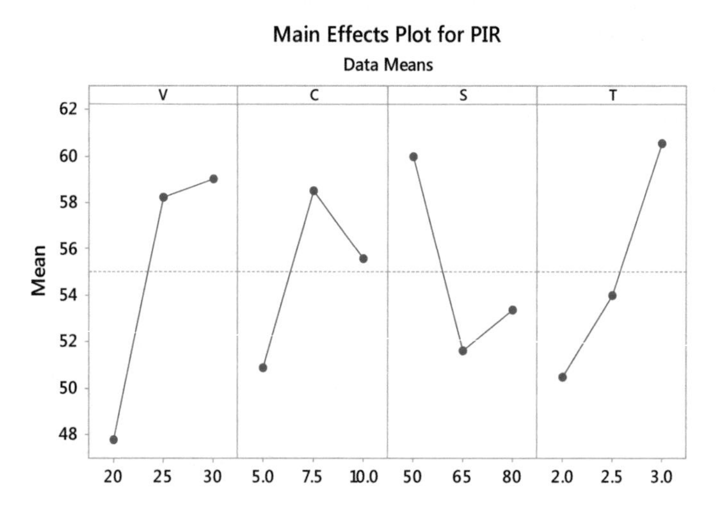

Fig. 4 Effect plot of all performance parameter on percent improvement on surface roughness

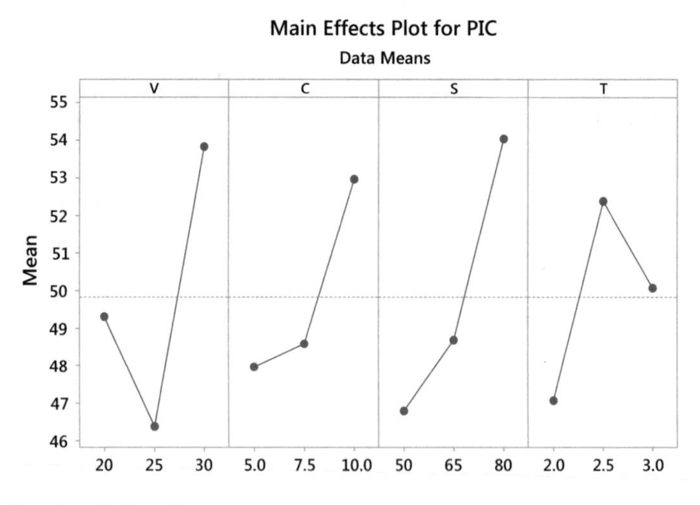

Fig. 5 Effect plot of all performance parameter on percent improvement on cylindricity

quickened anodic disintegration of the work material. In the meantime, an expanded current thickness additionally brings about a comparing increment in the rate of metal particle generation at the anode. The higher spots and tight ranges get a heavier electrochemical assault while the lower regions, still secured with the defensive film, have weaker electrochemical disintegration. The differential disintegration, along these lines made, is prominent to the point that the surface harshness as well as the swells are smoothed off quickly An adjustment in electrolyte fixation may likewise influence the defensive capacity and physical nature of metal oxide film. The net outcome is an enhanced differential disintegration at a higher electrolyte fixation as appeared in Figs. 4 and 5. The speed proportion and time fills in as a critical mechanical parameter in ECH in choosing the area of scouring of the passivating electrolytic metal oxide microfilm. Great outcomes on surface microgeometry manage the utilization of a legitimate blend of current force and turning-to-responding speed proportion, as reflected by the noteworthy commitment of C–SR association on PIRa. The procedure large-scale geometrical execution requires a particular coordination of mechanical cleaning and electrolytic disintegration because of the generally divided surface full scale geometrical.

4.2 ANOVA Analysis of Parameters

The experimental trial outcomes were dissected utilizing the examination of means (ANOM) and ANOVA. The ANOVA was performed on experimental information and the procedure execution attributes to distinguish the critical factors and measure their impacts on the reaction qualities as far as percent commitment to the aggregate variety, in this way building up the relative essentialness of the parameters. The pooled ANOVA outcomes for PIRa, PICyl, and DIR are outlined in Tables 4 and 5 individually. An essentialness level of 0.05 was chosen for this analysis. The examination of primary impacts 95 for every penny control points of confinement of means and pooled ANOVA show that the present power, electrolyte focus, electrolyte inlet temperature, stand-out weight, and stick coarseness measure have very extensive impacts when compared with the speed proportion and electrolyte flow rate (Fig. 6).

Table 4 ANOVA analysis and observation for PIR

Source	DF	Seq SS	Contribution (%)	Adj SS	Adj MS	F-value	P-value
V	1	189.77	31.73	189.8	189.77	4.79	0.094
C	1	33.46	5.59	33.46	33.46	0.84	0.41
S	1	63.13	10.89	65.13	65.13	1.64	0.269
T	1	151.23	25.28	151.2	151.23	3.82	0.122
e (pooled)	4	158.53	26.5	158.5	39.63		
Total	8	598.12	100				

Table 5 ANOVA analysis and observation for PIC

Source	DF	Seq SS	Contribution (%)	Adj SS	Adj MS	F-value	P-value
V	1	30.62	11.99	30.62	30.62	1.28	0.321
C	1	37.5	14.69	37.5	37.5	1.57	0.278
S	1	78.51	30.76	78.51	78.51	3.29	0.144
T	1	13.24	5.19	13.24	13.24	0.56	0.498
e (pooled)	4	95.39	37.37	95.39	23.85		
Total	8	255.26	100				

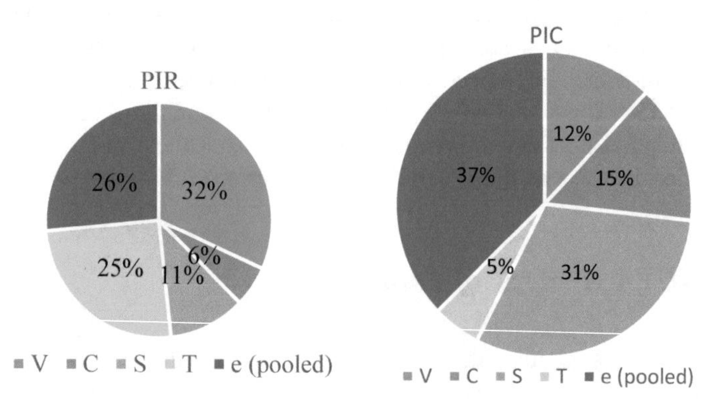

Fig. 6 Contribution of parameters in ECH process

5 Conclusion and Future Scope

ECH can be thought of as a communication procedure of controlled electrolytic disintegration and particular mechanical scouring. The productivity of rectifying the work surface smaller scale and large-scale geometrical mistakes requires legitimate coordination of the two activities. With the particular coordination of EC and mechanical activities, ECH can give an unmatched surface quality alongside decisively controlled size and work surface small-scale geometrical characters. The electrochemical sharpened surface is portrayed by great bearing and greasing up properties. Another gainful characteristic of ECH is the enhanced diametral increment rate (DIR) and process execution for work surface smaller scale and large-scale geometrical adjustments at higher current powers. In any case, appropriate determinations of electrolyte organization and fixation, rotating-to-responding speed proportion, and stick coarseness measure are significant all the while. Along these lines, ECH can be gainfully used as a remarkable exactness smaller scale completing innovation for basic segments of tribological hugeness. For future work the considered parameters can be upgraded by utilizing different procedures like 'utility-based Taguchi approach' 'LM organize demonstrating of electro-synthetic sharpening for process multi-execution control, NN-DFGA approach, GA-tuned fluffy Taguchi inference application, non-dominated sorting genetic algorithms and adaptive network-based fuzzy.

References

1. Merchant, M.E.: The manufacturing system concept in production engineering research. CIRP Ann. **10**, 77–83 (1961)
2. Rao, P.S., Jain, P.K., Dwivedi, D.K.: Electro chemical honing (ECH) of external cylindrical surfaces of titanium alloys. Procedia Eng. **100**, 936–945 (2015)
3. Shan, H.S.: Advanced Manufacturing Methods. Tata McGraw-Hill, New Delhi, India (2006)
4. Singh, H., Jain, P.K.: Remanufacturing with ECH—a concept. Procedia Eng. **69**, 1100–1104 (2014)
5. Jain, N.K., Naik, R.L., Dubey, A.K., Shan, S.: State-of-art-review of electrochemical honing of internal cylinders and gears. J. Eng. Manuf. **223**, 665–681 (2009)
6. Dubey, A.K.: Experimental investigations on electrochemical honing. Proc. IMechE Part B J. Eng. Manuf. **222**, 413–426 (2008)
7. Naik, R.L.: Investigation on precision finishing of gears by electrochemical honing. M. Tech dissertation, Mechanical and Industrial Engineering Department, I.I.T., Roorkee (2008)
8. Misra, J.P., Jain, P.K., Dwivedi, D.K.: Electrochemical honing—a novel technique for gear finishing. DAAAM Int. Sci. B **29**, 365–382 (2011)
9. Misra, J.P., Jain, P.K., Sevak, R.: ECH of spur gears-A step towards commercialization. DAAAM Int. Sci. B **17**, 197–212 (2012)
10. Shaikh, J.H., Jain, N.K.: High quality finishing of bevel gears by ECH. DAAAM Int. Sci. B **41**, 697–710 (2013)
11. Misra, J.P., Jain, P.K., Dwivedi, D.K., Mehta, N.K.: Study of time dependent behavior of electro chemical honing (ECH) of bevel gears, 24th DAAAM International symposium on intelligent manufacturing and automation 2013. J. Procedia Eng. **64**, 1259–1266 (2013)
12. Misra, J.P., Jain, P.K., Dwivedi, D.K., Mehta, N.K.: Study of electrochemical mechanical finishing of bevel gears. Int. J. Manuf. Technol. Manag. **27**(4–6), 154–169 (2013)
13. Chen, C.P., Liu, J., Wei, G.C., Wan, C.B., Wan, J.: Electrochemical honing of gears-a new method of gear finishing. Ann. CIRP **30**, 103–106 (1981)
14. Capello, G., Bertoglio, S.: A new approach by electrochemical finishing of hardened cylindrical gear tooth face. Ann. CIRP **28**(1), 103–107 (1979)
15. Misra, J.P., Jain, N.K., Jain P.K.: Investigations on precision finishing of helical gears by electrochemical honing process. Proc. IMechE Part B J. Eng. Manuf. **224**, 1817–1830 (2010)
16. Naik, L.R., Jain, N.K., Sharma, A.K.: Investigation on precision finishing of spur gears by electrochemical honing. In: Proceedings of the 2nd International and 23rd AIMTDR Conference, IIT Madras, India, pp. 509–514 (2008)
17. Chen, C.P., Liu, J., Wei, G.C., Wan, C.B., Wan, J.: Electrochemical honing of gears: a new method of gear finishing. Ann. CIRP **30**(1), 103–106 (1981)
18. Phadke, M.S.: Quality Engineering Using Robust Design. Prentice Hall, Englewood Cliffs, New Jersey (1989)

Prediction of the Material Removal Rate and Surface Roughness in Electrical Discharge Diamond Grinding Using Best-Suited Artificial Neural Network Training Technique

Siddharth Choudhary, Raveen Doon and S. K. Jha

Abstract This research revolves around proposing an Artificial Neural Network based model which can predict the Material Removal Rate and Surface Roughness with least permissible error. Here training method belonging to two different classes categorized as heuristic and numerical optimisation are compared for their convergence speed and the one with the fastest convergence is selected to train the network. To fortify the trained results, validation is done experimentally.

Keywords Electrical discharge diamond grinding (EDDG) · TOMAL-10 Levenberg–Marquardt · Cubic boron nitride (cBN) · Artificial neural network (ANN) · Box and Behnken model · Metal bounded diamond grinding wheel (MBDGW)

1 Introduction

In manufacturing science cost, reliability, durability, precision, and productivity plays a critical role, which has led to development of various super abrasives including diamond, Cubic Boron Nitride (cBN), and Borazon finished products. Design and manufacturing of such super abrasives based high-performance components are increasing, so we rely on advanced, nonconventional or hybrid machining processes to address this problem of complexity of the job with much ease in comparison to conventional machining processes [1–10]. Manufacturing

S. Choudhary (✉) · R. Doon
NSIT, Dwarka, Sector-3, New Delhi 110078, Delhi, India
e-mail: sid.choudhary007@gmail.com

R. Doon
e-mail: raveend06@gmail.com

S. K. Jha
MPAE Division, NSIT, Dwarka, Sector-3, New Delhi 110078, Delhi, India

© Springer Nature Singapore Pte Ltd. 2019

H. Malik et al. (eds.), *Applications of Artificial Intelligence Techniques in Engineering*, Advances in Intelligent Systems and Computing 697,
https://doi.org/10.1007/978-981-13-1822-1_45

products using these materials pose great challenges, one of which is the finishing aspect. Sensitivity to near surface damage plays a decisive role in a component's performance and its work-life is governed by surface finish.

Grinding is synonymous to finishing, which is vital for the industry due to its simplicity, efficiency, and effectiveness. To achieve desired surface finish, the choice of correct grinding wheel may prove to be critical and incorrect selection could be fatal as it may lead to reduced cutting ability, or the chips getting embedded in the wheel itself, high energy requirements and higher grinding force, high temperature at the tool–chip interface, low Material Removal Rate (MRR), low surface integrity.

MRR may be defined as ratio of the volume of material removed to the machining time. Surface roughness is one of the most critical elements of surface texture. It is often described by quantification of the variations in the normal vector direction of a real surface from its ideal shape (surface).

In this paper, ANN-based predictive model is introduced to quantify the extent of MRR and R_a obtained using EDDG after grinding of bi-metallic hexagonal TOMAL-10. The rest of the paper is organized as follows, Sect. 2 provides an overview of experimentation, Sect. 3 describes various training techniques available to train the neural network and Sect. 4 contains the results and Sect. 5 provides concluding remarks.

2 Experimentation

EDDG utilizes EDM's (Electrical discharge machining) principle for continuous in-process dressing of the diamond wheel. When a sufficient potential difference is attained between the MBDGW and the workpiece then electric field is generated. This electrostatic field of sufficient strength causes cold emission of electrons from the tool (cathode). Post cold emission, electrons accelerate towards the work piece (Anode). These liberated electrons after gaining sufficient velocity collide with the molecules of dielectric fluid. This results in knocking out of the electrons from dielectric fluid molecules resulting in positive ions, which is also known as dielectric dissociation. This process continues till a narrow column of ionized dielectric fluid is established between cathode and anode. Large conductivity of ionized column causes avalanche breakdown which is a spark. This spark produces a very high temperature in the order of (10,000–12,000 °C). Because of this high temperature, melting and vaporization of material takes place and evacuation of the metals is done by mechanical blast, which produces small crater's on both the electrode surfaces. The same cycle is repeated, and the next spark takes place at the next location of shortest gap between the electrodes.

Russian made model 3B642, a universal tool and cutter grinder was modified, and necessary changes were made to supply power by fixing brushes to the spindle head and the work table. Tomal-10 was grounded using the bronze alloy bounded diamond flaring cup type grinding wheels. A small-sized generator was particularly designed and built for the smooth performance of EDDG. Generator's characteristics could be easily adjusted for providing controlled electrical parameters. A SJ-210 portable surface roughness tester was used to measure the ground Tomal-10 specimens with each sampling length of 5 mm.

The work material chosen is TOMAL-10 which is basically bi-metallic hexagonal layer of Cubic Boron Nitride (cBN) on substrate of boron nitride along with copper and titanium in 1:1 ratio.

3 Prediction Techniques

Standard backpropagation is based on steepest gradient descent algorithm and network weight changes are aligned with the negative of the gradient of the performance function. Many fast training algorithms operating in batch mode are available to train the neural network, and a comparison is made in terms of convergence speed to select one with the best convergence time.

These training algorithms can be divided into two categories: Heuristic Techniques and Numerical Optimization Techniques. Heuristic Techniques include variable earning rate backpropagation algorithm and resilient backpropagation algorithm. Numerical optimization techniques include conjugate gradient, Quasi-newton and Levenberg–Marquardt training algorithm.

Conjugate gradient algorithm further has four variants

1. Fletcher–Reeves method
2. Polak–Ribiere method
3. Powell–beale restarts
4. Scaled conjugate gradient.

4 Results and Discussions

Following is the average time taken by all nine training algorithms to converge.

$$\mathrm{AvTime} = [0.1381 \quad 0.1148 \quad 0.1173 \quad 0.1186 \quad 0.1205 \quad 0.1163 \quad 0.1202 \quad 0.1161 \quad 0.1131]$$

The best convergence speed (least time) was given by Levenberg–Marquardt algorithm as 0.1131 s, therefore we will use this training technique to train our data (Tables 1 and 2).

4.1 Neural Network

Levenberg–Marquardt training technique is used to train the network. The graph of mean squared error with respect to number of iterations is given below for MRR (Figs. 1, 2, 3 and 4).

Table 1 Process parameters and their levels. Factor levels

Parameters	Symbols	Level 1 (−1)	Level 2 (0)	Level 3 (+1)	Level interval
Depth of cut	D	16	32	48	16
Wheel speed	V	15	25	35	10
Grit size	Z	40	120	200	80
Diamond concentration	K	50	100	150	50

Units Depth of cut (μm/double stroke)
Wheel speed (m/s)
Grit size (μm)
Diamond concentration (%)

Table 2 Design matrix

Run	D	V	Z	K	MRR	Ra
1	−1	−1	−1	−1	0.29	0.58
2	−1	−1	−1	−1	0.28	0.57
3	−1	−1	−1	1	0.22	0.33
4	−1	−1	1	−1	0.52	0.42
5	−1	−1	1	1	0.81	0.16
6	−1	1	−1	−1	0.67	0.2
7	−1	1	−1	1	0.28	0.15
8	−1	1	1	−1	0.76	0.23
9	−1	1	1	1	0.685	0.28
10	1	−1	−1	−1	0.17	0.26
11	1	−1	−1	1	0.13	0.1
12	1	−1	1	−1	0.4	0.14
13	1	−1	1	1	0.325	0.18
14	1	1	−1	−1	0.35	0.2
15	1	1	−1	1	0.26	0.22
16	1	1	1	−1	0.64	0.17
17	1	1	1	1	0.329	0.58
18	1	0	0	0	0.18	0.16
19	−1	0	0	0	0.32	0.13
20	0	1	0	0	0.37	0.19
21	0	−1	0	0	0.28	0.14
22	0	0	1	0	0.59	0.18
23	0	0	−1	0	0.22	0.17
24	0	0	0	1	0.3	0.16

MRR (mg/mm^3)
Ra (μm)
Inputs D, V, Z, K
Outputs MRR and Ra
Y_{MRR} represents output of trained network, these values are very good approximations of target MRR values

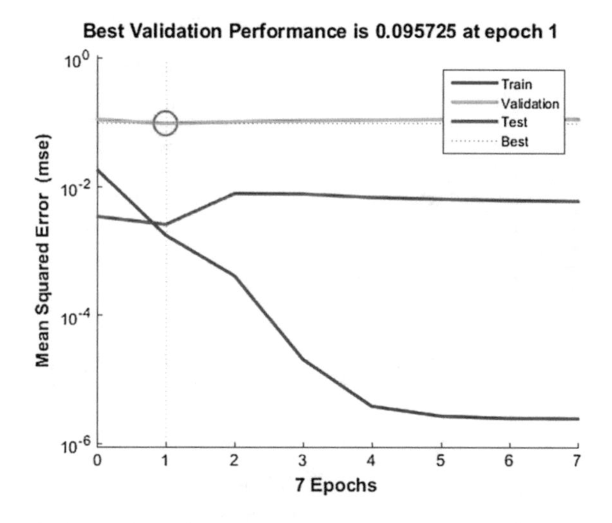

Fig. 1 The best performance occurs at epoch 1 and network at that configuration is 102 retained. Minimum MSE is 0.095725.

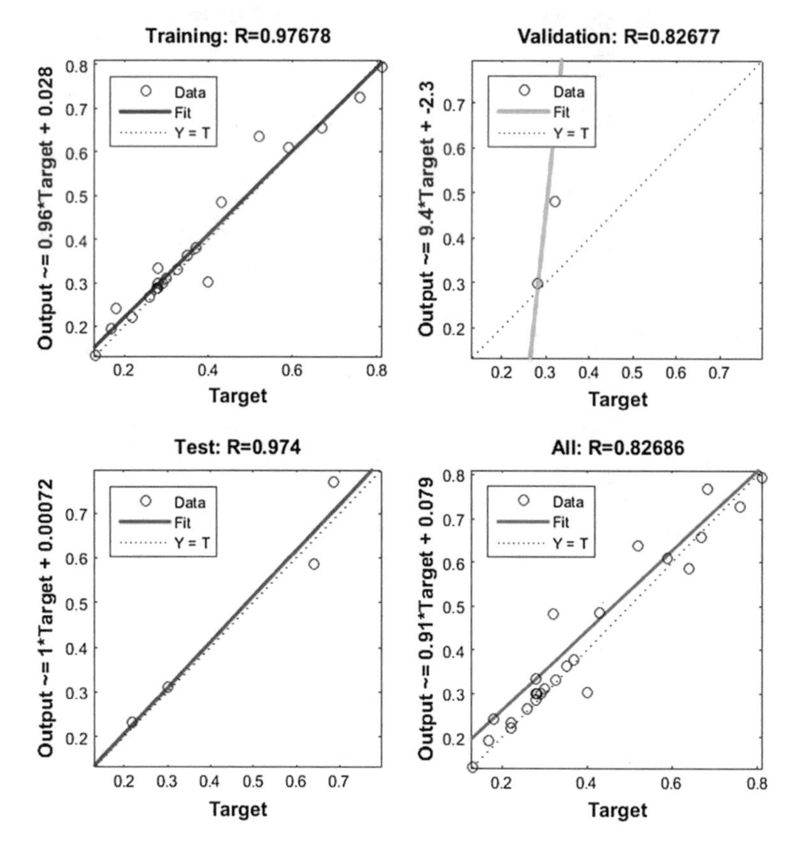

Fig. 2 Regression plot for MRR

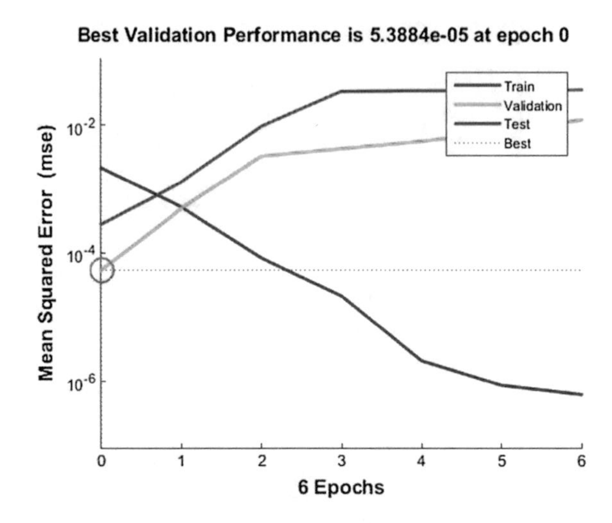

Fig. 3 Performance plot for surface roughness. Minimum MSE (epoch 0) is 5.3884e−05.

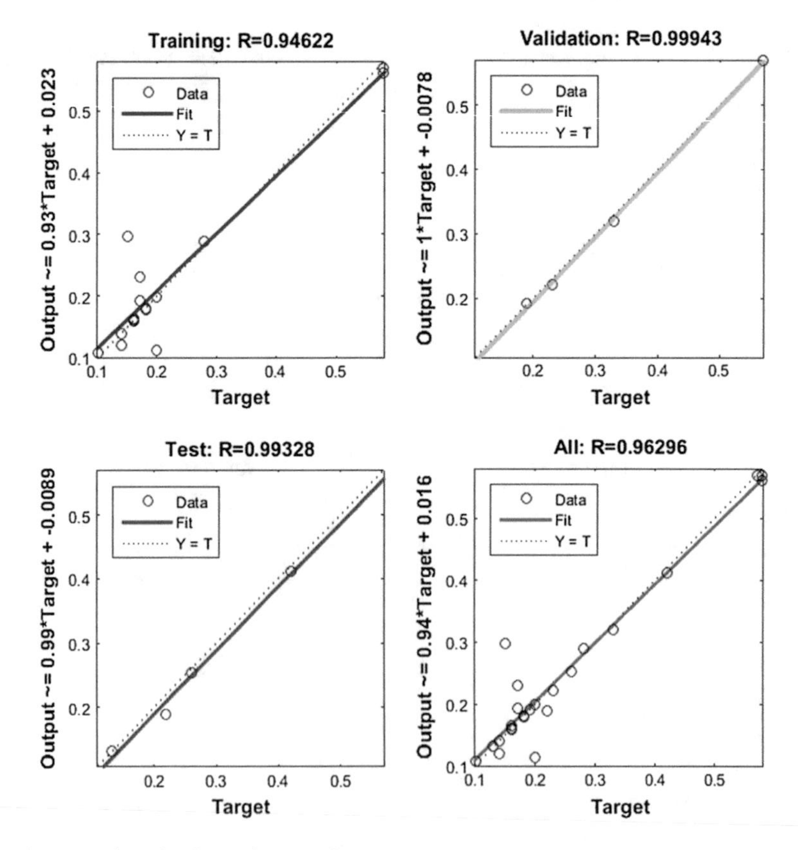

Fig. 4 Regression plot for surface roughness

S. no.	Test inputs				Predicted value	
	D	V	Z	K	Ymrr	Yr
1	0	0	0	−1	0.584	0.1591
2	1	1	1	1	0.3184	0.3439
3	−1	−1	−1	−1	0.2820	0.5616
4	−1	1	−1	1	0.2796	0.5693

Ymrr Simulated MRR output for test inputs
Yr Simulated surface roughness output for test inputs

Output of trained network		
S. no.	Ymrr	Yra
1	0.2985	0.5693
2	0.2985	0.5693
3	0.2198	0.3188
4	0.6367	0.4113
5	0.7950	0.1607
6	0.6574	0.1985
7	0.2855	0.2961
8	0.7268	0.2206
9	0.7688	0.2880
10	0.1926	0.2528
11	0.1328	0.1079
12	0.3013	0.1399
13	0.3310	0.1785
14	0.3617	0.1129
15	0.2658	0.1886
16	0.5854	0.1928
17	0.7430	0.5616
18	0.2414	0.1637
19	0.4802	0.1315
20	0.3788	0.1905
21	0.3340	0.1203
22	0.6087	0.1804
23	0.2313	0.2305
24	0.3114	0.1591

The regression plots show the relation between the output of the network and the target. For best training, R value would be close to 1.

Performance plot shows test curve and validation curve, which are very similar. An increase in test curve before validation curve increases would indicate overfitting. Hence similar test and validation curves are desirable.

The Levenberg–Marquardt algorithm is used to minimize an objective function of a data fitting problem. LM algorithm is a combination of other two minimization methods, the gradient descent method and Gauss–Newton method. It is an iterative method like other numerical techniques. The process starts with the parameters initialized to random values. After each iteration parameter vector is replaced by new estimate. By adding an extra term proportional to the identity matrix, I in the normal equation obtained in the Gauss–Newton method: $(J'J)\ \Delta a = J'(y - f(a))$.

Levenberg changed this equation by: $(J'J + \lambda I)\ \Delta a = J'\ (y - f(a))$.

Where λ is the nonnegative damping factor, which is adjusted at each iteration. J is jacobian matrix of derivatives of the residuals with respect to the parameters.

If the reduction of cost function is rapid a smaller value of λ can be used and the algorithm is like the Gauss–Newton algorithm. If the iteration is slow a larger value of λ can be used and steps are taken approximately in the direction of the gradient.

Marquardt improved the algorithm by scaling each component of the gradient according to the curvature, so that there is larger movement along the directions where the gradient is smaller. This avoids slow convergence in the direction of small gradient. A diagonal matrix replaces identity matrix,

$$(J'J + \lambda(J'J))\Delta a = J'(y - f(a))$$

This algorithm finds a local minimum and not global minimum.

5 Conclusion

Above results and discussions have led to following conclusions:

- Feedforward Artificial Neural network was able to generalize system characteristics by predicting output closer to actual values and as a result giving good prediction of MRR and Surface roughness.
- Levenberg–Marquardt training method had the least convergence time of all the techniques considered and hence was used for training feedforward neural network.

Acknowledgements I would like to thank Material Cutting Department, National Technical University "kharkiv Polytechnic Institute (KPI)" for their support and resources which made possible the availability of training data through experimentation.

References

1. McKie, A., Winzer, J., Sigalas, I., Herrmann, M., Weiler, L., Rödel, J., et al.: Mechanical properties of cBN–Al composite materials. Ceram. Int. **37**, 1–8 (2011)
2. Clark, T.J., DeVries, R.C.: Super Abrasives and Ultra-hard Tool Materials, ASM Handbook, ASM International, pp. 1008–1018 (1993)
3. Klocke, F., König, W.: Fertigungsverfahren: Drehen, Fräsen, Bohren, 8th edn. Springer, Berlin, Heidelberg (2008)

4. Denkena, B., Tönshoff, H.K.: Spanen—Grundlagen, 3rd edn. Springer, Berlin, Heidelberg (2011)
5. Harris, T.K., Brookes, E.J., Taylor, C.J.: The effect of temperature on the hardness of polycrystalline cubic boron nitride tool materials. Int. J. Refract. Met. Hard Mater. **22**, 105–110 (2004)
6. Kress, J.: Polykristallines kubischer Bornitrid—Einsatz in der Praxis. In: Weinert, K., Biermann, D., (eds.) Spanende Fertigung—Prozesse, Innovationen, Werkstoffe, Essen. Vulkan-Verlag GmbH, pp. 176–187 (2008)
7. Biermann, D., Baschin, A., Krebs, E., Schlenker, J.: Manufacturing of dies from hardened tool steels by 3 axis micromilling. Prod. Eng. Res. Dev. **5**, 209–217 (2011)
8. Sieben, B., Wagner, T., Biermann, D.: Empirical modeling of hard turning of AISI6150 steel using design and analysis of computer experiments. Prod. Eng. Res. Dev. **4**, 115–125 (2010)
9. Karpat, Y., Özel, T.: Mechanics of high speed cutting with curvilinear edge tools. Int. J. Mach. Tools Manuf. **48**, 195–208 (2008)
10. Astakhov, V.P.: Geometry of Single-Point Turning Tools and Drills. Springer, London (2010)

Step-Back Control of Pressurized Heavy Water Reactor by Infopid Using DA Optimization

Sonalika Mishra and Biplab Kumar Mohanty

Abstract It is a highly challenging task to control the nonlinear PHWR. In this paper a FOPID is designed using Dragonfly algorithm for a pressurized heavy water reactor. The objective is to control the highly nonlinear reactor control power, i.e., the step-back. The gains and other parameters of FOPID are found using Dragonfly algorithm by considering eight models of PHWR at eight operating points and also eight interval situations. Dragonfly is a new population based meta-heuristic algorithm based on swarming nature of dragonflies. By considering an objective function of minimizing integral absolute error the INFOPID is optimized. In simulation by using MATLAB 2016, it is observed that proposed DA optimized INFOPID gives better output then conventional INFOPID. System settling time and steady-state error is reduced. It is also able to track the set point efficiently for a variation of parameters. Swarming intelligence feature of DA and high degrees of freedom of FOPID is able to perform better work than other used controllers and currently operating Reactor Regulating System.

Keywords Pressurized heavy water reactor · Step-back · Fractional order proportional integral differential · Interval fractional order proportional integral differential · Dragonfly algorithm

1 Introduction

Nowadays in power industry sector Pressurized Heavy Water Reactor plays a vital role. It is a nuclear reactor. It uses uranium as fuel, heavy water as coolant. Coolant is pumped into the core where it is heated. This heated water is then used for steam generation to rotate the turbine, for electricity generation purpose. Depending on

S. Mishra (✉) · B. K. Mohanty
VSSUT, Burla, Odisha, India
e-mail: sonalikamishra146@gmail.com

B. K. Mohanty
e-mail: biplabkumar123@gmail.com

© Springer Nature Singapore Pte Ltd. 2019

H. Malik et al. (eds.), *Applications of Artificial Intelligence Techniques in Engineering*, Advances in Intelligent Systems and Computing 697,
https://doi.org/10.1007/978-981-13-1822-1_46

the load demand, if there is a need for reduction of reactor power, the universal power of water reactor can be decreased by shoving control rods into it to a pre-specified level. This phenomenon is known as STEP-BACK. An undesirable torpid response with power undershoot is resulted in controlling step-back by Indian reactor regulating system. An efficient robust controller is needed to be designed in order to control the step-back of this highly nonlinear PHWR at its ever changing load condition.t. Fuzzy PID controllers, PID tuned by LQR along with FO phase sharper [1] are developed for different linearized models of PHWR at a changing load condition to control this step-back condition. These all gave solution but did not give the efficient solution. Models of PHWR and linearized nonlinear equations over a steady-state operating range are developed by [2, 3]. In different methods FOPID are designed. Reduced fractional models are also obtained by [4] gives little modeling error. Using pole placement method PID controller was proposed. Fractional order controller gives better results than the PID controller. FOPID controller by [4], by frequency domain tuning FOPI controller by Bhase and Patre [5] using stability boundary method to control FO systems. The wide benefits of FO controllers have forced the researchers to implement this in different systems. As the parameters of PHWR are dynamic many controllers developed for fixed values were unable to cope with the changing environment of it. This difficulty inspired many scholars to consider the interval model of PHWR to be controlled by FOPID controller. Various methods QFT, Pole placement, LMI, etc., was proposed for formulation of controller for interval models [6]. However this technique gave not an optimum solution. To study the stability of FO interval models Edge theorem is used. FO controller has already been used for the interval systems [7–10]. Recently, FOPID is being developed for interval models of PHWR by using edge theorem and stability boundary locus [11]. It has also given a good result, but the result obtained after DA optimization is better than it. In this paper a FOPID is designed for interval models of PHWR using DRAGONFLY algorithm, here this algorithm is used to optimize integral absolute error. Eight models of water reactors are considered at eight operating points [4]. As it is an uncertain as well as an un-robust system, eight interval transfer functions are taken. Here interval forms are considered, as the parameters of reactor are dynamic, many controllers are developed for fixed values were unable to cope with the changing environment of it. The models of reactor are identified by using system identification technique. Controller designed for stabilizing the model can stabilize all the other models at different operating points as it has the feature that is large DC gain and least phase margin [2]. The subscript denotes the control rod drop level and superscript denotes the initial reactor power.

The paper is organized as follows. Section 2 fractional order models and interval models of the system are given in Sect. 3 dragonfly algorithm is analyzed. Section 4 is giving the details about best result obtained. Section 5 is all about the conclusion.

2 Fractional Order Models of PHWR

2.1 Tabulation

This interval form was obtained by taking ±50% higher and lower limit uncertainties of P^{30}_{100} model [1]. These eight interval transfer functions are considered to design the fractional order PID which will be in the form of

$$C(s) = K_p + \frac{K_I}{S^{\lambda}} + K_d S^{\mu} \tag{1}$$

K_p, K_i, K_d are proportional, integral and derivative gain. λ and μ are differential and integral orders of FOPID. For this FOPID have high degree of freedom which helps in better tuning. There are many designing techniques for FOPID. Here DRAGONFLY algorithm is used to get an optimized FOPID (Tables 1 and 2).

Table 1 Models of PHWR

Model identified	Reduced model(Fractional Order)
p^{30}_{100}	$\frac{1522.8947}{s^{2.0971}+8.1944s^{1.0036}+7.7684}e^{-2.0043\times 10^{-12}s}$
P^{90}_{30}	$\frac{1359.2345}{s^{2.0972}+8.1906s^{1.0036}+7.7075}e^{-1.5968\times 10^{-9}s}$
P^{30}_{80}	$\frac{1027.3027}{s^{2.0163}+6.7859s^{0.99388}+6.5268}e^{-2.5346\times 10^{-5}s}$
P^{30}_{70}	$\frac{1074.396}{s^{2.0961}+8.2663s^{1.0037}+7.8641}e^{-3.1431\times 10^{-10}s}$
P^{50}_{100}	$\frac{529.1365}{s^{2.1002}+7.1111s^{1.0002}+9.0873}e^{-1.6049\times 10^{-5}s}$
P^{50}_{90}	$\frac{604.2541}{s^{2.2986}+8.871s^{1.0321}+11.2993}e^{-6.5\times 10^{-6}s}$
p^{50}_{80}	$\frac{337.846}{s^{2.2038}+6.7453s^{1.0132}+7.4275}e^{-1.8479\times 10^{-7}s}$
p^{50}_{70}	$\frac{325.2142}{s^{2.1969}+7.1459s^{1.0113}+8.3167}e^{-2.343\times 10^{-7}s}$

Table 2 Interval form

Transfer function	Interval model of P^{30}_{100}
$G_1(s)$	$\frac{761.44735}{s^{2.0971}+4.0972s^{1.0036}+3.8842}$
$G_2(s)$	$\frac{761.44735}{s^{2.0971}+4.0972s^{1.0036}+11.6526}$
$G_3(s)$	$\frac{761.44735}{s^{2.0971}+12.2916s^{1.0036}+3.8842}$
$G_4(s)$	$\frac{761.44735}{s^{2.0971}+12.2916s^{1.0036}+11.6526}$
$G_5(s)$	$\frac{2284.3420}{s^{2.0971}+4.0972s^{1.0036}+3.8842}$
$G_6(s)$	$\frac{2284.3420}{s^{2.0971}+4.0972s^{1.0036}+11.6526}$
$G_7(s)$	$\frac{2284.3420}{s^{2.0971}+12.2916s^{1.0036}+3.8842}$
$G_8(s)$	$\frac{2284.3420}{s^{2.0971}+12.2916s^{1.0036}+11.6526}$

3 Dragonfly Algorithm

It is an essential task for controlling a system optimally, i.e., to find out the optimum system parameters. Many optimizing techniques are there for adjusting control or system parameters. Some of the algorithms are designed based on swarming feature of insects. DRAGONFLY algorithm is one of them. The Dragonfly algorithm for optimizing a problem is generated from the idea of this. Activities of Dragonflies can be studied in five steps separation, alignment, attraction, distraction [12].

3.1 Separation

Dragonflies have to get separated from each other in a distance for attacking a food. The distance from the nearer dragonfly denoted by vector

$$S_i = -\sum_{j=1}^{N} \left(x - x_j\right) \tag{2}$$

x = current position of individual, x_j = position of jth neighboring individual, N = number of neighboring individual.

3.2 Alignment

After certain separation dragonflies are aligned in a particular direction. This alignment can be denoted as vector An

$$A_i = \frac{\sum_{j=1}^{N} V_J}{N} \tag{3}$$

V_J = velocity of jth neighboring individual.

3.3 Cohesion

In order to hunt the food source, next to the alignment dragonflies gathers in the direction of food. The vector cohesion is indicated as C.

$$C_i = \frac{\sum_{j=1}^{N} x_j}{N} - x \tag{4}$$

3.4 Attraction

All dragonflies start to fly towards the food source in the purpose of hunting it. Attraction to the food source F_i

$$F_i = x^+ - x \tag{5}$$

x^+ = Position of food source.

3.5 Distraction

If any enemy exists in place of food, then the dragonflies get distracted towards outside in order to save themselves.

$$E_i = x^- + x \tag{6}$$

x^- = position of enemy.

For the updating of position of dragonflies and also for simulation of motion two vectors are used that are step vector and position vector. Step vector can be denoted as

$$\Delta x_{t+1} = (sS_i + aA_i + cC_i + fF_i + eE_i) + w\Delta x_i \tag{7}$$

where *s*, *a*, *c*, *f*, *e*, and *w* are weights specified to separation, alignment, cohesion, food attraction, enemy distraction, step vector, respectively.

Step vector is used to indicate the direction of motion of the insects. If one step count is over than next step position vector is as

$$x_{t+1} = x_t + \Delta x_{t+1} \tag{8}$$

In the absence of neighboring solution the position can be updated with random walk. We can write the position vector as

$$x_{t+1} = x_t + \text{Levy}(d) \times x_t \tag{9}$$

t = current iteration, d = dimension of position of the position vector.

$$\text{levy}(d) = 0.01 \times \frac{r_1 \times \sigma}{r_2^{1/\beta}} \tag{10}$$

r_1, r_2 are the random numbers in [0, 1], b is a constant.

Explorative and exploitative behaviors are achieved at the optimization when the separation, alignment, cohesion, food position, and enemy position are calculated. In every iteration, by using equation step and position vectors get updated. The updating of position is done until the desired criterion is achieved. Its weights are adaptive and are tuned in such a way that they first explore the search space and then get converge to hunt the optimal solution. This property makes this algorithm best. When iteration is increased DA explore the search space.

3.6 Steps of Execution of Dragonfly Algorithm

(1) Initialize Dragonfly population Xi and step vector ΔXi, (for $i = 1, 2, 3\ldots n$).
(2) Check for end condition. If satisfied then optimal values of desired parameters are obtained, otherwise execute steps 3, 4, 5.
(3) Update food source, enemy and s, a, c, f, e.
(4) Update neighboring radius by calculating S, A, C, F, and E.
(5) If Dragonfly has any neighboring Dragonfly update step and position vector by Eqs. (6 & 7), else update position vector by Eq. (8) and repeat the whole process from step (2).

4 Result and Discussion

C_1, C_2, C_3, C_4, C_5, C_6, C_7, C_8, are the controllers obtained after optimization by DA for G_1, G_2, G_3, G_4, G_5, G_6, G_7, G_8 respectively. At different iterations and search agent the optimized controller is obtained. After comparing all the eight controllers individually with each transfer function it is obtained that the C_1 controller is able to stabilize all the transfer function with in a small time as compared to the controller obtained using edge theorem and stability boundary locus [1] Taking 400 iterations this controllers are obtained (Table 3).

So the resulted controller is

$$C(s) = 0.0350 + \frac{0.0243}{s^{0.9360}} + 0.0326s^{0.9701} \tag{11}$$

Output Waveforms: Figs. 1, 2, 3, 4, 5, 6, 7, and 8

Table 3 Parameters of FOPID

	K_p	K_I	K_D	λ	μ	Search agent
C_1	0.0350	0.0243	0.0326	0.9360	0.9701	6
C_2	0.0481	0.0353	0.0386	0.8564	0.9102	4
C_3	0.0409	0.0169	0.0274	0.9031	0.7175	4
C_4	0.0390	0.0422	0.0198	0.9744	0.7078	4
C_5	0.0258	0.0113	0.0124	0.8371	0.9209	4
C_6	0.0217	0.0257	0.0110	0.9764	0.9766	4
C_7	0.0331	0.0243	0.0199	0.8360	0.6954	6
C_8	0.0343	0.0347	0.0081	0.9285	0.8103	4

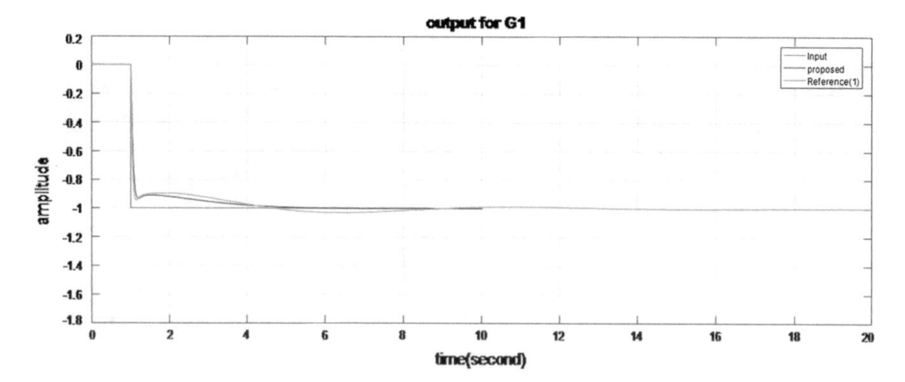

Fig. 1 Output waveform of optimized controller1

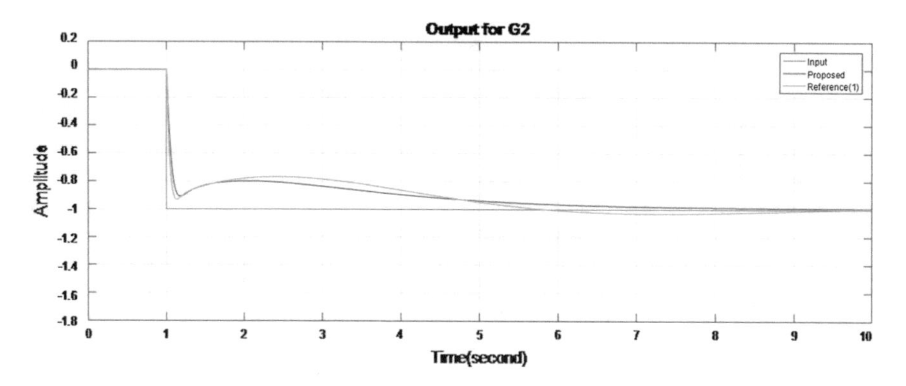

Fig. 2 Output waveform of optimized controller2

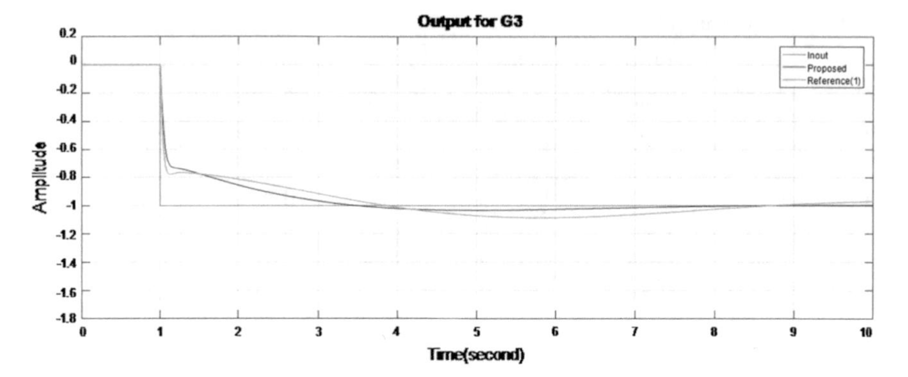

Fig. 3: Output waveform of optimized controller3

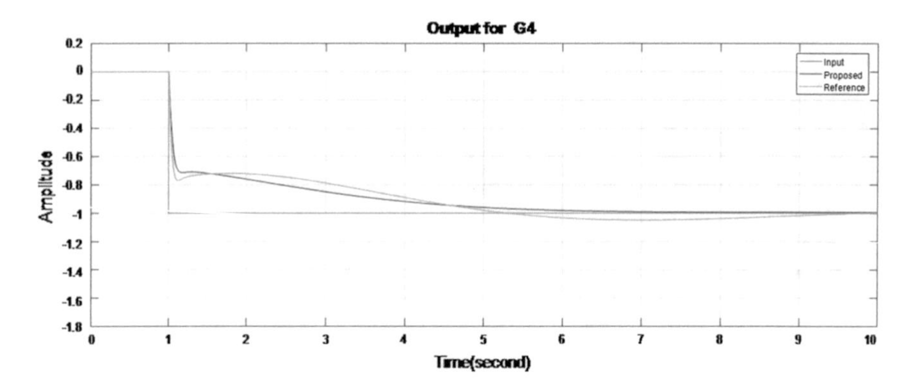

Fig. 4 Output waveform of optimized controller4

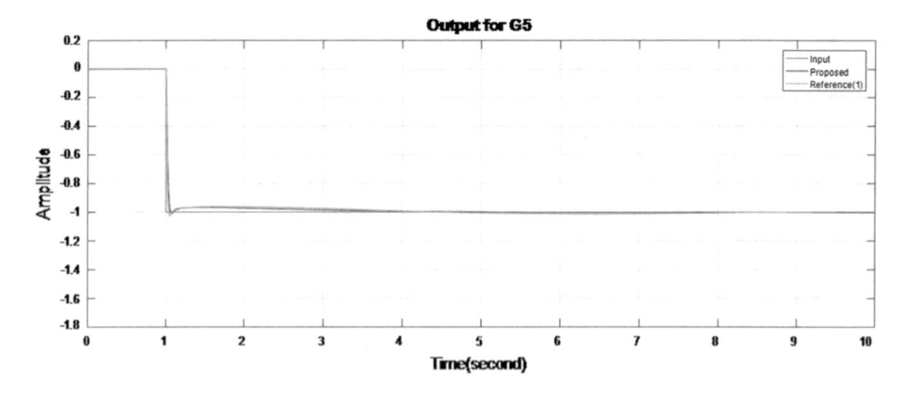

Fig. 5 Output waveform of optimized controller5

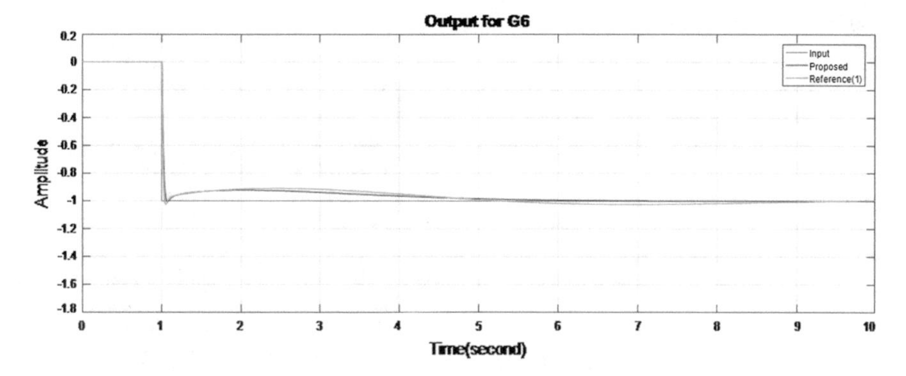

Fig. 6 Output waveform of optimized controller6

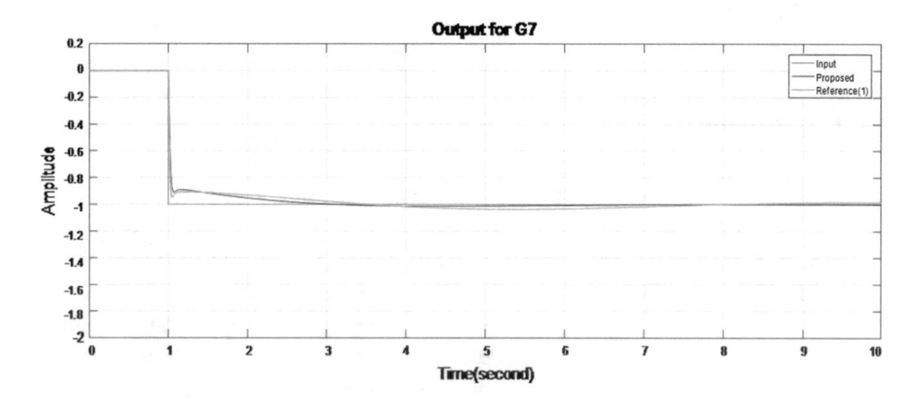

Fig. 7 Output waveform of optimized controller7

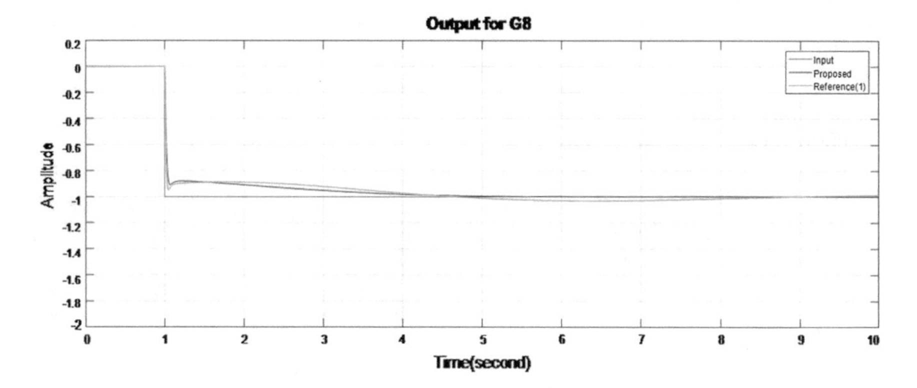

Fig. 8 Output waveform of optimized controller8

Table 4 (Result) reference [1] proposed result

G_1	IAE-0.3632 ITAE-1.201 ISE-0.04 Settling time-13.1 s	IAE-0.2459 ITAE-0.5638 ISE-0.04447 Settling time-5.2 s	G_5	IAE-0.1337 ITAE-0.4534 ISE-0.012 Settling time-8.3 s	IAE-0.0828 ITAE-0.1774 ISE-0.01439 Settling time-4.5 s
G_2	IAE-0.7842 ITAE-2.543 ISE-0.147 Settling time-15.8 s	IAE-0.722 ITAE-2.377 ISE-0.1228 Settling time-16 s	G_6	IAE-0.3231 ITAE-1.144 ISE-0.289 Settling time-13.3 s	IAE-0.2447 ITAE-0.7171 ISE-0.02513 Settling time-7 s
G_3	IAE-0.6988 ITAE-2.593 ISE-0.11 Settling time-18.2 s	IAE-0.4435 ITAE-1.1416 ISE-0.0912 Settling time-8.2 s	G_7	IAE-0.2715 ITAE-1.032 ISE-0.022 Settling time-12 s	IAE-0.1587 ITAE-0.3678 ISE-0.02121 Settling time-6.6 s
G_4	IAE-0.9048 ITAE-2.892 ISE-0.19 Settling time-15.9 s	IAE-0.7319 ITAE-1.985 ISE-0.1606 Settling time-15.7 s	G_8	IAE-0.3757 ITAE-1.279 ISE-0.037 Settling time-13.2 s	IAE-0.2458 ITAE-0.5626 ISE-0.03135 Settling time-5.3 s

5 Conclusion

An efficient algorithm is needed which can control step-back in PHWR for bigger range of operating condition, as PHWR plays a leading role in power generating industry. In this work a fractional order PID controller is designed for interval form of PHWR. Here DRAGONFLY algorithm is used to get the optimized FOPID. This method results in enhancement of robustness for a range of uncertainties instead of a fixed value. It is seen that this controller is able to stable more, faster than the reference controller. As observed from the Table 4 and all the waveforms proposed controller is better than the existing controller G_2 and G_4 are taking a little more to be stable but with less steady-state error. The system robustness can be studied by adding disturbance to the system. In future more advance controller can be designed to control this system and a comparative study can be done between the controllers.

References

1. Saha, S., Das, S., Ghosh, R., Goswami, B., Balasubramanian, R., Chandra, A.K., Das, S., Gupta, A.: Design of a fractional order phase sharper fpr iso-damped control of a PHWR under step-back condition. Nucl. Sci. **57**, 1602–1612 (2010)
2. Talange, D.B., Bandyopadhyay, B., Tiwari, A.P.: Spatial control of a large PHWR by decentralized periodic output feedback and model reduction techniques. Nucl. Sci. **53**, 2308–2317 (2006)
3. Sondhi, S., Hote, Y.V.: Fractional order controller and its application: a review. In: Proceedings AsiaMIC, Phuket, Thailand (2012)
4. Das, S., Gupta, A.: Fractional order modeling of a PHWR under step-back condition and control of its global power with a robust PI $I^{\lambda}D^{\mu}$ controller. Nucl. Sci. **58**, 2431–2441 (2011)

5. Bhase, S.S., Patre, B.M.: Robust FOPI Controller design for power control of PHWR under step-back condition. Nucl. Eng. **274**, 20–29 (2014). https://doi.org/10.1016/j.nucengdes.2014.03.041
6. Bhattacharya, S., Keel, L.H., Bhattacharya, S.P.: Robust stabilizer synthesis for interval plants using H-infinity methods. In: Decision and Control, pp. 3003–3008 (1993)
7. Saxena, S., Hote, Y.V., Sondhi, S.: Fractional order PI control of DC servo system using the stability boundary locus approach. In: Industrial and Information Systems, pp. 182–186 (2015)
8. Ostalczyk, P., Stolarski, M.: Fractional order PID controllers in a mobile robot control. IFAC Proceedings **42**(13), 268–271 (2009)
9. Sondhi, S., Hote, Y.V.: Fractional order PI controller with specific gain-phase margin for MABP control. IETE J. Res **61**, 142–153 (2015)
10. Carderon, A.J., Vingare, B.M., Feilu, V.: Fractional order control strategies for power electronic buck converters. Sig. Process **86**, 2803–2819 (2006)
11. Lamba, R., Singla, S.K., Sondhi, S.: Fractional order PID controller for power control in perturbed pressurized heavy water reactor. Nucl. Eng. Des. **323**, 84–94 (2017). https://doi.org/10.1016/j.nucengdes.2017.08.013
12. Pathania, A.K., Mehta, S., Rza, C.: Multi objective dispatch of thermal system dragonfly algorithm. Int. J. Eng. Res. **5**(11), 861–866 (2016). https://doi.org/10.17950/ijer/v5s11/1106

Implementation of Current Mode Integrator for Fractional Orders *n*/3

Divya Goyal and Pragya Varshney

Abstract This paper presents an implementation of Current Integrator for fractional orders (*n*/3). The circuit uses second-generation current conveyor (CCII) as the active element along with the passive elements based fractional-order element (FOE). The resistors and capacitors of FOE are in geometric progression and are arranged as RC ladder. The functionality of the fractional integrator has been verified for *n*, viz. $n = \{0.75, 1.0, 1.5, 2.0 \text{ and } 2.25\}$. The simulations results have been obtained in PSPICE using 0.25 μm CMOS technology parameters.

Keywords Current conveyor · Fractional · Fractional-order element
Current mode · Integrator

1 Introduction

The merits of current mode approach over the traditional voltage mode based circuits have evolved many new active devices [1]. These devices suffice most of the requirements of analog signal processing. They have the capability to operate on very low bias voltages, provide good performance, and high bandwidth. Additionally, the potential to act as a building block for both integer domain and fractional domain circuits have made these active elements widely implemented in the designs.

The fractional circuits have the competency to be highly accurate and precise as per the system's specifications. The realization of fractional-order circuits requires replacement of the true capacitor or the true inductor of integer-order circuits with an appropriate fractional-order element (FOE) [2]. Emulation of FOE is possible by

---

D. Goyal (✉)
Department of ECE, MAIT, New Delhi, India
e-mail: divyagoyal256@gmail.com

P. Varshney
Division of ICE, NSIT, New Delhi, India
e-mail: pragya.varshney1@gmail.com

© Springer Nature Singapore Pte Ltd. 2019

H. Malik et al. (eds.), *Applications of Artificial Intelligence Techniques in Engineering*, Advances in Intelligent Systems and Computing 697,
https://doi.org/10.1007/978-981-13-1822-1_47

using only passive elements or electro-chemically [3, 4].Nowadays, active elements such as OTA and CFOA have also been utilized to realize FOE. These active realizations have found to be useful in field of biomedicines as well as in representing complex biological processes [5, 6].

In this work, the authors have implemented the Current Mode fractional integrator of order $n/3$ for values of $n = \{0.75, 1.0, 1.5, 2.0, \text{ and } 2.25\}$. CCII, the active element of the fractional integrator, is realized using MOSFETs. And, FOE, the fractional entity of the Current Integrator is realized using resistors and capacitors. The magnitudes of all these passive elements are in geometric progression and are further arranged as series RC branches of a parallel ladder. Frequency responses have been obtained using 0.25 µm technology parameters. The OrCAD PSPICE simulation results for all fractional-orders are found to be in accordance with their theoretical values and hence, are satisfactory.

2 Circuit Description

2.1 Second Generation Current Conveyor (CCII)

CCII is a current mode device that operates as voltage and current follower. The device has three terminals: high impedance input terminal Y and output terminal Z and low impedance input terminal X. CCII exhibits the characteristics, as given by equations:

$$V_X = V_Y \tag{1a}$$

$$I_Y = 0 \tag{1b}$$

$$I_Z = I_X \tag{1c}$$

The applications of CCII are widely available in the literature [7–11]. CCII has been implemented to realize integrators, differentiators, universal filters, amplifiers, and PID controllers.

2.2 Fractional-Order Element (FOE) Emulation

The analog designing of a fractional device requires a FOE to be introduced in the realization. Generally, FOE comprises of passive elements that are arranged as a network and provide a fractional impedance of $Z(s) = s^r$, in the circuit. The value 'r' is the order of FOE incorporated and henceforth the order of the fractional circuit. 'A' is the magnitude of the impedance exhibited by FOE. Amongst the passive elements, the feasibility to implement resistors and capacitors (R & C) over resistors and inductors (R & L) makes former combination a dominant choice [2]. In literature, FOE is also termed constant phase element (CPE), as it displays a constant phase according to its order, r [12].

Various methods of designing an analog model of FOE have been illustrated in [12–19]. The arrangement of passive elements in different topologies such as: parallel RC ladder, domino ladder, cross ladder, tree structure, and H structure, are few of the emulation structures. Additionally, the methodology to calculate magnitude of these passive elements also differs. These emulations may utilize all constant-valued elements, or obtained them by implementing any of the rational approximation technique (which is a tri-step method), or calculate the value of the elements by a direct approach.

The technique comprising of only constant-valued R & C network is a simple way of emulating FOE as only the standard and easily available components are required. But, the inflexibility to design for any fractional value of 'r' has restricted their implementation in the fractional circuits [13].

The tri-step approach is an indirect method that comprises of approximation, decomposition, and arrangement as network [14].

- Approximation: In this first step, transfer function of an integer-order system corresponding to the transfer function of the fractional-order system is calculated using either of the rational approximation method illustrated in the literature.
- Decomposition: The approximated higher integer-order transfer function is decomposed into poles and zeros by partial fractions or by continued fraction expansion (CFE) into admittances and impedances.
- Arrangement: Magnitudes of R and C are thereafter calculated from the pole-zero-gain combinations or admittance-impedance pairs. Finally, all are suitably arranged as a network and then incorporated in a circuit with active elements to design a fractional-order system.

Whereas, in the direct methodology illustrated in [12], the consecutive passive elements are in geometric progression, i.e., they are multiplied by constant multipliers $(M_R$ and $M_C)$ to obtain their magnitude values, as

$$R_{a+1} = M_R \cdot R_a \tag{2}$$

$$C_{a+1} = M_C \cdot C_a \tag{3}$$

for $a = (1, 2, \ldots (A-1))$. This calculation of multipliers and magnitudes of the passive elements are dependent on the parameters: operating frequency range $(f_{\min}, f_{\max})$, the order of FOE (r), and the allowable error in phase (e_{p}). The final structure of FOE also includes a balancing resistor (R_{B}) and a balancing capacitor (C_{B}) to balance the requirement of this parallel RC ladder to be of infinite length. These two additional elements are obtained by

$$R_{\mathrm{B}} = R_1 \cdot \frac{1 - M_R}{M_R}$$
$$C_{\mathrm{B}} = C_1 \cdot \frac{(M_C)^A}{1 - M_C} \quad (4)$$

This approach has been implemented in [15, 16] to present the performance of various analog models of fractional-order systems.

2.3 Current Mode Fractional Integrator Realization

The design of Current Mode fractional integrator comprises of CCII and a grounded FOE, as shown in Fig. 1. The *r*-order fractional integrator comprises of a single *r*-order FOE. The FOE is a network of *R* & *C's*, each having their magnitudes in geometric progression. These elements are organized in parallel ladder with series RC branches. A direct methodology has been implemented to emulate the analog FOE of the fractional integrator.

The magnitude response of fractional integrator has slope of $(-20r)$ dB for every decade change and a constant phase of $(-90r)$ degrees. Both the constant slope and constant phase of this integrator are dependent on the fractional order, *r*.

3 Simulation Parameters and Results

The focus of this work is to implement the circuit of current mode fractional integrator for five values of *n*, viz. $n = \{0.75, 1.0, 1.5, 2.0, \text{ and } 2.25\}$. For the

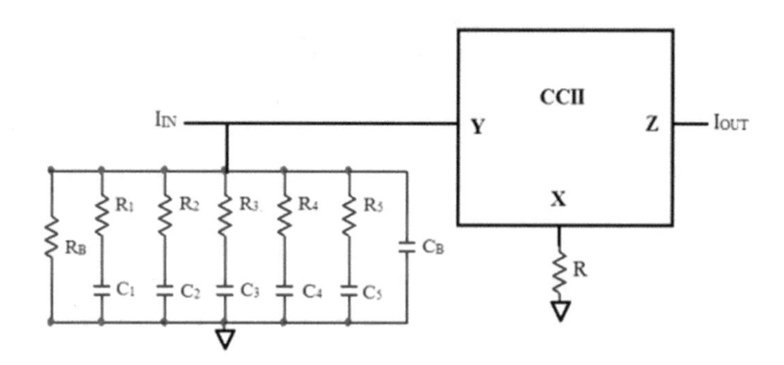

Fig. 1 Block diagram for current mode fractional integrator

chosen values of n, the order of fractional integrators become $r = n/3$, i.e., $r = \{0.25, 0.33, 0.5, 0.66 \text{ and } 0.75\}$.

The MOSFETs based schematic of CCII [20] is supplied with biased voltage of ± 1.25 V for the PSPICE simulations of all the five fractional integrators. 0.25 μm technology parameters have been used to simulate the designs. The analog model of FOE is designed using the direct method of [12]. The FOE has been designed for allowable error in phase, $e_p = 0.4°$ and in frequency range, $(f_{max}, f_{min}) = (10, 10^4)$ rad/s. The length of the ladder (A) is five and the initial values of resistance and capacitance is 10 kΩ and 10 μF, respectively. With the chosen parameters of FOE, the product of constant multipliers is obtained as $M_R \cdot M_C = 0.1714$. For the implemented FOEs, the value of all the passive elements and their constant multipliers are tabulated in Table 1.

Figures 2, 3, 4, 5 and 6 present the frequency responses for $n/3$ fractional current mode integrators. All the frequency responses were observed for the fractional integrators with the input current to be order μA. Ideally, these fractional integrators exhibit slope of $(-20r)$ dB per decade and a constant phase of $(-90r)$ degrees for all values of $r = n/3$, i.e. $r = \{0.25, 0.33, 0.5, 0.66 \text{ and } 0.75\}$.

- The response of integrator of order 0.25 has slope −5 dB per decade and phase −22.5 degrees and it is evident from Fig. 2 that the observed responses are in accordance with their ideal values. Although the phase response is seen to be constant at its ideal value near to 3.0 kHz, but the magnitude response is quite satisfactory for entire range of frequency.
- The response of integrator of order 0.33 has slope −6.67 dB per decade in the frequency range (3.0 Hz to10.0 kHz), as shown in Fig. 3. Moreover, the constant phase of −30° is only observed till 3.0 kHz.
- In Fig. 4, it can be seen that the current mode integrator of order 0.5 displays a constant slope decay of −10 dB per decade with phase remaining constant at −45°. Both the results closely match their respective ideal magnitudes.
- The responses for integrator of order 0.66 as presented in Fig. 5 shows that the slope magnitude and the phase are following their ideal values of −13.3 dB per decade and −60°, respectively. For the three decades of frequencies i.e. from 3.0 Hz to 3.0 kHz, the results are found to be satisfactory.
- It is visible from the Fig. 6 that the current mode integrator of fractional order 0.75 is presenting an ideal slope of −15 dB per decade till 10.0 kHz. But, the phase is appearing to be constant at a value of −67.5° near to 1.0 kHz.

Table 1 Value chart for passive elements and constant multipliers for FOE in fractional integrator

		$r = 0.25$	$r = 0.33$	$r = 0.5$	$r = 0.66$	$r = 0.75$
Constant multiplier	M_R	0.6435	0.5555	0.4140	0.3086	0.2664
	M_C	0.2664	0.3086	0.4140	0.5555	0.6435
Value of passive elements	R_1	10.0 K	10.0 K	10.0 K	10.0 K	10.0 K
	R_2	6434.6	5555.1	4140.39	3085.94	2664.17
	R_3	4140.3	3085.9	1714.2	952.3	709.7
	R_4	2664.1	1714.2	709.7	293.8	189.09
	R_5	1714.2	952.3	293.8	90.6	50.37
	C_1	10.0×10^{-06}	10.0×10^{-06}	10.0×10^{-06}	10.0×10^{-06}	10.0×10^{-06}
	C_2	2.66×10^{-06}	3.08×10^{-06}	4.14×10^{-06}	5.5×10^{-06}	6.43×10^{-06}
	C_3	7.097×10^{-07}	9.5×10^{-07}	1.71×10^{-06}	3.08×10^{-06}	4.1×10^{-06}
	C_4	1.89×10^{-07}	2.93×10^{-07}	7.09×10^{-07}	1.71×10^{-06}	2.66×10^{-06}
	C_5	5.037×10^{-08}	9.06×10^{-08}	2.93×10^{-07}	9.52×10^{-07}	1.7×10^{-06}
Value of Balancing element	R_B	5541.0	8001.4	14.15 K	22.405 K	27.53 K
	C_B	1.8296×10^{-08}	4.0477×10^{-08}	2.0765×10^{-07}	1.1902×10^{-06}	3.0938×10^{-06}

Unit of Resistance (R) is ohms, Capacitance (C) is Farad

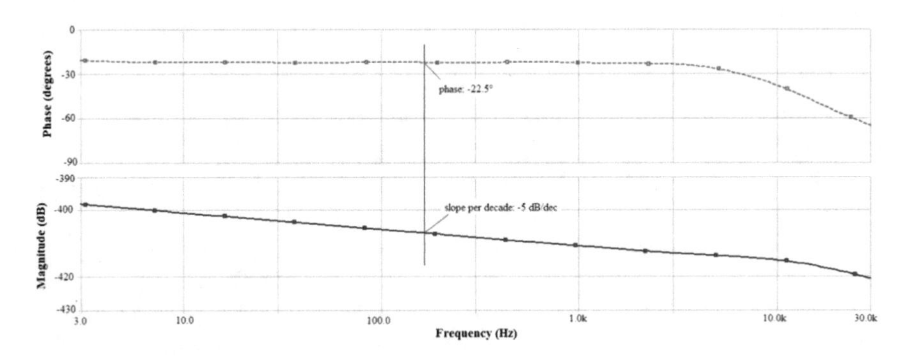

Fig. 2 Frequency response for current mode fractional integrator of order 0.25

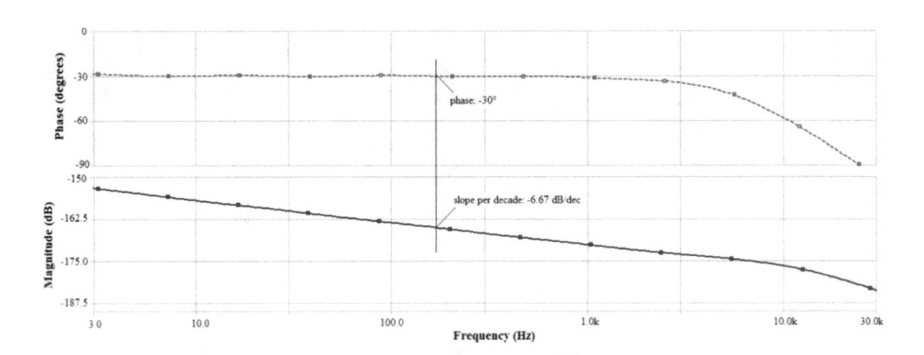

Fig. 3 Frequency response for current mode fractional integrator of order 0.33

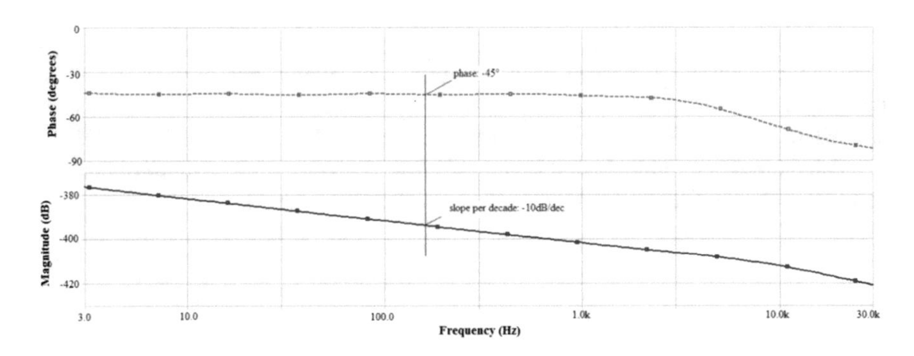

Fig. 4 Frequency response for current mode fractional integrator of order 0.5

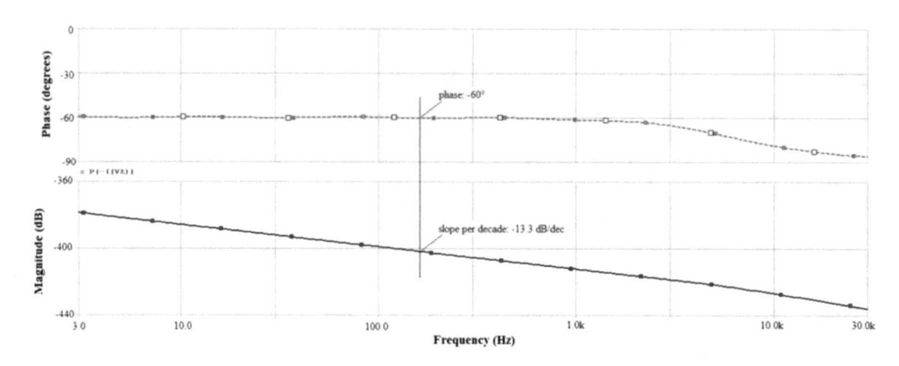

Fig. 5 Frequency response for current mode fractional integrator of order 0.66

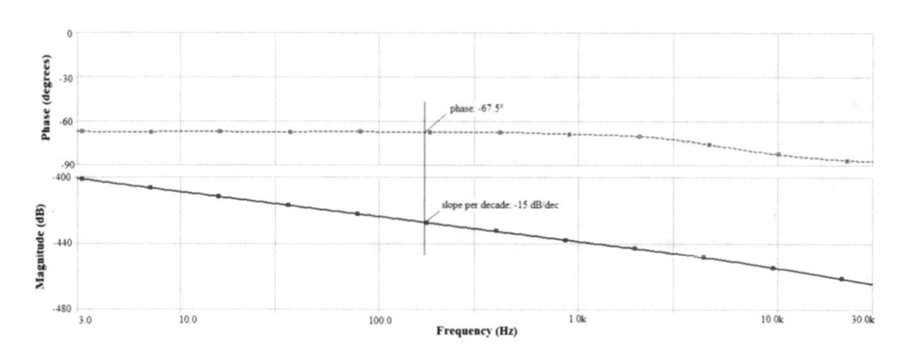

Fig. 6 Frequency response for current mode fractional integrator of order 0.75

4 Conclusion

The implementation of current mode fractional integrators has been presented in this paper. The integrators of orders 0.25, 0.33, 0.5, 0.66 and 0.75 have been simulated in PSPICE and their frequency response has been obtained. All the simulated circuits exhibited magnitude and phase response that are in accordance to their ideal values for nearly three decades of frequency range.

References

1. Biolek, D., Senani, R., Biolkova, V., Kolka, Z.: Active elements for analog signal processing: classification, review, and new proposals. Radioengineering **17**, 15–32 (2008)
2. Pu, Y.F.: Measurement units and physical dimensions of fractance-part I: position of purely ideal fractor in Chua's axiomatic circuit element system and fractional-order reactance of fractor in its natural implementation. IEEE Access **4**, 3379–3397 (2016)
3. Adhikary, A., Khanra, M., Pal, J., Biswas, K.: Realization of fractional order elements. INAE Lett. **2**, 41–47 (2017)

4. Khanra, M., Pal, J., Biswas, K.: Rational approximation and analog realization of fractional order differentiator. In: 2011 IEEE International Conference on Process Automation, Control and Computing (PACC), pp. 1–6 (July 2011)
5. Tsirimokou, G., Psychalinos, C.: Ultra-low voltage fractional-order circuits using current mirrors. Int. J. Circuit Theory Appl. **44**(1), 109–126 (2016)
6. Vastarouchas, C., Tsirimokou, G., Freeborn, T.J., Psychalinos, C.: Emulation of an electrical-analogue of a fractional-order human respiratory mechanical impedance model using OTA topologies. AEU-Int. J. Electron. Commun. **78**, 201–208 (2017)
7. Patranabis, D., Ghosh, D.: Integrators and differentiators with current conveyors. IEEE Trans. Circuits Syst. **31**, 567–569 (1984)
8. Lee, J.Y., Tsao, H.W.: True RC integrators based on current conveyors with tunable time constants using active control and modified loop technique. IEEE Trans. Instrum. Meas. **41**, 709–714 (1992)
9. Liu, S.I., Hwang, Y.S.: Dual-input differentiators and integrators with tunable time constants using current conveyors. IEEE Trans. Instrum. Meas. **43**(4), 650–654 (1994)
10. Erdal, C., Kuntman, H., Kafali, S.: A current controlled conveyor based proportional-integral-derivative (PID) controller. IU-J. Electr. Electron. Eng. **4**(2), 1243–1248 (2004)
11. Yuce, E., Minaei, S.: New CCII-based versatile structure for realizing PID controller and instrumentation amplifier. Microelectron. J. **41**(5), 311–316 (2010)
12. Valsa, J., Vlach, J.: RC models of a constant phase element. Int. J. Circuit Theory Appl. **41**(1), 59–67 (2013)
13. Yifei, P., Xiao, Y., Ke, L., Jiliu, Z., Ni, Z., Yi, Z., Xiaoxian, P.: Structuring analog fractance circuit for 1/2 order fractional calculus. In 2005 IEEE 6th International Conference on ASICON, 2005, pp. 1136–1139 (2005)
14. Charef, A.: Analogue realisation of fractional-order integrator, differentiator and fractional $PI^{\lambda}D^{\mu}$ controller. IEE Proc. Control Theory Appl. **153**(6), 714–720 (2006)
15. Abulencia, G.L., Abad, A.C.: Analog realization of a low-voltage two-order selectable fractional-order differentiator in a 0.35 um CMOS technology. In 2015 IEEE International Conference on Humanoid, Nanotechnology, Information Technology, Communication and Control, Environment and Management (HNICEM), pp. 1–6 (Dec 2015)
16. Gonzalez, E., Dorčák, L., Monje, C., Valsa, J., Caluyo, F., Petráš, I.: Conceptual design of a selectable fractional-order differentiator for industrial applications. Fract. Calc. Appl. Anal. **17**(3), 697–716 (2014)
17. Podlubny, I., Petraš, I., Vinagre, B.M., O'leary, P., Dorčák, L.: Analogue realizations of fractional-order controllers. Nonlinear Dyn. **29**(1), 281–296 (2002)
18. Djouambi, A., Charef, A., Voda, A.: Numerical simulation and identification of fractional systems using digital adjustable fractional order integrator. In: 2013 IEEE European Conference on Control (ECC), 2615–2620 (July 2013)
19. Gonzalez, E.A., Petráš, L.: Advances in fractional calculus: control and signal processing applications. In: 2015 IEEE 16th International Conference on Carpathian Control (ICCC), 147–152 (May 2015)
20. Tangsrirat, W.: Floating simulator with a single DVCCTA. Indian J. Eng. Mater. Sci. **20**, 79–86 (2013)

CSTR Control Using IMC-PID, PSO-PID, and Hybrid BBO-FF-PID Controller

Neha Khanduja and Bharat Bhushan

Abstract A hybrid control approach (BBO-FF) based PID is used for optimal control of a non linear system, i.e., CSTR and this hybrid approach is compared with conventional Z-N tuned PID, IMC-PID, and another optimal control method PSO-PID. The aim of this study is to find an optimal controller for nonlinear process control systems and to achieve best desired output.

Keywords CSTR · PSO · IMC · BBO · FFA

1 Introduction

Operational conditions of dynamical systems are monitored and controlled by controllers [1]. IMC is a controller which is model based. It has a single tuning parameter, i.e., IMC filter coefficient. PID controller parameters are function of this filter coefficient. The proper selection of this filter coefficient is must for robustness of the closed loop system [2].

Particle Swarm Optimization, i.e., PSO is a modern heuristic optimization algorithm which is simple and computationally efficient and was first developed by James Kennedy and Russell Eberhart. Solution of continuous nonlinear optimization problems can be found in robust way. The main advantages with PSO are that it has stable convergence characteristics [3].

Biogeography-based optimization, i.e., BBO is a modern metaheuristic optimization algorithm which was first introduced by Simon in 2008. In BBO the movement of species within different islands depends on water resources, temperature, the diversity of vegetation and land area. These all features are known as suitability index variable, i.e., SIV [4].

The firefly algorithm was developed by Yang X.S. The working principle is based upon the flashing phenomenon of fireflies [5]. The nature of independent

N. Khanduja · B. Bhushan (✉)
Department of EE, Delhi Technological University, New Delhi, Delhi, India
e-mail: bharat@dce.ac.in

© Springer Nature Singapore Pte Ltd. 2019

H. Malik et al. (eds.), *Applications of Artificial Intelligence Techniques in Engineering*, Advances in Intelligent Systems and Computing 697,
https://doi.org/10.1007/978-981-13-1822-1_48

working and close aggregate around the optima values of flies is basic consideration. It helps in finding efficient global optima and local optima simultaneously.

Stability Analysis of CSTR has been explained in Sect. 2. Section 3 explains the PID Controllers. Section 4 describes IMC-based PID controller. Section 5 explains PSO and PSO based PID controller. Sections 6 and 7 explains the concept of FA and BBO. The hybrid of BBO and FF algorithm is explained in Sect. 8. Simulation results are shown in Sect. 9 along with a comparative study.

2 Mathematical Modeling of CSTR

Continuously stirred tank reactors play a most important role in process industry. First-order reaction is considered and it is assumed that reaction is exothermic. It is assumed that CSTR is perfectly mixed, has constant volume and constant parameter values. CSTR has a surrounding jacket and jacket also has in feed and out streams. It is assumed that temperature of jacket is lower than the reactor and it is perfectly mixed (Figs. 1, 2, 3 and 4) [2, 6].

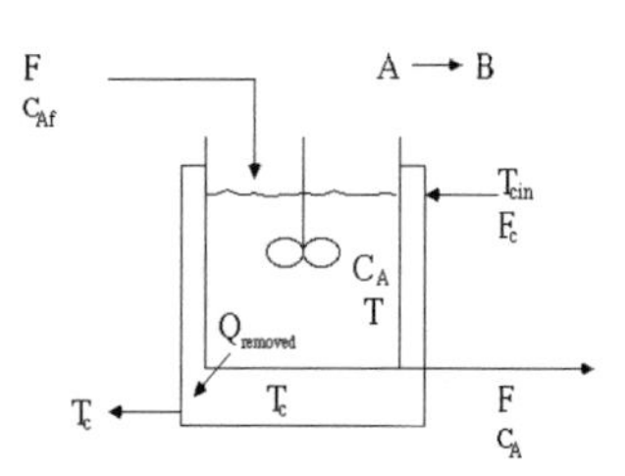

Fig. 1 CSTR state variable representation of dynamic equations

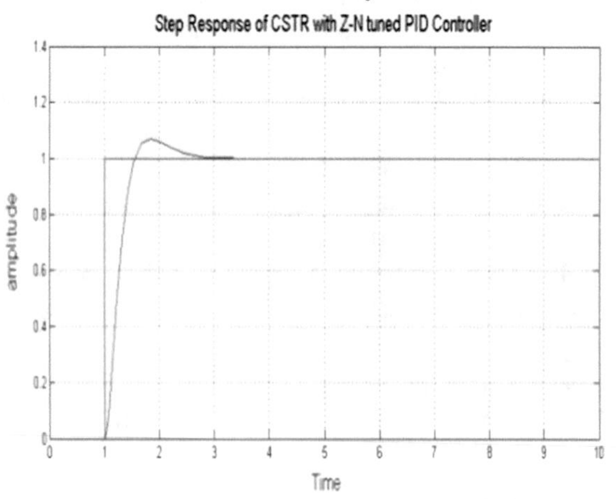

Fig. 2 Step response with ZN-PID controller

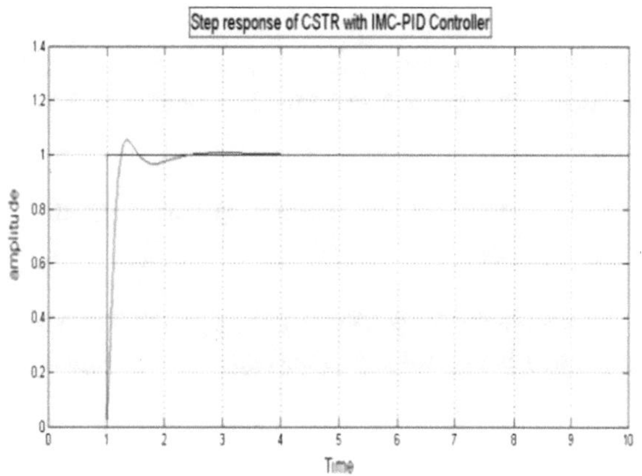

Fig. 3 Step response of PID controller tuned by IMC

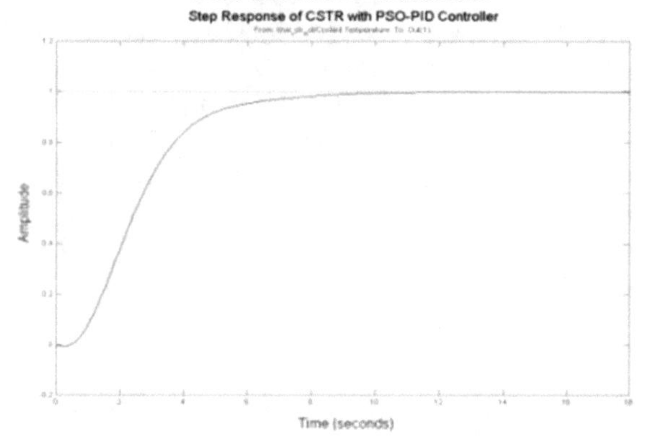

Fig. 4 STEP response of PSO based PID controller

$$f_1(C_A, T) = \frac{dC_A}{dt} = \frac{F}{V}(C_{Af} - C_A) - r \quad (1)$$

$$f_2(C_A, T) = \frac{dT}{dt} = \frac{F}{V}(T_f - T) + \left(\frac{-\Delta H}{\rho c_p}\right) r - \frac{UA}{V\rho c_p}(T - T_j) = 0 \quad (2)$$

State space matrix of CSTR is

$$A = \begin{bmatrix} -\frac{F}{V} - k_s & -C_{As}k_s^{'} \\ \frac{-\Delta H}{\rho c_p} k_s & \left(\frac{-\Delta H}{\rho c_p}\right) C_{As}k_s^{'} - \frac{F}{V} - \frac{UA}{V\rho c_p} \end{bmatrix}$$

3 PID Controller

The PID controller is simple and easy to implement, that is why majorly used in industry. PID controller output is given by following equation:

$$u(t) = K_{\mathrm{p}}e(t) + K_{\mathrm{i}} \int_{0}^{t} e(t)\mathrm{d}t + K_{\mathrm{d}}\frac{\mathrm{d}e(t)}{\mathrm{d}t}, \tag{3}$$

where K_{p} is the proportional gain, K_{i} is the integration gain, and K_{d} is the derivative gain, $u(t)$ is control signal and $e(t)$ is error signal [1].

4 Internal Model Controller (IMC)

IMC is a model-based procedure in which process model and controller are embedded. It was developed by Garcia and Morari. To get a desired temperature trajectory how much heat flow is to be added to the process—this decision is taken by process model and is explained by the set point [1–9].

IMC-based PID controller is used to handle uncertainty and instability in an open-loop control system. Invertible part of transfer (process) acts as controller. As there is a single tuning parameter, i.e., filter tuning parameter which makes IMC an easy and simple controller. Parameters of PID controller are function of filter coefficient [7].

5 Particle Swarm Optimization (PSO)

PSO is a swarm-based optimization technique which was developed by James Kennedy and Russell Eberhart. Efficiency of PSO lies in the concept of exchange of information within the group [8] present velocity, previous position, and the position of its neighbors are the parameters which are used to attain optimum position by individual particle [3].

Velocity update equation is shown in Eq. 4.

$$v_{id}^{n+1} = w \cdot v_{id}^{n} + a_1 \cdot r \cdot \left(p_{id}^{n} - x_{id}^{n}\right) + a_2 \cdot r\left(p_{gd}^{n} - x_{id}^{n}\right) \tag{4}$$

and position update equation is shown in Eq. 5.

$$x_{id}^{n+1} = x_{id}^{n} + v_{id}^{n+1}, \tag{5}$$

$i = 1,2\ldots\ldots p$
$d = 1,2,\ldots\ldots m$
where n—generation pointer; p—total no. of members in a particle; m—total no. of $x_{id}^{(n)}$ current position of particle j at generation n; v_n^{id}—velocity of particle i at generation n; $p\text{best}_j$—personal best position of particle i; $g$best—global best position of the group; r ()—random number whose value lies in between 0 and 1. c1, c2-acceleration constant.

$$W = \frac{w_{\max} - w_{\min}}{t_{\max}} * t \tag{6}$$

Inertia weight factor W is given by Eq. 6 its function is to maintain a balance between global and local search [10, 12].

6 Firefly Algorithm

It is a nature-based metaheuristic algorithm. The main assumptions of FA are

1. Fireflies are attracted by each other and they are unisex in nature.
2. Brightness and attractiveness are directly proportional.
3. Distance and level of brightness has inverse relation.

Value of brightness represents the objective function. Fireflies can communicate to each other because most of the fireflies are visible at limited distance in night [9].

Inverse relation of distance and attractiveness is shown in Eq. 7.

$$I(r) = I_0 e^{-\gamma r^2} \tag{7}$$

I—current intensity of light; I_0—initial intensity of light; γ—absorption coefficient of light; r—distance between two fireflies

β is attractiveness which is given by

$$\beta = \beta_0 e^{-\gamma r^2} \tag{8}$$

$\beta_0 =$ Attractiveness when distance between two fireflies is 0

It is the tendency that firefly i move towards the more attractive firefly j. Mathematically it can be defined as

$$\Delta x_i = \beta_0 e^{-\gamma r_{ij}^2}\left(x_j^t - x_i^t\right) + \alpha\varepsilon_i, \quad x_i^{t+1} + \Delta x_i \tag{9}$$

If $\gamma \to 0$ the brightness and attractiveness are continuous, if $\gamma \to \infty$ the attractiveness and brightness decreases. α is a random parameter whose value lies between 0 and 1 [5].

7 Biogeography-Based Optimization (BBO)

BBO algorithm tells about migration phenomenon of species from one island to another. This algorithm also describes that how new species come into existence and how old species get vanished. Suitability index variables (SIV) are an important parameter of BBO because it characterizes the habitability [4].

Immigration rate λ is given by following equation:

$$\lambda = I_{\max}(1 - R(s)/n) \tag{10}$$

$I_{\max}$ is the maximum immigration rate, $R(s)$ is the fitness rank of solution, and n is number of species

Emigration rate μ is defined by following equation:

$$\mu = E_{\max}(R(s)/n) \tag{11}$$

$E_{\max}$ is the maximum emigration rate [7].

8 Hybrid BBO-FF Algorithm

The main reason to hybridize two or more algorithm is to overcome the limitations of individual algorithm and to get a better and improved version of algorithm. It is also required to find the strength of the hybrid algorithm so that global optima can be achieved in a shorter span of time. A new hybrid algorithm is proposed in this paper by combining firefly and biogeography-based optimization. This algorithm not only utilizes the benefits of both the algorithms but avoid their weakness too. FA is employed to find global optima while BBO has good convergence characteristic and also it has an elitism process which retains the best solution. In this hybrid algorithm process begins with FFA and computation continues with FFA for certain number of iterations so that global best position can be found in search space; then search process is switched to BBO to speed up the convergence characteristic and best solution can be retained.

9 Simulation Results

Z-N method gives high overshoot and high ISE.

Figure 5 shows better step response is obtained by IMC-PID controller.

Hybrid FF-BBO based PID Controller (Tables 1 and 2).

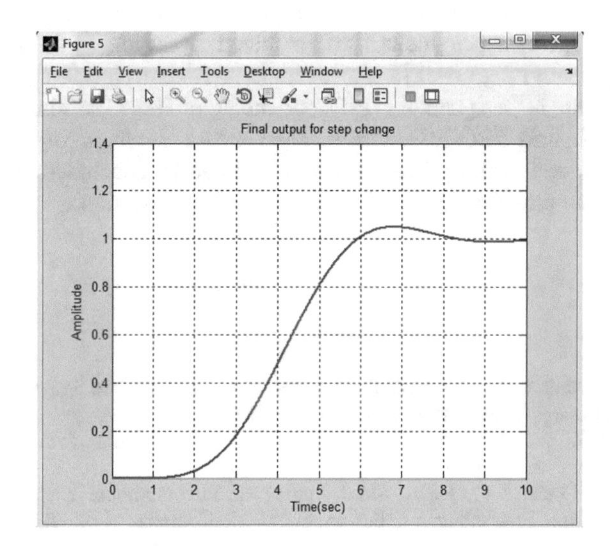

Fig. 5 Step response of hybrid controller

Table 1 Hybrid controller parameters

Parameter	Values	Parameter	Values
BBO algorithm		FF algorithm	
No. of generations (maximum)	200	Lb (matrix)	[0.05 0.25 2.0]
Population size	10	Ub (matrix)	[2.0 1.3 15.0]
Keep rate	0.2	D	15
P-mutation	0.1	Number of decision variables	5
		No. of decision variables Lb	10
		No. of decision variables Ub	10

Table 2 Comparative analysis of different controllers

Performance criterion	Z-N	IMC-PID	PSO-PID	Hybrid BBO-FF PID
Settling time	>20	>15	8.65 s	7.8 s
Maximum overshoot	6.7%	5.34%	0	0.05%
Rise time	1.5 s	1.20 s	4.51 s	3.0 s
P.I.	0.1691 (ISE)	0.06748 (ISE)	0.03 (ISE)	1.93

10 Conclusion

IMC-PID gives satisfactory set point response compared to ZN-tuned PID controller. Better performance indices are obtained by using PSO-PID controller. PSO-PID has main effect on ISE and overshoot. A new algorithm, i.e., BBO-FF hybrid algorithm is designed for PID parameter tuning. The simulation results obtained from hybrid BBO-FF algorithm have shown that main impact of BBO-FF is on overshoot and settling time. In addition, it can be concluded that the hybrid controller gives optimal results and minimize the cost function.

References

1. Cominos, P., Munro, N.: PID controllers: recent tuning methods and design to specification. IEEE Proc. Control Theory Appl. **149**(1), 46–53 (2002)
2. Bequette, B.W.: Process Control, Modeling, Design and Simulation. Prentice-Hall of India Private Limited, India (2003)
3. Solihin, M.I., Tack, L.F., Kean, M.L.: Tuning of PID controller using particle swarm optimization. In: Proceeding of the International Conference on Advanced Science, Engineering and Information Technology, pp. 458–461, Putrajaya, Malaysia (2011)
4. Bhattacharya, A., Chattopadhyay, P.K.: Biogeography-based optimization for different economic load dispatch problems. IEEE Trans. Power Syst. **25**, 1064–1077 (2010)
5. Yang, X.S.: Firefly algorithm, stochastic test functions and design optimization. Int. J. Bio-Inspired Comput. **2**, 78–84 (2010)
6. Glandevadhas, G., Pushpakumar, S.: Robust temperature controller design for a chemical process. Int. J. Eng. Sci. Technol. **2**(10), 5831–5837 (2010)
7. Jain, J., Singh, R.: Biogeographic-based optimization algorithm for load dispatch in power system. Int. J. Emerg. Technol. Adv. Eng. **3**(7), (July 2013)
8. Kennedy, J., Eberhart, R.C.: Particle swarm optimization. In: IEEE Proceedings International Conference on Neural Networks, vol. 4, pp. 1942–1948. Perth, Australia (1995)
9. Naidua, K., Mokhlisa, H., Bakarb, A.H.A.: Application of firefly algorithm (FA) based optimization in load frequency control for interconnected reheat thermal power system. In: IEEE Jordan Conference on Applied Electrical Engineering and Computing Technologies (AEECT) (2013)

Analysis of EEG Signals for Emotion Recognition Using Different Computational Intelligence Techniques

Papia Ray and Debani Prasad Mishra

Abstract Emotion is a very essential aspect in day-to-day life. Emotion can be analyzed by using facial expression, gesture, verbally, and many other ways. But there are some demerits in this technique, so Electroencephalography (EEG) signal is used for recognition of emotion. The most important role of wavelet transform is to remove the noise from the biomedical signals. Analysis of EEG signals using computational intelligence technique like discrete wavelet transform and bionic wavelet transform is presented in this paper. A new modified wavelet transform called Bionic Wavelet Transform (BWT) has been applied here for analysis of biomedical signals. By adapting value of scales, T-function of bionic wavelet transform is varied and its effects on the value of the threshold are noticed. This is called the BWT which is used for emotion recognition using EEG signals. For classification purposes, different classifiers, i.e., Artificial neural network (ANN), k-nearest neighbor (K-NN), Naïve Bayes, and support vector machine (SVM) are presented in this paper. From the proposed algorithm, i.e., with BWT, it is observed that the emotion is better classified than with WT. EEG data are taken from enterface06_emobrain dataset in which there is having the dataset of two subjects which are applied to evaluate the performance of the proposed classifier. In order to find the best method for denoising, signal-to-noise ratio is calculated for different emotions of EEG signal and it is observed that BWT removes the noise better from EEG signal than WT.

Keywords EEG signals · Emotion recognition · Computational intelligence Classification · Bionic wavelet transform · Support vector machine Classifier · Denoising

P. Ray (✉)
Department of Electrical Engineering, VSSUT, Burla, Odisha, India
e-mail: papia_ray@yahoo.co.in

D. P. Mishra
Department of Electrical Engineering, IIIT, Bhubaneswar, Odisha, India
e-mail: debani@iiit-bh.ac.in

© Springer Nature Singapore Pte Ltd. 2019

H. Malik et al. (eds.), *Applications of Artificial Intelligence Techniques in Engineering*, Advances in Intelligent Systems and Computing 697,
https://doi.org/10.1007/978-981-13-1822-1_49

1 Introduction

Biomedical signal processing is an interdisciplinary field that studies extract significant information from biomedical signals. EEG is a biomedical complex signal, i.e., used for analysis of brain irregularities. Here bionic wavelet transform plays an important role in emotion recognition, which contains both time and frequency content of signals. EEG has frequency band spectrum, i.e., five frequency bands for making relationship between EEG and their behaviors. Different methods, i.e., facial declaration [1], speech recognition [2], motion through content [3] are used for emotion recognition. Emotion recognition using independent component analysis (ICA) and Granger causality analysis (GCA) is described in [1]. Based on different emotional condition, ICA was used to decay independent component. To perceive the interrelationship between each independent component, GCA is used in [1]. It is based on graph theory. Features extraction for emotion recognition using DWT is described in [2]. Recognition of emotions using text is described in [3]. There are six emotions, i.e., happiness, disgust, sadness, fear, surprise, anger which is in text format. Here corpus-based syntactic bigrams is used for the improvement of the performance of baseline system. The introduction part of wavelet is described in [4]. Here one- and two-dimensional data have been used for analysis purpose. Fourier transform and wavelet transform are described in [5]. In Fourier transforms, the only frequency component is there. But in wavelet transform both time and frequency components of the signals is present. So wavelet transform is better than Fourier transform. Bionic wavelet transform (BWT) is instigated by Yao and Zhang in [6] and it is mainly used in the human biosystem and has given promising results about speech processing. From speech signal processing, it is concluded that the performance of the BWT is better than WT for cochlear implant. Denoising of ECG signals has been done by multiadaptive bionic wavelet transform (MABWT) [7]. By adapting the values of T-function of the mother wavelet gives accurate results than other techniques. Baseline wander of ECG signals is reduced by using MABWT technique is described in [7]. In multiadaption technique both hard and soft thresholding techniques are used.

Probabilistic neural network (PNN) is used as a classifier; K-NN gives better result than PNN [8]. With lesser computational multifaced nature, K-NN gives 91.33% accuracy on beta band [8]. Artificial neural network (ANN) is used as a classifier for emotion recognition in [9]. As a feature, event-related brain potential (ERP) is used and those features are P100, N100, P200, and P300. ANN and feedforward classifier are used as a classifier in [9]. Seven electrodes, i.e., used in [9] are FZ, FC2, CZ, FC1, PZ, F1, and F2. For recognition of depression state, alpha band wave in EEG signal is analyzed in [10]. Alpha band wave describes about frontal brain asymmetry which is used for diagnosis of depression. Discrete wavelet transform coefficients are used as the features of EEG signals which is used for emotion recognition [11].

The remaining parts of the paper are prepared as follows. Section 2, illuminates computational intelligence techniques alike discrete wavelet transform, bionic wavelet transform. In Sect. 3, focus is made on classification methods. Section 4 discusses simulation outcome and Sect. 5 draws the conclusion.

2 Computational Intelligence Techniques

2.1 Discrete Wavelet Transform (DWT)

Wavelet takes in two arguments: time and scale. The original wavelet transform is called mother wavelet $h(t)$ which used for generating all basic functions. The impulse component and the unwanted high frequency can be eliminated using WT technique. Pyramid algorithm, i.e., in the forward algorithm is used, the signal is processed by WT and in the backward algorithm, and the signal is processed by inverse discrete wavelet transform (IDWT). Wavelet transform of the sampled waveform can be obtained by implementing DWT. For emotion recognition, frequency band partition is necessarily required. DWT decompose the EEG signals to delta, theta, alpha, beta, gamma with sampling frequency 256 Hz. The frequency band, i.e., disintegrated are given in Table 1.

After the first level of decomposition, the signal is decomposed into low and high determination. Then the low determination is further decomposed into low and high determination. Bionic wavelet transform (BWT), which is made known by Yao and Zhang, i.e., used in the human bio system and gives necessary results about speech processing [11].

3 Classification Techniques

In machine learning technique, classification is used for classified the data using different classifiers. The data is divided into two set, i.e., training class data and testing class data in 70–30 proportion. As a classifier, artificial neural network (ANN), K-nearest algorithm (K-NN), Naïve Bayes and support vector machine (SVM) has been used here.

Table 1 Decomposition of EEG signals into different frequency bands

Frequency angle (Hz)	Frequency bands	Decomposition level	Frequency bandwidth (Hz)
0–4	DELTA	CA5	4
4–8	THETA	CD5	4
8–16	ALPHA	CD4	8
16–32	BETA	CD3	16
32–64	GAMMA	CD2	32
64–128	NOISE	CD1	64

3.1 Artificial Neural Network (ANN)

Artificial neural network (ANN) is based on human's central nervous system. In human's central nervous system, many neurons are interconnected with each other. The artificial neural network is a combination of many characteristics. The input signal should be of discreteness and dimensionality type. Adaptive weights are present in this network which is adaptively tuned by many learning algorithm. It is used to increase the speed of the system. Due to ANN, the performance of the system is in time.

3.2 K-Nearest Neighbor (K-NN)

K-nearest neighbor is a nonparametric method which is used for classification purpose. The classification of data is done by the nearest training dataset. It is also used for regression purpose.

If K-NN is used for classification purpose, than the object is classified and the value of that object is assigned to that k-nearest neighbor value. If k value is 1, then the value of object is assigned to single nearest neighbor value. If K-NN is for regression purpose than the value of the object will be the average of k-nearest neighbor.

3.3 Support Vector Machine (SVM)

Support vector machine (SVM) is a machine learning algorithm which gives the maximum classification accuracy as compared to other algorithm. In SVM classifier, Q-dimensional vector is there and it is required to separate that vector from (Q-1) dimensional vector. So it requires a hyperplane. It is called linear classifier. To make the largest separation between two classes, a hyperplane is chosen which makes the two classes more separable. So the distance from the hyperplane to its nearest data is maximized. To decrease the empirical classification error and increase the geometric margin, SVM is used.

3.4 Naïve Bayes

They are basically Bayesian Networks and searches for the class which is most likely with a probabilistic summary. Here each has a variable as sole parent and the variable has no parent.

4 Simulation Results

The performance evaluation is done by using MATLABR2016a. The performance of WT, BWT techniques is analyzed and the SNR values of EEG signals and accuracy are also discussed here. All the five data are taken here from enterface06_emobrain. The different EEG signals are shown here for an emotion recognition purpose in case wise manner (Fig. 1).

4.1 Case-1: Results of Wavelet Transform of Subject1

The data of subject1 of the wavelet transform is fed into four classifiers and they gave different outputs. Some data are trained and the other is given for tests and Table 2 shows how many samples it has given back depending on which accuracy is calculated. Table 2 shows samples for classifying sad and neutral. The bold part shows that K-NN classifier is a better one than others.

Table 3 shows samples for classifying sad and happy of subject1 and the data of wavelet transform is fed into these four classifiers. The bold part shows that k-NN classifies better than others.

Fig. 1 Sad neutral and happy

Table 2 Sample set details of WT for sad versus neutral of subject1

Samples no.	Artificial neural network	K-nearest neighbor	Naive-bayes	Support vector machine
Training	13,975	**13,975**	13,975	13,975
Test	5990	**5990**	5990	5990
Classification	5227	**5475**	3999	2882

Table 3 Sample set details of WT for sad versus happy of subject1

Samples no.	Artificial neural network	K-nearest neighbor	Naive-bayes	Support vector machine
Training	13,975	**13,975**	13,975	13,975
Test	5990	**5990**	5990	5990
Classification	5586	**5754**	4143	2733

Table 4 shows samples for classifying neutral and happy. It has been seen here that K-NN classifier gives better output than other classifiers.

Table 5 shows accuracy rate of four different classifiers used and the difference in accuracy between them. The bold part shows that in k-NN classifier, the accuracy rate of classification is more than others.

4.2 Results of Bionic Wavelet Transform of Subject1

The same representation of classification has been done for BWT and it has been noticed that the samples received are better than Wavelet Transform. Table 6 shows the classification of sad and neutral. The bold part shows that k-NN classifies more data than others.

Table 7 shows the classification for sad and happy emotions. The bold part shows that k-NN classifies more data than others.

Table 8 shows the classification between neutral and happy emotions and it is noticed that K-NN classifies better than others.

The accuracy rate determined from the above samples is shown here in Table 9 above. The bold part shows that in k-NN classifier, the accuracy rate is more than

Table 4 Sample set details of WT for neutral versus happy

Samples no.	Artificial neural network	K-nearest neighbor	Naive-bayes	Support vector machine
Training	13,975	**13,975**	13,975	13,975
Test	5990	**5990**	5990	5990
Classification	5122	**5344**	4375	3082

Table 5 Accuracy rate by WT of subject1

Emotions	Artificial neural network (%)	K-nearest neighbor (%)	Naive-bayes (%)	Support vector machine (%)
Sad	87.25	**91.39**	66.76	48.10
Neutral	93.25	**96.06**	69.15	45.62
Happy	85.50	**89.20**	73.03	51.45

Table 6 Sample set details of BWT for sad versus neutral of subject1

Samples no.	Artificial neural network	K-nearest neighbor	Naive-bayes	Support vector machine
Training	13,975	**13,975**	13,975	13,975
Test	5990	**5990**	5990	5990
Classification	5547	**5950**	5654	3782

Table 7 Sample set details of BWT for sad versus happy of subject1

Samples no.	Artificial neural network	K-nearest neighbor	Naive-bayes	Support vector machine
Training	13,975	**13,975**	13,975	13,975
Test	5990	**5990**	5990	5990
Classification	5820	**5928**	5221	3165

Table 8 Sample set details of BWT for neutral versus happy of subject1

Samples no.	Artificial neural network	K-nearest neighbor	Naive-bayes	Support vector machine
Training	13,975	**13,975**	13,975	13,975
Test	5990	**5990**	5990	5990
Classification	5616	**5949**	5314	4036

Table 9 Accuracy rate by BWT of subject1

Emotions	Artificial neural network (%)	K-nearest neighbor (%)	Naive-bayes (%)	Support vector machine (%)
Sad	92.6	**99.33**	94.39	63.13
Neutral	97.15	**98.95**	87.15	52.83
Happy	93.75	**99.30**	88.71	67.38

others. So it is finally concluded from Tables 5, 6, 7, 8 and 9 that the classification in BWT is better than WT.

4.3 CASE-2: Results of Wavelet Transform of Subject2

The data of the wavelet transform is fed into four classifiers and they gave different outputs. Some data are trained and the other is given for tests and Table 10 shows how many samples it has given back depending on which accuracy is calculated. Table 10 shows samples for classifying sad and neutral. It can be observed from Table 10 that K-NN classifies more samples than other ones.

Table 11 shows samples for classifying sad and happy and the data of wavelet transform is fed into these four classifiers. It can be observed from Table 11 that K-NN classifies more samples than other ones.

Table 12 shows samples for classifying neutral and happy. It can be said here that K-NN classifier gives better output than other classifiers.

Table 13 shows accuracy rate of four different classifiers used and the difference of accuracy between them. It can be noticed from Table 13 that K-NN is better classifier than others.

Table 10 Samples set the details of WT for sad versus neutral of subject2

Samples no.	Artificial neural network	K-nearest neighbor	Naive-bayes	Support vector machine
Training	12,323	**12,323**	12,323	12,323
Test	5282	**5282**	5282	5282
Classification	2631	**4991**	4910	2648

Table 11 Samples set the details of WT for sad versus happy of subject2

Samples no.	Artificial neural network	K-nearest neighbor	Naive-bayes	Support vector machine
Training	12,323	**12,323**	12,323	12,323
Test	5282	**5282**	5282	5282
Classification	2846	**5104**	3825	2434

Table 12 Samples set the details of WT for neutral versus happy of subject2

Samples no.	Artificial neural network	K-nearest neighbor	Naive-bayes	Support vector machine
Training	12,323	**12,323**	12,323	12,323
Test	5282	**5282**	5282	5282
Classification	2425	**4366**	3851	2766

Table 13 Accuracy rate by WT of subject2

Emotions	Artificial neural network (%)	K-nearest neighbor (%)	Naive-bayes (%)	Support vector machine (%)
Sad	49.8	**94.49**	92.95	50.13
Neutral	53.9	**96.63**	72.42	46.07
Happy	45.9	**82.66**	72.91	52.37

4.4 Results of Bionic Wavelet Transform of Subject2

The same representation of classification has been done for BWT and we see that the samples received are better than in case of Wavelet Transform. Table 14 shows the classification of sad and neutral and noticed that K-NN classifies more samples than others.

Table 15 shows the classification of sad and happy emotions. It can be noticed that K-NN classifies more samples than others.

Table 16 shows samples for classifying neutral and happy. It has been seen here that K-NN classifier gives better output than other classifiers.

Depending on the classification, accuracy percentage of the samples is calculated. Table 17 shows accuracy rate of four different classifiers used and can be observed that K-NN is a better classifier than others.

Table 14 Samples set the details of BWT for sad versus neutral of subject2

Samples no.	Artificial neural network	K-nearest neighbor	Naive-bayes	Support vector machine
Training	12,323	**12,323**	12,323	12,323
Test	5282	**5282**	5282	5282
Classification	2641	**5281**	3683	2818

Table 15 Samples set the details of BWT for sad versus happy of subject2

Samples no.	Artificial neural network	K-nearest neighbor	Naive-bayes	Support vector machine
Training	12,323	**12,323**	12,323	12,323
Test	5282	**5282**	5282	5282
Classification	2895	**5282**	4098	2642

Table 16 Samples set the details of WT for neutral versus happy of subject2

Samples no.	Artificial neural network	K-nearest neighbor	Naive-bayes	Support vector machine
Training	12,323	**12,323**	12,323	12,323
Test	5282	**5282**	5282	5282
Classification	2429	**4742**	3913	3806

Table 17 Accuracy rate by BWT of subject2

Emotions	Artificial neural network (%)	K-nearest neighbor (%)	Naive-bayes (%)	Support vector machine (%)
Sad	50.0	**99.94**	69.72	53.34
Neutral	54.8	**100**	77.58	50.02
Happy	46.0	**89.78**	74.07	72.06

Table 18 SNR values of EEG signals

Subject	Sad		Neutral		Happy	
Techniques	WT (dB)	BWT (dB)	WT (dB)	BWT (dB)	WT (dB)	BWT (dB)
Subject1	0.43	21	14.41	25.8	0.36	32
Subject2	0.87	25.6	2.65	29	1.67	26

Table 18 shows the SNR values of EEG signals using WT and BWT techniques. It can be noticed that BWT has better SNR than WT, so is a better noise remover from the signal than WT.

5 Conclusion

In this paper, the wavelet transform is used for emotion recognition purpose using EEG signals, but a new technique, i.e., BWT is used for emotion recognition process. The major difference between WT and BWT is that the mother wavelet function is fixed in WT, but in BWT it is not fixed. WT and BWT is both used to extract the frequency component, i.e., delta, beta, alpha, gamma, and theta wave. For classification purposes, different classifiers, i.e., ANN, K-NN, Naïve Bayes, and SVM are used in this paper. From the comparative analysis, it is observed that the accuracy of classification in BWT technique is more than the other computational intelligence technique, i.e., wavelet transform. A K-nearest algorithm with BWT technique gives maximum accuracy than others. It almost gives 100% accuracy, i.e., the data are accurately classified. This proposed technique, i.e., BWT is reliable and simple as compared to other computational intelligence techniques. Also BWT is a better noise remover from the EEG signal than WT which is observed from the SNR values.

References

1. Dongwei, C., Fang, W., Wang, Z., Haifang, L., Junjie, C.: EEG-based Emotion recognition with Brain Network using Independent Components Analysis and Granger Causality, in: Computer Medical Applications, ICCMA 2013, 1-6
2. Singh, M., Singh, M., Gangwar, S.: Feature extraction from EEG for emotion classification. Int. J. Inf. Technol. Knowl, Manag. **7**(1), 6–10 (December, 2013)
3. Aman, S.: Recognizing emotions in text, M.sc. thesis, University of Ottawa, Canada, pp. 1–97 (2007)
4. Williamsand, J.R., Amaratungay, K.: Introduction to Wavelets in Engineering, pp. 2365–2388. Intelligent Engineering Systems Laboratory, Massachusetts Institute of Technology, MA, USA
5. Sifuzzaman, M., Islam, M.R., Ali, M.Z.: Application of wavelet transform and its advantages compared to fourier transform. J. Phys. Sci. **13**, 121–134 (2009)
6. Yao, J., Zhang, Y.T.: Bionic wavelet transform: a new timefrequency method based on an auditory model. IEEE Trans. Biomed. Eng. **48**(8), 856–863 (2001)
7. Sayadi, O., Shamsollahi, M.B.: Multiadaptive bionic wavelet transform: application to ECG denoising and baseline wandering reduction. EURASIP J. Adv. Signal Process. **1**, 1–11 (2007)
8. Murugappan, M.: Human emotion recognition through short time electroencephalogram (EEG) signals using fast fourier transform (FFT). In: IEEE 9th International Colloquium on Signal Processing and its Applications, Kuala Lumpur, Malaysia, Mac, 2013, pp. 8–10
9. Singh, M., Singh, M.M., Singhal, N.: ANN based emotion recognition along valence axis using EEG. Int. J. Inf. Technol. Knowl. Manag. **7**(1), 56–10 (December, 2013)
10. Niemiec, A.J., Lithgow, B.J.: AIM Alpha-band characteristics in EEG spectrum indicate reliability of frontal brain asymmetry measures in diagnosis of depression. In: Engineering in Medicine and Biology 27th Annual Conference Shanghai, China, September 2005, pp. 1–4
11. Rendi, E., Yohanes, J.: Discrete wavelet transform coefficients for emotion recognition from EEG Signals. In: 34th Annual International Conference of the IEEE EMBS San Diego, California USA, 28 Aug–1 Sept 2012, pp. 2251–2254

Improvization of Arrhythmia Detection Using Machine Learning and Preprocessing Techniques

Sarthak Babbar, Sudhanshu Kulshrestha, Kartik Shangle, Navroz Dewan and Saommya Kesarwani

Abstract Cardiac arrhythmia is a medical condition in which heart beats irregularly. Even with the state-of-the-art equipment, it can be hard to detect. It has to be caught right when it happens. The aim of this paper is to detect and classify this medical condition using machine learning. For this, a variety of sophisticated classifiers such as Random Forest, Logistic Regression (LR), Support Vector Machine (SVM), Naive Bayes, SVM + LR and SVM + RFM have been used to train and test the data. A new approach combining the above models in an exhaustive fashion with Synthetic Minority Over-Sampling Technique (SMOTE) and extremely randomized tree-based feature selection has been used to achieve far better test accuracies than the paper we surveyed.

Keywords Supervised machine learning · Support vector machine
Random forest · Synthetic minority over-sampling technique · Cardiac arrhythmia

S. Babbar · S. Kulshrestha (✉) · K. Shangle · N. Dewan · S. Kesarwani
Jaypee Institute of Information Technology, Noida, India
e-mail: sudhanshu.kulshrestha@jiit.ac.in

S. Babbar
e-mail: sarthakbabbar3@gmail.com

K. Shangle
e-mail: kartikshangle123@gmail.com

N. Dewan
e-mail: navrozdewan@gmail.com

S. Kesarwani
e-mail: saommya@gmail.com

© Springer Nature Singapore Pte Ltd. 2019

H. Malik et al. (eds.), *Applications of Artificial Intelligence Techniques in Engineering*, Advances in Intelligent Systems and Computing 697,
https://doi.org/10.1007/978-981-13-1822-1_50

1 Introduction

Anomaly from the normal sequence of electrical impulses resulting in extremely fast and disorganized rhythm is known as 'Arrhythmia'. Lungs, brain, and other organs fail to work properly and may shut down or get damaged because in arrhythmia, the lower chambers quaver and the heart is unable to pump any blood. Given, the high error rates of computerized interpretation, detecting irregular heartbeat is quite challenging. Arrhythmia detection is usually performed by expert technicians and cardiologists using electrocardiogram. Dissimilarity amongst various types of arrhythmia only makes it harder to detect. It is found that only about 50% of all the computer predictions were correct for non-sinus rhythms [1]. Due to the variability in wave morphology between patients and also because of the presence of noise, this is quite difficult to judge. Various supervised classifications are available and the ones used here are Naive Bayes, Logistic Regression, Support Vector Machine, Random Forest, and further the combination of Logistic Regression and Support Vector Machine, Logistic Regression and Random Forest. The primary aim of our research is to improve the existing research of detection and classification of arrhythmia. We need to achieve better accuracies than the existing ones to practically use software detection of diseases in hospitals. A new approach combining the above-mentioned models in an exhaustive fashion with Synthetic Minority Over-Sampling Technique (SMOTE) and extremely randomized tree-based feature selection have been used. The paper has been organized as follows: Sect. 2 provides the details about previous studies carried out on cardiac arrhythmia classification. In Sect. 3, the detailed dataset is explained. Section 4 explains the methodology and preprocessing. In Sect. 5, results and discussions are summarized to show the performance and comparison of the applied classifiers. Finally, the paper is concluded in Sect. 6.

2 Background Study

Out of all the industries in which technology prevails, health care has been the most paramount. Not only does it save countless lives, machine learning can now accurately predict when the disease will occur. Various researches have been conducted using an approach similar to the one delineated here. One of them includes the research by Gupta et al. [2] in which they have used the machine learning techniques to derive the correlation between 11 chronic diseases and the tests involved in diagnosing them. The research conducted by Nguyen et al. [3] proposes a consolidation of fuzzy Standard Additive Model (SAM) with Genetic Algorithm (GA) to counter with the computational challenges. Wavelet transformation is utilized to extract the important features from high-dimensional datasets. Research by Nivetha and Samundeeswari [4] finds that fuzzy-based classification yields better results than the SVM model. A new feature known as amplitude difference for detection of arrhythmia is studied using the random forest classifier in the research by Park et al. [5]. Research by Kalidas and Tamil [6] uses a combination of LR- and SVM-based machine learning techniques for the classification of each arrhythmia that has been achieved to reduce the occurrence of false alarms in the intensive care unit during the detection of various types of arrhythmia. Research by Rajpurkar et al. [7] develops a 34-layer convolutional neural network for detection of arrhythmia from electrocardiograms. The authors in Kourou et al. [8] have presented a review of the latest ML techniques employed in the modeling of cancer progression. The prognostic models are described based on numerous supervised ML techniques and on various input features and samples of data. Research by Asl et al. [9], explains the benefits of using HRV signal instead of ECG and predicts six types of arrhythmia, which are classified by using generalized discriminant feature reduction scheme with SVM classifier. Another research conducted by Uyar and Gurgen [10] suggests that using a combination of techniques is a good way to overcome the limitations of individual classifiers. It proposes the use of a serial fusion of SVM and LR giving the accuracy of 80%. The feature selection was one of our main concerns with the dataset. The research by Hall and Smith [11] compares two major techniques—wrapper and Correlation-based Feature Selection (CFS) for feature selection. It evaluates these using two algorithms and CFS turns out to be faster and more accurate than the usually better wrapper. Another research by Cruz and Wishart [12] uses MATLAB techniques for feature selection and achieve a maximum accuracy of 77.4%. The following are the techniques we have used in our research.

2.1 *Naive Bayes*

Naive Bayes algorithm is useful for classification of problems. It belongs to the family of probabilistic classifiers, which is based on Bayes theorem. It assumes that the existence of a specific feature is not dependent on the occurrence of other features. It is measured by calculating conditional probability which is given by Bayes' theorem:

$$P(X|Y) = \frac{P(Y|X)P(X)}{P(Y)}, \tag{1}$$

where $P(X|Y)$ is the conditional probability of event X assuming that the event Y is also true, $P(X)$ and $P(Y)$ are probabilities of the occurrence of events X and Y, respectively.

2.2 Random Forest

Random forest is an ensemble algorithm which is used for the classification and regression that will construct a number of decision trees at the time of training and outputting the class that is the mode of the output of individual trees. The drawback of single decision tree is that it is prone to noise. Random forest tries to reduce the noise as it can aggregate many decision trees to give more accurate results.

2.3 Logistic Regression

Logistic regression (LR) is one of the most popular machine learning algorithms for binary classification. It is an easy algorithm for performing very fine on a wide range of problems. Taking $\times 1$ and $\times 2$ as input variables at the training instance and B0, B1 and B2 as coefficients that are updated implementing (3), the output of the model used for making 'prediction' is computed as given by (2)

$$\text{Prediction} = 1/\left(1 + e^{(-(B_0 + B_1 * x_1 + B_2 * x_2))}\right) \quad (2)$$

Let 'B' be the coefficient to be updated, 'y' be the output variable, 'x' be the input value for the coefficient and 'a' be the parameter known as 'learning rate', which is specified at the beginning of the training run to control the amount of changes in the coefficients, and hence in the model. The coefficients B_0, B_1 and B_2 are updated to new coefficient values as given by (3)

$$B = B + a * (y - \text{Prediction}) * \text{Prediction} * (1 - \text{Prediction}) * x. \quad (3)$$

2.4 Support Vector Machine

Support Vector Machine (SVM) is a good supervised technique and is powerful towards outliers as well. It has the capability of dealing with high-dimensional data. It has a unique technique called the kernel trick, which basically converts not distinguishable problem to distinguishable problem, and kernels are the name of these functions. In SVM, each data item is plotted as a point in x-dimensional space, where number of features your dataset has is equal to x. Then, we carry-out by finding a hyperplane, which segregates the two classes well. We will be using SVM with three different kernels as follows:

- Linear—A hyperplane is written as the set of points $\vec{y}$ which will. Where in the normal vector to the hyperplane is $\vec{r}$. The parameter which defines the offset of the hyperplane from the origin along the normal vector $\vec{r}$ is $u/\|\vec{r}\|$.

$$\vec{r} \cdot \vec{y} - u = 0. \tag{4}$$

- Polynomial—polynomial kernel is defined by a degree one. We obtain from training or test samples, feature vectors z and m and $q \geq 0$ are parameters responsible for trading off the effect of lower order terms with higher order terms.

$$k(z, m) = (z^T m + q)^1. \tag{5}$$

- RBF—Radial Basis Function kernel. Here, we have the distance (squared Euclidean) between the two vectors that are equal to $|n{-}n'||^2$. The free parameter is represented by σ.

$$k(n, n') = e\left(-||n - n'||^2\right)/2\sigma^2. \tag{6}$$

2.5 *Fusion of Above Techniques*

Ensembles are one of the ways to improve your accuracies on the dataset. Here, a voting method is being used which combines different types of techniques and their predictions to give an improved and better result. Here, a combination of SVM (linear classifier) with logistic regression and a combination of SVM with random forest is implemented using an ensemble. Combining two techniques helps to overcome the shortcomings of individual technique.

3 Data Collection

The dataset which we are working on has been taken from UCI machine learning repository [13]. The title of the dataset is Cardiac Arrhythmia Database. The original owners of the database are Dr. H. Altay Guvenir, Burak Acar and Dr. Haldun Muderrisoglu. The dataset contains 452 instances and 279 features. Table 1 mentions some of the features of the dataset. The dataset contains 11 classes of arrhythmia, one normal class and one unlabeled as shown in Table 2.

Table 1 Features description

S. no.	Feature	Type of data
1.	Age	Linear
2.	Sex	Nominal
3	Height	Linear
4	Weight	Linear
5	QRS duration	Linear
6	P-R interval	Linear
7	Q-T interval	Linear
8	T interval	Linear
9	P interval	Linear
10	QRS	Linear
11	T	Linear
12	P	Linear
13	QRST	Linear
14	J	Linear
15	Heart rate	Linear
16	Q wave	Linear
17	R wave	Linear
18	S wave	Linear
19	R′ wave	Linear
20	S′ wave	Linear

Table 2 Class distribution

Label code	Output label	No. of rows
01	Normal	245
02	Ischemic changes (coronary artery disease)	44
03	Old anterior myocardial infarction	15
04	Old inferior myocardial infarction	15
05	Sinus tachycardia	13
06	Sinus bradycardia	25
07	Ventricular premature contraction (PVC)	3
08	Supraventricular premature contraction	2
09	Left bundle branch block	9
10	Right bundle branch block	50
11	First-degree Atrioventricular block	0
12	Second-degree AV block	0
13	Third-degree AV block	0
14	Left ventricular hypertrophy	4
15	Atrial fibrillation	5
16	Others	22

4 Methodology

The diagnosis of cardiac arrhythmia involves an ECG test that records the heart's electrical impulses. The parameters produced by the test along with other features like age, height and weight are used to construct the dataset. Using various supervised machine learning models, the dataset is trained and the classification is done into one of the 13 classes.

4.1 Data Preprocessing

It is critical to preprocess our data since it can be noisy, incomplete, missing attribute values, errors, and outliers and discrepancies in codes and names. Following are the techniques used:

- *Missing Values*: Inbuilt feature in Pandas is used to delete columns with missing values using a function called dropna() and to replace missing values using a function called fillna().
- *Imbalanced Dataset*: SMOTE generates synthetic samples of the minor class instead of creating copies (oversampling) with the help of similar instances, as shown in Fig. 1. SMOTE samples are chosen on the basis of the following equation:

$$s = z + y \times (z_R - z), \tag{7}$$

  with $0 \leq y \leq 1$; z_R is selected among the five nearest minority classes. SMOTE produces better results than simple oversampling and SMOTE samples are not redundant. Figure 1 shows the comparison of the initial and SMOTE datasets. A comparison of the age data (one of the features) before and after applying SMOTE is shown in Figs. 2 and 3.
- *High Dimensionality*: The dataset is highly dimensional with 280 features, a lot of them being redundant. Use of extremely randomized tree-based feature

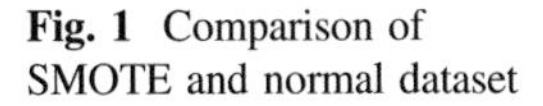

Fig. 1 Comparison of SMOTE and normal dataset

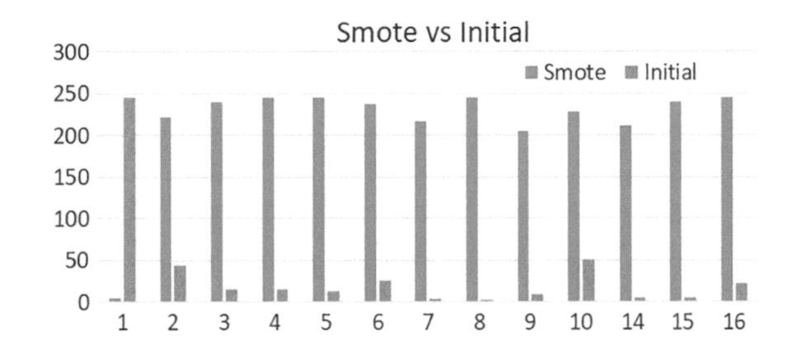

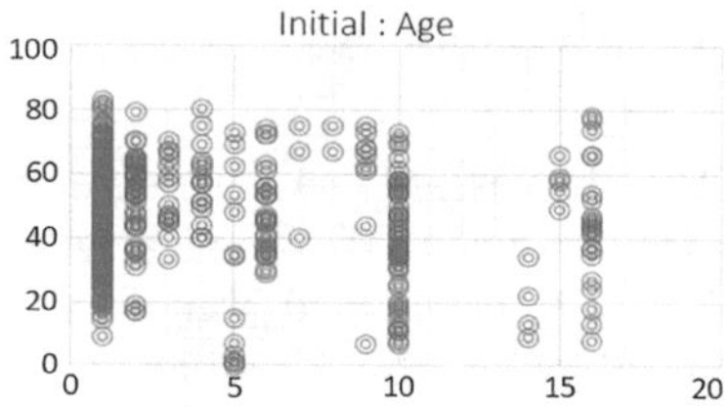

Fig. 2 Before SMOTE applied on age attribute

Initial : Age

(X-axis:Class no. Y-axis:age in years)

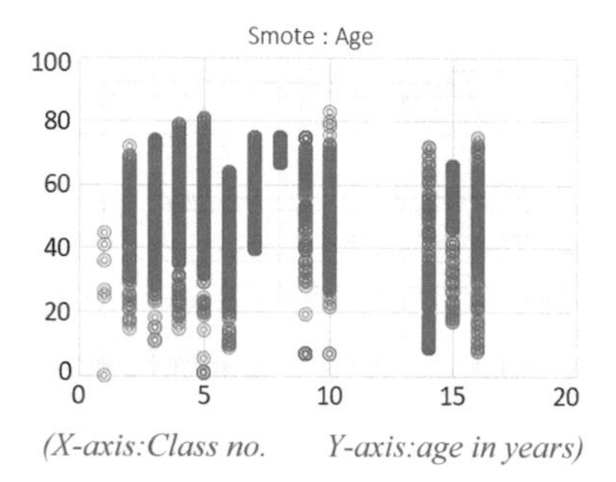

Fig. 3 After SMOTE applied on age attribute

selection has been made. Just like in random forest, a random subset of features is selected, but instead of choosing the best split among the subset of features, it draws thresholds at random for all candidates and chooses the best one out of these for the splitting. It reduces the variance of the model but increases the bias.

Figure 4 shows the importance of individual features. The above classifier provided the importance of all the features in a numerical form from which we have further selected the top 100 features of the 280 total features. This reduces the computation time as well as increases the accuracy of techniques at the same time. For combining these two algorithms, SMOTE is applied over extremely randomized tree-based feature selected dataset. The result will have only 100 columns but a lot more instances due to SMOTE oversampling.

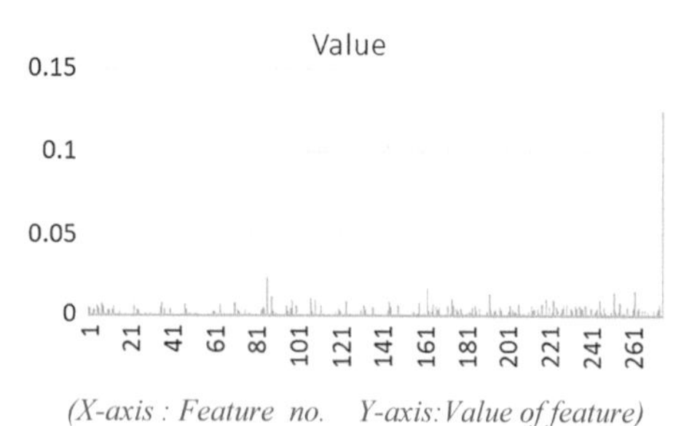

Fig. 4 Importance of features

4.2 Model Selection

The choice of selecting a perfect model depends on the dataset available and the size of the dataset, whether we have high-dimensional or low-dimensional dataset. A rigorous testing approach is followed to get the maximum possible accuracy by testing as many models as possible. Naive Bayes, Random Forests, SVM, LR, SVM + LR and SVM + RFM are the models which were trained and tested.

The accuracies of the above-mentioned models are measured. Comparing them we get a prominent and stupendous picture of selecting the best model possible for our dataset.

5 Result Analysis

The research carried out implements a total of six models and evaluates them for four different kinds of preprocessed datasets. Naive Bayes was removed for testing with the dataset other than the original one because it gave highly variable accuracies each time. The results for the original dataset show SVM + LR outperforming all other models (Table 3, Fig. 5).

Tables 4 and 5 give the metric scores of LR and SVM, where p and f are the number of true positives and false positives, respectively. Precision is the ratio which is given as follows:

$$p/(p+f). \tag{8}$$

The precision describes the ability of not classifying a label as positive for a sample that is negative. The recall score is given by the ratio:

$$p/(p+f). \tag{9}$$

Table 3 Accuracy for original dataset

Classifiers	Accuracy
RFM + SVM	72.0821256
RFM	69.19323672
LR	75.16778523
SVM + LR	78.5024
Naïve Bayes	59.33
SVM (poly)	70.58
SVM (linear)	67.64
SVM (RBF)	50

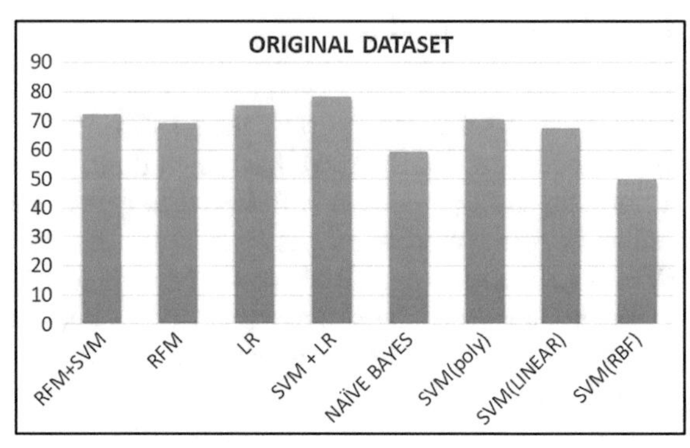

(X-axis:Name of Technique Y-axis:Accuracy in %)

Fig. 5 Accuracy of different classifiers on normal dataset

Table 4 Metrics scores of logistic regression

Dataset	Precision	Recall	F1-score	Support
SMOTE	0.99	0.99	0.99	936
Tree-based	0.76	0.74	0.73	149
Original	0.76	0.75	0.74	149
SMOTE and tree	0.99	0.99	0.99	971

Table 5 Metrics scores of support vector machine

Dataset	Precision	Recall	F1-score	Support
SMOTE	0.93	0.52	0.58	426
Tree-based	0.3	0.54	0.38	68
Original	0.3	0.54	0.38	68
SMOTE and tree	0.99	0.99	0.99	971

It defines the classifier's power of finding the positive samples. The F1 score is the harmonic average of recall scores and precision scores. The support defines the average occurrence of each class in the target array (class feature).

With the tree-based feature selection as shown in (Table 6, Fig. 6), SVM + LR continues to give the highest accuracy that is 81.16%. The tree-based feature selection improves almost all the models except SVM (poly), which reduces by about 1%. This is due to the fact that the irrelevant features are discarded. Training models were also faster on this dataset as it has only 100 features compared to the original dataset.

It can be seen in Table 7 and Fig. 7 that the SMOTE drastically improves the accuracy of SVM (poly) from 70.58 to 99.76%, which is also the highest. While SMOTE does not improve the accuracy of all the models, LR, SVM (poly) and SVM (linear) reach near perfect accuracies of about 99%. This improvement was expected as SMOTE has balanced the dataset, so that there is no bias towards majority classes.

Table 6 Comparison of techniques with Tree-Based Feature Selection

Classifier	Normal (%)	Tree-based (%)
RFM + SVM	72.08	77.16
RFM	69.19	75.38
LR	75.17	73.83
SVM + LR	78.50	81.16
SVM (poly)	70.58	69.11
SVM (linear)	67.64	76.47
SVM (RBF)	50	54.41

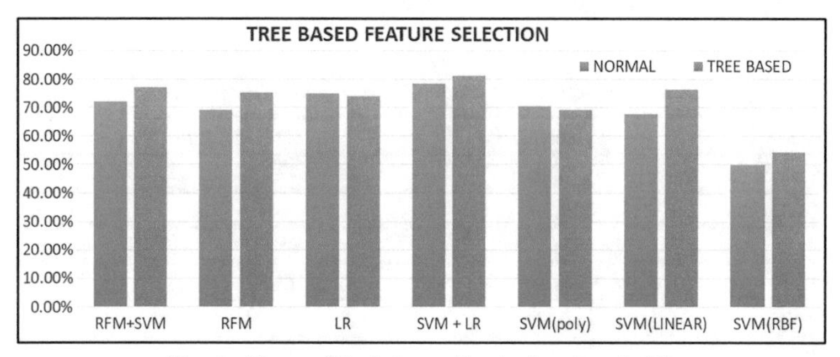

(X-axis: Name of Technique Y-axis:Accuracy in %)

Fig. 6 Comparison of model accuracy of normal and tree-based dataset

Table 7 Comparison of techniques with SMOTE

Classifiers	Normal (%)	SMOTE (%)
RFM + SVM	72.08	57.70
RFM	69.19	70.31
LR	75.17	99.14
SVM + LR	78.28	58.29
SVM (poly)	70.58	99.76
SVM (linear)	67.64	99
SVM (RBF)	50	52.11

We can note that unlike SMOTE, in which RFM + SVM and SVM + LR suffer a huge downgrade in accuracies, the combination of SMOTE and tree (Table 8; Fig. 8) offers a great improvement in accuracy without having a drastic negative effect. It is interesting to note in Fig. 9 that both SMOTE and SMOTE and Tree-based datasets show a similar improvement trend throughout denoting an '*M*' shape. The tree-based dataset, on the other hand follows a trend similar to the initial dataset. The SVM (poly) with the SMOTE and tree dataset gives the highest accuracy of 99.77%. The accuracy of LR, SVM (poly) and SVM (linear) stand out from the rest of the models for the SMOTE and the combination of SMOTE and tree-based dataset.

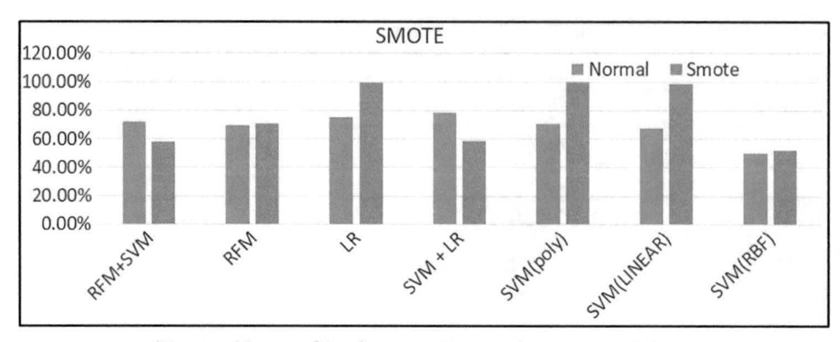

(X-axis: Name of Technique Y-axis:Accuracy in %)

Fig. 7 Comparison of techniques with SMOTE dataset

Table 8 Comparison of techniques with combination of SMOTE and tree

Classifiers	Normal (%)	SMOTE and tree (%)
RFM + SVM	72.08	73.22
RFM	69.19	73.46
LR	75.17	98.97
SVM + LR	78.28	73.87
SVM (poly)	70.58	99.77
SVM (linear)	67.64	99
SVM (RBF)	50	59.50

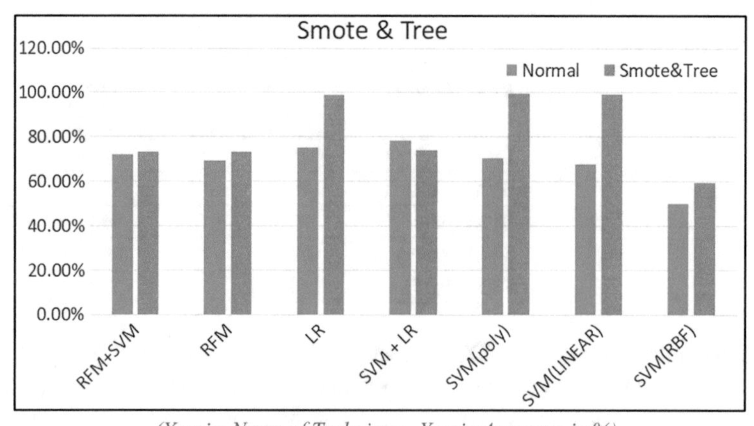

(X-axis: Name of Technique Y-axis:Accuracy in %)

Fig. 8 Comparison of techniques with SMOTE and tree-based dataset

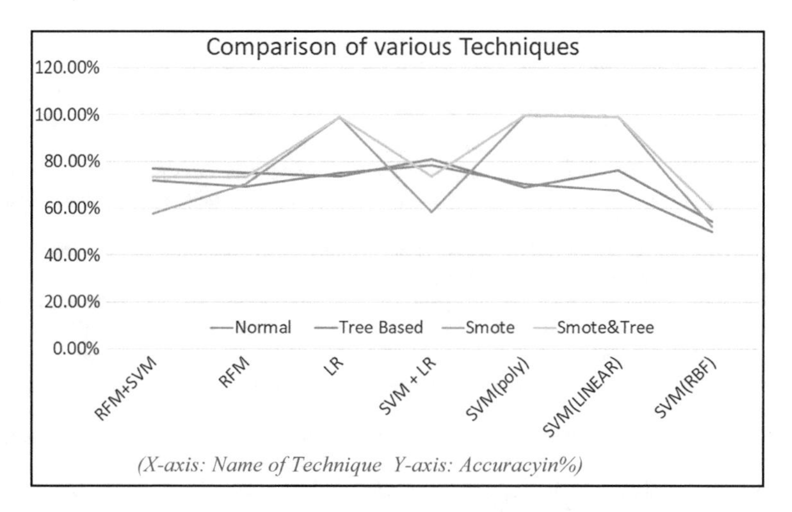

Fig. 9 Comparison of all techniques

6 Conclusion

The previous researches by Gupta et al. [14] conducted on cardiac arrhythmia show the accuracies of models as Naïve Bayes: 47%, Random Forests: 72%, SVM with RF: 77.4%, SVM with polynomial degree 2: 66%, Pattern net: 69% and TWO-level RF: 70%. The study suggested here achieves better accuracies: SVM (poly): 99.77%, SVM (Linear): 99% and LR: 99.14%. This has been achieved using the new and unique combination of preprocessing techniques such as SMOTE and extremely randomized trees along with the above-mentioned models. A larger dataset, with the already impeccable accuracy of this proposed model, could help to introduce such software for actual disease detection in hospitals. Recording the data from the ever popular digital watches with heart rate monitors could help increase the dataset. It will help doctors in better understanding and detection of cardiac arrhythmia. Whether or not the software will altogether overtake doctors in the future still remains a question.

References

1. Shah, Atman P., Rubin, Stanley A.: Errors in the computerized electrocardiogram interpretation of cardiac rhythm. J. Electrocardiol. **40**(5), 385–390 (2007)
2. Gupta, D., Sangita K., Ashish A.: A method to predict diagnostic codes for chronic diseases using machine learning techniques. In: 2016 International Conference on Computing, Communication and Automation (ICCCA), IEEE (2016)
3. Nguyen, T., et al.: Classification of healthcare data using genetic fuzzy logic system and wavelets. Expert Syst. with Appl. **42**(4), 2184–2197 (2015)

4. Nivetha, S., Samundeeswari, E.S.: Predicting survival of breast cancer patients using fuzzy rule based system. Int. Res. J. Eng. Technol. **3**(3), 962–969 (2016)
5. Park, J., Lee, S., Kang, K.: Arrhythmia detection using amplitude difference features based on random forest. In: 2015 37th Annual International Conference of the IEEE Engineering in Medicine and Biology Society (EMBC), IEEE (2015)
6. Kalidas, V., Tamil, L.S.: Enhancing accuracy of arrhythmia classification by combining logical and machine learning techniques. In: Computing in Cardiology Conference (CinC), 2015, IEEE (2015)
7. Rajpurkar, P., et al.: Cardiologist-level arrhythmia detection with convolutional neural networks. arXiv preprint arXiv:1707.01836 (2017)
8. Kourou, K., et al.: Machine learning applications in cancer prognosis and prediction. Comput. Struct. Biotechnol. J. **13**, 8–17 (2015). https://doi.org/10.1016/j.csbj.2014.11.005
9. Asl, B.M., Setarehdan, S.K., Mohebbi, M.: Supportvector machine-based arrhythmia classification using reduced features of heart rate variability signal. Artif. Intell. Med. **44** (1), 51–64 (2008)
10. Uyar, A., Gurgen, F.: Arrhythmia classification using serial fusion of support vector machines and logistic regression. In: 4th IEEE Workshop on Intelligent Data Acquisition and Advanced Computing Systems: Technology and Applications, IDAACS 2007, IEEE (2007)
11. Hall, M.A., Smith, L.A.: Feature selection for machine learning: comparing a correlation-based filter approach to the wrapper. In: FLAIRS Conference, vol. 1999 (1999)
12. Cruz, J.A., Wishart, D.S.: Applications of machine learning in cancer prediction and prognosis. Cancer Inf. **2**, 59 (2006)
13. UCI Machine Learning Repository: Arrhythmia data set https://archive.ics.uci.edu/ml/datasets/arrhythmia
14. Gupta, V., Srinivasan, S., Kudli, S.S.: Prediction and classification of cardiac arrhythmia (n.d.)

Identification of Hidden Information Using Brain Signals

Navjot Saini, Saurabh Bhardwaj and Ravinder Agarwal

Abstract A lot of information can be extracted and understood from brain signals. The brain gives an insight into the various processes underlying our behavior and responses. If an individual is hiding information in the brain, this information can be detected using brain signals. In the present work, electroencephalogram (EEG) signals of the subjects are analyzed to detect hidden information in the individuals. After preprocessing the data, wavelet features are extracted. Initially, classification algorithm based on support vector machine (SVM) has been used to identify the subjects hiding the information. Afterwards, k-nearest neighbor (kNN) algorithm has been used to distinguish between the guilty and innocent participants. SVM performs better than kNN and achieves a significant classification accuracy of 80% in identification of the concealed information.

Keywords EEG · SVM · Wavelet · ERP · kNN

1 Introduction

An individual can hide some information for any reason. The individual can be a culprit or may be hiding information to protect another individual. Wide research is taking place around the world to detect concealed information using various techniques. Some of the methods being used are electroencephalography (EEG) [1], electrocardiography (ECG), skin conductance (SC) measurement and functional magnetic resonance imaging (fMRI) [2]. P300, a component of event related

N. Saini (✉) · S. Bhardwaj · R. Agarwal
Electrical and Instrumentation Engineering Department, Thapar Institute of Engineering & Technology, Patiala, India
e-mail: navjot.saini@thapar.edu

S. Bhardwaj
e-mail: saurabh.bhardwaj@thapar.edu

R. Agarwal
e-mail: ravinder_eeed@thapar.edu

© Springer Nature Singapore Pte Ltd. 2019

H. Malik et al. (eds.), *Applications of Artificial Intelligence Techniques in Engineering*, Advances in Intelligent Systems and Computing 697,
https://doi.org/10.1007/978-981-13-1822-1_51

potential (ERP), is generated in the brain [3] when meaningful information is presented to an individual. In the present work, EEG data has been analyzed. EEG is the recording of the activity of the neurons in the brain of a person.

There are generally three kinds of stimuli presented on a computer screen, probe stimuli, target stimuli, and irrelevant stimuli. Probe stimuli are related to the information which is known to a person. Irrelevant stimuli are related to information which is unknown to a person. The target stimuli are used to recognize whether a person is attentive during testing session or not. Various time, frequency, and wavelet-based features can be extracted from the acquired physiological data. If features are extracted properly, then it leads to better classification of the data.

The remaining paper is prepared such that the description of EEG data, preprocessing, extraction of features, and classification methodology have been explained in part 2. Afterwards, the results have been deliberated upon in part 3. Finally, the concluding information is presented in part 4.

2 Materials and Methods

In the present work, EEG data has been obtained from one of the studies [4]. In that study, 12 electrodes were used to acquire the data of the participants. The vertical electrooculogram (VEOG) and the horizontal electrooculogram (HEOG) signals were also recorded. The sampling rate was 500 Hz. Ten subjects' data have been analyzed in this work. Five subjects belong to the guilty group and five subjects belong to the innocent group. Each subject's probe response in one session has been examined.

2.1 Preprocessing of EEG Data

Preprocessing of the brain signals has been performed using EEGLAB toolbox [5]. The data has been filtered using a bandpass filter in the frequency range of 0.1–30 Hz. Various artifacts like eye blinks, muscle artifacts, and power line noise have been removed to obtain artifact free EEG data. A flowchart depicting the processing of the signals is presented in Fig. 1.

2.2 Features Extraction

A channel of EEG signals after removal of artifacts is depicted in Fig. 2 for a guilty and an innocent subject. Wavelets [6] are being used in many applications. Other types of features such as the recurrence plot features [7] and autonomic features [8] have been studied to analyze brain signals. Wavelet features have been extracted in

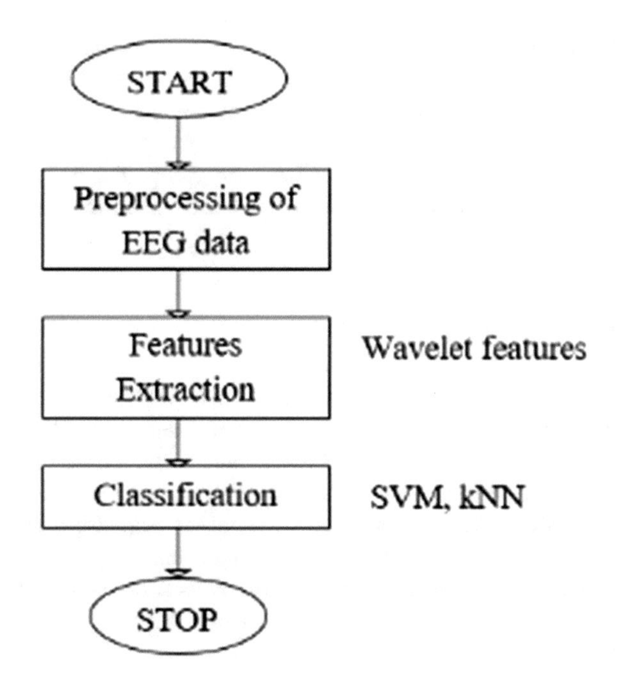

Fig. 1 General signal processing flowchart

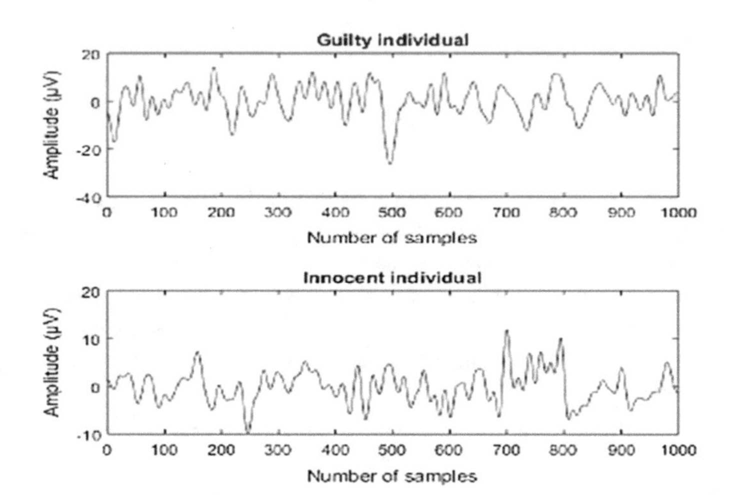

Fig. 2 A preprocessed channel of EEG data corresponding to a guilty individual (top row) and an innocent individual (bottom row)

this study. Symlet (sym4) wavelet has been used to extract wavelet coefficients in the delta frequency band (0.1–4 Hz) of the EEG signals. Each individual's EEG data is decomposed into seven pairs of wavelet coefficients related to different frequency ranges. In this way, features based on two groups are obtained. One group is the guilty group and other is the innocent group. The guilty group data is given the class label 1 and the innocent group data is given the class label −1.

2.3 Classification

Machine learning techniques are being used for analyzing EEG signals [9]. Earlier, unsupervised classifier [10] has been used in P300 detection. However, support vector machine (SVM) classifier is efficient and widely used in classification applications. In SVM, a hyperplane as a decision boundary is obtained, such that the separation between the target classes is maximum. Suppose training data consists of data points x_i and the corresponding target values y_i, $i = 1,2,\ldots,k$, where k denotes the total instances of data points. y_i can consist of values such as [0, 1] or [− 1, 1]. In SVM, following optimization problem is solved

$$\begin{aligned} &\min_{w,b,\xi} \quad \frac{1}{2} w^{\mathrm{T}} w + \mathrm{C} \sum_{i=1}^{k} \xi_i \\ &\text{such that} \quad y_i\left(w^{\mathrm{T}} \varnothing(x_i) + b\right) \geq 1-, \xi_i, \\ &\xi_i \geq 0 \end{aligned} \tag{1}$$

where, $C > 0$ is regularization parameter.

Radial basis function (RBF) kernel is applied, which is specified as: $K(x, y) = \mathrm{e}^{-\gamma||x-y||^2}, \gamma > 0$. 10-fold cross validation is used to select the optimum values of the classifier parameters which are further used for training and testing of data. In SVM classification, classifier parameter C and kernel parameter γ are tuned.

The classifier based on k-nearest neighbor (kNN) has been additionally used for comparison with SVM. The kNN method classifies the test samples based on the class values of the nearest training samples. In kNN, no model is obtained using the training samples.

Suppose, there are m number of training samples, $d_1, d_2, \ldots, d_m$ and T is the test sample to be classified. If d_l is the most similar sample out of the m training samples, to the test sample T, then, T is assigned the class of d_l. In order to determine the similarity between T and d_l, distance between them is evaluated as

$$d(T, d_l) = \min\{d(T, dj)\}, \tag{2}$$

where, $j = 1,2,\ldots,m$ and $d(T,dj)$ represents the distance between T and each of the m training samples.

T is assigned the class of d_l, due to the minimum distance between T and d_l out of all the other distance values. Here, the value of k has been set to 2, since the training data is small.

3 Experimental Results

3.1 Wavelet Features Extraction

The Symlet (sym4) wavelet coefficients extracted in the delta frequency band (0.1–4 Hz) for the three subjects, each in the guilty and innocent groups, are presented in Fig. 3.

3.2 Classification

Classification accuracy of 80% has been achieved using SVM-based classification algorithm. During classification, 50% data is used for finding the optimum parameter values and training. The remaining 50% data is used for testing.

The kNN classifier classifies the test data with an accuracy of 60%. Similar to SVM, 50% data has been utilized as training data and 50% data is utilized for testing.

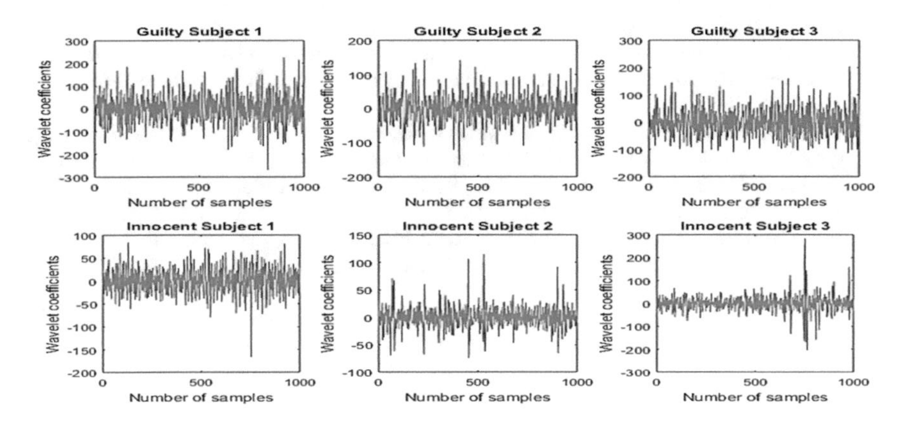

Fig. 3 sym4 wavelet coefficients extracted for guilty subjects (top row) and innocent subjects (bottom row)

4 Conclusion

It has been observed that the developed signal processing procedure has extracted the underlying hidden information in the brain of an individual to a large extent. Using wavelet features and SVM-based classification, classification accuracy of 80% has been achieved. SVM outperforms the kNN classifier in accurately distinguishing between the guilty and innocent participants. The obtained testing accuracy value holds significance, considering that only 10 subjects' data have been considered in this work. In future, it is proposed to analyze more data and features to improve the results.

References

1. Zahhad, M.A., Ahmed, S.M., Abbas, S.N.: State-of-the-art methods and future perspectives for personal recognition based on electroencephalogram signals. IET Biometrics **4**(3), 179–190 (2015)
2. Gao, J., Yang, Y., Huang, W., Lin, P., Ge, S., Zheng, H., Gu, L., Zhou, H., Li, C., Rao, N.: Exploring time- and frequency dependent functional connectivity and brain networks during deception with single-trial event related potentials. Sci. Rep. **6**, 37065 (2016). https://doi.org/10.1038/srep37065
3. Demiralp, T., Ademoglu, A., Comerchero, M., Polich, J.: Wavelet analysis of P3a and P3b. Brain Topogr. **13**(4), 251–267 (2001)
4. Gao, J., Tian, H., Yang, Y., Yu, X., Li, C., Rao, N.: A novel algorithm to enhance P300 in single trials: application to lie detection using F-score and SVM. PLoS ONE **9**(11), e109700 (2014). https://doi.org/10.1371/journal.pone.0109700
5. Delorme, A., Makeig, S.: EEGLAB: an open source toolbox for analysis of single-trial EEG dynamics including independent component analysis. J. Neurosci. Methods **134**, 9–21 (2004)
6. Mehrnam, A.H., Nasrabadi, A.M., Ghodousi, M., Mohammadian, A., Torabi, S.: A new approach to analyze data from EEG-based concealed face recognition system. Int. J. Psychophysiol. **116**, 1–8 (2017)
7. Akhavan, A., Moradi, M.H., Vand, S.R.: Subject-based discriminative sparse representation model for detection of concealed information. Comput. Methods Progr. Biomed. **143**, 25–33 (2017)
8. Farahani, E.D., Moradi, M.H.: Multimodal detection of concealed information using genetic-SVM classifier with strict validation structure. Inf. Med. Unlocked **9**, 58–67 (2017)
9. Coelho, V.N., Coelho, I.M., Coelho, B.N., Souza, M.J.F., Guimar˜aes, F.G., da S. Luz, E.J., Barbosa, A.C., Coelho, M.N., Netto, G.G., Costa, R.C., Pinto, A.A., de P. Figueiredo, A., Elias, M.E.V., Filho, D.C.O.G., Oliveira, T.A.: EEG time series learning and classification using a hybrid forecasting model calibratedwith GVNS. Elect. Notes Discrete Math. **58**, 79–86 (2017)
10. Lafuente, V., Gorriz, J.M., Ramirez, J., Gonzalez, E.: *P*300 brainwave extraction from EEG signals: an unsupervised approach. Expert Syst. with Appl. **74**, 1–10 (2017)

Raw Sequence to Target Gene Prediction: An Integrated Inference Pipeline for ChIP-Seq and RNA-Seq Datasets

Nisar Wani and Khalid Raza

Abstract Next Generation Sequencing technologies, such as ChIP-seq and RNA-seq are increasingly being used to understand the genomic mechanisms whereby the interaction between transcription factors (TFs) and their target genes give rise to different gene expression patterns, thereby regulating various phenotypes and controlling diverse disease mechanisms. TFs play a pivotal role in gene regulation by binding to different locations on the genome and influencing the expression of their target genes. In this paper, we propose an inference pipeline for predicting target genes and their regulatory effects for a specific TF using next generation data analysis tools. Compared to other bioinformatics tools that use processed genome-wide protein–DNA-binding information for identifying regulatory targets, our approach provides an end-to-end solution, from preprocessing raw sequences to generating regulatory targets for transcription factors. Besides this, we also integrate transcriptomic data to evaluate whether the direct effect of TF binding leads to activating or repressive expression of target genes.

Keywords NGS · Pipeline · Target gene · Prediction · ChIP-seq
RNA-seq · Regulatory potential

1 Introduction

Omics technologies are key drivers of the data revolution that has taken place in the life sciences domain from last few decades. These technologies enable unbiased investigation of biological systems at genomic scales. Using high throughput Next Generation Sequencing (NGS) methods, genome-wide data is collected from cells,

N. Wani · K. Raza (✉)
Department of Computer Science, Jamia Millia Islamia, New Delhi, India
e-mail: kraza@jmi.ac.in

N. Wani
Govt. Degree College Baramulla, University of Kashmir, Srinagar, Jammu and Kashmir, India

© Springer Nature Singapore Pte Ltd. 2019

H. Malik et al. (eds.), *Applications of Artificial Intelligence Techniques in Engineering*, Advances in Intelligent Systems and Computing 697,
https://doi.org/10.1007/978-981-13-1822-1_52

tissues, and model organisms [30]. These data are key to investigate biological phenomena governing different cellular functions and also help biomedical researchers to better understand the disease etiologies which have not been previously explored. NGS protocols such as, ChIP-seq and RNA-seq are to generate datasets from where we can obtain genome-wide binding map of TFs and epigenetic signatures [16, 26] and can also measure the gene expression abundance within the cell for the entire genome [11, 37].

Numerous efforts have been put forth to uncover the interplay between genomic datasets obtained from ChIP-seq and RNA-seq for gene regulation studies of individual TFs [36] or mapping Transcription Regulatory Networks as in [35]. Revealing such interaction between these data has significant biomedical implications in various pathological states as well as in normal physiological processes [39]. Therefore, there is a compelling need to integrate these data to predict the pattern of gene expression during cell differentiation [20] and development [9] and to study human diseases such as, cancer as outlined in [28].

The aim of this study is to integrate genome-wide protein–DNA interaction (ChIP-seq) and transcriptomic data (RNA-seq) using a multi-step bioinformatics pipeline to infer the gene targets of a TF which serve as building blocks of a transcriptional regulatory network. We have developed a Perl script that implements this multistage pipeline by integrating tools in the same order as depicted in Fig. 1, the choice of tools for each stage is a consequence of thorough literature study among the set of tools available in their respective domains. Our implementation is a partially automated system that requires supervision at the time of quality control of raw reads, but progresses smoothly onwards without any manual intervention to integrate the two datasets and generate TF-specific gene targets.

This paper is organized into multiple sections and subsections. Following the introduction is the related literature section covering the studies related to this domain. A detailed description about datasets being used for this study and steps involved in generating TF targets is covered in materials and methods section. The results section presents the inference pipeline output followed by discussion and conclusion.

2 Related Literature

Software tools and computational methods exist that predict and analyze gene targets by processing ChIP-seq data. A distinguishable group of peak callers such as, CisGenome [19], BayesPeak [34], and Model-based Analysis of ChIP-seq (MACS), Peakseq [32], SICER [40] are some of the widely used tools that identify TF-binding sites. These peak callers identify the target genes either by looking for peaks in promoter region or assign a proximal nearest gene in the vicinity of peaks. However, with most TFs ChIP-seq data having peaks in and around the promoter

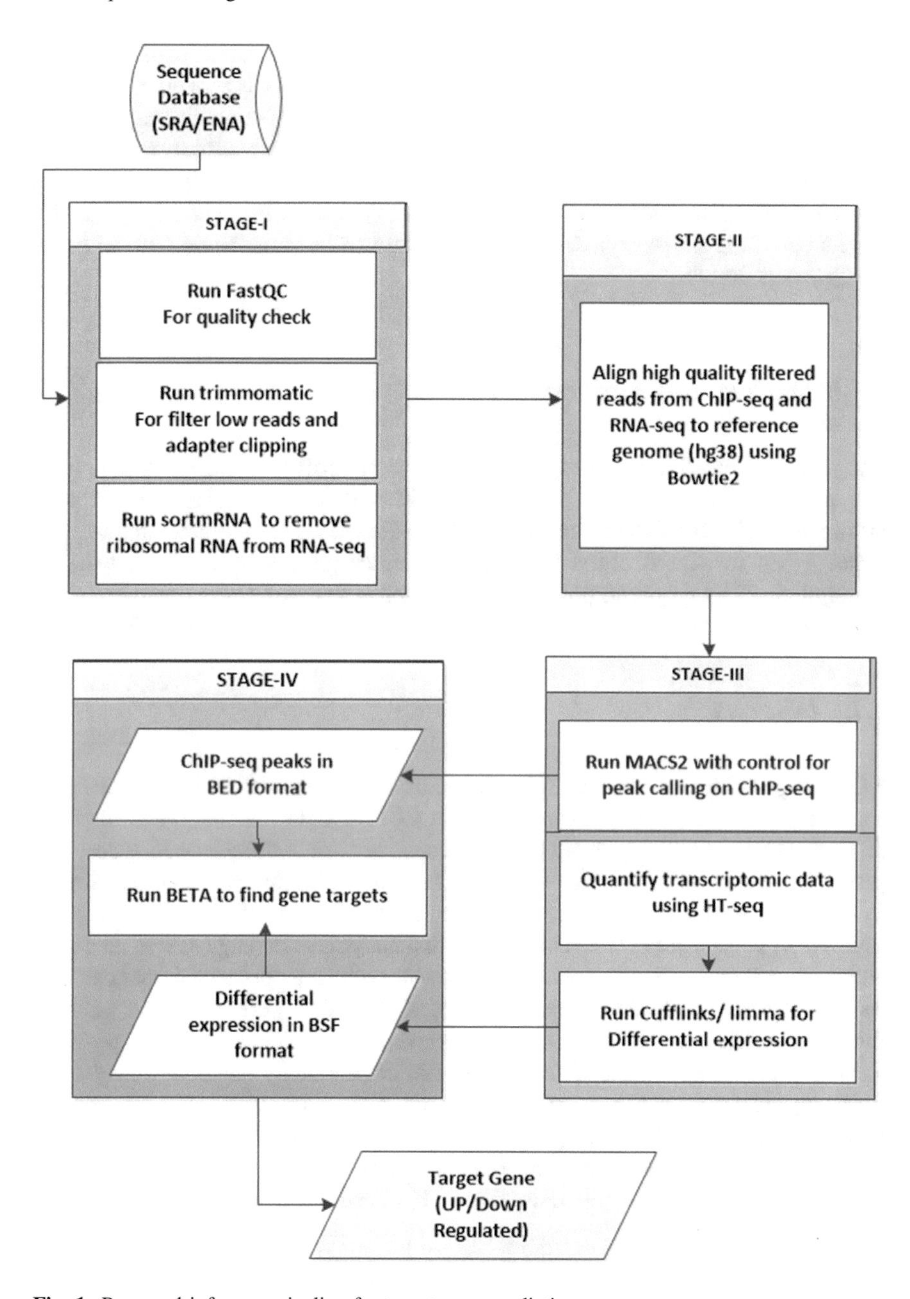

Fig. 1 Proposed inference pipeline for target gene prediction

regions is very less. Also predicting targets using nearest peak is not always reliable. TIP [7] is another tool that builds a probabilistic model to predict gene targets, but does not take into account gene expression data. Certain databases such as

JASPAR [33], TRED [18], etc., identify target genes for a selected set of TFs based on the motif analysis of the promoter regions using Position Weight Matrices (PWMs). A recent study [13] combines multiple approaches to predict target genes.

On the contrary, some earlier studies used only gene expression data for predicting target genes. For example, authors in [29] use Support Vector Machines (SVM) to discover relationships between the TFs and their targets; [17] identify targets with time-series expression data by creating a linear activation model based on Gaussian process.

3 Materials and Methods

The proposed pipeline operates on raw NGS data, ChIP-seq, and RNA-seq. After preprocessing the raw sequences it yields differential gene expression and peak information of genes from these datasets. Both these datasets are integrated to yield target genes for the ChIPred (Chromatin immunoprecipitated) TF. A working description of the proposed pipeline is presented below.

3.1 Datasets

NGS data is primarily accessed from Sequence Read Archive (SRA) at NCBI [24] and European Nucleotide Archive (ENA) at EBI [23]. Raw sequences in the form of FASTQ files are freely available for download for a variety of cell types, diseases, treatments, and conditions. Besides the public databases, a number of projects and consortia offer public access to their data repositories. For example, ENCODE [12] is a publicly funded project that has generated large sets of data for a variety of cell lines, tissues, and organs. Raw as well as preprocessed data can also be accessed and freely downloaded from ENCODE data portal. Another publicly funded research project, The Cancer Genome Atlas (TCGA) [38] also provides datasets for a variety of cancer types. For the current study we have downloaded ChIP-seq data of MCF7 breast cancer cell line from ENCODE experiment *ENCFF580EKN* and RNA-Seq data of transcriptomic study *PRJNA312817* from European Nucleotide Archive. The RNA-seq experiment contains 30 samples of time course gene expression data from MCF7 cell line subjected to estrogen stimulation.

3.2 Pipeline Workflow

Molecular measurements within the NGS data exist in the form of millions of reads and are stored as FASTQ files. Information within the raw files is hardly of any value and needs extensive preprocessing before this data can be analyzed. The preprocessing task is a multi-step process and involves the application of a number of software tools. In this section, we present a detailed NGS pipeline that describes necessary steps from preprocessing of RNA-Seq and ChIP-seq data to target genes regulated by ChIPred TF. The pipeline has been implemented using a Perl script by integrating various NGS data processing and analysis tools. Figure 1 is a graphical depiction of the proposed inference pipeline.

3.3 Quality Control

Almost all sequencing technologies produce their outputs in FSATQ files. FASTQ has emerged as the de facto file format for data exchange between various bioinformatics tools that handle NGS data. FASTQ [8] format is a simple extension to existing FASTA format; the files are plain ASCII text files with the ability to store both nucleotide sequences along with a corresponding quality score for each nucleotide call.

Base sequence qualities are usually interpreted in terms of Phred quality scores. Phred quality scores Q are defined as a property which is logarithmically linked to error probabilities P of called bases and can be computed as shown in Eq. (1).

$$Q = -10 \log_{10} P \tag{1}$$

Phred's error probabilities have been shown to be very accurate [14], e.g. if Phred assigns a quality score of 10 means that 1 out of 10 base calls is incorrect, a score of 20 depicts that 1 in 100 bases has been called incorrectly. Usually, a Phred score ≥ 20 is considered as acceptable read quality, otherwise read quality improvement is required. If read quality is not improved by trimming, filtering, and cropping, there may be some error during library preparation and sequencing. FASTQC [1] is a Java based tool that is used to assess the quality of the reads produced by Next Generation DNA Sequencers. Low-quality reads are excised from the FASTQ files to improve the quality of the reads. Various tools are available that can be used to trim bases with poor Phred scores, i.e., Phred score less than 20.

Trimmomatic [4] is a Java based open source tool used for trimming illumina FASTQ data and removing adapters. Additionally RNA extracted using NGS does include non-coding RNA molecules besides the coding ones. These non-coding RNAs are usually the ribosomal RNA. For quantifying the gene expression patterns using RNA-seq, it is essential that these non-coding RNAs be filtered from the existing reads. SortMeRNA [21] is a very efficient and accurate tool that is used to filter out the ribosomal RNA from the metatranscriptomic data. A diagrammatic flow of quality check and trimming is shown in Fig. 2.

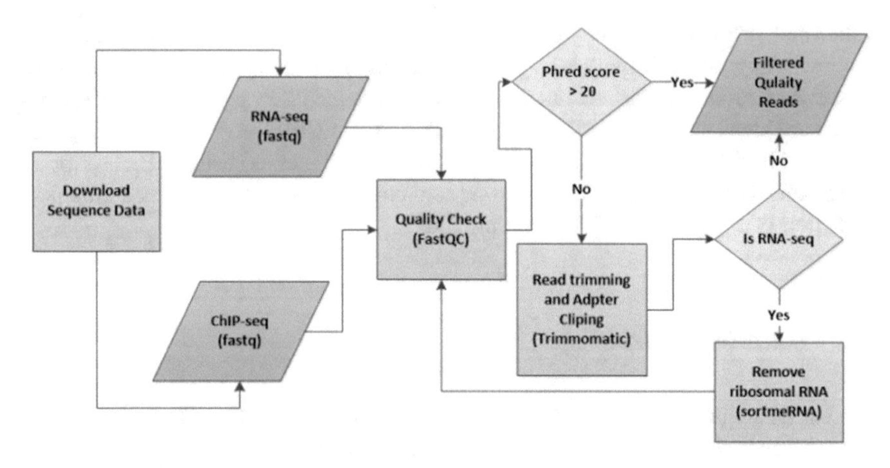

Fig. 2 Flowchart showing quality control check, adapter clipping and trimming of raw reads

3.4 Mapping Sequence Reads to a Reference Genome

With required read quality achieved after trimming and adapter clipping, the next step is to align the short reads to a reference sequence. The reference sequence is our case is human genome assembly hg38/hg19 but it can also be a reference transcriptome, or a de novo assembly incase a reference sequence is not available. There are numerous software tools that have been developed to map reads to a reference sequence. Besides the common goal of mapping, these tools vary considerably from each other both in algorithmic implementation and speed. A brief account can be found in [15].

In this pipeline we used Bowtie2 genome aligner [22], because it is a memory efficient and an ultrafast tool for aligning sequencing reads. Bowtie2 performs optimally with read lengths longer than 50 bp or beyond 1000 bp (e.g., mammalian) genomes. It builds an FM Index while mapping reads to keep its memory footprint small and for the human genome it is typically around 3.2 GB of RAM.

3.5 Expression Quantification of RNA-Seq

Although raw RNA-seq reads do not directly correspond to the gene expression, but we can infer the expression profiles from the sequence coverage or the mapping reads that map to a particular area of the transcriptome. A number of computational tools are available to quantify the gene expression profiles from RNA-seq data. Software tools, such as Cufflinks, HTSeq, IsoEM, and RSEM are freely available. A comparative study of these tools is presented in [5].

Despite clear advantages over microarrays, there are still certain sources of systematic variations that should be removed from RNA-seq data before performing any downstream analysis. These variations include between sample differences, such as sequencing depth and within sample differences, e.g., gene length, GC content, etc. In order to circumvent these issues and exploit the advantages offered by RNA-seq technology, the reads/kilobase of transcript per million mapped reads (RPKM) normalizes a transcript read count by both its length and the total number of reads in the sample [25, 27]. For data that has originated from the paired-end sequencing, a similar normalization metric called FPKM (fragments per kilobase of transcript per million mapped reads) is used. Both RPKM and FPKM use similar operations for normalizing single end and paired-end reads [10].

Counts per million (CPM) is another important metric provided by limma package to normalize gene expression data. Once the normalized expression estimates are available, we can obtain a differential expression of gene lists across the samples or conditions using limma voom [31] provided by R bioconductor.

3.6 Peak Calling

Early preprocessing steps of ChIP-seq data resemble that of RNA-seq. Beginning with the quality check of raw reads, read trimming and adapter elimination, filtered high-quality reads are then mapped to reference genome using Bowtie2 as described above.

In order to identify the genomic locations where the protein of Interest (POI) has attached itself to DNA sequences, the aligned reads are subjected to a process known as Peak Calling. Software tools that predict the binding sites where this protein has bound itself by identifying location within the genome with significant number of mapped reads (peaks) are called Peak Callers. A detailed description of various ChIP-seq peak callers is presented in [27]. Although a number of tools are available, but for this study have used the most efficient and open source tool called Model-based Analysis of ChIP-seq (MACS). Nowadays an upgraded version of MACS called MACS2 is commonly used for this purpose. A MACS2 algorithm does process aligned ChIP-seq bam files both with control and without control samples. Mapped reads are modeled as sequence tags (an integer count of genomic locations mappable under the chosen algorithm). Depending upon the type of protein being ChIPred, different types of peaks are observed when viewing the information in a genome browser, such as Integrated Genome Viewer (IGV). Most of the TFs act as point-source factors and result in narrow peaks, factors such as histone marks generate broader peaks, and proteins such as RNA (Fig. 3).

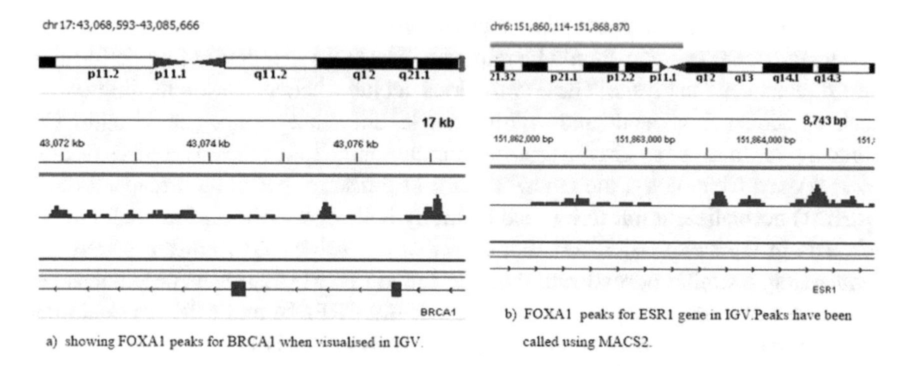

Fig. 3 Integrated Genome Browser view of FOXA1 peaks for BRCA1 and ESR1 targets generated by MACS2 peak caller

4 Results

The proposed pipeline first generates a list of differentially expressed genes (DEGs) from the normalized expression profiles, thereby identifying the gene activity for both factor-bound and factor-unbound conditions. These DEGs are then integrated with the binding information from stage-III of the pipeline. Both these intermediate data sets are passed as input to stage-IV that employs Binding and expression target analysis (BETA) for prediction process; it calculates binding potential derived from the distance between transcription start site and the TF-binding site, thereby modeling the manner in which the expression of genes is being influenced by TF-binding sites. Using contributions from the individual TF-binding sites, we obtain a cumulative score of overall regulatory potential (probability of a gene being regulated by a factor) of a gene. The percentage of up- and down-regulated genes is shown in Fig. 4.

During the process of target prediction each gene receives two ranks, one from the binding potential R_{bp} and other from differential expression R_{de}. Both these ranks are multiplied to obtain rank product $R_{\mathrm{p}} = R_{\mathrm{bp}} \times R_{\mathrm{de}}$. Genes with more regulatory potential and more differential expression are more likely to be as real targets. Table 1 shows a list of up and down regulated genes.

Predicted targets from the inference pipeline results have been widely reported in literature, e.g., FOXA1 up regulated BRCA1 and down regulated ESR1, GATA3 and ZNF217 have been reported in [2]. Similarly evidence regarding the role of multiple loci on TERT gene are related to ER(−ve) breast cancer [3]. Many of these predicted targets are well-known prognostic biomarkers whose role has been established well in the scientific literature. Once we have a set of target genes, a further downstream analysis of these genes can be done by using gene ontology-based tools such as DAVID to map them with their corresponding biological functions.

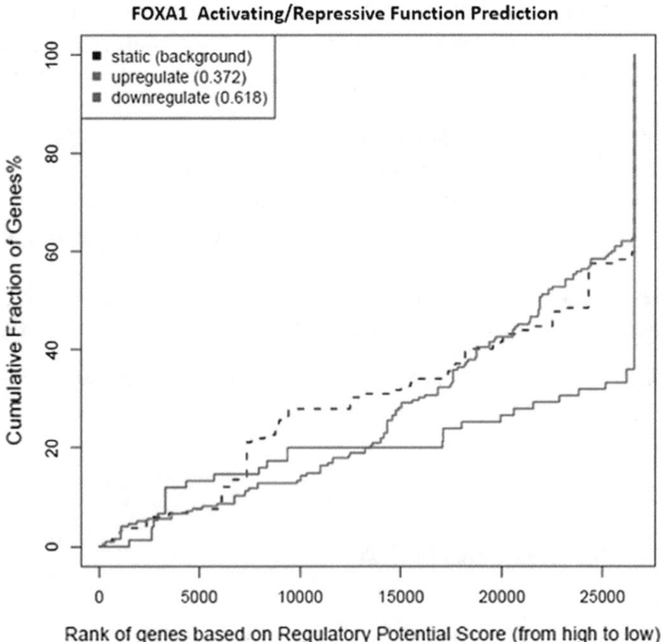

Fig. 4 FOXA1 predicted genes activation and repression function

Table 1 Target genes predicted by inference pipeline

Chromosome	RefSeqID	Rank product gene symbol		Regulation
chr17	NM 007298	2.19E−04	BRCA1	Up
chr12	NM 004064 s	2.96E−04	CDKN1B	Up
chr5	NM 001193376	6.16E−04	TERT	Up
chr7	NM 000492	9.59E−04	CFTR	Up
chr2	NM 001204109	9.77E−04	2 BCL2	Up
chr10	NM 001002295	4.39E−04	GATA3	Down
chr14	NM 138420	7.89E−04	AHNAK	Down
chr20	NM 006526	2.72E−03	ZNF217	Down
chr6	NM 001122741	5.79E−03	ESR1	Down
chr1	NM 001878	5.79E−03	CRABP2	Down

As there is hardly any ground truth data of target genes available when comparing with different computational methods. We compare our proposed pipeline based on various in built characteristics with an existing method, such as Target identification from profiles (TIP) [7]. Given below are some important observations:-

1. Whereas the TIP uses a probabilistic model to capture the binding signals from raw ChIP-seq files, out method employs refined peak information obtained from state-of-the art peak calling program (MACS) integrated in our proposed pipeline.
2. TIP operates on binding signals and their distance from the transcription start site (TSS) to generate confidence scores for targets, whereas our method uses narrow peaks from MACS to calculate a weighted measure to gauge regulatory potential of a TF for all the binding sites on the whole genome to identify targets.
3. TIP does not support integration of transcriptomic data for identifying up/down regulation of the target genes compared to the proposed inference pipeline.

More recently tools, such as CHIP-BIT [6], etc., that are used for identifying target genes of TFs have been developed in this domain, but their core functionality is peak calling and only use ChIP-seq data for this purpose.

5 Discussion and Conclusion

The availability and expansion of ChIP-seq and RNA-seq datasets is fueling an exponential rise in the number of studies being conducted in the area of integrated computational analysis. The motive behind these research endeavors is to address basic questions about how multiple factor binding is related to transcriptional output within in vivo DNA. The proposed inference pipeline is used to decipher the regulatory relationship between TFs that bind to DNA and their corresponding target genes that they influence resulting in their activation/repression. From RNA-seq and ChIP-seq reads, the pipeline generates one file containing differential expression of genes and the other DNA-binding events in the form of peaks. Both these files are integrated in the final stage to yield targets for the TF/TFs whose peaks files were used.

The inference pipeline presented in this paper extracts target genes and hence the regulatory network for a specific TF that has been ChIPred. In case we are required to build a regulatory network for a set of TFs, we need to input new peak files for every new TF in a loop and record the target genes of this TF and its regulatory influence in a separate file. In the current study, we considered only TFs and their influence on gene expression. However, a wider study can include multiple TFs, methylation data, histone marks, and polymerase loading to improve the efficiency of the proposed pipeline.

Deciphering the transcriptional regulatory relationships and understanding the elements of regulatory mechanisms that control gene expression is a key research area of regulatory biology. Therefore computational integration of factor binding and other genome-wide data, such as gene expression will be sought after to extract functionally important connections of a working regulatory code.

Acknowledgements The author Nisar Wani acknowledges Teacher Fellowship received from University Grants Commission, Ministry of Human Resources Development, Govt. of India vide letter No. F.BNo. 27-(TF-45)/2015 under Faculty Development Programme.

References

1. Andrews, S., et al.: Fastqc: a quality control tool for high throughput sequence data (2010)
2. Baran-Gale, J., Purvis, J.E., Sethupathy, P.: An integrative transcriptomics approach identifies mir-503 as a candidate master regulator of the estrogen response in mcf-7 breast cancer cells. RNA **22**(10), 1592–1603 (2016)
3. Bojesen, S.E., Pooley, K.A., Johnatty, S.E., Beesley, J., Michailidou, K., Tyrer, J.P., Edwards, S.L., Pickett, H.A., Shen, H.C., Smart, C.E., et al.: Multiple independent variants at the tert locus are associated with telomere length and risks of breast and ovarian cancer. Nat. Gene. **45**(4), 371–384 (2013)
4. Bolger, A.M., Lohse, M., Usadel, B.: Trimmomatic: a flexible trimmerfor illumina sequence data. Bioinformatics **30**(15), 2114–2120 (2014)
5. Chandramohan, R., Wu, P.-Y., Phan, J.H., Wang, M.W.: Benchmarking RNA-seq quantification tools. In: Engineering in Medicine and Biology Society (EMBC). 2013 35th Annual International Conference of the IEEE, pp. 647–650. IEEE (2013)
6. Chen, X., Jung, J.-G., Shajahan-Haq, A.N., Clarke, R., Shih, I.-M., Wang, Y., Magnani, L., Wang, T.-L., Xuan, J.: Chip-bit: Bayesian inference of target genes using a novel joint probabilistic model of chip-seq profiles. Nucl. Acids Res. **44**(7), e65 (2016)
7. Cheng, C., Min, R., Gerstein, M.: Tip: a probabilistic method foridentifying transcription factor target genes from chip-seq binding profiles. Bioinformatics **27**(23), 3221–3227 (2011)
8. Cock, P.J.A., Fields, C.J., Goto, N., Heuer, M.L., Rice, P.M.: The sanger fastq file format for sequences with quality scores, and the solexa/illumina fastq variants. Nucl. Acids Res. **38**(6), 1767–1771 (2009)
9. Comes, S., Gagliardi, M., Laprano, N., Fico, A., Cimmino, A., Palamidessi, A., De Cesare, D., De Falco, S., Angelini, C., Scita, G., et al.: L-proline induces a mesenchymal-like invasive program in embryonic stem cells by remodeling h3k9 and h3k36 methylation. Stem Cell Rep. **1**(4), 307–321 (2013)
10. Conesa, A., Madrigal, P., Tarazona, S., Gomez-Cabrero, D., Cervera, A., McPherson, A., Szcześniak, Gaffney, D.J., Elo, L.L., Zhang, X., et al.: A survey of best practices for RNA-seq data analysis. Genome Biol. **17**(1), 13 (2016)
11. Costa, V., Angelini, C., De Feis, I., Ciccodicola, A.: Uncovering the complexity of transcriptomes with RNA-seq. BioMed Res. Int. (2010)
12. ENCODE Project Consortium, et al.: The encode (encyclopedia of dna elements) project. Science **306**(5696), 636–640 (2004)
13. Essebier, A., Lamprecht, M., Piper, M., Boden, M.: Bioinformatics approaches to predict target genes from transcription factor binding data. Methods (2017)
14. Ewing, B., Hillier, L., Wendl, M.C., Green, P.: Base-calling of automated sequencertraces using phred. I. Accuracy assessment. Genome Biol. (1998)
15. Flicek, P., Birney, E.: Sense from sequence reads: methods for alignment and assembly. Nat. Methods **6**, S6–S12 (2009)
16. Furey, T.S.: Chip-seq and beyond: new and improved methodologies to detect and characterize protein-dna interactions. Nat. Rev. Genet. **13**(12), 840 (2012)
17. Honkela, A., Girardot, C., Gustafson, E.H., Liu, Y.-H., Furlong, E.E.M., Lawrence, N.D., Rattray, M.: Model-based method for transcription factor target identification with limited data. Proc. Natl. Acad. Sci. **107**(17), 7793–7798 (2010)
18. Jiang, C., Xuan, A., Zhao, F., Zhang, M.Q.: Tred: a transcriptional regulatory element database, new entries and other development. Nucl. Acids Res. **35**(suppl 1), D137–D140 (2007)

19. Jiang, H., Wang, F., Dyer, N.P., Wong, W.H.: Cisgenome browser: a flexible tool for genomic data visualization. Bioinformatics **26**(14), 1781–1782 (2010)
20. Kadaja, M., Keyes, B.E., Lin, M., Pasolli, H.A., Genander, M., Polak, L., Stokes, N., Zheng, D., Fuchs, E.: Sox9: a stem cell transcriptional regulator of secreted niche signaling factors. Genes Develop. **28**(4), 328–341 (2014)
21. Kopylova, Evguenia, Noé, L., Touzet, H.: Sortmerna: fast and accuratefiltering of ribosomal RNAs in metatranscriptomic data. Bioinformatics **28**(24), 3211–3217 (2012)
22. Langmead, B., Salzberg, S.L.: Fast gapped-read alignment with bowtie 2. Nat. Methods, **9**(4), 357–359 (2012)
23. Leinonen, R., Akhtar, R., Birney, E., Bower, L., Cerdeno-Tárraga, A., Cheng, Y., Cleland, I., Faruque, N., Goodgame, N., Gibson, R., et al.: The European nucleotide archive. Nucl. Acids Res. **39**(suppl 1), D28–D31 (2010)
24. Leinonen, R., Sugawara, H.: The sequence read archive. Nucl. Acids Res. **39**(suppl 1), D19–D21 (2010)
25. Park, P.J.: Chip-seq: advantages and challenges of a maturing technology. Nat. Rev. Genet. **10**(10), 669 (2009)
26. Pepke, S., Wold, B., Mortazavi, A.: Computation for chip-seq andrna-seq studies. Nat. Methods **6**, S22–S32 (2009)
27. Portela, A., Esteller, M.: Epigenetic modifications and human disease. Nat. Biotechnol. **28**(10), 1057–1068 (2010)
28. Qian, J., Lin, J., Luscombe, N.M., Yu, H., Gerstein, M.: Prediction of regulatory networks: genome-wide identification of transcription factor targets from gene expression data. Bioinformatics **19**(15), 1917–1926 (2003)
29. Raza, K., Ahmad, S.: Recent advancement in Next Generation Sequencing techniques and its computational analysis. arXiv preprint arXiv:1606.05254 (2016)
30. Ritchie, M.E., Phipson, B., Wu, D., Hu, Y., Law, C.W., Shi, W., Smyth, G.K.: Limma powers differential expression analyses for RNA sequencing and microarray studies. Nucl. Acids Res. **43**(7), e47 (2015)
31. Rozowsky, J., Euskirchen, G., Auerbach, R.K., Zhang, Z.D., Gibson, T., Bjornson, R., Carriero, N., Snyder, M., Gerstein, M.B.: Peakseq enables systematic scoring of chip-seq experiments relative to controls. Nat. Biotechnol. **27**(1), 66–75 (2009)
32. Sandelin, A., Alkema, W., Pär Engström, Wasserman, W.W., Lenhard, B.: Jaspar: an open-access database for eukaryotic transcription factor binding profiles. Nucl. Acids Res. **32**(suppl 1), D91–D94 (2004)
33. Spyrou, C., Stark, R., Lynch, A.G., Tavaré, A.G.: Bayespeak: Bayesian analysis of chip-seq data. BMC Bioinform. **10**(1), 299 (2009)
34. Wade, J.T.: Mapping transcription regulatory networks with chip-seq and RNAseq. In: Prokaryotic Systems Biology, pp. 119–134. Springer, Berlin (2015)
35. Wang, S., Sun, H., Ma, J., Zang, C., Wang, C., Wang, J., Tang, Q., Meyer, C.A., Zhang, Y., Liu, X.S.: Target analysis by integration of transcriptome and chip-seq data with beta. Nat. Protocols **8**(12), 2502 (2013)
36. Wang, Zhong, Gerstein, Mark, Snyder, Michael: Rna-seq: a revolutionary tool fortranscriptomics. Nat. Rev. Genet. **10**(1), 57–63 (2009)
37. Weinstein, J.N., Collisson, E.A., Mills, G.B., Mills Shaw, K.R., Ozenberger, B.A., Ellrott, K., Shmulevich, I., Sander, C., Stuart, J.M.,Cancer Genome Atlas Research Network, et al.: The cancer genome atlas pan-cancer analysis project. Nat. Gene. **45**(10), 1113–1120 (2013)
38. Yue, F., Cheng, Y., Breschi, A., Vierstra, J., Wu, W., Ryba, T., Sandstrom, R., Ma, Z., Davis, C., Pope, B.D., et al.: A comparative encyclopedia of dna elements in the mouse genome. Nature **515**(7527), 355 (2014)
39. Zang, C., Schones, D.E., Zeng, C., Cui, K., Zhao, K., Peng, W.: A clustering approach for identification of enriched domains from histone modification chip-seq data. Bioinformatics **25**(15), 1952–1958 (2009)
40. Zhang, Y., Liu, T., Meyer, C.A., Eeckhoute, J., Johnson, D.S., Bernstein, B.E., Nusbaum, C., Myers, R.M., Brown, M., Li, W., et al.: Model-based analysis of chip-seq (macs). Genome Biol. **9**(9), R137 (2008)

An Improvement in DSR Routing Protocol of MANETs Using ANFIS

Vivek Sharma, Bashir Alam and M. N. Doja

Abstract In mobile ad hoc networks(MANET's), dynamically changing topology, constraints over energy, breakage of links and providing security to network results in the generation of multiple routes from source to destination. One of the well-known reactive routing protocol of MANET is named as Dynamic Source Routing (DSR) which stored the value of previously determined path to be used in route cache of intermediate nodes. Therefore, while selection from multiple routes this cache information is used to determine new route. Since the amount of cache is limited, therefore it is impossible to store all the information over cache. One of the solutions to this problem is to have cache information not only on the basis of hop count as done by DSR. Hence in the proposed algorithm, the output parameter "Anfis cost (Anf)" is computed on the basis of input parameters "hop count", "energy", and "delay" to select the optimal route in cache. The proposed routing algorithm uses adaptive fuzzy inference system (ANFIS) to improve the performance of DSR protocol hence named as A-DSR. The simulation results proved that the proposed A-DSR protocol have substantial high value of packet delivery ratio compared to state-of-art soft computing-based protocol and the conventional DSR.

Keywords ANFIS · A-DSR · Cache · MANET

V. Sharma (✉) · B. Alam · M. N. Doja
Department of Computer Engineering, Jamia Millia Islamia, New Delhi, India
e-mail: Vivek2015@gmail.com

B. Alam
e-mail: balam2@jmi.ac.in

M. N. Doja
e-mail: mdoja@jmi.ac.in

© Springer Nature Singapore Pte Ltd. 2019

H. Malik et al. (eds.), *Applications of Artificial Intelligence Techniques in Engineering*, Advances in Intelligent Systems and Computing 697,
https://doi.org/10.1007/978-981-13-1822-1_53

1 Introduction

Mobile ad hoc networks have several characteristics like mobility, dynamic topology, bandwidth, security. For applications in the MANET's design of routing protocol are important issues. There are thousands of routing protocol has been proposed by the researcher in last three decades. As there is no routing protocol fit in every scenario. The MANETs on demand reactive routing protocol is dynamic source routing(DSR) [1–4]. It uses the route cache to store all the possible information about the route stored in data packet. The intermediate node uses this cache information to reply a route request message generated by the source, if they have the route information.

Earlier routing protocol used one metric to optimize the routing protocols of the mobile ad hoc networks. The DSR uses the hop count metrics to select the shortest route. To provide stable path the Associativity-Based Long-Lived routing protocol [5] selects the paths that show the greater time connectivity among the nodes. Current routing protocol uses more than one metric to improve the performance of routing protocol. To improve the cache decision [6] presents the fuzzy logic system based DSR protocol [7] that select the route path by considering the combined effect of metrics link strength, energy and hop count. To prevent the DSR from denial of service attack, the energy metric has been considered by authors [8]. To achieve the route balance in the data flow, [9] modified DSR in IPv6 network. The author in [10] predicts the delay in MANET's by using regression and fuzzy time series. For improving the performance of DSR routing protocol, author in [11] presents routing protocol that forwards its RREQ packet based on its own residual battery, received signal strength and speed.

The structured overview of the paper has been presented in which Sect. 2 reports the ANFIS. Section 3 represents the proposed description of A-DSR routing protocol. Section 4 shows the performance evaluation of A-DSR routing protocol which is followed by conclusions of Sect. 5.

2 ANFIS

ANFIS is composed of fuzzy logic [12] and neural networks as presented by Jang [13]. Fuzzy inference system is constructed by the ANFIS in which the given training data set has been used. The membership function parameters of the above ANFIS are changed to make better by back propagation algorithm or in combination with least square type of method. The five layers are depicted in Fig. 1. The fist layer represents the inputs variables hop count, delay, and energy. The fuzzy rules are denoted by the intermediate (hide), and the outputs Variable is represented by the third layer represent. The weight links are encoded with fuzzy sets.

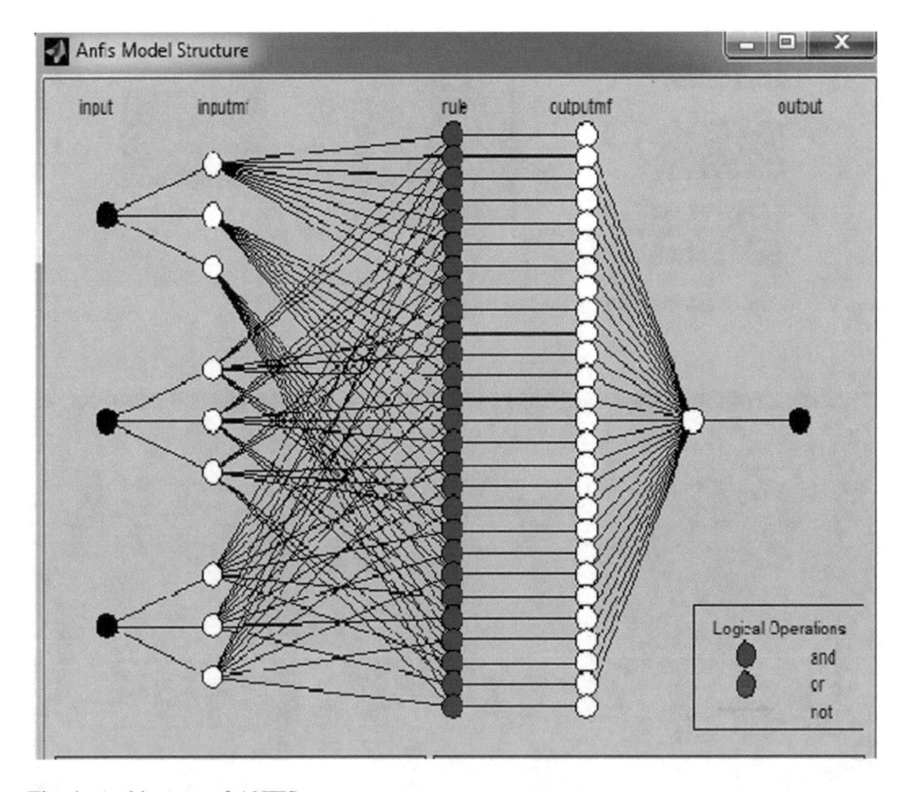

Fig. 1 Architecture of ANFIS

3 Proposed ANFIS DSR (A-DSR)

In DSR routing protocol each node has route cache that contained the route information. The route cache is continuously updated. It has two phases: (1) route discovery and (2) route maintenances. If a node wishes to communicate with other nodes, it checks cache whether the route is available. If route is available in cache it transfers the packet otherwise route discovery takes place. During the route discovery phase, intermediate node receives the hop count, energy and delay parameter information and these metrics are processed as input to ANFIS system and output metric Anf cost is calculated. The process of calculating Anf cost is depicted in Fig. 2.

The membership function used for input and output metric are shown in Fig. 3a–d respectively. The rule base of ANFIS system contains 27 rules as shown in Table 1 and Data set for the training has 1000 point. From its 300 is used for training and rest of it are used for testing. This cost is assigned to the path while creating the route as a response to a route request. Each intermediate node execute this process is until it meets to the destination. A source node receives the route reply message produced from the destination through the path kept in the route record.

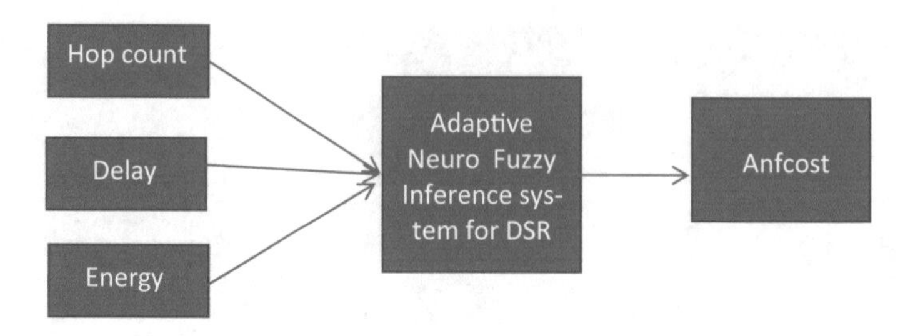

Fig. 2 Process of calculating Anf cost for proposed A-DSR

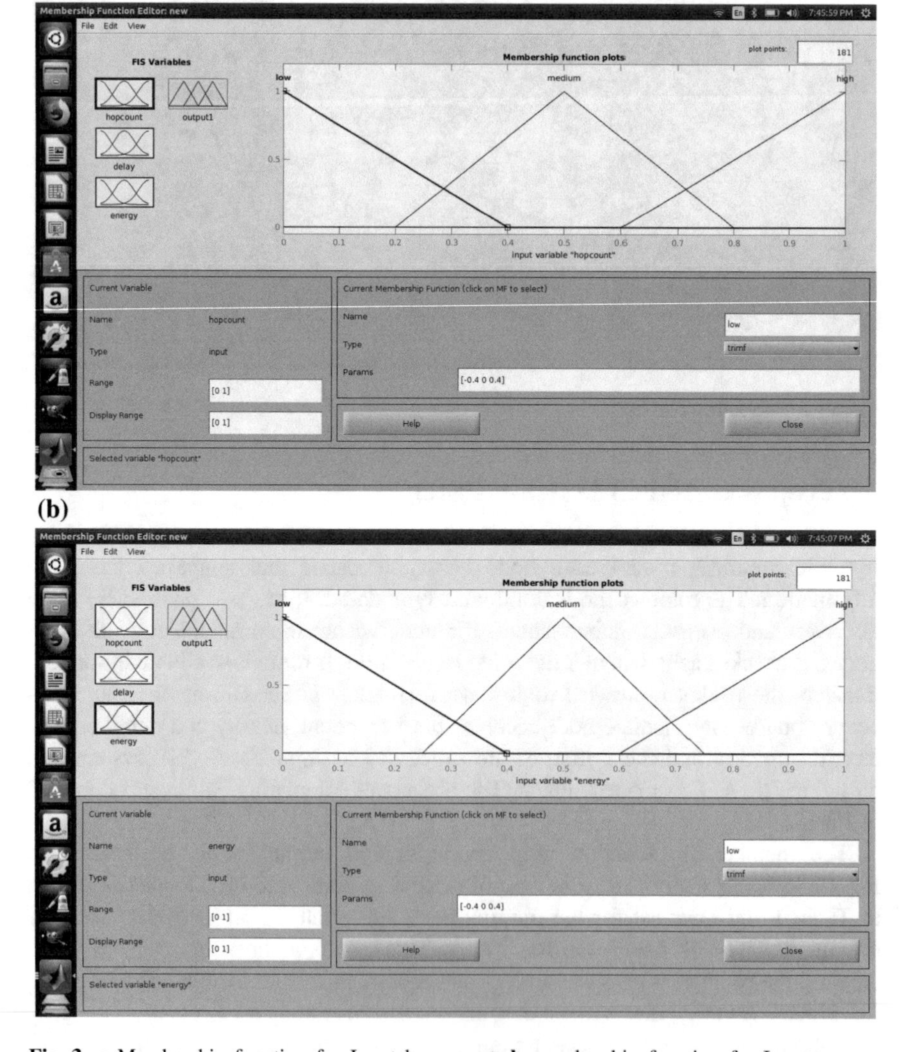

Fig. 3 **a** Membership function for Input hop count; **b** membership function for Input energy; **c** membership function for input delay; **d** membership function for output anfcost

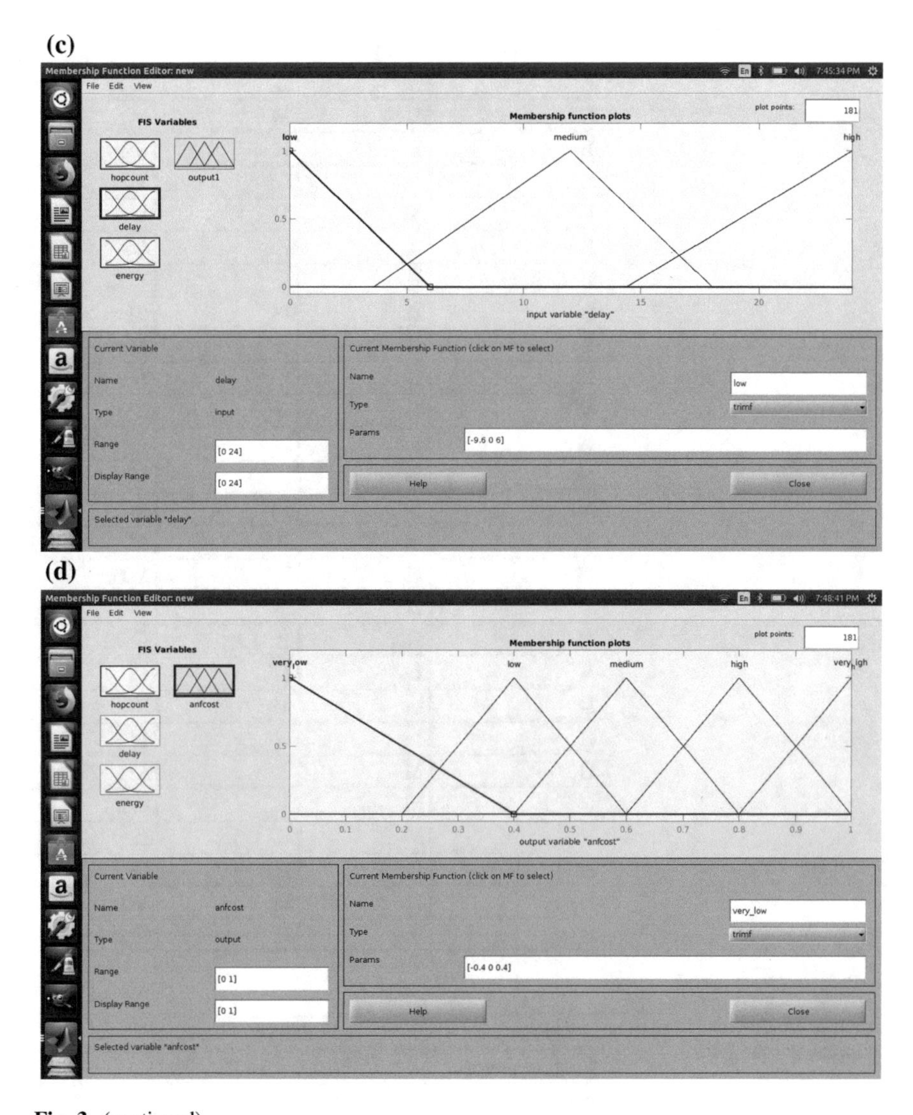

Fig. 3 (continued)

4 Simulation Model

The simulation environment consists of 30 nodes and placed within 1200 m 500 m area of simulation. The network range of each node is 250 m. The random way point model is used for Mobility. The range of speed attains maximum value of 20 m/s. The traffic generator Constant Bit Rate (CBR) is used. The size of data payload was 512 bytes. The simulation conducted in network simulator tool NS2.35 for 130 s for DSR routing protocol and proposed A-DSR (Table 2).

Table 1 Rule value for ANFIS

Hop count	Delay	Energy	Output
L	L	L	V.H
L	L	M	V.L
L	L	H	V.L.
L	M	L	V.H.
L	M	M	L
L	M	H	L
L	H	L	V.H.
L	H	M	M
L	H	H	M
M	L	L	H
M	L	M	L
M	L	H	L
M	M	L	H
M	M	M	M
M	M	H	M
M	H	L	V.H.
M	H	M	H
M	H	H	H
H	L	L	M
H	L	M	M
H	L	H	M
H	M	L	M
H	M	M	H
H	M	H	H
H	H	L	H
H	H	M	V.H
H	H	H	V.H

Table 2 Simulation environment variables

Parameter	Values
Routing protocol	DSR, A-DSR
No. of nodes	30,
Area	1200 * 500
Channel capacity	54 Mbps
Traffic type	CBR
Mobility model	Random way point
Pause time	50, 100, 150, 200, 250

4.1 Results for Packet Delivery Ratio (PDR)

In Fig. 4, packet delivery ratio is shown for variation in pause time that is calculated by ratio of the summation of all the received packets to the summation of all the generated packets that have been sent. The PDR of proposed algorithm is higher than that of DSR protocol and M-DSR protocol based on Fuzzy Inference System due to efficient route selection algorithm that consider hop count, energy, and delay of link node.

4.2 Results for Average End to End Delay

Figure 5 represents the Average End to End Delay with variation in pause time and it is calculated to measure the average delay between the time taken by a packet from its origin to reach its destination in the network. The Average End-to-End Delay of proposed algorithm is lower than that of DSR protocol and M-DSR protocol based on Fuzzy Inference System.

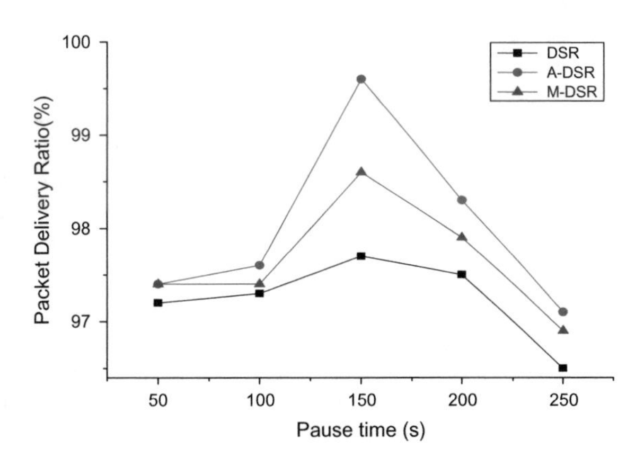

Fig. 4 Packet delivery ratio versus pause time

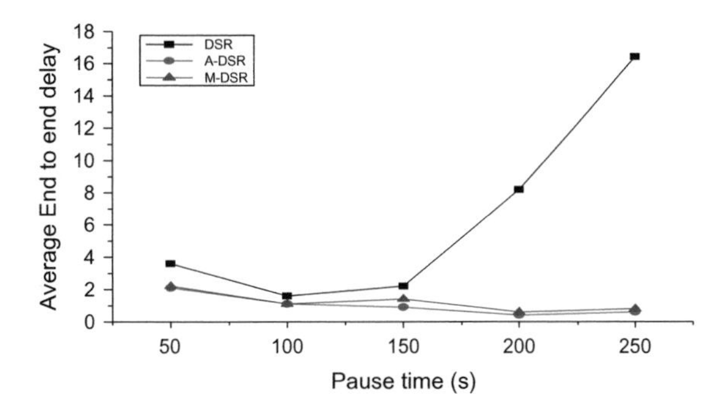

Fig. 5 Average end to end delay versus pause time

5 Conclusion

The proposed algorithm A-DSR employs ANFIS method to select the route from cache. ANFIS is applied for processing intermediate node energy, hop count, and delay to compute the node cost (Anf cost). The simulated results show that A-DSR algorithm have significantly high packet delivery ratio than the DSR routing protocol and recently developed fuzzy-based routing protocol. The simulation results also show that the proposed A-DSR algorithm has the low average value of end-to-end delay.

References

1. Murthy, C.S.R., Manoj, B.S.: Ad hoc wireless networks: architecture and protocols. Pearson Ltd., London (2004)
2. Sharma, V., Alam, B.: Unicaste routing protocols in mobile ad hoc networks: a survey. Int J Comput Appl USA **51**, 148–153 (2012)
3. Johnson, D.B., Maltz, D.A., Yih-Chun, H.: The dynamic source routing protocol for mobile ad hoc networks (DSR). IETF Internet Draft (2007). http://www.ietf.org/internet-drafts/draft-ietf-manet-dsr-08.txt
4. Toh, C.K.: Ad Hoc Mobile Wireless Networks protocols and Systems. Prentice Hall, New Jersey (2002)
5. Sridhar, K.N., Chan,M.C.: Stability and hop-count based approach for route computation in MANET. In: 14th International Conference on Computer Communications and Networks (2005). https://doi.org/10.1109/icccn.2005.1523800
6. Rea, S., Pesch, D.: Multi-metric routing decisions for ad hoc networks using fuzzy logic. In: 1st International Symposium on Wireless Communication Systems (2004)
7. Zhang, X., Cheng, S., Feng, M., Ding, W.: Fuzzy logic QoS dynamic source routing for mobile ad hoc networks. In: The Fourth International Conference on Computer and Information Technology (2004)
8. Upadhyay, R., Bhatt, U. R., Tripathi, H.: DDOS attack aware DSR routing protocol in WSN. In: International Conference on Information Security & Privacy (ICISP2015)
9. Liu,Q., Qin, H.: Implementation and Improvement of DSR in Ipv6. In: International Workshop on Information and Electronics Engineering (IWIEE) (2012)
10. Singh, J.P., Dutta, P., Chakrabarti, A.: Weighted delay prediction in mobile ad hoc network using fuzzy time series. Egypt. Inform. J. **15**, 105–114 (2014)
11. Bhatt, U.R., Nema, N., Upadhyay, R.: Enhanced DSR: an efficient routing protocol for MANET. In: IEEE International Conference on Issues and Challenges in Intelligent Computing Techniques (ICICT) (2014)..https://doi.org/10.1109/icicict.2014.6781282
12. Zuo, J., Ng, S.X., Hanzo, L.: Fuzzy logic aided dynamic source routing in cross-layer operation assisted ad hoc networks. In: IEEE 72nd Vehicular Technology Conference Fall (VTC 2010-Fall) (2010)
13. Jang, J.-S.R.: ANFIS: adaptive-network-based fuzzy inference system. IEEE Trans. Syst. Man Cybern. **23**(3), 665–685 (1993)

Building Domain Familiarity for Elicitation: A Conceptual Framework Based on NLP Techniques

Syed Taha Owais, Tanvir Ahmad and Md. Abuzar

Abstract Elicitation is the entry point of any digital transformation project. Domain knowledge is a critical factor for successful elicitation. Hence, elicitation is usually executed by subject matter experts (SME), who have fair amount of domain knowledge required for the project. However, such SME are scare and costly resources. Hence there is an urgent need to design a framework that can help build domain knowledge of the business analyst (BA) involved in elicitation in order to lower dependence on SME. Such framework should be programmatically implementable and should be able to work seamlessly with the existing techniques of elicitation. Several solutions based on organizational theories, cognitive models, and strategic frameworks have been proposed with varying results. This work is in progress. As elicitation generates a web of interactions in natural language, natural language processing (NLP) techniques can be utilized to extract and build domain knowledge. This paper investigates the application of NLP to extract working domain knowledge from the available organizational documents to assist analyst during elicitation phase. The paper proposes a conceptual framework to extract and build domain familiarity. This paper also analyzes the gaps and new areas of research in this direction.

Keywords Building · Elicitation · NLP techniques

1 Introduction

Any software development project, typically, begins with the elicitation of requirement [1–3]. The requirement elicitation is usually accomplished by one or more persons known as business analyst [4]. They analyze and synthesize

S. T. Owais (✉) · T. Ahmad · Md. Abuzar
NIC, Lodhi Road, New Delhi 10003, India
e-mail: owais@nic.in

T. Ahmad
e-mail: tahmad2@jmi.ac.in

© Springer Nature Singapore Pte Ltd. 2019

H. Malik et al. (eds.), *Applications of Artificial Intelligence Techniques in Engineering*, Advances in Intelligent Systems and Computing 697,
https://doi.org/10.1007/978-981-13-1822-1_54

information from various sources and liaison between user of the system and its developer. The job of the analyst is to elicit correct information which will lead to right solutions delivered within schedule and budget so that it meets the vested interest of the stakeholders and deliver value [5]. This is easier said than done.

Elicitation is a long and arduous journey to identify the need for solution (how, why, when, by whom, for whom, etc.) which can deliver value in a context [3]. It is exploration of concepts, investigation of relationships, and discovery of new ideas. As a result, elicitation generates complex web of interactions and negotiations with steep learning curve [3]. As such, the team of analyst is expected to possess analytical thinking capabilities, business knowledge, communication skills, computer application well versed with latest tools and techniques along with behavioral characteristics like ethical, honesty and other professional attributes [5, 6]. But this is too much to ask for.

Domain familiarity can be referred to a set of valid and strategic knowledge about an area of human endeavor [7]. The persons owning such knowledge are known as Subject Matter Expert (SME). They are aware about the dominant concepts, nitty-gritty of complex inter-relationships, terms of common parlance, jargons, trends and techniques, etc. They are also aware about the current pain points have general idea about users and industry, legislation, compliances, stories, and news. They have the ability to know (acquire and transfer knowledge), propose (generate as well consume new ideas) and justify (action and implementation) [4]. However such resources are costly and may not always be available.

The history of software development is replete with broken codes, abandoned software, fuming stakeholders, and wasted resources time and money [8]. This has resulted in lost opportunities and the blame game.

Standish Group International [2] has surveyed 30,000 IT projects in USA and concluded that factors contributing to failures of software projects (both failed and challenged projects) are requirement planning (19.9%), requirement elicitation (25.4%) and user communication (24.8%). Similar conclusions have been reported from studies in UK and Australia. In Indian context not many studies have been conducted on failures of software. However, public service computerizations are regularly audited by CAG (Comptroller Accountant General). CAG has done audit of many projects done by government bodies with similar conclusions [9].

Hence poor elicitation will result in challenged projects and can happen due to several reasons like lack of domain familiarity, poor communication, planning, etc. Either costly SMEs are deployed as analyst or we find a way to develop domain familiarity for analysts who are not SMEs. This paper uncovers the importance of domain familiarity among the analysts in Sect. 2. Section 3 describes the research methodology and knowledge diffusion of application of natural language processing (NLP) in the area of software engineering. Section 4 shows that using NLP techniques it is possible to develop working knowledge to assist the analyst during elicitation phase. Finally analyzes gap and share new direction for research into elicitation.

2 Related Work

2.1 Elicitation Scenario

Figure 1 represents a standard scenario of requirements elicitation. The job of analyst is to understand the objective of the software development endeavor and document HOW, WHY, WHEN, WHERE, FOR WHOM, BY WHOM, etc. This is done through an intense communication process involving the stakeholders.

Here stakeholders are all those entities who as impacted (negatively or positively) by the outcome of the project and have vested interest in success or failure of the project. The stakeholder may represent the whole canvas of the organization from top management (patron) to specialists (SME) to normal end users. They are guided by the organizational philosophy, power and politics, custom and culture, shared values, etc. Besides this, the organization generally owns a repository of data and documents in various formats, some existing legacy systems and its manuals, etc. On the other hand, analyst will do all activities to draw out actual requirements from the stakeholder and the organization repository. He is equipped with many elicitation techniques like interview, observation JAD/RAD, etc., and perhaps brings in his experience of similar project.

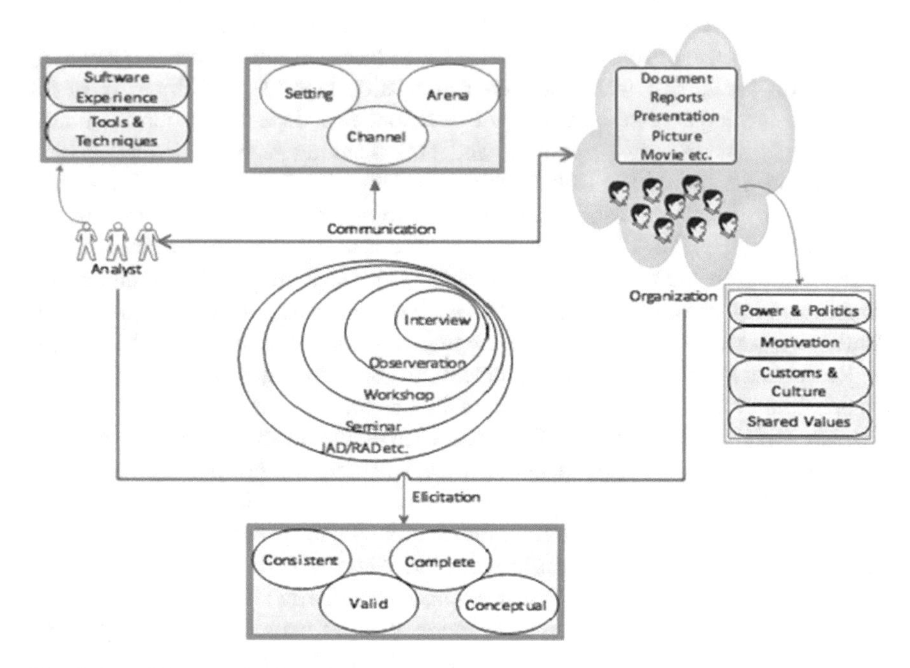

Fig. 1 Broad framework of requirement elicitation

The communication that ensues is a key to the success of the project [10]. The communication happens in an arena with some setting using a channel (face-to-face, telephone, email, etc.). Each channel has bandwidth and noise associated with it [11]. This results in requirement elicitation, which should ideally be consistent, conceptual, valid, and implementable.

2.2 Elicitation Issues

Despite best effort by all, requirement elicitations are hardly consistent, keep changing, and never complete [2]. Lack of domain knowledge, large volume of organizational documents, non-familiarity with jargons and parlance of the domain, nitty-gritty of the domain area, misinformation, faulty assumptions, etc., can derail the communication process for the purpose of requirement elicitation [12].

2.3 Significance of Domain Familiarity

Researchers and practitioners have acknowledged the importance of domain familiarity. Conversation cannot be considered as a black box [5]. Low on domain knowledge creates gap in perception [5, 6]; induces misinformation [13]; wrong assumption [5, 8, 13]; weak interpersonal relationships [3]. Effectiveness of all elicitation techniques like interview, observation, etc., depends on domain familiarity [2, 3]. Asking right questions and understanding reply in correct context depends on the higher level of domain knowledge [2, 3, 5, 8, 13, 14]. Domain familiarity by analyst enhances stakeholder participation [6]; helps to distinguish needs from desires [12, 15, 16]; helps to set priorities and provide direction else the analyst gropes in the darkness of ignorance [15]. Davis model of communication challenge [11] shows that when both analyst and stakeholder possess domain familiarity, it results in shared understanding and concepts involved in the project.

Researchers have attempted to model the domain familiarity and designed models/frameworks to incorporate domain knowledge. These are implemented through tools and techniques. Tacit Knowledge Framework, which is used to identify and articulate requirement can work when analyst has domain familiarity [12]. Clark's theory for common ground enhances communication when domain familiarity is present [11]. Organizations are dynamic, ever looking for new opportunities [2]. Socio-technical theories like Maslow's Motivational theory can be effectively applied when analyst has domain familiarity [11]. Recent researches stress the fact that elicitation is best represented in natural language [17, 18]. Concepts of ontology [11], semantic business vocabulary and rule [19], text mining techniques [21] have been proposed. Common domain knowledge can be extended

[11]. Cognition models have also been supported, which help analyst to interpret the non-verbal cues and body language to extract tacit information. This will be effective when analyst is equipped with domain familiarity. Similarly, elicitation was can developed used the ethnographical frameworks [6]. Elicitation is a learning experience for all involved [1–3, 8, 13, 14]. Education theory when applied properly will enhance domain knowledge, harness negotiation skills, and mental mapping [10]. However, it is also possible that domain knowledge can lead to unintentional omission of some critical tacit information. Hence smart (with domain knowledge) but stupid analyst (who asks more questions rather than providing more answers) stands more chances of getting elicitation right [13].

There are several ways to build domain knowledges. Summarizing the existing knowledge sources [7], working with domain expert [14], data mining and text mining [15], rule mining from structured and unstructured sources [15], by exploiting NLP tools and techniques [20], by building semantic business vocabulary and rules (SBVR) [19] etc.

Domain familiarity is one of the pillars on which elicitation rests. Domain ignorance can lead to incomplete understanding, overlooking of tacit assumptions, misinterpretation of synonyms, over or under commitment, lack of correlation between different views expressed on same topic [13].

3 Research Method

Domain knowledge with the analyst can get right elicitation and there are different ways to build domain knowledge. As majority of the elicitation exists in natural language, it is pertinent to ask whether NLP can be used to build domain familiarity that can assist analyst during elicitation in framing right question and eliciting right responses.

3.1 Research Question

RQ: How to develop domain familiarity of an analyst involved in elicitation from existing documents by using NLP techniques.

In other words, by applying NLP techniques on the existing document, a body of knowledge can be built which an analyst can use to plan the elicitation process, enhance interaction with stakeholders, understand explicit and implicit requirements, etc. The main objective of this paper is to propose a framework that can utilize the existing data and document repository in the organization, where analyst is involved in requirement elicitation, to develop a systematic working knowledge. Such processed knowledge will help analyst to move in correct direction.

3.2 Data Sources

The primary sources of the data are the technical papers on the requirement engineering topics. The technical papers were searched on Google(TM) Scholar and CiteSeerx (TM) using appropriate keywords. The sample search phrase included ("NLP in RE", "How to develop domain familiarity", "Extract Domain knowledge", "Automatic RE", "NLP tools for RE", "NLP techniques for RE"). The search results of papers were added to the online library and were labeled for later reference.

The search string returned several papers. The list of relevant open access papers was prepared based on their abstract, number of citations, etc. Only journals and conference papers were considered. The final list of papers consisted of three categories namely elicitation as a communication problem, domain knowledge for requirement engineering, and use of NLP in software engineering.

3.3 Diffusion of Knowledge

All the authors have highlighted the limitation of conventional requirement elicitation techniques, which is still prevalent in this era. This often results in the gap between the requirement expressed in natural language and requirement specification in some formal representation. As a result around 42% of the original proposed functionality is delivered. Also the conventional elicitation techniques assume that analysts are domain expert of the problem under study, which may not always be true.

Describing the system in the language of the stakeholder is the most effective way of capturing the requirements. As per one estimate 80% of the elicitation of requirement is expressed in natural language [18].

Various techniques of NLP have been described like tokenization, parts-of-speech (PoS) tagging, parsing, rule-based natural language processing, statistical NLP, classification and clustering techniques, etc.

Review of many techniques like heuristics, contextual NLP, use of glossary and terms to convert natural language description into formal specification [20]. The author has also proposed three-stage model—namely tokenization, parsing and term manager, by using dictionary of 32,000 terms and 79 rules. The same was implemented using LISP IDE for simple unambiguous descriptions.

Several tools like ReBuilder, GOAL, NL-OOML have also been developed using NLP concepts [18], UML Generators which can automatically generate UML document from requirement document expressed in natural language. The author proposes OpenNLP (Apache Foundation Project) to transform the specification in natural language into object model. This has been done by leveraging the SBVR (Semantics Business Vocabulary and Rules approved by OMG).

Attempts have been made to solve the problem of disambiguation [21]. The templates are designed in such a way that the requirements expressed are true representation of the proposed system. Here authors have used Ruth and EARS template. The template based requirement elicitation document is fed to an NLP engine for tokenization. The resultant tokens are then subjected to chunker, which generates noun phrase and verb phrase. Using BNP grammar full parse tree is generated. Using the pattern matching language JAPE, the authors transform their approach into a tool named as RETA (REquirement Template Analyzer). The tool has been evaluated under various test cases like absence of glossary, effectiveness of approach in identifying non-conformance defects.

4 Proposed Framework

The proposed framework assumes that analyst does not have working knowledge of the area of elicitation (it may be his first project) but he has access to all the documents which describe about the processes. As an analyst he needs to interact with various stakeholders with the purpose of elicitation, but his effort is limited by his knowledge about the project. The proposed conceptual frame attempts to fill the gaps as illustrated in Fig. 2.

The analyst has access to all the documents (both structured and unstructured) that describe the domain of the problem. The analyst has to interact with the stakeholder. In the above system, the framework will consume documents and generate a working knowledge on the basis of which the analyst will communicate with the stakeholder. Hence the proposed framework is built upon the conventional elicitation process as depicted in Fig. 1.

4.1 Component of Framework

The proposed conceptual model primarily consists of a NLP engine that consumes the different sources of data and produces a set of objects and relationships. The second component is the SBVR builder. The concept of SBVR is approved by OMG (Object Management Group). The output of NLP is used to build the business vocabulary and rules. These two components generate the third component working domain knowledge. The working domain knowledge is used by analyst to design the elicitation process.

4.2 Working of Framework

The collection of document is processed by NLP engine. The document is parsed, tokenized to identify noun and verb phrases. The noun phrases will be clustered into

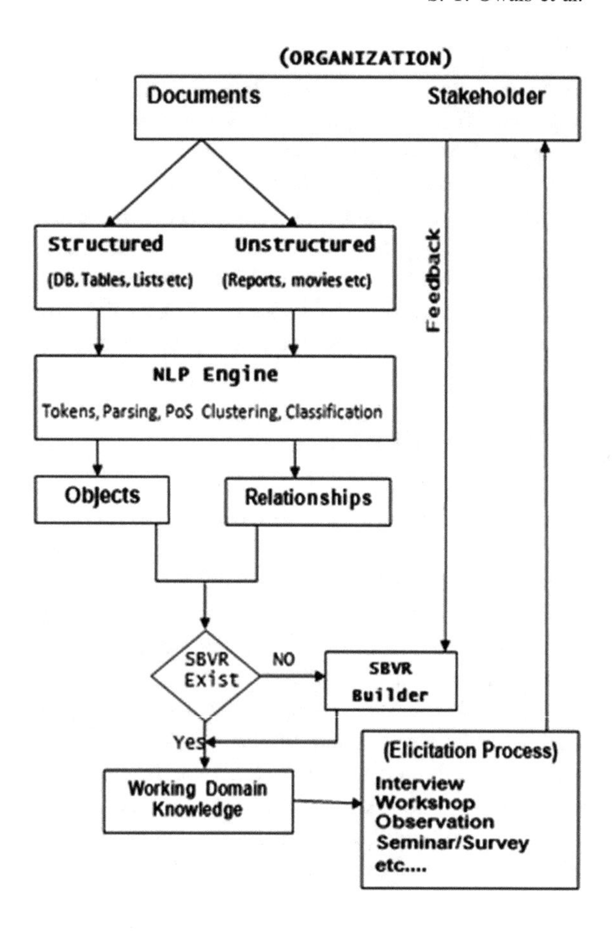

Fig. 2 Conceptual framework for domain familiarity extraction

similar concepts and further classified to identify the broad modules. The database of dictionary (for synonym and antonym) will be looked up and a glossary of terms will be created. Parts of speech tagging will be done to establish relationships. These relations and objects will used to build semantic business vocabulary and rules. This will result in working domain knowledge. Using this domain familiarity, the analyst will design the plan for interview, observation and other elicitation methods. This will enable the analyst to conduct the communication process with the stakeholders in more effective and informed manner. There will be feedback mechanism which will further enrich the domain knowledge of the analyst (Fig. 3).

Programming languages are Java and Python support OpenNLP and NLP-Kit respectively. These provide necessary functionalities for tokenization, parsing, classification, and clustering. Besides acquisition of information can be done by using number of boilerplate solution from Google™ and Yahoo™.

There are solutions available for visualization and statistical processing of terms and entities extracted.

Hence the proposed framework will enable cost and time saving on SME required for elicitation. This is very true for large eService projects. It will also lead to development of systematic body of knowledge for a particular domain, which can be updated, reviewed, and validated. Such sources can also be used for other purposes like risk assessment, code testing, bug fixing, and future enhancement and these can also be used for training, negotiation and for reviving challenged projects. By systematic synthesis of various domains, new domain areas can be identified or created leading to development of new product—for example consider two seemingly disparate domains namely game of soccer and music. These two domains can be synergized to create new domain like music based pain management system, etc.

The major challenging factors for implementation of this framework will be the level of digitization and organization of documents, volume, and categories of information. If the majority of document is organized and structured like database, tables, lists, websites, design documents, etc., then the framework can be easily implemented.

4.3 Gap and Research Direction

NLP is primarily tokenization, PoS tagging and parsing of information expressed in natural language. However NLP is an evolving technology and many new ways are being proposed to handle complex and non-formal sources of information expressed in natural language. Disambiguation is a challenging area of research.

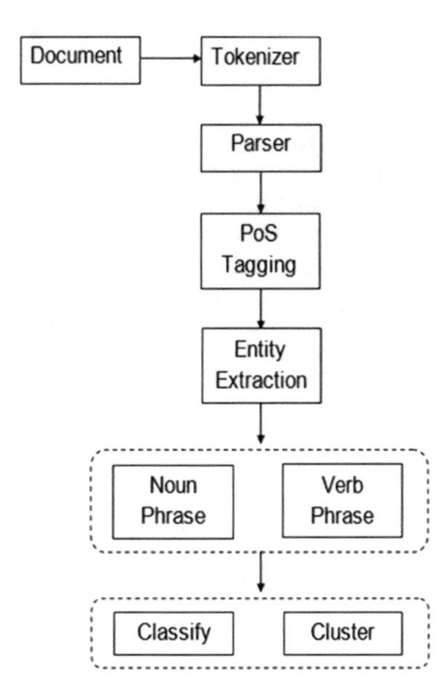

Fig. 3 NLP based process flow

Building new business semantics and rules, strengthening the existing one can greatly help in automatic conversion of requirement documents to formal computer models. It can also identify and extract new domain areas. Validation of model generated automatically is another area of research. e-Services are specialized IT-based service offering. The public service delivery can take the benefit of NLP during requirement elicitation as public service delivery require large array of e-services, which are owned by many and used by large number of citizens. Not much work is reported in this direction. Researchers and practitioners can exploit NLP to generate validated design for e-services for eGovernment.

References

1. Coughlan, J., Robert Macredie, D.: Effective Communication in Requirement Elicitation: A Comparison of Methodologies. Springer, London (2002). https://doi.org/10.1007/s007660200004
2. Davey, B., Cope, C.: Requirements elicitation—what's missing? Issues Inf. Sci. Inform. Technol. **5** (2008)
3. Teven, H., Houn-Gee, C., Gary, K.: Communication skills of IS providers: an expectation gap analysis from three stakeholders perspectives. IEEE Trans. Profess. Commun. **46**(1), 17–34. https://doi.org/10.1109/tpc.2002.808351
4. Business Analyst Book of Knowledge from International Institute of Business Analysis
5. Davey, B., Parker, K.: Requirements elicitation problems: a literature analysis. Issues Inf. Sci. Inform. Technol. **12**, 71–82 (2015)
6. Thew, S., Sutcliffe, A.: Investigating the role of 'soft issues' in RE process. In: 19th IEEE International Conference on Requirements Engineering Conference, pp. 63–66. https://doi.org/10.1109/re.2008.35
7. Herlea Damian, E.: Challenges in requirements engineering
8. Davis, C.J., Trimblay, M.C., Fuller, R., Berndt, D.J.: Communication challenges in requirements elicitation and the use of repertory grid technique. J. Comput. Inform. Syst. **46**(5) (2006)
9. CAG Office of India: 2016. CAG audit report on communication and IT sector. Presented Nov 2016 (source website)
10. Hada, I., Soffer, P., Kenzi, K.: The Role of Domain Knowledge in Requirements Elicitation Via Interviews: An Exploratory Study. Springer, London (2012)
11. Business Analyst Book of Knowledge from International Institute of Business Analysis
12. Friedich, W.R., Van Der Poll, J.A.: Towards a methodology to elicit tacit domain knowledge from users. Interdisc. J. Inform. Knowl. Manage. **2**(1), 179–193 (2007)
13. Appan, R., Browne, G.J.: The impact of analyst-induced misinformation on the requirements of elicitation process. MIS Q. **36**(1), 85–106 (2012)
14. Milne. A., Maiden, N.: Power and politics in requirements engineering: a proposed research agenda. In: 19th IEEE International Conference on Requirements Engineering Conference, pp. 187–196 (2011). https://doi.org/10.1109/re.2011.6051646
15. Kaya, H., Shimizu, Y., Yasui, H., et. al.: Enhancing domain knowledge for requirements elicitation with web mining. In: 17th Asia Pacific IEEE Software Engineering Conference (APSEC), pp. 3–12 (2010). https://doi.org/10.1109/apsec.2010.11
16. Aranda, G.N., Vizcaino, A., Paittini, M.: A Framework to Improve Communication During the Requirements Elicitation in GSD Projects. Requirement Engineering, pp. 397–417. Springer, Berlin (2010)

17. Sedelmaier, Y., Landes, D.: Using business process models to foster competencies in requirements engineering. In: CSEE&T 2014, Klagenfurt, Austria (2014)
18. Mohanan, M., Samuel, P.: Software Requirement Elicitation Using Natural Language Processing. Springer International Publishing, Switzerland (2016). https://doi.org/10.1007/978-3-319-28031-8_17
19. The Object Management Group: Semantics of business vocabulary and business rules (SBVR). Formal specification, v1.0 (2008)
20. MacDonell, S.G., Min, K., Connor, A.M.:.Autonomous requirements specification processing using natural language processing. Technical Report (2014)
21. Arora, C., Sabetzadeh, M., Briand, L., Zimmer, F.: Automated Checking of conformance to requirements templates using natural language processing. IEEE Trans. Softw. Eng. (2015). https://doi.org/10.1109/tse.2015.2428709

Optical Character Recognition on Bank Cheques Using 2D Convolution Neural Network

Shriansh Srivastava, J. Priyadarshini, Sachin Gopal, Sanchay Gupta and Har Shobhit Dayal

Abstract Banking system worldwide suffers from huge dependencies upon man-power and written documents thus making conventional banking processes tedious and time-consuming. Existing methods for processing transactions made through cheques causes a delay in the processing as the details have to be manually entered. Optical Character Recognition (OCR) finds usage in various fields of data entry and identification purposes. The aim of this work is to incorporate machine learning techniques to automate and improve the existing banking processes, which can be achieved through automatic cheque processing. The method used is Handwritten Character Recognition where pattern recognition is clubbed with machine learning to design an Optical Character Recognizer for digits and capital alphabets which can be both printed and handwritten. The Extension of Modified National Institute of Standards and Technology (EMNIST) dataset, a standard dataset for alphabets and digits is used for training the machine learning model. The machine learning model used is 2D Convolution Neural Network which fetched a training accuracy of 98% for digits and 97% for letters. Image processing techniques such as segmentation and extraction are applied for cheque processing. Otsu thresholding, a type of global thresholding is applied on the processed output. The processed segmented image of each character is fed to the trained model and the predicted results are obtained. From a pool of sample cheques that were used for testing an accuracy of 95.71% was achieved. The idea of combining convolution neural

S. Srivastava (✉) · J. Priyadarshini · S. Gopal · S. Gupta · H. S. Dayal
Vellore Institute of Technology, Chennai, India
e-mail: shriansh.srivastava2015@vit.azc.in

J. Priyadarshini
e-mail: priyadarshini.j@vit.ac.in

S. Gopal
e-mail: sachin.gopal2015@vit.ac.in

S. Gupta
e-mail: sanchay.gupta2015@vit.ac.in

H. S. Dayal
e-mail: harshobhit.dayal2015@vit.ac.in

© Springer Nature Singapore Pte Ltd. 2019

H. Malik et al. (eds.), *Applications of Artificial Intelligence Techniques in Engineering*, Advances in Intelligent Systems and Computing 697,
https://doi.org/10.1007/978-981-13-1822-1_55

network with image processing techniques on bank cheques is novel and can be deployed in banking sectors.

1 Introduction

Handwritten character recognition is a research field in machine learning, pattern recognition, and computer vision. A handwriting recognizer acquires and detects characters in documents, images, and other sources and converts them into machine-readable form. This technology finds its application in fields like optical character recognition (or optical character reader, OCR) and other character recognition systems. The implementation of an OCR via machine learning mechanisms is an emerging field of research. The machine learning algorithms that are best suited for these purposes are neural networks, multi-class Support Vector Machines (SVM) and a few unsupervised paradigms of machine learning.

Machine learning is a research field in computer science that deals with learning from a dataset which can be used in practically any real-time application such as value prediction, classification, and decision-making. A supervised machine learning model is input with instances of data specific to a problem. The algorithm has to learn from the input data and provide a universal hypothesis that conforms to the data as effectively as possible. An important aspect of a successful machine learning algorithm is that it is not only able to provide answers to the training data but also performs significantly well when tested on a fresh dataset of completely different instances [1].

Neural Network is a set of mathematical models that are inspired by adaptive biological learning. Neural networks can be applied in various fields such as OCR pattern recognition, classification, regression, and language translation. A neural network is a collection of perceptron learners brought together and arranged in layers thereby referred to by the name multi-layered perceptron. A neural network is a supervised learning algorithm where learning typically occurs through training of data [2]. An advanced version of the neural networks, the convolution neural network which is known to extract minute details from the input images to provide a high degree of accuracy, has been used in the proposed work [3].

In banking applications, neural networks can be used to recognize characters on bank cheques. The important text details on the cheque are segmented out and input to the machine learning OCR that stores the output in the text format. The cheque is scanned and the required information from the cheque is obtained as text that can be fed to the banking software.

The contents of the paper have been distributed as follows. A literature survey of related works has been done in Sect. 2. The related works, their pros, and cons have been brought out and results from those have been used for comparative analysis in Sect. 3.4. Section 3 contains the proposed work where Sect. 3.1 tells about the algorithm. Section 3.2 contains the information about the architectural design. Further, the paper contains important tables and plots which give an insight into the proposed work. Finally, conclusion and future scope have been added in Sect. 4.

2 Literature Survey

Uhliarik [1] primarily focuses on Handwritten Character Recognition using Backpropogation Algorithm and resilient backpropagation (RPROP). In this work median filter is used in order to remove any noise in the image. Binary Thresholding is applied and contours are created. A bounding box is created around the character and then the character is segmented to obtain an image of size 20 × 20. Projection Histograms are used as feature vectors and then fed to the Neural Network. Backpropogation took 100 optimization epochs and the accuracy obtained was 96.25%. The main disadvantage of this approach is that using contours to find the characters can lead to failure in case of digits like 0 where both inner and outer circle will be counted as contour.

Shih and Wei [4] primarily recognize handwritten Sanskrit characters using two-stage multi-classifier. In the first stage kNN is applied to group the data into k classes and a multilayer classifier of *k* classes to label the data. The key aspect of this work is that the handwriting recognition is done for a language other than English. However its limitation is that the accuracy was 85% which could still be improved.

LeCun et al. [5] demonstrates how learning networks such as back propagation network can be applied for recognizing handwritten zip codes. The key feature of this work is the application of constrained back propagation which helps in shifting invariance but also vastly reduces the entropy. However, overlapping characters were not recognized by the image segmentation method used in the work.

Coates et al. [6] focuses on character recognition using Unsupervised feature Learning. Here, a variant of *K* Means clustering is preferred over the likes of auto-encoders and sparse encoding which makes this method faster and simpler. The key aspect apart from faster computation and simplicity is usage of large-scale algorithm for learning from unlabeled data automatically. However, improvements could be made in image preprocessing for even better results.

Alam Miah et al. [7] focuses on using Segmentation-based recognition strategy and further using Neural Network with Backpropagation. 26 features are extracted from the obtained segmented image (20 × 15) and then fed into the Neural Network for the handwritten digit recognition. 500 iterations were required for prediction which suggests that the system takes time to converge. Moreover, the data used for training is very less as compared to EMNIST database as only 50 samples are there for each digit.

Suresh Kumar Raju and Mayuri [8] uses hybrid feature extraction technique that consists of Feed forward Neural Network, and radial basis function Neural Network and nearest neighbor network for recognizing handwritten English alphabets. The scanned bank cheque is converted to gray-scale and then edge detection is applied in order to segment the characters from the cropped image. Image dilation and filtering are applied in order to remove any noise and to smoothen the image. Blob analysis is used for getting the image of each character. The images obtained are resized to 6 × 8 and fed into the neural network for classification. The main disadvantage of this system is that the dataset used by them is self-created which signifies that the model can be biased towards some samples.

Mehta et al. [9] focuses on the courtesy, amount and the signature, while the courtesy amount payable amount is sent to the character segmentation the signature is authenticated using sum graph and Hidden Markov Model (HMM). The character segmentation uses median filtering and then binarization of the cheque image before segmentation using sum graph method on the negative of the binarized image and segmented on the minima. The work resulted in character segmentation accuracy is found to be 95%, character recognition efficiency 83%, Digit recognition efficiency is 91% and system detects forgeries with an accuracy of 80% and can detect the signatures with 91% accuracy.

Narkhede and Patil [10] describe the process for the authentication of the signature of a person. This work used the basic preprocessing techniques such as thinning, binarization, and rotation and then calculates the shape context score by calculating the best image of a point p_i to a point $q_{j.}$ on the template signature. These scores are then fed to the KNN which assigns the most common class out of the neighbor for predicting whether the signature is authentic or not.

3 Proposed Work

The objective of the machine learning model is to provide an efficient system for recognizing characters on bank cheques to do away with the manual data entering methods prevalent in banks. The presence of neural networks ensures high accuracy and efficiency in finding out the handwritten characters. The image decomposes into segments of fixed pixel values and these segmented characters are fed to the neural network which predicts the characters from the cheque and stores it in the text format. Verification of amount in words can be done with amount in digit. This ensures that the two fields match with each other and the cheque is valid. In case there is an empty field, the output is an error message is displayed where the user is told which field(s) is missing.

3.1 Algorithm

Input: Bank Cheque Image
Step 1: Segmentation: It involves segmenting the text part of the bank cheque such as name, the amount in words, the amount in digits, date, etc., by cropping or segmenting those parts from the cheque image.
Step 2: Character Extraction: Segmenting each word or numbers obtained from step 1 into individual characters.
Step 3: Prediction: Feeding the segmented characters into the convolution 2D neural networks for character recognition.
Output: The result obtained from the neural network is concatenated and stored in a text file from where it can be fed.

3.2 Architectural Design

The EMNIST dataset is a set of handwritten character digits derived from the NIST Special Database 19 and converted to a 28 × 28 pixel image format and dataset structure that directly matches the MNIST dataset.

It is important in image processing to select an adequate threshold of gray level for extracting objects from their background. Thresholding is required in order to create a contrast between the written characters from the background and to also convert gray-scale images to a binary image. This paper makes use of Otsu thresholding which is an automatic threshold selection region based segmentation method. Otsu method was proposed by Scholar Otsu in 1979 which is widely used because it is simple and effective [11]. The proposed work involves usage of 2D Convolution Neural Network which is an important operation in signal and image processing (Tables 1 and 2).

Table 1 Description of architectural model

Architecture	Description
Input layer	A bank cheque image is capture at 1601 × 719 pixels
Segmentation and extraction layer	Cropping: the input image is cropped into various rectangular field that capture the payable amount (words and digits), account number, date, cheque number, signature, recipient's name. Fixed pixel values are cropped to match the exact region of interest. Missing parts of the cheque can be detected and marked invalid
	Otsu thresholding: it is used for background removal of the above regions
	Connected component labeling: it is applied on outputted regions and the boundary location of the component are extracted
	Component extraction: the character are extracted from the original image form the boundary location obtained in the previous step and is given to the CNN for prediction
Prediction layer	A 2 dimensional CNN is used to recognize the character extracted at the end of the previous step and output the character
Output	The characters recognized by the prediction layer are stored into a text file for further use

Table 2 Training results

Trained data	Accuracy in %
Alphabets [A–Z]	97.45
Digits [0–9]	98.87
Average	98.16

3.3 Performance Analysis

The convolution model employed for the character recognition task was trained on the EMNIST dataset that consists of: Capital Letters: 145,600 characters; 26 balanced classes, Digits: 280,000 characters. 10 balanced classes. Images shown in Table 3 have been created by the author own self for the study purpose. These images are not collected from anywhere.

Table 3 Experimental results of the proposed system

S. No.	Image	Output obtained in text format	Accuracy in %
1	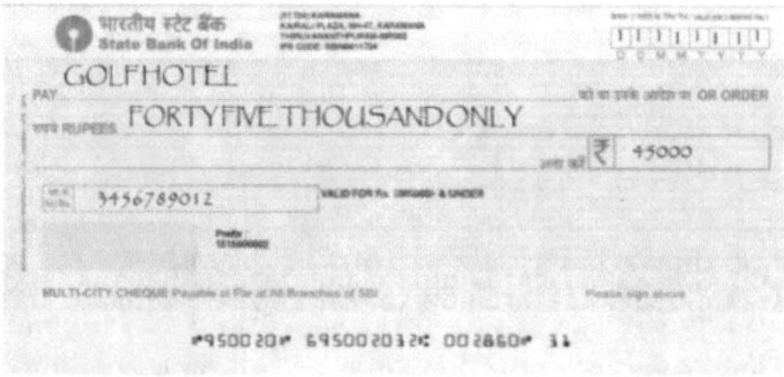	Name—GOLF HOTEL Amount in words—FORTY FIVE THOUSAND ONLY Amount in digits—45,000 Account number—3956789012 Date—11/11/1111	97.77
2	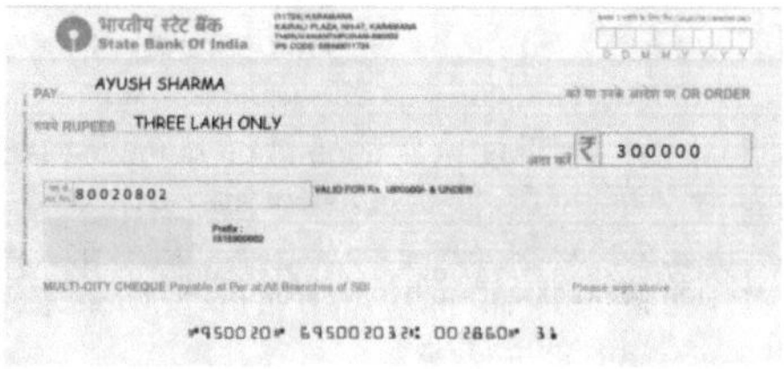	Name—AYUSH SHARMA Amount in words—THREE LAKH ONLY Amount in digits—300,000 Account number—80020802 Date—_______ (invalid cheque)	100
3	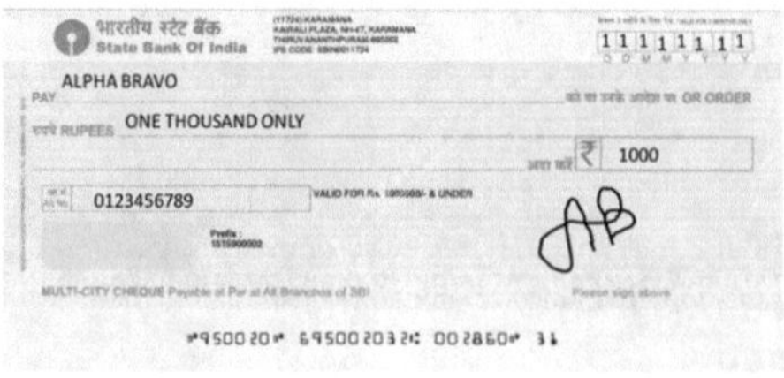	Name—ALPHA BRAVO Amount in words—ONE THOUSAND ONLY Amount in digits—2000 Account number—0223955789 Date—11/11/1111	89.74

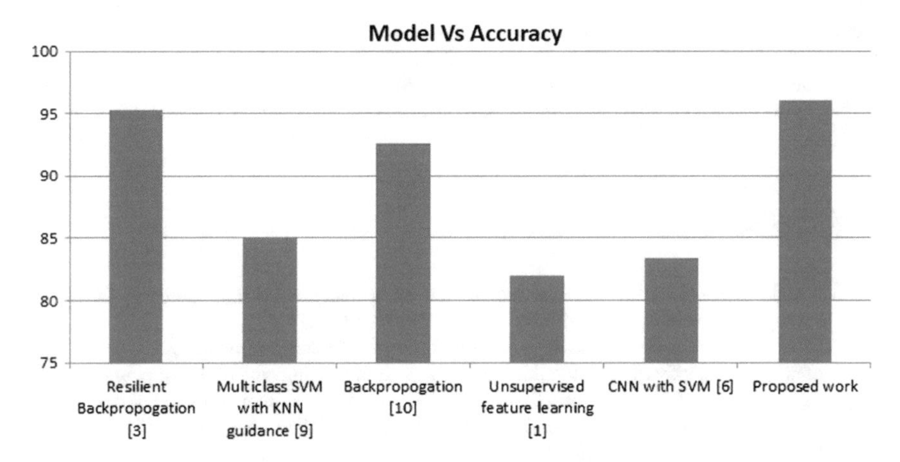

Fig. 1 Model versus accuracy (in %) comparison graph

3.4 Comparative Study

See Fig. 1.

4 Conclusion and Future Scope

The convolution neural network currently used in the system deploys handwriting character recognition for block letters and digits using pattern recognition. This gives an optical character recognition (OCR) tool built using machine learning. This OCR can be used in banking applications to automate the whole banking transaction procedure. The accuracy of OCR may depend on external noisy factors such as damage (which may affect the written part) on the cheque surface. However, the model has been developed in such a manner that the place (lighting, orientation) do not affect the accuracy of the OCR. While OCR is used effectively in other domains, it is still finding its way in the real-time banking sector today. This method can be very successful as it reduces human effort and can speed up the entire banking process. The proposed work produced an accuracy of about 95.71% when tested with the bank cheques as shown in Table 3, however the model's validation accuracy while training was 98% for letters and 97% for alphabets.

Future scope of the proposed work includes increasing the accuracy of the character recognition tool. Additions such as support for cursive handwriting, punctuations can be further incorporated with better image segmentation. The system can also be scaled up to other applications that require reading of the text as their main operation in other sectors as well. The similar work can be done for small letters and signature verification can also be added for better cheque authenticity.

Acknowledgements Authors would like to thank to the bank accounts holder of SBI bank account number 80020802, 3956789012 and 0223955789. The images shown in Table 3 have been created by the authors own self with the help of these account numbers for the study purpose. Authors also thank to the publisher to provide the enlighten platform for this research work.

References

1. Uhliarik, I.: Handwritten character recognition using machine learning methods. Bachelor's Thesis, Department of Applied Informatics, Bratislava (2013)
2. Vision Tech: OCR neural networks and other machine learning techniques
3. Fanany, M.I.: Handwriting recognition on form document using convolutional neural network and support vector machines (CNN-SVM). In: 2017 5th International Conference on Information and Communication Technology (ICoIC7), Melaka, pp. 1–6 (2017). https://doi.org/10.1109/icoict.2017.8074699
4. Shih, Y., Wei, D.: Handwritten Sanskrit recognition using a multi-class SVM with K-NN guidance. MIT
5. LeCun, Y., Boser, B., Denker, J.S., et al.: Backpropagation applied to handwritten zip code recognition. Neural Comput. **1**, 541–551 (1989). https://doi.org/10.1162/neco.1989.1.4.541
6. Coates, A., Carpenter, B., Case, C., et al.: Text detection and character recognition in scene images with unsupervised feature learning. In: ICDAR 2011, Computer Science Department, Stanford University, Stanford, CA, USA. https://doi.org/10.1109/icdar.2011.95, Nov 2011
7. Alam Miah, M.B., Yousuf, M.A., et al.: Handwritten courtesy amount and signature recognition on bank cheque using neural network. Int. J. Comput. Appl.(0975–0887) **118**(5) (2015)
8. Suresh Kumar Raju, C., Mayuri, K.: A new approach to handwritten character recognition. Int. J. Innov. Res. Comput. Commun. Eng. **5**(4), 9003–9012 (2017). https://doi.org/10.15680/IJIRCCE.2017.0504354
9. Mehta, M., Sanchat, R., et al.: Automatic cheque processing system. Int. J. Comput. Electr. Eng. **2**(4) (2010). https://doi.org/10.17706/ijcee
10. Narkhede, S.G., Patil, D.D.: Signature verification for automated cheque authentication system based on shape contexts. Int. J. Comput. Sci. Inform. Technol. **5**(3), 3297–3300 (2014)
11. Qu, Z., Hang, L.: Research on image segmentation based on the improved Otsu algorithm. Int. J. Innov. Res. Comput. Commun. Eng. **5**(6) (2017)

An Approach to Color Image Coding Based on Adaptive Multilevel Block Truncation Coding

Nadeem Ahmad, Zainul Abdin Jaffery and Irshad

Abstract Block Truncation Coding (BTC) has been considered as a prime choice in many instances because it is the simplest to implement and is quite satisfactory from the viewpoint of the resulting fidelity and compression ratio. To effectively compute multidimensional color data, an efficient expression of color data is necessary, therefore this paper presents a novel BTC algorithm called Adaptive Multilevel Block Truncation Coding (AMBTC) to achieve the optimal coding performance by adaptively choosing two-level or four-level BTC according to the edge property of the coding block. The AMBTC uses an adaptive selector level based on the sum of absolute difference (SAD) to get optimal coding performance. In order to reduce the bit rate, we utilize luminance bitmap that represents the three color bitmaps. This method improves the two-level and four-level BTC used in AMBTC. The simulation result is compared with existing conventional BTC and it is found that the proposed BTC outperforms in PSNR and compression ratio and achieves better image quality of reconstructed images.

Keywords BTC · AMBTC · SAD · PSNR · Compression ratio

1 Introduction

During the past decade, the use of Internet and Multimedia technologies is growing exponentially. This poses a greater stress on computer hardware for information transmission, storage, and management. Digital images usually require a large

N. Ahmad (✉) · Z. A. Jaffery · Irshad
Department of Electrical Engineering, Jamia Millia Islamia, New Delhi, India
e-mail: nadeemiete@gmail.com

Z. A. Jaffery
e-mail: zjaffery@jmi.ac.in

Irshad
e-mail: irshad.jmi@gmail.com

© Springer Nature Singapore Pte Ltd. 2019

H. Malik et al. (eds.), *Applications of Artificial Intelligence Techniques in Engineering*, Advances in Intelligent Systems and Computing 697,
https://doi.org/10.1007/978-981-13-1822-1_56

number of bits for its representation; it causes a serious problem for image data transmission and storage. A true color digital image is an array of three (R, G, and B) color components, each component is quantized to 8-bits, so to represent a color image, it requires 24 bits per pixel. A similar case is found in color images. One approach is to apply the grayscale compression technique on each color plane to compress color data. However, the correlation found between the color planes is not utilized by this method. To overcome this problem, a more advanced approach is utilized by first converting color image planes to less correlated YCbCr, YUV, YIQ, etc., format, and then further use a grayscale image coding on each converted color component. In all color image representation of Y** format, the first component, Y stands for the intensity of the pixel while remaining other two components represent the color statistics. In the color image coding, more weight is given to color components in comparison to the corresponding luminance component.

This paper proposed a novel color image coding technique based on block truncation coding (BTC). Several algorithms and conceptual modification to the BTC have been reported in literature, viz. Delp and Mitchell [1] presented a grayscale image coding approach based on some analytical aspect of the moment-preserving quantizers, representing the generalization of the two-level BTC techniques to multilevels. Also for image enhancement by upper and lower means BTC, absolute moment BTC, and adaptive BTC are discussed in references [2–4]. Based on predictive BTC, a low bit rate, good visual quality, and low computational cost techniques in encoding and decoding operation are presented in [5]. A method of getting more reduction in bit rate is achieved using pattern fitting method in [6]. However, all above-discussed BTC methods are based on two-level bit-pattern. To enhance the visual quality of reconstructed images, multilevel BTC method is also discussed in many kinds of literature. Reference [7] proposed a classified BTC based on vector quantization and equispaced three-level BTC is discussed in [8]. Wu and Coll [9] reported their investigations on the application of BTC techniques to color images. Color images, in general, can be looked upon as consisting of three monochromatic image planes—red, green, and blue (R/G/B). One can, therefore, achieve an additional compression ratio of approximately 3:1, if one could use a single bitmap for all of the three primary color planes, except the quantization values, would be different for each color plane. A scheme that joins the three planes through a bit-by-bit majority rule (among the R, G, and B bits in each pixel) into a single fused bitmap is suggested in [9]. Quad-tree segmentation BTC method is presented in [10] to reduce the bit rate in color image coding. Hu et al. [11] presented block prediction coding and AMBTC (AMBTC-BPC) approach for color image coding.

This paper is further divided into four sections. Conventional BTC and its limitations are discussed in Sect. 2, AMBTC details in Sect. 3 and simulation results in Sect. 4, and the last Sect. 5 discussed the conclusion of the overall paper.

2 BTC Overview

Block truncation coding techniques fall within the category of what has since been generalized as moment-preserving quantization methods of data compression. In its original form, the central idea was to preserve the first two sample moments of a small block (of size, let us say, $n \times n$) within the image under compression-oriented processing. Let $\{x_i, i = 1, \ldots, n^2\}$ be the n^2 intensity values of the pixel set in this sub-image block. Then, the first two moments $\bar{x}$ and $\overline{x^2}$ the standard deviation σ, a function of these two moments, can be written as

$$\bar{x} = \frac{1}{n^2}\sum_{i=1}^{n^2} x_i \tag{1}$$

$$\overline{x^2} = \frac{1}{n^2}\sum_{i=1}^{n^2} x_i^2 \tag{2}$$

$$\sigma = \left(\overline{x^2} - \bar{x}^2\right)^{1/2} \tag{3}$$

The concept central to this analysis is to define a 1-bit, that is, two-level quantizer, as shown in Fig. 1. With a threshold x_{thr} and two output levels, such that

$$\left.\begin{array}{l} x_i \geq x_{\text{thr}} \Rightarrow y_i = x^+ \\ x_i \leq x_{\text{thr}} \Rightarrow y_i = x^- \end{array}\right\} \forall i = 1, \ldots, n^2 \tag{4}$$

An intuitively obvious threshold in this context would be the mean value itself. The unknown output values x^+ andx^- can be determined by setting up the expressions that equate (preserve) the moments before and after quantization, as follows:

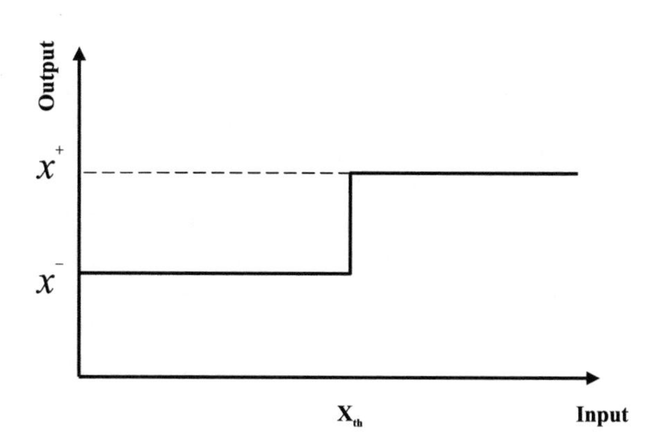

Fig. 1 Input–output relationship of a binary quantizer

$$n^2 \cdot \bar{x} = n^- x^- - n^+ x^+ \quad (5)$$

$$n^2 \overline{x^2} = n^- [x^-]^2 - n^+ [x^+]^2 \quad (6)$$

where n^+ and n^- are, respectively, the number of pixels above and below the threshold (mean) value, which together adds up to n^2 (the total number of pixels in the block). Now, in Eqs. (5) and (6), the only unknowns are x^+ and x^-, since n^+ and n^- can be counted once the mean value is determined. Solving the simultaneous nonlinear Eqs. (5) and (6) for x^+ and $x-$, we have

$$x^- = \bar{x} - \sigma \sqrt{\frac{n^+}{n^-}} \quad (7)$$

$$x^+ = \bar{x} + \sigma \sqrt{\frac{n^-}{n^+}} \quad (8)$$

Thus the output levels, x^+ and x^-, are biased symmetrically about the mean value. Both the positive and negative biases are proportional to the standard deviation.

However, the above BTC method has some limitations when applied to color image coding. If gradual variation is observed in edge information of the block than two-level BTC is not suitable for color image coding. The performance of the reconstructed image is adversely affected, when the different location of a coding block uses the similar edge information.

3 Proposed AMBTC

This paper proposed a unique color image coding techniques known as adaptive multilevel BTC (AMBTC) that reduces the drawbacks of two-level BTC. In this way, AMBTC provides optimal coding performance. Figure 2 shows the detailed block diagram of AMBTC. The implementation of two-level and four-level BTC are executed in parallel on each block by dividing the input color image into nonoverlapping blocks of 4 $\times$ 4 size. The preferences of choosing the level of BTC solely depended on the edge detection of input blocks and it is decided by computing the sum of the absolute difference (SAD) value between the original input data and the reconstructed data for each level. The minimum value of SAD results in selecting the optimal coding level of AMBTC, in this way it increases the accuracy of edge detection and minimize memory cost requirement. Although this method helps in the selection of optimal coding level, on the other hand, it increases hardware cost due to parallel processing utilized in two-level and four-level BTC, pixel reconstruction and SAD value calculation. Based on the edge characteristics of Image coding blocks, reference [12] suggested, a bit rate of 6 bpp in conventional two-level and 9 bpp in four-level BTC, respectively. To design a codec for

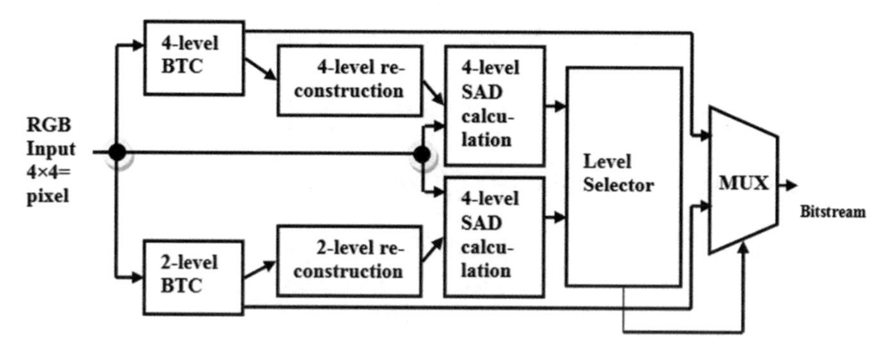

Fig. 2 Block diagram of AMBTC

color images, it is desirable to fix the bit rate to 4 bpp, and it is very difficult to design a codec which maintains the bit rates of two-level and four-level BTC in (4 × 4) block to 4 bpp. To solve this issue, the three color bitmaps are represented by only luminance bitmaps. This paper proposed a method to enhance the two-level and four-level BTC and detail of these methods is discussed in the Sects. 3.1 and 3.2.

3.1 Improved Two-Level BTC

In order to compress the color images using the proposed BTC approach, three bitmaps (bR, bG, and bB) are required to be generated. Since image block consists of edge characteristics, the bitmap is used in place of three bitmaps to represent the color image edge information. One of the critical problems arises when required edge information is more precisely represented by a single bitmap. This issue is more precisely treated by considering that R, G, and B components are represented by luminance and therefore, to generate a single bitmap, a method based on luminance called BTC-Y is adopted for image coding. The computed bitmap represented by luminance is given as

$$
\begin{gathered}
b^{2L}(i) = \begin{cases} 1 & \text{if } Y(i) > T \\ 0 & \text{otherwise} \end{cases} \\
Y = \alpha \cdot R + \beta \cdot G + \gamma \cdot B
\end{gathered}
\tag{9}
$$

where coding block luminance is represented by $Y(i)$ and is used as conversion parameters. The RVs of R, G, and B represented by high representative value $\left(\text{hrv}^{R,G,B}\right)$ and low representative value can be computed using luminance bitmaps as

$$\mathrm{hrv}^{R,G,B} = \sum P^{R,G,B}(i) b^{2L}(i) / \sum b^{2L}(i) \tag{10}$$

$$\mathrm{lrv}^{R,G,B} = \sum P^{R,G,B}(i)\left[1 - b^{2L}(i)\right] / \sum \left[1 - b^{2L}(i)\right] \tag{11}$$

Moreover, the reconstructed pixel $Q_{2L}^{R,G,B}(i)$ is computed as

$$Q_{2L}^{R,G,B}(i) = \begin{cases} \mathrm{hrv}^{R,G,B} & \text{if } b^{2L}(i) = 1 \\ \mathrm{lrv}^{R,G,B} & \text{otherwise} \end{cases} \tag{12}$$

Two other parameters are also used to generate the single bitmap. One is represented by BTC-A and used as an average the *R*, *G* and *B* values of each pixel as

$$A = \frac{(R + G + B)}{3} \tag{13}$$

For each pixel to find the average (minimum and maximum) values of *R*, *G*, and *B*, we used the other parameters represented by BTC-M and it is computed as

$$M = \frac{[\mathrm{Min}(R, G, B) + \mathrm{Max}(R, G, B)]}{2} \tag{14}$$

The above two parameters, generate bitmaps same as generated by the proposed BTC-Y method.

While in the proposed method, the RVs of each color are computed with the help of *R*, *G*, and *B* bitmap only and luminance bitmap is not required. Due to this reason, this approach is represented by BTC-Ync.

3.2 Improved Four-Level BTC

If the edge information of color image block consists of multiple level of intensity variations than two-level BTC is not suitable for image coding application. To reduce the bit rate, the same procedure is utilized in four-level BTC as used in two-level BTC. The parameter luminance bitmap is used here to compute *R*, *G*, and *B* bitmaps (four-level to 2 bpp). The reference [12] depicts the procedure to compute four-level BTC for luminance, it first finds out the highest value $Y_{\max}$ and the lowest value $Y_{\min}$ of image block and further computes the threshold value, "TH" and "TL" from the value of $Y_{\max}$ and $Y_{\min}$ as

$$\mathrm{TH} = 3Y_{\max} + Y_{\min} \tag{15}$$

$$\mathrm{TL} = Y_{\max} + 3Y_{\min} \tag{16}$$

With the help of threshold (TH and TL) average value of one-block M^Y, and dynamic range D^Y is calculated as

$$M^Y = \frac{\left[\frac{1}{n}\sum_{Y(i)\geq \mathrm{TH}}^{n} Y(i) + \frac{1}{m}\sum_{Y(i)\leq \mathrm{TL}}^{m} Y(i)\right]}{2} \tag{17}$$

$$D^Y = \frac{1}{n}\sum_{Y(i)\geq \mathrm{TH}}^{n} Y(i) - \frac{1}{m}\sum_{Y(i)\leq \mathrm{TL}}^{m} Y(i) \tag{18}$$

where coding block luminance value is represented by $Y(i)$. From RVs, the Eq. (19) is used to compute the threshold value that quantizes each pixel as mentioned below

$$t_N = M^Y + \frac{N-2}{3} D^Y, \quad N \in \{1, 2, 3\} \tag{19}$$

Using 4-level BTC, we compute the luminance bitmap as

$$\left(b^{4L}(i) = \begin{cases} 3 & \text{if } Y(i) \geq t_3 \\ 2 & \text{if } t_2 \leq Y(i) < t_3 \\ 1 & \text{if } t_1 < Y(i) < t_2 \\ 0 & \text{if } Y(i) < t_1 \end{cases} \right) \tag{20}$$

Then, reconstructed pixels are computed as

$$Q_{4L}^{R,G,B}(i) = M^{R,G,B} + \left[2 \cdot b^{4L}(i) - 3\right] D^{\frac{R,G,B}{6}} \tag{21}$$

The same principle is used in computing $M^{R,G,B}$ and $D^{R,G,B}$ as used in M^Y and D^Y

4 Simulation Results

In this paper, we used still color images of (size 512 × 512, Lake, Airplane, Splash, Sea boat, Lena, and Peppers) to analyze the performance of the proposed AMBTC. Since variation of intensity is found in edge block information in color images, and in some cases, two-level and four-level BTC methods are suitable to provide

Table 1 Image coding results in PSNR

Test images 512 × 512	Image coding performance (dB)		
	AMBTC		Conventional BTC
	2-level BTC	4-level BTC	
Lake	31.64	35.41	29.92
Airplane	32.44	35.05	32.22
Lena	32.74	33.23	31.47
Splash	36.22	31.26	30.05
Seaboat	29.28	28.83	27.52
Peppers	32.73	29.54	28.12

efficient coding results while in other cases, it failed to provide optimum coding results therefore in this paper, an algorithm is developed called AMBTC that adaptively selects the BTC level, if there is slight variation found in edges, then it selects two-level BTC, and if there are edges that consists of multilevel variation, then it adopts four-level BTC. (PSNR) and (CR) which are the basic parameters are used here to evaluate the coding performance of color images. The simulation results are compared between the proposed 4 × 4 AMBTC and conventional 4 × 4 BTC which yields the bit rate of 6 bpp and 4 bpp. The coding performance for still images for two-level and four-level BTC are tabulated in the Table 1. The proposed AMBTC achieves 32.50 dB on average for two-level BTC and 30.58 dB on average for four-level BTC when compared with conventional BTC, obtain a gain of 2.96 dB with two-level and 1.03 dB with four-level BTC, respectively. According to edge property, the proposed AMBTC selects adaptively the two-level or four-level BTC and outperforms the other method used in previous literature and improve the coding performance. A comparative analysis of reconstructed images is shown in Figs. 3, 4 and 5 between the proposed method and conventional BTC clearly show that proposed AMBTC outperform in edges due to multilevel bit allocation approach.

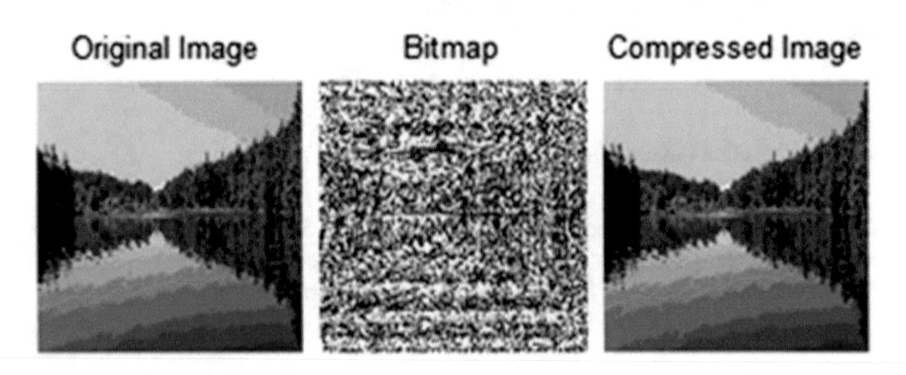

Fig. 3 Improved four-level BTC result with compression ratio 6 **a** Lake image

Fig. 4 Improved two-level BTC with compression ratio 6 **a** Lake image

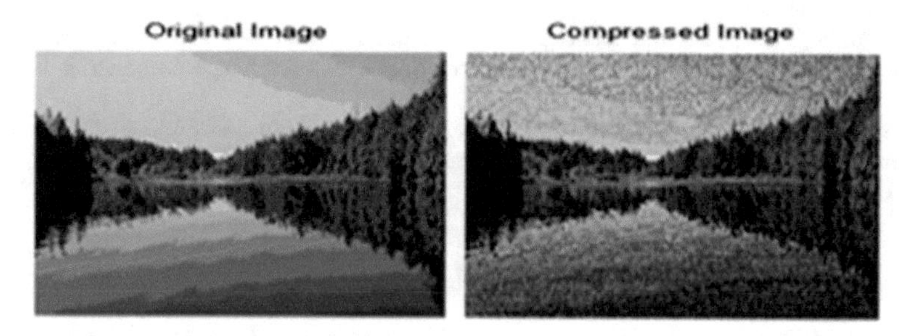

Fig. 5 Conventional BTC with compression ratio 4 **a** Lake image

5 Conclusion

In this paper, we developed a novel color image coding algorithm namely AMBTC to minimize the bandwidth and memory cost. The proposed method adapts (two-level/four-level) for efficient compression of color image. In order to reduce the correlation between the color components, the AMBTC first transforms the RGB space into less correlated YCbCr, YUV, and YIQ, etc., space. AMBTC works by adaptively decomposing the image pixel block in four-level and two-level BTCs as per the color value variation found in the pixel blocks. Simulation results demonstrate that the proposed method outperforms in compassion ratio and PSNR value as compared to conventional BTC approach. We also achieve a bit rate reduction in AMBTC algorithm. Further progress in this work will include developing a codec that allows reducing the bit rate while increasing the CR by 10:1 or greater and cut the cost for multimedia applications.

References

1. Delp, E.J., Mitchell, O.R.: Image compression using block truncation coding. IEEE Trans. Commun. **27**, 1335–1341 (1979)
2. Udpikar, V.R., Raina, J.P.: A modified algorithm for block truncation coding of monochrome images. Electron. Lett. **21**, 900–902 (1987)
3. Lema, M.D., Mitchell, O.R.: Absolute moment block truncation coding and applications to color images. IEEE Trans. Commun. **32**, 1148–1157 (1984)
4. Hui, L.: An adaptive block truncation coding algorithm for image compression. In: Proceeding of ICASSP90, vol. 4, pp. 2233–2236 (1990)
5. Hu, Y.C.: Prediction moment preserving block truncation coding for gray-level image compression. J. Electron. Imaging **13**, 871–877 (2004)
6. Dhara, B.C., Chandra, B.: Block truncation coding using pattern fitting. Pattern Recogn. **37**, 2131–2139 (2004)
7. Efrati, N., Licztin, H., Mitchell, H.B.: Classified block truncation coding vector quantization: an edge sensitive image compression algorithm. Sig. Process. Image Commun. **3**, 275–283 (1991)
8. Mor, I., Swissa, Y., Mitchell, H.B.: A fast nearly optimum equi-spaced 3-level block truncation coding. Sig. Process. Image Commun. **6**, 397–404 (1994)
9. Wu, Y., Coll, D.C.: Single bitmap block truncation coding for color image. IEEE Trans. Commun. **35**, 352–356 (1987)
10. Chen, W.L., Hu, Y.C., Liu, K., Lo, C.C., Wen, C.H.: Variable–rate Quad tree-segmented block truncation coding for color image compression. Int. J. Sig. Process. Image Process. Pattern Recogn. **7**(1), 65–76 (2014)
11. Hu, Y.C., Su, B.H., Tsai, P.Y.: Color image coding scheme using absolute moment block truncation coding and block prediction technique. Imaging Sci. J. **56**(5), 254–270 (2008)
12. Han, J.W., Hwang, M.C., Kim, S.G., You, T.H.: Vector quantizer based block truncation coding for color image compression in LCD overdrive. IEEE Trans. Consum. Electron. **54**, 1839–1845 (2008)

Extreme Learning Machine-Based Image Classification Model Using Handwritten Digit Database

Hasmat Malik and Nilanjan Roy

Abstract The collection of different types of data leads to the problem of its classification. There have been different techniques presented by different authors and researchers to improve the classification process, some of which can be named like spectral–spatial approach where the spectral and spatial information are taken care of individually for hyperspectral image classification (HSI). This paper concentrates on the novel method called Extreme Learning Machine and the approach is tested on MNIST data set. Results are obtained using ANN and ELM, thereby leading to a comparison between the two techniques which is the sole motto behind this paper.

Keywords Modified national institute of standards and technology (MNIST) Extreme learning machine (ELM) · Hyperspectral image (HSI) and artificial neural network (ANN)

1 Introduction

Different types of methods have been tried and experimented for image classification process. Some proposed techniques create hierarchical architecture with the help of sparse coding to learn hyperspectral image classification [1] while some techniques use end-to-end and pixel-to-pixel classification using convolutional networks [2]. Techniques such as Supervised Latent Linear Gaussian Process Latent Variable Model (LLGPLVM) utilizes the simultaneous process of extraction and classification requiring smaller training sets and providing with higher accuracy [3]. The process of extracting information from a multiband image and thereby processing it

H. Malik (✉)
Electrical Engineering Department, IIT Delhi, New Delhi 110016, India
e-mail: hmalik.iitd@gmail.com

N. Roy
ICE Department, NSIT Delhi, New Delhi 110078, India
e-mail: nilanjan25roy@gmail.com

© Springer Nature Singapore Pte Ltd. 2019

H. Malik et al. (eds.), *Applications of Artificial Intelligence Techniques in Engineering*, Advances in Intelligent Systems and Computing 697,
https://doi.org/10.1007/978-981-13-1822-1_57

to create maps is what is known as image classification. The most common type of classification stands out to be the unsupervised classification as it does not require samples for classification process. ANN requires a huge pool of samples along with large time consumption and is sensitive to overfitting; backpropagation neural network (BPNN) is time-consuming, complex and is sensitive to number of neurons present in hidden layer; Support Vector approach suffers from the disadvantage of choosing the kernel along with parameter selection problems [4]. In the present day, CNN-based contextual classification exploits the spatial–spectral characteristics using multi-scale convolutional filter bank which provides the baseline for CNN-based approach [5]. The abovementioned problems can be taken care by the proposed approach known as Extreme Learning Machine (ELM). Results are obtained in a much quicker time and accuracy obtained is also quite significant.

2 Methodology

2.1 Data Set Used

The data set used is MNIST (Modified National Institute of Standards and Technology). Various systems containing images are trained by this large database of handwritten digits which was created from remixing the samples of MNIST data sets. This data set consists of 60,000 training samples and 10,000 testing samples [6]. 70% of the training data set (42,000 samples) is treated as training samples and 10% (9000 samples) used as validation and testing purpose each.

2.2 Proposed Approach

The step-wise-step procedure of the proposed approach for image classification has been presented in Fig. 1, which includes seven steps training and testing phases. In step 1, data set has been collected and divided into two categories (70% data for training and 30% data for testing purposes) in step 2. In step 3, ELM model has been designed and trained in step 4. The testing and error evaluation have been presented in step 5 to 6. Finally, model has been saved for future prospective in step 7.

2.3 Extreme Learning Machine (ELM)

Extreme Learning Machine uses feedforward neural networks and is used for the purpose of regression and classification. They contain a single layer or multilayer of hidden nodes where the parameters of hidden nodes need not be tuned. The name

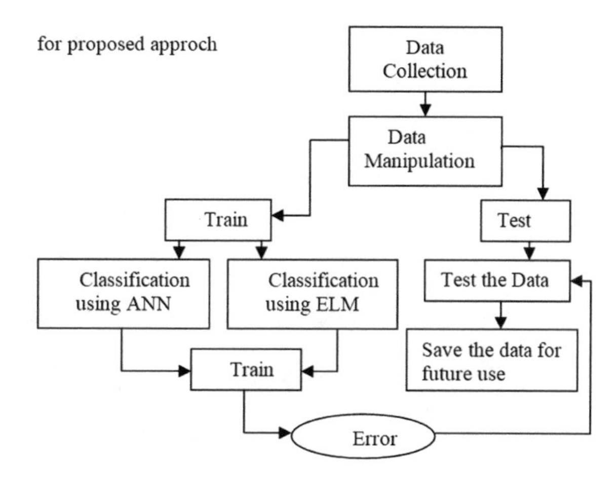

Fig. 1 Flowchart for the proposed approach

was given by Guang-Bin Huang who claimed that such models are able to provide a much more generalized performance and are 1000 times faster than models using backpropagation. The models are much better than many state-of-the-art models with accuracies outperforming Support Vector Machines.

2.3.1 Mathematical Modeling of ELM [4, 7–9]

The input weight vector between hidden and all input nodes is denoted by "*m*". Number of hidden nodes can be denoted by "*N*". The activation function is symbolized as "*k*" and the input feature vector "*x*" has dimension "*D*". The connectivity between input and hidden layers is a function of feature mapping from a *D* dimensional space to *N* dimensional space. Henceforth, the mapping vector can be denoted as

$$g(x) = [k(x; m_1, b_1)\ldots, k(x; m_N, b_N)] \quad (1)$$

Supposing the activation function to be a sigmoidal function, we can write

$$k(x;\ m_i,\ b_i) = \frac{1}{1 + e^{-(x * m_i + b_i)}} \quad (2)$$

Let the number of output nodes be denoted by *O*. The output weight between the *i*th hidden node and the *j*th output node can be denoted by γ_{ij}, where $j = 1, 2, 3\ldots, O$.

The value of an output node *j* can be calculated as

$$e_j(x) = \sum_{i=1}^{N} \gamma_{ij} * k(x; m_i, b_i) \quad (3)$$

$$\text{i.e.} \quad e(x) = [e_1(x), e_2(x) \ldots e_o(x)] \tag{4}$$

$$g(x) * \gamma; \text{where}, \gamma = \gamma = \begin{pmatrix} \gamma_1 \\ \gamma_2 \\ \vdots \\ \gamma_N \end{pmatrix} = \begin{pmatrix} \gamma_{1,1} & \cdots & \gamma_{1,O} \\ \vdots & \ddots & \vdots \\ \gamma_{N,1} & \cdots & \gamma_{N,O} \end{pmatrix} \tag{5}$$

The class level of x during recognition can be determined as

$$\text{Label}(x) = \arg\max e_i(x); \tag{6}$$

$$\text{i} \in \{1 \ldots o\} \tag{7}$$

Let n be the number of training sample. Now, for linear representation

$$G\gamma = Y \tag{8}$$

where Y is the output vector and

$$G = \begin{pmatrix} g(x_1) \\ \vdots \\ g(x_n) \end{pmatrix} = \begin{pmatrix} k(x_1; m_1, b_1) & \ldots & k(x_1; m_N, b_N) \\ \vdots & \ddots & \vdots \\ k(x_n; m_1, b_1) & \cdots & k(x_n; m_N, b_N) \end{pmatrix} \tag{9}$$

$$Y = \begin{pmatrix} y_1 \\ \vdots \\ y_n \end{pmatrix} = \begin{pmatrix} y_{1,1} & \cdots & y_{1,O} \\ \vdots & \ddots & \vdots \\ y_{n,1} & \cdots & y_{n,O} \end{pmatrix} \tag{10}$$

Training error minimizes the difference: $\|T - G\gamma\|^2$ and norm of output $\|\gamma\|$.

2.4 Artificial Neural Network (ANN) [10–17]

A neural network can be described like (A, B, and c) where A is set of neurons and B $\{(i, j) \mid i, j \in \text{A}\}$ provides connectivity between neuron i and neuron j. The function c defines the weights, where c (i, j), can be written like $c_{i,j}$. A learning machine which has approximation property engulfed in its domain along with the capability to approximate mathematical models to a prespecified degree is what we call Artificial Neural Network (ANN). The advantage of ANN lies in the fact that it can very easily predict the nonlinear bonding between input and output for any system model.

Table 1 Classification accuracy

Technique used	Training phase accuracy (%)	Testing phase accuracy (%)
ANN	98.4	96.6
ELM (highest value in %)	99.0	98.4

3 Results and Discussions

The comparative results for image classification using ANN and ELM are given in Table 1, where ELM is trained and tested each time by changing the hidden layer neurons from 1 to 500 and achieved highest classification accuracy is 99% for training phase and for testing phase is 98.4% which is comparatively higher than ANN results.

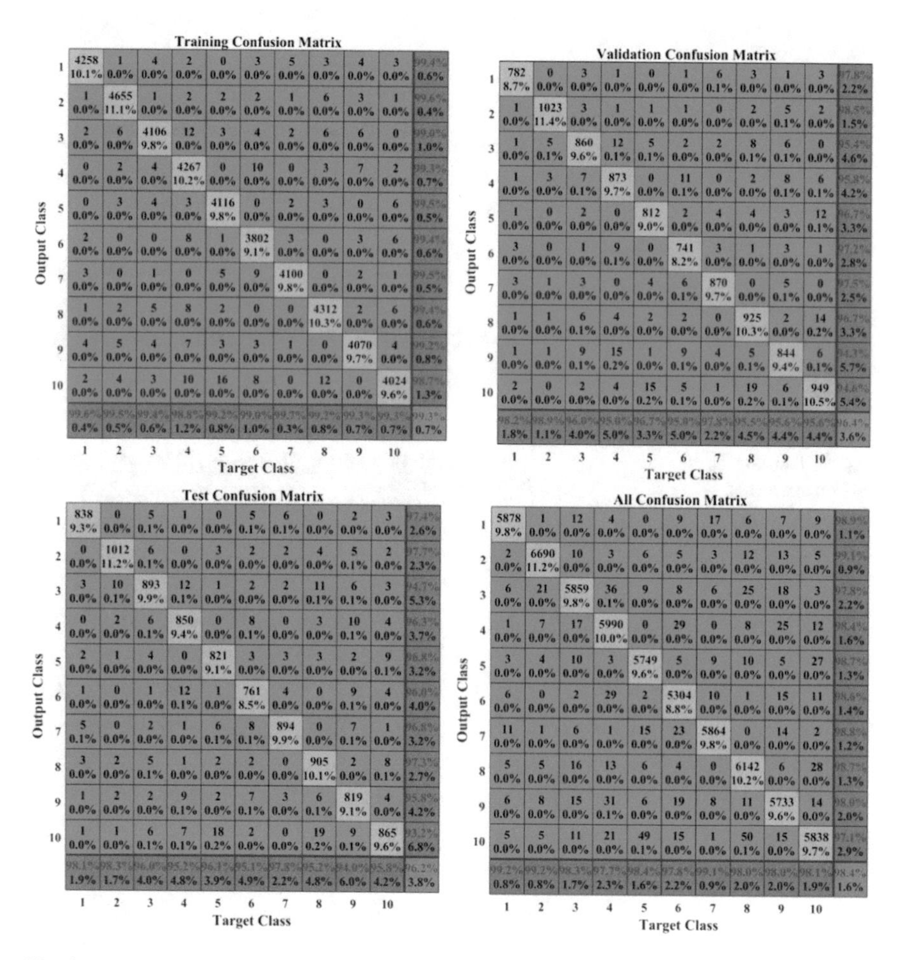

Fig. 2 Training phase confusion matrix

3.1 ANN-Based Image Classification

The diagonal elements of the confusion matrix depict the correct classification pertaining to ten classes, whereas the other elements show misclassified data. Overall training accuracy comes out to be 98.4% and testing accuracy comes out to be 96.6%. The training state and performance plots should decrease with increase in the number of epochs for best classification results. Figures 3 and 4 show gradient

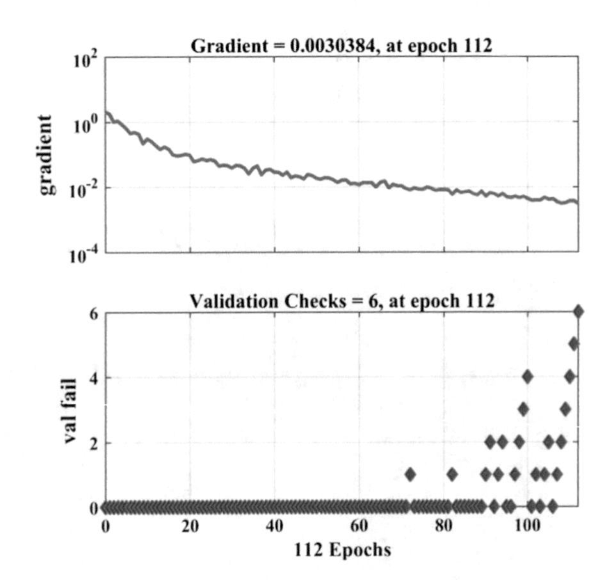

Fig. 3 Training phase training state plot

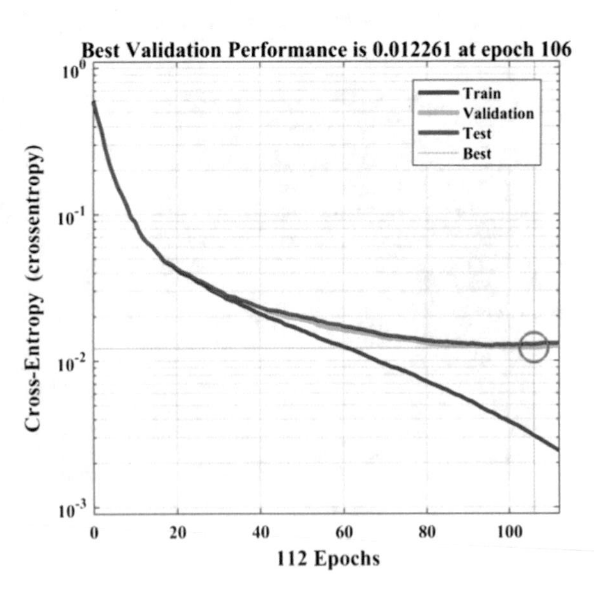

Fig. 4 Training phase performance plot

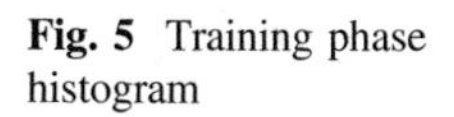

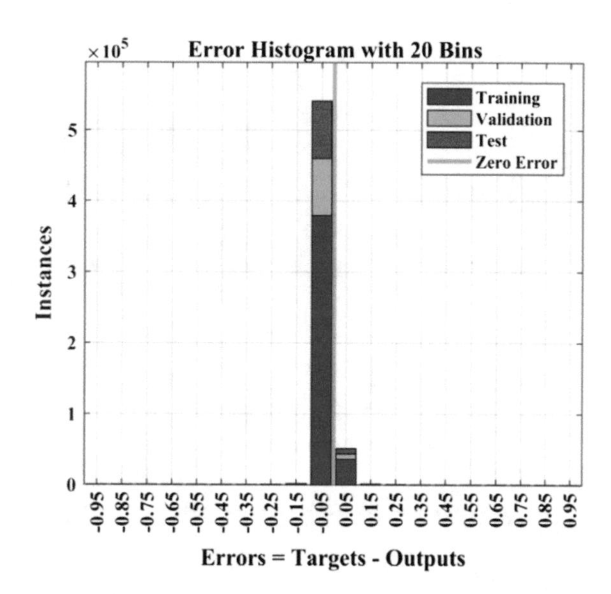

Fig. 5 Training phase histogram

value of 0.0030384 and 0.012261 at epochs 112 and 106, respectively. The ROC plot shows classification of ten different classes ranging from 0 to 9 and the error histogram shows 70% data as training and 15% each for validation and testing. The training phase results are presented in Figs. 2, 3, 4, 5 and 6, whereas in Figs. 7, 8 testing phase results are represented in (Fig. 5).

The confusion matrix shows the correlation between correct classified samples with respect to the total samples. Here, confusion matrix for training and testing phases has been represented along with performance plot. The performance plot further shows the validation of the ANN model. The ROC for each training and testing case has also been represented, which shows the model performance of train and test data set.

3.2 ELM-Based Image Classification

Learning curve for the training and testing sets shows classification of data ranging the neurons from 1 to 500. An increasing curve shows accurate classification procedure. The highest training and testing accuracies come out to be 99.0% and 98.4%, respectively. The model's learning curve for training and testing phases are represented in Figs. 9 and 10 where hidden layer neurons are changed from 1 to 500.

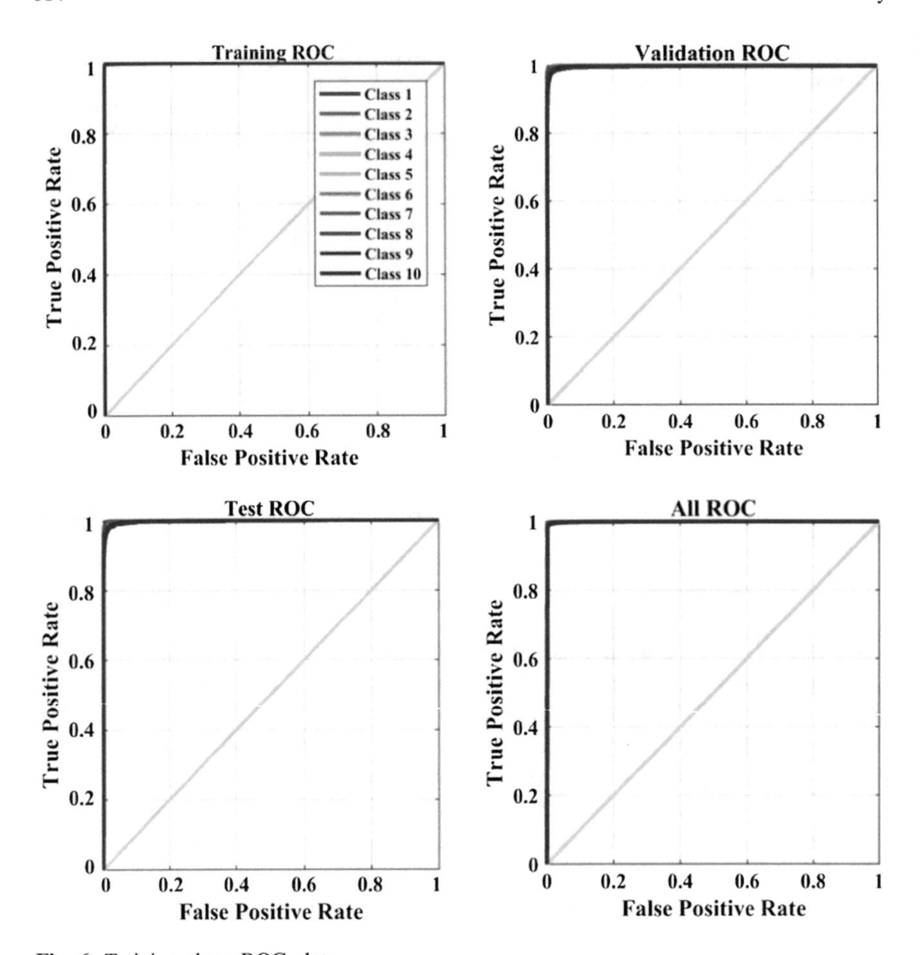

Fig. 6 Training phase ROC plot

The obtained results from both ANN and ELM technique have been compared, and we found that ELM gives better performance in a short operating time period. Moreover, ELM has another beauty that it will handle large size data set in an easy way which is not possible in ANN.

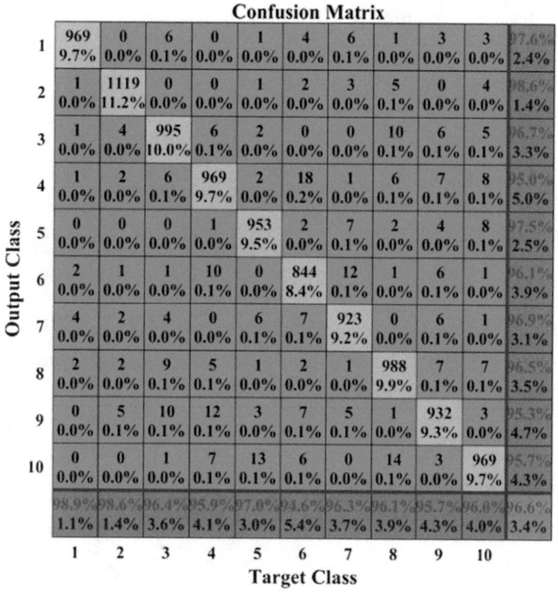

Fig. 7 Testing phase confusion matrix

Confusion Matrix

Output Class \ Target Class	1	2	3	4	5	6	7	8	9	10	
1	969 9.7%	0 0.0%	6 0.1%	0 0.0%	1 0.0%	4 0.0%	6 0.1%	1 0.0%	3 0.0%	3 0.0%	97.6% 2.4%
2	1 0.0%	1119 11.2%	0 0.0%	0 0.0%	1 0.0%	2 0.0%	3 0.0%	5 0.1%	0 0.0%	4 0.0%	98.6% 1.4%
3	1 0.0%	4 0.0%	995 10.0%	6 0.1%	2 0.0%	0 0.0%	0 0.0%	10 0.1%	6 0.1%	5 0.1%	96.7% 3.3%
4	1 0.0%	2 0.0%	6 0.1%	969 9.7%	2 0.0%	18 0.2%	1 0.0%	6 0.1%	7 0.1%	8 0.1%	95.0% 5.0%
5	0 0.0%	0 0.0%	0 0.0%	1 0.0%	953 9.5%	2 0.0%	7 0.1%	2 0.0%	4 0.0%	8 0.1%	97.5% 2.5%
6	2 0.0%	1 0.0%	1 0.0%	10 0.1%	0 0.0%	844 8.4%	12 0.1%	1 0.0%	6 0.1%	1 0.0%	96.1% 3.9%
7	4 0.0%	2 0.0%	4 0.0%	0 0.0%	6 0.1%	7 0.1%	923 9.2%	0 0.0%	6 0.1%	1 0.0%	96.9% 3.1%
8	2 0.0%	2 0.0%	9 0.1%	5 0.1%	1 0.0%	2 0.0%	1 0.0%	988 9.9%	7 0.1%	7 0.1%	96.5% 3.5%
9	0 0.0%	5 0.1%	10 0.1%	12 0.1%	3 0.0%	7 0.1%	5 0.1%	1 0.0%	932 9.3%	3 0.0%	95.3% 4.7%
10	0 0.0%	0 0.0%	1 0.0%	7 0.1%	13 0.1%	6 0.1%	0 0.0%	14 0.1%	3 0.0%	969 9.7%	95.7% 4.3%
	98.9% 1.1%	98.6% 1.4%	96.4% 3.6%	95.9% 4.1%	97.0% 3.0%	94.6% 5.4%	96.3% 3.7%	96.1% 3.9%	95.7% 4.3%	96.0% 4.0%	96.6% 3.4%

Fig. 8 Testing phase ROC plot

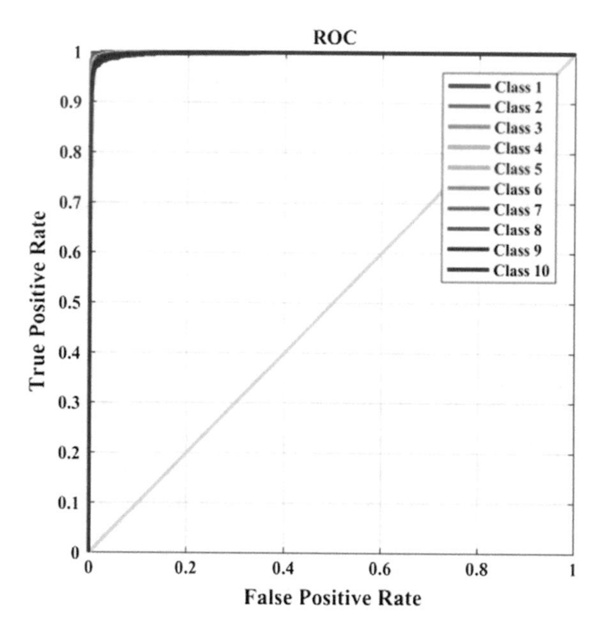

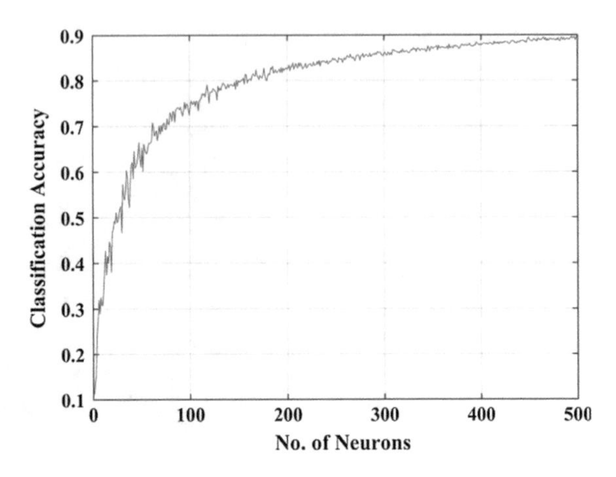

Fig. 9 Plot for training phase accuracy

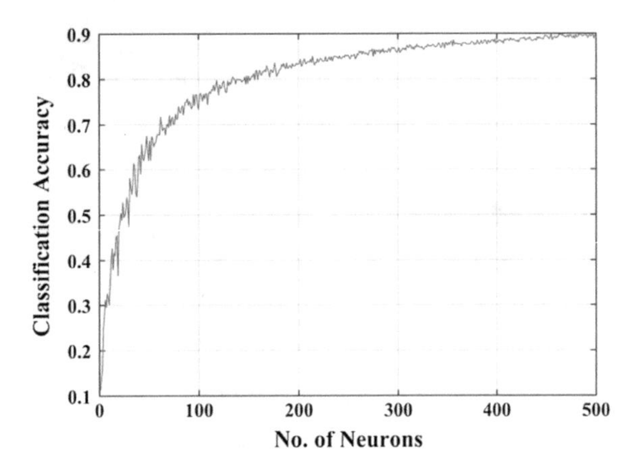

Fig. 10 Plot for testing phase accuracy

4 Conclusion

In this paper, a comparative study of ANN and ELM is presented for image classification of handwritten images using publicly available MNIST image data set. The classification accuracy of ELM is higher than ANN method in term of highest classification accuracy at a particular number of hidden layer neuron. Whereas, the average value of ANN is higher than ELM. Moreover, the processing time of ELM is very less as compared with ANN.

Future work is focused on improving the efficiency of the system using ELM and reaching better classification accuracies.

References

1. Fan, J., Chen, T., Lu, S.: Superpixel guided deep-sparse-representation learning for hyperspectral image classification. IEEE Trans. Circ. Sys. Video Tech. **PP**(99), 1 (2017). https://doi.org/10.1109/tcsvt.2017.2746684
2. Tao, Y., Xu, M., Zhang, F., Du, B., Zhang, L.: Unsupervised-Restricted Deconvolutional Neural Network for Very High Resolution Remote-Sensing Image Classification. IEEE Transactions on Geoscience and Remote Sensing **55**(12) (2017). https://doi.org/10.1109/tgrs.2017.2734697
3. Jiang, X., Fang, X., Chen, Z., Gao, J., Jiang, J., Cai, Z.: Supervised gaussian process latent variable model for hyperspectral image classification. IEEE Geosci. Remote Sens. Lett. **14** (10), 1760–1764 (2017). https://doi.org/10.1109/lgrs.2017.2734680
4. Malik, H., Mishra, S.: Selection of most relevant input parameters using principle component analysis for extreme learning machine based power transformer fault diagnosis model. Int. J. Electr. Power Compon. Sys. **45**(12), 1339–1352 (2017). https://doi.org/10.1080/15325008.2017.1338794
5. Lee, H., Kwon, H.: Going deeper with contextual cnn for hyperspectral image classification. IEEE Trans. Image Process. **26**(10), 4843–4855 (2017). https://doi.org/10.1109/tip.2017.2725580
6. Schaetti, N., Salomon, M., Couturier, R.: Echo state networks-based reservoir computing for mnist handwritten digits recognition. In: Computational Science and Engineering (CSE) and IEEE International Conference on Embedded and Ubiquitous Computing (EUC) and 15th International Symposium on Distributed Computing and Applications for Business Engineering (DCABES), 2016 IEEE International Conference on 24–26 Aug 2016. https://doi.org/10.1109/cse-euc-dcabes.2016.229
7. Sharma, S., Malik, H., Khatri, A.: External fault classification experienced by three-phase induction motor based on multi-class ELM. Elsevier Procedia Comput. Sci. **70**, 814–820 (2015). https://doi.org/10.1016/j.procs.2015.10.122
8. Malik, H., Mishra, S.: Application of extreme learning machine (ELM) in paper insulation deterioration estimation of power transformer. Proc. Int. Conf. Nanotech. Better Living **3**(1), 209 (2016). https://doi.org/10.3850/978-981-09-7519-7nbl16-rps-209
9. Malik, H., Mishra, S.: Extreme learning machine based fault diagnosis of power transformer using IEC TC10 and its related data. In Proceedings IEEE India Annual Conference (INDICON), pp. 1–5 (2015). https://doi.org/10.1109/indicon.2015.7443245
10. Yadav, A.K., Malik, H., Chandel, S.S.: Application of rapid miner in ANN based prediction of solar radiation for assessment of solar energy resource potential of 76 sites in Northwestern India. Renew. Sustain. Energy Rev. **52**, 1093–1106 (2015). https://doi.org/10.1016/j.rser.2015.07.156
11. Yadav, A.K., Malik, H., Chandel, S.S.: Selection of most relevant input parameters using WEKA for artificial neural network based solar radiation prediction models. Renew. Sustain. Energy Rev. **31**, 509–519 (2014). https://doi.org/10.1016/j.rser.2013.12.008
12. Yadav, A.K., Sharma, V., Malik, H., Chandel, S.S.: Daily array yield prediction of grid-interactive photovoltaic plant using relief attribute evaluator based radial basis function neural network. Renew. Sustain. Energ. Rev. **81**(2), 2115–2127 (2018). https://doi.org/10.1016/j.rser.2017.06.023
13. Malik, H., Yadav, A.K., Mishra, S., Mehto, T.: Application of neuro-fuzzy scheme to investigate the winding insulation paper deterioration in oil-immersed power transformer. Electr. Power Energ. Syst. **53**, 256–271 (2013). https://doi.org/10.1016/j.ijepes.2013.04.023
14. Malik, H.: Application of artificial neural network for long term wind speed prediction. In: Proceedings IEEE CASP-2016, pp. 217–222, 9–11 June 2016. https://doi.org/10.1109/casp.2016.7746168

15. Azeem, A., Kumar, G., Malik, H.: Artificial neural network based intelligent model for wind power assessment in India. In: Proceedings IEEE PIICON-2016, pp. 1–6, 25–27 Nov 2016. https://doi.org/10.1109/poweri.2016.8077305
16. Saad, S., Malik, H.: Selection of most relevant input parameters using WEKA for artificial neural network based concrete compressive strength prediction model. In: Proceedings IEEE PIICON-2016, pp. 1–6, 25–27 Nov 2016. https://doi.org/10.1109/poweri.2016.8077368
17. Azeem, A., Kumar, G., Malik, H.: Application of waikato environment for knowledge analysis based artificial neural network models for wind speed forecasting. In: Proceedings IEEE PIICON-2016, pp. 1–6, 25–27 Nov 2016. https://doi.org/10.1109/poweri.2016.8077352

Tracking Control of Robot Using Intelligent-Computed Torque Control

Manish Kumar, Ashish Gupta and Neelu Nagpal

Abstract Tracking control of nonlinear and uncertain system like robot is always a challenge for the designer. In this paper, the simulation study of trajectory tracking control of 3-degree of freedom (DOF) robotic manipulator is presented. In this, the control structure is based on computed torque control method. The control objective is to make three joints of robot to trace the desired time-varying trajectory. The trajectory tracking is done with different approaches using computed torque proportional–derivative (CTC-PD) control, computed torque fuzzy logic control (CTC-FLC) and computed torque adaptive network fuzzy inference system (CTC-ANFIS). The Simulink model of Phantom Omni^{TM} Bundle robot is used. The comparative performance analysis of the controllers shows that the joints of robot track the trajectory with small tracking error in case of CTC-ANFIS.

Keywords Computed torque control · Trajectory control · Proportional–derivative controller · Fuzzy logic controller · ANFIS

1 Introduction

A robotic manipulator is a complex dynamic system, which suffers from coupled, nonlinearity and uncertainty effects [1]. Thus, the control algorithm should tackle these undesirable effects when employed for tracking control. Due to simple structure and cost-effectiveness, conventional proportional–integral–derivative (PID) controllers have been preferred but fail to give an effective result in case of a nonlinear and uncertain process. For nonlinear robotic system, CTC approach is applied for the reason of feedback linearization ability and can control each joint independently but require exact dynamics of the robot [2]. Alternative solutions

---

M. Kumar (✉) · A. Gupta · N. Nagpal
EEE Department, Maharaja Agrasen Institute of Technology, New Delhi, India
e-mail: manishkumar251195@gmail.com

N. Nagpal
e-mail: nagpalneelu1971@gmail.com

© Springer Nature Singapore Pte Ltd. 2019

H. Malik et al. (eds.), *Applications of Artificial Intelligence Techniques in Engineering*, Advances in Intelligent Systems and Computing 697,
https://doi.org/10.1007/978-981-13-1822-1_58

have been found which can improve the result. As the popularity of PID/PD controller is more, attempts have been made to introduce some intelligence in these controllers. Fuzzy logic controller has self-tuning capabilities to provide better results than PID controller [3]. It is an intelligent control technique with the benefits of flexibility and optimization space. Also, the capability of the expert knowledge and tuning can make a FLC as indispensable control. The self-tuning of FLC depends on the rule base, parameters of membership function (MF) that include shape, width and spacing of each MF. Further, it has been shown that asymmetrical spacing of MF gives better results [4] as it fine tunes the controller. The type and values of MF are decided by the system designer and can be adjusted according to the need of the system.

Nature-inspired algorithms find application in optimal tuning of FLC [5]. Neural network has self-learning capability and when coupled with FLC, it provides a possible solution for the tuning problem [6]. Neural network can be used to extract the fuzzy rules from the input–output sets, to obtain the input–output function of the system and to construct the controllers. Neural networks when used as controller, are able to realize the dynamics of the mode [1]. In this paper, neuro-fuzzy technique is used as the prime tool, which is known as adaptive network-based fuzzy inference system (ANFIS). ANFIS is the mixing of the Fuzzy Logic techniques with Artificial Neural Network in which a fuzzy inference system is made by using neural network to determine its parameters (mainly its membership functions) [4, 6, 7]. In this system, fixed membership functions which are chosen erratically and a rule base which is predetermined by the users according to the characteristics of the variables is used. ANFIS controller has been used in many applications [8–16].

In this paper, the Simulink model of Phantom Omni™ Bundle robot is developed. A reference sinusoidal trajectory is given to its three active joints. The controllers, i.e. CTC-PD, CTC-FLC and CTC-ANFIS are designed so that each joint follow reference trajectory. FLC is designed with five triangular membership functions and the rule base for each joint. ANFIS controller is developed using Gaussian MF in fuzzy model. For learning fuzzy parameters back propagation method is used. It is used in two different modes, i.e. constant and linear mode. For each controller, simulations are used to evaluate the performance of the system.

In this paper, following sections are described as: Sect. 1 describes the mathematical modelling of robot used. Design of different control structures are explained in Sect. 2. Section 3 provides the simulation results and its discussion and finally Sect. 4 concludes the paper.

2 Robot Model

Phantom *Omni*™ Bundle is an electromechanical device having the facility of kinaesthetic feedback. It has 6-DOF out of which 3-DOF are driven by DC motors as shown in Fig. 1. It can be used for testing controller developed in MATLAB.

Fig. 1 Phantom Omni™ Bundle

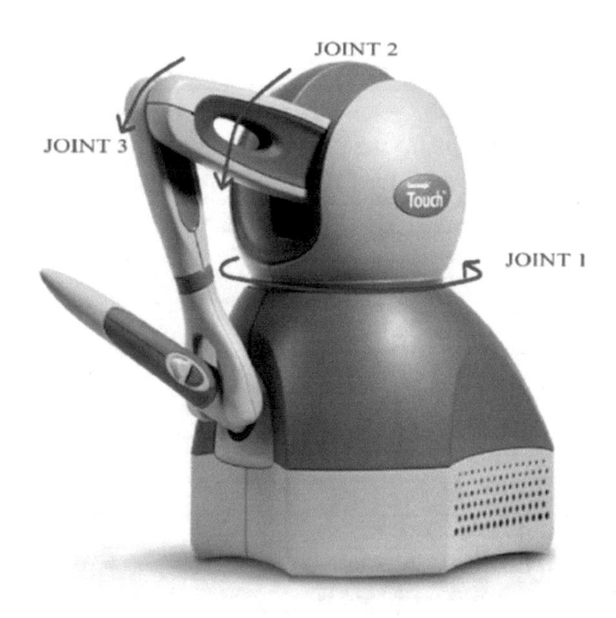

It uses the interface IEEE-1394 Fire wire 6 pin port for communication. In this work, we have developed the Simulink model of Phantom robot for the trajectory control of 3 joints considering the dynamics referred from [1]

The inertia matrix of robot is

$$M(f) = \begin{bmatrix} m_{11} & m_{12} & m_{13} \\ m_{21} & m_{22} & m_{23} \\ m_{31} & m_{32} & m_{33} \end{bmatrix} \quad (1)$$

where

$m_{11} = c_1 + c_2 \cos(2f_2) + c_3 \cos(2f_3) + c_4 \cos(f_2) \sin(f_3)$	$m_{12} = c_3 \sin(f_2)$	$m_{13} = 0$
$m_{21} = c_5 \sin(f_2)$	$m_{22} = k_6$	$m_{23} = -0.5c_4 \sin(f_2 - f_3)$
$m_{31} = 0$	$m_{32} = -0.5c_4 \sin(f_2 - f_3)$	$m_{33} = c_7$

$D(f,\dot{f})$ represents the Coriolis and Centrifugal forces.

$$D(f,\dot{f}) = \begin{bmatrix} D_{11} & D_{12} & D_{13} \\ D_{21} & D_{22} & D_{23} \\ D_{31} & D_{32} & D_{33} \end{bmatrix} \quad (2)$$

where

$D_{11} = c_2\dot{f}\sin(2f_2) - c_3\dot{f}_3\sin(2f_3)$ $- 0.5c_4\dot{f}_2\sin(f_2)\sin(f3)$ $+ 0.5c_4f_3\cos(f_2)\cos(f_3)$	$D_{12} = -c_2\dot{f}_1\sin(2f_2)$ $- 0.5c_4\dot{f}_1\sin(f_2)\sin(f_3)$ $+ c_5\dot{f}_2(f_2)$	$D_{13} = -c_3\dot{f}_1\sin(2f_3) +$ $0.5c_4f_1\cos(f_2)\cos(f_3)$
$D_{21} = c_2\dot{f}_1\sin(2f_2)$ $+ 0.5c_4f_1\sin(f_2)\sin(f_3)$	$D_{22} = 0$	$D_{23} = 0.5c_4\dot{f}_3\cos(f_2 - f_3)$
$D_{31} = c_3\dot{f}_3\sin(2f_3) +$ $0.5c_4\dot{f}_1\cos(f_2)\cos(f_3)$	$D_{32} = 0.5c_4\dot{f}_2\cos(f_2 - f_3)$	$D_{33} = 0$

The values of c_1 to c_{10} is given by the values

$c_1 = 0.00179819707554751$	$c_2 = 0.000864793119787878$
$c_3 = 0.000486674040957256$	$c_4 = 0.00276612067958414$
$c_5 = 0.00308649491069651$	$c_6 = 0.00252639617221043$
$c_7 = 0.000652944405770658$	$c_8 = 0.164158326503058$
$c_9 = 0.0940502380783103$	$c_{10} = 0.117294768011206$

3 Controller Design

In this section, CTC-PD controller, CTC-FLC and an optimized CTC-ANFIS have been applied on robot.

The dynamics of robot is represented as

$$M(f)\ddot{f} + D(f,\dot{f}) = \tau \quad (3)$$

where

$fn \times 1$	Joint displacement vector
$\dot{f}n \times 1$	Joint velocities vector
$\tau n \times 1$	Torque matrix applied by the actuators
$\mathrm{M(f)}n \times n$	Inertia matrix
$\mathrm{D}(\mathrm{f},\dot{f})\mathrm{n} \times 1$	Centripetal and Coriolis torques matrix.

The different controllers are described as follows in the following subsections.

A. *CTC-PD*

CTC is a control method used for linearizing robot dynamics. It uses dynamical model of robots in such manner that motion control of all the joints can be done individually. The model consists of inner loop which compensates nonlinearity and outer loop which provides control signal as shown in the equation below.

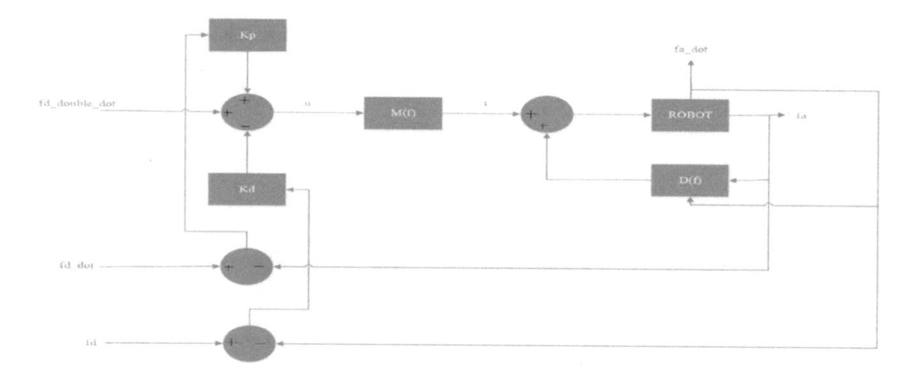

Fig. 2 Diagram of CTC-PD

$$T_c = M(f)\left[\ddot{f} + k_d\left(\dot{f}_d - \dot{f}_a\right) + k_p(f_d - f_a)\right] + D\left(f,\dot{f}\right) \tag{4}$$

This control input is given to each joint independently to track the desired trajectory. Figure 2 shows diagram of CTC-PD

B. *CTC-FLC Controller*

The process of FLC is shown in Fig. 3. The first block (fuzzifier) is the part of the controller, which scales the input crisp values into normalized universe of discourse U then converts the inputs to degrees of membership function (MF).

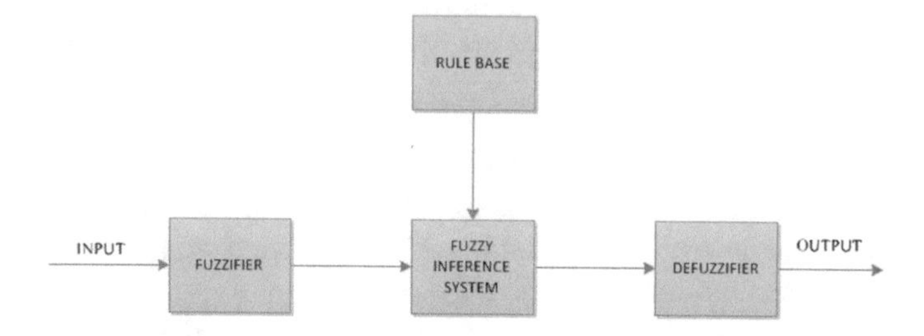

Fig. 3 Block diagram of FLC

Table 1 Rule base

$\dot{e}$/e	LN	SN	Z	SP	LP
LN	LN	LN	SN	SN	Z
SN	LN	LN	SN	SN	Z
Z	LN	SN	Z	SP	LP
SP	Z	Z	SP	SP	LP
LP	Z	Z	SP	SP	LP

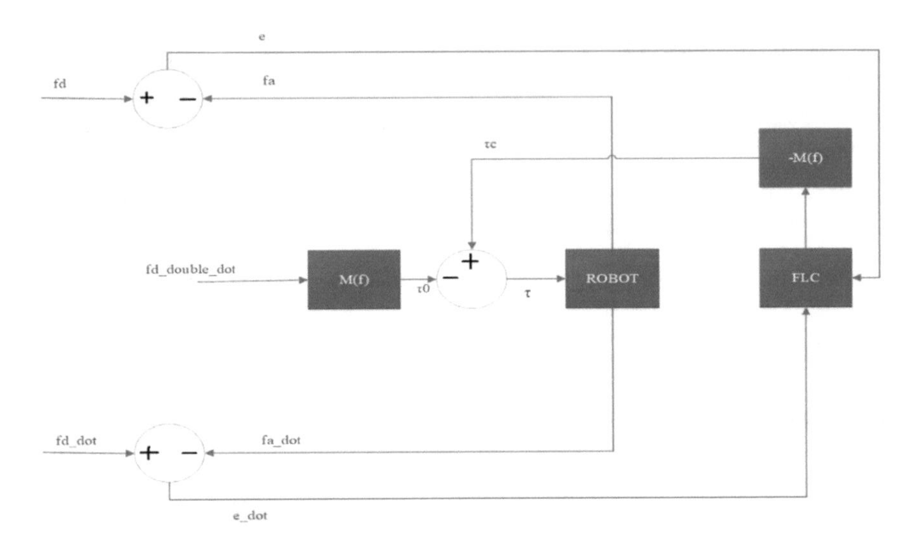

Fig. 4 Block diagram of CTC-FLC

In this paper, error signal (e) and the derivative of error signal ($\dot{e}$) have been used in fuzzy logic controller to generate the output. Each membership functions (input and output) are defined by five linguistic variables which are taken as large negative (LN), small negative (SN), zero (Z), small positive (SP), large positive (LP). The rule base of FLC is shown in the Table 1. The block diagram of CTC-FLC is referred from [2, 12] as shown in Fig. 4

where $\dot{e}$ represents the difference of the actual and desired velocities and e represents the difference of actual and desired positions.

C. *CTC-ANFIS*

A neural network and the fuzzy interference system are collectively referred as Adaptive network-based fuzzy inference system (ANFIS). The estimation of parameters is done in such a way that the ANFIS architecture represents both the Sugeno and Tsukamoto fuzzy models [13].

Initially, a fuzzy model is formulated in which the fuzzy rules are obtained from the input–output data set. Further the rules of fuzzy model are tuned by the neural network. The network is tuned using ANFIS methodology. By applying an optimal data selection criterion, the number of training data used is reduced. Figure 5 shows the ANFIS structure.

In this paper, we have used two modes for tuning the membership function of FLC, i.e. constant and linear mode. In the constant mode, we will have symmetrical membership function while in the linear mode asymmetrical membership functions are observed.

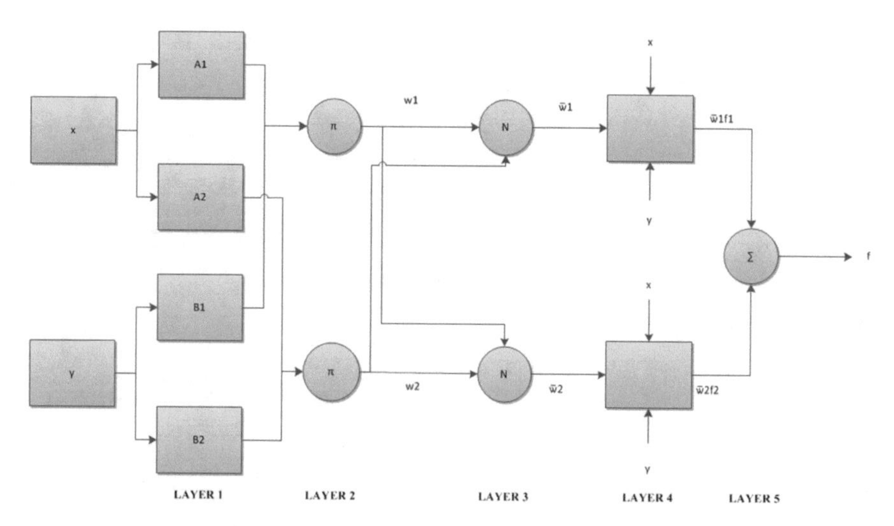

Fig. 5 ANFIS structure

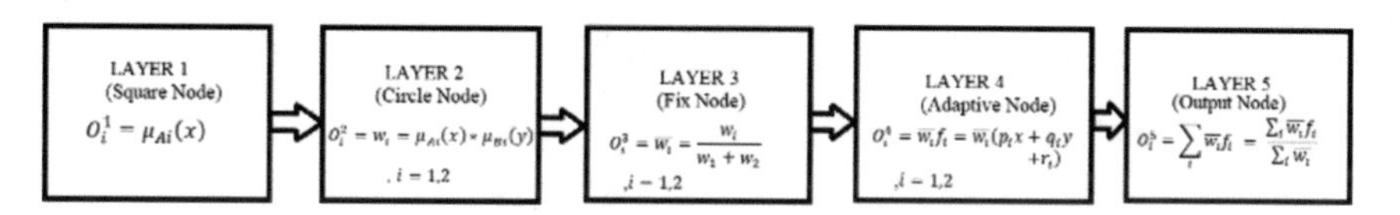

Fig. 6 Layer description

A and B are the fuzzy sets. x and y are the input variables while $f(x, y)$ is the polynomial function of x and y.

The output node O_i^n corresponding to ith node of each layer is shown in Fig. 6 where $n = 1, 2 \ldots 5$

1. *SIMULATION RESULTS*

The Simulink model of Phantom Omni robot for each type of controller is developed and the simulations have been performed. The reference trajectory is given and the trajectory tracking of three joints is observed. Figures 7, 8 and 9 describe the tracking of different joints with CTC-PD, CTC-FLC and CTC-ANFIS with constant and linear mode.

From Table 2, it is observed that RMS error of CTC-ANFIS with linear mode is lesser than other two (CTC-PD and CTC-FLC) controller.

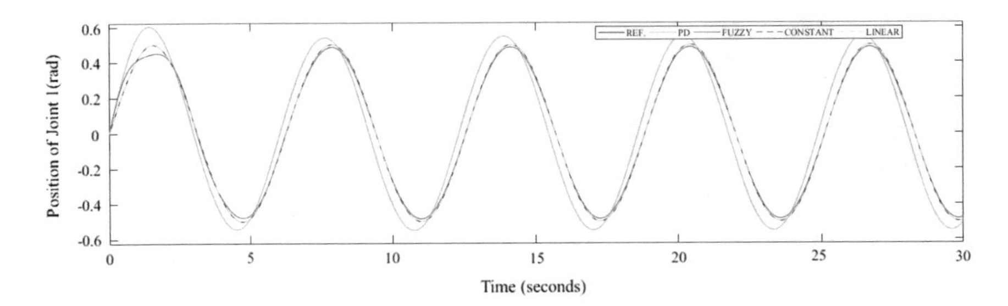

Fig. 7 Tracking of joint 1

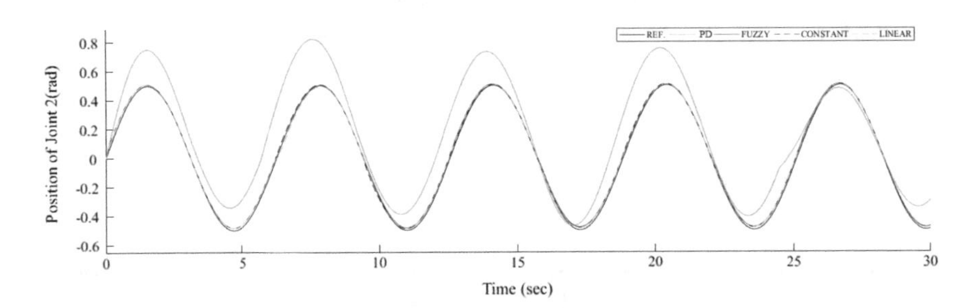

Fig. 8 Tracking of joint 2

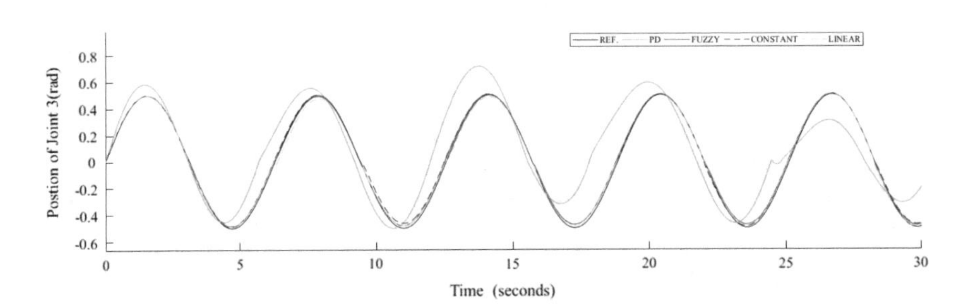

Fig. 9 Tracking of joint 3

Table 2 Comparison results of controllers

Joints	RMS error (10^{-3})			
	CTC-PD	CTC-FLC	CTC-ANFIS	
			Constant	Linear
Joint 1	17.48	11.98	10.409	1.46
Joint 2	21.34	17.50	8.13	0.73
Joint 3	26.83	22.78	4.61	1.02

4 Conclusion

The mathematical model for 3-DOF Omni robot was developed. The trajectory control of Omni robot using different controllers, i.e. CTC-FLC, CTC-ANFIS was studied. Conventional controllers are not capable of providing effective tracking results in case of nonlinear systems like robot. The simulation results of trajectory using the ANFIS-CTC with linear mode and Gaussian membership function shows better results than other methods used. Simulation results on robotic joints are provided for the tracking control proficiency and the efficacy of the proposed design. We have implemented CTC-ANFIS as an optimization technique to tune the CTC-FLC for motion control strategy of robotic arm. In future, there is a scope of using any advance optimization techniques like BFOA, PSO, etc., to tune the controller.

References

1. Song, Z., Yi, J., Zhao, D., Li, X.: A computed torque controller for uncertain robotic manipulator system: fuzzy approach. Fuzzy Sets Syst. **154**(2), 208–226 (2005)
2. Kapoor, N., Ohri, J.: Fuzzy sliding mode controller (FSMC) with global stabilization and saturation function for tracking control of a robotic manipulator. J. Control Sys. Eng. **1**(2), 50–56 (2013)
3. Nagpal, N., Bhushan, B., Agarwal, V.: Intelligent control of four DOF robotic arm. In: 1st IEEE International Conference on Power Electronics (ICPEICES), 2016
4. Precup, R.E., Borza, A.L., Radac, M.B, Petriu, E.M.: Bacterial foraging optimization approach to the controller tuning for automotive torque motors. In: IEEE 23rd International Symposium on Industrial Electronics (ISIE), pp. 972–977, doi:https://doi.org/10.1109/ISIE.2014.6864744 (2014)
5. Benmiloud, T.: Improved adaptive neuro-fuzzy inference system. Neural Comput. Appl. **21**(3), 575–582 (2012)
6. Chatterjee, A., Chatterjee, R., Matsuno, F., Endo, T.: Augmented stable fuzzy control for flexible robotic arm using LMI approach and neuro-fuzzy state space modeling. IEEE Trans. Industrial Electr. **55**(3), 1256–1270 (2008)
7. Nugroho, P.W., Du, H., Li, W., Alici, G.: Implementation of adaptive neuro fuzzy inference system controller on magneto rheological damper suspension. In: 2013 IEEE/ASME International Conference on Advanced Intelligent Mechatronics (AIM), Wollongong, Australia, 9–12 July (2013)
8. Solatian, P., Abbasi, S.H., Shabaninia, F.: Simulation study flow control based on PID ANFIS controller for non-linear process plants. American J. Intell. Sys. **2**(5), 104–110 (2012)
9. Alavandar, S., Nigam, M.J.: Inverse kinematics solution of 3DOF planar robot using ANFIS. Int. J. Comput. Commun. Control., ISSN 1841-9836, E-ISSN 1841-9844, vol 3, Suppl. Issue: Proceedings of ICCCC 2008, pp. 150–155 (2008)
10. Elyazed, M.M.A., Mabrouk, M.H., Elnor, M.E.A., Mahgoub, H.M.: Trajectory planning of five DOF manipulator: dynamics feed forward controller over computed torque controller. Int. J. Eng. Res. Tech. **4**(9), 401–406 (2015)
11. Tseng, C.S., Chen, B.S., Uang, H.J.: Fuzzy tracking control design for nonlinear dynamic systems via T-S fuzzy model. IEEE Trans. Fuzzy Sys. **9**(3), 381–392 (2001)

12. Navaneethakkannam, C., Sudha, M.: Analysis and implementation of ANFIS-based rotor position controller for BLDC motors. J. Power Electr. **16**(2), 564–571 (2016)
13. Mala, C.S., Ramachandran, S.: Design of PID controller for directon control of robotic vehicle. Glob. J. Res. Eng. Electr. Electr. Eng. **14**(3), 2014
14. Manjaree, S., Agarwal, V., Nakra, B.C.: Kinematic analysis using neuro fuzzy intelligent technique for robotic manipulator. Int. J. Eng. Res. Tech. **6**(4), 557–562 (2013)
15. Yi, S.Y., Chung, M.J.: A robust fuzzy logic controller for robot manipulators with uncertainties. IEEE Trans. Syst. Man Cybern. B Cybern. **27**(4), 706–713 (1997)
16. Vick, A., Cohen, K.: Genetic fuzzy controller for a gas turbine fuel system. In: Advances in Intelligent and Autonomous Aerospace Systems, Chapter 6. Progress in Astronautics and Aeronautics, vol. 241, pp. 173–198. American Institute of Aeronautics and Astronautics Inc., Virginia (2012)

Tracking Control of 4-DOF Robotic Arm Using Krill Herd-Optimized Fuzzy Logic Controller

Neelu Nagpal, Hardik Nagpal, Bharat Bhushan and Vijyant Agarwal

Abstract Robots have become an integral part of the automated world. The motion control in joint space is a challenging problem with the objective of the controller is to force the joints of the manipulator to track the desired trajectory. In this paper, Krill Herd-Optimized Fuzzy Logic Controller (KHO-FLC) is designed, tested, and implemented for the first time for the tracking control of a 4-degree of freedom (DOF) robotic arm. The algorithm of Krill Herd (KH) is used to tune the parameters of Fuzzy Logic Controller (FLC). The proposed work is validated by simulation study of a robotic manipulator for the tracking of joint motion control. A comparison of performance analysis of the system using KHO-FLC and that of Fuzzy-PD, and Particle Swarm Optimized Fuzzy Logic Controller (PSO-FLC) controller demonstrates the effectiveness of the proposed algorithm.

Keywords Robotic arm · Krill Herd · Fuzzy logic control · Particle swarm optimization · Degree of freedom

N. Nagpal (✉) · B. Bhushan
Electrical Deapartment, Delhi Technological University, New Delhi, India
e-mail: nagpalneelu1971@gmail.com

B. Bhushan
e-mail: bharat@dce.ac.in

H. Nagpal
MAIT, Delhi, India
e-mail: nagpal.hardik5@gmail.com

V. Agarwal
Division of MPEA, Netaji Subhas Institute of Technology, New Delhi, India
e-mail: vijaynt.agarwal@gmail.com

© Springer Nature Singapore Pte Ltd. 2019

H. Malik et al. (eds.), *Applications of Artificial Intelligence Techniques in Engineering*, Advances in Intelligent Systems and Computing 697,
https://doi.org/10.1007/978-981-13-1822-1_59

1 Introduction

The widespread applications of robotic arm require interminable developing control techniques to control the movements of its joints and that is of end-effector. Different control strategies have generally been developed for processing the error signal generated by comparing the actual joint position and the desired joint position, thereby feeding a control voltage signal to the actuator. Despite simplicity and robustness, conventional Proportional–Integral–Derivative (PID) controllers lack in performance when applied to nonlinear, uncertain, and coupled robotic system. To enhance the performance of PID controller, versions of modified PID controller [1–3] and intelligent controllers [4, 5] have been proposed. Hybridization of Fuzzy logic with PID control creates a more appropriate solution to control robot manipulator. In FLC, selection and tuning of membership function along with the framing of rules are done heuristically. For better performance, determination of scaling factors and membership functions plays an important role. The fine tuning of FLC is achieved using asymmetrical membership function [6]. Apart from the conventional PID controllers, fractional-order proportional and derivative (FOPD) controllers have shown has a superior performance as compared to the traditional PID/PD controllers in terms of dynamic performance and robustness [7]. Also, Fractional-Order Fuzzy Controllers are also popular for their performance [8]. The theory of the evolutionary process has been exploited for global search optimization. PSO [9] has been used to optimize the spacing and width of the membership functions of FLC to get better tracking [10]. Also, many other nature-inspired algorithms are employed for the optimal tuning of the controller [11, 12]. Krill Herd algorithm [13] has been employed in the applications of general FLC [14], electrical and power system [15], wireless and network system [16] and for neural network training [17]. The comparison of KH and PSO has been done in [18]. In this present work, this algorithm is exploited to tune the scaling factor of the fuzzy membership function, optimizing rule base, and finding optimal gain parameters namely 'K_P' and 'K_D' and 'K_O'. This algorithm is designed, implemented, and tested for the tracking control of multi-degree robotic arm. The present work is associated with the recently developed KH based, fuzzy-PD control scheme. Each joint of 4-DOF robotic arm is required to track the predefined trajectory. Further, simulations are carried out for the comparative analysis of different control algorithms resulting in minimum tracking error of each joint of the robotic arm. The Simulink model for the control of 4-DOF robotic arm has been developed using MATLAB/Simulink environment. Further, simulations are carried out for the comparative analysis of different control algorithms resulting in minimum tracking error of each joint of the robotic arm. The Simulink model for the control of 4-DOF robotic arm has been developed using MATLAB/Simulink environment. Each joint is required to follow the time-varying trajectory and the effectiveness of this approach is established by comparing the simulation results with FLC and PSO-FLC. The performance of the proposed approach is better than the other two approaches with the minimum root mean square error.

The paper has been organized as follows: A brief overview of the 4-DOF robot model is given in Sect. 2. Different controller and optimization schemes are explained in Sect. 3. In Sect. 4, the fitness function for the optimization algorithm is given. Simulation results and the comparison in terms of root mean square error are given in Sect. 5. Finally, in Sect. 6 the conclusion is drawn based on the results.

2 Model of 4-DOF Robotic Arm

In practice, the exact mathematical model of the physical system is not known. The control designer aims that the ability of the control calculation to coordinate the exact model can be effectively demonstrated. SimMechanics is an advanced tool of Simulink that allows the possibility to verify model-based control algorithm. In this, a mechanical framework can be represented by a connected block diagram similar to the Simulink model. The complex architectures of robotic manipulators with any degree of freedom can be developed using the enriched features of this software. The Simulink model of 4-DOF has been created utilizing SimMechanics. The model of the robotic arm is a subsystem of the whole system. The robotic arm is further composed of DC motor subsystem and Robotic mechanics subsystem. The arm is developed by serially connected rigid links with four joints named as Turntable, Bicep, Forearm, and Wrist joint. Each joint is powered by DC motor that causes rotational motion to it. The mechanical and electrical specifications of DC motor as per the desired rating are shown in Tables 1 and 2, respectively. In order

Table 1 DC motor electrical parameter

Armature inductance	1.2×10^{-5} H
Stall torque	2.4×10^{-4} N m
No-load speed	1.91×10^{4} rpm
Rated DC supply voltage	6 V
No-load current	9.1 mA

Table 2 DC motor mechanical parameter

Rotor inertia	0 kg m^2
Initial rotor speed	0 rpm
Damping coefficient	0.001 N m/rad/s

Table 3 Robot parameters

Joint	Mass (kg)	Link length (m)
Turntable	m1 = 0.181	a = 0.067
Bicep	m2 = 0.049	b = 0.152
Forearm	m3 = 0.161	c = 0.185
Wrist	m4 = 0.173	d = 0.071

to amplify or attenuate the incoming signals to the desired rating levels of the DC motor, a Power Amplifier is used. The link length and mass of each joint are given in Table 3.

3 Controllers

The objective of a positional controller is to drive each joint motor so that the actual joint angular displacement can track a desired angular displacement decided by a predetermined trajectory. Thus, the control technique is based on utilizing the error signal from the desired and the actual angular positions of the joint. Initially, the robotic arm is considered to be in a fully extended vertical position setting all joint angles to be zero. The function of a controller is to generate the control signals, i.e., voltages to DC motor for each joint, using the difference between the reference joint angles and the actual joint angles as the input. Different controllers used are elaborated in subsections given below.

3.1 FLC

The PD-type fuzzy logic controller (FLC) is shown in the Fig. 1. Error (e) and change in error (ce) act as inputs and single output (cout), designed with triangular membership functions. The number of fuzzy sets should be an odd integer greater than one. More precision can be obtained using more number of membership functions for various inputs and outputs in fuzzy logic control. Selection of proper membership function from the available data affects the system response and generally, attempt has been made for selecting optimal value heuristically. In FLC, the range for the error and change in error is [−1.57, 1.57]. In the membership functions, the second point of each MF coincides with the third point of the left one and the first point of the right one.

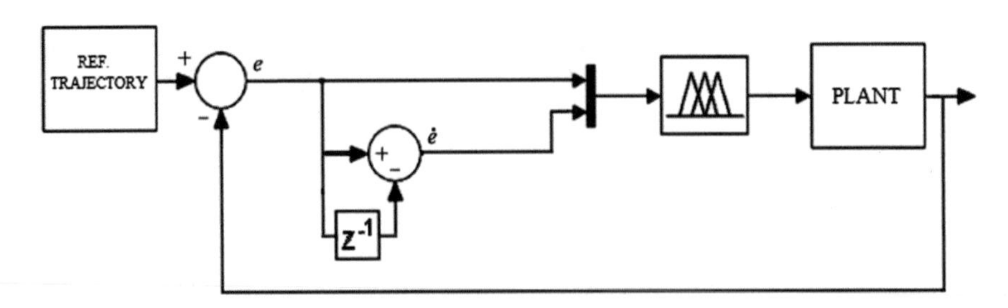

Fig. 1 Fuzzy logic controller

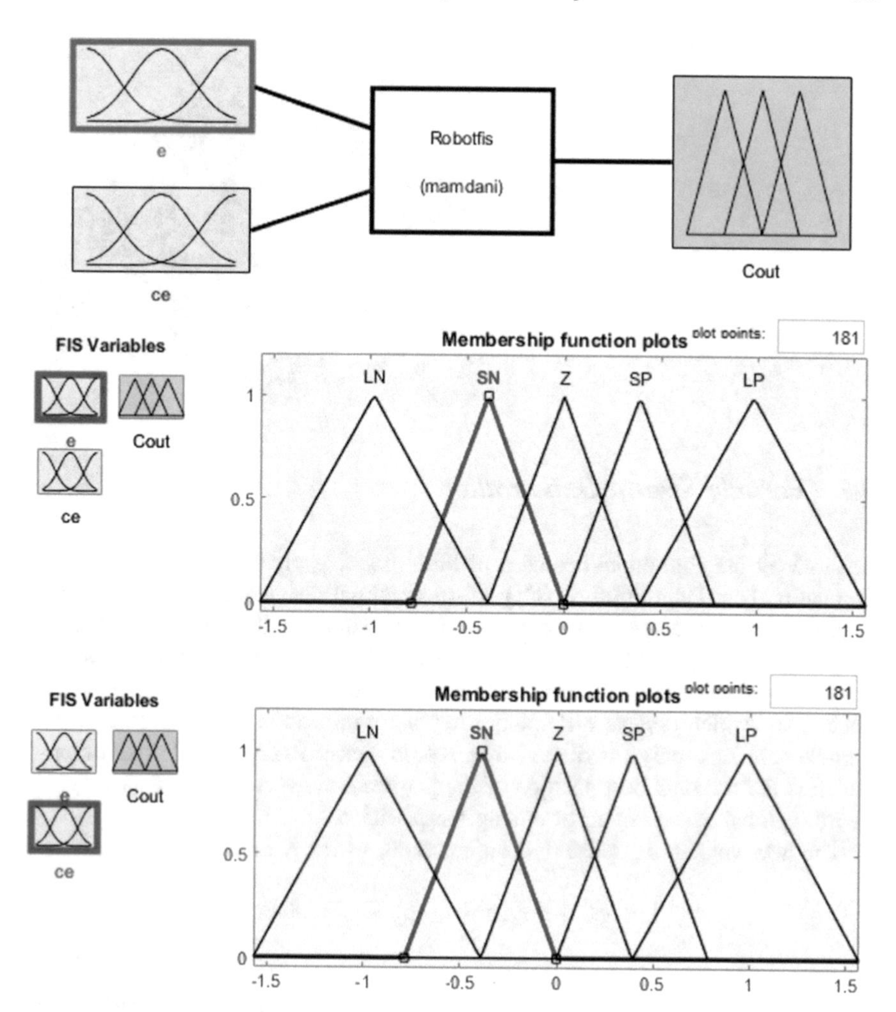

Fig. 2 Input membership functions for error and change in error

In this work, the membership function (MF) chosen for error, change in error and volt output are shown in Fig. 2. And, the associated surface relationship for Turntable joint is shown in Fig. 3, where the linguistic labels LN, SN, Z, SP, and LP are Large Negative, Small Negative, Zero, Small Positive, and Large Positive, respectively. The major drawback of FLC is that the fine tuning of the membership function which adapt the nonlinear model is time consuming and not accurate.

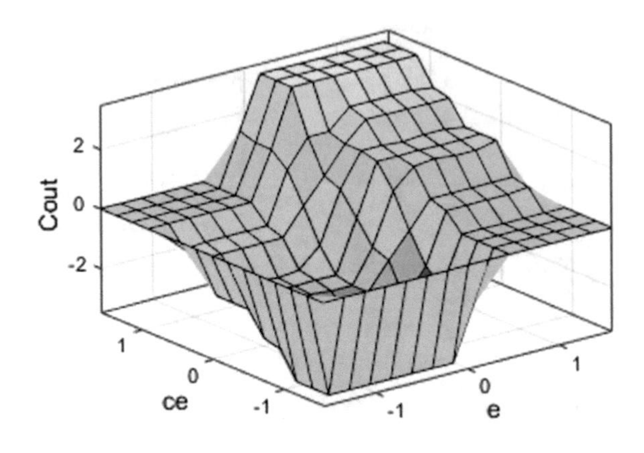

Fig. 3 Surface relationship

3.2 *Particle Swarm Optimization*

Inspired by the communal behavior of birds flocking or fish schooling, PSO is a population-based algorithm, which performs optimal search using the population of particles. The implementation of the algorithm is simple with only a few parameters to modify as compared to some other metaheuristic optimization algorithms such as genetic algorithm, etc. Every particle has a velocity vector, 'v_i' and a position vector, 'x_i' which depicts a d solution to the optimization problem. Initially, the particles are randomly distributed in a search space. Every particle has its *pbest* which is the personal best position of the particle and every particle follows *gbest,* i.e., the global best position of among the particles.

The new velocity is updated as an equation, which is follows:

$$v_n^{t+1} = wv_n^t + c_1r_1(pbest_d - x_d^t + c_2r_2\left(gbest_d - x_d^t\right)$$

where

w, $c_{1,}$ and c_2 are called the coefficients of inertia, cognitive, and society study, respectively. r_1 and r_2 are uniformly distributed random numbers ranging [0, 1].

With a change in the velocity of particles brings a change in the position toward individual best and global best position. Thus, the equation for the position update is stated as

$$x_n^{t+1} = x_n^t + \alpha v_n^t$$

Figure 4 shows the implementation of PSO-FLC in the form of a block diagram. The arrow crossing the FLC block shows that PSO is optimizing the FLC membership functions.

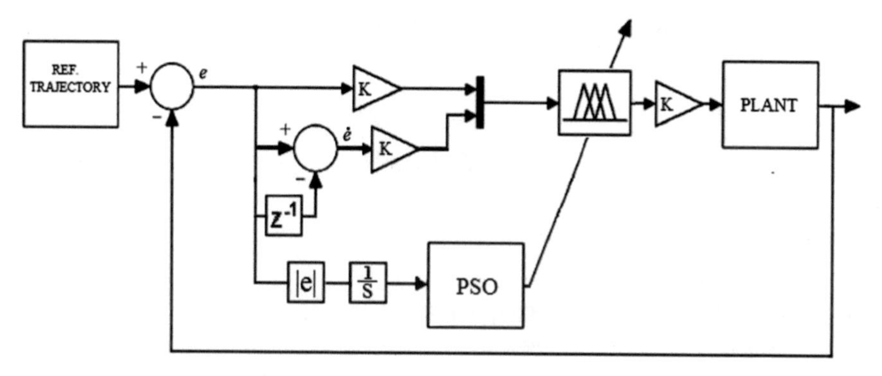

Fig. 4 PSO-FLC block diagram

3.3 *Krill Herd Algorithm*

Inspired by the foraging strategy of an organism, KH is the latest algorithm and finds application for solving engineering optimization problems. In the process of searching for food, Krill's get divided into two groups keeping the minimum distance between each individual krill from the food and from highest density of the heard krill. This results in the computation of an objective function with highest density krill groups. The basic steps involved in this algorithm are Induced motion (N_i), Foraging motion (F_i), and Random diffusion (D_i). The Lagrangian model is designed as

$$\frac{\mathrm{d}X_i}{\mathrm{d}t} = N_i + F_i + D_i \tag{1}$$

Equation (2) is shown for the induced motion for krill individual

$$\begin{aligned} N_i^{\text{new}} &= N^{\max}\alpha_i + w_n N_i^{\text{old}} \\ \alpha_i &= \alpha_i^{\text{local}} + \alpha_i^{\text{target}} \end{aligned} \tag{2}$$

where

$N^{\max}$ is the maximum induced speed
ω_n is inertia weight in motion induced ranging between [0, 1]
N_i^{old} is the previous motion induced
α_i^{local} is the local effect due to neighbors
α_i^{target} is the target direction effect of the best krill individuals.

Similarly, Eq. (3) gives a foraging motion that includes the location of the food and its experience

$$\begin{aligned} F_i &= V_f \beta_i + w_f F_i^{\text{old}} \\ \beta_i &= \beta_i^{\text{food}} + \beta_i^{\text{best}} \end{aligned} \tag{3}$$

where

V_f	is the foraging speed
ω_{f}	is inertia weight for foraging motion with range [0, 1]
F_i^{old}	is the last foraging motion
β_i^{food}	is the food attractive
β_i^{best}	is the effect of the best fitness of the ith krill till every iteration.

The physical diffusion of the krill individuals is treated to be a random process, and is given by (4)

$$D_i = D_{\max}\left(1 + \frac{\text{itr}}{\text{itr}_{\max}}\right)\delta \tag{4}$$

where

$D_{\max}$	is the maximum diffusion speed,
δ	is a random directional vector ranging [−1, 1].

At last, the position vector X_i is updated as given in (5)

$$X_i^{(t+\Delta t)} = X_i^{(t)} + \Delta t \frac{\mathrm{d}X_i}{\mathrm{d}t} \tag{5}$$

where

Δt is an important parameter either randomly chosen according to the optimization problem or depends upon the search space as shown in (6)

$$\Delta t = C_t \sum_{i=0}^{n} (\text{UB} - \text{LB}) \tag{6}$$

where

LB and UB	are lower and upper bounds of Krill's position in search space.
C_t	is a constant ranging [0, 2].

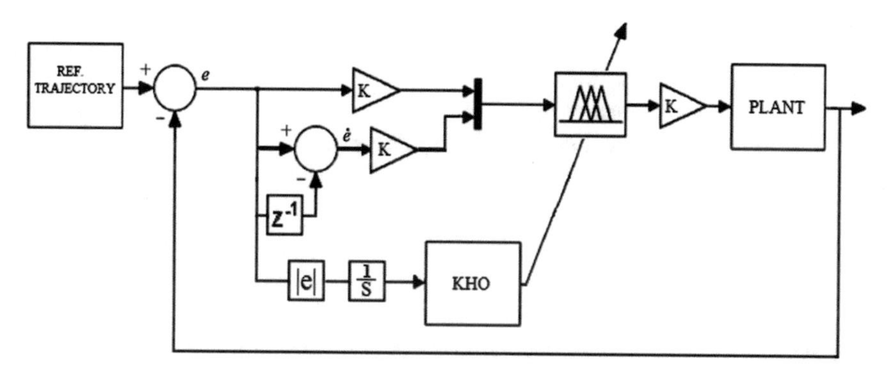

Fig. 5 KHO-FLC block diagram

Pseudo Code of Krill Herd Algorithm

Start
Step 1. Initialize parameters: $itr = 1$, V_f, D_{max}, N_{max},
Step 2 Opposition-based population initialization
Step 3. For $itr < itr_{max}$
- ➢ Objective Function evaluation
- ➢ Movement induced
- ➢ Foraging motion
- ➢ Physical diffusion
- ➢ Crossover and mutation

Step 4. Update krill's position
Step 5. Update the current best
End

Figure 5 shows the implementation of KHO-FLC in the form of a block diagram. The arrow crossing the FLC block shows that KH is tuning the FLC membership functions.

4 Objective Function

In this paper, spacing parameter for MFs of input/output variables, optimal rule base, and optimal gain in the input of FLCs are determined with KH algorithm. Figures 4 and 5 show the block diagram of the two methods with PSO-FLC and KHO-FLC. This method of tuning parameters is based upon minimizing the absolute time integral error of joints.

$$e(t) = q_d(t) - q(t)$$

$$\mathrm{ITAE} = \int_{t=0}^{t} t.\left|e_j(t)\right|$$

where

q_d is the desired trajectory
q is the output trajectory then,
$e_j(t)$ is the error at sampling time 't' for joint 'j'.

5 Simulation Results

The evaluation of the proposed controller designed for the robotic model is done through a computer simulation for the arm movement as shown in Fig. 6 and an internal structure of the controller is shown in Fig. 7. A time-varying trajectory is given as a reference to which each joint of the arm has to be followed. The model is tested with each type of controller, i.e., Fuzzy-PD, PSO-FLC, and KHO-FLC. The results for the Turntable, Bicep, Forearm, and Wrist are shown in Figs. 8, 9, 10 and 11, respectively. The following observations have been noticed from the simulation results:

- For Turntable (Joint-1) as in Fig. 8, the steady-state response of the PSO-FLC and KHO-FLC is much faster than FLC but initially there are some oscillations that settle down later. Also, KH shows better results as it shows less tracking error for the same number of iterations.
- For Bicep (Joint-2) in Fig. 9, the FLC is unable to track the desired trajectory but after optimizing with PSO and KH, it is able to track. PSO-FLC shows more

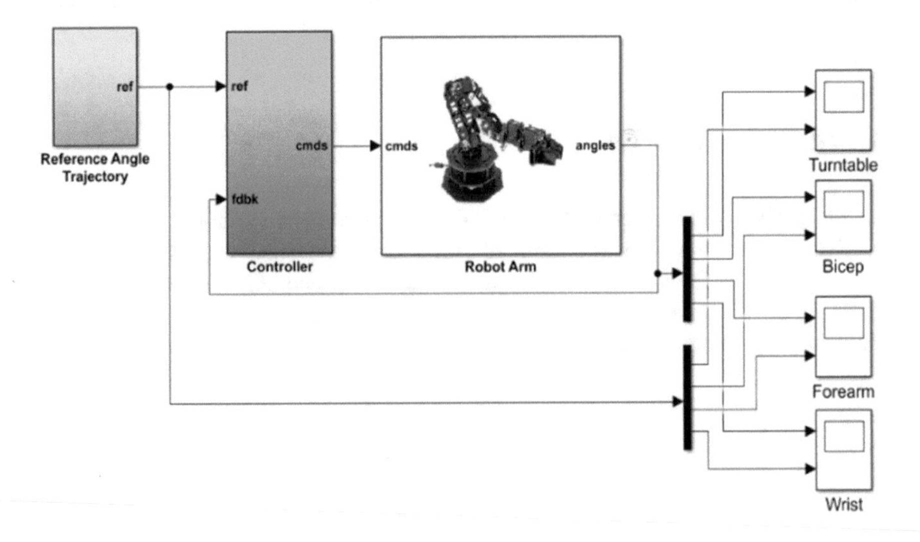

Fig. 6 Simulink model

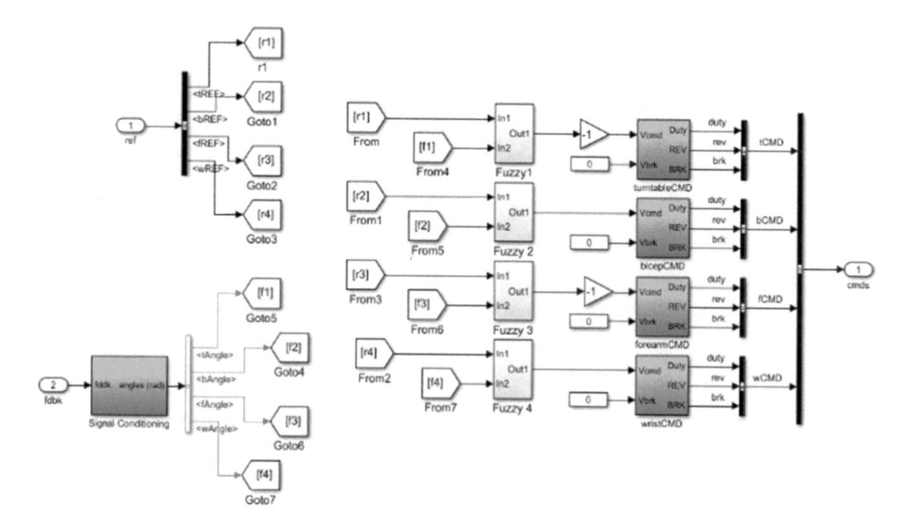

Fig. 7 Controller simulink model

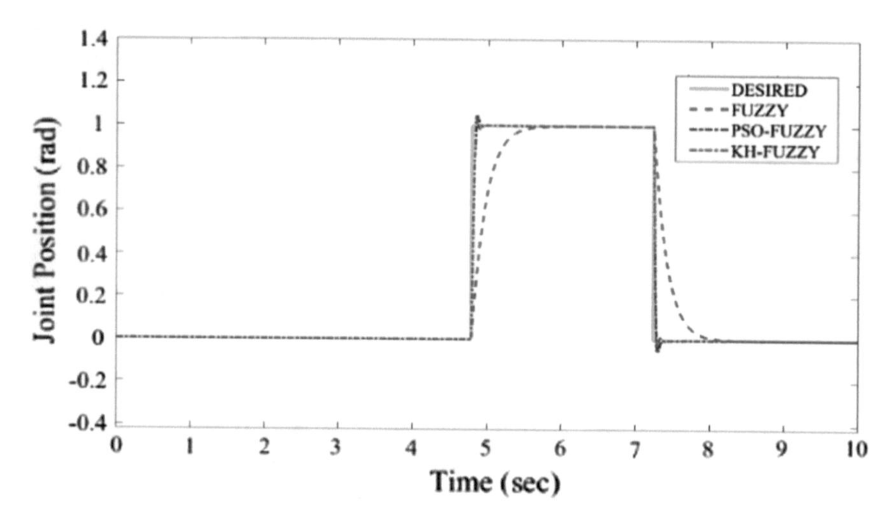

Fig. 8 Trajectory tracking of turntable (joint 1)

transient error with higher peak oscillations as compared to KHO-FLC. Thus, KHO-FLC shows better results with less tracking error.

- For Forearm (Joint-3) in Fig. 10, due to the coupling effect with Bicep, transient response is not so good but is able to track desired trajectory for KHO-FLC and PSO-FLC. Also, with KHO-FLC shows less tracking error.
- For Wrist (Joint-4) in Fig. 11, a similar response is obtained with KHO-FLC as shown in the above joints. But, PSO-FLC is unable to track the desired

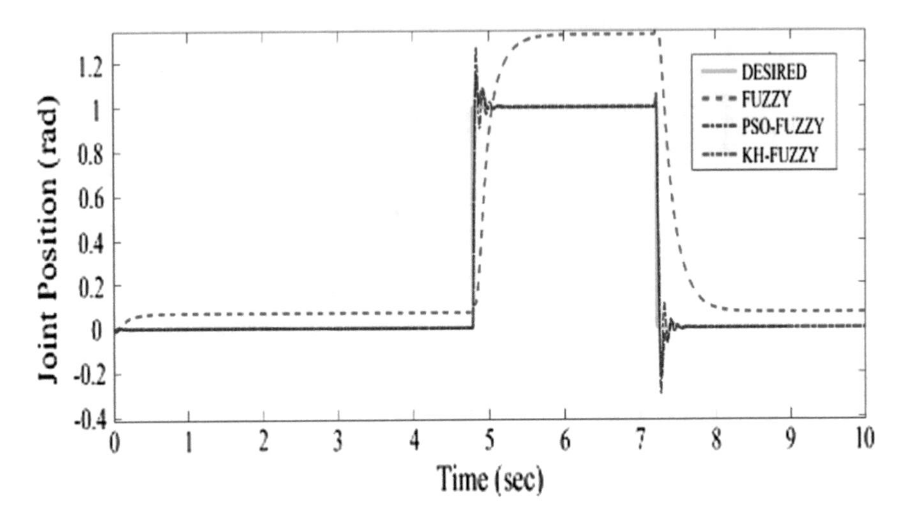

Fig. 9 Trajectory tracking of bicep (joint 2)

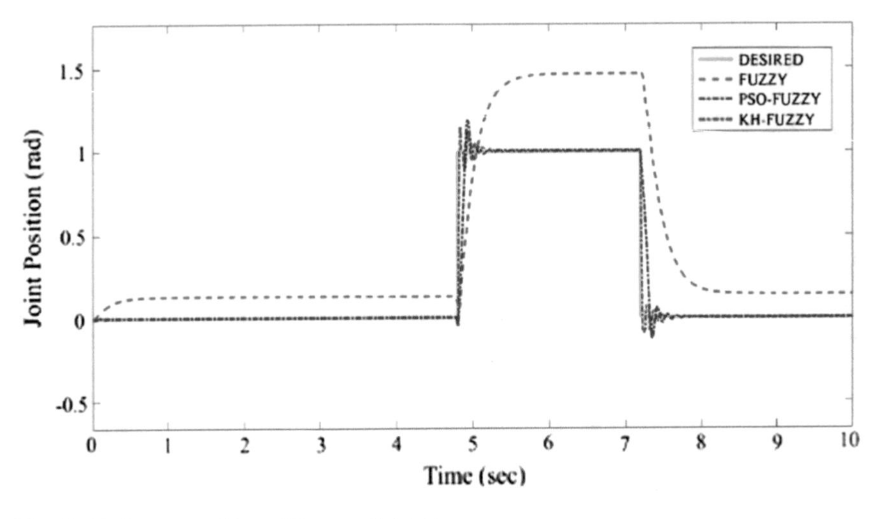

Fig. 10 Trajectory tracking of forearm (joint 3)

trajectory or not able to optimize FLC within a set of number of iterations as compared to KHO-FLC.

- Among the different control structures, the performance of KHO-FLC is the best compared to other techniques applied. The root mean square tracking error is compared in Table 4.
- After each iteration, the tracking error is reduced as KH finds the optimal solution of the parameters. Figure 12 shows the reduction in error for Turntable.

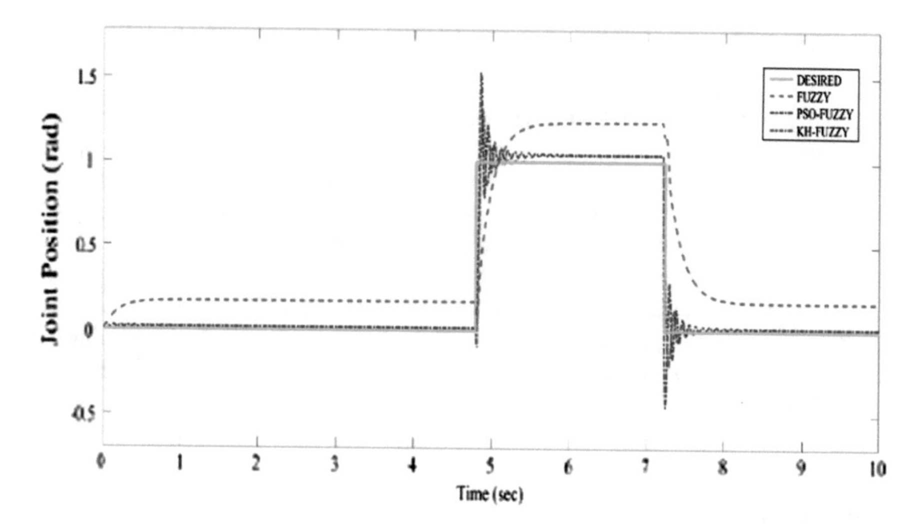

Fig. 11 Trajectory tracking wrist (joint 4)

Table 4 Root mean square error

	Turntable	Bicep	Forearm	Wrist
FLC	0.0937	0.282	0.305	0.262
PSO-FLC	0.0351	0.086	0.082	0.107
KHO-FLC	0.0349	0.073	0.060	0.076

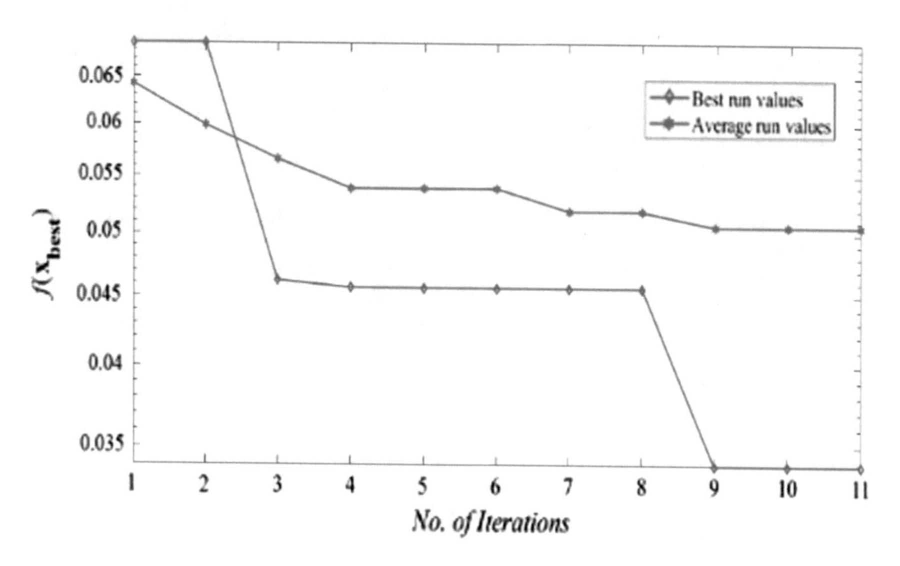

Fig. 12 Tracking error for turntable after every iteration

6 Conclusions

While designing FLC, the choice of the membership function, their spacing, and rule base play an important role to control the system effectively. In this paper, successfully implementation of Krill herd algorithm to optimal tune the fuzzy logic controller for the tracking of 4-DOF robotic arm is done. Through MATLAB simulations, the actual trajectory of the proposed controller is compared with the conventional FLC and PSO-FLC. The comparative results confirmed the improved performance of the proposed controller in terms of transient response and RMS tracking error. Further, the work encouraged the implementation of an evolutionary algorithm to optimize intelligent controller for a system when the mathematical model is not provided and depends only on input–output data.

References

1. Radaideh, S.M., Hayajneh, M.T.: A modified PID controller. J. Franklin Inst. **339**, 543–553 (2002)
2. Shakibjoo, A.D., Shakibjoo, M.D.: 2-DOF PID with re-set controller for 4-DOF robot arm manipulator. In: International Conference on Advanced Robotics and Intelligent Systems, pp. 1–6 (2015)
3. Nagpal, N., Bhushan, B., Agarwal, V.: Estimation of stochastic environment force for master–slave robotic system. Sadhana **42**(6), 889–899 (2017)
4. Nagpal, N., Bhushan, B., Agarwal, V.: Intelligent control of four DOF robotic arm. In: IEEE International Conference on Power Electronics, Intelligent Control and Energy Systems (2016)
5. Manjaree, S., Nakra, B.C., Agarwal, V.: Comparative analysis for kinematics of 5-DOF industrial robotic manipulator. Acta Mechanica et Automatica **9**(4), 229–240 (2015)
6. Gupta, N., Garg, R., Kumar, P.: Asymmetrical fuzzy logic control to PV module connected micro-grid. In: IEEE India International Conference JMI (2015)
7. Chen, Y.Q., Petras, I., Xue, D.: Fractional order control—a tutorial. In: American Control Conference, June 2016
8. Sharma, R., Gaur, P., Mittal, A.P.: Optimum design of fractional-order hybrid fuzzy logic controller for a robotic manipulator. Arabian J. Sci. Eng. **42**(2), 739–750 (2017)
9. Kennedy, J., Eberhart, R.: Particle swarm optimization. In: IEEE International Conference on Neural Networks, vol. 4, Perth, pp. 1942–1948 (1995)
10. Bingul, Z., Karahan, O.: A fuzzy logic controller tuned with PSO for 2 DOF robot trajectory control. Expert Syst. Appl. **38**(1), 1017–1031 (2011)
11. Bhushan, B., Singh, M.: Adaptive control of nonlinear systems using bacterial foraging algorithm. Int. J. Comput. Electr. Eng. **3**(3), 335–342 (2011)
12. Aghajarian, M., Kiani, K., Fateh, MdM: Design of fuzzy controller for robot manipulators using bacterial foraging optimization algorithm. J. Intell. Learn. Syst. Appl. **4**, 53–58 (2012)
13. Gandomi, A.H., Alavi, A.H.: Krill Herd: a new bioinspired optimization algorithm. Commun. Nonlinear Sci. Numer. Simul. **17**, 4831–4845 (2012)
14. Fattahi, E., Bidar, M., Kanan, H.R.: Fuzzy Krill Herd optimization algorithm. In: First International Conference on Networks & Soft Computing (2014)
15. Shayanfar, H.A., Shayeghi, H., Younesi, A.: Fuzzy logic-based load frequency control in a nonlinear deregulated multi-source power system. In: International Conference on Artificial Intelligence, pp. 74–80 (2017)

16. Shopon, Md., Adnan, Md.A., Mridha, Md.F.: Krill herd based clustering algorithm for wireless sensor networks. In: International Workshop on Computational Intelligence (IWCI), pp. 96–100 (2016)
17. Kowalski, P.A., Łukasik, S.: Training neural networks with krill herd algorithm. Neural Process Lett **44**, 5–17 (2015)
18. Chaturvedi, S., Pragya, P., Verma, H.K.: Comparative analysis of particle swarm optimization, genetic algorithm and krill herd algorithm. In: IEEE International Conference on Computer, Communication and Control (2015)

Author Index

© Springer Nature Singapore Pte Ltd. 2019
H. Malik et al. (eds.), *Applications of Artificial Intelligence Techniques in Engineering*, Advances in Intelligent Systems and Computing 697,
https://doi.org/10.1007/978-981-13-1822-1

Printed in the United States
By Bookmasters